The Chemical Biology of Carbon

Chemical Biology

Titles in the series:

1: High Throughput Screening Methods: Evolution and Refinement
2: Chemical Biology of Glycoproteins
3: Computational Tools for Chemical Biology
4: Mass Spectrometry in Chemical Biology: Evolving Applications
5: Mechanisms of Primary Energy Transduction in Biology
6: Cyclic Peptides: From Bioorganic Synthesis to Applications
7: DNA-targeting Molecules as Therapeutic Agents
8: Protein Crystallography: Challenges and Practical Solutions
9: Oxidative Folding of Proteins: Principles, Biological Regulation and Design
10: Chemical and Biological Synthesis: Enabling Approaches for Understanding Biology
11: Synthetic Glycomes
12: The Chemical Biology of Sulfur
13: The Chemical Biology of Phosphorus
14: DNA Damage, DNA Repair and Disease: Volume 1
15: DNA Damage, DNA Repair and Disease: Volume 2
16: The Discovery and Utility of Chemical Probes in Target Discovery
17: Inhibitors of Protein–Protein Interactions
18: The Chemical Biology of Nitrogen
19: RNA Polymerases as Molecular Motors: On the Road
20: Prebiotic Chemistry and Life's Origin
21: The Chemical Biology of Carbon

How to obtain future titles on publication:
A standing order plan is available for this series. A standing order will bring delivery of each new volume immediately on publication.

For further information please contact:
Book Sales Department, Royal Society of Chemistry, Thomas Graham House, Science Park, Milton Road, Cambridge, CB4 0WF, UK
Telephone: +44 (0)1223 420066, Fax: +44 (0)1223 420247,
Email: booksales@rsc.org
Visit our website at www.rsc.org/books

The Chemical Biology of Carbon

By

Christopher T. Walsh
Stanford University, USA

ROYAL SOCIETY OF CHEMISTRY

Chemical Biology No. 21

Print ISBN: 978-1-83916-802-4
PDF ISBN: 978-1-83916-950-2
EPUB ISBN: 978-1-83916-951-9
Print ISSN: 2055-1975
Electronic ISSN: 2055-1983

A catalogue record for this book is available from the British Library

The Royal Society of Chemistry is a charity, registered in England and Wales, Number 207890, and a company incorporated in England by Royal Charter (Registered No. RC000524), registered office: Burlington House, Piccadilly, London W1J 0BA, UK, Telephone: +44 (0) 20 7437 8656.

For further information see our website at www.rsc.org

Printed in the United Kingdom by CPI Group (UK) Ltd, Croydon, CR0 4YY, UK

Dedication

This work is dedicated to:

Diana
Allison
Thomas
Sean

Chemical Biology No. 21
The Chemical Biology of Carbon
By Christopher T. Walsh

Published by the Royal Society of Chemistry, www.rsc.org

Preface

The chemical biology of carbon can be recast as the organic chemistry of life. The rules of functional group reactivity, electronics, and stereochemistry that govern abiotic organic chemistry are carried over without exception to the organic chemical transformations, metabolism, that occur in living organisms. Readers with only one to two semesters of organic chemistry background will readily navigate familiar concepts to analyze the organic chemistry of life. Yet, this approach to carbon chemical biology should also be instructive to graduate and postgraduate students, medicinal chemists, organic chemists, and chemical biologists in industry and academic research settings.

There is an underlying chemical and molecular logic that organisms, including humans, employ to interconvert hundreds of thousands of organic reactants onto products. That dynamic organic chemical inventory is bent to two major metabolic purposes. One is to extract energy from dietary intake into chemically useful molecules. The second purpose is to use those molecules, a core group of thermodynamically active but kinetically stable molecules, for all the biosynthetic pathways that sustain growth, enable information storage, and transfer in biopolymers such as DNA, RNA, and proteins.

We approach the complexities of the organic chemistry of life in Section II from a few principles of molecular structure and reactivity that govern abiotic organic chemistry. A first theme, encompassing Section II, Chapters 2–5, is to analyze the major classes of organic metabolites according to their functional groups since those condition possible reactivities.

These start with the simplest functional group, alkene formation and reactivity, then proceed to C–O hetero functional groups (*e.g.* alcohols, aldehydes, acids, carbon dioxide), C–S functional groups (thiols, thioethers, thioesters, disulfides), and C–N functional groups (amines, imines, amides). Because the five elements C, H, O, N, and S make up 98% of the mass of

Chemical Biology No. 21
The Chemical Biology of Carbon
By Christopher T. Walsh

Published by the Royal Society of Chemistry, www.rsc.org

human bodies, this approach catalogs essentially all the common functional group chemistry available to carbon in biologic systems. Carbon–sulfur chemistry in particular, not usually taught in introductory organic chemistry, is uniquely placed in life's chemical firmament. So is phosphate ester and phosphate anhydride chemical biology.

A second theme is oxidation–reduction chemistry that relates most of the inventory of organic metabolite functional groups to each other. Alkane to alcohol, alcohol to aldehyde or ketone, aldehyde to acid and acid to one carbon shorter alkane and CO_2, amine to imine to amide, dithiol to disulfide are all related by net two electron transfer steps. This lens is not only useful for understanding the flow of organic substrates through metabolic pathways. It also illuminates the chemical logic for such energy-yielding pathways as glycolysis, fatty acid oxidations and the citric acid cycle. This integrated redox and functional group-centered approach similarly demystifies the chemical steps and strategies in fatty acid biosynthesis and all the many variations of isoprenoid and steroid assemblies.

A third principle that enables prediction of the reactivity of the great bulk of organic metabolites is the chemical principle of nucleophilicity (the reaction of electron rich functional groups) and its opposite, electrophilicity (the reaction behavior of electron poor functional groups). All C–C bonds made or broken *in vivo* in two electron reaction manifolds are explicable by nucleophiles capturing electrophiles.

Section III of this book works off another central premise: that nature has espoused heterocycles. Nature apparently needs rigid, constrained molecular structures in general, introducing regio- and stereocontrol for high affinity interactions with biologic targets (enzymes, receptors, RNA, DNA). Chapters 6–9 start with a chapter devoted to purines and pyrimidines, nitrogen-containing heterocyclic building blocks for RNA and DNA that may have been among the earliest heterocycles to emerge during molecular evolution. Chapter 7 summarizes major classes of natural carbacycles, from three to eight carbons, examining assembly principles. These include the assembly logic for bioconstruction of single and fused aromatic ring systems.

Chapters 8 and 9 examine two other aspects of nitrogen heterocycles. The first deals with additional metabolically central low molecular weight nitrogen metabolites including, pyrrole rings, the pyrrolidine ring of proline, the 1,3-bis nitrogen imidazole side chain of histidine and the bicyclic indole ring of tryptophan. How they form and how they react are key features of biological catalysis. Chapter 9 turns to the vitamin heterocycles noting that 9 of 13 human vitamins function as chemo-specific coenzymes to use their distinct heterocycles to enable life-sustaining human chemistry.

Among the array of nitrogen heterocycles brought to bear are the thiazolium ring of thiamin, the pyridinium ring of nicotinamides, the tricyclic isoalloxazine ring system of riboflavin, the pyridine aldehyde coenzyme form of pyridoxal-phosphate, the bicyclic pterin ring of tetrahydrofolate coenzymes, and an oxygen heterocycle, the radical scavenging cyclic enediolate ring of vitamin C. Nature has conscripted a chemical menagerie of heterocycles essential for the reactions of life where the chemistry accessible to each different heterocycle has been evolved to fit essential biologic function.

From these fundamental precepts of chemical reactivity and the inventory of organic functional groups in nature, we turn in Section IV, Chapters 10–13, to the very few types of reactions that organic molecules can undergo in living organisms. The reactions of carbon functional groups in biologic systems are essentially limited to four mechanistic manifolds.

One can exemplify the possibilities for a substrate which undergoes a C–H bond cleavage in the active site of an enzyme catalyst. The C–H bond can only break heterolytically or homolytically. Two heterolytic possibilities exist: the electrons in the bond, as it breaks, stay with the carbon (an incipient carbanion) as the hydrogen departs as a proton (H^+). At the other extreme, the two electrons depart with the hydrogen as a hydride ion (H^-), leaving an incipient electron deficient carbocation atom. There is only one homolytic (equal-breaking) C–H bond cleavage outcome. One electron remains with carbon as an incipient carbon radical while one electron departs with the hydrogen as a hydrogen atom ($H^\bullet$).

The two heterolytic and one homolytic routes then generate three types of carbon-centered transition states, or, if they have finite lifetimes beyond the 10^{-13} s^{-1} half life of a molecular vibration, yield carbon-centered unstable intermediates. Proton transfers yield carbanions. Hydride transfers yield carbocations. Hydrogen atom transfers yield carbon radicals. These are the three possible carbon-centered intermediates in reactions, whether in abiotic organic chemistry or the organic chemistry of life. (Carbenoid pathways are not generally operant.)

The fourth reaction mode possibility is the lack of any reaction intermediates, with a smooth energy surface between reactant(s) and product(s). In abiotic chemistry these are concerted, pericyclic reactions, collated in the 1960s by the Woodward Hoffmann rules that explained allowed reactions pathway based on orbital symmetry principles. In the past two decades a variety of biological pericyclic reactions have been discovered and validated, bringing into view diverse families of pericyclase enzymes that catalyze four types of possible cyclic reaction manifolds in carbon chemical biology.

Section V, the concluding section, consists of Chapters 14–16. They examine three distinct iconic organic metabolites, the polyhydroxy-aldehyde glucose, the linear hexa-olefinic hydrocarbon squalene, and the amino-methylpyrrole nitrogen heterocycle porphobilinogen. Each metabolite is examined in the context of metabolic pathways where key chemical steps utilize the specific functional groups available to those substrates.

Glucose, as equilibrating linear aldehyde with its predominant cyclic hemiacetal forms, reveals the diversity of carbonyl and redox chemistry in three pathways: glycolysis, the pentose-phosphate pathway, and polysaccharide chain elongations (Chapter 14).

Squalene is a key isoprenoid C_{30} hydrocarbon built from head to tail five carbon alkylations and then a head to head self-alkylation of two C_{15} units, before undergoing directed cyclization cascades to the tetra- and pentacyclic frameworks of sterols (Chapter 15).

Finally, Chapter 16 focuses on the pyrrolic nitrogen heterocyclic porphobilinogen. Its chemical capacity to serve as precursor to both carbon nucleophile and carbon electrophile is the underlying chemical tenet to tetrapyrrole macrocycle biosynthetic logic. Prophobilinogen's pyrrole ring-based chemical versatility to form C–C bonds in linear tetrapyrroles and then undergo directed cyclization to the tetrapyrrole macrocycle framework is the key assembly principle for the pigments of life.

These few organic chemical principles anchor the organic chemistry of life. They deconvolute the complexities of metabolite dynamics, illuminate metabolic pathways that intersect at certain nodal points, and address issues of carbon economy. The principles explain control of reactivity of such promiscuous molecules as NH_3, H_2S, $CH_2=O$, validate the value of a few core C=C, C–O, C–S, and C–N functional groups among the myriad organic metabolites, and detail the life giving chemical features of heterocyclic metabolites and coenzymes. This approach to the chemical biology of carbon uncovers and illustrates the underlying chemical logic for major pathways that utilize the rules and constraints of abiotic organic chemistry in the organic chemistry of life. The chemical biology of carbon subsumes much of the chemical biology of hydrogen and oxygen, and integrates and builds on the chemical biology of sulfur, nitrogen, and phosphorus. Together, these six elements account for 99% of the mass of the human body.

Christopher T. Walsh

Contents

Chemical Biology No. 21
The Chemical Biology of Carbon
By Christopher T. Walsh

Published by the Royal Society of Chemistry, www.rsc.org

Section I

Section I, comprising Chapter 1, is the shortest of the five sections in this monograph on the chemical biology of carbon. The operating premise is that carbon chemical biology is nothing more or less than the organic chemistry of life. To that end Chapter 1 starts with the persistent tetravalency of carbon and the proposal that the inventory of tens of thousands of organic molecules (metabolites) can be parsed into a limited set of functional groups and straightforward chemical routes for their interconversion.

The four key general categories of functional groups for *in vivo* chemistry are alkenes (olefins), C–O, C–S, and C–N functional groups. Within the C–O, C–S, and C–N groups many are interconnected by two electron redox steps. A second general tenet of this book is that carbacycles and heterocycles (oxygen-, nitrogen-, and sulfur-heterocycles), their formation, reactivity, and accessible chemistry, are key determinants that enable much of the primary metabolic functional group transformations.

For the predominant two electron reaction pathways for metabolite interconversions, the concept of carbon nucleophiles as electron rich species that attack electron poor electrophilic carbons puts a logical framework on *in vivo* organic chemistry.

These above three convergent approaches to the kinds of organic functional groups available in carbon chemical biology set the stage for a third major precept: that there are only four reaction types that make or break C–H and C–C bonds in both abiotic and biotic organic chemistry. These are carbanion-driven reactions, carbocation-driven reactions, carbon-radical driven-reactions and pericyclic reactions (concerted *via* cyclic transition states).

Our approach to carbon chemical biology is then fourfold: (1) available carbon-based functional groups; (2) the most common carabacyles and heterocycles (including 10 of 13 human vitamins); (3) nucleophilic carbons attack electrophilic carbons to make new C–C bonds; and (4) there are few mechanistic routes to make and break C–H and C–C bonds. These four reactivity parameters explicitly encompass the underlying logic of metabolic pathways that constitute the organic chemistry of life.

CHAPTER 1

Introduction to Carbon Chemical Biology

1.1 The Chemical Biology of Carbon Is Nothing More or Less Than the Organic Chemistry of Life

The chemistry of carbon, organic chemistry, encompasses the structure and reactivity of all molecules that contain one or more carbon atoms. The word organic can be defined as "relating to living matter". In a real sense then, organic chemistry has its roots in the carbon chemistry of living organisms. Over the past three centuries the field of organic chemistry has expanded dramatically to include carbon chemistry far outside of the molecules of life. Indeed, it has been estimated that some 10 million carbon compounds can be computed,[1] taking the subject 100 fold and more beyond the inventory of molecules found in living organisms. There are more carbon-containing molecular scaffolds than for any other set of elements, attesting to the dominant place of carbon in the discipline of chemistry.

The chemical biology of carbon reaches back to the origins of organic chemistry[2] as the chemistry of living organisms. Carbon is the second most abundant element in humans, after oxygen and before hydrogen and nitrogen. Seven elements, H, C, N, O, S, P, and Ca, make up 99% of the biomass of organisms. While abiotic organic chemistry courses take up C–O functional groups early, C–N, C–S, and $C–OPO_3^{2-}$ functional groups are central to carbon chemical biology. Their biosynthetic logic and chemical reactivity in metabolism are key attributes of carbon chemical biology. Abiotic organic chemistry rarely delves into C–N, C–S, and $C–OPO_3^{2-}$ functional group chemistry in general courses.

We define carbon chemical biology as the organic chemistry of life, comprising all the reactions of organic molecules in the cells and tissues of living

Chemical Biology No. 21
The Chemical Biology of Carbon
By Christopher T. Walsh

Published by the Royal Society of Chemistry, www.rsc.org

organisms. We explore the types of carbon-based functional groups that are commonly present and account for the interconversions of reactants (substrates) into products by the catalytic action of thousands of enzymes *in vivo*. The organic molecules of life are often categorized as metabolites, molecules that partake in metabolism, comprised of the plethora of metabolic pathways that operate in intracellular and extracellular chemistry of life.

Metabolites are typically small organic molecules. Some metabolites are obtained from diet and transformed into other metabolite frameworks in hundreds of biosynthetic and degradative pathways in cells and tissues. Most such molecules have one or more carbon atoms with one or more attached hydrogens in C–H bonds. The number of known metabolites has increased markedly over the period from 2009, when the first release of the Human Metabolite Data Base (HMDB) was compiled, to the 2018 release #4 that lists 114 000 identified metabolites.[3] The increase of up to 100 fold over a decade in some metabolite arenas is a testament to improved sensitivity in analytical detection and separation methods that have enabled the rise of metabolomics science.[4]

The rules of organic chemistry, accrued and promulgated over many decades, apply to all the carbon-containing molecules, metabolites, in all living organisms.[5] In turn, life cannot be understood in chemical terms without parsing the structure and reactivity of carbon atoms in the constituent metabolites. The two most relevant attributes are the tetravalent nature of carbon atoms and the array of functional groups that arise from the combination of carbon atoms with hydrogen, nitrogen, oxygen, sulfur, and phosphorus in biologic metabolites.

The guiding precepts of organic chemistry for structure and reactivity of carbon-based functional groups that apply (without exception) to the molecules of life are prerequisites and predictors of the molecular logic of metabolic pathways. These pathways and their underlying chemical logic encompass both the breakdown of ingested foodstuffs and biosynthesis, including biopolymers for informational, structural, and catalytic purposes.

The definition we use in this book is that the chemical biology of carbon is nothing more or less than the organic chemistry of life. The level of organic chemical knowledge assumed in this volume is a one or two semester introductory organic chemistry course. Yet, the level of presentation will be useful to all practitioners from undergraduate, graduate, and postgraduate students to medicinal chemists, industrial and academic researchers, and professors.

The discipline of biochemistry, over the full arc of the 20th century, would be the predominant historical pillar for much of the terminology of metabolites and metabolism, providing a logical framework of biosynthetic and degradative metabolic pathways.[6] The historical subset of chemically oriented biochemists have been precursors to the current broad cohort of chemical biologists who can look more deeply into the molecular and chemical features of biology in the postgenomic era. In sum, this volume parses implementation of the reactivity precepts of organic chemistry in the carbon-containing functional groups that comprise the molecules of life, many of them heterocycles.[7]

1.2 Features of Carbon the Element

Many authors have stated that carbon scaffolds form the basis of all life on earth, perhaps the primary historical origin of the term "organic chemistry". Carbon, the element has been known for some six thousand years of human history, back to the times of the Sumerian and Egyptian empires, in the form of soot (carbon black) and charcoal. Carbon on planet earth, summed over the core, mantle and crust, is estimated to comprise some 4 trillion tonnes, with about 0.025% of the total in the upper crust[7] (and presumably available to life forms throughout evolution). At the level of an individual adult human, the 18.5% average abundance of carbon in a 70 kg adult amounts to ~24 pounds of carbon in the molecules of life. Carbon the element is proposed to arise by triple fusion events from helium atoms in the blazing interiors of white dwarf stars.

The atomic number of carbon, a first row element of the periodic table sandwiched between boron and nitrogen, is six. It has six electrons. An atomic orbital diagram for filling of those six electrons, shown in Figure 1.1, is in the order $1s^22s^22p^2$. In addition to the filled 1s shell, there are four valence electrons ($2s^22p^2$). Promotion of the 2s electrons up to the 2p shell energy gives four valence electrons to be paired up. Carbon is therefore always tetravalent,[5] needing another four electrons to attain a filled $2p^6$ shell. Hybridization of the atomic orbitals of carbon to molecular orbitals results in four sp^3 orbitals of equal energy. The tetravalency of carbon is the fundamental attribute of carbon-based bonding and chemistry. Monomeric atomic carbon is an evanescent species, not found in biologic systems. Stable, accessible carbon compounds always have four bonds to substituents.

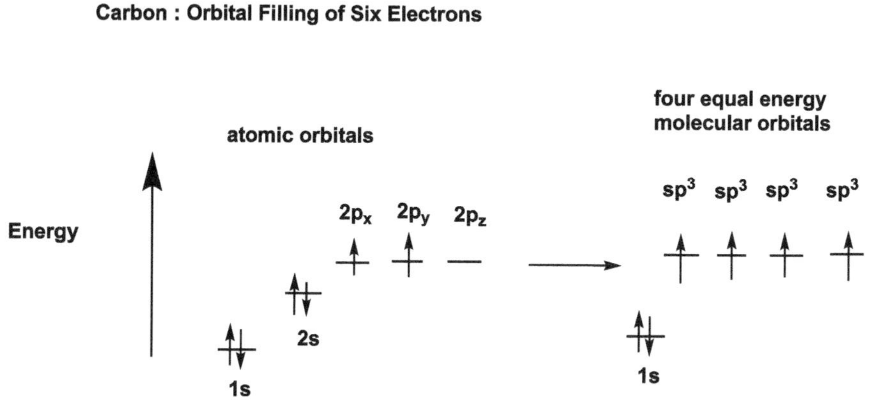

Figure 1.1 Carbon orbital filling diagram for six electrons: atomic orbitals to molecular orbitals.

1.2.1 Common Carbon Allotropes and Isotopes

Although there are many distinct geometrical arrangements (allotropes) of all carbon arrays,[8] three are most readily observed (Figure 1.2). The most stable form thermodynamically is graphite in which six carbon atoms connect to each other in planar hexagonal arrays. The layers stack vertically over and below each other, allowing the layers to slide past one another. For that reason, graphite is soft. It is also opaque and black in color. Graphene arrays represent a single layer peeled away from layers of graphite, with conductance and other desirable properties still under exploration.

A second, highly valued carbon allotrope is diamond, occupying the other end of the hardness scale for materials. Both hard and transparent, diamonds crystallize in the indicated face-centered tetrahedra in cubic unit cells (Figure 1.2). Much more recently appreciated are the fullerenes (sometimes termed "Bucky Balls"), characterized in the 1980s with some representing the geodesic domes of the architect Buckminster Fuller.[8] Free carbon atoms are not found in biologic systems, only tetravalent compounds of carbon with hydrogen, oxygen, sulfur, and nitrogen (and phosphorus) bonds.

Two radioactive isotopes have different biomedical research utilities: ^{11}C has a half-life of only 20 minutes, while ^{14}C has a half-life of 5200 years for decay. Carbon chemical biology is essentially the universe of the carbon 12 isotope. The ^{11}C and ^{14}C isotopomers of carbon are not generally involved in the organic chemistry of life. For those biotic roles, carbon always combines with some other element, hydrogen, nitrogen, oxygen or sulfur.

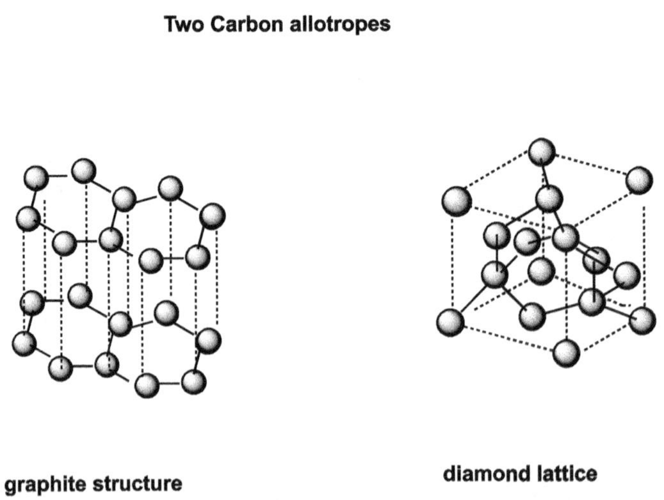

Figure 1.2 Allotropes of carbon. Graphite *vs.* diamond lattices.

1.3 Features of Carbon-based Common Functional Groups in Chemical Biology

While C–C bonds are central to much of carbon chemical biology, and biologic routes to C–C bond formation and cleavage are explored in detail in this volume, C–H, C–O, C–N, C–S, and even C–P bonds, connect carbon to the other five elements that constitute 98% of human body mass. These carbon-based functional groups are determinative for life to proceed. Indeed, organic chemistry texts, after introducing orbital diagrams and some stereochemical definitions, often launch into functional group chemistry to begin the journey through reactivity patterns of organic compounds.[5,9]

In carbon chemical biology, in addition to C–C bonds, the formation and reactivity of C–H, C–O, C–N, and C–S bonds comprise most of the carbon metabolic universe.

1.3.1 Hydrocarbons: C–H Bonds in Alkanes

The simplest forms of carbon compounds that occur in biological systems, current and ancient, are hydrocarbons. The simplest hydrocarbon of all is methane, CH_4, with four C–H bonds, at angles of 109.5°. This is a classic sp^3-hybridized tetrahedral carbon atom. Methane is a major component of natural gas deposits and is generated by one carbon reductive metabolism from carbon dioxide, by fastidiously anaerobic archaeal strains (microbes) known as methanogens.[10] Likewise, in mixed microbial consortia, methane oxidizers take methane in the opposite direction, oxidizing it to methanol.[11]

A second component of natural gas is the two carbon alkane ethane (CH_3–CH_3). Two sp^3 tetrahedral CH_3 groups are joined by a C–C bond. In oil deposits,[12] representing buried layers of long-ago organic metabolites largely from anaerobic microbial metabolic pathways,[11,13] four gases may be present as volatile alkanes, along with the much longer hydrocarbon chains that have higher melting points. In addition to methane and ethane one may find the C_3 alkane propane and the C_4 alkane butane. If these are linear hydrocarbons they are known as *n*-alkanes, *n*-propane and *n*-butane (Figure 1.3). The internal carbons (C_2 in propane and C_2, C_3 in butane) are CH_2 groups (methylene groups), each connected to two other carbon atoms.

Oil deposits around the world are typically mixtures of alkanes, residual reduced metabolites from ancient microbial metabolism buried and compressed by accumulating layers over the geologic eons. The sets of constituent alkanes in crude oils are typically fractionated by melting point differences, where short chain alkanes volatilize starting at ~30 °C all the way up to a bitumen residue that remains a solid mass at temperatures of 500 °C.

The methane to butane components come off initially as gases, followed by a C_5–C_{12} petrol fraction used largely for motor fuel. The kerosene fraction boils between 180–300 °C and contains C_{16}–C_{18} hydrocarbons, of use in jet fuels. At the high temperature end, one finds paraffin waxes, solids at room

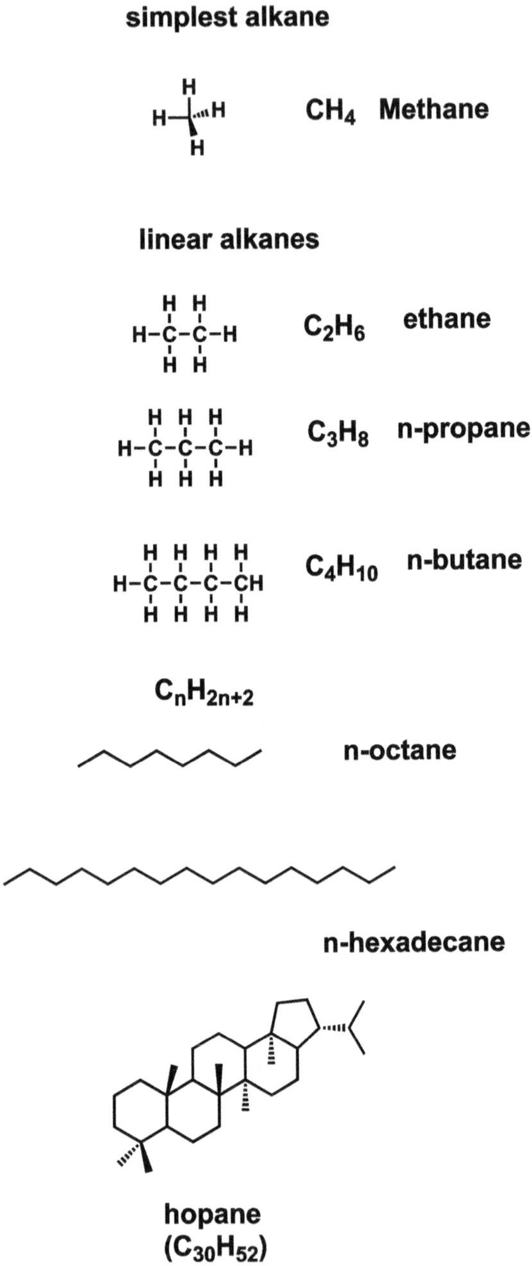

Figure 1.3 Some alkanes found in organismal metabolism.

temperature with long chain hydrocarbons of typical C_nH_{2n+2} composition (candles and waxes). The final, nonvolatile residue of bitumen contains the asphalt mixtures used in roof tars and road construction.

Also, among the alkane hydrocarbon components of crude oils are branched chain hydrocarbons built up of reduced isoprene units such as the linear C_{20} phytanes and the tetracyclic C_{30} bacterial hopane metabolites, telltale remnant molecules of isoprenoid and steroid metabolism (Figure 1.4). We will take up both isoprenoid and steroid biosynthetic metabolic logic in Chapters 2 and 15, detailing a few distinct patterns of C–C bond formation mechanisms. For now, we note that the hopanes introduce the concept of cyclic alkane scaffolds, with three fused six-membered cyclohexane rings and one fused cyclopentane ring in the characteristic ABCD ring scaffold of steroid metabolites. All the carbons in the linear, methyl branched, and cyclic hopanes are sp^3 tetrahedral carbon centers,

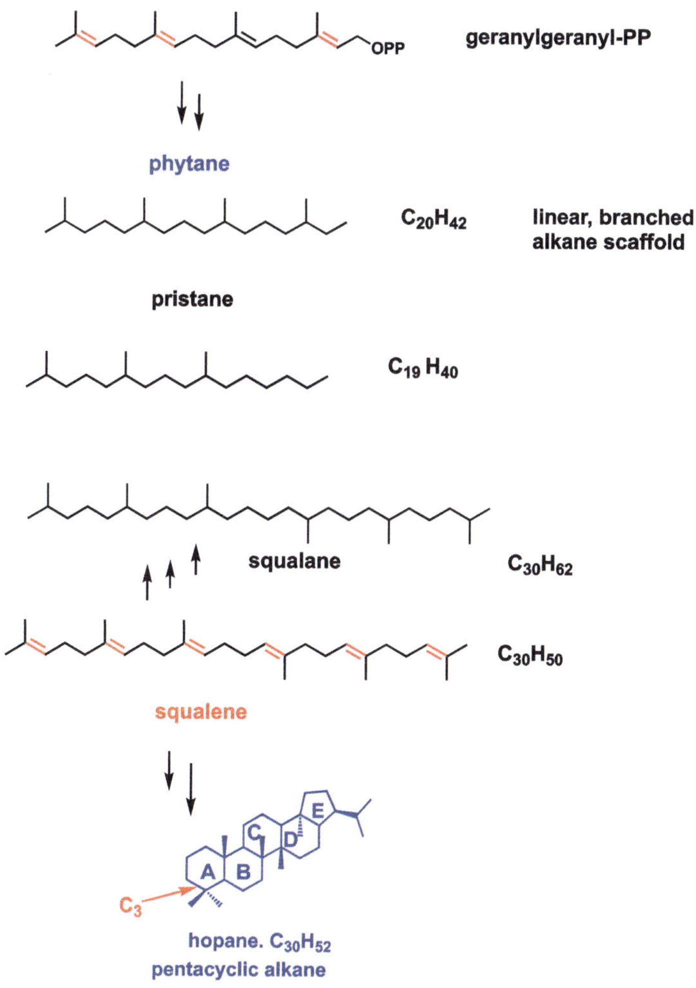

Figure 1.4 Precursors to phytane, pristane and squalane branched hydrocarbons. Hopane is a pentacyclic bacterial metabolite.

even when a carbon atom, such as C_3 in the A ring of hopanes has no C–H bonds, instead connected to four adjacent carbon centers.

Hydrocarbons as central components of fossil fuels are most probably chemically reduced remnants of organic metabolites, reduced functional groups by anaerobic organisms over geologic eons. Without functional groups such simple hydrocarbons are utilizable as growth sources by microbes but not typically by higher, multicellular organisms. Microbes that can use alkanes as energy sources typically carry out oxidations to introduce organic functional groups that allow subsequent breakdown into one to three carbon fragments such as the simple carboxylic acid anions formate, acetate, and propionate.

Alkane degradation is therefore almost exclusively the province of microbial metabolism, with bacteria and archaeal species as leading players, from oxidation of methane to the longer alkane chains. Many bacteria grow anaerobically on hydrocarbons. These include C_{13}–C_{18} alkane hydrocarbon degradation (ultimately to molecules of CO_2) by anaerobic sulfate-reducing bacteria.[14,15] The electrons removed from the alkanes, *e.g.* the C_{16} hexadecane, are funneled into sulfate as it undergoes eight electron reduction to H_2S (Figure 1.5A). Concomitantly, a terminal methyl group undergoes eight electron oxidation to CO_2 as the C_1–C_2 bond in the C16 hydrocarbon is cleaved to form the one carbon shorter C15 hydrocarbon product.

Among the hydrocarbons in contemporary organisms are two metabolically central polyolefins: C30 linear hexaene squalene in animals and the linear C40 nonaene phytoene in plants. Squalene is a precursor in bacteria to the tetracyclic hopene (Figure 1.5B). Squalene and phytoene have methyl branches, spaced every four carbons, indicating they arise by alkylative extension of five carbon isoprenoid building blocks (see Chapter 2). Downstream metabolism of these long chain branched hydrocarbons typically involves oxidative conversion to the arrays of organic functional groups noted below and in subsequent chapters.

1.3.2 Alcohols from Alkanes

In addition to anaerobic bacterial metabolism on hydrocarbons such as hexadecane noted above, obligate aerobic bacteria such as *Pseudomonas oliverans* use O_2 as cosubstrate *via* iron-based monooxygenases (Chapter 12) to attack the quite unreactive carbon atoms in substrates such as *n*-octane.[16] The enzymatic products are monooxygenated alkanes (alcohols), either at the terminal methyls C_1 or C_8 to generate 1-octanol, or at any of the internal methylene carbons C_2–C_7 (sub-terminal = secondary hydroxylation) (Figure 1.6).

The introduction of an oxygen atom has created an alcohol, one of a set of possible chemical functional groups in metabolism. As depicted in Figure 1.6, subsequent oxidative metabolic processing of 1-octanol will go through the aldehyde (octanal), and then to the acid octanoic acid, as a cascade of carbon–oxygen functional groups, each oxidized by two electrons compared to its precursor. We examine this oxidation (reduction in the

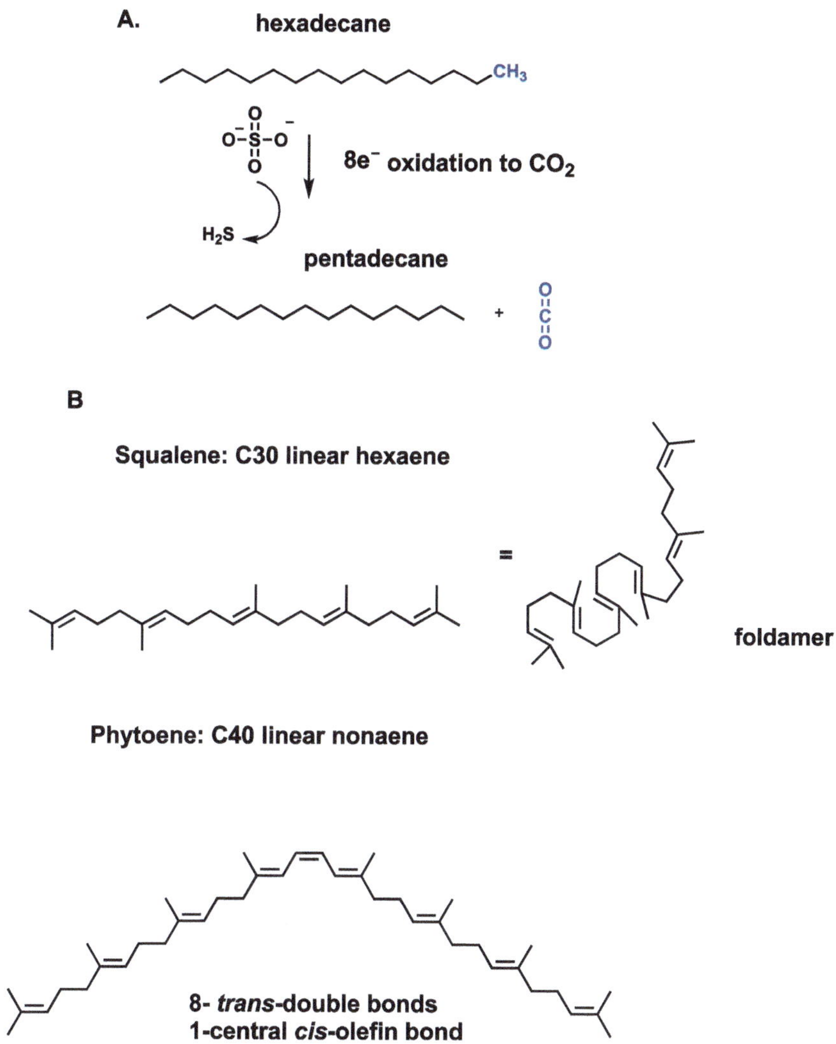

Figure 1.5 (A) Metabolic oxidation of 16 carbon hexadecane by sulfate-reducing bacteria to the C_{15} pentadecane and CO_2. (B) Squalene is a C30 hydrocarbon with six alkene groups. Phytoene is a C40 nonaene plant precursor of carotenes.

back direction) strategy in the context of methane to CO_2 in Section 1.5 below. When the initial alcohol metabolite is at an internal carbon, for example 4-octanol, subsequent metabolism involves two electron oxidation to a carbonyl group (4-octanone) that is now definitionally a ketone, not an aldehyde) (Figure 1.6). To get to a carboxylic acid downstream metabolite, the bacteria introduce another O_2-derived oxygen atom to produce the internal oxoester (also known as a lactone) by action of an enzyme that carries

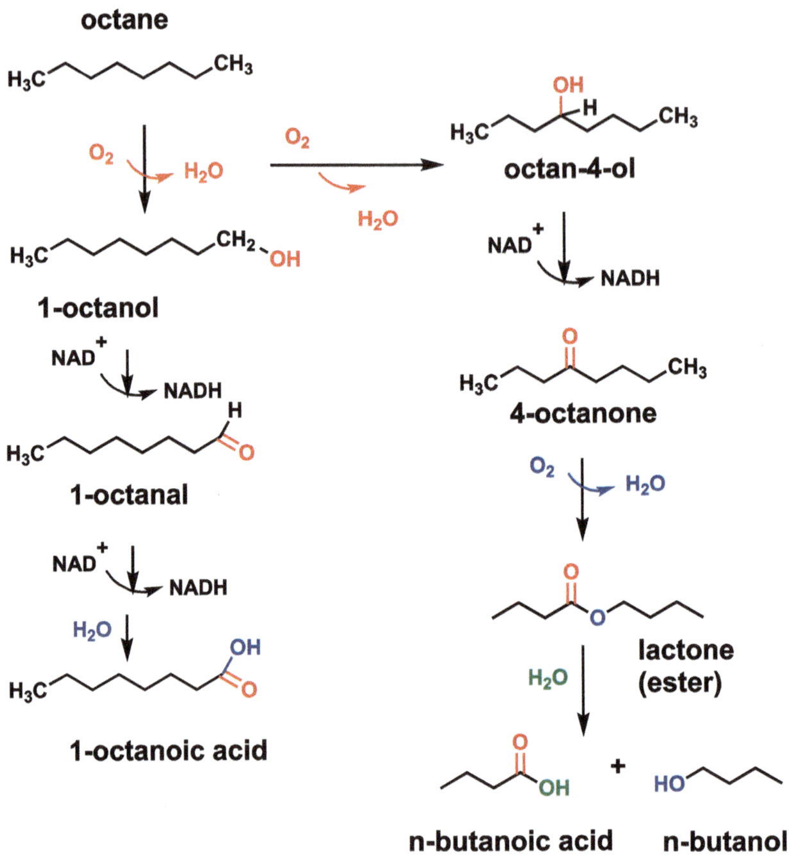

Figure 1.6 Schematic for oxygenative bacterial metabolism of octane by either primary carbon or secondary carbon oxygenation to aldehyde or ketone. The 4-octanone route involves a Baeyer–Villiger type oxygen insertion to form the lactone.

out Baeyer–Villiger type chemistry. Catalyzed hydrolysis of the ester yields two fragments, the butanoic acid and butanol metabolites, respectively.

Thus, aerobic bacteria run cascades of both oxygenative (incorporation of one atom from O_2) and oxidative reactions (removal of pairs of electrons but formally without the participation of O_2, indicating some other partner molecule is accepting the electrons directly). The general logic is to use enzymes to functionalize inert hydrocarbon scaffolds by introduction of a bevy of carbon–oxygen functional groups, elaborated below.

1.3.3 Hydrocarbons: Alkenes and Alkynes as Functional Groups

The biologic alkanes are viewed as highly inert molecules. The sp^3 carbon atoms with their C–H bonds at CH_3 (methyl), CH_2 (methylene), C–H (methine at branchpoints) carbons are not readily reactive, despite the metabolic

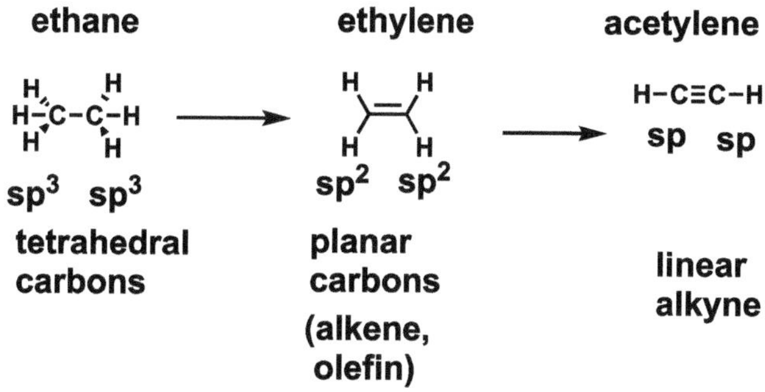

Figure 1.7 Alkane to alkene to alkyne: ethane to ethylene to acetylene.

strategies outlined above. On the other hand, desaturation of paired methylene groups, CH_2–CH_2, to CH=CH products create an alkene, simplest rendition as ethylene. Further desaturation of the alkene creates an alkyne linkage in acetylene (Figure 1.7). Compared to alkanes, the double bond in alkanes and the triple bond in alkynes are reactive functional groups. Alkenes are also known as olefins and we use the terms interchangeably.

In the olefin = alkene connectivity both carbons are now sp^2 hybridized, with trigonal not tetrahedral geometry. The C=C double bond has both a σ and π bond component. We will note that the π electrons can participate in a number of C–C bond-forming enzyme-catalyzed reactions. Analogously, the C–C triple bond in acetylene (and generically in other alkyne structures) has one σ and two π bonds. The two π bonds are in orthogonal planes.

Both the olefin/alkene and the alkyne groups are termed (potentially) reactive functional groups in contrast to the sp^3 hybridized alkane linkages. These two functional groups are different from the C–O, C–N, and C–S functional groups, polarized with electrons drawn to the heteroatoms, in that the alkene and alkyne groups involve only C–H bonded scaffold elements. The alkenes may qualify as the simplest functional group: we examine their biologic formation and reactivity in Chapter 2.

We will see many metabolic examples of generation and utilization of alkene functional groups, *e.g.* in fatty acid and isoprenoid metabolism. The alkene group in the biologic isopentenyl building blocks, Δ^2-isopentenyl diphosphate and Δ^3-isopentenyl diphosphate, is the key functional group that shows nucleophilic behavior of the π electrons of the alkene that captures an accessible allylic cation in the Δ^2-prenyl partner molecule for tens of thousands of C–C bond forming reactions. The squalene and phytoene structures in Figure 1.6 have six and nine double bonds, respectively.

Alkene double bonds can also participate as electrophiles, the electron deficient components, for example in conjugate addition reactions by nucleophiles (Figure 1.8A). A typical configuration for alkenes in biological metabolites is to be in conjugation with carbonyl groups, as in the thioester 2-butenyl *S*-CoA, or the dicarboxylate fumarate. In those cases the double bond acts as an electrophile for enzyme-mediated water addition

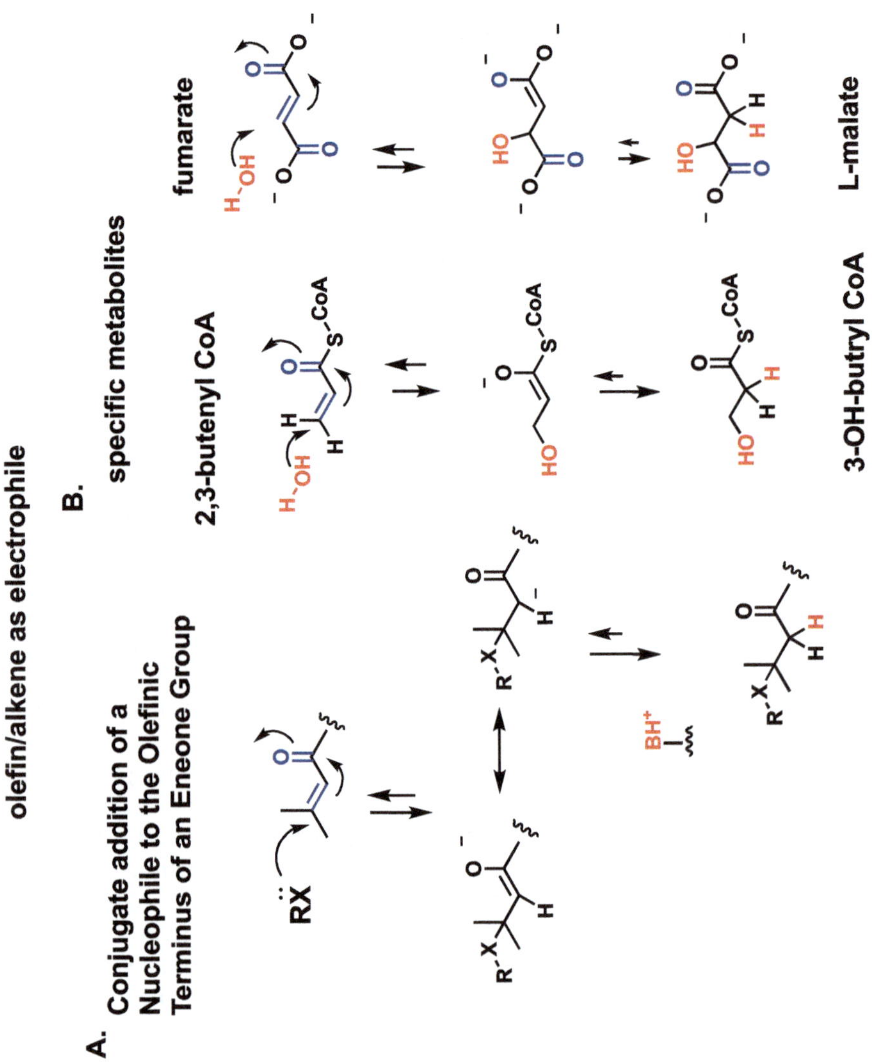

Figure 1.8 Conjugate addition of nucleophiles to conjugated olefins.

reactions (Figure 1.8B). By contrast, alkyne metabolism is very rare in biologic systems,[17] although the alkyne group as a probe in click chemistry applied to biologic systems has been a widely useful application.[18]

1.4 Common Functional Groups of Carbon Involving Oxygen, Nitrogen and/or Sulfur Atoms

In organic chemistry the concept of functional groups is one of the broad organizing principles that parse the observed reactivities of different organic and organometallic functionalities based on the behavior of a small number of molecular arrays that show up over and over in myriad possible reactions. In addition to the two C–H-only alkene and alkyne functional groups, there are three other sets of common carbon-centered functional groups that account for the great bulk of chemical transformations in the major metabolic pathways of life.

Figure 1.9 collects such common functional groups, according to C–O, C–N, and C–S chemical bond types. We have noted in passing above that the common aerobic strategy for microbes dealing with alkane hydrocarbons in their immediate neighborhoods is oxygenative.[19] This array of functional groups are the subjects of Chapters 3–5.

We have seen alcohol, aldehyde/ketone, and acid functional groups in the context of metabolism of the octane scaffold. These represent the three successive two electron oxidation states during traverse from a CH_3 group to a CH_2OH, to a CHO to a COOH group, of which more in Section 1.5 below on redox states.

Analogously, in Figure 1.9 the formal replacement of a C–H bond by a C–N bond, where nitrogen is trivalent, turns a methyl group to a primary amine. Further two electron oxidation of the amine generates the imine (either aldimine (from aldehyde) or ketimine (from ketone)) linkage that is central to the multiple strategies of amino acid metabolism. Yet, a second two electron oxidation (now of an aldimine) formally yields the amide derivative of a carboxylic acid. As in the oxygen-based functional groups, all three C–N bond functional groups are pervasive in primary metabolism[20] and provide low energy entries into distinct types of reactivities. Note that the amide functional group has two types of heteroatoms (O and N) attached to the same carbon.

Correspondingly, carbon–sulfur bonds in turn create functional groups[21] that lie at the core of many transformations in carbon chemical biology (Figure 1.9). Formal replacement of a C–H bond by a C–S bond, where sulfur is divalent, yields a C–SH bond, a thiol group. Because the pK_a of the thiol groups in biology are between pH 8 and 9 (pK_a = pH where a species is half ionized, half neutral), there is substantial thiolate ion chemistry available to organisms.

Thioethers have a C–S–C linkage and are central in metabolism of the proteinogenic building block L-methionine. They are analogous to the C–O–C connectivity in oxoethers. Oxidation of that thioether to the corresponding sulfonium cation in *S*-adenosyl methionine provides the key functional

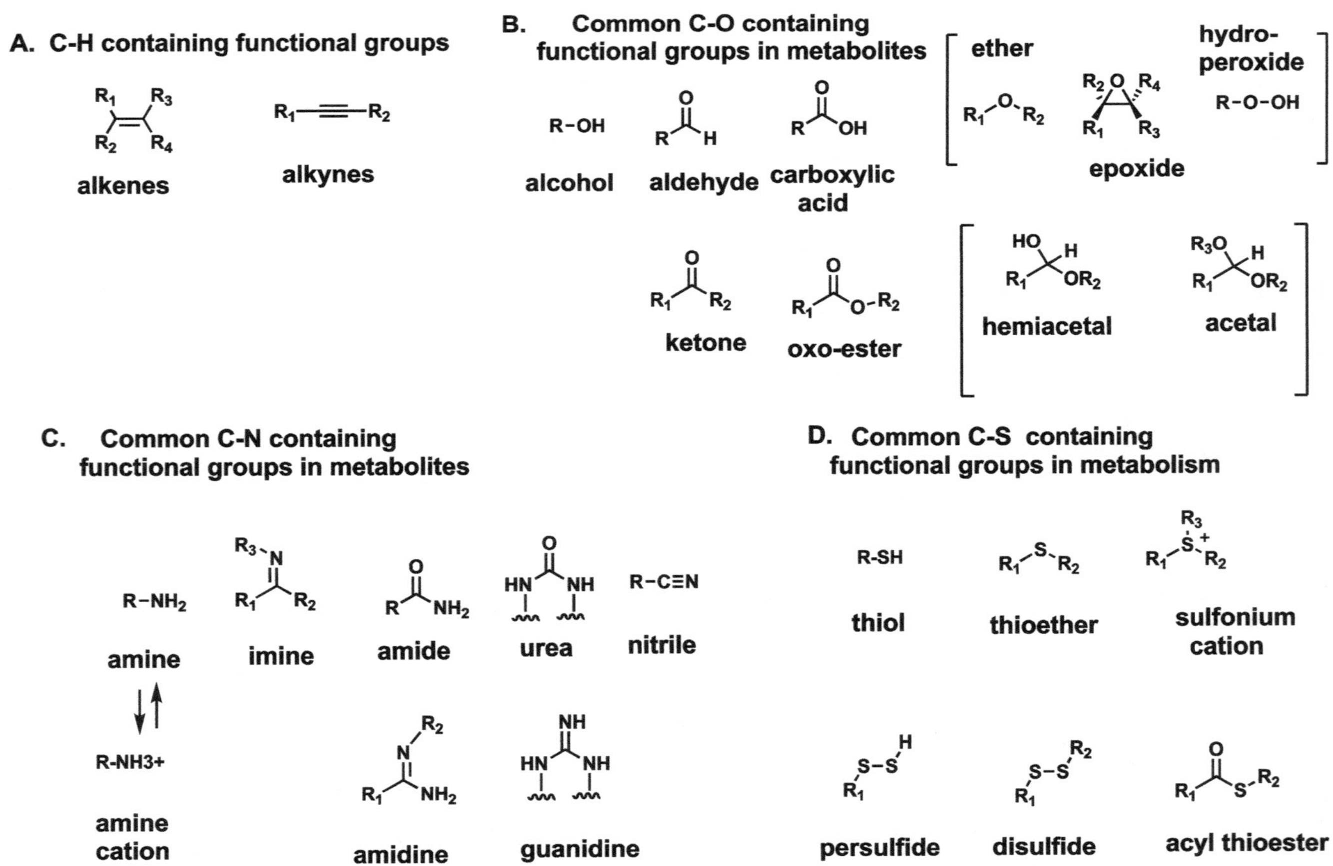

Figure 1.9 (A)–(D) Some common carbon-based functional groups in biologic systems. C–H, C–O, C–N, and C–S functional groups.

group for tens to hundreds of millions of methyl group transfers in every cell cycle. A related C–S oxidation state occurs in the C–S–S–C disulfide functional groups that crosslink and stabilize protein conformations in extracellular spaces of the body. Thioaldehydes are rare in primary metabolism, whereas thioesters (at the acyl group oxidation state) are central to hundreds of acyl transfer reactions in organisms. We examine the parallel and divergent properties of alcohol, amine, and thiol functional groups in major metabolic classes in Chapters 3–5. The central involvement of organosulfur functional group chemistry in carbon chemical biology is a feature distinct from abiotic organic chemistry syllabuses.

1.4.1 Phosphate Ester C–OPO_3^{2-} Functional Group Chemical Biology

A particularly notable variation of the alcohol functional group in organismal metabolism is enzymatic conversion to phosphate esters by phosphoryl group transfers[22] (Figure 1.10). The formal combination is of an alcohol with a phosphoric acid, generating the mixed ester. In a comparable combination of an alcohol group with a carboxylic acid, the result would be a classic (oxo) ester.

We take up the phosphate ester strategy and the multiple consequences in Chapters 2, 3, and 14. From trapping of glucose inside cells as impermeant

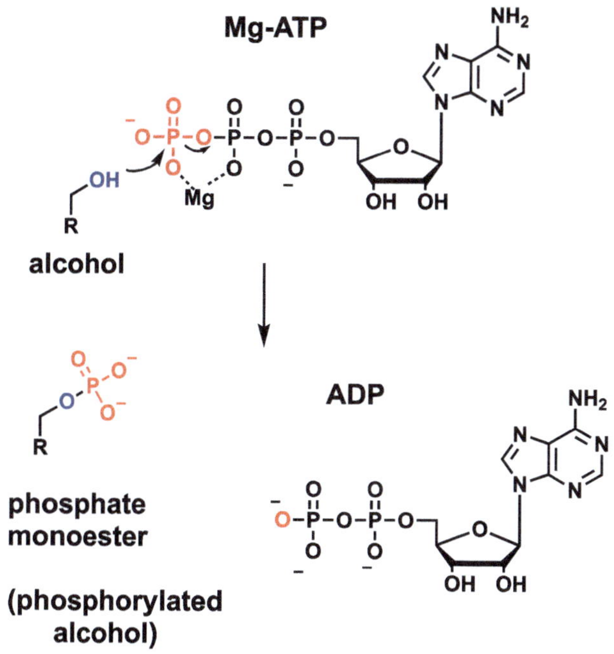

Figure 1.10 Cleavage of ATP to ADP by alcohol attack on Pγ in the ATP side chain yields phosphorylated alcohol and ADP.

glucose-6-phosphate to an estimated 270 000 protein side chain phosphorylations in signal transduction biology, alcohol group-targeted enzymatic phosphorylations are key chemical derivatizations of neutral alcohol groups to tetrahedral phosphate ester dianions. This broad use of phosphate in chemical biology is in contrast to abiotic organic chemistry where phosphate esters are not mainstream activating/leaving groups.

1.5 Two to Four to Six to Eight Electron Traverses of Carbon Functional Groups

The sets of functional groups listed in Figure 1.9 offer carbons at different hybridization states: alkynes have sp-hybridized carbons, alkenes have sp^2 and alkanes have sp^3 carbons. Those carbons are thereby at different formal oxidation states for chemical reactions, both abiotically and in cellular systems. As illustrated for the two simplest alkanes, methane and ethane, in Figure 1.11A, the gap from the alkane (most reduced) to the fully oxidized CO_2 involves an eight electron traverse, with stable intermediates separated by the removal of two electrons. Six electrons are removed as the system proceeds through three of the common C–O functional groups: from alkane to alcohol, to aldehyde, and then to carboxylic acid. Likewise, the conversion of formate to carbon dioxide is a two electron oxidation at carbon (the enzyme formate dehydrogenases generates CO_2 as the C–H bond is broken with the H moving with two electrons transferred as a hydride equivalent (H^-) to cosubstrate oxidized nicotinamide adenine dinucleotide (NAD^+; see Chapters 2, 3, and 9 among others).

An analogous eight electron traverse, redox chemistry, is shown for the two carbon alkane ethane on the way to ethanol, acetaldehyde, and acetic acid, six electrons oxidized from ethane. The last pair of electrons is removed as the carboxylate carbon is oxidized to CO_2. This assignment of redox state for C–O, C–N, and C–S metabolites holds generally for more complex scaffolds.

There are several schemes for determining the oxidation state of reactants and products. One simple approach suitable for chemical biology is: in any reaction sequence carbons with more hydrogens are in a more reduced state compared to carbons with fewer hydrogen substituents. The formal oxidation state of $H = +1$ (in most cases, $H^+ =$ protons, although we shall note that the most abundant redox cofactors in cellular metabolism, nicotinamide adenine dinucleotides, carry out apparent hydride transfers with hydrogen at the $H = -1$ oxidation state).

In terms of electronegativity scales, carbon is more electronegative than hydrogen. Thus, in methane (CH_4) the formal oxidation state of each H is $+1$ and that of the single carbon is -4. Tetravalent carbon cannot have more than four hydrogen substituents so this alkane is the most reduced form of one carbon molecule and is the end product of reductive methanogenic metabolism by anaerobic archaeal strains that make a living by dumping electrons into dissolved CO_2. We noted above in Figure 1.11A that the conversion of alkane to alcohol involves removal of two electrons from the alkane. As shown in Figure 1.11B, the formal oxidation state of the carbon in

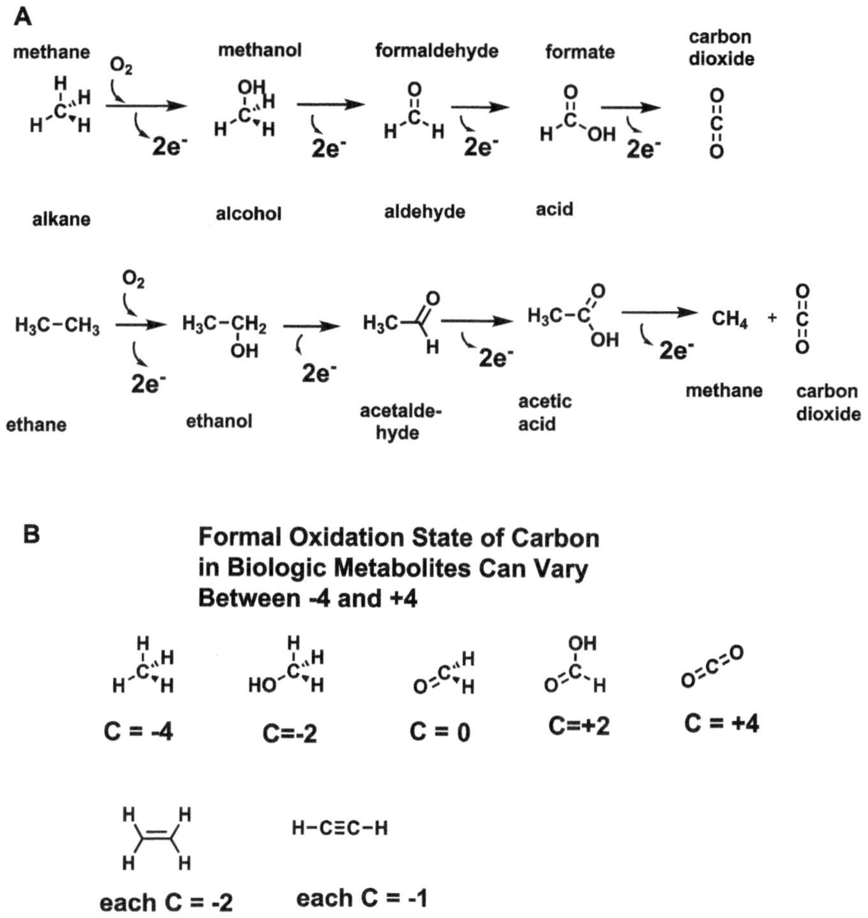

Figure 1.11 (A) Eight electron span between alkanes and carbon dioxide. Methane and ethane as examples. (B) Formal oxidation state of carbon can vary from −4 (CH_4) to +4 (CO_2).

methanol has now changed to C = −2, consistent with the notion of a two electron oxidation (two electrons removed from carbon). Likewise, formaldehyde had the sole carbon at C=O, formic acid has C = + 2, and CO_2 (with no hydrogen substituents and two oxygen substituents (each O at −2) has the carbon at a formal oxidation state of +4. This is a complementary metric to show that methane is the most reduced and CO_2 the most oxidized of the one carbon molecules. Each of these is a major player in certain aspects of carbon chemical biology.

Inter alia, the realization that alkanes, alcohols, aldehydes (and ketones), and carboxylic acids are related by two electron transfers, out for oxidation, or two electrons in for reductive interconversions, gives explicit recognition to the role of reduction/oxidation (= redox) chemistry in functional group

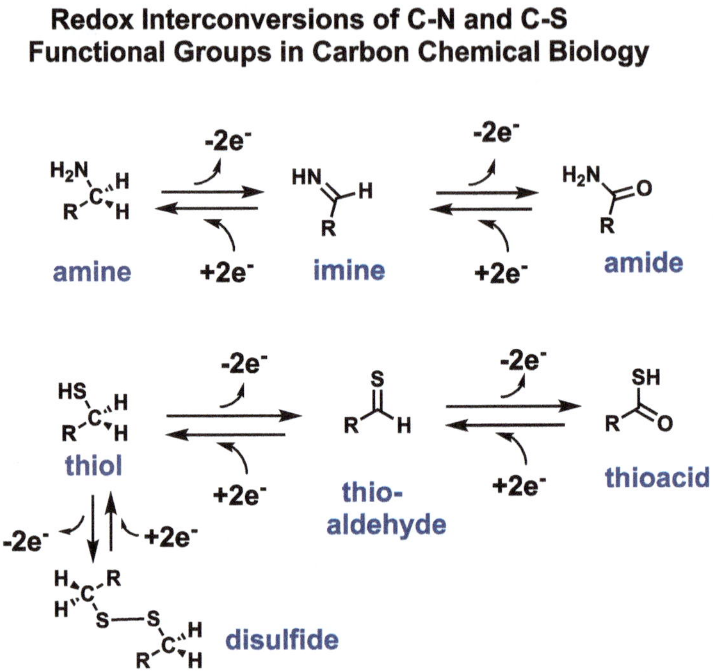

Figure 1.12 Common C–S and C–N functional groups related by two electron redox reactions.

transformations. This underlying principle holds both for abiotic organic chemistry and for metabolism in cellular biologic systems.

One can apply similar logic to analyze the interconversion of C–N[20] and C–S[21] functional groups that are common in the chemical biology of carbon frameworks (Figure 1.12). There is a two electron redox traverse between amine and imine and another two electron traverse between imine and amide. A similar four electron redox traverse exists between thiol and thio acid, with the intermediate thioaldehyde. Both the thioaldehyde and thio acid redox states are known, but relatively rare in chemical biology. We will note below that the much more common acyl thioester functional group typically arises by attack of a thiol group on an electrophilic acyl group. (In general, in this book we will not analyze carbon-based transformations in terms of the formal oxidation states at carbon.)

1.6 Enzymatic Catalysts

The great bulk of coding genes (in humans around 23 000 genes) within an organism's genome are translated *via* intermediate messenger RNA molecules into proteins. A substantial subset (several thousands) of those proteins have catalytic activity, accelerating chemical transformations of a specific set of reactant metabolites (substrates) into products by lowering the energy barrier(s) between substrates and products. Like all catalysts, abiotic or biotic, enzymes cannot affect the free energy differences between substrates and products, the equilibrium positions, but can accelerate rates. Among the largest described rate accelerations are ratios of 10^{24} for enzymatic over nonenzymatic rates. Some enzymes operate in the realm of 10^5 catalytic events per second. If such an enzyme accelerated a nonenzymatic reaction by 3×10^{11} fold, an enzymatic turnover would occur in 10 microseconds, while the nonenzymatic reaction would occur once in a year. Average turnovers of 10–100 per second for enzymes involved in primary metabolic pathways in organisms with rate enhancements of 10^7–10^{10} over nonenzymatic rates are essential for life. The difference between a reaction being complete in 10 milliseconds, compared to 1–10 years in the uncatalyzed case, not only emphasizes the dramatic rate accelerations carried out by enzymes but also emphasizes the chemical stability of the thousands of low molecular weight organic molecules that are the metabolic inventory in chemical biology.

Proteins with enzymatic activity may be as small as linear chains of 120–140 amino acid residues (14 000–16 000 Daltons in molecular weight) to dodecamers of subunits of 50 000 Daltons (12 mers) to hetero-oligomeric catalysts of unequal stoichiometries (*e.g.* $\alpha_2\beta_4\gamma_2$). These macromolecular catalysts provide chiral active sites for recognition and catalysis of specific substrates and impose regiochemical and stereochemical control on both substrate recognition and product outcome. Chirality control is a given for organic substrate transformations in chemical biology. Nineteen of the twenty amino acids that are the backbones of proteins are all L-isomers at C_2 (the twentieth, glycine, has a prochiral CH_2 group at C_2). Sugars related to glucose are D-hexoses, unless epimerase enzymes have been at work.

The principles by which the folded conformations of protein enzymes achieve such stupendous rate accelerations with consummate product regiochemical and stereo-chemical specificity continue to be examined. Enzyme names typically have the suffix -ase to indicate catalytic activity, such as oxygenases, racemases, reductases, polymerases, *etc.*

1.7 Electrophilic Group Transfer Reactions: Thermodynamic Activation and Kinetic Stability

An orthogonal reactivity axis governs a large subset of carbon-based functional group chemical biology. Much of the flux in metabolic pathways is driven by spending a small set of organic metabolites that have the twin properties of thermodynamic activation coupled to kinetic stability under physiologic reaction conditions.[23] The select set of ATP, NAD(P)H, 3′-phosphoadenosine 5′-phosphosulfate (PAPS), UDP-glucose, molecular oxygen, carbamyl phosphate, *S*-adenosyl methionine, acetyl-coenzyme A, and Δ^2-isopentenyl diphosphate (Figure 1.13) reflect nature's different chemical strategies to build molecules capable of diverse, largely electrophilic, group transfer reactions to cellular nucleophiles. ATP molecules transfer either phosphoryl, pyrophosphoryl or adenylyl groups to drive myriad otherwise unfavorable equilibria in macromolecular dehydrative biosynthesis of DNA,

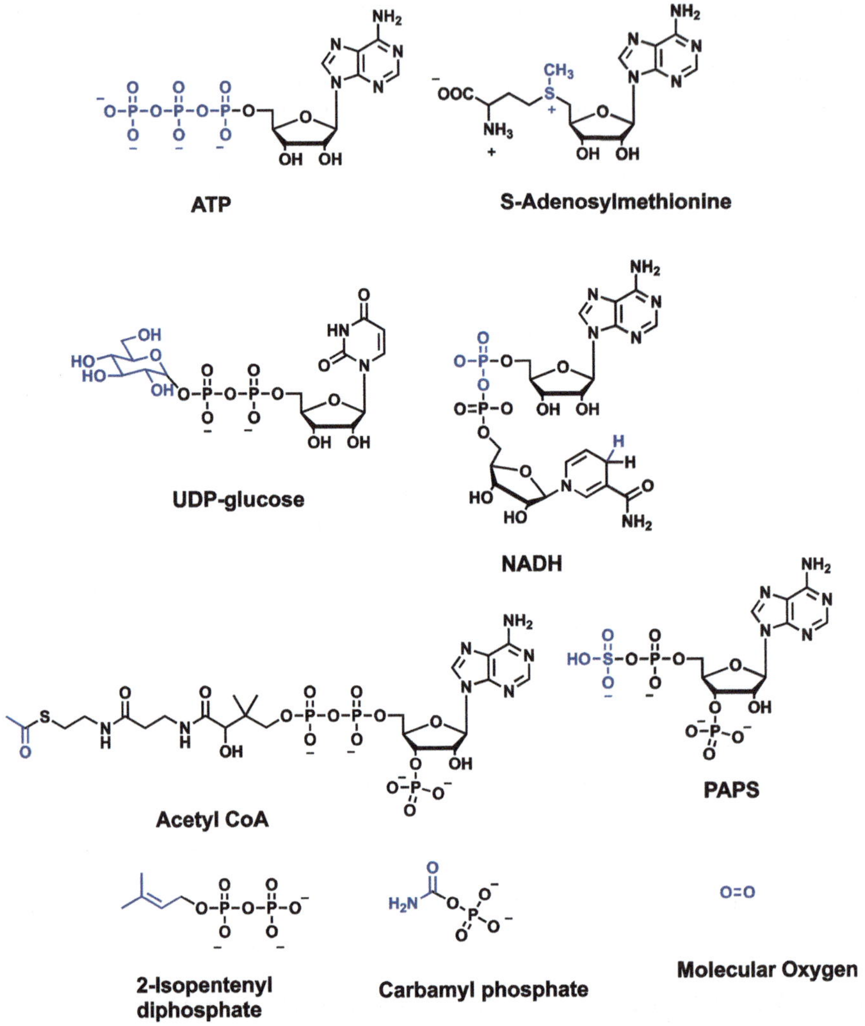

Figure 1.13 Thermodynamically activated, kinetically stable metabolites that can drive coupled equilibria *in vivo*.

RNA, proteins, and polysaccharides (Figure 1.14). Analogously, PAPS is the cellular reagent for sulfuryl group transfers to nucleophiles (Figure 1.14). O_2 at 21% of earth's atmosphere is reductively activated for a wide range of carbon centered hydroxylations, as noted in Chapter 12.

The remaining set of core metabolites NAD(P)H, acetyl CoA, Δ^2-IPP, SAM, UDP-glucose, and carbamyl phosphate are involved in carbon-based functional group interconversions and will be examined in following chapters. NAD(P)H is the cellular hydride donor and $NAD(P)^+$ the cellular hydride acceptor in the opposite direction, most often in alcohol to ketone/aldehyde and aldehyde to carboxylic acid redox transformations (Figure 1.14).

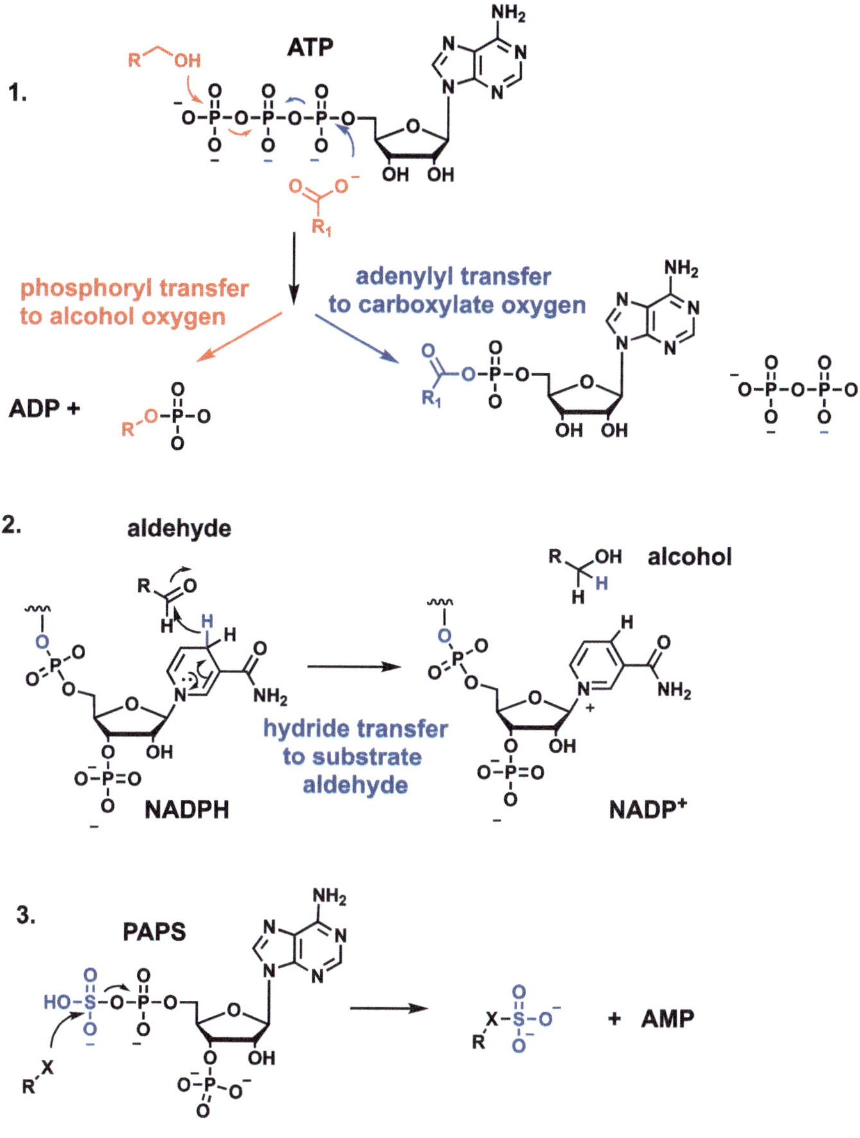

Figure 1.14 ATP undergoes phosphoryl or nucleotidyl transfer to alcohol or carboxylate oxygens; aldehydes undergo hydride ion transfer from NADH during reduction to alcohols; PAPS (3′-phospho-adenosine-5′-phosphosulfate undergoes sulfuryl group transfer to nucleophiles.

Acetyl CoA and longer acyl CoA homologs are the premier acyl transfer reagents to alcohols, amines and thiols, respectively, generating acyl-oxo esters, acyl amides, and acyl thioesters, three of the central acylated functional groups of chemical biology (Figure 1.15).

The remaining three agents, Δ^2-IPP, SAM, and UDP-glucose, generate carbenium ion transition states to deliver, respectively, allyl cation

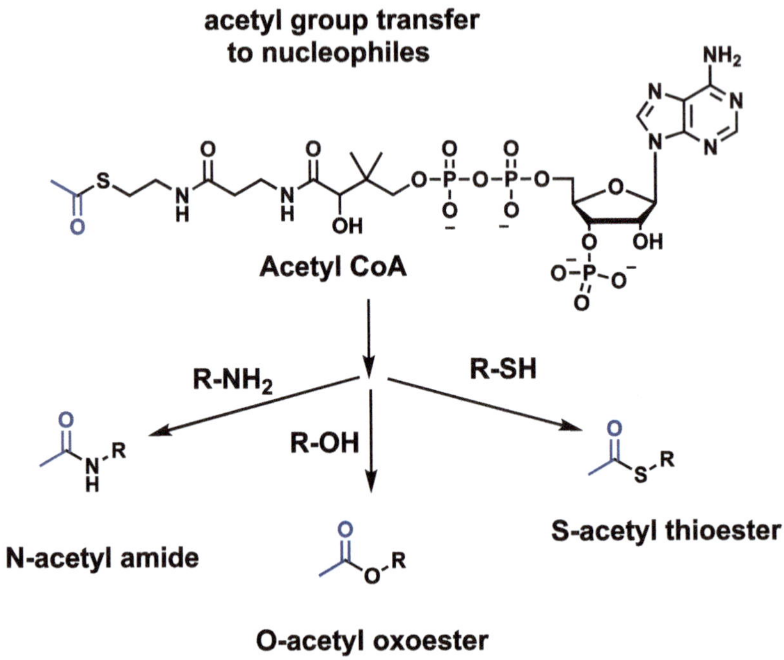

Figure 1.15 The acetyl group in the thioester acetyl CoA undergoes acetyl group transfer to nitrogen, oxygen, and sulfur nucleophiles.

equivalents, methyl cation equivalents, and glucosyl C_1–O_5 oxocarbenium ion equivalents to nucleophilic atoms in cosubstrates, Figure 1.16.

These three distinct chemical strategies for transient generation of carbocationic species are a segue to the next section on the nature of carbon-centered reaction intermediates. They also highlight three diverse forms of functional group logic to deliver electron deficient (electrophilic) fragments to electron rich (nucleophilic) cosubstrate atoms that include O, N, S, and even accessible carbon nucleophiles. For Δ^2-IPP the reactive functional group in the starting material is the allyl-pyrophosphate ester set up for low energy C_1–O bond cleavage. For *S*-adenosyl methionine the trivalent cationic sulfonium center is the basis for low energy transfer of a $[CH_3^+]$ equivalent in S_N2 reaction sequences. The potential electrophile in UDP-glucose and other NDP hexose congeners is the acetal linkage (see Figure 1.9 and Chapter 14) between C_1 of the glucopyranose ring and the pyrophosphate bridge of the UDP portion. Nature has conscripted distinct chemical types of thermodynamically labile but kinetically stable functional groups to power different kinds of chemical transformations in organisms.

1.8 Major Reaction Types That Make and Break C–C Bonds

The brief introduction to the metabolic centrality of the five thermodynamically activated kinetically stable donors of electrophilic groups in the

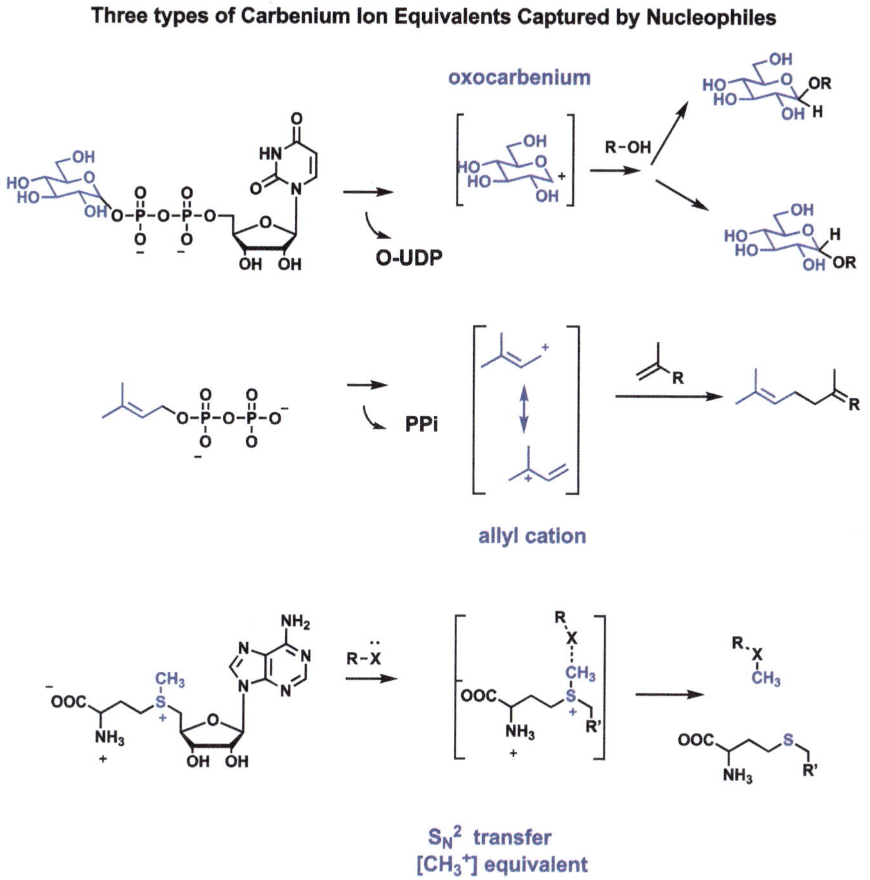

Figure 1.16 Three types of electrophilic groups captured by cellular nucleophiles are glucosyl C_1 oxocarbenium ions, allyl cations from Δ^2-isopenetenyl-PP, and methyl groups from *S*-adenosylmethionine.

above section raises the question about what types of carbon-centered intermediates occur in chemical biology reactions. Section IV, representing Chapters 10–13, takes up enzyme- and vitamin-mediated examples in some detail. We note that organisms have evolved enzymes and coenzymes to accelerate reaction pathways, by stabilizing three different types of otherwise unstable carbon-centered intermediates along with a set of concerted, pericyclic transition states.

There are four common routes to C–X (including C–C) bond formation used by chemists in total synthesis of complex scaffolds[24] (Figure 1.17). Two involve heterolytic coupling of carbanion or carbenium ion equivalents. A third involves homolytic coupling where each partner brings one electron to the new C–X bond formation. This route involves carbons with unpaired electrons, carbon radicals. The fourth route encompasses pericyclic

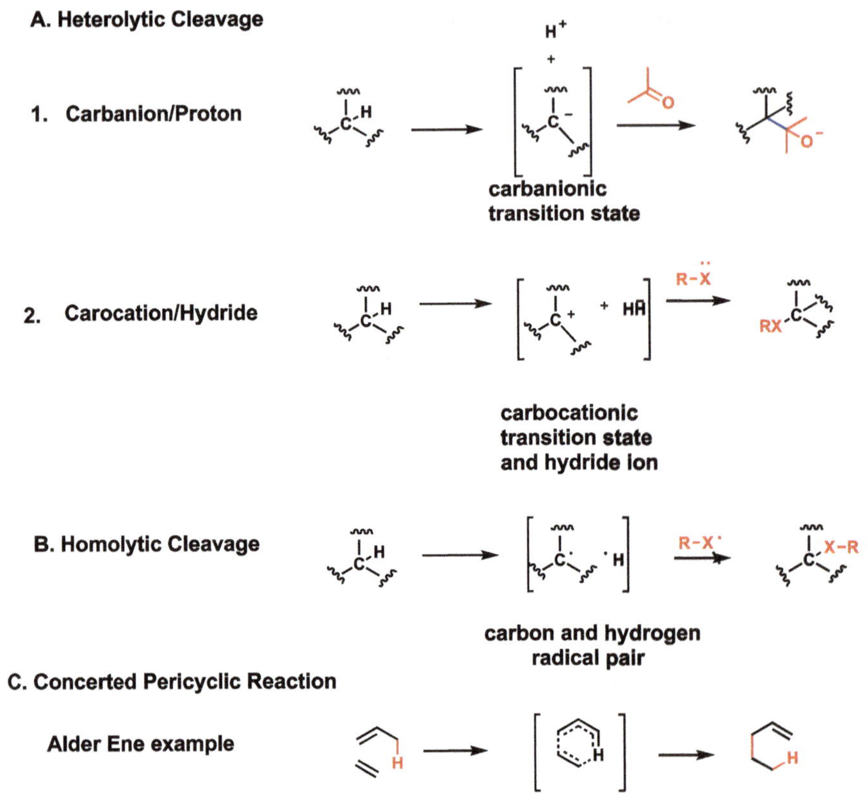

Figure 1.17 Four pathways to C–H and/or C–C bond cleavages and formations.

reactions, where there is concerted bond breaking and bond making in cyclic transition states.[25]

All four of these routes are relevant for carbon-based transformations in chemical biology. Each depends on the energy barriers for various mechanisms, involving both substrate structure and the architecture and catalytic inventory of the biologic catalysts, the enzymes, (and coenzymes, see below) that carry out the specific rate accelerations.

1.8.1 Carbanion Equivalents

Carbanion equivalents can be kinetically accessible in biologic systems when a given substrate can stabilize the incipient negative charge on carbon by delocalization into a neighboring, more electronegative substituent. In functional group terminology three to four types of functional groups in substrates can act as carbanion/carbon nucleophile progenitors. The first involves (enolizable) aldehydes and ketones (Figure 1.18) where an adjacent α-carbon has at least one hydrogen substituent. Enolization can occur by abstraction of that Cα–H as a proton. The resultant negative charge on carbon can be shared/delocalized in a resonance contributor that has the

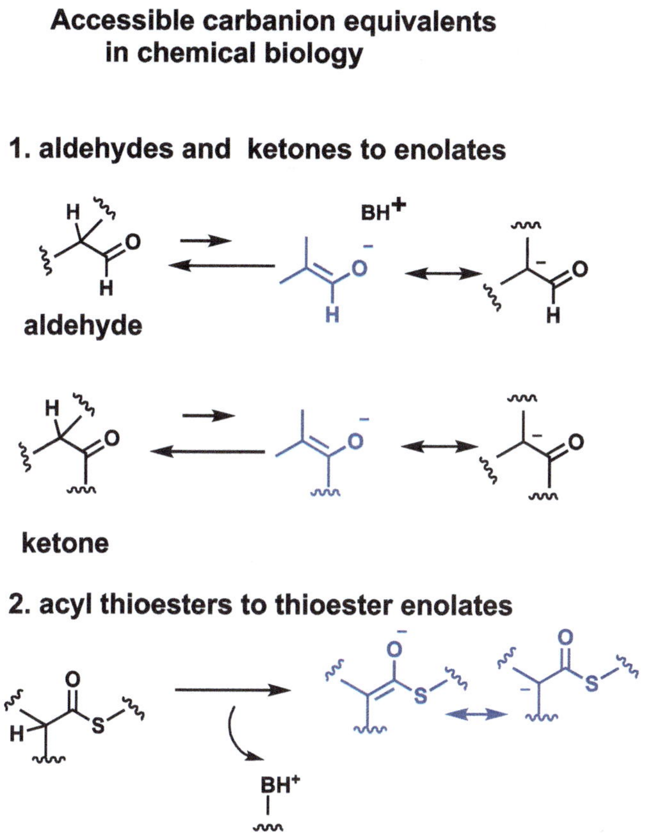

Figure 1.18 Accessible carbanion equivalents in chemical biology; enolates and thioester enolates.

negative charge on oxygen. This is an enolate anion. Oxygen is a more electronegative element than carbon, so the lower energy and more highly populated resonance contributor has the negative charge density higher on oxygen. However, there is enough contribution from the ensemble with the negative density on Cα to act as a carbon nucleophile, *e.g.* in C–C bond formation. The three carbon 2-ketoacid pyruvate is a common metabolic precursor to the pyruvate enolate ion.

A second metabolically central set of carbanion equivalents are acyl thioester enolates, arising by equivalent proton abstraction from C_2 of the acyl moiety, *e.g.* in acetyl CoA (Figure 1.18). The C_2 carbanion equivalent from acetyl CoA is a key intermediate in both fatty acid chain elongation and early in isoprenoid biology, as examined in Chapter 11.

Comparable logic is at work in the chemical biology of imine/iminium ion functional groups (Figure 1.19). Cα–H abstraction from iminium functional groups leads to enamines as the neutral resonance contributors to carbanion structures. In many instances enamines are lower energy, more stable

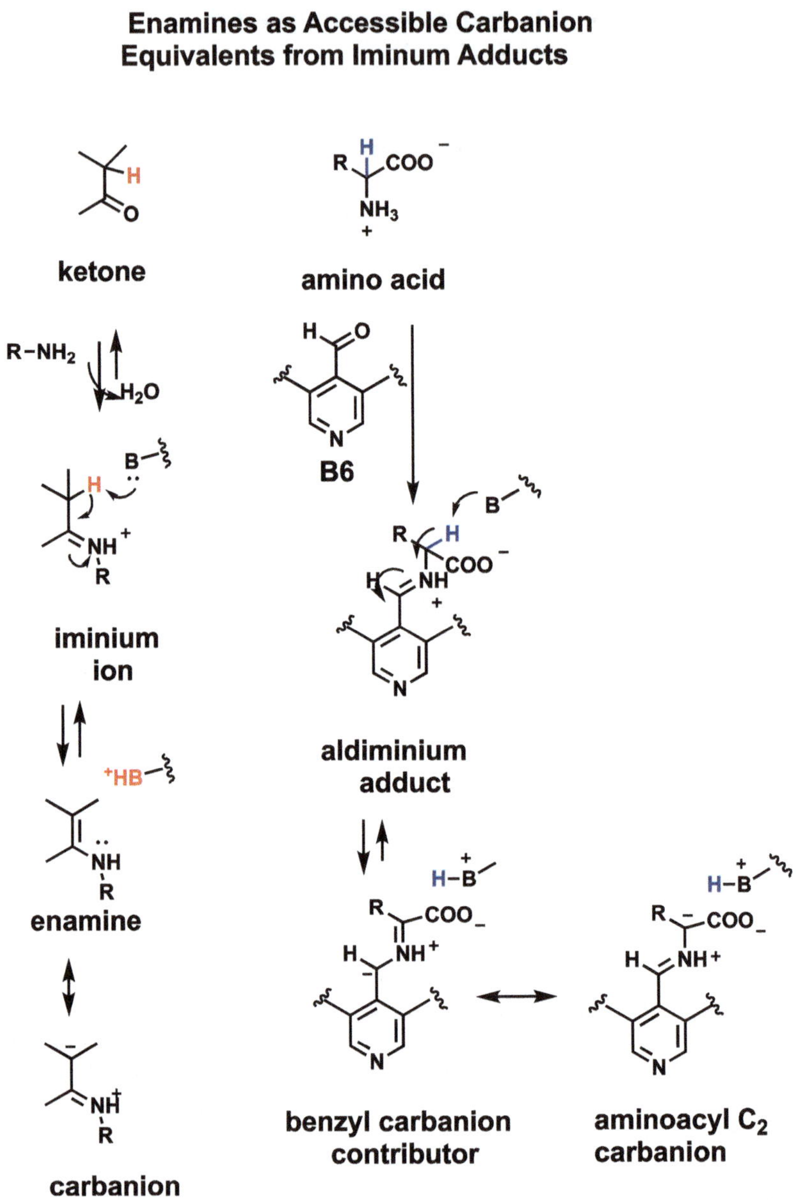

Figure 1.19 Enamines from imines as resonance contributors to carbanion structures.

forms of substrate carbanions than the corresponding enolate ions.[26] Amino acid chemical biology mediated by the coenzyme form of vitamin B6, pyridoxal-phosphate, is catalyzed in dozens to hundreds of enzymes that generate C_2-carbanions from amino acid-PLP aldimine adducts (Chapter 10)[27] (Figure 1.19).

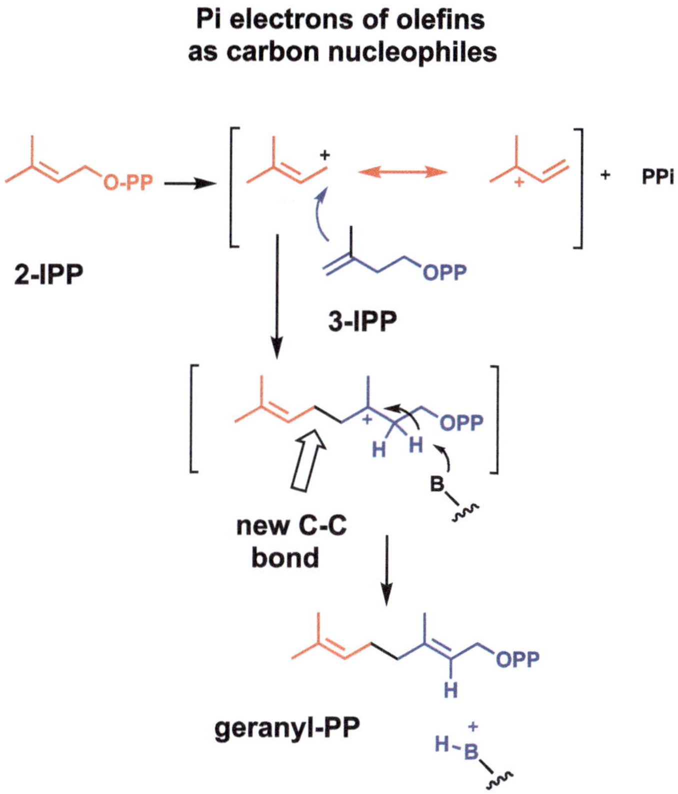

Figure 1.20 π-Electrons of olefins as carbon nucleophiles: Δ^3-IPP.

The π-electrons of alkenes (olefins) can be put to use as carbon nucleophiles in a variety of chemical biology contexts. Perhaps most pervasive is the use of the terminal olefin of Δ^3-isopentenyl-PP on the allylic carbocation arising from early C_1–O bond cleavage in the Δ^2-IPP partner by chain-elongating prenyltransferases. Such enzymes operate in the isoprenoid chain elongation phases of pathways leading to >70 000 known isoprenoid scaffolds (Figure 1.20).[28]

1.8.2 Carbocation Equivalents

In the prevalent routes to C–C bond formation in carbon chemical biology, substrate-derived carbanion equivalents need electrophilic carbon centers with which to react. The two most common electrophilic partners in C–C bond formations may be (1) aldehyde/ketone functional groups and (2) carbonyl groups of acyl thioesters. In aldehydes and ketones, the C=O double bonds are polarized, with partial positive charge on carbon and partial negative charge on oxygen. Thus, these carbonyl groups are obvious partners for enzymatic condensation with carbanion equivalents (Figure 1.21). The related imine functional groups similarly react as

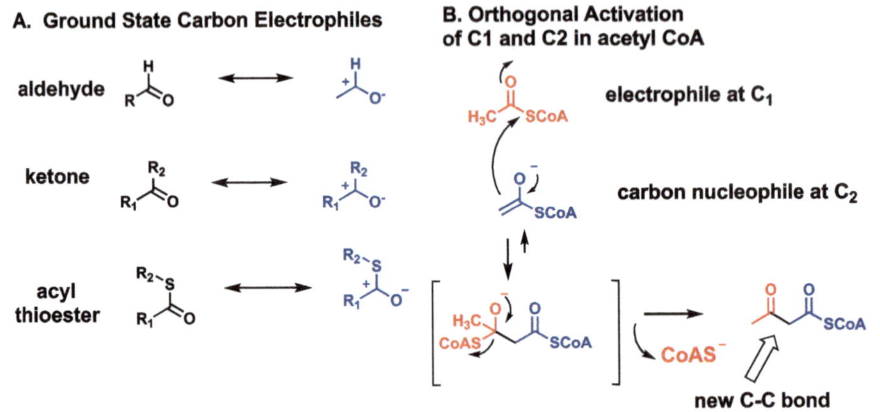

Figure 1.21 (A) Ground state carbon electrophiles. (B) Orthogonal activation of carbons 1 and 2 in acyl CoAs as carbon electrophiles and carbon nucleophiles.

electrophiles at the carbon atom. We will examine physiologic examples of these metabolically central aldehyde/ketone/imine capture in C–C bond formations in Chapter 10.

The second major source of substrate electrophiles for biosynthetic C-C bond formations are the carbonyl groups of acyl thioesters.[29] Compared to acyl-oxoesters, the C–S bond is longer than the C–O bond and the carbonyl group behaves more "ketone-like" in its polarization. A classic and central example of acetyl CoA serving both as electrophile at the C_1 acetyl carbonyl and nucleophilic carbon at C_2 is in the very first step of isoprenoid biosynthesis. The enzyme generates the C_2 carbanionic thioester enolate in its active site and uses it to attack the electrophilic C_1 carbonyl for C–C bond formation. Hydrolysis of one of the two C_1–SCoA bonds drives the equilibrium towards C–C bond formation, albeit the reverse direction C–C bond thiolytic cleavage is also observed (Figure 1.21).

In Section 1.6, on strategies for electrophilic group transfers in chemical biology, we noted briefly that three different thermodynamically activated scaffolds have been evolved to serve as carbocation equivalents in metabolically central reactions: the acetal linkage in UDP-glucose, the allylic double bond in Δ^2- but not Δ^3-isopentenyl-PP, and the sulfonium cation linkage in *S*-adenosylmethionine (where all three carbon substituents on that trigonal sulfur are activated for electrophilic transfers).[23,30,31] Most of the time those group transfers occur in nucleophilic oxygen, nitrogen, or sulfur atoms in cosubstrates. However, Figure 1.22 shows that phenolate anions can act as cyclohexadienyl carbanion resonance structures for capture of the electrophilic C_1 of the glucosyl group of UDP-glucose to form a rare *C*-glucoside.

Analogously, *C*-methylation during novobiocin biosynthesis involves capture of the SAM-derived $[CH_3^+]$ transition state by the indicated phenolate

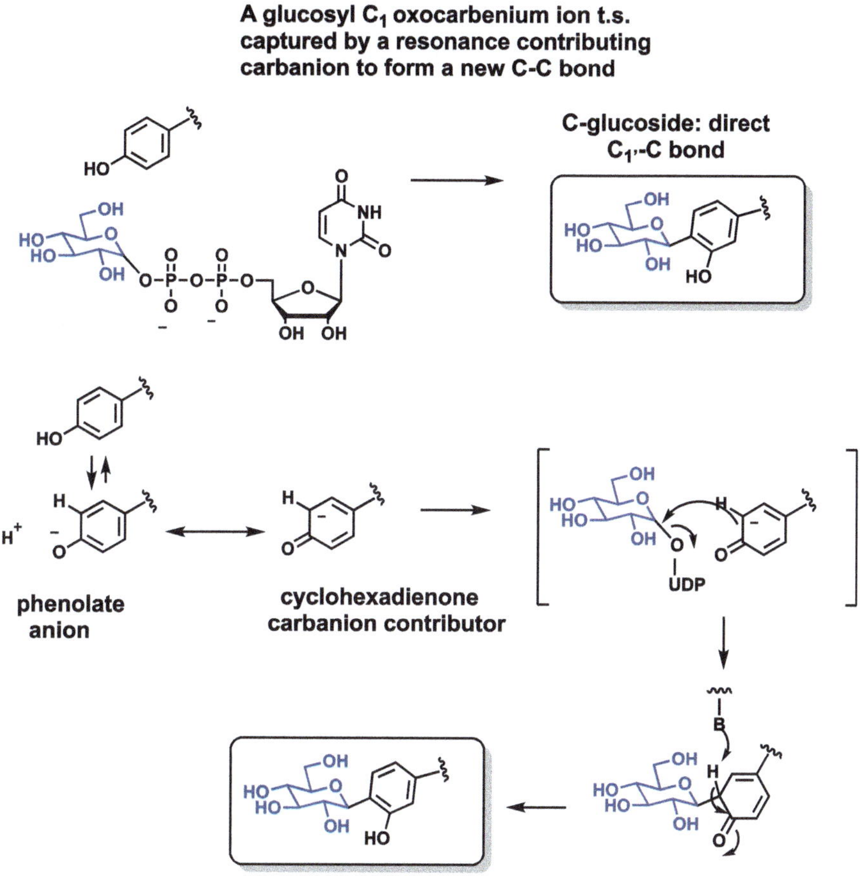

Figure 1.22 Capture of an electrophilic glucosyl C_1-oxocarbenium ion transition state by a carbon nucleophile.

anion resonance contributor.[31] Finally, the resonance-delocalized allyl cation derived from Δ^2- IPP is capturable both at C_1 (in the prenyl transfer chain elongations) and at C_3 of the allyl scaffold, as shown for capture by the C_3 enamine of tryptamine. That result establishes partitioning proof of the delocalized allyl cation's structure (Figure 1.23).

1.8.3 Carbon Radical Equivalents

For most of the 20th century it was assumed that homolytic carbon chemistry was rare in chemical biology, relegated to a few anaerobic metabolic pathways, such as in coenzyme B12-mediated rearrangements practiced by bacteria. As it turns out, carbon-centered radicals, as intermediates or evanescent odd electron transition states, are quite common, one set under anaerobic conditions, the second set requiring molecular oxygen as coreactant.

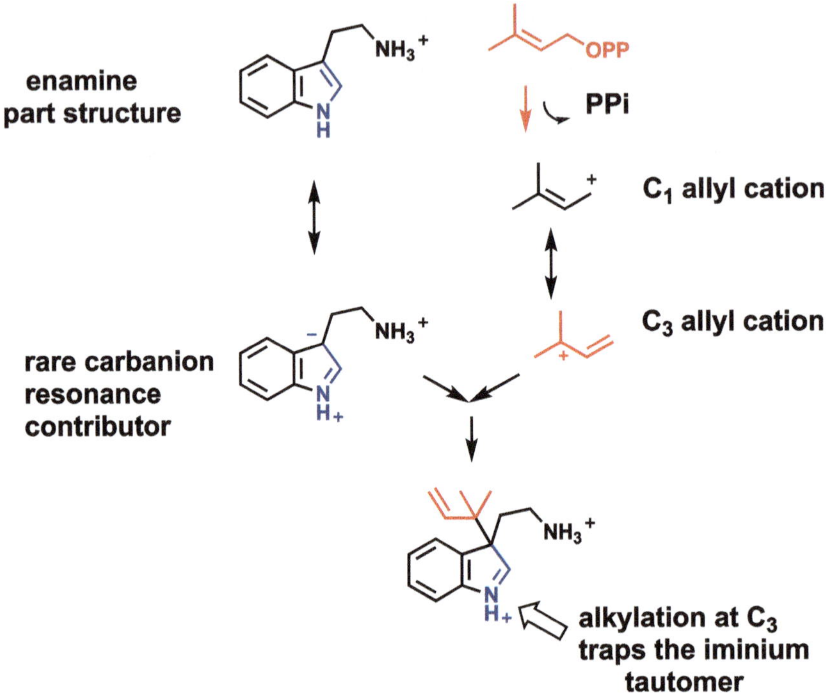

Figure 1.23 C–C bond formation between a C_3 indole enamine carbon nucleophile and a delocalized allyl carbocation.

On the highly oxygenated end, at an average 240 µm [O_2] cellular concentration, molecular oxygen is reductively activated by one electron transfer pathway.[32] Flavin-dependent monooxygenases typically make covalent flavin peroxides while iron-based oxygenases create high valent oxo-iron intermediates as robust oxidants.[33] The iron-based enzymes, in particular, go on to homolyze C–H bonds of bound substrates (Figure 1.24), creating carbon-based radicals.

A typical fate for an sp^3 carbon radical is [$OH^•$] transfer in a radical rebound step to create a hydroxyl group in an iron-based oxygenase active site. This is the classic alkane to alcohol transformation shown in Figure 1.7. For an alkene functional group in a bound substrate, the net outcome is instead epoxidation (Figure 1.24), accounting for the epoxide functional groups shown in Figure 1.9. An alternative fate for some sp^3-carbon radicals is internal radical rearrangement, kinetically outcompeting the intermolecular [$OH^•$] transfer. That "thwarted oxygenase" outcome[34] happens twice in the fungal construction of the 4,5-bicyclic ring system of penicillins (see Chapter 12).

At the other end of the aerobic/anaerobic spectrum, a family of radical *S*-adenosylmethionine enzymes[35] now numbers 700 000 entries as (largely) predicted catalysts that generate carbon-centered radicals in bound

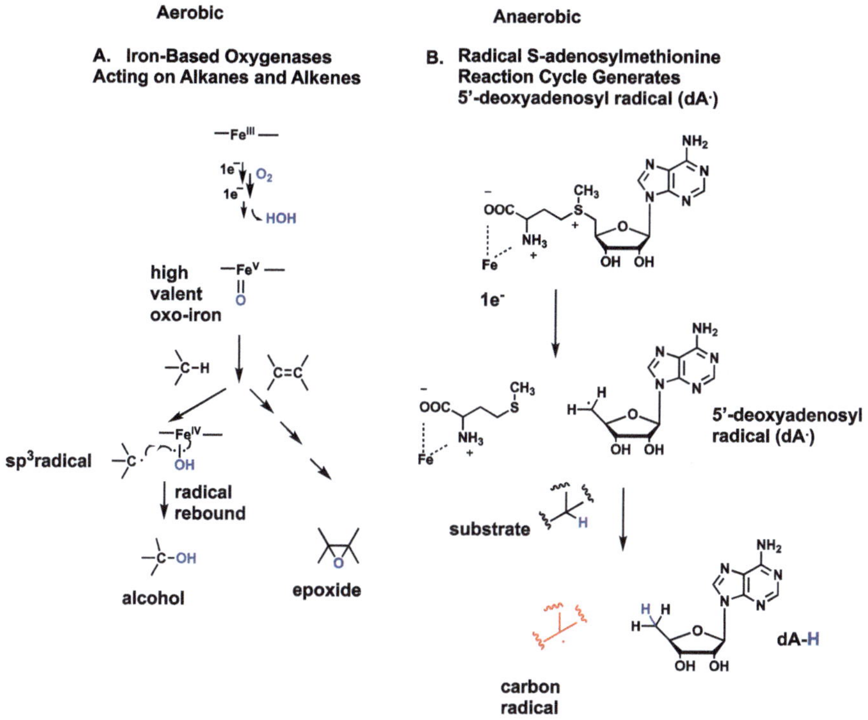

Figure 1.24 Two paths to carbon-centered radicals. (A) Aerobic from one electron reductions of O_2. (B) Anaerobic from one electron-mediated homolysis of *S*-adenosylmethionine into methionine and the 5′-deoxyadenosyl radical.

substrates. The radical generator in these >half million cases is largely the 5′-deoxyadenosyl radical (Figure 1.24), arising by one electron input and homolysis of the deoxyadenosyl-$C_{5'}$-sulfur bond in SAM. The dA$^{\bullet}$ radical is sufficiently robust to abstract a hydrogen atom from nearby bound substrate C–H bonds and thereby initiate carbon radical-based rearrangements on the way to bound product radicals.[36] Those nascent product radicals are often quenched by re-abstraction of a hydrogen atom back from the nascent 5′-deoxyadenosine terminal methyl group. Radical rearrangements, still to be examined in over 700 000 predicted pathways, offer the prospect of novel chemical pathways in carbon chemical biology.

1.8.4 Pericyclic Reactions in Carbon Chemical Biology

Pericyclic reactions come in several subclasses united by the criterion that there are no intermediates. The making and breaking of a set of bonds is concerted (the specific bond-forming or bond cleavage steps may be asynchronous, with one or more lagging behind early bond completions). From the initial discovery in 1928 of Diels and Alder of the 4+2 pericyclic cycloaddition reactions that bear their name,[37] the arenas of pericyclic organic chemistry have been dynamic ones. Predictive understanding may have culminated in the 1965–1970 formulation of the Woodward–Hoffman rules[38] to predict and explain allowed thermal *vs.* photochemical pericyclic reactions according to principles of orbital symmetry. Pericyclic reactions have been employed in a myriad of synthetic steps to introduce scaffold complexity and regio- and stereochemical control of the products,[39] often in one-pot cascade reactions.

The study of pericyclases in carbon chemical biology has become an explosive arena in the postgenomic era where hundreds of thousands of microbial genome sequences are now catalogued, biosynthetic gene clusters identified, and likely pericyclases exported to surrogate hosts for successful expression and characterization and validation of the pericyclic nature of the catalyzed reactions.[40]

As adumbrated by Woodward and Hoffman,[38] four of several types of pericyclic reactions are cycloadditions, electrocyclizations, sigmatropic rearrangements, and Alder ene reactions (Figure 1.25). Over the past two decades, a number of potential Diels Alderases have been examined and some have stood the test of time.[41] They seem to be authentic pericyclases for cycloaddition reactions, particularly in the construction of both decalin ring systems and spirotetronates during introduction of complexity in

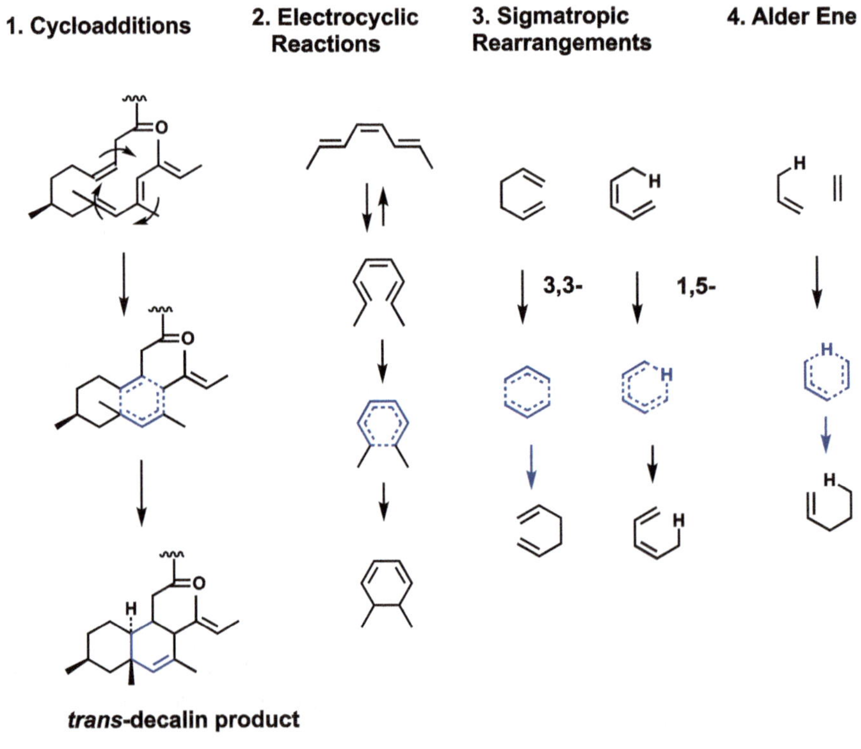

Figure 1.25 Four types of pericyclic reactions in chemical biology.

natural product pathways. Similarly, there are now examples of biologic electrocyclizations, sigmatropic rearrangements (Figure 1.25), and Alder-ene reactions, as described in Chapter 13.

1.9 Nature Embraces the Biosynthesis and Utilization of Heterocycles

Organic chemistry classes, up through the first three semesters, typically go light on heterocycles. Although specific ring systems may be encountered, such as pyridines or pyrroles, heterocyclic chemistry is often viewed as a specialized topic, with each ring system showing some idiosyncratic reactivity features. Natural product chemists and medicinal chemists may be the ones who encounter heterocycles most frequently, on one hand in isolation and structure determination and, on the other hand, as synthetic isosteres for natural therapeutic scaffolds.

In contrast, nature seems to have been enthusiastic about both carbacyclic and heterocyclic chemistry from the earliest days of life forward. Heterocycles from three atoms to eight atoms are found throughout life forms, with differing connectivities as well as distinct carbon, nitrogen, oxygen, and sulfur atom compositions (Chapters 7 and 8). Biologic information storage and transfer in DNA and RNA molecules are built on the interactions of stacked purine and pyrimidine monomers (Chapter 6). The part structures of amides, amidines, ureas, and guanidines bring heterocyclic carbon–nitrogen functional groups to the forefront. In proteins the five-membered ring imidazole side chain of histidine and the bicyclic 5,6-indole ring of tryptophan are key heterocycles. Hundreds to thousands of alkaloid natural products are constructed by enzymes using indole ring chemistry from tryptamine, the decarboxylation product of tryptophan.

Seven of the eight B vitamins contain heterocycles that add to the protein catalysts' inventory of functional groups[42] (Chapter 9). B1 and B3 vitamins use thiazolium and pyridinium cations as key functional units. B6 displays a pyridine aldehyde in the coenzymatically active form. A bicyclic tetrahydropterin ring is the business end of folate (vitamin B9). A tricyclic isoalloaxine ring system is the key to riboflavin (B2) chemical biology. The fused tetrapyrrolic corrin ring system of B12 enables its radical rearrangement roles in chemical biology. No examination of carbon chemical biology would be complete without analysis of the roles of these heterocyclic systems in the chemistry of life.

1.10 One Piece of Terminology, One Piece of Graphics

In this volume we repeatedly use two concepts from abiotic organic chemistry to illustrate *in vivo* carbon chemistry. The first is the concept of nucleophilicity and electrophilicity. Electron rich functional groups behave as attacking species, they are nucleophiles. Examples of biologic nucleophiles include C–S bonds where sulfur is divalent (*e.g.* thiolate anions), C–O bonds (*e.g.* carboxylate anions but also transient alcoholate anions) and C–N bonds (deprotonated basic forms of amines) as electron rich functional groups. We spend much of the book also detailing the types of carbon nucleophiles, often stabilized carbanion equivalents) that can attack electron deficient target atoms (Figure 1.26). Indeed, a main theme for the bulk of organic chemical transformations in organismal metabolism is to identify carbon-based functional groups that qualify as nucleophiles *vs.* electrophiles and how they react with each other.

Electron deficient (electrophilic) targets in carbon chemical biology can include carbon electrophilic C=O functional groups such as aldehydes, ketones, and acyl thioesters. They also include C=N– electrophiles such as iminium ions (Figure 1.26) and C–S electrophiles in the form of disulfide bonds (–C–S–S–C–).

We formulate reaction mechanisms where electron rich nucleophiles attack and form bonds to electron poor nucleophiles (Figure 1.26). Many of the figures that illustrate points in the text of each chapter show diagrams of such reactions. The direction of electron flow is always shown proceeding from an electron rich nucleophile to an electron poor electrophile.

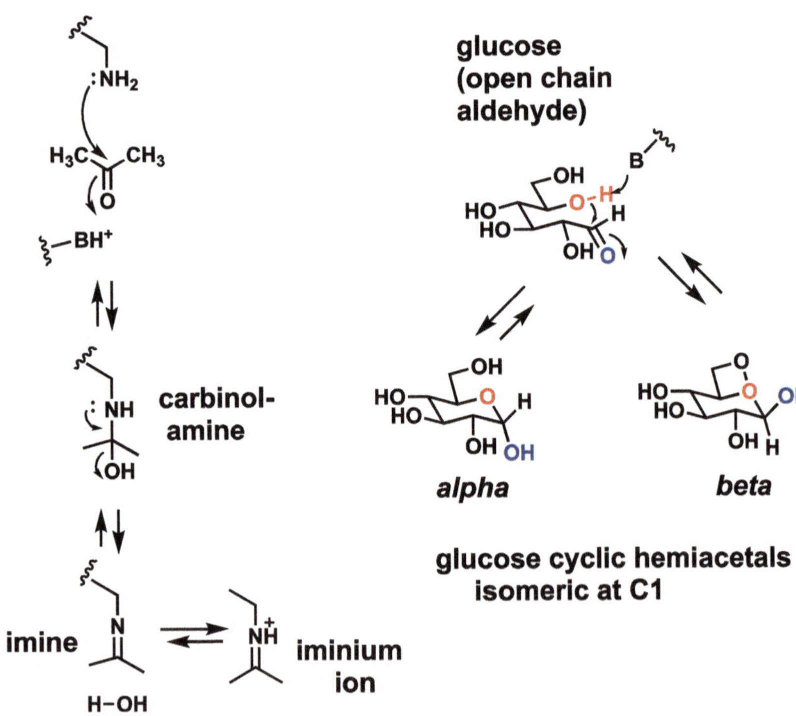

Figure 1.26 Double headed curved arrows used as standard organic chemistry shorthand for flow of two electrons from a nucleophilic to an electrophilic center. The formation of an intermolecular iminium ion *vs.* an intramolecular hemiacetal in glucose are illustrations of the curvy arrow usage.

The shorthand device that indicates electron flow is the use of curved arrows. Figure 1.27 shows examples of amine addition to a ketone carbonyl to produce the iminium ion and the intramolecular cyclization of the minor open chain aldehyde form of the sugar glucose to two cyclic hemiacetal forms, isomeric at C1 for the -OH (below = α) or above the plane (above = β). These types of electron flows by arrow pushing are schematized in standard organic chemistry texts and utilized in this volume. An ancillary feature is that curved arrows with standard, two barbed arrow heads refer to movement of two electrons at a time.

In Chapter 12 we take up one electron transfer reactions in chemical biology, we employ single headed arrows for reactions that form or consume odd electron species, *e.g.* carbon and oxygen radicals. Figure 1.27 depicts the quenching of the 5′-deoxyadenosyl radical (dA•), see Chapter 12) by hydrogen atom (H•) transfer from a bound substrate C–CH_3 group. Each one electron transfer is shown by single headed curvy arrows. One of the electrons in the C–H bond being cleaved remains at the substrate carbon as a transient -CH_2• radical equivalent.

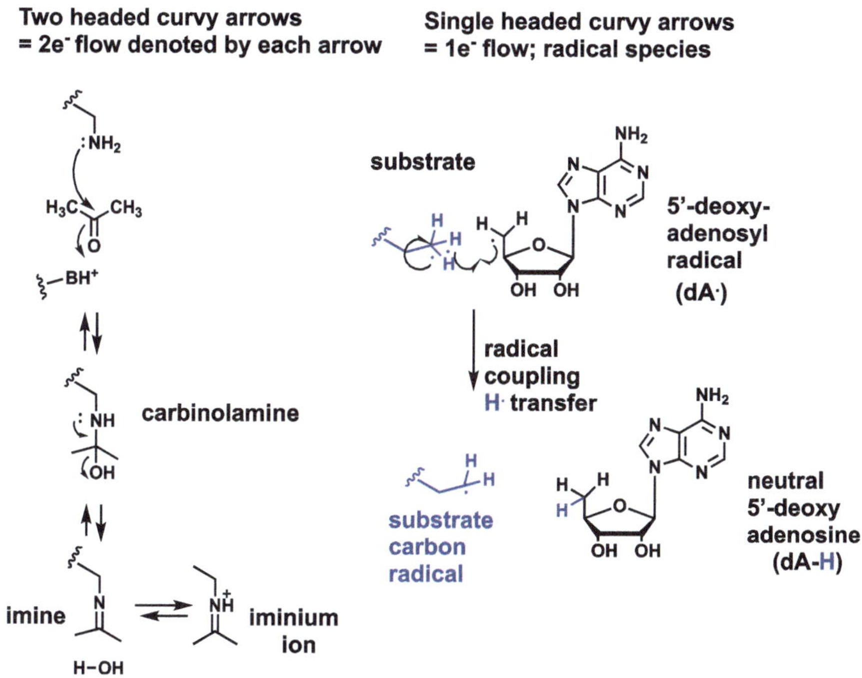

Figure 1.27 Single headed curvy arrows show the direction of flow of single electrons in reactions where one electron transfers occur. The specific case of hydrogen atom transfer from a bound substrate C–CH_3 group represents a hydrogen atom transfer to quench the 5′-deoxyadenosyl radical while creating a transient carbon-centered substrate radical.

1.11 Organization of This Volume

The premise of this book is that there is underlying logic to carbon chemical biology, the organic chemistry of life, that can be approached and appreciated by analysis of a few core principles. Those principles of carbon-based functional group identity and rules of reactivity carry over fully from abiotic to biotic organic chemistry. This premise leads to a straightforward but nonetheless unusual approach to coming to grips with the myriad kinds of biologic metabolites, their transformations, their interconversions at metabolic nodal points and the myriad individual pathways that lead to an estimated 100 000 low molecular weight metabolites.

To that end, after an initial introductory chapter there are four additional sections to the book. Section II examines the functional group inventory of small organic metabolites, starting with the simplest functional group, alkenes (–C=C–), and then proceeding to common C–O, C–S, and C–N functional groups. This survey identifies the common patterns of functional groups in small organic metabolites that populate the many primary metabolic pathways. This section also focuses on the central concepts of redox transformations that interrelate functional groups and explain their types of transformations.

The third concept that simplifies the reactivity patterns of most two electron reaction manifolds connecting substrates and products is the division into electron rich attacking nucleophiles and electron poor electrophilic groups that are captured by nucleophiles. The fourth principle of *in vivo* organic metabolism is that nature uses a core set of thermodynamically activated but kinetically stable molecules that drive coupled equilibria in execution of electrophilic group transfers to cosubstrate nucleophiles.

Section III is centered around the premise that nature makes and needs heterocyclic ring systems as functional groups to conduct the enzyme-catalyzed reactions of life. (It is too anthropomorphic to write that nature loves heterocycles, but the anthropomorphizing has utility.) Oxygen- and nitrogen-containing heterocycles from three to six atoms are most abundant. The monocyclic pyrimidine and bicyclic purine building blocks for RNA and DNA serve as exhibit A for the essentiality of nitrogen-containing heterocycles for life. Analysis of key nitrogen heterocycles is divided between the ring systems both in free substrates and as side chains of proteins (pyrroles, imidazoles, indoles) in Chapter 8. In Chapter 9 we summarize the remarkable proliferation of distinct natural nitrogen heterocycles in the coenzyme forms of the human B vitamins, including thiazolium, pyridinium, pyridine aldehyde, pterin and flavin ring systems. These heterocyclic coenzymes are necessary for enabling central transformations of substrate functional groups and for lowering energy barriers to allow enzyme partners to carry out *in vivo* organic chemistry on the millisecond time scale.

Section IV, Chapters 10–13, then examines the types of reaction mechanisms available to C–H and C–C bonds in organic substrates as they are transformed in carbon chemical biology reactions. Four alternatives, examined in each of the four chapters of this section, stem from the different ways to cleave or form a bond to carbon. Chapter 10 focuses on energetically accessible carbon nucleophiles. These include not only the alkene groups acting as nucleophiles, but also structurally stabilized carbanions such as enolate anions and thioester enolate anions. Chapter 11 focuses on the other partner for C–C bond formations, carbon electrophiles. In addition to the ground state carbonyl groups of aldehydes, ketones, and acyl thioesters, the carbocation inventory available in biologic systems notably includes transient allyl carbocations, and glycosyl C_1 oxocarbenium ions as well as all three carbon substituents attached to the sulfonium cation of *S*-adenosylmethionine.

Chapter 12 examines the consequences of homolytic bond breakage at carbon centers and the fate of transiently generated carbon radicals. These occur in anaerobic radical SAM enzymes and also in aerobic iron-based oxygenase catalysis. Chapter 13 concludes the analysis by examining concerted reactions where bonds break and form simultaneously. There are no intermediates. Four categories of such pericyclic reactions in biologic systems have come into focus over the past two decades and we examine recently characterized classes of pericyclases.

Section V, the concluding section of three chapters, examines the chemical biology of three paradigmatic metabolites, central to distinct metabolic pathways to understand how their carbon-based functional groups react according to the underlying chemical principles of the previous sections. In turn, analysis of three metabolites with their distinct functional groups reveals the chemical logic for five metabolic pathways. Glucose, with its hydroxyl, aldehyde, and hemiacetal functional groups, enables three different metabolic pathways, one degradative, two biosynthetic (Chapter 14). Chapter 15 examines the formation and then cascade cyclization of the linear hexaene squalene to the tetracyclic framework of sterols. Squalene is midway in functional group complexity as a key intermediate in the biosynthetic pathway from 18 acetyl CoA molecules to cholesterol. The third paradigmatic metabolite illuminates the chemical virtuosity of the nitrogen heterocyclic aminomethylpyrrole porphobilinogen. This dicarboxy-aminomethylpyrrole is progenitor to both carbon nucleophile and carbon electrophile in creation of, first the linear, then the macrocyclic tetrapyrrole pigments of life (Chapter 16).

Readers can apply this carbon-centered functional group chemical logic, with a core appreciation of how nature has invented its inventory of heterocycles, to analyze (and predict) reactivity patterns of metabolites in primary and secondary metabolic pathways, both degradative (for energy storage) and biosynthetic. These few principles for biotic organic chemistry provide a foundation for deciphering the molecular complexities of life.

This volume on carbon chemical biology can be viewed as an integrative volume that picks up threads of and is complementary to three previous volumes that examine the chemical biology of nitrogen, the chemical biology of phosphorus, and the chemical biology of sulfur. This volume on carbon chemical biology also deals extensively with hydrogen (hydride ions, hydrogen atoms, protons) and oxygen chemical biology (*e.g.* Chapter 12).

It thereby integrates the chemical biology of the six elements (C, H, N, O, P, S) that comprise 99% of the body mass of organisms.

A parting word in this introductory chapter about references. The decision has been to err on the side of fewer references for the general concepts of reactivity and functional group identity that can be found in standard texts on abiotic organic chemistry and even biochemistry. When specific enzymatic examples of functional group transformations of particular metabolites are analyzed, references are added to key original research papers or reviews.

References

1. Laboratory LAN, Periodic Table of Elements, LANL, https://periodic.lanl.gov.
2. H. M. Leicester and H. S. Klickstein, *A Source Book in Chemistry 1400–1900*, Harvard University Press, 1951.
3. D. S. Wishart, Y. D. Feunang, A. Marcu, A. C. Guo, K. Liang and R. Vazquez-Fresno, *et al.*, HMDB 4.0: the human metabolome database for 2018, *Nucleic Acids Res.*, 2018, **46**(D1), D608–D617.
4. D. S. Wishart, C. Knox, A. C. Guo, R. Eisner, N. Young and B. Gautam, *et al.*, HMDB: a knowledgebase for the human metabolome, *Nucleic Acids Res.*, 2009, **37**(Database issue), D603–D610.
5. J. Clayden, N. Greeves and S. Warren, *Organic Chemistry*, Oxford University Press, 2012.
6. J. M. Berg, J. L. Tymoczko and L. Stryer, *Biochemistry*, WH Freeman, San Francisco, 6th edn, 2007.
7. Editors, Carbon, Chemical Element, Encyclopedia Britannica (retrieved 4-28-2021).
8. https://courses.lumenlearning.com, LL, Carbon Boundless Chemistry (retrieved 01-20-2021).
9. R. J. Ouellette and J. D. Rawn, *Organic Chemistry, key concepts, problems and solutions*, Elsevier, 2015.
10. R. K. Thauer, Biochemistry of methanogenesis: a tribute to Marjory Stephenson, *Microbiology*, 1988, **144**, 2377–2406.
11. A. J. Holmes, P. Roslev, I. R. McDonald, N. Iversen, K. Henriksen and J. C. Murrell, Characterization of methanotrophic bacterial populations in soils showing atmospheric methane uptake, *Appl. Environ. Microbiol.*, 1999, **65**(8), 3312–3318.
12. chem.libretexts.org/Courses CLT. 4.12: Uses and Sources of Alkanes (retreived 02-15-2021).
13. Y. Wang, G. Wegener, J. Hou, F. Wang and X. Xiao, Expanding anaerobic alkane metabolism in the domain of Archaea, *Nat. Microbiol.*, 2019, **4**(4), 595–602.
14. C. M. So and L. Y. Young, Initial reactions in anaerobic alkane degradation by a sulfate reducer, strain AK-01, *Appl. Environ. Microbiol.*, 1999, **65**(12), 5532–5540.

15. C. M. So and L. Y. Young, Isolation and characterization of a sulfate-reducing bacterium that anaerobically degrades alkanes, *Appl. Environ. Microbiol.*, 1999, **65**(7), 2969–2976.
16. J. B. van Beilen, M. G. Wubbolts and B. Witholt, Genetics of alkane oxidation by Pseudomonas oleovorans, *Biodegradation*, 1994, **5**(3–4), 161–174.
17. J. A. Marchand, M. E. Neugebauer, M. C. Ing, C. I. Lin, J. G. Pelton and M. C. Y. Chang, Discovery of a pathway for terminal-alkyne amino acid biosynthesis, *Nature*, 2019, **567**(7748), 420–424.
18. J. E. Moses and A. D. Moorhouse, The growing applications of click chemistry, *Chem. Soc. Rev.*, 2007, **36**(8), 1249–1262.
19. Y. Ji, G. Mao, Y. Wang and M. Bartlam, Structural insights into diversity and *n*-alkane biodegradation mechanisms of alkane hydroxylases, *Front. Microbiol.*, 2013, **4**, 58.
20. C. T. Walsh, *The Chemical Biology of Nitrogen*, Royal Society of Chemistry, London, 2021.
21. C. T. Walsh, *The Chemnical Biology of Sulfur*, Royal Society of Chemistry Press, London, 2020.
22. C. T. Walsh, *The Chemical Biology of Phosphorus*, Royal Society of Chemistry Press, London UK, 2021.
23. C. T. Walsh, B. P. Tu and Y. Tang, Eight Kinetically Stable but Thermodynamically Activated Molecules that Power Cell Metabolism, *Chem. Rev.*, 2018, **118**(4), 1460–1494.
24. K. C. Nicolaou, D. J. Edmonds and P. G. Bulger, Cascade reactions in total synthesis, *Angew. Chem., Int. Ed.*, 2006, **45**(43), 7134–7186.
25. S. Arns and L. Barriault, Cascading pericyclic reactions: building complex carbon frameworks for natural product synthesis, *Chem. Commun.*, 2007, **22**, 2211–2221.
26. B. List, Enamine catalysis is a powerful strategy for the catalytic generation and use of carbanion equivalents, *Acc. Chem. Res.*, 2004, **37**(8), 548–557.
27. Y.-L. Du and K. S. Ryan, Pyridoxal phosphate-dependent reactions in the biosynthesis of natural products, *Nat. Prod. Rep.*, 2019, **36**, 430–457.
28. C. T. Walsh and Y. Tang, *Biosyntheis of Natural Products*, Royal Society of Chemistry Press, London UK, 2019.
29. W. P. Jencks, *Catalysis in Chemistry and Bilogy*, McGraw Hill, New York, 1979.
30. C. T. Walsh, B. P. Tu and Y. Tang, Eight kinetically stable but thermodynamically activated molecules that power cell metabolism, *Chem. Rev.*, 2018, **118**, 1460–1494.
31. M. Pacholec, J. Tao and C. T. Walsh, CouO and NovO: *C*-methyltransferases for tailoring the aminocoumarin scaffold in coumermycin and novobiocin antibiotic biosynthesis, *Biochemistry*, 2005, **44**(45), 14969–14976.
32. S. Harayama, M. Kok and E. L. Neidle, Functional and evolutionary relationships among diverse oxygenases, *Annu. Rev. Microbiol.*, 1992, **46**, 565–601.

33. C. T. Walsh and T. A. Wencewicz, Flavoenzymes: versatile catalysts in biosynthetic pathways, *Nat. Prod. Rep.*, 2013, **30**, 175–200.
34. C. T. Walsh and B. S. Moore, Enzymatic Cascade Reactions in Biosynthesis, *Angew. Chem., Int. Ed.*, 2019, **58**(21), 6846–6879.
35. H. J. Sofia, G. Chen, B. G. Hetzler, J. F. Reyes-Spindla and N. E. Miller, Radical SAM, a novel protein superfamily linking unresolved steps in familiar biosynthetic pathways with radical mechanisms: functional characterization using new analysis and information visualization methods, *Nucleic Acids Res.*, 2001, **29**, 1097–1106.
36. P. A. Frey, A. D. Hegeman and F. J. Ruzicka, The Radical SAM Superfamily, *Crit. Rev. Biochem. Mol. Biol.*, 2008, **43**(1), 63–88.
37. O. Diels and K. Alder, Synthesen in der hydroaromatiche Reihe, *Justus Liebigs Ann. Chem.*, 1928, **460**, 98.
38. R. Hoffmann and R. B. Woodward, Orbital symmetry control of chemical reactions, *Science*, 1970, **167**(3919), 825–831.
39. K. C. Nicolaou, S. A. Snyder, T. Montagnon and G. Vassilikogiannakis, The Diels–Alder reaction in total synthesis, *Angew. Chem., Int. Ed.*, 2002, **41**(10), 1668–1698.
40. M. Ohashi, F. Liu, Y. Hai, M. Chen, M. C. Tang and Z. Yang, *et al.*, SAM-dependent enzyme-catalysed pericyclic reactions in natural product biosynthesis, *Nature*, 2017, **549**(7673), 502–506.
41. C. S. Jamieson, M. Ohashi, F. Liu, Y. Tang and K. N. Houk, The expanding world of biosynthetic pericyclases: cooperation of experiment and theory for discovery, *Nat. Prod. Rep.*, 2019, **36**(5), 698–713.
42. C. T. Walsh and Y. Tang, *The Chemical Biology of Human Vitamins*, Royal Society of Chemistry, London, 2019, p. 446.

Section II

In this section of the book we analyze the most common types of carbon-based functional groups in carbon chemical biology, encompassing C=C, C–O, C–S, and C–N bonds. The elements C, H, O, N, P, and S make up 99% of human body mass. While there are a few hundred metabolites with direct C–P bonds in microbes, there are none in humans. Thus, analysis of the above four types of above functional groups account for the great bulk of functional group inventory of organic molecules that are acted on *in vivo*.

We start in Chapter 2 with alkenes (olefins) and to a lesser degree the rare involvement of alkynes in natural molecules. We examine common routes to alkene formation and then their accessible reactivities under physiologic conditions. The C=C double bond constitutes the simplest functional group type. Depending on structural context, most notably when alkenes are conjugated to carbonyl groups, C=C bonds can react as electrophiles, *e.g.* for hydride reduction to the saturated alkane ($-CH_2-CH_2-$) or by water addition to an alcohol. More typically, olefins can react *via* their π-electrons as carbon nucleophiles. The remarkable cyclization of the polyolefin squalene to the bacterial tetracyclic hopene metabolites can be viewed as a nucleophilic cascade of olefin reactivity. Alkene groups can also act as carbon nucleophiles towards enzyme-bound flavin peroxides for transfer of electrophilic oxygen [OH^+] for epoxide ring formation. Alternatively, iron-based oxygenases instead use the C=C double bond of substrates as a radical generator and transfer [$OH^\bullet$] equivalents. Thus, olefins can behave in enzyme active sites as electrophiles, nucleophiles, or carbon radicals.

Chapter 3 is the first of three chapters that summarize covalent bonds of carbons to heteroatoms, the functional groups thereby created, their reactivity, and their interrelationships. This chapter takes up the common C–O (and C=O) bonds found in organic molecules in biology (= metabolites). Enzymatic formation and interconversions of familiar alcohols, aldehydes or ketones, acyl groups, and carbon dioxide (CO_2) encompass much of the functional group transformations in biology. They also remind us that each of these groups is two electrons away from each other, in both the oxidative and reductive directions. The acyl groups can be present as the free carboxylic acids (carboxylate anions at physiologic pH), as linear oxoesters, or as cyclic oxoesters known as lactones. The acyl groups can involve a second heteroatom, sulfur or nitrogen, to give acyl thioesters or amides, entries to Chapters 4 and 5. An additional set of C–O groups of biologic relevance are cyclic ethers, most commonly at the epoxide, tetrahydrofuran, and tetrahydropyran ring sizes.

Chapter 4 turns to C–S bonds, examining the enzymatic step in which inorganic H_2S (from bacterial reduction of sulfate anions) crosses over to become an organic thiol group. The chapter examines the reactivity of thiols, compared to alcohols, and also the rare thioaldehydes and thioketones along with thiamides *vs.* their oxygen analogs. Four additional C–S groups of metabolic importance, not mimicked by C–O group chemical biology, are the thioether (= sulfide group, *e.g.* methionine), the trivalent sulfonium cationic group (*S*-adenosylmethionine), the disulfide group (R–S–S–R), *e.g.* between cysteine side chains in proteins, and acyl thioesters. The acyl coenzyme A thioesters are key thermodynamically activated, kinetically stable forms of acyl groups that facilitate nucleophilic carbon behavior at C_2 of the acyl group and electrophilic behavior at C_1.

Chapter 5 examines the common types of C–N functional groups in organic metabolites, starting with the crossover of inorganic NH_3 into organic frameworks. NH_3 forms imines with carbonyl groups, such as ketones. The enzyme glutamate dehydrogenase can run in the reductive direction, using a hydride donated by NADH to trap the hydrolytically labile iminium ion as the 2-amino group of L-glutamate. The second route of ammonia incorporation is mediated by glutamine synthetase which converts free ammonia into the C_5 carboxamide of glutamine.

These three functional groups, amine, imine, and amide, account for the bulk of C–N chemical biology. The coenzyme form of vitamin B_6 mediates amino group transfers between amino acid donors and keto acid acceptors by reversible imine–amine redox chemistry. Three additional C–N functional groups of biologic relevance are ureas, gunanidines, and amidines. The amidines are the stable imines of amides, while guanidines are the comparable stable imines of ureas.

CHAPTER 2

Olefin and Alkyne Functional Groups

Perhaps the simplest functional groups in carbon scaffolds are alkenes (=olefins) and alkynes (acetylenes), involving only carbon and hydrogen atoms without involvement of nitrogen, oxygen, phosphorus, or sulfur atoms. The alkenes are two electron oxidized from the parent alkanes while the alkynes are a further two electrons oxidized from the alkenes[1] (Figure 2.1). The C–C double bond in alkenes and the much rarer C–C triple bond in alkynes offer potentially reactive functional groups compared to the typically unreactive adjacent pair of CH_2–CH_2 groups in alkanes. We take up alkene formation and reactivity first, then turn briefly to the strategies organisms employ for conversion of alkene linkages to alkynes. In a sense the alkenes are the simplest type of organic functional groups found in natural metabolites as they do not involve a heteroatom (oxygen, nitrogen or sulfur) to create reactive centers. The main heteroatom-containing functional groups of chemical biology are addressed in the next three chapters.

2.1 Alkenes

2.1.1 Alkanes to Alkenes and Back to Alkanes

The enzymatic conversion of a pair of adjacent alkane methylene groups to the alkene product is a formal two electron oxidation (Figure 2.2). The paired CH_2 groups in the alkane have been converted to the adjacent CH=CH pair in the alkene. Depending on the other substituents on the starting carbon atoms, the geometry of the product olefin can be *cis* or *trans*.

The removal of two hydrogens and two electrons could in principle occur by two routes. One mechanism would be loss of each hydrogen with one of the electrons in each C–H bond, that is a pair of hydrogen atoms in flight,

Chemical Biology No. 21
The Chemical Biology of Carbon
By Christopher T. Walsh

Published by the Royal Society of Chemistry, www.rsc.org

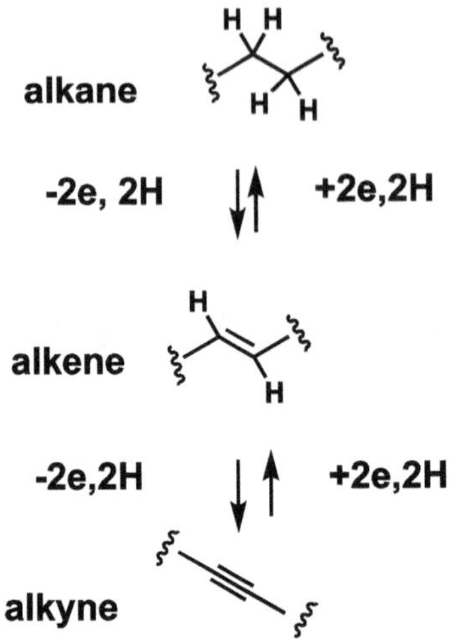

Figure 2.1 Alkanes, alkenes, and alkynes are separated by two electrons in redox interconversions. While alkanes are generally unreactive, both alkenes and alkynes are biologically useful functional groups that can react in enzyme active sites.

with a pair of carbon radicals (as transition states or high energy intermediates). The electrons remaining on the two carbons could form the new carbon–carbon bond in the product olefin. Alternatively, the polarity for cleavage of each of the two C–H alkane bonds could be distinct. One could depart as a proton, the other as a hydrogen with two electrons, a hydride ion. In either route there must be a cosubstrate acceptor that can accept the two electrons being removed from the adjacent alkane carbons, either as two hydrogen atoms (two $H^•$) or as the hydride ion with its associated pair of electrons (Figure 2.2).

The proton/hydride route is operant when the paired methylene groups of an alkyl chain are adjacent to a carbonyl group. The radical route appears to be operant when the alkane methylenes are isolated, not activated by an adjacent carbonyl group. We take up the formation and reduction of conjugated alkenes first then turn to isolated alkene formation *in vivo*. Desaturation of a pair of unactivated methylene groups in an alkane to alkene product is difficult and typically requires high valent oxoiron enzyme oxidants to carry out radical chemistry on the way to olefins, typically with *cis*-geometry.

For a pair of methylene carbons adjacent to a carbonyl group (aldehyde, ketone, thioester, even free acid), families of enzymes can catalyze a low energy pathway to generate a product olefin conjugated to the carbonyl group. That is typically a desaturase acting by a proton–hydride route. If the carbon α- to the carbonyl has at least one C–H bond, that is the hydrogen removable as a proton. The incipient carbanion can be delocalized and stabilized as the enolate/thioester enolate (Figure 2.3) and is thereby kinetically accessible in the active site of dehydratases.

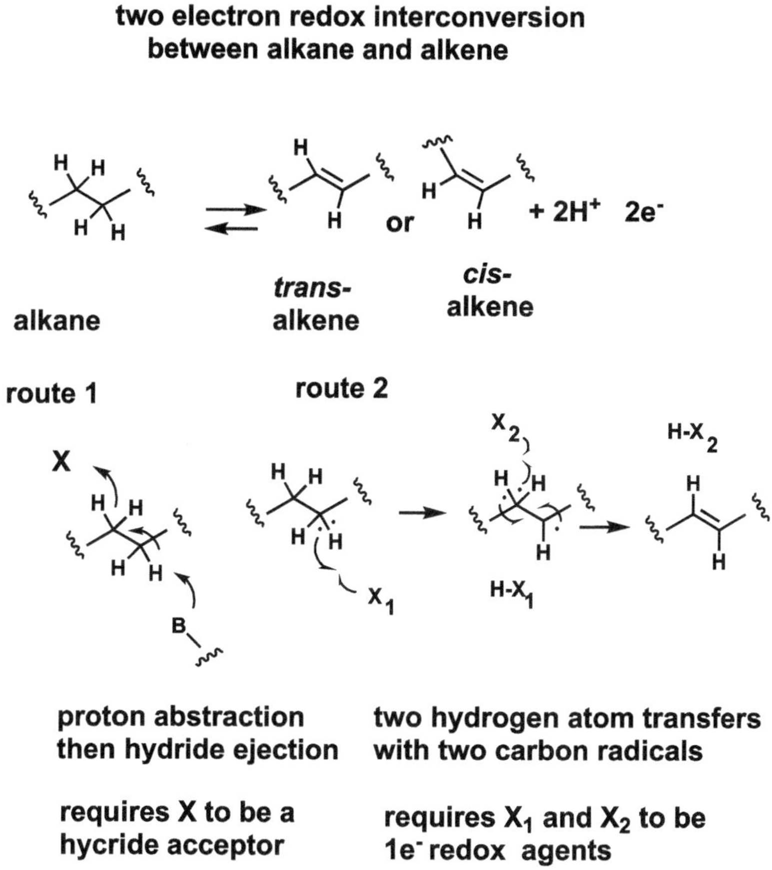

Figure 2.2 Two electron redox interconversion between alkane and alkene: two mechanistic possibilities: route 1: proton/hydride $2e^-$ mechanism; route 2; two $1e^-$ steps involving two hydrogen atoms and two carbon radicals (transition states). Route 1 requires cosubstrate X to be a hydride acceptor. Route 2 requires X_1 and X_2 to be one electron redox active.

2.1.2 Two Common Hydride Transfer Coenzymes

There are two types of heterocyclic organic molecules that serve as hydride acceptors in functional group redox reactions: nicotinamides and flavins, the coenzyme forms of vitamin B3 and B2, respectively.[2–4] We will iterate in later chapters that the coenzyme forms of vitamin B3, the nicotinamide adenine dinucleotides, NAD^+ and $NADP^+$ (Figure 2.4) are obligate hydride transfer cofactors while the coenzyme forms of vitamin B2, riboflavin, use their tricyclic isoalloxazine heterocycle as permissive for either two electron hydride transfer or one electron transfers, depending on cosubstrate redox potentials. Hydride transfers to $NAD(P)^+$ in enzymatic transfers are directed to the C_4 position of the N_1 quaternized-cationic pyridinium ring, quenching the positive charge on N_1 as the 1,4-dihydopyridine ring of NAD(P)H forms.

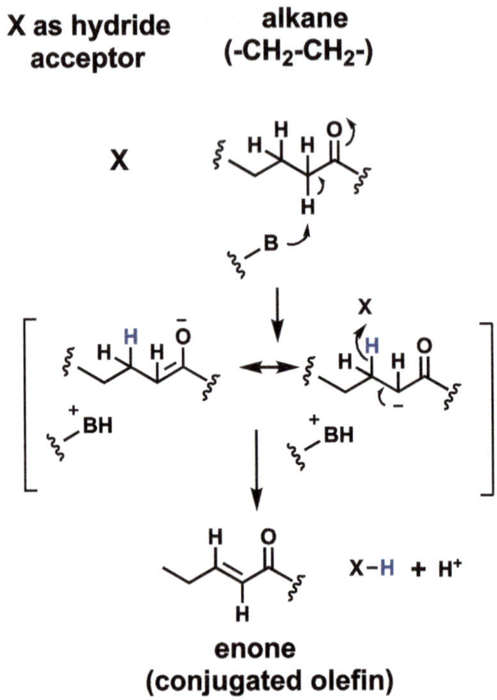

Figure 2.3 The hydride transfer route suggests initial abstraction of a C_2–H as a proton, to give the stabilized C_2 carbanion/thioester enolate anion as a species that can eject a C_3–H as a hydride ion.

For comparison hydride transfer from NADH to a tightly bound oxidized coenzyme in a flavin-dependent oxidoreductase active site goes to N_5 in the central ring of the isoalloxazine cofactor, generating the N_1, N_5 dihydroflavin regioisomer (Figure 2.4). These two central redox coenzymes emphasize early in this book two features of carbon chemical biology: the use of hydride transfers for much of *in vivo* redox chemistry and the key roles of nitrogen heterocyclic rings in the organic chemistry of life.

With these two hydride transfer coenzymes identified, we can turn back to how alkanes are dehydrogenated by hydride mechanisms to alkenes *in vivo*. The first step in mobilization of energy, *via* harvesting electrons from the fatty acid oxidation, is enzymatic desaturation of alkanyl thioesters to the conjugated 2,3-enoyl thioesters[5] (Figure 2.5). Correspondingly, the final step in fatty acid biosynthesis involves the two electron reduction of 2,3-enoyl thioesters in conjugate hydride ion additions, as a vehicle for storing electrons. We observe utilization of oxidized flavin coenzymes as hydride acceptors in the alkane to alkene direction. However, NAD(P)H is the specific hydride donor in the reverse direction, alkene thioester to alkane thioester reductions.[6]

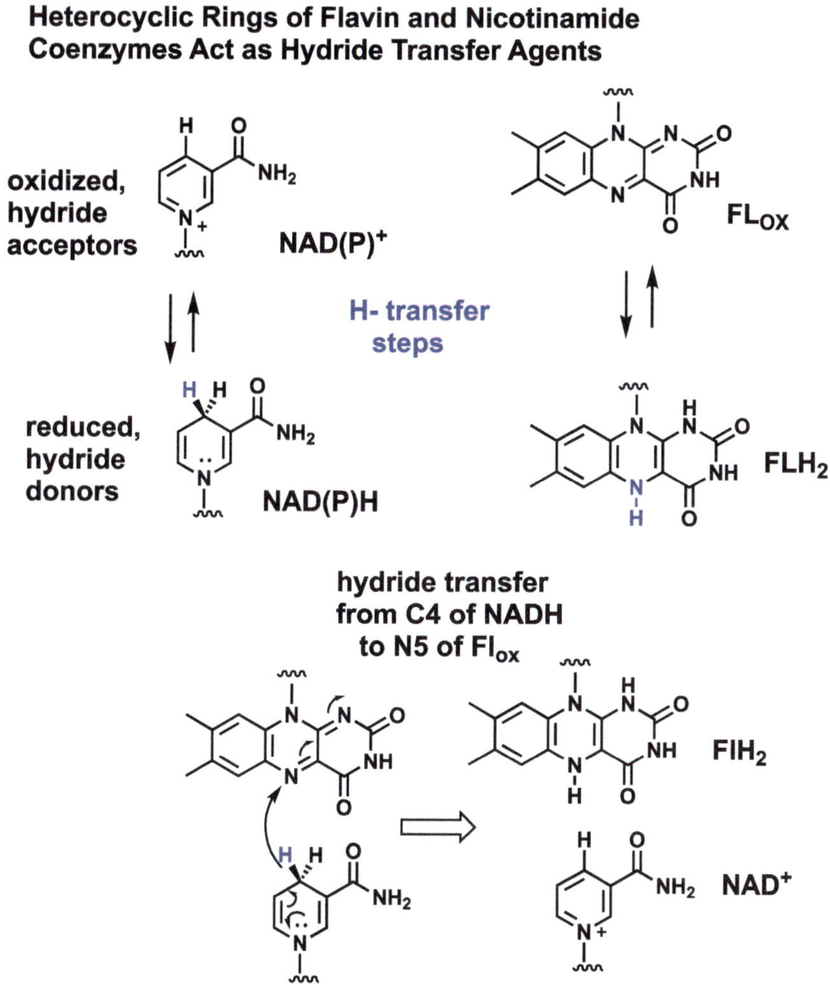

Figure 2.4 The heterocyclic rings of flavin and nicotinamide coenzymes can act as hydride transfer agents. In its reduced NAD(P)H state the dihydropyridine ring is a hydride donor of one of its two C_4 hydrogens. In the oxidized state flavin coenzymes (FMN, FAD) use their tricyclic isoalloxazine ring as hydride acceptor at central ring N_5. The oxidized form of $NAD(P)^+$ has the heteroaromatic pyridinium ion as electron sink. In the FlH_2 oxidation state the dihydroflavin can function as a hydride donor from N_5 of the dihydroflavin.

2.1.3 Acyl CoA Dehydrogenases

Three sets of FAD-containing acyl CoA dehydrogenases work selectively on long chain (C_{12}–C_{18}) acyl CoA thioesters, medium chain (C_8–C_{12}) acyl CoAs, and short chain (C_4–C_8) acyl CoAs to introduce conjugated *trans*-2,3-double bonds.[5] This is the first step in the oxidative pathway, followed by hydration

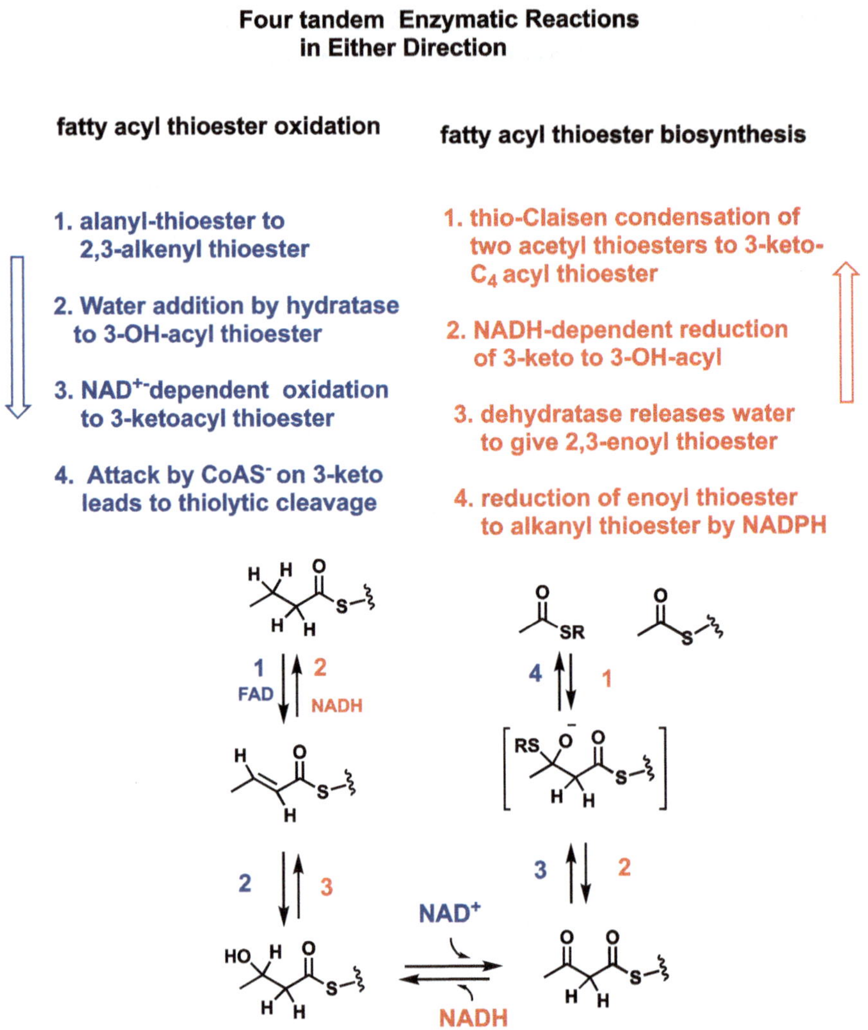

Figure 2.5 Four chemical steps that operate in reverse for fatty acid oxidation *vs.* fatty acid biosynthesis. Hydride transfer is the first step in fatty acid oxidation and the last step in fatty acid reductive biosynthesis. FAD-containing acyl CoA dehydrogenases are the hydride acceptors in fatty acid oxidation. NADPH is the hydride donor in enoyl thioester reduction. NAD^+/NADH is the hydride transfer coenzyme in the 3-OH- to 3-keto redox step in both pathways. Both pathways encompass four electron transfers in or out of the acyl scaffolds by a pair of hydride transfer events.

to the 3-hydroxy-acyl CoAs (nucleophilic addition of water OH group to electrophilic C_3 of the enoyl thioester), oxidation of alcohol to ketone, and then thiolytic cleavage of an acetyl CoA equivalent as the acyl chain is shortened by two carbons (Figure 2.5).

Acyl CoA dehydrogenases catalyze the net dehydrogenation to conjugated enoyl thioester products by the proton, hydride route (Figure 2.6). The directionality is set by the fact that the C_2–H_s of acyl chains in acyl CoA can be cleaved in the enzymes' active sites to the C_2-carbanion by an active site basic group. The incipient carbanion can be delocalized and stabilized by the resonance contributor that has the negative charge on oxygen, the thioester enolate contributor (Figure 2.6). That C_2 carbanionic transition state, or high energy short-lived intermediate, can use that delocalized electron pair to assist in expulsion of one of the two C_3–H substituents as a hydride ion to N_5 of oxidized FAD as acceptor. This generates the *trans*-2,3-enoyl CoA and the dihydro $FADH_2$.

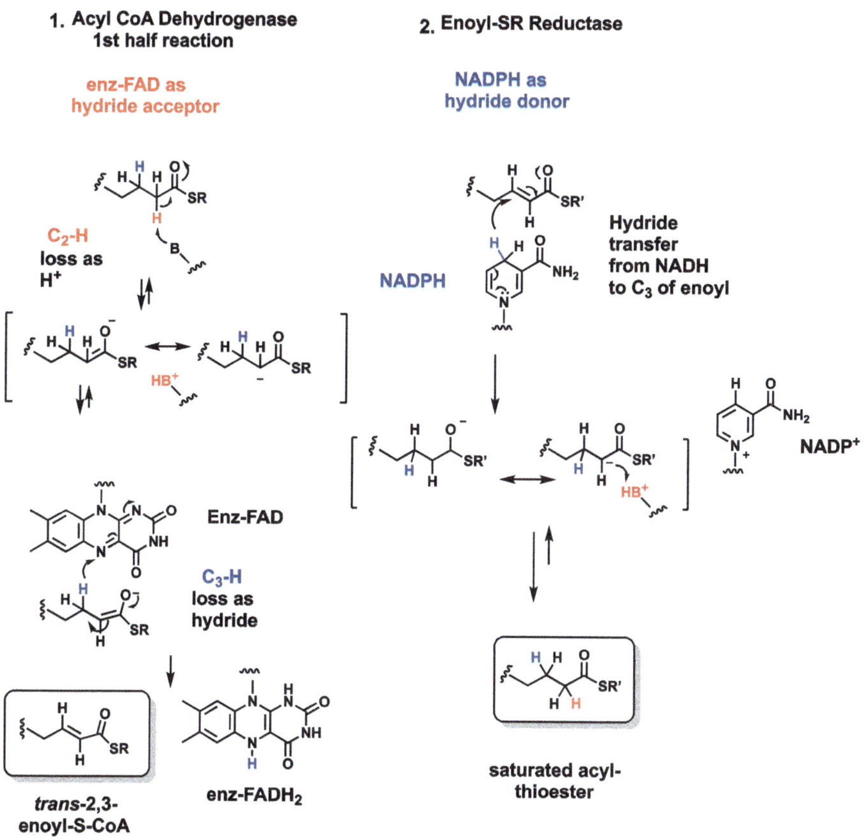

Figure 2.6 Parallel but distinct strategies for hydride transfer by FAD-dependent acyl CoA dehydrogenases *vs.* NADPH-dependent enoyl thioester reductases. The acyl CoA dehydrogenases function in the alkanyl to 2,3-enoyl-SR direction while the enoyl reductases run in the opposite direction with conjugate addition of a hydride into C_3 of the enoyl thioester cosubstrate. In both directions a thioester enolate is a resonance contributor of a stabilized C_2 carbanion, albeit at different stages of the reaction cycles.

Because FAD and its cognate FMN coenzyme are tightly held in enzyme active sites and do not freely dissociate during or at the end of a catalytic cycle,[4] a reoxidative half reaction must occur by which two electrons are transferred from $FADH_2$ back out to a second substrate (Figure 2.6). We will note in subsequent chapters that in such second half reactions many $FADH_2$ groups undergo two single electron transfer steps in reoxidation. For oxidases and oxygenases, the cosubstrate is O_2 for one electron reductive steps. For dehydrogenases, other flavoproteins and/or iron-containing proteins are the reducible cosubstrates.

From the point of view of the acyl CoA substrate, the first half reaction has generated the conjugated 2,3-enoyl thioester product. Both the proton abstraction step at C_2 and the subsequent hydride ejection step at C_3 are stereospecific (see Section 3.4.2). The newly generated alkene group is polarized, by conjugation to the thioester carbonyl, with the olefinic terminus at C_3 functioning as an electrophilic carbon. Evidence for conjugated olefins acting as electrophiles is provided by the next enzyme set that hydrate the 2,3-alkenyl thioesters specifically to the 3-hydroxyacyl thioesters, never to the 2-hydroxyacyl products (Figure 2.7).

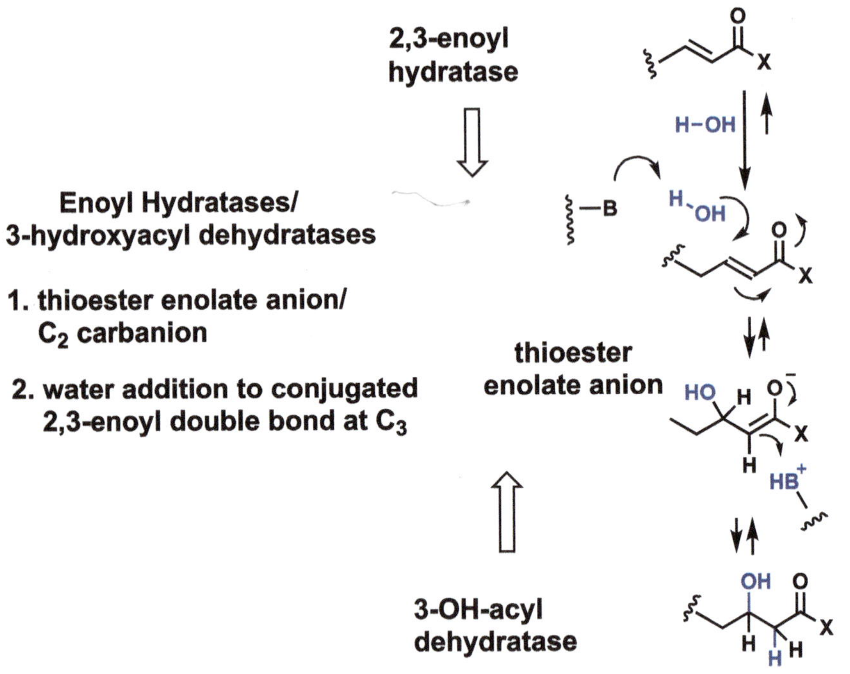

Figure 2.7 Enoyl CoA hydratases and the cognate 3-OH-acyl thioester dehydrases use equivalent chemical strategies, in the reverse physiologic directions. The two steps, occurring in different orders, are (1) formation of a stabilized, delocalized C_2 carbanion equivalent and (2) water addition to a conjugated 2,3-enoyl thioester C=C double bond.

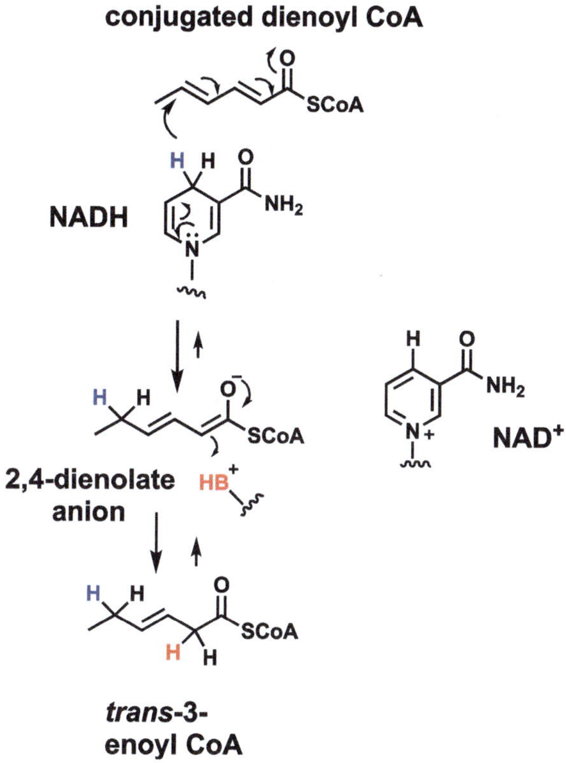

Figure 2.8 Reduction of 2,4-dienoyl CoA substrates by NADH-utilizing reductases is directed to hydride transfer to the C_5-olefinic terminus of the diene. Subsequent protonation of the thioester dienolate anion at C_2 yields the unconjugated *trans*-3-enoyl CoA.

The proton, hydride strategy, likely stepwise *via* thioester enolate chemistry (C_2-carbanions), is generalized to other flavin-linked alkane to alkene desaturases. One central metabolic example is succinate dehydrogenase, functioning in the citric acid cycle. Succinate is desaturated to the *trans*-olefin fumarate as the enzyme-$FADH_2$ oxidation state is formed by a comparable proton/hydride redox mechanism (Figure 2.8). Dihydroflavin reoxidation proceeds by one electron transfer paths to membrane soluble quinones and then the heme B cofactor.

2.1.4 Enoyl Reductases Uses NADPH as Hydride Donor to C_3 of the 2,3-Enoyl-thioester Substrates

The physiologic reverse of fatty acid β-oxidation is deposition of saturated long chain fatty acyl groups during fatty acid biosynthesis. In times of excess reducing power in cells (high $NADH/NAD^+$, $NADPH/NADP^+$ ratios), cells and organisms store reducing electrons in the alkanyl chains of fatty acyl groups.[6] When chemical energy needs to be mobilized, fatty acids are instead oxidized to provide reducing equivalent in the forms of $FADH_2$ and NADH to be used as hydride transfer agents for other metabolic reactions.

Cells have various strategies for keeping the oxidative pathways separate from the reductive laying down of fatty acyl chain reserves. One is to use soluble coenzyme A thioesters for fatty acid oixidation, but protein-bound thiols for fatty acid reductive biosynthesis. A second distinction is to use FAD in the oxidative route, as noted above, to go from alkane to conjugated alkene, but to use NADPH as a proximal hydride donor in the reverse direction from conjugated alkenyl thioester to alkanyl thioester (Figure 2.6).

A reaction scheme for a prototypic enoyl reductase (Figure 2.6) uses NADPH, bound in the active site, as hydride donor to C_3 of the 2,3-enoyl thioester substrate. The reaction features N_1 of the dihydronicotinamide ring as a neighboring group participant to facilitate expulsion of one of the two C_4–H as a hydride ion. That two electron transfer step converts NADPH to the oxidized heteroaromatic pyridinium cation of $NADP^+$. Concomitantly, the nascent acyl thioester is the C_2-carbanionic/thioester enolate resonance hybrid structure (lower energy to form that carbanion due to its delocalization). Chiral protonation at C_2 completes the conversion of conjugated alkenyl to alkanyl thioesters.

An extension of the use of NAD(P)H as hydride donor to reduce 2,3-enoyl thioesters is observed in the action of 2,4-dienoyl CoA reductases.[7] These reductases transfer a hydride ion from NADH specifically to the diene terminus at C_5. Regiospecific protonation of the thioester dienolate anion gives the *trans*-3-enoyl CoA, a net 1,5-reduction by a doubly vinylogous conjugated alkene participation. This 3-*trans*-double bond is out of conjugation (Figure 2.8). Also of interest are enzymes that fix CO_2 as a carboxylate group while reducing a conjugated double bond *via* hydride addition from NADH. An example is the reductive crotonyl CoA carboxylase (Figure 2.9). The requisite carbon nucleophile at C_2 to capture the electrophilic carbon in CO_2 arises from prior conjugate addition of the hydride ion from NADH to C_3 of crotonyl CoA.[8]

The acyl CoA dehydrogenases and the enoyl thioester reductases run in opposite directions physiologically (Figures 2.5 and 2.6). Both use hydride ion transfers to carry out their respective two electron redox functions. These two central metabolic redox interconversions showcase the surprising reliance of nature on hydride transfers for creation and reduction of alkene functional groups. To implement such hydride transfer strategy, nature has invented two heterocyclic redox active coenzymes (see Chapter 9), nicotinamide and flavin coenzymes, obligate *vs.* permissive hydride transfer reagents. Indeed, hydride transfers account for all 24 electrons (12 pairs) removed during complete catabolism of glucose to six CO_2 molecules (see Chapter 14).[9]

Both the creation of conjugated olefins and their reduction back to alkanes utilize hydride transfer strategies. In subsequent chapters, *e.g.* Chapter 6 on cycloalkanes, we will contrast this strategy with successive dehydrogenation by oxygen functional group elimination routes for C=C double bond creation, from cyclohexyl to cyclohexenyl to cyclohexadienyl to

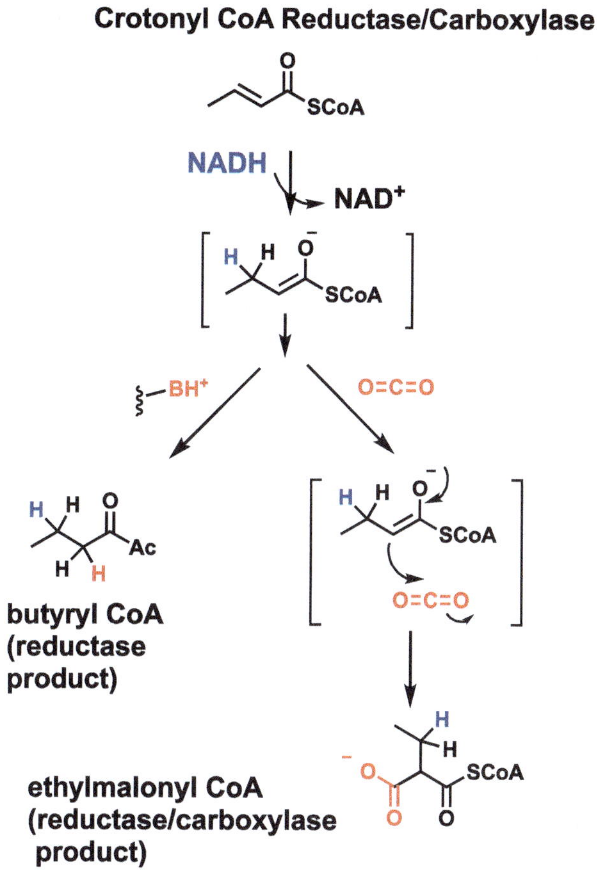

Figure 2.9 The crotonyl CoA reductase/carboxylase is an unusual catalyst for fixing CO_2 as a carboxylate group. The strategy is to employ initial conjugate addition of a hydride ion from NADH to C_3 to give the C_2-carbanion/thioester enolate. That carbanion can be reprotonated to the saturated butyryl CoA product in competition with capture of electrophilic CO_2 by the carbanion to give ethylmalonyl CoA.

phenyl rings in the chorismate pathway. In Section 2.2, we also contrast these nicotinamide/flavin redox mechanistic axes with the introduction of isolated, unactivated olefinic bonds in unsaturated fatty acids.

2.1.5 Olefin Hydration/Dehydration Is Nonoxidative

Note that, by contrast, dehydration of 3-hydroxy acyl thioesters, intermediates in both the fatty acid oxidation and fatty acid reductive synthetic pathways are nonoxidative (Figure 2.7). Addition of water to an 2,3-enoyl thioester in fatty acid oxidation or the reverse, dehydration of the 3-hydryacyl thioester to the conjugated 2,3-enoyl thioester, do not involve any change in oxidation state. No hydride transfers or redox coenzymes are required.

What is required is the ability of a dehydratase to generate a C_2-carbanion/thioester enolate. Expulsion of the OH^- from C_3 is formally equivalent to H^- expulsion but OH^- is instantly protonated to water. In the reverse, hydratase direction water must be activated in an active site to function as oxygen nucleophile at electrophilic C_3 of the 2,3-enoylthioester substrate. The immediate addition product is the stabilized C_2 carbanion/thioester enolate, readily protonatable with chirality control in that active site.

2.2 *De Novo* Olefin Formation in Isoprenoid Pathways

A second distinct strategy for *de novo* olefin/alkene bond formation is pervasive in isoprenoid and steroid biosynthetic pathways. While the best studied isoprenoid biosynthetic route, operant in animals and plants, goes through hydroxymethylglutaryl CoA and mevalonate,[10] some bacteria, including human pathogens instead generate the two isomers of isopentenyAl-PP *via* methylerythritol phosphate (MEP) and radical enzyme chemistry to create the double bonds.[11] We focus here only on the mevalonate route but briefly analyze the logic underpinning the MEP route in Chapter 15.

As depicted in Figure 2.10, three molecules of acetyl CoA are used to initiate the steroid biosynthetic pathway by action of a tandem pair of thio-Claisen condensation enzymes to yield the branched six carbon hydroxymethylglutaryl CoA, with one of the two carboxylates activated as the electrophilic thioester. That thioester is reduced by the rate-limiting enzyme HMG CoA reductase, first to the aldehyde mevaldate and then on to the alcohol mevalonate by a tandem pair of NADPH-dependent hydride transfers (Figure 2.10B). *Inter alia*, the four electron reduction of acyl CoA to alcohol differentiates the two functional end groups (carboxylate anion *vs.* alcohol).

The next three enzymatic steps are tandem phosphorylations of the two alcoholic groups in mevalonate by ATPs (Figure 2.11). First the mevalonate 5-pyrophosphate is generated by tandem phosphorylations. Then, the third ATP is attacked by the 3-OH to give the mevalonate 3-PO_3^{2-}-5-pyrophosphate. This newly added phosphoryl group activates the carboxylate for low energy decarboxylation with C_3–O bond cleavage. The production of CO_2 and inorganic phosphate is accompanied by olefin formation at the new terminus of the resultant Δ^3-isopentenyl-PP (3-IPP). The ejection of the oxygen at carbon 3 is enabled by the prior phosphorylation. From a pK_a comparison, ejection of a phosphate ion (p$K_{a2} = 6$) compared to hydroxide ion (pK_a for HOH = 14) could be up to 10^8 fold faster (lower in energy).

This decarboxylation/C–O bond cleavage route is a completely distinct mode of alkene generation from the proton/hydride pathway noted above in Section 2.1.1. The chemical utility of this alkene as a functional group is manifested in several kinds of metabolic transformations, the first of which is a protonation/deprotonation sequence that generates the Δ^2-IPP isomer that then functions as a key allylic carbocation generator (Figure 2.12).

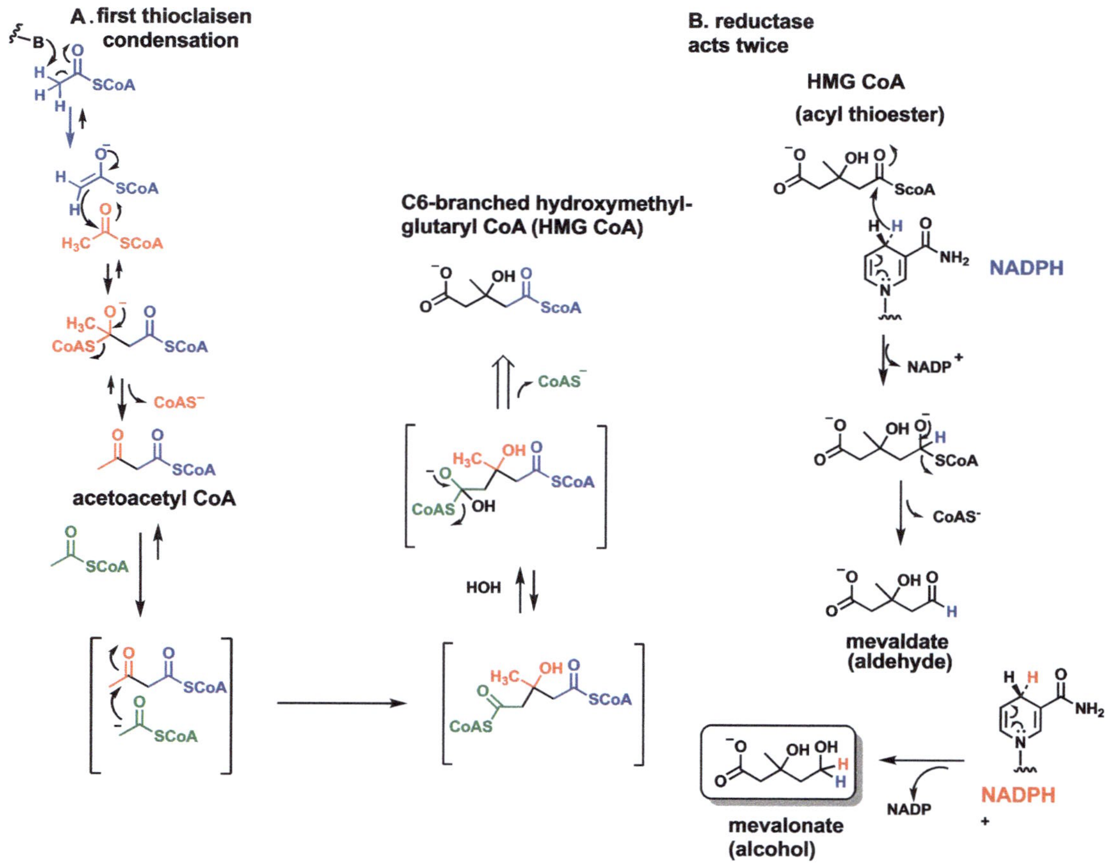

Figure 2.10 Three thio-Claisen condensations by the first two enzymes in the isoprenoid biosynthetic pathway build a six carbon branched 3-hydroxyglutaryl acyl chain. The first enzyme generates the linear four carbon 3-ketoacyl CoA, acetoacetyl CoA. Directed attack by the next enzyme on that 3-keto electrophilic carbonyl creates the branched chain hydroxyglutaryl group.

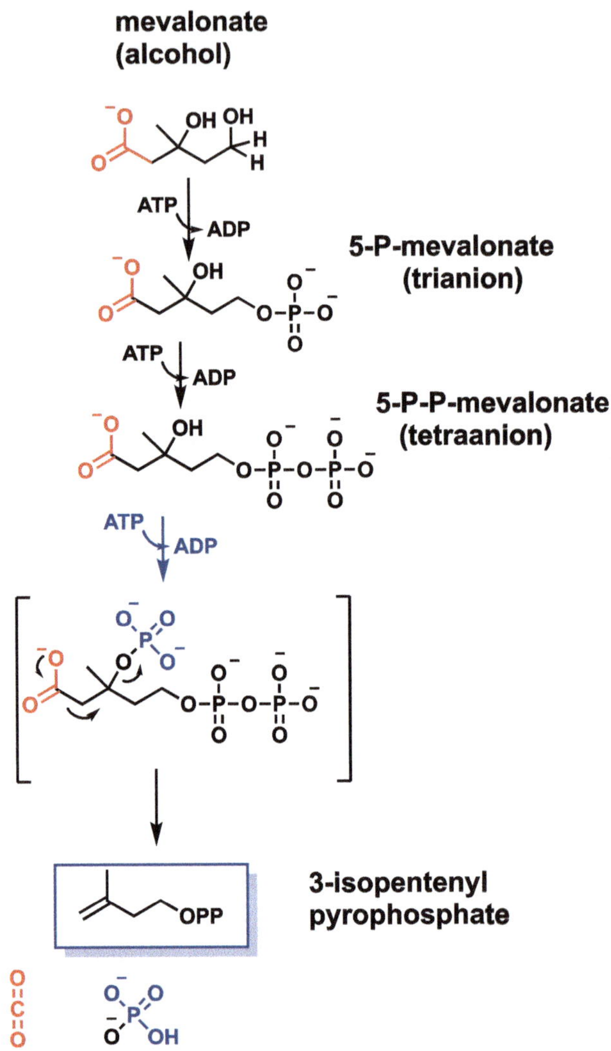

Figure 2.11 The olefin in 3-isopentenyl-PP forms by coupled (concerted?) enzymatic decarboxylation and -phosphate elimination from -3P-mevalonate-5-PP.

2.3 Reversible Enzymatic Protonation of Alkenes by Isoprenoid Isomerases

Once generated enzymatically, C=C double bonds are employed as highly versatile functional groups in a plethora of metabolic pathways, both as carbon electrophiles, carbon nucleophiles, carbon bases, even as carbon radical equivalents. The alkene of Δ^3-IPP can act as an electron rich species, reacting with a proton as a carbon base, or with a carbon center as a carbon nucleophile. Among these possibilities, catalyzed protonation of Δ^3-IPP is a

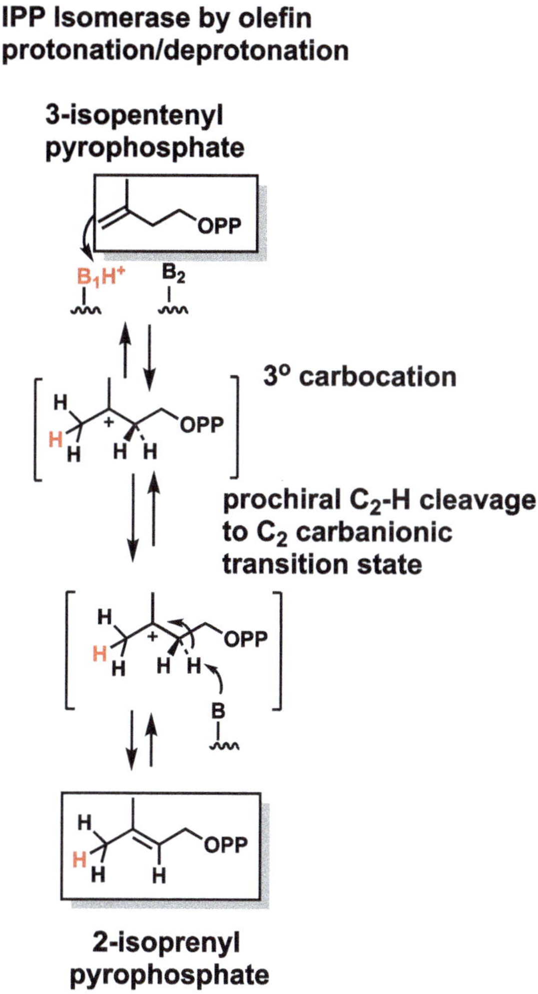

Figure 2.12 Olefin isomerization by isopentenyl-PP by IPP isomerase involving reversible protonation of either double bond *via* the tertiary carbocation.

likely route to Δ^2-IPP, the key other half of the isoprenoid building block inventory, by a suite of IPP isomerase enzymes (Figure 2.12). Reversible protonation of the terminal olefin in an IPP isomerase active site would generate a stabilized tertiary carbocation as shown.[12] Such a species, transition state or high energy intermediate, could partition by loss of one of two protons. Loss of one of the three chemically equivalent methyl hydrogens as a proton gives back the starting Δ^3-IPP. Loss of one of the two prochiral hydrogens at C_2 as a proton would quench the charge on the C_3 cation as the Δ^2-IPP isomer forms. Reversible alkene protonation with partition of two possible proton removals accounts for isomerization. In this case the read out is a net migration of the C–C double bond from the original terminal position to an internal locus. That Δ^2-IPP isomer is essential for electronic labilization of the C_1–OPP bond to facilitate early C–O bond cleavage in thousands of isoprenoid/steroid metabolic transformations across microbial, plant, and animal kingdoms.

2.4 Differential Reactivity of the Two Olefins in Δ^3- *vs.* Δ^2-IPP

The terminal C=C double bond in Δ^3-IPP and the internal C=C double bond in the Δ^2-IPP isomer allow complementary chemistry for C–C bond formation by large families of prenyltransferases. The Δ^2-IPP partner undergoes early C_1–OPP bond cleavage to create a stabilized, delocalized allyl cation in the transferase active sites (Figure 2.13). The double bond at C_2 of Δ^2-IPP has enabled C_1 to function as a carbocation precursor.

The π-electrons in the terminal alkene in the Δ^3-IPP partner then act as electron rich carbon nucleophile and attack the allyl cation at C_1 of the Δ^2-IPP derived partner. As denoted in Figure 2.13, this leads to a new C_1–C_4 C–C single bond as a stabilized tertiary carbocation is formed.[13] Analogous to decomposition of the IPP isomerase intermediates in the front direction, enzymatic cleavage of one of the two prochiral C_2–H bonds in that adduct will lead to the linear C_{10} product geranyl-PP. The chirality of most prenyltransferases is that the *E* (*entgegen*)-configuration of the elongated olefinic product is generated as product.[14]

That stereoselectivity and regioselectivity creates a product Δ^2-prenyl-PP that can react again as the electrophile in the next five carbon prenyl group extensions. Such iterative actions of prenyltransferases generate C15, C20 and C25 sequiterpene, diterpene, and sesquiditerpene frameworks (Figure 2.13). Clearly the alkenes act as differential functional groups for alkylative C–C bond formation, a mainstay reaction in chemical biology.

2.5 Intramolecular Capture of the Alkene-enabled Allylic Carbocations in Isoprenoid Metabolism: Alkenes as Carbon Nucleophiles

The chain-elongating prenyltransferases noted above catalyze intermolecular head to tail alkylations between a Δ^3-IPP molecule acting as carbon nucleophile and the elongating Δ^2-prenyl pyrophosphates acting as allyl cation generators (Figure 2.14). Chain elongation is an iterative process of C5 alkylations by the Δ^3-IPP acting as carbon nucleophile. Δ^2-IPP metabolites at the C10, C15, and C20 level have rich intramolecular chemical alternatives that are widespread in plant metabolism.[15–18]

For example, the 10 carbon monoterpene metabolite geranyl-PP has two double bonds. Once the C_{1-3}-allyl cation is generated in an enzyme active site, the *E*-allyl cation could undergo catalyzed isomerization to the *Z*-allyl

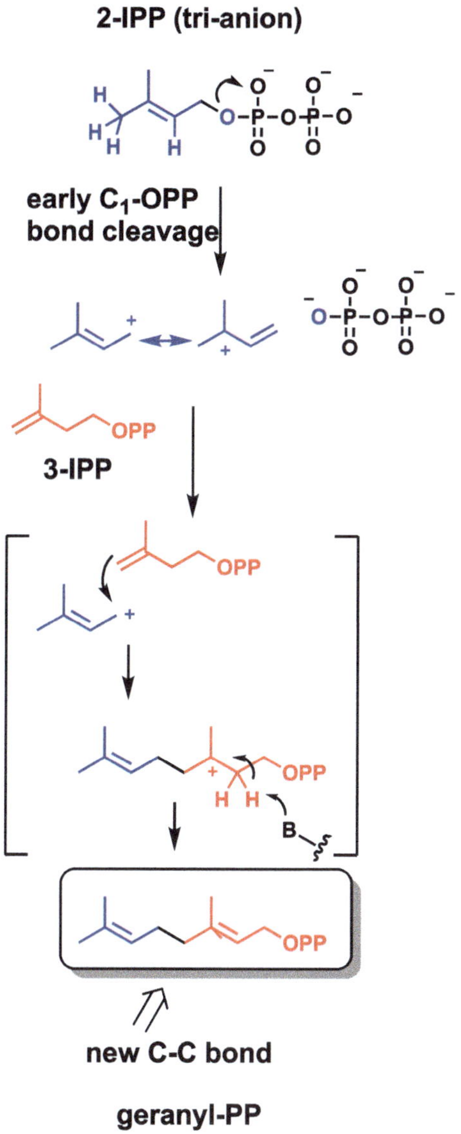

Figure 2.13 Mechanism of prenyltransferase C–C bond formation. 3-IPP as olefin nucleophile; 2-IPP as allyl cation precursor. C–C bond formation generates a tertiary carbocation. Deprotonation at C2 leads to the 2-prenyl-PP. The C10 product is geranyl-PP with two *E*-double bonds.

cation by way of nerolidol-PP (Figure 2.15). If the linear substrate is folded such that the terminal double bond can approach the *Z*-allyl cation geometric isomer, and react as a carbon nucleophile, intramolecular C–C bond formation ensues to form the cyclic α-terpinyl cation (Figure 2.15). That carbocation can be quenched by the three routes suggested in Figure 2.15 to

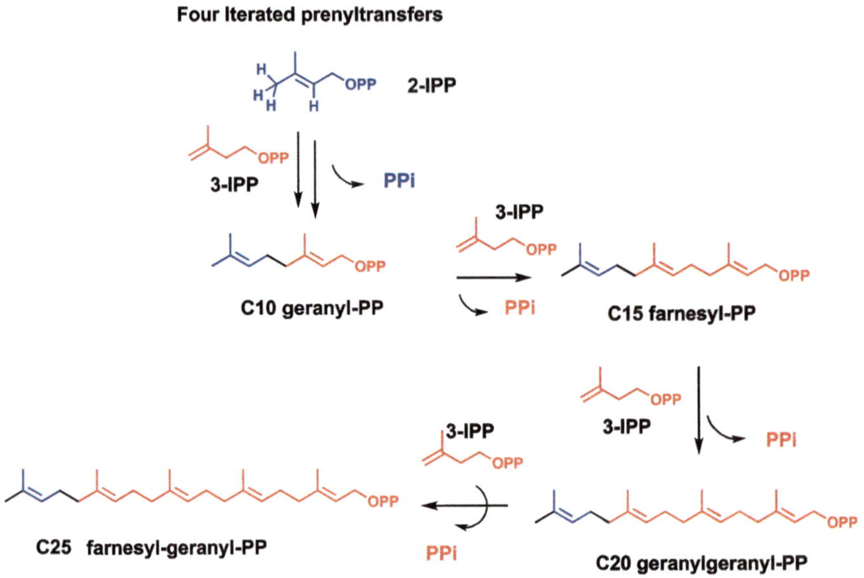

Figure 2.14 Iterative five carbon alkylations of growing 2-prenyl chains by addition of 3-IPP monomers.

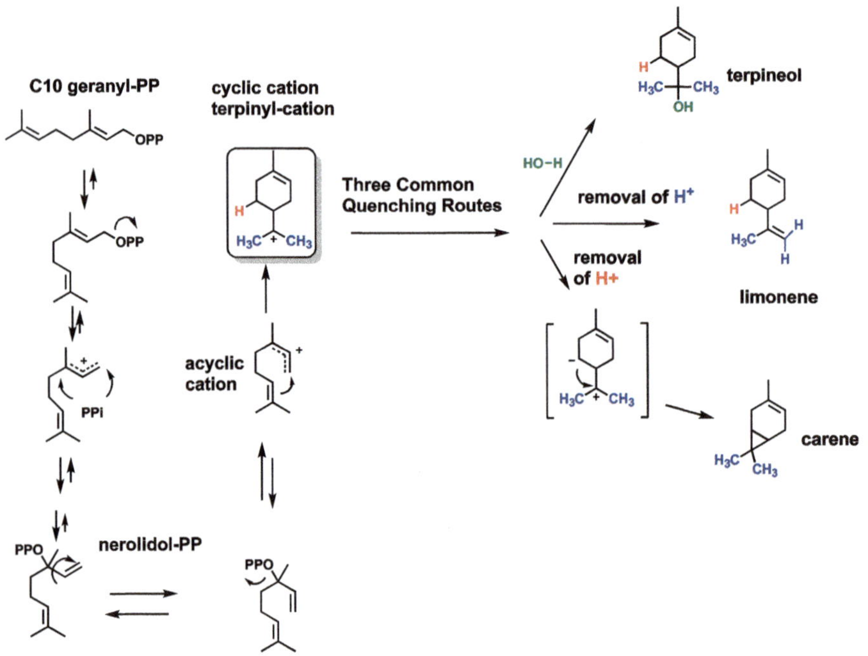

Figure 2.15 Carbocation-based rearrangement of linear geranyl-PP *via* nerolidol-PP to the cyclic terpinyl cation. Three common quenching routes for isoprenoid cations involve: water addition to give an alcohol; neighboring C–H bond cleavage by proton abstraction to generate an olefin; C–C single bond migration to generate a cyclopropane.

yield monocyclic alcohol α-terpineol, monocyclic bis-olefin, or the fused 3,6-bicyclic system of carene (a trimethyl-bicycloheptane framework). All three products prefigure the cycloalkane discussion in Chapter 7.

Analogously, Figure 2.16 shows how the C15 farnesyl-PP as a Δ^2-prenyl-PP can react to form humulenes by intramolecular catalysis.[19] Similarly, at the C20-diterpene level, geranylgeranyl-PP can be processed enzymatically by such intramolecular olefin/cation capture routes to both *ent*-kaurene[20] or to the 6-8-6 tricyclic scaffold of taxadiene,[21] under the direction of different diterpene cyclases (Figure 2.17), a remarkable plasticity to fused tri-carbacyclic frameworks.

These are the tips of veritable metabolite icebergs available for allyl cation species reacting intramolecularly. All exemplify the behavior of one or more alkene linkages, within the mono-, sesqui-, and diterpene scaffolds, as carbon nucleophiles that capture carbocations.

Enzymes have evolved three major routes to quench and terminate reaction manifolds mediated by intramolecular cationic species (Figure 2.15). One route is abstraction of a proton from a C–H bond adjacent to a carbocation. This route generates an alkene in the (rearranged) product. A second route is enzymatic addition of a (nucleophilic) OH group from a cosubstrate water molecule to form an alcohol product at the carbon site of the captured

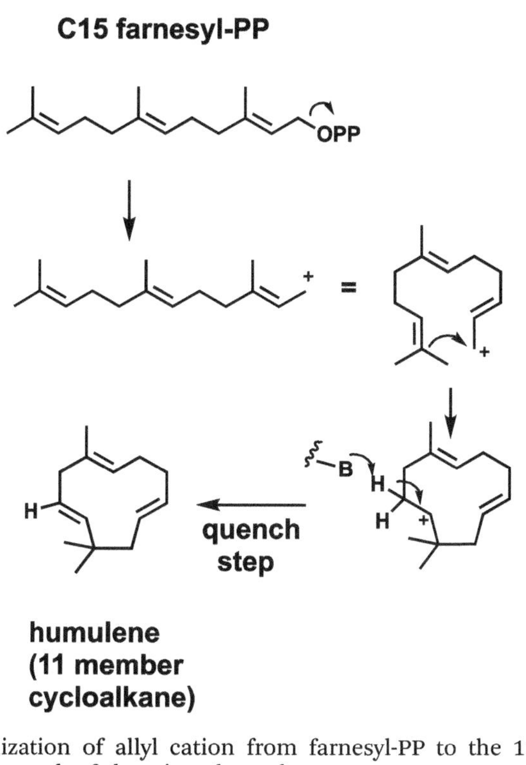

Figure 2.16 Cyclization of allyl cation from farnesyl-PP to the 11-member cyclic framework of the triene humulene.

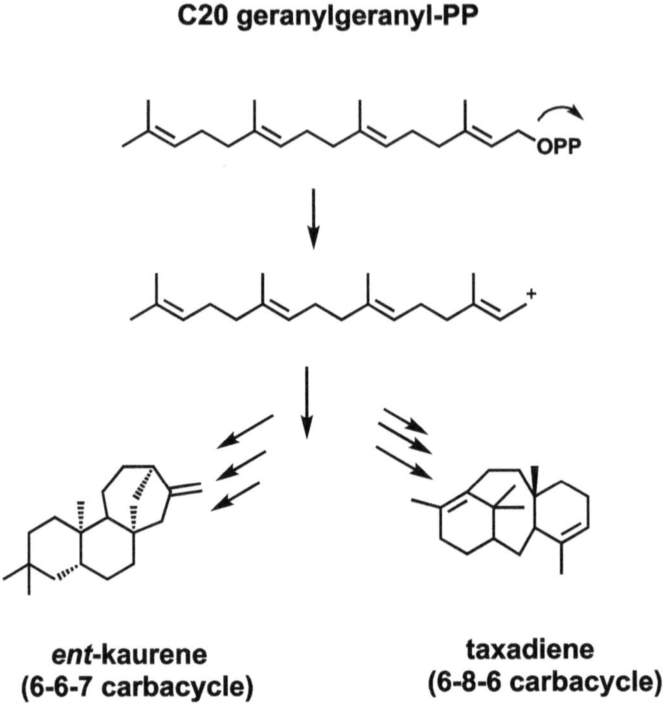

Figure 2.17 The C_{1-3} allyl cation from the linear diterpene geranylgeranyl-PP can undergo cation-olefin cyclizations to the 6-6-7 tricyclic scaffold of *ent*-kaurene and also to the 6-8-6 carbacyclic taxadiene.

carbocation. The third quenching strategy is by C–C bond migration. This last route leads to cyclopropane rings as in the plant metabolite carenes. Figure 2.15 illustrates how plants employ these three strategies in differentially quenching the α-terpinyl cation that arises from intramolecular capture of the geranyl-PP allylic cation by the terminal alkene.

2.6 Olefins as Carbon Nucleophiles in Squalene Cyclization Cascades

The linear C30 hexaene squalene, first isolated from shark liver, is a key biosynthetic intermediate in biosynthesis of tetracyclic and pentacyclic triterpenes. Squalene is formed in a head to head (C_1 to C_1) joining by squalene synthases, where cyclopropane-containing presqualene pyrophosphate arising from two molecules of farnesyl-PP are coupled enzymatically[20,22,23] (elaborated in more mechanistic detail in Chapter 15) (Figure 2.18). We will note that squalene itself can undergo enzymatic cyclization, as evidenced by hopene and hopanol formation in bacteria (Figure 2.19). In the next chapter we discuss that a more extensive metabolic cyclization flux occurs from squalene-2,3-oxide after metabolic epoxidation of one of the terminal double bonds in squalene.

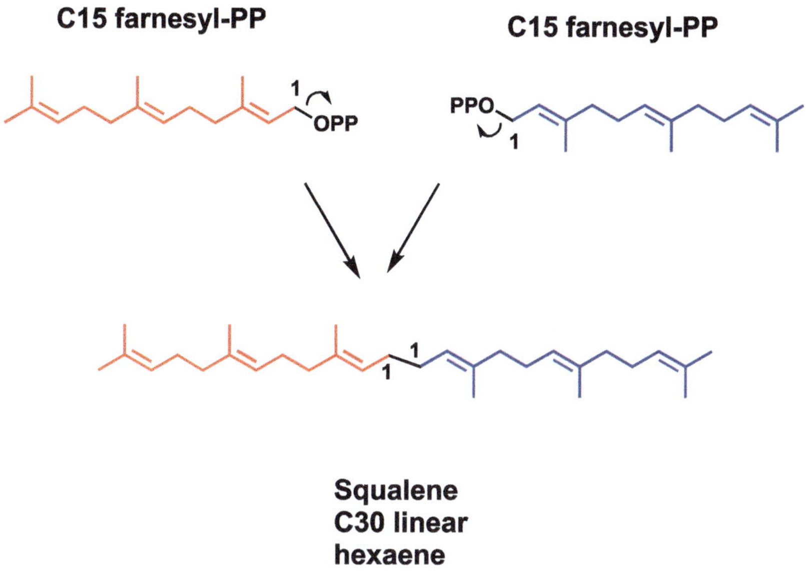

Figure 2.18 Head to head coupling of two C15 farnesyl-PPs yield the linear C30 hexaene squalene.

While squalene 2,3-epoxide is a stronger electrophile than the terminal double bond in squalene itself, the squalene cyclases can use unmodified squalene for protonation of one of the two terminal double bonds to initiate the cyclization cascade. As depicted in Figure 2.19, the product hopane scaffold is pentacyclic, with a 6, 6, 6, 6, 5 fused ring system. The linear hexaene framework of substrate squalene must be folded in the cyclase active site such that five of the six olefins participate as carbon nucleophiles in the cyclization cascade (Figure 2.19). The cyclization route forms four cyclohexanes and one cyclopentane ring with a presumed bound product-like hopenyl carbocation. We noted in Figure 2.15 two routes for quenching enzyme-bound isoprenoid carbocations are by: (1) adjacent C–H bond cleavage to turn the cationic intermediate into a product olefin or (2) by capture by a water molecule to give the alcohol. Indeed, both hopene and hopanol are products of the bacterial squalene/hopene cyclase (Figure 2.19).

Whether this cyclase and related cyclase family members carry out concerted or stepwise catalysis (*e.g.* forming some rings first *via* series of intermediate carbocations with finite lifetimes) may vary for different triterpene cyclases. Some triterpene cyclases leak minor amounts of partially cyclized products, suggesting stepwise interception of regiospecific carbocations.[24] Whatever the mechanistic continuum, the triterpene cyclases build remarkable carbacyclic structural complexity from a linear polyene scaffold *via* participation of multiple alkenes as carbon nucleophiles that capture transient carbocations.

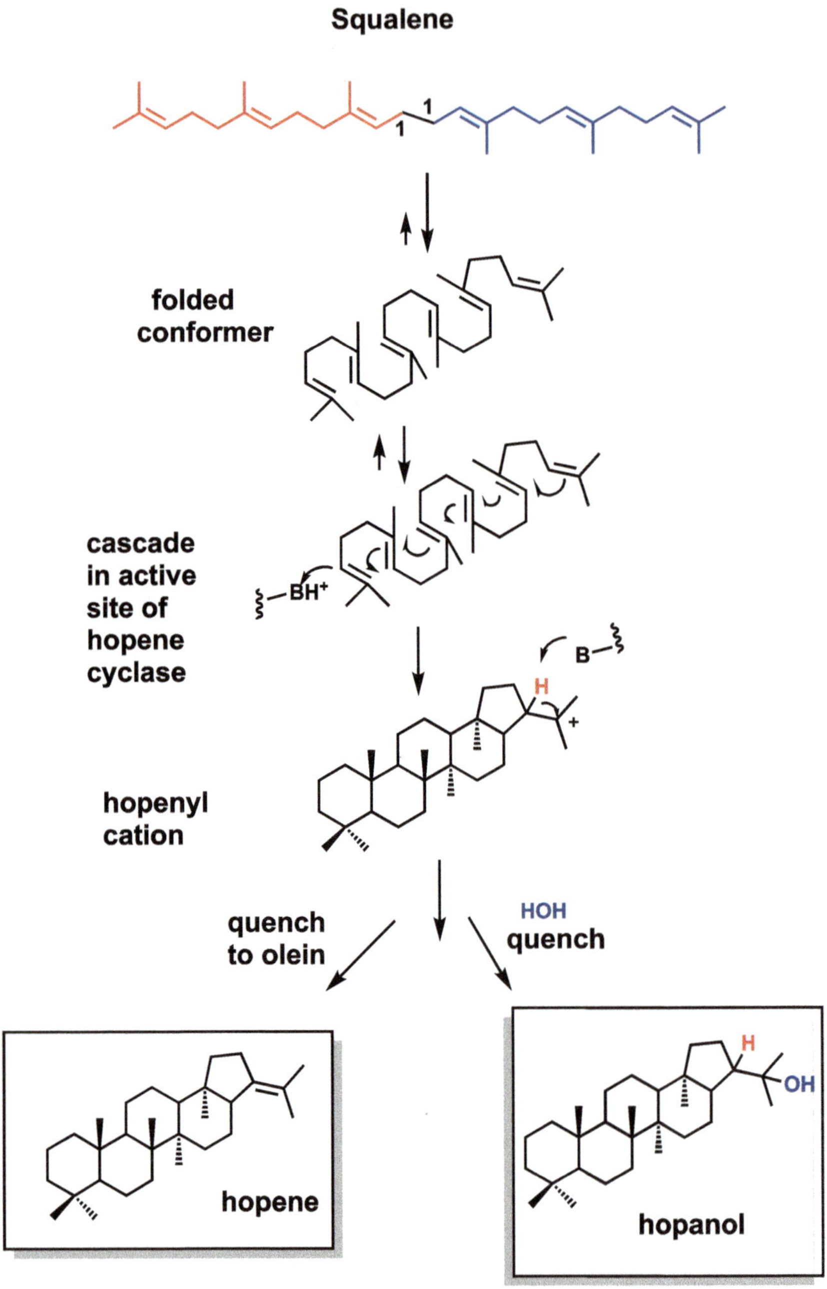

Figure 2.19 Hopene cyclase protonates a folded conformer of squalene to set in motion a cyclization cascade leading to a pentacyclic cation. Quenching by water give hopanol. Quenching by adjacent proton abstraction gives the exocyclic olefin in hopene.

2.7 *cis*-(*Z*-)-Geometry Imposed by Some Prenyltransferases

The large family of intermolecular C–C bond forming, head to tail isoprenoid chain-extending prenyl transferases quench their cationic intermediates to give *E*-configured (*trans*)-alkene functional groups. However, there are a smaller number of isoprenoid metabolites with *Z* (zussamen)-(*cis*) double bonds, interspersed among the *E*-olefinic linkages.[25]

One example is found in the 55 carbon bactoprenol-PP where 10 alkylative chain extension steps have produced the long chain lipid carrier molecule that moves growing muramyl peptide units from the inside to the outside of the bacterial cytoplasmic membrane during cell wall biosynthesis (Figure 2.20). Inspection of the bactoprenol pyrophosphate structure from *Lactobacilli* shows that the olefin in the first five carbon unit has been saturated, while the subsequent seven olefins have *Z*-geometry. The C-terminal three olefins have the anticipated *E*-geometry and are added by a distinct prenyltransferase. The seven *Z* double bonds impose a kinked structure on this lipid carrier.

Two sets of prenyltransferases work *ad seriatim* during oligomerization of the IPP building blocks, one Δ^2-IPP and ten Δ^3-IPPs. The first two elongations,

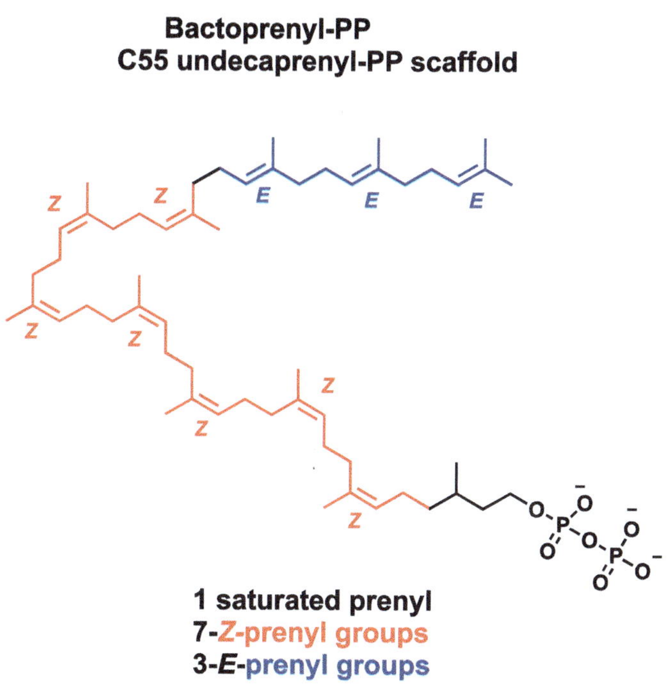

Figure 2.20 Bactoprenol-PP a C55 undecaprenyl-PP arises by 10 five-carbon alkylation steps. The first three olefins are established by an *E*-specific prenyltransferase. The next seven double bonds are created by a *Z*-specific prenyltransferase.

to the 15 carbon farnesyl-PP, arise from quenching of the tertiary carbocation intermediate by stereospecific removal of one of the two prochiral hydrogens at C_2 as a proton to set *E*-olefin geometry (Figure 2.21). Then, the related prenyltransferase, undecaprenyl-PP synthase, takes the all *E*-farnesyl-PP onto bactoprenyl-PP in eight consecutive extensions using eight molecules of Δ^3-IPP as carbon nucleophile.[26] This second, undecaprenyl-PP synthase must remove the other prochiral hydrogen at C_2 of the tertiary carbocations, generating *cis*-(*Z*)-double bonds each time. The last to be added prenyl group must undergo a net reduction of the olefin proximal to the pyrophosphate group. Isoprenoid double bond geometry is controlled by which proton is removed to quench the tertiary carbocation during alkylative chain elongations. The shapes of elongated *E*- *vs.* *Z*-polyolefins are dramatically different and enable distinct biologic and physiologic functions.

2.8 Olefins as Carbon Nucleophiles or Carbon Radical Equivalents

A brief set of comments about olefin epoxidation before considering the heterocyclic epoxide rings in the context of C–O bonds in Chapter 3 and in heterocyclic chemical biology in Chapter 8. The flavin-containing squalene epoxidase creates the 2,3-β-epoxy-squalene by delivery of an electrophilic oxygen equivalent [OH^+] from a covalent flavin hydroperoxide intermediate.[27] The olefin has behaved as carbon nucleophile, leaving an adjacent transient carbocation to be captured in the second C–O bond-forming step that generates the epoxide (Figure 2.22).

By contrast, in Chapter 12, carbon radicals generated by high valent oxoiron species in iron oxygenase active sites can be featured as occurring by formal one electron transfer processes. Such a route would generate a transient carbon radical that could collapse to the epoxide by radical recombination as an [$OH^{\bullet}$] equivalent is transferred from the oxoiron species[28] (Figure 2.23). Two distinct epoxidation routes for alkenes depend on the chemistry available to the proximal oxygen transfer agents.

Further indication for substrate olefins acting as carbon radical generators can be seen in some radical SAM enzyme catalyses. As we will summarize in Chapter 12, that enzyme superfamily generates 5′-deoxyadenosyl radicals by homolytic cleavage of *S*-adenosylmethionine in enzyme active sites. Among the fates for quenching the carbon-centered 5′-deoxyadenosyl radical is capture by a cosubstrate olefin, creating an adenosyl–carbon bond and an adjacent nascent carbon radical. An example of this reactivity pattern of a cosubstrate olefin as a radical trap for the deoxyadenosyl radical occurs in capture of an enolpyruvyl-benzoyl-ether metabolite derived from chorismate in a bacterial route to menaquinone[29] (Figure 2.24). The nascent product radical undergoes substantial rearrangement before release.

Alkene groups in substates, influenced by other electronic and structural functional groups, can be coaxed in enzyme active sites to react as carbon electrophiles, carbon nucleophiles, or as one-electron carbon radical traps.

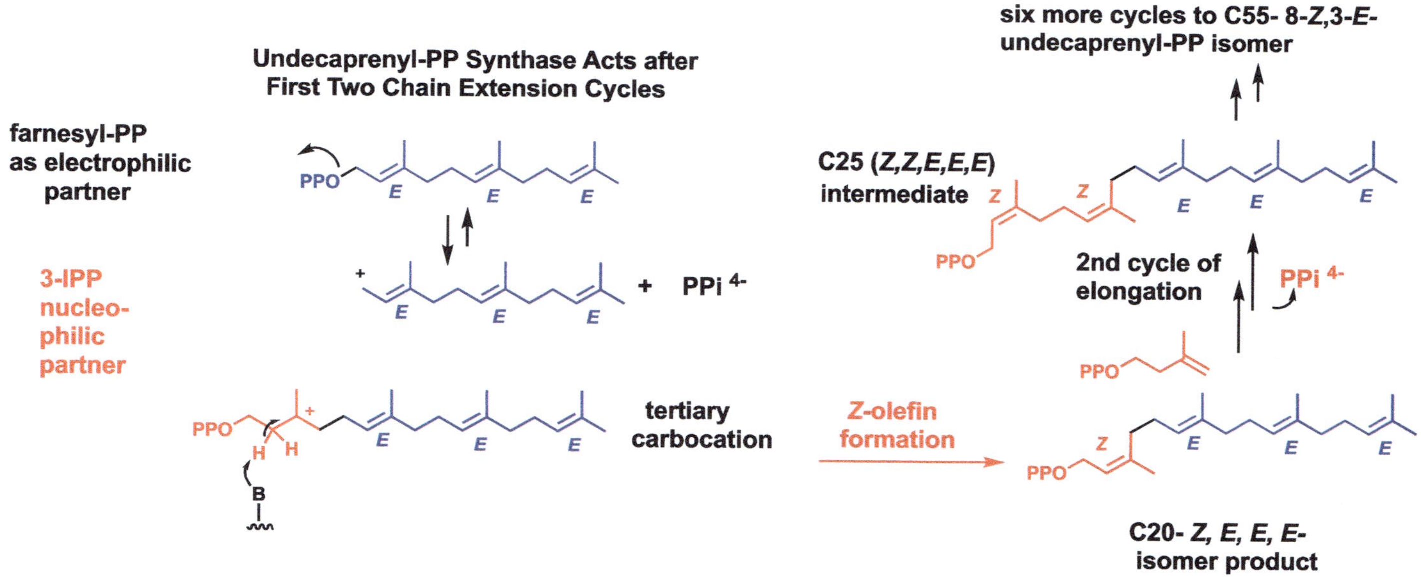

Figure 2.21 Undecaprenyl-PP synthase elongates C15 farnesyl-PP by eight five-carbon alkylation cycles with 3-IPP as monomer. The quenching of the initial tertiary cation in every chain elongation gives the 2-prenyl chain in every cycle but that newly formed olefin has *Z*-geometry.

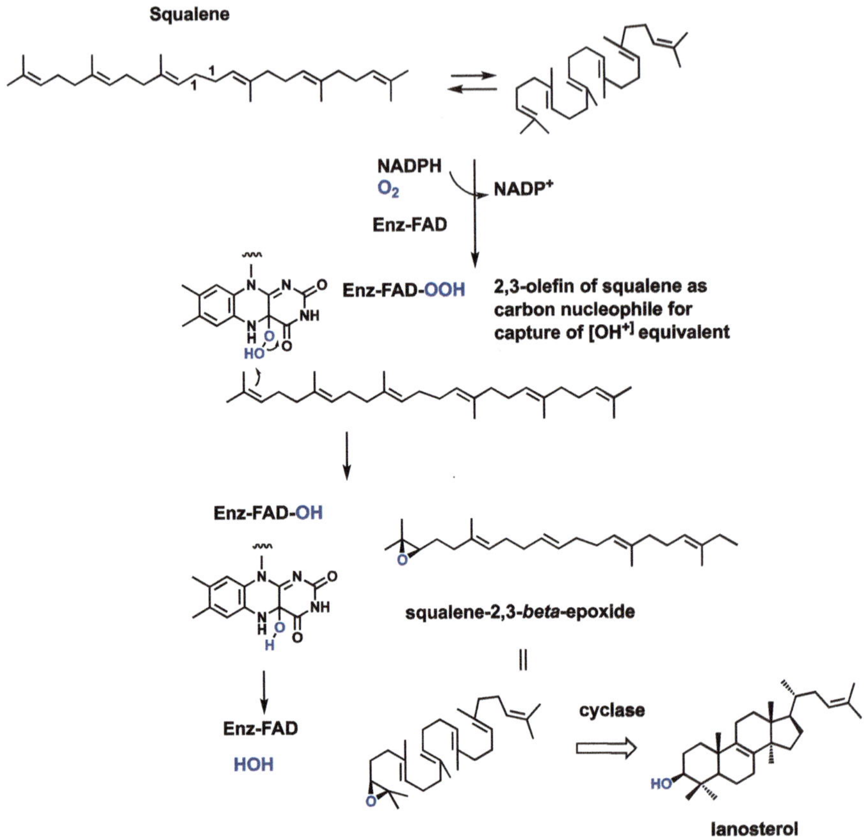

Figure 2.22 Squalene epoxidase is a flavoprotein monooxygenase using the terminal 2-olefin as nucleophile to form squalene 2,3-beta-epoxide. The proximal donor of an electrophilic oxygen to the squalene double bond is FAD–O–OH.

2.9 *De Novo cis*-Alkene Formation Strategies: Bacterial Case

Both bacteria and higher organisms contain a large range of long chain fatty acid molecules, differing in lengths and degrees and position of alkene linkages. Typical unsaturated fatty acids in bacteria are the Δ^9 palmitoleic and the Δ^{11} oleic acids (Figure 2.25). The single olefinic group has the unusual *cis*-(*Z*)-geometry. Plants also make a C18:2 linoleic acid, and C18:3 α-linolenate ($\Delta^{9,12,15}$). Animals make the C20:4 eicosanoate known as arachidonate ($\Delta^{5,8,11,14}$). All of the olefins have *cis*-geometry. The *cis*-olefins cause kinking in the acyl chains such that the acyl chains pack less well in membranes, increasing membrane fluidity in those local regions, presumptively critical for creating heterogeneous membrane microenvironments.

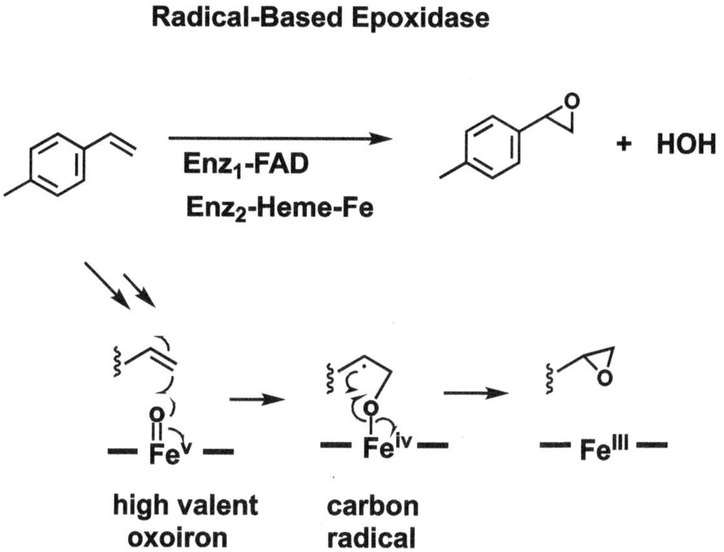

Figure 2.23 By contrast to squalene epoxidase that delivers an [OH^+] equivalent to cosubstrate olefin, iron-based oxygenases use high valent oxoiron species to generate radical-based pathways.

The strategies employed by bacteria to create isolated *cis*-alkene groups differ from that used by the eukaryotic organisms. As noted below, during fatty acid chain elongation cycles bacteria dehydrate the β-hydroxydecanoyl thioester intermediate to the normal α-olefin intermediate but then isomerize it to an unreducible γ-olefin.[30] That γδ-*cis*-olefin then persists during normal chain elongations. By contrast, eukaryotic plants and animals use O_2-reducing fatty acid desaturases for *de novo* introduction of *cis*-double bonds.

The bacterial strategy to the accumulation of oleate (*cis*-18:1) with the double bond at carbons 11–12 focuses around a dehydratase/isomerase that operates at the level of the C_{10} acyl group. During every two carbon chain elongation cycle of fatty acid biosynthesis, the initial 3-ketoacyl CoA condensation product is reduced to the 3-OH acyl CoA by hydride transfer from NADPH. The 3-OH acyl CoA suffers enzymatic dehydration to the 2-enoyl CoA and that is reduced to the pair of methylene groups from an NADPH cosubstrate molecule (Figure 2.5). The three enzymes, keto reductase, dehydratase, enoyl reductase, accomplish a four electron reduction (3-keto to 3-CH_2) in every elongation cycle, highlighting the storage of electrons in long chain fatty acids and their role as mobilizable energy-rich molecules (Figure 2.5).

Bacteria express a fatty acyl thioester dehydratase/isomerase that operates specifically on the C_{10} 2,3-enoyl thioester intermediate.[30] The isomerase catalyzes the αβ-*trans*-double bond isomerization from the αβ-substrate to a γδ-*cis*-olefin (Figure 2.26). This *cis*-olefin is now out of conjugation with the thioester carbonyl. It is higher in energy and might be the minor component at isomerase equilibrium. The isomerization is reminiscent of the double

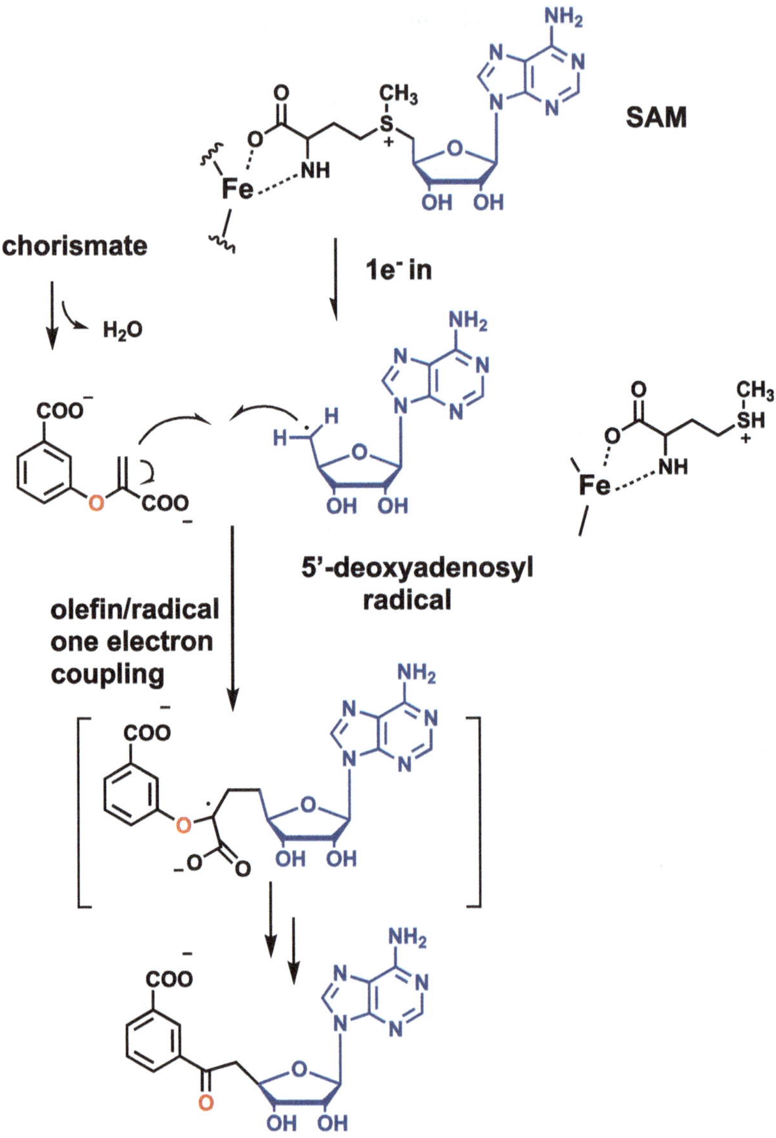

Figure 2.24 An example of a substrate olefin reacting with a carbon-based radical by one electron pathways is found in the net addition of the C=C double bond of enolpyruvyl benzoate to the 5′ deoxyadenosyl radical in an anaerobic pathway to menaquinone.

bond isomerization of Δ^3-IPP to Δ^2-IPP but this reaction is featured as proceeding by prochiral C_4–H proton abstraction to give the dienolate anion. That anion can be partitioned in the isomerase active site to reprotonate back at C_4 or instead reprotonate at C_2 which is the isomerization outcome (Figure 2.27).

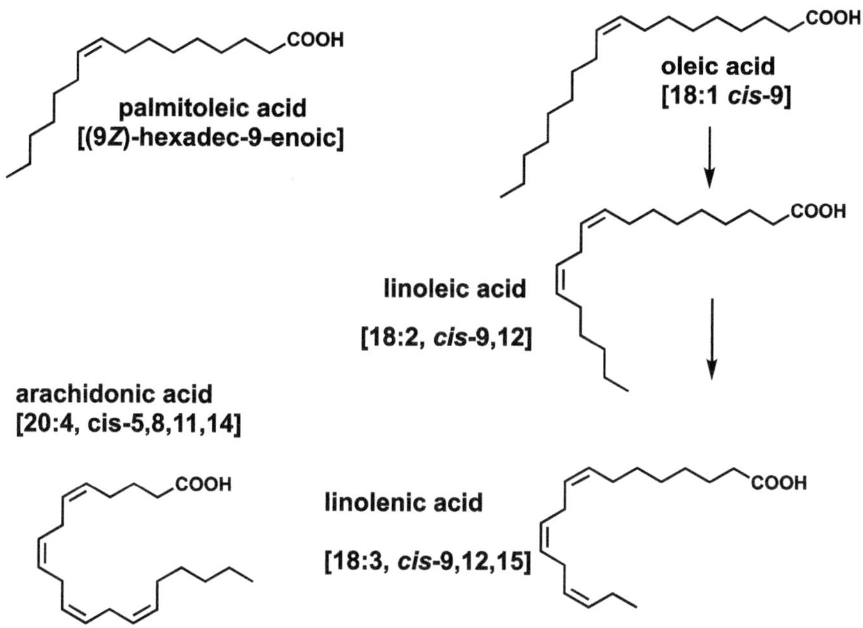

Figure 2.25 Naturally occurring *cis*-double bonds in fatty acids impose bent geometries on the alkyl chains.

The γδ-*cis* olefin, removed from electronic conjugation to the electrophilic thioester carbonyl, is not polarized to accept a hydride ion. It is unreducible because it is not activated to behave as an electrophile by an adjacent carbonyl group. The unconjugated double bond persists. The next cycle of two carbon chain elongation will generate a C12 acyl thioester with the double bond now at $C_{5,6}$. Three more chain elongations move that *cis*-double bond to $C_{7,8}$, then $C_{9,10}$, and then $C_{11,12}$ in the accumulating oleyl CoA product of bacterial fatty acid synthesis (Figure 2.27).

2.10 *De Novo cis*-Alkene Formation: Animal Case

Higher organisms utilize a distinct strategy. They use a desaturase enzyme with O_2 as cosubstrate to accept electrons from the pair of adjacent methylene groups that undergo oxidative removal of two electrons to a *cis*-olefin[31,32] (Figure 2.28). We defer discussion of the orthogonal reactivity of O_2 with spin-paired organic molecules to Chapter 12. Here we note that the fatty acid desaturases of plants and animal have two atoms of iron in the active site that bind O_2 and undergo net two-electron transfer to the bound O_2 to generate an oxoiron species that acts as a radical generator towards specific methylene groups of bound fatty acyl CoA cosubstrate (Figure 2.28). The external source of two electron reductant is NADPH, *via* a stepdown $2e^-/1e^-$ flavoenzyme cytochrome b5 reductase to deliver two electrons one at time to the diiron stearoyl CoA desaturase.

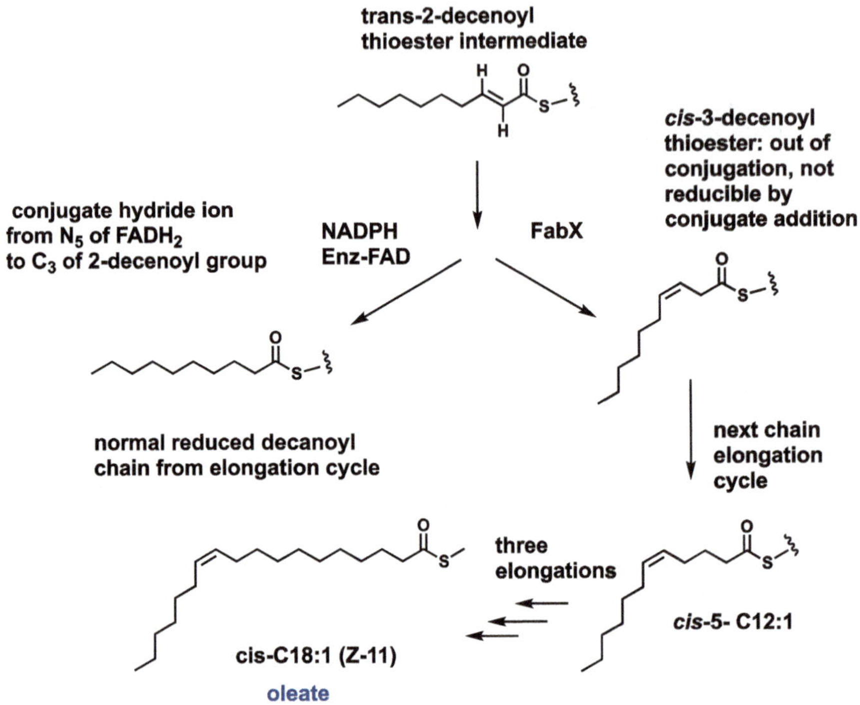

Figure 2.26 Bacterial monounsaturated *cis*-olefinic fatty acids: the key step occurs at the C10 acyl chain level. An isomerase can act to isomerize the conjugated *trans*-2-enoyl-thioester to the unconjugated 3-*cis*-enoyl thioester. That unconjugated double bond is not electrophilic enough to be reduced by hydride addition. The *cis*-olefin persists during subsequent cycles of chain elongation, for example all the way to the *Z*-11-oleate.

Successive cleavage of a C_{11}–H then a C_{12}–H bond in stearoyl CoA by radical paths yields a *cis*-11,12 olefin in product oleyl CoA. The two electrons removed from stearoyl CoA plus the electrons removed from each of the Fe^{II} atoms in the active site constitute four electrons funneled into the O_2 substrate, yielding two water molecules as products (Figure 2.28). Palmitoyl CoA is also a substrate, for desaturation to palmitoleyl CoA. These O_2-dependent desaturases, in generating substrate carbon radicals, but failing to incorporate an oxygen atom into products, qualify as "thwarted oxygenases" as discussed in Chapter 12.

Fatty acid desaturases can act iteratively. Thus, linoleate bears two *cis*-olefins, introduced stepwise *de novo* from the saturated stearoyl CoA. Linoleate is precursor to the 18:3 linolenate (Figure 2.25). The use of O_2 as cosubstrate for radical-based alkane desaturation to alkenes with regiochemical and stereochemical control presumably evolved quite late, once global oxygen levels rose to provide sufficient molecular oxygen as coreactant. The bacterial route to unsaturated fatty acids is oxygen independent and does

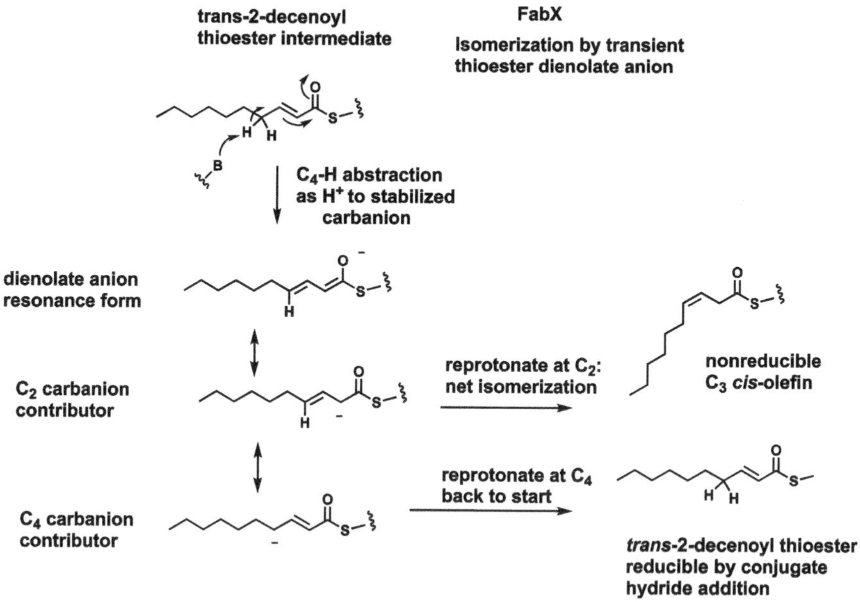

Figure 2.27 The mechanism of the FabX isomerase at the C10 acyl chain level to move the double bond out of conjugation involves the dienolate anion as a contributor to both a C_2 and a C_4 carbanion.

not use radical intermediates. In either case the utility of the alkene functional group led to its independent invention anaerobically and aerobically.

2.11 Alkene Groups as Carbon Electrophiles

To this point, in enumeration of the potential reactivity of alkene groups in biological molecules, we have called out reactivity of the π-electrons of the terminal double bond in Δ^3-IPP and in the linear hexaene squalene to function as carbon nucleophiles. We also have commented on the ability of IPP isomerase to protonate the double bond of Δ^3-IPP to allow isomerization to its partner Δ^2-IPP, uncovering the role of the 2-olefin isomer to stabilize an allylic cation, setting up the head to tail chemistry explored by C–C bond forming intramolecular and intermolecular cyclases and prenyltransferases. The IPP isomerase is fully reversible. Likewise, the *E*- to *Z*-isopentenyl olefin geometry in undecaprenyl-PP synthase is proposed to use a similar reversible olefin protonation route.

We also noted in Section 2.1.4 that the conjunction of olefin with carbonyl groups in both fatty acid degradation and fatty acid biosynthesis polarizes the olefin to act instead as electrophile for hydride transfers to the conjugated olefin in the reductive direction (Figure 2.5) and from the corresponding C_3 of the saturated acyl CoAs in the oxidative direction. That is reactivity of the conjugated olefin as an electrophile, not as a nucleophile.

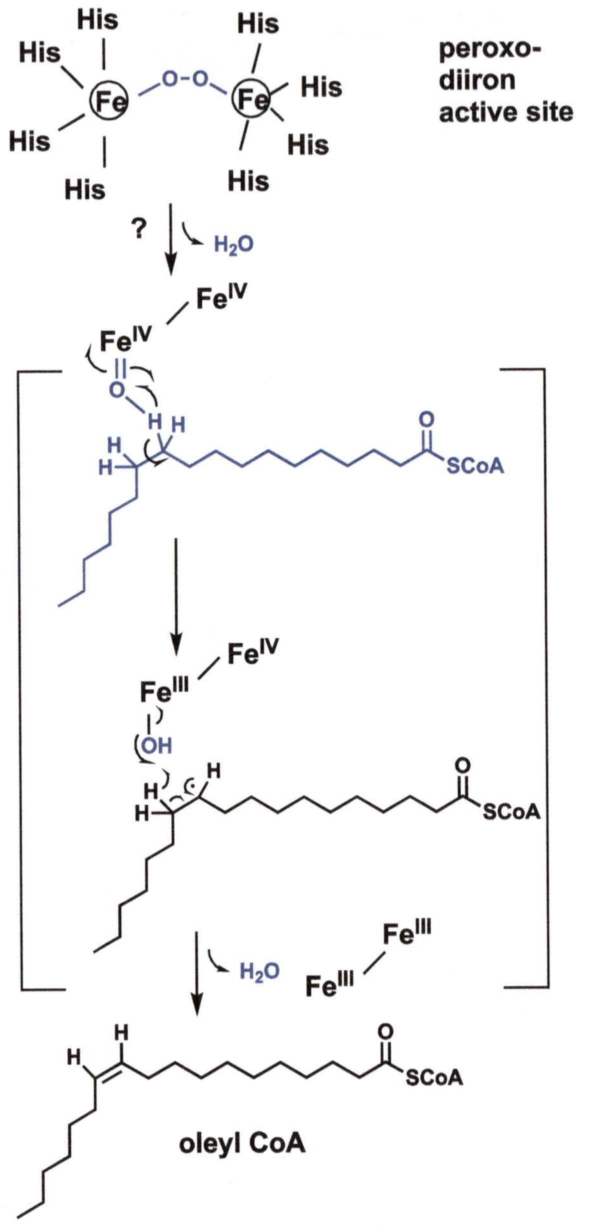

Figure 2.28 Animal cells introduce the 11,12-*cis*-olefin *de novo* by action of a diiron oxygenase that likely involves carbon radical species as the internal CH_2–CH_2 pair at $C_{11,12}$ is regiospecifically and stereospecifically oxidized to the *cis*-olefin.

Further evidence for the electrophilic reactivity of such conjugated olefins is manifested by the hydratase/dehydratase couples both in fatty acid metabolism and in the fumarate to malate step in the citric acid cycle

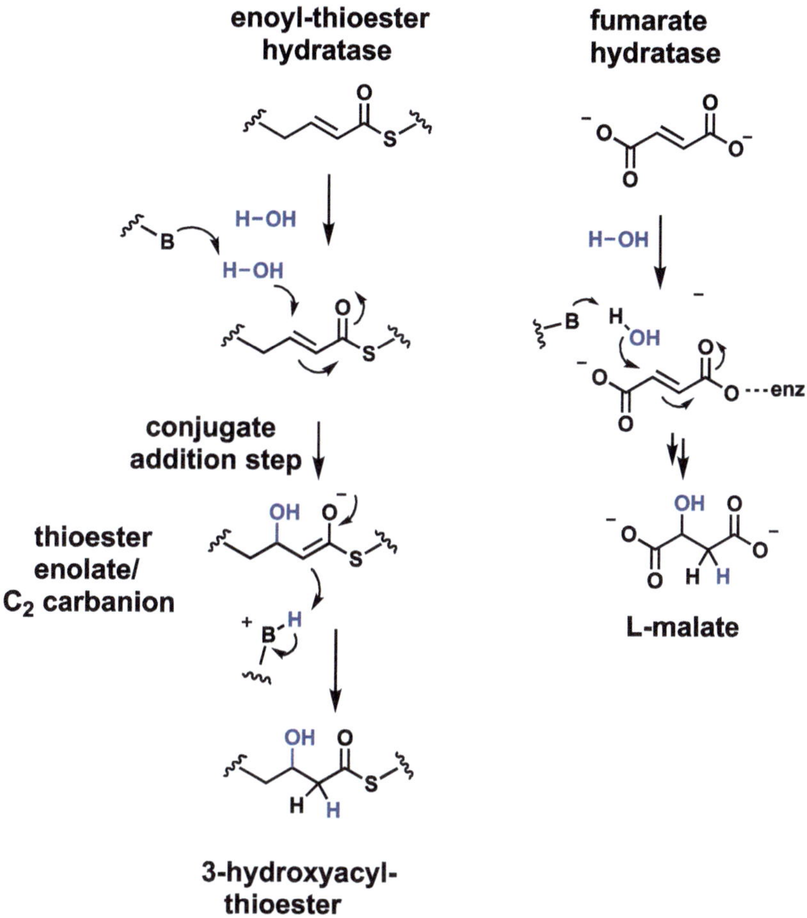

Figure 2.29 The enzymatic hydration of conjugated enoyl groups utilizes the substrate's double bond as an electrophile. This polarity persists in the enzymatic hydration of fumarate to L-malate.

(Figure 2.29). Both those pathways, in different metabolic arenas, carry out three enzyme, four electron oxidations of a CH_2 group in a substrate to the C=O group in product, by way of the CH–OH intermediates. The fumarate to malate conversion catalyzed by fumarate hydratase is a classic addition of water to the central olefin followed by rapid protonation of a carbanionic transition state/intermediate.[33]

Incidentally, fumarate hydratase (fumarase) will also recognize the alkyne link in the surrogate substrate acetylene dicarboxylate as an electrophilic functional group, hydrating it to the enol.[34] The enol spontaneously ketonizes to oxaloacetate. The alkyne link (see Section 2.14) can thus also function biologically as an electrophile when conjugated to a carbonyl group (Figure 2.30).

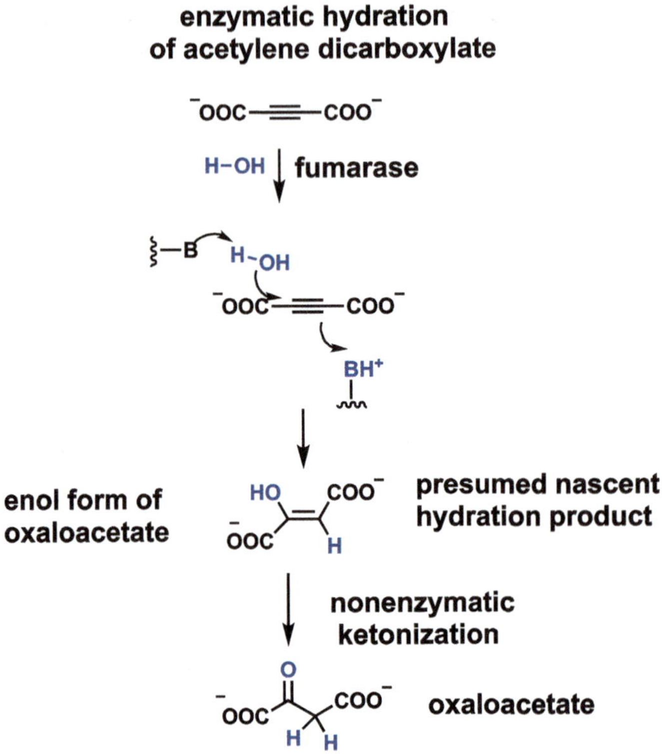

Figure 2.30 Enzymatic hydration of the C–C triple bond in acetylene dicarboxylate generates the enol of oxaloacetate as initial product.

2.12 Cyclopropanations of Alkenes

One more metabolic example of alkene functionalization in metabolism without involvement of heteroatoms is enzyme-catalyzed conversion to cyclopropane rings, the smallest of the cycloalkanes. Depending on molecular context, there are several biologic routes to cyclopropanes, some avoiding olefins altogether. Here, we focus on fatty acid pathways in bacteria and plants where an olefin is starting substrate and *S*-adenosylmethionine provides the transferring CH_2 unit that derives from the CH_3 attached to the sulfonium sulfur in SAM (Figure 2.31). Only two of the three hydrogens in the original CH_3 group of cosubstrate SAM end up in the product cyclopropane ring.

An intriguing variation occurs in some plant seed oils where an SAM-dependent cyclopropane synthase acts on the *cis*-double bond of oleate to produce the cyclopropane-containing dihydrosterculic acid.[35] Then, a flavoenzyme desaturase removes two electrons from the strained cyclopropane ring to produce the even more strained cyclopropene ring in sterculic acid, often the dominant lipid in the oils from *Malves* plant species (Figure 2.31).

For the conversion of *cis*-olefin to cyclopropane, the olefin could act as carbon nucleophile (*e.g.* at C_{10}) on the CH_3 of SAM to give a cationic species

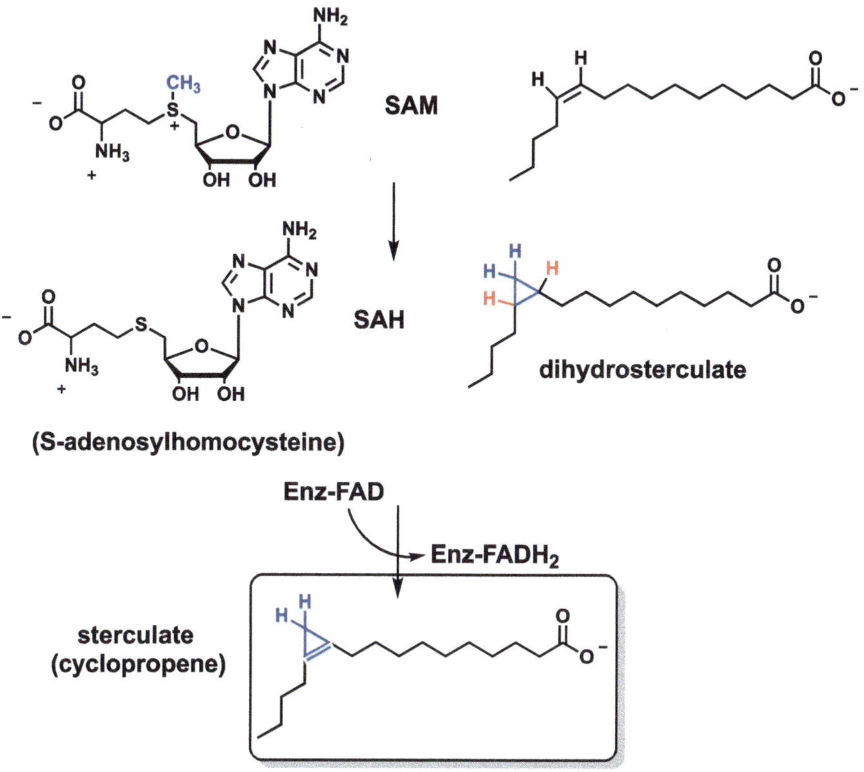

Figure 2.31 Cyclopropanation of oleate to dihydrosterculate utilizes *S*-adenosylmethionine (SAM) as CH_2 donor. Subsequent desaturation by a flavoenzyme gives the cyclopropene ring of sterculic acid anion.

at C_9, that then closes to the cyclopropane (Figure 2.32). Alternatively, the enzyme could abstract one of the hydrogens from the SAM-CH_3 group to create the ylid that is a formal carbanion, attacking the olefin as an electrophile (Figure 2.32), a reverse reactivity profile.[36]

Finally, in addition to classic utilization of SAM by organisms in S_N2-like two-electron reaction routes (net $[CH_3^+]$ transfer), SAM is also reversibly fragmented to the 5′-deoxyadenosyl radical that fosters one electron routes in enzyme active sites (see Chapter 12) Such radical fragmentative routes are thought to be the way that a polyolefinic (octenoyl) precursor is converted to the five cyclopropane rings in the C18 acyl chain of jawsamycin[37] (Figure 2.32.), mechanism not fully elucidated.

2.13 Epoxidations on the Way to Further C–O Functionalities

Another classic reactivity profile for olefins, especially in secondary metabolism is conversion to the comparable three member epoxide rings. Unlike the cyclopropanes and all the alkene reactions noted to date in this chapter, epoxides

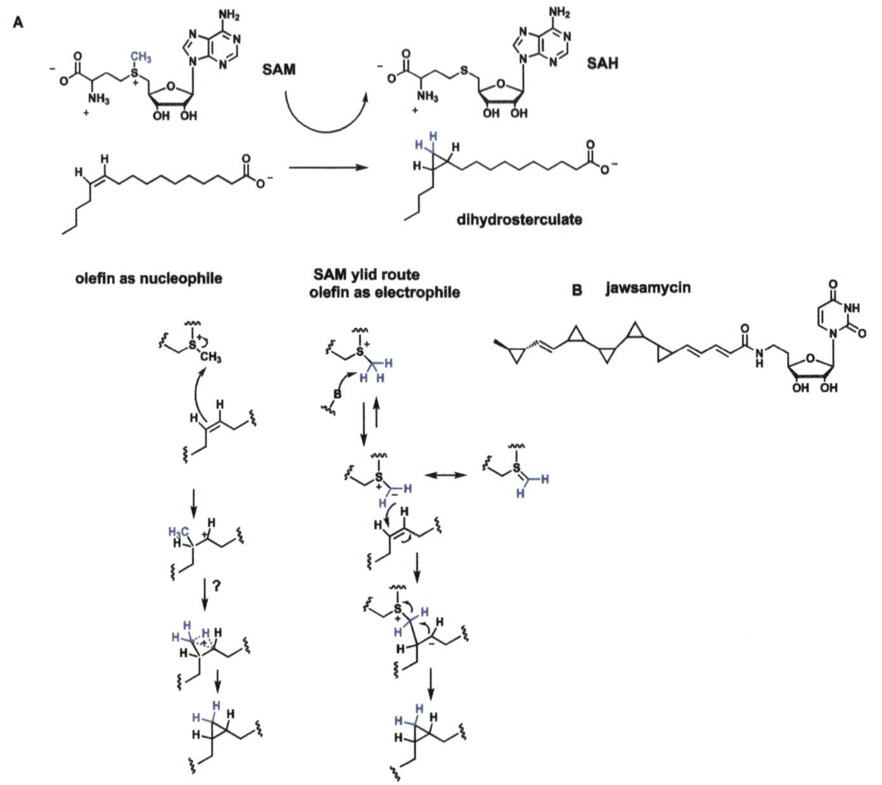

Figure 2.32 Two mechanistic possibilities for cyclopropanation of oleate feature the olefin either as nucleophile or propose a SAM ylid as carbanion equivalent to capture the olefin as electrophile.

introduce heteroatom enzyme chemistry. Epoxide formation and reactivity serve as a good segue to the common C–O functional groups in the next chapter.

We have noted above that two types of oxygenases convert olefins to epoxides-iron-based oxygenases and flavin coenzyme-based oxygenases (Figures 2.20 and 2.21). The flavoenzyme epoxidases transfer an electrophilic oxygen equivalent from a flavin peroxide intermediate so the substrate olefins behave as nucleophiles.[38] This is the mechanism for the flavoenzyme squalene epoxidase[27] converting squalene to its beta-2,3-epoxide, the proximal substrate for oxidosqualene cyclases[39] (Figure 2.22), examined further in Chapter 12. By contrast, high valent oxoiron species in iron oxygenase active sites may proceed to oxygen transfer by one electron, carbon radical routes, emphasizing that olefins may function in both two-electron and one-electron modes for net epoxidations (Figure 2.23).

2.14 Alkenes to Alkynes

Alkynes, both in internal positions in fatty acyl chains and as terminal functional groups are natural metabolites in bacteria, fungi, insects, and plants.[40]

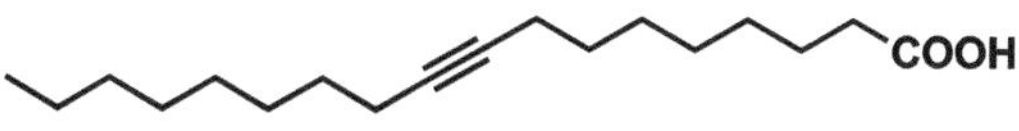

stearolic acid
[9-octadeynoic acid]

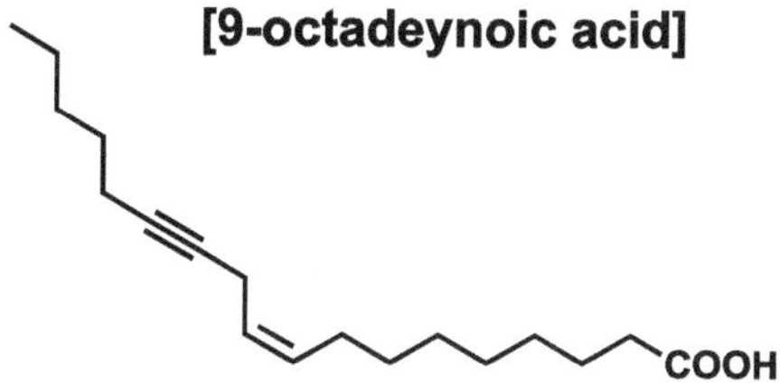

crepenynic acid
[9Z-octacecen—9-en-12-ynoic]

Figure 2.33 Structures of stearolic acid and crepenynic acid.

Examples include the acetylenic C18 fatty acids stearolic acid with one Δ^9-alkyne, and crepenynoic acid with a 9-*cis*-olefin and a Δ^{12}-alkyne group (Figure 2.33). Studies of crepenynoic acid in tomato plants have demonstrated that the fatty acid desaturase Crep1 will take oleic acid and convert it to the 9,12-diolefinic linoleic acid (Figure 2.34).

That same desaturase will then process the 12-*cis*-olefin in linoleate to the Δ^{12}-alkyne in the crepenynoate product. Each desaturation at $C_{12,13}$ is a two-electron oxidation: methylene pairs to *cis*-alkene and then *cis*-alkene to alkyne. The alkyne-forming subset of fatty acid desaturase are also known as acetylenases. They are members of the diiron oxygenase family noted in Section 2.9 and use oxoiron-directed radical chemistry, presumably in both alkene and then alkyne formation (Figure 2.35). The alkane to alkene to alkyne functional group transformation is a net four electron oxidation that is shape-altering along the way.

Additional categories of secondary metabolites where alkenes and alkyne groups have formed include the enediyne antitumor agents[40] and the mixed polyketide-nonribosomal peptides such as jamaicamide with a terminal alkyne in a 5-hexynoyl chain, presumably derived by tandem fatty acid desaturase actions.

2.14.1 Alkynyl Amino Acid Biosynthesis from Alkenes and Allenes

Two natural amino acids with terminal alkyne groups are L-propargylglycine and beta-ethynyl-L-serine (Figure 2.36). The biosynthetic route has been deciphered in *Streptomyces cattleya*, establishing that propargylglycine is a precursor to ethynylserine.[41] Unexpectedly, L-lysine is the starting substrate that undergoes chlorination at C_4 by an O_2-dependent iron halogenase (BesD)

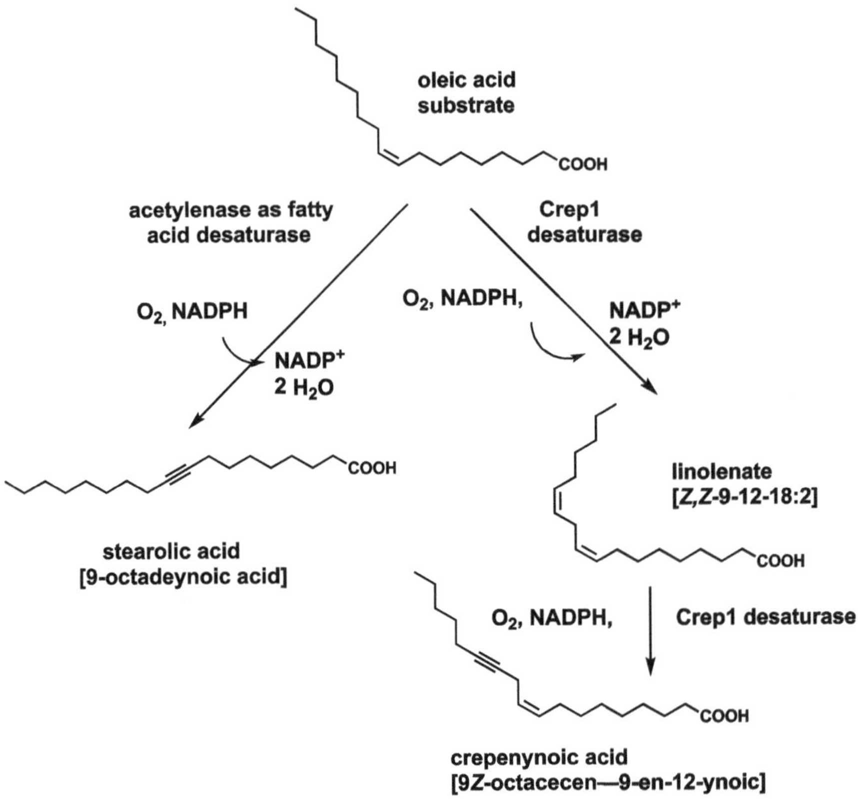

Figure 2.34 Acetylenases act as a subset of O_2-dependent fatty acid desaturases.

that introduces a chlorine atom by radical chemistry at C4 of lysine. Then, BesC is an oxygenase that cleaves the chloro-lys side chain to somehow liberate C_6 and the amino group to produce 4-Cl-allylglycine. BesB is a pyridoxal-P enzyme that then eliminates the 4-chloro substituent to yield L-propargylglycine.

This net conversion of a chloroalkene to an alkyne group is unusual and probably proceeds *vis* an allene intermediate that is isomerized to the terminal alkyne (Figure 2.36). Propargylglycine to ethynylserine involves enzymatic protection of the amino group as an *N*-glutamyl amide followed by oxygenase action at the C_3–CH_2 group to install the OH group (Figure 2.36). Amidase action then liberates the free ethynylserine product. The alkyne in propargylglycine can be converted by PLP enzymes into a conjugated allene that can inactivate target PLP enzymes, accounting for one natural use of the alkyne functional group.

2.15 Summary of Alkenes as Functional Groups

Alkene/olefin groups are generated and consumed *in vivo* by several routes. Containing only C–H and C=C bonds, alkenes have the virtue of being the

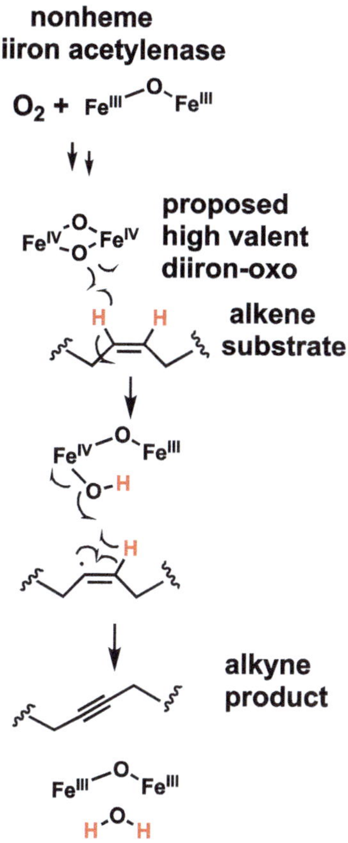

Figure 2.35 O_2-dependent fatty acyl desaturases take alkane groups to alkenes and alkenes to alkynes by radical routes.

simplest of the carbon-based functional groups in metabolism. They offer reactivity tailoring based on structural context.

Alkenes can participate in two electron reaction manifolds, either as electrophile or nucleophile. They also engage in one electron reaction manifolds *via* transient carbon radical species. Internal olefins also offer geometric variation as *trans* (*E*)-isomers of *cis*-(*Z*)-isomers, with *trans*-geometries predominating except in some isoprenoid and unsaturated fatty acyl chains.

The metabolism of alkanes to alkenes as first step in fatty acid oxidation and the corresponding saturation of the conjugated 1,3-enoyl thioesters in fatty acid biosynthesis reflect ongoing mobilization *vs.* storage of reducing equivalents in cellular energy metabolism. In both physiologic directions these redox steps utilize hydride transfers. In the oxidative direction enzyme-bound FAD is the hydride acceptor at N_5. In the reductive direction NADPH is the hydride donor from C_4. Thus, these coenzyme forms of vitamin B2 and B3, with their nitrogen heterocyclic ring systems, are central hydride transfer

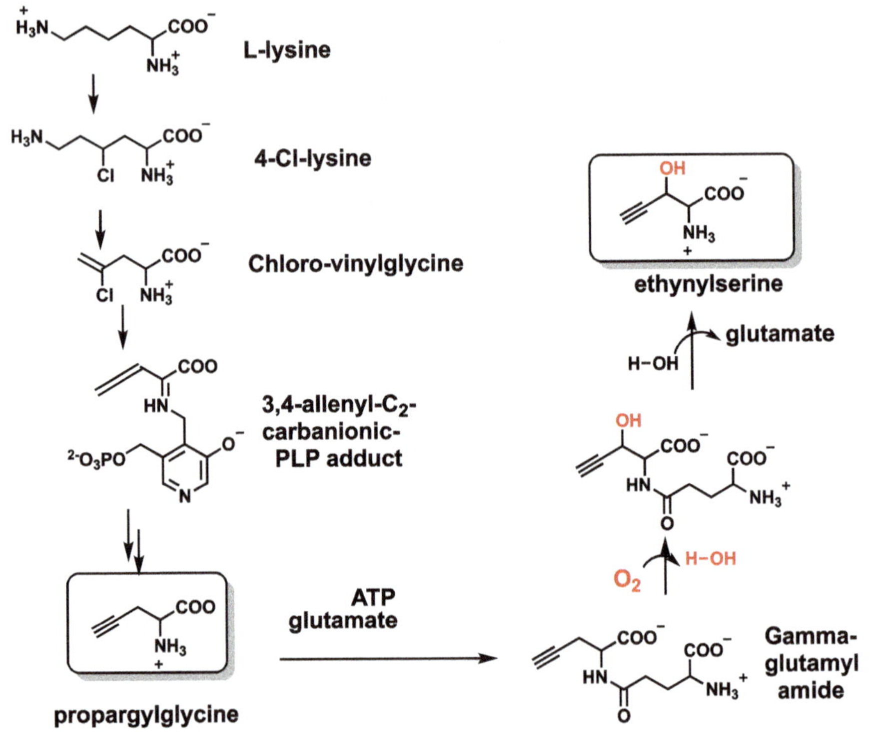

Figure 2.36 Lysine to propargylglycine and beta-ethynylserine: chlorovinylglycine as an intermediate.

reagents for much of *in vivo* redox chemistry, including reversible alkane to alkene interconversions.

The alkylative chain elongations catalyzed by prenyltransferases between Δ^3-IPP monomers and elongating Δ^2-IPP isomers illuminate a high flux C–C bond forming strategy between olefinic isomers. This coupling strategy leads to the largest class of natural metabolites, the isoprenoid/terpenoid collection of organic molecules. The use of 1,4-connected polyolefins as carbon nucleophiles in C–C bond-forming cascades to sterol frameworks are striking examples of how nature uses olefins as tandem carbon nucleophiles.

Olefins also are utilized in nature to create two types of small cyclic ring metabolites, cyclopropane and epoxide groups. The methyl carbon at the cationic sulfonium center of *S*-adenosylmethionine is the source of the CH_2 group in cyclopropanes, while an oxygen atom from O_2 is the proximal oxygen donor in many epoxides. Both strained three membered rings can serve as electrophiles. Cyclopropanes are intermediates in the head to head joining of a pair of C15 farnesyl-PP substrates to generate squalene. The cyclopropane can also serve as electrophile in some DNA alkylation reactions.[42] Epoxides can persist in natural product frameworks or serve as selectively reactive electrophiles. The formation and reactivity of epoxide rings is a good transition to C–O functional groups in Chapter 3.

References

1. J. Ashenhurst, *Oxidation Ladders: alkanes, alkenes, alkynes*, 2021.
2. C. T. Walsh and Y. Tang, *The Chemical Biology of Human Vitamins*, Royal Society of Chemistry, London, 2018.
3. N. Pollack, C. Dolle and M. Ziegler, The poer to reduce: pyridine nucleotides-small molecules with a multitude of functions, *Biochem. J.*, 2007, **402**, 205–218.
4. C. Walsh, Flavin coenzymes: at the crossorads of biological redox chemistry, *Acc. Chem. Res.*, 1980, **13**, 148–155.
5. S. Ghisla and C. Thorpe, Acyl CoA dehydrogenases a mechanistic overview, *Eur. J. Biochem.*, 2004, **271**, 494–508.
6. R. P. Massengo-Tiasse and J. E. Cronan, Diversity in enoyl-acyl carrier protein reductases, *Cell. Mol. Life Sci.*, 2009, **66**, 1507–1517.
7. K. L. Fillgrove and V. E. Anderson, The mechanism of dienyl-CoA reduction by 2,4-dienyl-CoA reductase is stepwise: observation of a dienolate intermediate, *Biochemistry*, 2001, 12412–12421.
8. T. J. Erb, V. Brecht, G. Fuchs, M. Muller and B. E. Alber, Carboxylation mechnaism and stereochemistry of crotonyl-CoA carboxylasze/reductase, a carboxylating enoyl-thioester reductase, *Proc. Nat. Acad. Sci. USA*, 2009, **106**, 8871–8876.
9. C. T. Walsh, B. P. Tu and Y. Tang, Eight kinetically stable but thermodynamically activated molecules that power cel metabolism, *Chem. Rev.*, 2018, **118**, 1460–1494.
10. H. M. Miziorko, Enzymes of the mevalonate pathway of soprenoid biosynthesis, *Arch. Biochem. Biophys.*, 2011, **505**, 131–143.
11. A. Frank and M. Groll, The methylerythritol pathway to isoprenoids, *Chem. Rev.*, 2017, **117**, 5675–5703.
12. J. E. Reardon and R. H. Abeles, Mechanism of action of isopentenyl pyrophosphate isomerase: evidence for a carbonium ion intermediate, *Biochemistry*, 1986, **25**, 5609–5616.
13. C. D. Poulter and H. C. Rilling, Prenyltransferase: the mechanism of the reaction, *Biochemistry*, 1976, **15**, 1079–1083.
14. H.-Y. Chang, T.-H. Cheng and A. H.-J. Wang, Structure, catalysis, and inhibition mechanism of prenyltransferase, *IUBMB Life.*, 2020, DOI: 10.1002/iub.2418.
15. B. Tian, C. D. Poulter and M. P. Jacobson, Defining the chemical space of monoterpenoid synthases, *PLOSA Comput. Biol.*, 2016, **12**, e1005053.
16. D. J. Tantillo, Importance of inherent reactivity in enzyme-promoted carbocation cyclizations/rearrangements, *Angew. Chem., Int. Ed.*, 2017, **56**, 10040–10045.
17. D. W. Christianson, Structure and chemical biology of terpenoid cyclases, *Chem. Rev.*, 2017, **117**, 1157–11648.

18. C.-C. Chen, S. R. Malwal, X. Huan, W. Liu, L. Ma, C. Zhai, L. Dai, J.-W. Huang, A. Shillo, J. Desai, X. Ma, Y. Zhang, R.-T. Guo and E. Oldfield, Terpene cyclases and prenyltransferases: structures and mechnaisms of action, *ACS Catal.*, 2021, **11**, 290–303.
19. C. Schotte, P. Lukat, A. Deuschmann and W. Blankenfeldt, Understanding and engineering the stereoselectivity of humulene synthase, *Angew. Chem., Int. Ed.*, 2021, **60**, 20308–20312.
20. W. Liu, X. Feng, Y. Zheng, C.-H. Huyang, C. Nakano, T. Hoshino, S. Bogue, T.-P. Ko, C.-C. Chen, Y. Cui, J. Li, I. Wang, S.-T. D. Hsu, E. Oldfield and R.-T Guo, Structure, function and inhibition of ent-kaurene synthase from Bradyrhizobium japanicum, *Sci. Rep.*, 2014, **4**, 6214.
21. M. Kokasl, Y. Jin, R. M. Coates, R. Croteau and D. W. Christianson, Taxadiene synthase structure and evolution of modular architecture in terpene biosynthesis, *Nature*, 2010, **469**, 116–120.
22. B. S. J. Blagg, M. B. Jarstfer, D. H. Rogers and C. D. Poulter, Recombinant squalene synthase. A mechanism for the rearrangement of presqualene diphsphate to squalene, *J. Am. Chem. Soc.*, 2002, **124**, 8846–8853.
23. C.-I. Liu, W.-Y. Jeng, W.-J. Chang, M.-F. Shih, T.-P. Ko and A. J. H. Wang, Structural insights into the catalytic mechanism of hiuman squalene synthase, *Acta Crystallogr., Sect. D: Biol. Crystallogr.*, 2014, **70**, 231–241.
24. S. C. Hammer, P.-O. Syren, M. Seitz, B. M. Nestl and B. Hauer, Squalene hopene cyclases: highly promiscuous and evolvable catalysts for stereoselective CC and CX bond formation, *Curr. Opin. Chem. Biol.*, 2013, **17**, 293–300.
25. J. S. Dickschat, Isoprenoids in three-dimensional space: the stereochemistry of terpene biosynthesis, *Nat. Prod. Rep.*, 2011, **28**, 1917–1936.
26. R.-T. Guo, T.-P. Ko, A. P.-C. Chen, C.-J. Kuo, A. J.-P. Wang and P.-H. Linag, Crystal structure of undecaprenyl pyrophodsphate synthase in complex with magnesium, isopentenyl pyrophosphate and farnesyl thiopyrophosphate: roile s of the metal ion and conserved ersidues in catalysis, *J. Biol. Chem.*, 2005, **280**, 20762–20774.
27. A. K. Padyana, S. Gross, L. Jin, G. Cianchetta, R. Narayanaswamy and F. Wang, *et al.*, Structure and inhibition mechanism of the catalytic domain of human squalene epoxidase, *Nat. Commun.*, 2019, **10**(1), 97.
28. J. Li, H.-J. Liao, Y. Tang, J.-L. Huang, L. Cha, T.-S. Lin, J. L. Lee, I. V. Kurnikov, M. G. Kurnikova, W.-C. Chang, N.-L. Chan and Y. Guo, Epoxidation catalyzed by the noinheme iron(II)-and 2-oxoglutarate-dependent oxygenase, AsqJ: mechanistic elucidation of oxygen atom transfer by a ferryl intermediate, *J. Am. Chem. Soc.*, 2020, **142**, 6268–6284.
29. N. Mahanta, D. Fedoseyenko, T. Dairi and T. P. Begley, Menaquinone biosynthesis: formation of aminofutalosine requires a unique radical SAM enzyme, *J. Am. Chem. Soc.*, 2013, **135**(41), 15318–15321.
30. M. Leesong, B. S. Henderson, J. R. Gillig, J. M. Schwab and J. L. Smith, Structure of a dehydratase-isomerase from the bacterial pathway for biosynthesis of unsaturated fatty acids: two catalytic activities in one active site, *Structure*, 1996, **4**(3), 253–264.

31. J. Shanklin, J. E. Guy, G. Mishra and Y. Lindqvist, Desaturases: emerging models for understanding functional diversification of diiron-containing enzymes, *J. Biol. Chem.*, 2009, **284**(28), 18559–18563.
32. P. Sperling, P. Ternes, T. K. Zank and E. Heinz, The evolution of desaturases, *Prostaglandins, Leukotrienes Essent. Fatty Acids*, 2003, **68**(2), 73–95.
33. M. Mescam, K. C. Vinnakota and D. A. Beard, Identification of the catalytic mechanism and estimation of kinetic parameters for fumarase, *J. Biol. Chem.*, 2011, **286**(24), 21100–21109.
34. D. H. Flint and R. G. McKay, Escherischia coli fumarase A catalyzed transfer of 18O from C-2 and 2H from C3 of malate to acetylene dicarboxylate to form 18O and 2H-labeled oxaloacetate, *J. Am. Chem. Soc.*, 1994, **116**, 5534–5539.
35. X. Bao, S. Katz, M. Pollard and J. Ohlrogge, Carbocyclic fatty acids in plants: biochmeical and molecular genetic characterization of cyclopropane fatty acid synthesis in Sterculia foetida, *Proc. Nat. Acd. Sci. U. S. A.*, 2002, **99**, 7172–7177.
36. B. J. Landgraf and S. J. Booker, The ylide has landed, *Nature*, 2013, **498**, 45–47.
37. T. Hiratsuka, H. Suzuki, A. Minami and H. Oikawa, Stepwise cyclopropanation on the polycyclopropanated polyketide formation in jawsamycin biosynthesis, *Org. Biomol. Chem.*, 2017, **15**, 1076–1079.
38. T. A. Wenciewicz and C. T. Walsh, Flavoenzymes versatile catalysts in biosynthetic pathways, *Nat. Prod. Rep.*, 2013, **30**, 175–200.
39. K. U. Wendt, G. E. Schulz, E. J. Corey and D. R. Liu, Enzyme mechanisms for polycyclic triterpene formation, *Angew. Chem., Int. Ed.*, 2000, **39**, 2812–2833.
40. X. Li, J.-M. Lv, D. Hu and I. Abe, Biosynthesis of alkyne-containing natural products, *RSC Chem. Biol.*, 2021, **2**, 166–180.
41. J. A. Marchand, M. E. Neugebauer, M. C. Ing, C.-I. Lin, J. G. Pelton and M. C. Y. Chang, Discovery of a pathway for terminal-alkyne amino acid biosynthesis, *Nature*, 2019, **567**, 420–424.
42. M. R. Wilson, Y. Jiang, P. W. Villalta, A. Stornetta, P. D. Boudreau, A. Carra, C. A. Brennan, E. Chun, L. Ngo, L. D. Samson, B. P. Engelward, W. S. Garrett, S. Balbo and E. P. Balskus, The human gut bacterial genotoxin colibactin alkylates DNA, *Science*, 2019, **363**, eaar7785.

CHAPTER 3

C–O Bond Formation and Reactivity: A Family of Related Functional Groups

The six elements that make up ~99% of human body mass are hydrogen, oxygen, carbon, nitrogen phosphorus and sulfur. Carbon engages in covalent bonds to all the other five elements, although direct carbon–phosphorus bonds (phosphonates and phosphinates) are quite rare in biology. The introductory chapter noted that organic molecules are transformed by enzymatic catalysis in chemical biology by way of functional group interconversions. Chapter 2 introduced alkenes and alkynes as carbon based functional groups lacking any heteroatom bonds and summarized common modes of formation and reactivity. The common carbon-based functional groups with heteroatoms attached to carbon for *in vivo* metabolic chemistry have carbon–oxygen, carbon–nitrogen, and/or carbon–sulfur bonds as depicted back in Figure 1.10.

This chapter takes up the most abundant carbon–oxygen functional groups in the organic chemistry of life, how they are formed, and how they react. One can argue that the common C–O functional groups, alcohols, aldehydes and ketones, carboxylate anions and other acyl derivatives, constitute the heart of functional group transformations in metabolism. This chapter also serves as a companion to Chapter 4 on C–S functional groups and Chapter 5 on the metabolic origin and roles of corresponding C–N functional groups.

3.1 Sources of Inorganic Oxygen

A first question in C–O functional group chemical biology might be: where do the oxygen atoms come from? The two most abundant sources of inorganic oxygen on earth are water and molecular oxygen. O_2 constitutes 21% of earth's

Chemical Biology No. 21
The Chemical Biology of Carbon
By Christopher T. Walsh

Published by the Royal Society of Chemistry, www.rsc.org

atmosphere and water covers two-thirds of the globe's surface. Water can serve as a hydration reagent in some enzyme-catalyzed conjugate addition reactions to olefins (depicted in Chapter 2, *e.g.* in Figure 2.7, or in (enzymatic) hydrolysis of the major classes of biopolymers: nucleic acids, proteins, and polysaccharides (Figure 3.1). Hydratases typically produce alcohols,[1] while hydrolases[2] incorporate one oxygen atom from water into carboxylate oxygens.

Both water as universal solvent and O_2 as major atmospheric gas are kinetically stable in the presence of organic metabolites except in particular enzyme active sites of hydratases and hydrolases (water as reactant), *vs.* oxidases and mono- and dioxygenases (O_2 as reactant) (Figure 3.2). Aspects of the metabolic strategies that aerobic organisms have evolved to reductively activate O_2 for the one electron oxygenations in Figure 3.2 are covered in Chapter 12 on carbon-based radical intermediates.

3.2 C–O Functional Group Chemical Biology

The enzymatic conversion of C–H bonds to C–O bonds results in the creation of new functional groups that are reactivity handles for the chemical transformations of substrates into metabolic products. We illustrated the traverse from alkane to alcohol to aldehyde/ketone to carboxylic acid anion for both one carbon and two carbon frameworks in Chapter 1 to illustrate the redox relationships between the four functional groups, each separated from the other by two electron removal. The redox relationships of these central carbon–oxygen functional groups are further emphasized in this chapter (Figure 3.3).

The interconvertible troika of C–O functional groups – alcohols, aldehydes and ketones, and carboxylate anions – in scaffolds ranging from one to six carbons dominate the organic chemistry of hundreds to thousands of low molecular weight metabolites.[3] Mixing of the three functional groups in the same organic framework are explored in almost every combination in metabolic inventories. Figure 3.4 shows a subset of three carbon metabolites with alcohols, aldehydes/ketones, and carboxylate groups. They are all known natural compounds. The trihydroxy glycerol is the backbone of most membrane phospholipids. Oxidation of one of the terminal-OH groups gives glyceraldehyde, a key metabolite in catabolism of glucose. Further oxidation to the carboxylate gives glycerate. Pyruvate and lactate are related by reversible interconversion of the 2-keto and 2-OH group. Hydroxypyruvate, which can arise from oxidation of the C_2–OH of glycerate to the ketone, has all three C–O functional groups in its three carbon framework, enabling a wide range of *in vivo* chemistry. Reductive amination (taken up in Chapter 5) converts hydroxypyruvate to the key amino acid metabolite serine. Note that each of the three carbons of hydroxypyruvate is at a different C–O group oxidation state (alcohol, ketone, carboxylate), enabling multiple modes of reaction for such a branchpoint metabolite.

Common four carbon metabolites include the dicarboxy scaffolds of succinate, fumarate, malate, and oxaloacetate (Figure 3.5), interrelated by

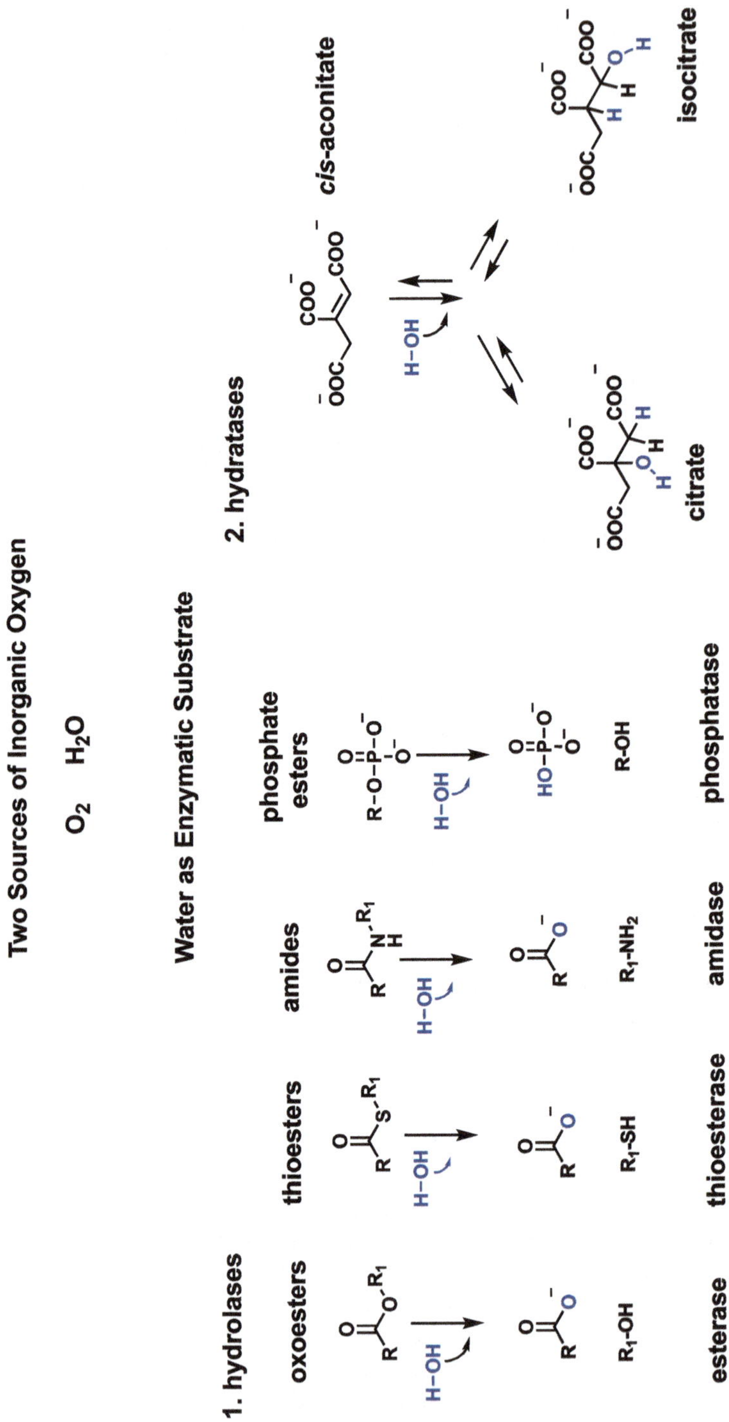

Figure 3.1 Water and molecular oxygen: two sources of inorganic oxygen.

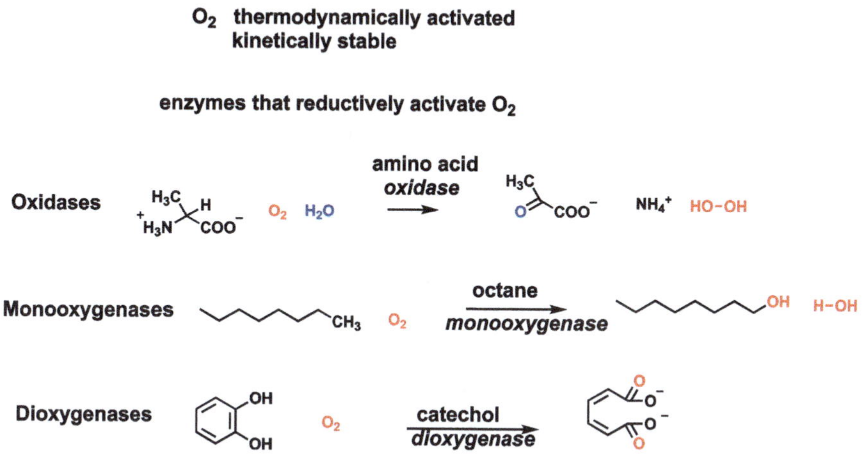

Figure 3.2 O_2 is a thermodynamically activated but kinetically stable substrate for oxidases, monooxygenases or dioxygenases.

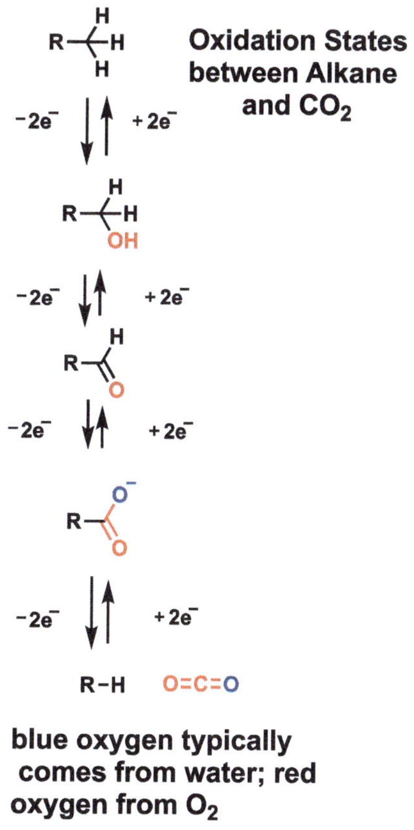

Figure 3.3 Oxidation states between an alkane and CO_2: an eight-electron traverse.

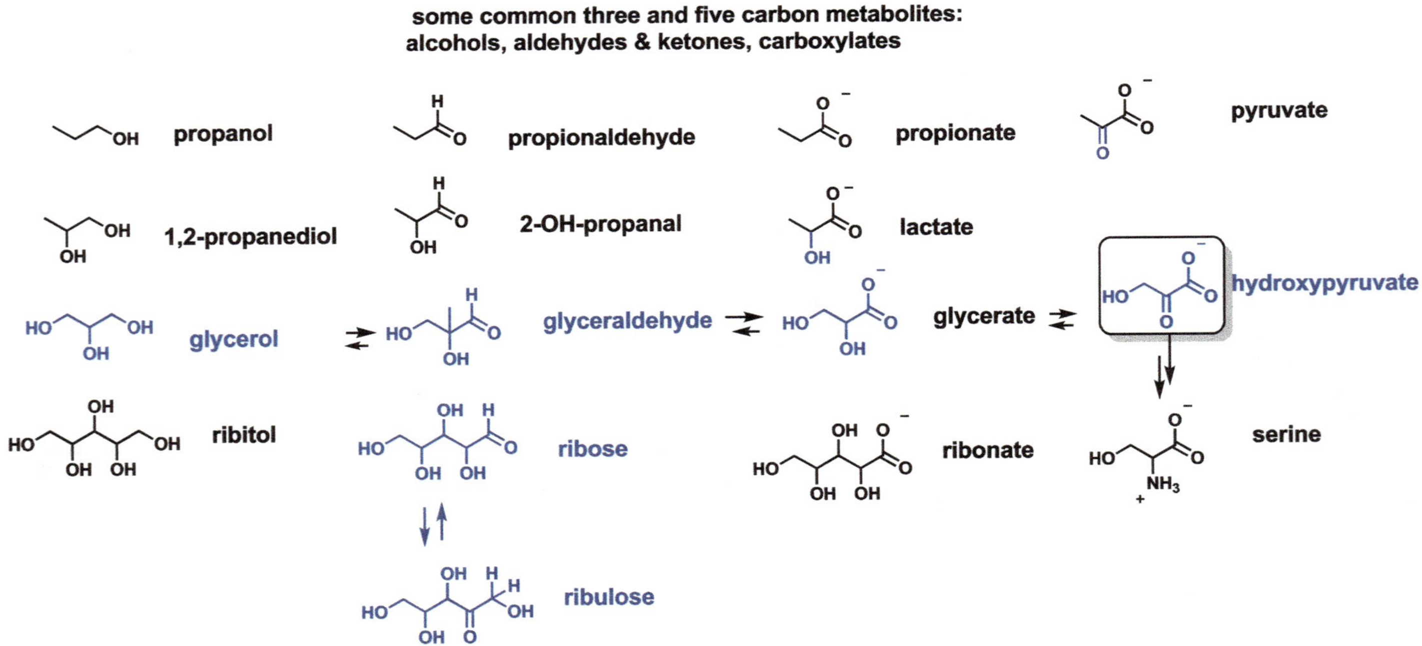

Figure 3.4 Some three carbon metabolites with mixtures of alcohol, aldehyde and ketone, and carboxylate groups. The metabolic progression from glycerol to hydroxypyruvate to serine is highlighted, as is the aldehyde to ketone isomerizatoin between ribose and ribulose.

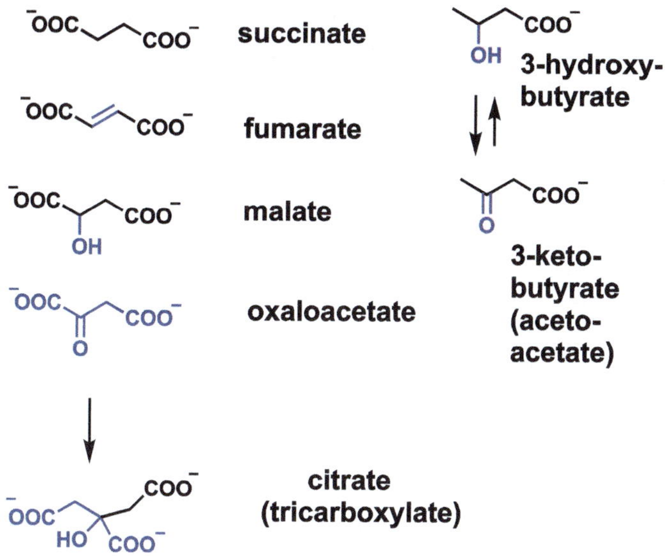

Figure 3.5 Two nested sets of four carbon metabolites: succinate, fumarate, malate, and oxaloacetate; 3-hydroxy- and 3-ketobutyrate; citrate is a six carbon tricarboxylate metabolite by addition of a two carbon moiety to the ketone of oxaloacetate.

alkane/alkene/alcohol/ketone functionality sandwiched between the two carboxylates (citrate, the metabolite named for the citric acid cycle has one –OH group and three carboxylates). Other common four carbon metabolites include the redox pair 3-hydroxybutyrate and 3-ketobutyrate (aka acetoacetate). Among the key five carbon metabolites with multiple C–O bonds are ribose with its terminal aldehyde and the isomeric ribulose with a ketone at C_2.

Much of the rest of this chapter examines the (often reversible) routes of interconversion between alcohols and aldehydes/ketones and between aldehydes and carboxylic acids. In turn, we then analyze the chemistry available to alcohols as oxygen nucleophiles, aldehydes and ketones as carbon electrophiles, and the activation strategies to convert nucleophilic carboxylates to electrophilic acyl derivatives set up for acyl group transfers.

3.3 Enzymatic Alcohol Formation

Alcohols are typically formed by one of three routes in chemical biology, depending on the hybridization state of the carbon atom being converted to alcohol. For sp^2 carbons in olefin linkage, most often conjugated to a carbonyl group, water is added across the olefinic terminus of the conjugated system by hydratases,[1] as discussed back in Chapter 2. Hydration of fumarate to 2*S*-malate or hydration of 2-butenyl-*S*-CoA to 3-hydroxybutyryl CoA are two examples (Figure 2.29). The conjugation of the olefin to the carbonyl group polarizes the olefin such that C_4 is relatively electron deficient and subject to addition of an OH^- equivalent from cosubstrate water in the requisite enzyme active site. The olefins are typically stable indefinitely in the absence of enzyme. We have observed that water addition to olefins is nonoxidative.

For isolated olefinic groups (unconjugated to electrophilic carbonyls), oxygenation to epoxides is noted in Chapter 2. O_2 is the source of oxygen in the epoxide product. While many epoxides are kinetically stable in organic metabolites, others can be opened enzymatically. For example, a partner epoxide hydrolase (epoxidase) can subsequently open the electrophilic epoxide by water addition to generate the *trans*-diol in a net two-step process[4] (Figure 3.6). One oxygen in the diol is from O_2 by action of the first enzyme; the second oxygen derives from water, reflecting the tandem use of both natural inorganic oxygen sources to build alcohol groups.

For sp^3 carbon atoms, *e.g.* at methyl or methylene groups, often distant from any activating functional group, the source of oxygen in the alcohol product is O_2. As elaborated in Chapter 12, the preponderance of oxygenating enzymes use one-electron chemistry to reductively activate and split O_2, creating high valent oxoiron oxidants that abstract hydrogen atoms ($H^•$) from such C–H bonds of substrates. The incipient carbon radical is typically captured in a radical rebound step by net [$OH^•$] transfer from the resultant $Fe^{III/IV}$–OH species. This is the pattern for oxygenative carving out of angular methyl groups and nine additional carbon fragments as the C_{30} alcohol

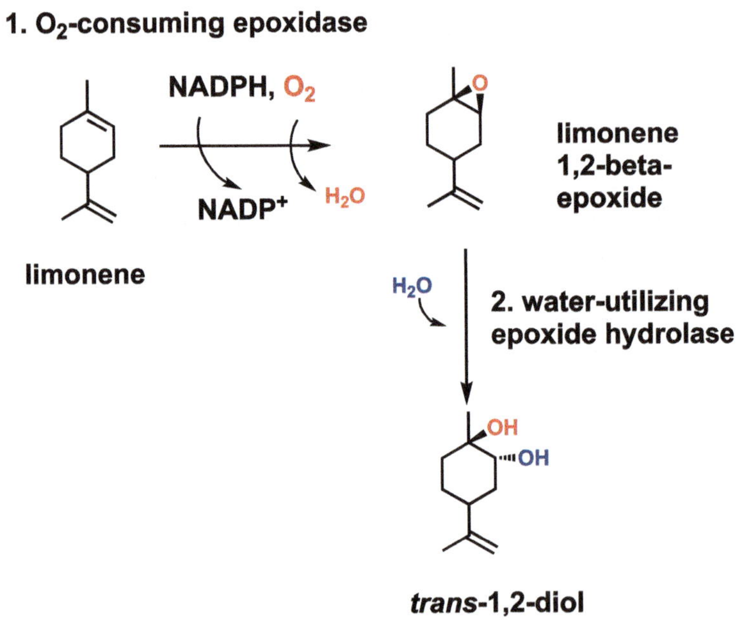

Figure 3.6 O_2-consuming limonene epoxidase followed by an epoxide hydrolase generates a *trans*-1,2-diol.

lanosterol, the first biosynthetic steroid metabolite, is converted to the C_{19} testosterone, the human male sex hormone, by a bevy of P450 oxygenase steps (see Chapter 15) (Figure 3.7). More on this strategy in Chapter 15.

The third route for alcohol generation is typically enzymatic reduction of aldehydes (to primary alcohols) or of ketones (to secondary alcohols) (Figures 3.4 and 3.8). That is, a reductive reaction, the reverse of the oxidative direction. These are independent of both O_2 and H_2O as cosubstrates. These reductions are typically mediated by hydride transfers to and from nicotinamide-dependent oxidoreductases that function reversibly. The Uniprot database (https://www.uniprot.org) lists some 500 000 members of the super family of aldehyde/ketone oxidoreductases that use nicotinamides as hydride acceptors,[5] broadening the scope of nicotinamide-based hydride transfers from the olefin-alkane universe noted in Chapter 2.

The thermodynamic basis for hydride transfer, from dihydronicotinamides and to oxidized nicotinamide coenzymes in the reverse direction, is taken up below[6] and in Chapter 9. The enzymatic conversion of D-glyceraldehyde-3-P to glycerol-3-P is such an example, providing the three carbon backbone for shunting glyceraldehyde-3-P flux from glycolysis (see Chapter 13) to glycerol-based phospholipids in cell membranes (Figure 3.9).

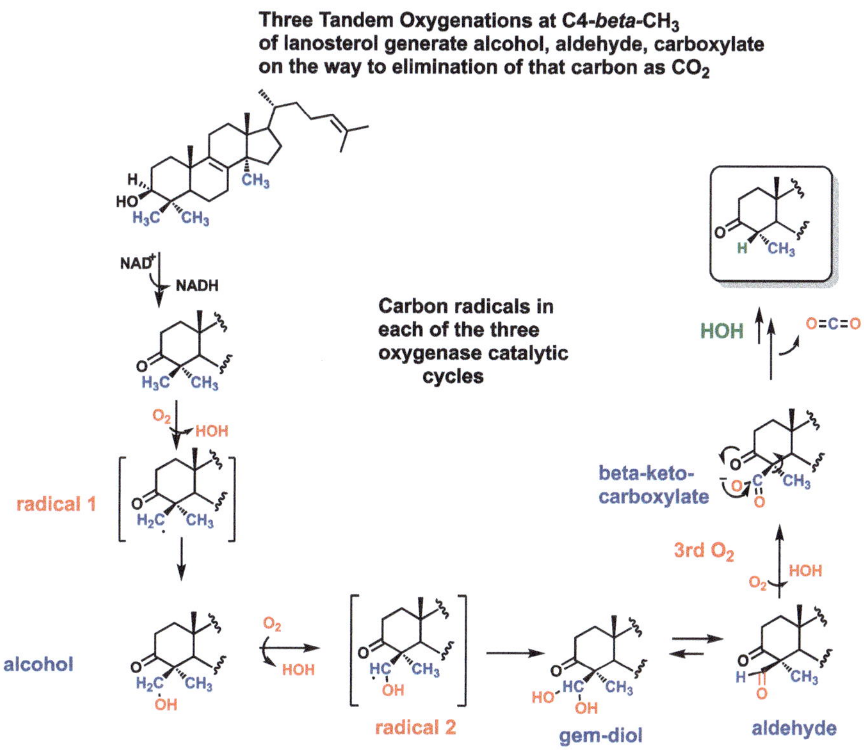

Figure 3.7 Redox traverse by oxygenase activity in conversion of C–CH_3 in lanosterol progressively to –C–CH_2OH, to C–CH=O, to C–COO^-, to C–H and CO_2.

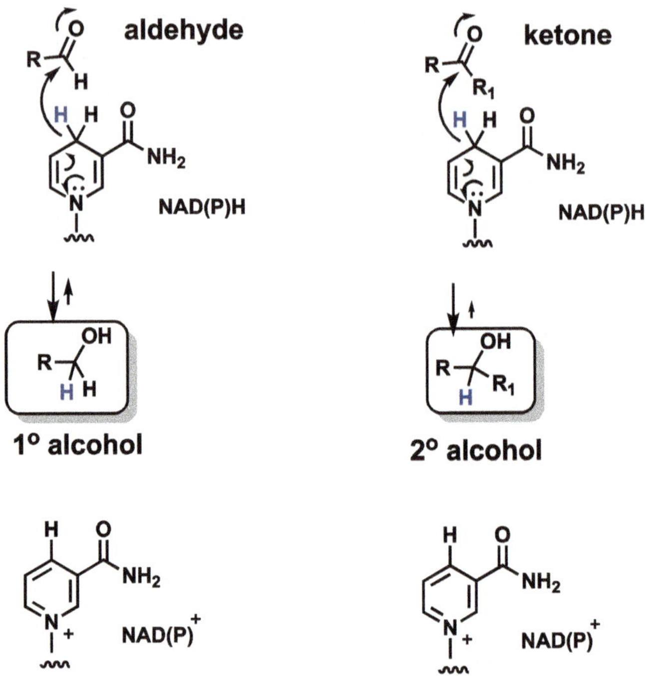

Figure 3.8 Redox interconversion between aldehyde and primary alcohol or ketone and secondary alcohol mediated by hydride transfer to and from nicotinamide coenzymes.

3.4 Alcohols to/from Aldehydes or Ketones: Nicotinamide Coenzymes Dominate

Reversible enzymatic oxidations of primary alcohols to aldehydes and secondary alcohols to ketones can proceed by standard $NAD(P)^+$-utilizing oxidoreductases as exemplified below. They can, in specific cases, also be part of the consecutive action of cytochrome P450 oxygenases as in the removal of the 4α-CH_3 and the 4β-CH_3 groups in the lanosterol framework (see Chapter 15) (Figure 3.8). This consecutive triple oxygenation strategy is used elsewhere in chemical biology for sequential conversion of CH_3 to COO^- groups, for example in removal of C_5–CH_3 groups on methyl-cytosine residues[7] (Figure 3.10). Both in the double demethylation of the C_4 pair of methyl groups in lanosterol and in the conversion of 5-methyl C residues in DNA to 5-hydroxymethyl C, the first step in these oxygenative cascades are the enzymatic oxygenations of alkanes (methyl groups) to alcohols. The subsequent conversions to aldehydes (*via* gemdiol tetrahedral adducts) and then to carboxylates are also O_2-consuming.

3.4.1 Nicotinamide Coenzymes as Hydride Transfer Reagents

The strategy of enzymatic oxygenation of alcohols to the gem diol form of aldehydes or ketones by the oxygenase family noted above is dramatically

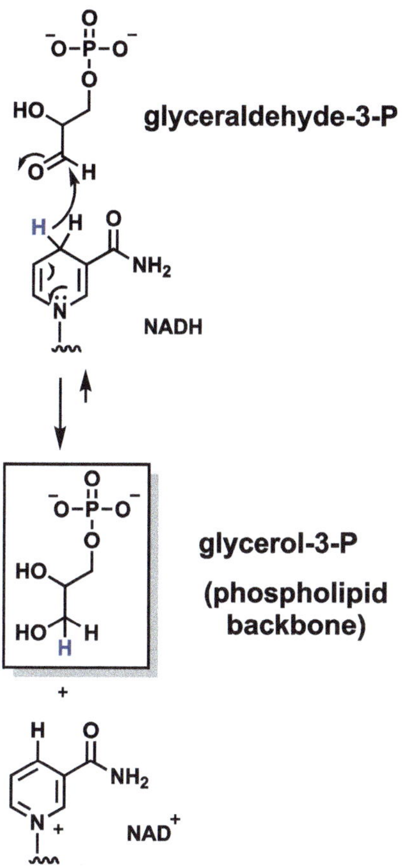

Figure 3.9 Reversible reduction of glyceraldehyde-3-P to glycerol-3-P by hydride transfer provides the three carbon triol backbone form glycerol-based lipids.

outweighed by an oxygen-independent oxidation strategy that doubtless harks back to biologic eons when global [O_2] levels where negligible. The major route from alcohols to aldehydes or ketones and back, a reversible mode not readily available in the monooxygenase strategy, utilizes nicotinamides, the coenzyme forms of vitamin B3[6] (see Chapters 2 and 9). Three nicotinamide-dependent oxidoreductases examples are shown in Figure 3.11: alcohol dehydrogenase, 3-hydroxybutyrate dehydrogenase (secondary alcohol to ketone), and the first of two two-electron reductive steps catalyzed by HMGCoA reductase: the reduction of the thioester acyl group to the aldehyde group in mevaldate. As named, this third enzyme functions physiologically in the reductase direction. In fermenting yeast, where extracellular ethanol levels can reach 5% by volume, the alcohol dehydrogenase can clearly function as an acetaldehyde reductase. The three enzymes in Figure 3.11 thus can act to oxidize a primary alcohol to an aldehyde (reversibly), a secondary alcohol to a ketone, and to reduce a thioester to an

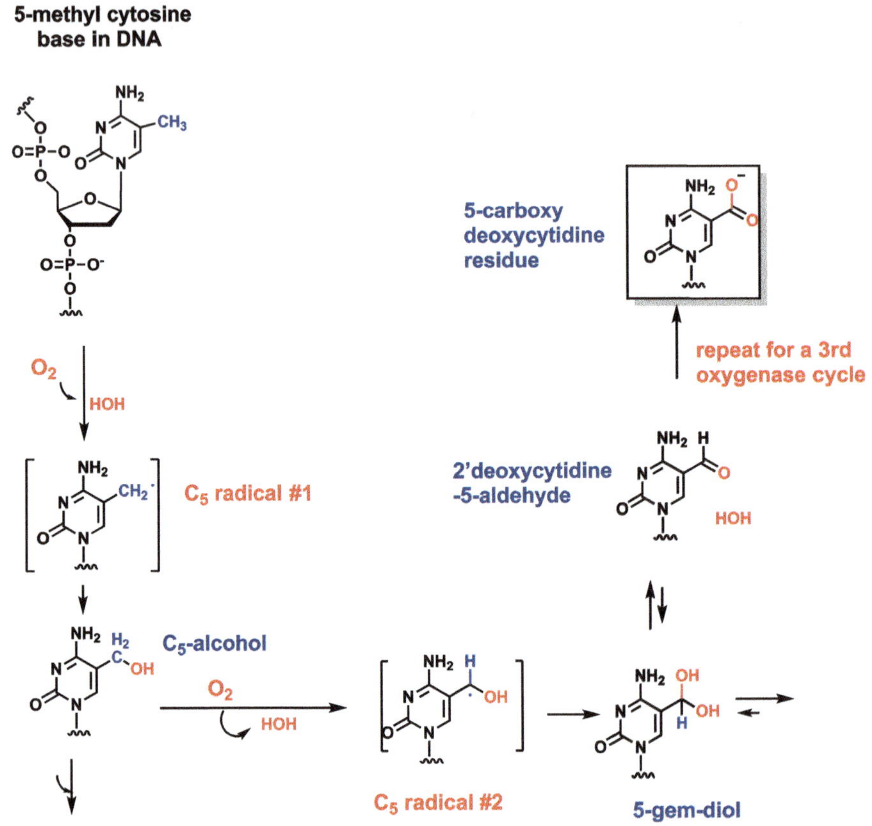

Figure 3.10 Oxidative transformation of 5-methylcytosine base in DNA to the 5-alcohol, the 5-aldehyde, and the 5-carboxylate by three successive oxygenations.

aldehyde, all *via* hydride transfers to and from the appropriate redox states of nicotinamides acting as consumable cosubstrates.

The two forms of nicotinamide coenzymes, NAD^+ and $NADP^+$ in oxidized state, and NADH and NADPH in two-electron reduced state are shown in Figure 3.9 for the redox active heterocyclic ring moiety only, as was the operant presentation mode back in Chapter 2. Both, the oxidized and reduced heterocycle are the key redox cofactors. The nicotinamides function as thermodynamically activated but kinetically stable molecules that transfer hydrogen and two electrons as a hydride ion. NADPH differs from NADH by a phosphate ester group at the 2′-OH of the ribose moiety (Chapter 9). Enzymes can distinguish the $NADP^+$ forms from NAD^+ forms by virtue of that dianionic phosphate group and tend to use NADPH for biosynthetic redox reactions (*e.g.* the HMGCoA reductase).

The reduced forms NADH/NADPH transfer a hydride ion to the electrophilic carbon of the carbonyl carbon of the cosubstrate. Their thermodynamic activation for hydride transfer arises from the fact that the reduced forms have

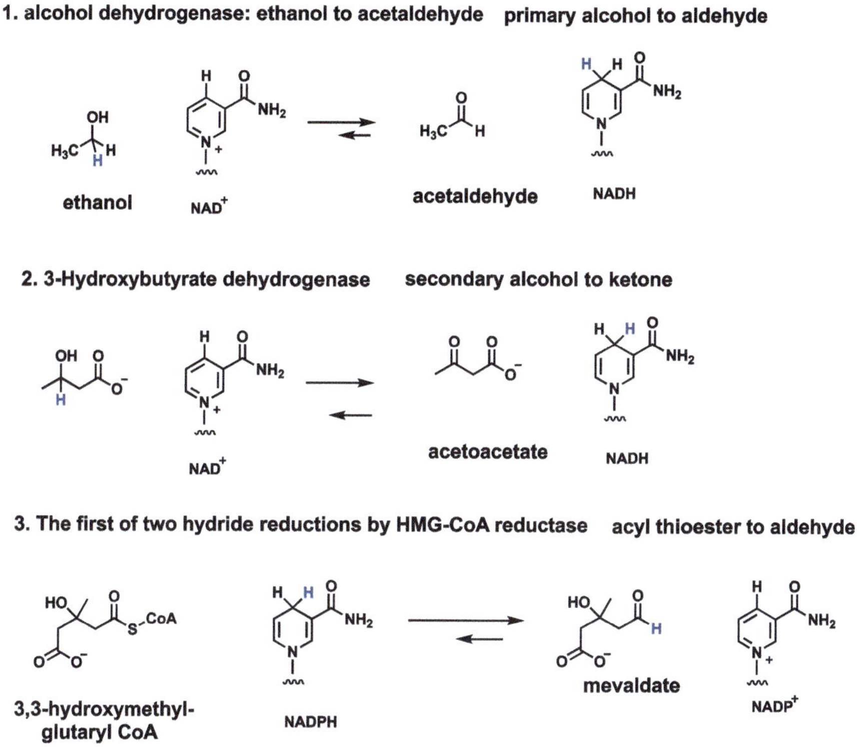

Figure 3.11 Three nicotinamide-dependent oxidoreductases: alcohol dehydrogenase, 3-hydroxybutyrate dehydrogenase, and HMGCoA reductase, first reductive step.

a dihydropyridine ring that is returned to aromaticity on ejection of a hydride ion. Complementarily, in the alcohol to ketone oxidative direction, the oxidized NAD^+/$NADP^+$ coenzymes have a permanently quaternized pyridinium ion that acts as an electron sink for hydride addition at the *para*-position to quench the positive charge on that N_1 cationic nitrogen (Figure 3.12). The dihydro form of nicotinamide coenzymes are sufficiently stable in oxygenated solutions to serve as the most commonly used, diffusible source of cellular reducing equivalents, an almost perfectly evolved heterocyclic molecule. The basis of the kinetic stability in the presence of O_2 is mechanistic incompatibility for electron transfer between obligate two-electron-only nicotinamides and obligate one-electron reactivity of O_2 (see Chapter 12).

3.4.2 Enzymes Act Stereospecifically at Prochiral Carbon Centers

Classic alcohol dehydrogenases involve the oxidation of ethanol to acetaldehyde as NAD^+ is reduced to NADH or *vice versa*, the reduction of aldehyde to alcohol as reduced nicotinamide is oxidized (Figures 3.11 and 3.12). These enzymes, chiral three dimensional structures, show the anticipated stereoselectivity in distinguishing between the two prochiral hydrogens at C_1

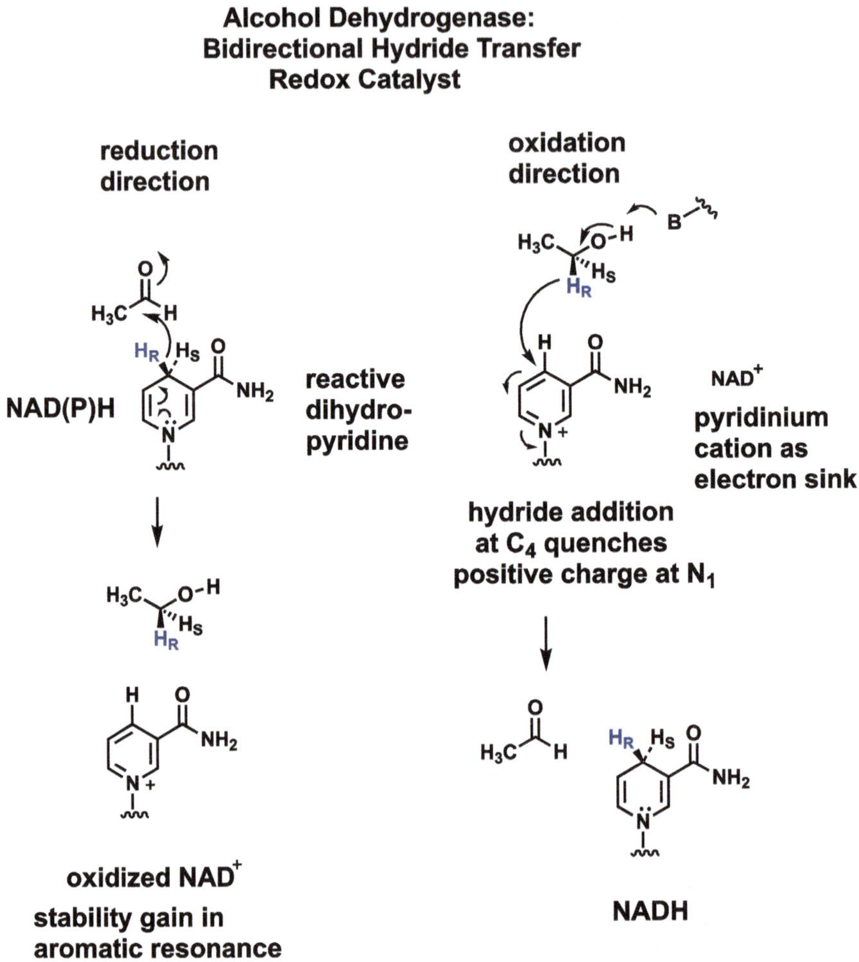

Figure 3.12 The mechanism of alcohol dehydrogenase, in each direction.

of ethanol and the two faces of NAD^+ at the trigonal, planar C_4 of the pyridinium ring (Figure 3.12). Both of those carbons have two identical and two nonidentical substituents. They are describable as C_{aabc} carbon atoms and are termed prochiral carbon centers.[8]

The word chiral is a synonym for handed or asymmetric. Prochiral carbon centers can give rise to chiral carbon centers. Replacement of one of the identical a groups by a d group would generate C_{abcd} centers with four distinct substituents, chiral carbon centers (Figure 3.13). One can assign the chirality of the two C_{abcd} chiral centers and from them the prochiral carbon centers that generated them by a stereochemical nomenclature termed the *R* and *S* system. *R* is from the latin word *rectus*, right-handed. *S* is from the latin word *sinister*, meaning left-handed. *R*-isomers have a clockwise order of substituents around the carbon in descending priority by atomic mass or atomic number.

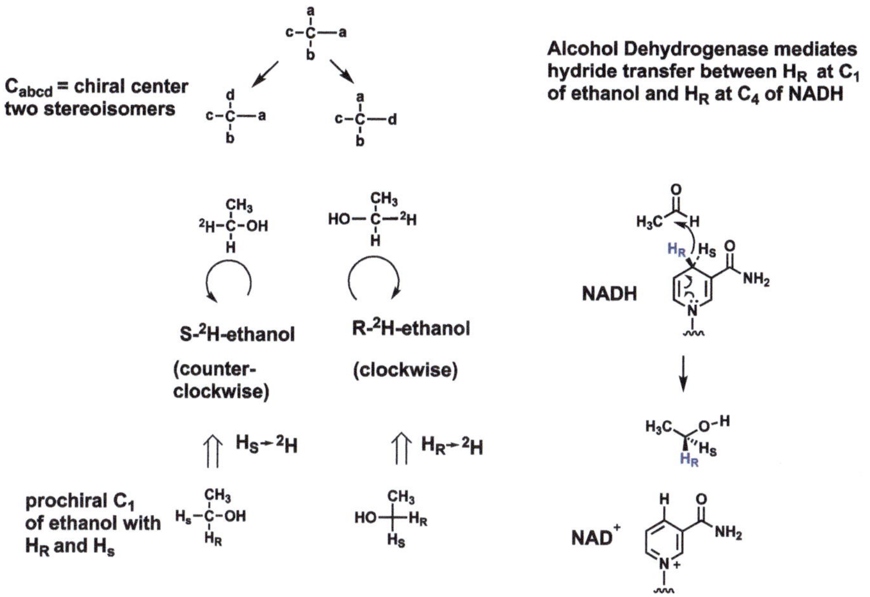

Figure 3.13 Prochiral and chiral carbon centers; stereochemistry of alcohol dehydrogenase action at C_1 of ethanol and C_4 of NADH.

Hydrogen is always the lowest priority substituent. Setting that C–H bond at the bottom of a two dimensional projection, if the descending priority substituents have a clockwise order, that is an *R*-stereochemistry. If anti-clockwise, that is an *S*-stereochemical chiral center.[8]

In Figure 3.13, C_1 of the ethanol has one hydrogen replaced with a deuteron (^{2}H), giving either *R*-deutero-ethanol or *S*-deutero-ethanol. In turn, one can see that in the prochiral C_{aabc} C_1 of ethanol, replacement of the proR hydrogen (H_R) by deuterium gives *R*-deutero-ethanol, while replacement of the proS hydrogen (H_s) gives the *S*-deutero-ethanol. Hence the prochiral designation of the two prochiral hydrogens at any CH_2XY center as H_R or H_s, for example the C_4 locus of hydride transfer at NAD(P)H (Figure 3.13). One can describe the observed stereochemical outcome of alcohol dehydrogenase catalysis as transfer of a hydride in the reductive direction from H_R at C_4 of NADH to the carbonyl carbon of acetaldehyde. That hydride ends up in the H_R position at C_1 of ethanol.

All such prochiral centers can be recognized stereospecifically in the chiral active sites of oxidoreductase enzyme catalysts. One can often find family members that have evolved either to be proR or proS specific.[9] All the enzymes of Chapter 2 that make C_2 carbanions/thioester enolate anions are either H_R or H_s specific.

3.4.3 Regiochemical Selectivity in Oxidoreductases

Enzymatic oxidation of secondary alcohols rather than primary alcohols by hydride transfer to NAD^+ gives ketone products in place of aldehydes.

In addition to the 3-hydroxybutyrate dehydrogenase reaction in Figure 3.11, another classic example is the action of isocitrate dehydrogenase that catalyzes hydride transfer from the C_2–OH group of isocitrate to give NADH and nascent bound oxalosuccinate, on the way to released 2-ketoglutarate. As shown in Figure 3.14, the red carboxylate is β- to that newly introduced ketone while the blue carboxylate is α-. The newly formed ketone group in oxalosuccinate can act as an electron sink for decarboxylation of red carboxylate by stabilization of the incipient C_3 carbanion of oxalosuccinate as the resonance contributing enolate anion. Protonation at C_3 of the product enolate gives the observed products CO_2 and 2-ketoglutarate.[10]

The nicotinamide coenzyme NAD^+ is the acceptor for 20 of the 24 electrons removed as glucose is processed *in vivo* to 6 CO_2 (and 12 H_2O) molecules in the form of 10 NADH molecules. The other four electrons removed are stored as hydride equivalents in reduced flavin $FADH_2$. Thus, all 24 electrons have been captured as 12 hydride ions by oxidoreductases. It is not clear that one would have thought a priori that nature would have invested so heavily in hydride transfer enzymology as the key strategy in energy metabolism.

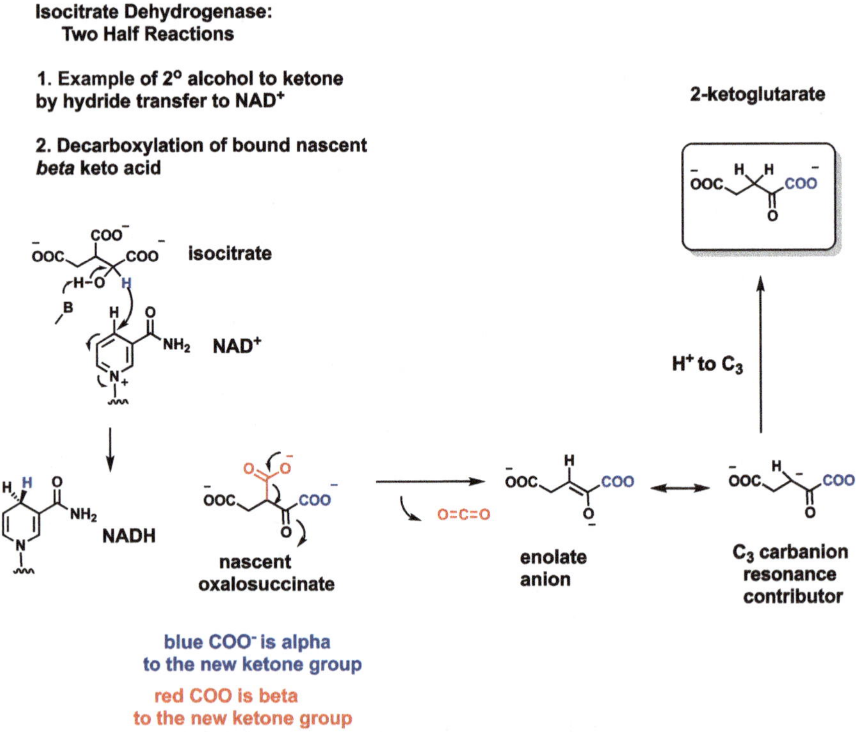

Figure 3.14 Isocitrate dehydrogenase two half reactions: oxidation of 2-OH to 2-keto in bound oxalosuccinate precedes decarboxylation of the carboxylate β- to the newly formed ketone.

3.5 Aldehydes as Geminal Diols and Hemiacetals

Aldehydes and ketones are two electrons more oxidized than the corresponding alcohols. Ketones lack any C–H bonds and are not oxidized further by direct enzymatic routes. (We look at an oxygen-dependent insertion that converts ketones to lactones in a subsequent section below). Aldehydes are more reactive carbonyl electrophiles than ketones. One piece of evidence is a greater propensity for aldehyde carbonyl groups to add water and generate some accumulation of the tetrahedral carbonyl hydrate, also known as a geminal diol (gem-diol)[11] (Figure 3.15).

Correspondingly, hydroxy aldehydes with hydroxy group 4 to 5 atoms distant from the aldehyde can readily cyclize intramolecularly to the gem diols, in these cases known as hemiacetals. Such aldehydes, hexoses such as glucose among them, can form strain free six membered cyclic internal addition products, hemiacetals[12] (Figure 3.15).

Typically, such hemiacetals are in rapid equilibrium with the acyclic open chain aldehyde form. The relative proportions of aldehyde to hemiacetal can vary. In neutral aqueous solutions of the ur-metabolite glucose the two hemiacetal isomers, isomeric at carbon 1, predominate to greater than 99% of the population, with the open chain aldehyde a very minor form (Figure 3.15) (see Chapter 14 for metabolic consequences).

An additional biologically relevant case of aldehyde in reversible formation of tetrahedral addition adducts involves thiolate anions as nucleophiles. As depicted in Figure 3.15, the corresponding tetrahedral adduct is a thiohemiacetal.[13] Thiols are stronger nucleophiles than hydroxyl groups so the equilibria may favor thiohemiacetals over corresponding hemiacetals. Thiohemiacetal formation comes up immediately below in the case of aldehyde dehydrogenase catalytic logic.

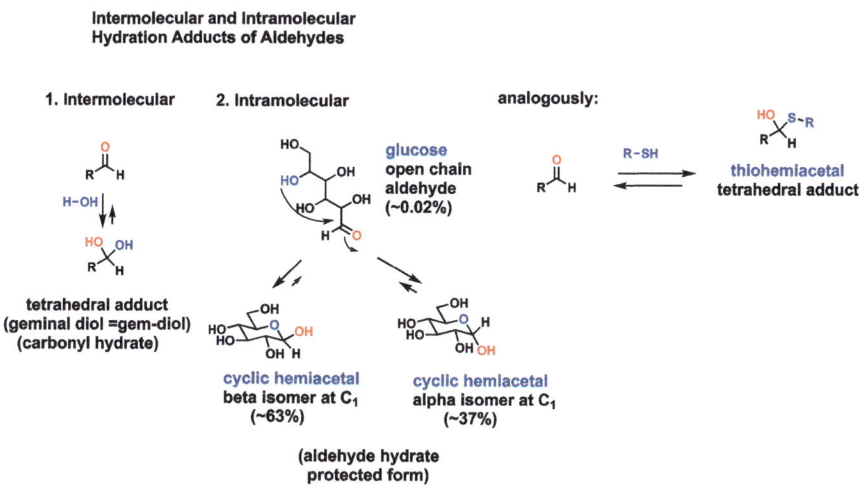

Figure 3.15 Intramolecular and intermolecular hydration adducts of aldehydes. Geminal diols and hemiacetals.

3.6 1,2- and 1,3-Hydroxy Carbonyl Group Chemistry

The combination of hydroxyl and carbonyl groups, both intramolecularly and intermolecularly, can lead to a series of chemical transformations set in motion by the apposition of the two types of C–O functional groups. These are characteristic reactivities of oxygen nucleophiles, electrophilic carbonyl carbons and accessible carbanionic transition states *via* enolizations at carbons adjacent to the ketone, aldehyde, or even thioester carbonyl group. The mechanisms for 1,2-isomerizations and 1,3-aldol cleavages are examined in more detail in the context of biologically accessible carbanions in Chapter 10, and also in Section 3.12 on generation of an enediolate anion in ribulose-1,5-P_2 during catalytic cycles of ribulose-1,5-bisphosphate carboxylase.

Compounds with 1,2-hydroxy carbonyl groups can undergo facile enzyme-catalyzed isomerizations. This is particularly relevant in both the polyhydroxy aldehyde framework of aldose sugars and also in the corresponding polyhydroxy ketose sugars. Figure 3.5 depicts the isomerization of ribose to ribulose, while (Figure 3.16) (right hand column) shows the comparable reaction for phosphoglucose isomerase where the open chain aldehyde form of 6-P-glucose undergoes 1,2-isomerization to fructose-6-P. The 1-,2-hydroxyaldehyde group at $C_{1,2}$ of the glucose moiety is isomerized to the 2-,1-keto-alcohol of fructose. Structural studies on phosphoglucoisomerase indicate a *cis*-geometry for the enediolate and a glutamyl side chain as the active site base transferring the proton between the two oxygens of the pair of enediolate isomers.[14] (By contrast, divalent cation-dependent xylose isomerase[15] appears to catalyze a direct-1,2-hydride transfer.[14])

Both of the reversible ribose–ribulose and glucose–fructose isomerizations start by formation of the C_2 carbanion, stabilized by resonance contribution from the C_1-aldehyde carbonyl. Inspection indicates that these are enediolate anions (Figure 3.16). That initial enediolate can be reversibly isomerized to a second enediolate by ready switching of the position of protonation on the diolate oxygen moiety of bound substrate in the isomerase active sites. The isomeric enediolate has a C_1-carbanion resonance contributor rather than the starting C_2 carbanionic character. Reprotonation on carbon will now occur at C_1 rather than C_2. The product is the ketose sugar. Chapter 14 examines the physiologic consequences of both the phosphoglucose isomerase action for the glycolysis pathway and the phosphoribose isomerase action in the pentose phosphate pathway for nucleic acid biosynthesis.

In addition to enzyme-catalyzed enediolate anion-based isomerization of 1,2-hydroxy-carbonyl groups, organic metabolites with 1,3 carbonyl groups can undergo classic carbanion-assisted aldol cleavages. In the reverse direction, 1,3-hydroxycarbonyl groups in a product indicate that the responsible enzyme may have performed an aldol condensation from two fragments. The reversible aldolase fragmentation of fructrose-1,6-P_2 into two three-carbon products, glyceraldehyde-3-P and dihydroxyacetone-3-P (an aldehyde and a ketone) by fructose-bis-phosphate aldolase[16] (shown in the right hand column of Figure 3.16) is a metabolically central example of

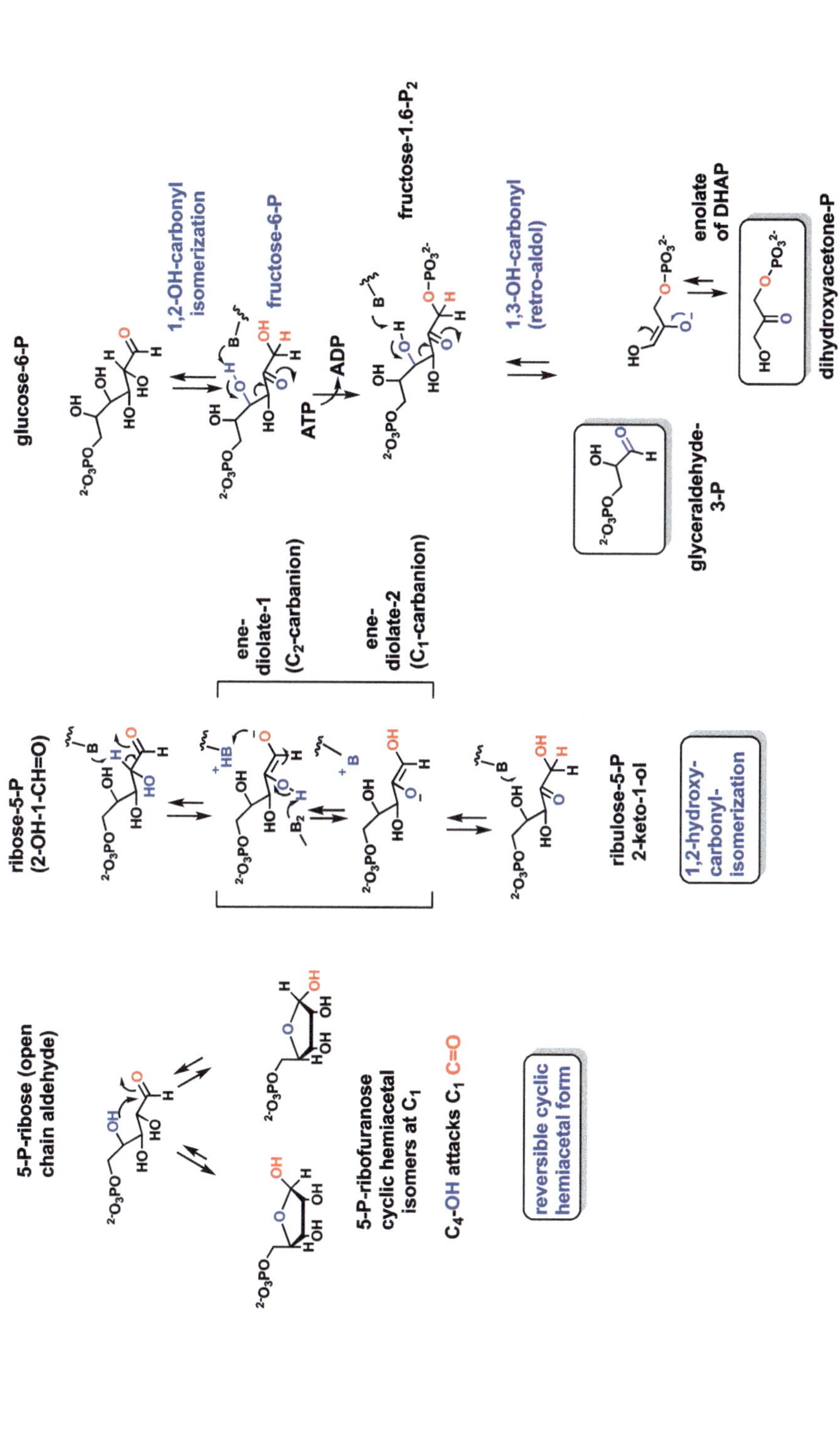

Figure 3.16 1,2-Hydroxy-carbonyl, 1,3-hydroxycarbonyl- and 1,4-hydroxycarbarbonyl enzyme-catalyzed reactions.

carbanion chemistry, accessible by enolization adjacent to the ketone carbonyl group (also see Chapters 10 and 14).

Indeed, that column shows that the first three enzymes of the glycolytic pathway carry out (1) a 1,2-hydroxy-carbonyl mediated isomerization of aldose to ketose, (2) phosphorylation of the newly created C_1-alcohol and then: (3) a 1,3-hydroxy-carbonyl retroaldol cleavage to give the two three carbon fragments glyceraldehyde-3-P and dihydroxyacetone-P. Carbons 2 and 4 of fructose-1,6-P_2 constitute the paired C–O groups that enable low energy aldol cleavage of the six carbon fructose-scaffold. We will observe a tandem pair of catalyzed aldol condensations, one intermolecular, one intramolecular in the biosynthetic pathway to the phenyl ring of phenylalanine in Chapter 7.

Finally, the left hand column in Figure 3.16 shows that C_4 of ribose-5-P engages in intramolecular attack on the C_1-aldehyde, creating a pair of C_1 isomers of the cyclic hemiacetals, reversible ribofuranose adducts. The predominant cyclic hemiacetal form of glucose-6-P occurs from intramolecular attack of the C_5–OH on the C_1 aldehyde. The C_4–OH can also attack C_1 to give a five member hemiacetal, a furanose, but that is a very minor component compared to the glucopyranose form of the glucose hemiacetals.

3.7 Aldehydes to Acids, Hemiacetals to Lactones, Thiohemiacetals to Acyl Thioesters

Aldehydes have carbonyl groups derived from oxidation of primary alcohols ($-CH_2OH$). Therefore, the aldehyde carbonyl groups have one remining C–H bond and can be oxidized by another two electrons to the oxidation state of carboxylic acids (Figure 3.17). The cosubstrate is typically the oxidized nicotinamide NAD^+, accepting the aldehyde C_1–H as a hydride ion during the two electron oxidation.[17] While the aldehyde could be the proximal form of enzyme-bound substrate oxidized directly to the carboxylic acid anion, several enzyme families make use of the tendency of aldehydes to form intermolecular and intramolecular tetrahedral adducts (gem diols, hemiacetals, thiohemiacetals) in preequilibrium. Then, the enzymes oxidize those tetrahedral adduct forms.

Figure 3.17 notes two enzymes that illustrate such a strategy of NAD^+-dependent aldehyde oxidoreductases: to oxidize enzyme-tethered thiohemiacetal adducts rather than the free aldehyde. These are, respectively, the aldehyde dehydrogenase superfamily members *vs.* the specific family member glyceraldehyde-3-P-dehydrogenase.

The aldehyde dehydrogenase enzymes have an active site cysteine residue that acts as a thiolate anion nucleophile, adding into the carbonyl group of aldehyde substrates to pre-form tetrahedral thiohemiacetal covalent enzyme adducts (Figure 3.17). Such tetrahedral adducts are then, and only then, oxidized by hydride transfer to cosubstrate NAD^+ to produce the two-electron-reduced NADH and the two-electron-oxidized acyl-thioester enzyme adduct. Catalyzed addition of water into the electrophilic carbonyl of the acyl-*S*-cys-enzyme then leads to hydrolytic release of the carboxylate anion

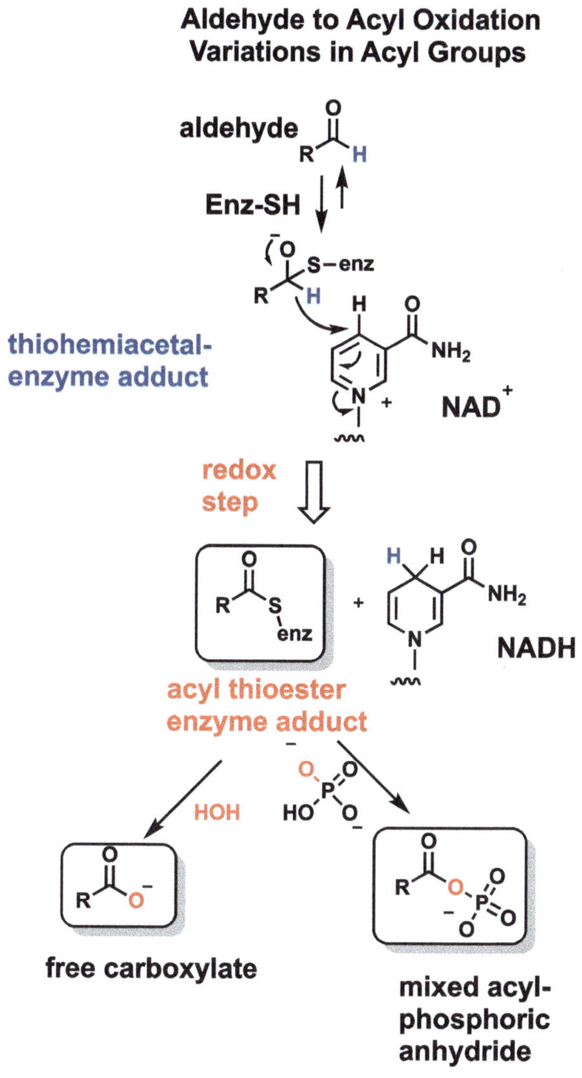

Figure 3.17 Aldehyde to acyl group oxidations. Mechanistic variants lead to alternate acyl group products.

product (Figure 3.16, left arm of partition). Water is a key, ubiquitous cosubstrate to release the nascent tethered acyl group product from its thioester linkage to the enzyme.

In the specific case of glyceraldehyde-3-P dehydrogenase, the fate of the acyl-*S*-thioester enzyme adduct is distinct.[18] The electrophilic acyl group is captured by a nucleophilic oxygen of cosubstrate inorganic phosphate instead of a water hydroxyl oxygen. The released product is the mixed acyl-phosphoric anhydride, an acyl-phosphate. Its thermodynamic potential is illustrated by the next enzyme that transfers the phosphoryl group to ADP to generate ATP, with a K_{eq} of ~3000 in the direction of ATP synthesis. We take up the acyl group transfer potential of acyl phosphates in Section 3.7.

A distinct strategic variant is shown in (Figure 3.18), as the two electron oxidation of a tetrahedral hemiacetal adduct rather than the free aldehyde form. Glucose-6-P dehydrogenase oxidzes the β-hemiacetal C_1 isomer to the corresponding 6-P-gluconolactone.[19] Oxidation of the very minor (<0.01%) free aldehyde form of glucose-6-P would have given 6-P-gluconate directly. Perhaps the enzyme is operating on the major accumulating hemiacetal form of 6-P-glucose to increase the bulk flux of glucose carbons through this first enzymatic step of the pentose phosphate pathway (ultimately generating D-ribose-5-P for nucleic acid biosynthesis).

Oxidation at C_1 occurs by enzyme-accelerated removal of the C_1–H as a hydride ion to nicotinamide cosubstrate $NADP^+$ so the reduced product is NADPH. The hemiacetal can be viewed as a reversibly protected form, not generally electrophilic, of the possibly promiscuously reactive C_1-aldehyde of 6-P-glucose. Only in the enzyme active site is the tetrahedral hemiacetal carbon directly oxidized to the lactone. The product lactone is clearly at a formal acyl group oxidation state. It is also a cyclic form of an oxoester,

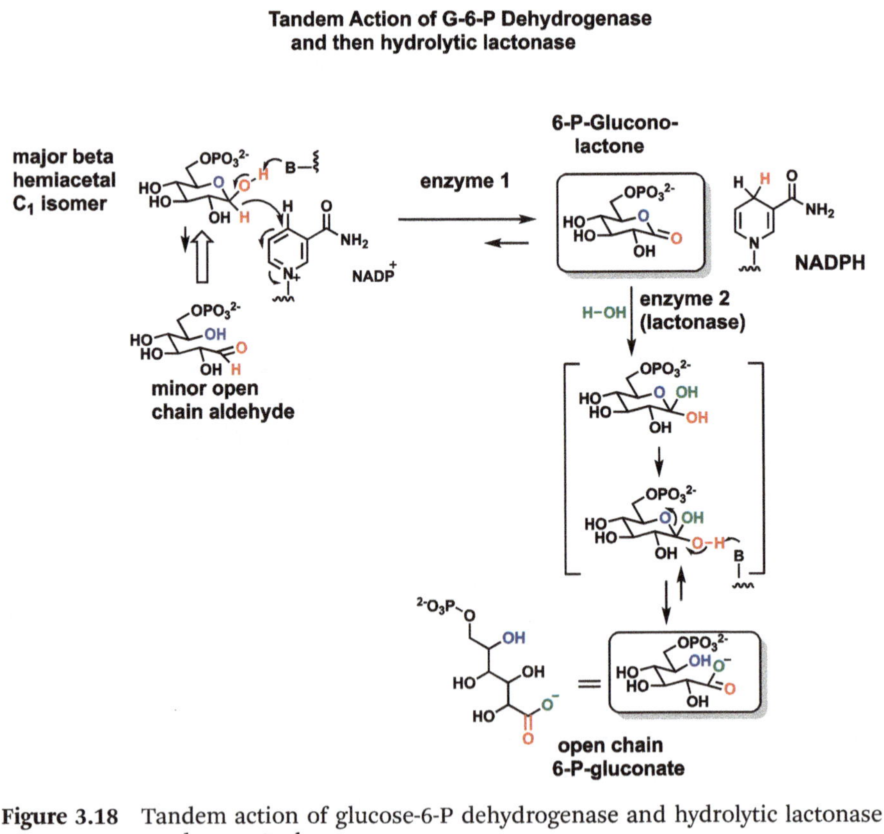

Figure 3.18 Tandem action of glucose-6-P dehydrogenase and hydrolytic lactonase produces 6-P-gluconate.

reminding that oxoesters and lactones are at the same formal oxidation state as acids.

3.7.1 Acyl Group Diversity from Aldehyde Oxidations

The acyl groups encountered thus far, in the action of oxidoreductases on aldehyde substrates or on the hemiacetal and thioacetal adducts in equilibrium with the aldehyde, are (1) the resonance stabilized carboxylate anions, (2) the thermodynamically activated acyl thioesters, (3) acyl phosphates, and (4) acyl lactones as cyclic forms of acyl esters (Figure 3.19). The only major acyl group missing from this biologic inventory are the amide groups taken up in Chapter 5 on carbon–nitrogen bonds. In terms of activation of the acyl groups for transfer as electrophilic group fragments to cellular nucleophiles, the resonance-stabilized carboxylates are lowest in (electrophilic) group transfer potential. Then the amides, up to the lactones and oxoesters on the way to acyl thioesters. The mixed acyl-phosphoric anhydrides define the upper end of increasing thermodynamic potential for transfer as activated, electrophilic forms of acyl groups (Figure 3.19).

A final comment on the metabolic fate of the 6-P-gluconolactone product from G-6-P dehydrogenase action. Organisms that employ this pentose phosphate pathway for flux of carbons from glucose-6-P also elaborate a tandemly-acting lactonase to accelerate the hydrolytic opening of the lactone to 6-P-gluconate[20] (Figure 3.18). This use of a separate enzyme is in contrast to the aldehyde dehydrogenase family above where the hydrolase activity was part of its intrinsic catalytic cycle. Here, a separate hydrolase has been encoded in the genome to get to the free 6-P-gluconic acid.

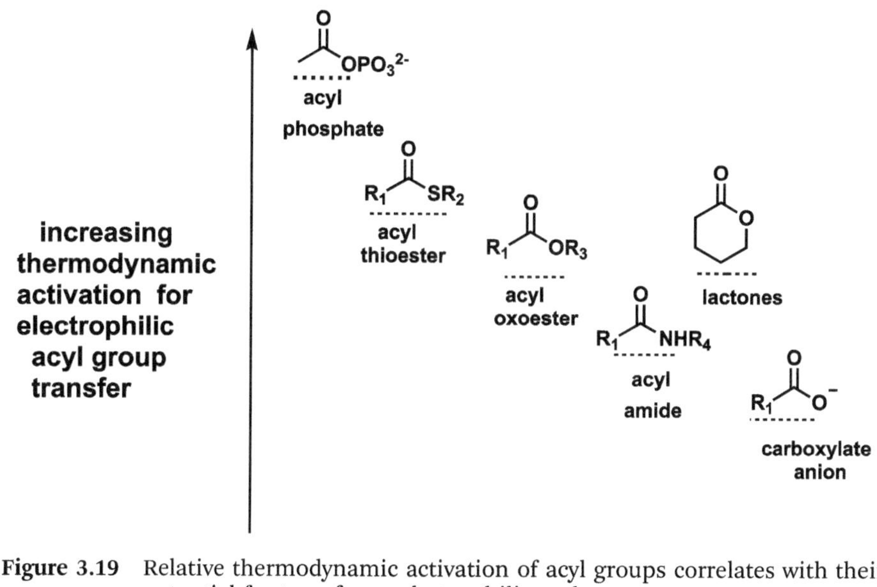

Figure 3.19 Relative thermodynamic activation of acyl groups correlates with their potential for transfer as electrophilic acyl groups.

We will note in Chapter 14 that the very next enzyme in the pathway is another $NADP^+$-reducing oxidoreductase, oxidizing the C_3–OH (secondary alcohol) to the C_3-ketone in 3-keto-6-phosphogluconate (a β-keto acid), set up for low energy C_1 decarboxylation *via* the 2,3-enediolate form of the C_2 carbanion.

3.8 Acyl Groups as Activated Electrophilic Fragments

Carboxylic acid metabolites generally have pK_a values for dissociation to a proton and the carboxylate anion in the range of 4–5. At pH 7 they are 99–99.9% ionized as the carboxylate anions. With the delocalized negative charge, the free carboxylates are unreactive as electrophilic carbonyl groups. Instead, the carboxylate can actually act with reversed polarity, as oxyanion nucleophiles, *e.g.* towards the α- or γ-phosphoryl groups of ATP.[21]

Consistent with the ranking of acyl groups in Figure 3.19, nature can generate three of those acyl derivatives by formal replacement of the oxyanion of carboxylates with: (1) an alcoholic oxygen to form oxoesters, (2) an amine to form an amide, or (3) with a thiol/thiolate anion to form the acyl thioester. To get to any of the three acyl derivatives, the carboxylate anions must be converted uphill to mixed acyl phosphoric anhydrides, either acyl phosphates or acyl-AMPs, from attack of the carboxylate oxygen on Pγ or Pα respectively of ATP. Then, these mixed acyl-phosphoric anhydrides are convertible to the (energetically) downhill thioesters, oxoesters, and amides (Figure 3.20). In terms of relative energetics, the acyl thioesters are most thermodynamically activated for transfer as electrophilic groups, then the oxoesters, then the amides, and finally the carboxylates (Figure 3.17). The acyl thioesters and acyl amides prefigure the content of Chapters 4 and 5, respectively. Note the expenditure of ATP, cleavage of a sidechain phosphoric anhydride bond, provides the energy to drive accumulation of the activated acyl-phosphoric anhydrides.

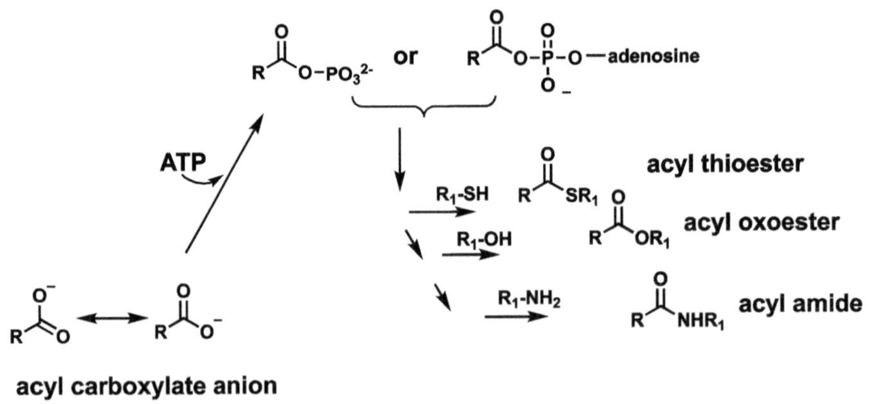

Figure 3.20 Schematic for conversion of resonance-stabilized carboxylate anions to mixed acyl-phosphoric anhydrides on the way to acyl thioesters, acyl oxoesters, and acyl amides.

3.8.1 Acyl Thioester Reduction to Aldehyde to Alcohol

Consistent with the thermodynamic activation and kinetic stability of acyl thioesters, they can serve as diffusible chemical packets of electrophilic acyl groups. Reductive capture of the acyl group occurs in the first step[22] of catalytic cycles of 3-hydroxymethyl-glutaryl CoA (HMGCoA) reductase[23] as noted back in Figure 3.9. This enzyme catalyzes the rate-limiting step in cholesterol biosynthesis in human liver.

In fact, that enzyme catalyzes two tandem hydride ion reductive steps from NADPH (The first hydride transfer reaction, shown back in Figure 3.11, reduces the acyl thioester by two electrons to the aldehyde (mevaldate). The second hydride transfer then more conventionally reduces the aldehyde to the alcohol mevalonate. This tandem action highlights the four electron interconversion of acyl group, aldehyde, and alcohol. The expenditure of 2 NADPH molecules as thermodynamically activated hydride transfer agents drives the reactions in the forward, reductive direction. More on placement of this double reduction step in the context of the isoprenoid pathway to steroids can be found in Chapter 15.

3.9 Alcohols to Oxoesters: Acyl Transfers to Oxygen: Diacyl Glycerol-phospholipids

Alcoholic groups are one half of ester functional groups (Figures 3.19 and 2.17). The second half is provided by an acid component. The alcohol oxygen behaves as nucleophile. The acyl group of a carboxylate must first be converted to an electrophilic acid derivative in order to be captured by the alcoholic oxygen.

When the acid is an inorganic phosphoric acid (electrophilic at the central phosphorus atom), the product is a phosphoester, a widespread functional group in both carbohydrate metabolism and in phosphoproteomics (Figure 3.21). When the acid component is an activated form of a carboxylate anion as depicted in Figure 3.20, the product is an acyl oxygen ester, often simply termed an ester (Figure 3.21). There is no change in oxidation state at the alcoholic carbon as it goes from free alcohol to oxoester, unlike the alcohol to aldehyde/ketone transformations.

3.9.1 Acetylcholine as an Oxoester Neurotransmitter

In primary metabolism thioester chemistry is much more prevalent than oxoester chemistry, as will be detailed in the next chapter.[24] However, at least three examples of acyl oxoester strategies come readily to mind for metabolites. The first is acetylcholine (Figure 3.22) the major transmitter at neuromuscular junctions, generated enzymatically by choline *O*-acetyl transferase from acetylation of the choline alcohol group as it attacks the electrophilic acetyl group of acetyl CoA.[25] This is a central metabolic example

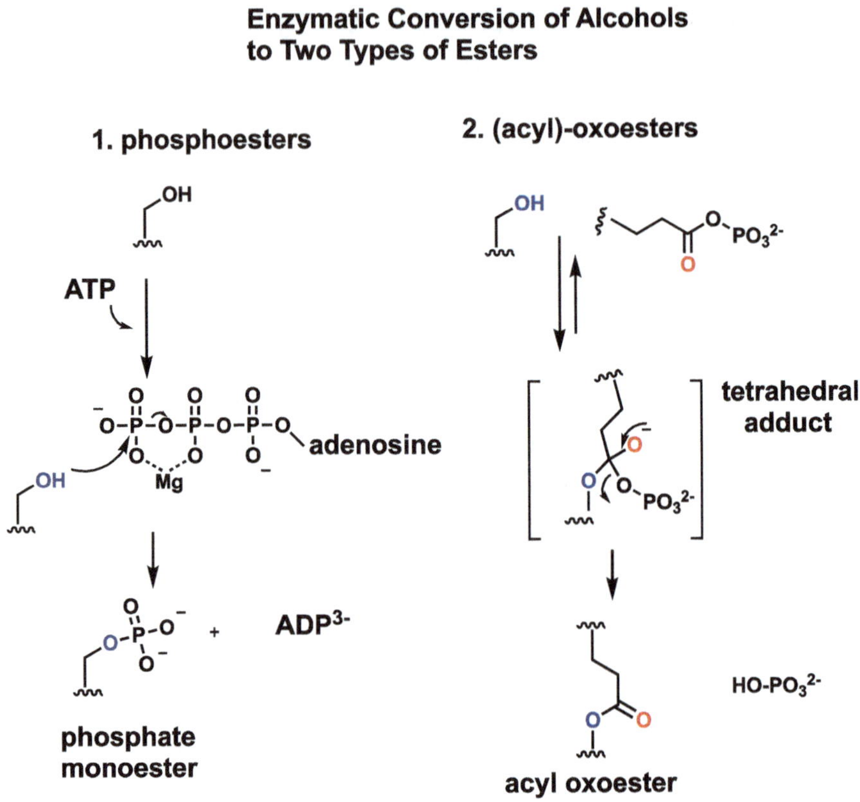

Figure 3.21 Enzymatic conversion of alcohols to phosphate esters or to acyl-oxoesters.

of the acetyl transfer potential of the cellular acetyl transfer metabolic currency acetyl CoA. Whereas choline is inactive as transmitter molecules at neuromuscular junctions, the acetyl choline oxoester is the active transmitter.

Control of the lifetime is important for nerve transmission and the lifetime of the oxoester can be readily mediated by a second enzyme, acetylcholinesterase. As an acyl oxoester, acetylcholine is chemically labile thermodynamically to hydrolysis back to the inactive alcohol and acid components. Indeed, the neurotransmitter lifetime is regulated by the hydrolytic enzyme acetylcholinesterase.[26] Note that in the two enzyme pathway the starting acetyl CoA thioester linkage has been converted to oxoester in acetyl choline and then on hydrolysis to the resonance-stabilized acetate anion (Figure 3.22). Hydrolysis of the neurotransmitter is thermodynamically favored (Figure 3.19). The acetylcholine is stable enough kinetically to form and diffuse in synaptic clefts. It is also labile to enzyme-accelerated hydrolysis, making the acyl oxoester linkage well-suited to a time-limited biologic signaling molecule.

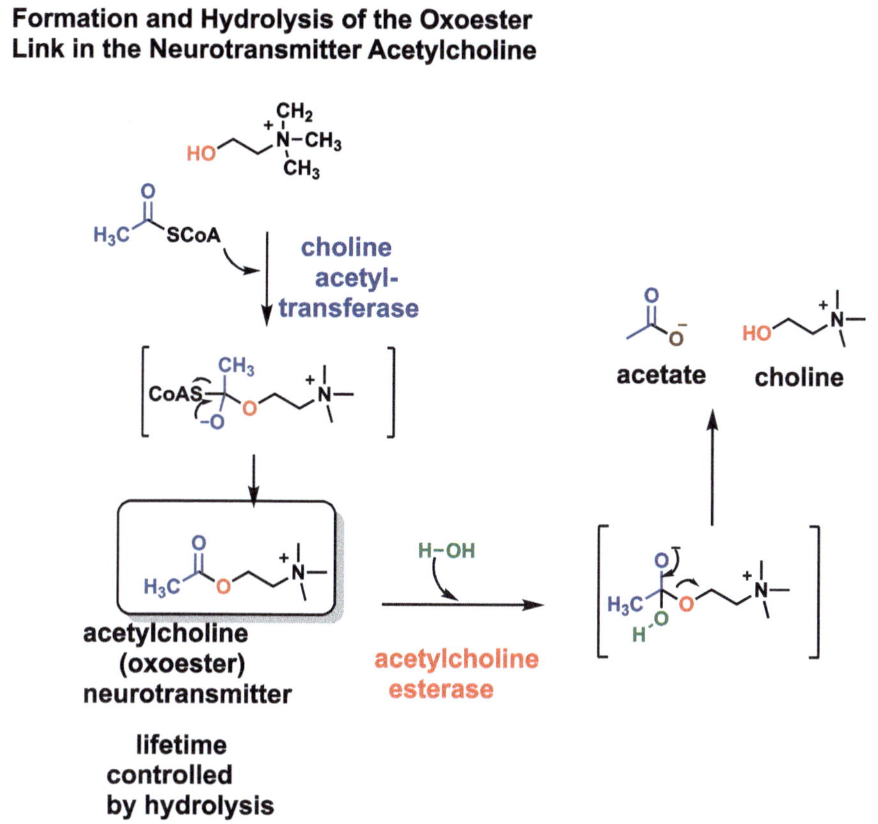

Figure 3.22 Formation and hydrolysis of the oxoester link in the neurotransmitter acetylcholine.

3.9.2 Acyl Group Activation Strategy: Aminoacyl-AMP, Aminoacyl-*O*-tRNA, Aminoacyl-Amide Bonds in Protein Biosynthesis

The second common metabolic locus of coupled oxoester formation and utilization are the chemical linkages between aminoacyl groups and the 3′-OH of the 3′-terminal adenosine in transfer RNAs at the ribosomes in cells (Figure 3.23). This discussion also prefigures C–N bond chemical biology of Chapter 5 because both (amino)acyl transfers to an alcohol (oxoester formation) and then to amines (amide group formation as the oxoester is consumed) are in play. Overall, as the carboxylate anions of amino acid building blocks are incorporated into proteins, aminoacyl esters are formed as intermediates and then used to make the more stable amide bonds in proteins.

The logic is to convert the resonance-stabilized carboxylate anions of proteinogenic amino acids into activated mixed aminoacyl-AMP anhydrides.[21] Then, react them with specific alcohol group partners

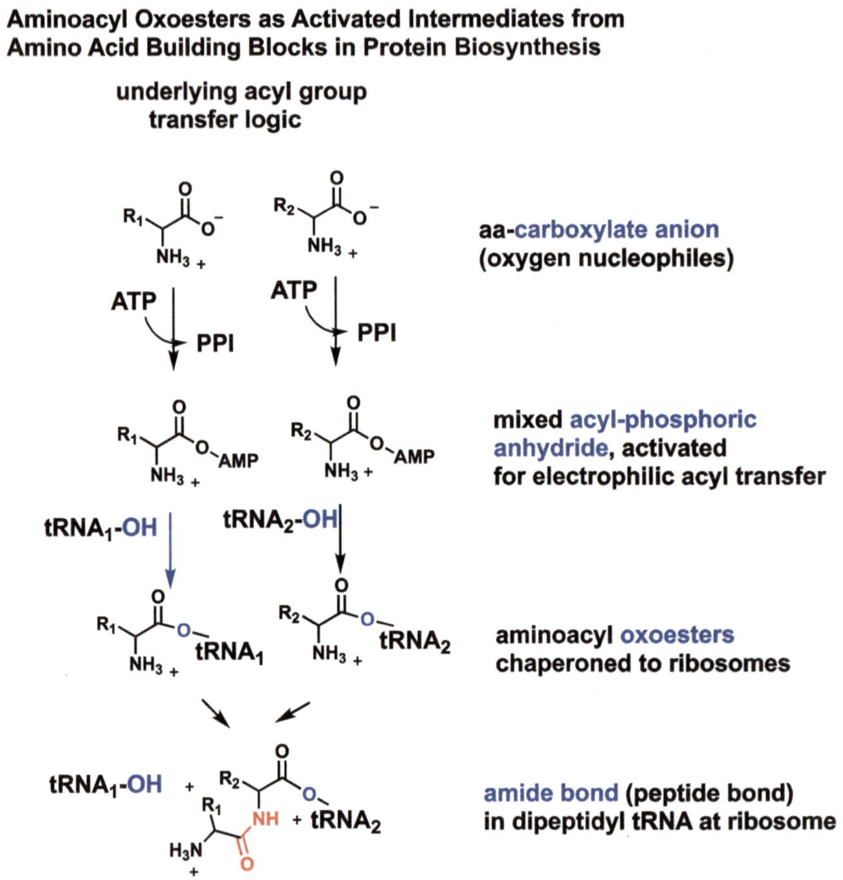

Figure 3.23 Aminoacyl oxoesters (aminoacyl tRNAs) as activated intermediates from amino acid building blocks used in protein biosynthesis.

(the terminal 3′-OH of AMP residues in cognate tRNAs) to create the oxoester linkages in aminoacyl tRNAs. The balanced thermodynamic activation coupled with sufficient kinetic stability allows all the aminoacyl tRNAs to be chaperoned intact to the peptidyl transferase center of the ribosomes. There, the oxoester group of the growing peptidyl chain reacts as electrophile with the free base form of the incoming aminoacyl tRNA (Figure 3.24). That generates the amide bond, characteristic coupling chemistry in all proteins. The biologic controls on selectivity and regulation at many stages, including the RNA-based peptidyl transferase center itself, can overwhelm appreciation of the simplicity of the chemical strategy for activation of amino acid monomers and then coupling them. They go from free carboxylate to aminoacyl-AMP. From there, the aminoacyl groups undergo transfer to aminoacyl oxoesters in aminoacyl tRNAs and finally to aminoacyl amides (peptide bonds) in nascent proteins (Figure 3.20). This strategy, practiced

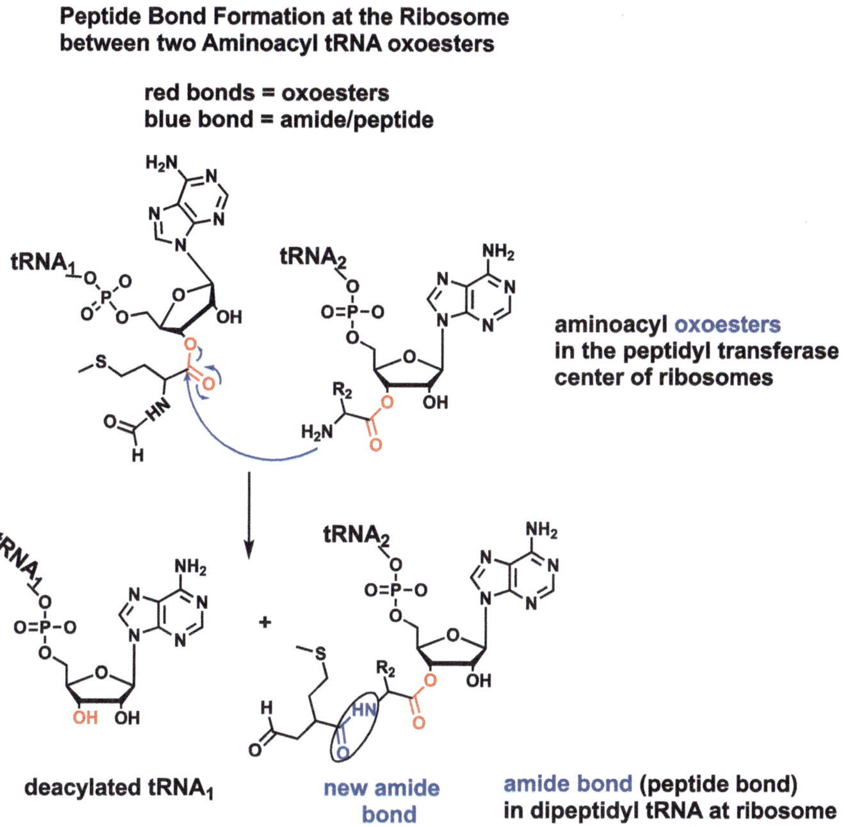

Figure 3.24 Peptide bond formation between two aminoacyl tRNAs (oxoesters) in the peptidyltransferase center of the ribosome.

millions of times per cell cycle *in vivo*, follows the simple acyl group chemistry charts of Figures 3.19 and 3.20.

3.9.3 Glycerol-based Diacyl Phospholipids Make Up Cell Membranes

A third locus of oxoester bonds are found in the pair of long chain acyl groups attached to the hydroxyl groups at C_1 and C_2 of glycerol-3-phosphate (Figure 3.25). The C_2-ketone of the glycolytic intermediate dihydroxyacetone-P is reduced to the C_2-secondary alcohol in glycerol-3-P in an NADPH-dependent hydride transfer step. There is a second oxidoreductase, one that functions irreversibly in the direction of glycerol-3-P oxidation, in mitochondria. Unlike the soluble cytoplasmic enzyme functioning in the DHAP reductase direction for membrane lipid production, the mitochondrial membrane-associated, flavin-dependent glycerol-3-P dehydrogenase provides two electrons to the membrane respiratory chain. Together, the two enzymes provide a glycerol-3-P

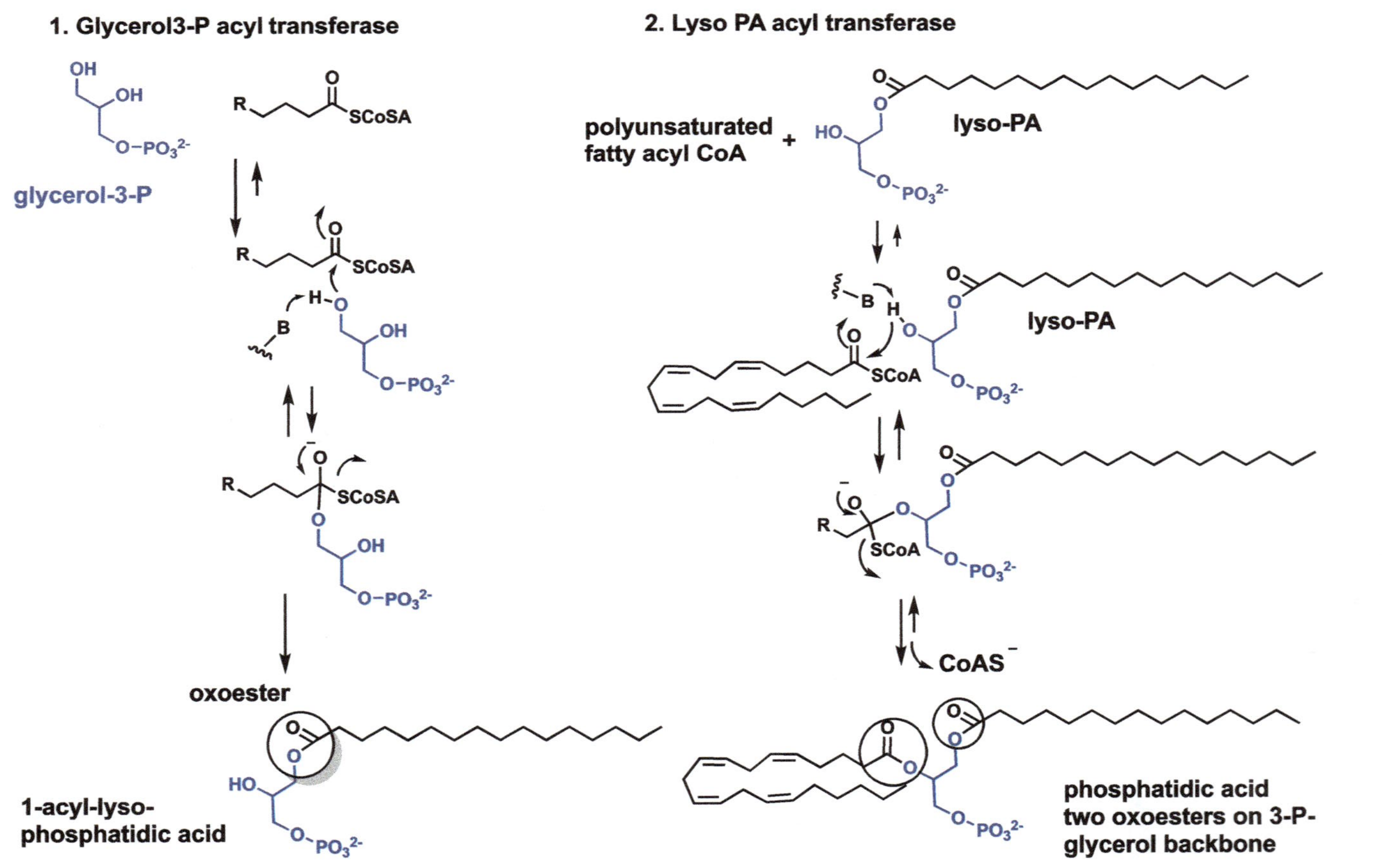

Figure 3.25 Tandem enzymatic conversion of two alcohol groups in glycerol-3-P to oxoesters in lysophosphatidic acids and then diacyl phosphatidic acids for membrane lipid biogenesis.

redox shuttle between pools of DHAP and glycerol-3-P in cytoplasm and mitochondrial compartments of cells. There is a third aldehyde oxidoreductase, one that uses O_2 as cosubstrate. That is a molybdenum-dependent oxidase, removing two electrons from aldehydes and nitrogen heterocycles such as xanthine. (Those are molybdopterin oxidases containing a molybdenum^{VI} ion coordinated to an ene dithiol as part of the core of a molybdenum cofactor and are not discussed further here.)

Two ester bonds are formed on the glycerol-3-P backbone sequentially from attack of the C_1–OH and C_2–OH oxygens on the activated carbonyl groups of acyl CoA thioester cosubstrates. The first enzyme is a glycerol-3-P acyltransferase, typically adding a saturated fatty acyl group, such as the C_{16} palmitoyl chain from the long chain fatty acyl CoA, thermodynamically activated as an electrophilic acyl group.[27] The attacking nucleophile is the C_1–OH of glycerol-3-P (Figure 3.23). This attack generates an oxoester linkage in a molecule of lysophosphatidic acid, (lyso-PA) with detergent like properties, arising from hydrophilic and hydrophobic regions within the framework of a lyso-PA.

The lyso-PAs are then substrates for a second acyltransferase that shows regiospecificity for using the C_2–OH as oxygen nucleophile and typically a polyunsaturated fatty acyl CoA as electrophilic partner.[28] Figure 3.25 shows the tetra-unsaturated ($\Delta^{5,8,11,14}$) arachidonic acid chain (see Chapter 2 for polyunsaturated fatty acids) incorporated at the C_2–oxygen of the glycerol backbone. These diacylated glycerol-3-P products are known as phosphatidic acids, precursors to the bulk phospholipid constituents of cell membranes that occur in 10^7 to 10^8 copies in every human cell. Given 10^{14} cells per person this is a serious commitment to acyl-*O*-ester lipid chemistry. Every phosphatidic acid has two acyl-oxoester bonds to long chain acyl groups (Figure 3.25).

The phosphatidic acids in eukaryotic cells can be partitioned down one of two further biosynthetic pathways as schematized in Figure 3.26. One path is phosphatase-mediated to release inorganic phosphate and a diacylgycerol (DAG). These DAGs are themselves lipid signaling molecules as well as being penultimate intermediates on the way to such membrane phospholipids as phosphatidylethanolamine and phosphatidylinositol as well as energy-storing triacylglycerols (triglycerides) (Figure 3.26). The second route is catalyzed by cytidylyltransferases that use CTP as electrophilic substrate to partner with a phosphatidic acid, in a classic cytidylyl (nucleotidyl) transfer of the CMP moiety of CTP to the attacking phosphate anion of a phosphatidic acid. The resultant CDP-diacylglycerols are immediate precursors to different membrane phospholipids. For example, attack by the OH group of ethanolamine or serine on the proximal phosphate of CDP-diglycerides yield phosphatidylethanolamines or phosphatidylserines, respectively, with expulsion of CMP[29] (Figure 3.26).

These oxoester linkages, two in membrane phospholipids, three in triacylglycerol lipid droplets, are kinetically stable as chemical groups embedded in hydrophobic scaffolds, under various physiologic conditions. Perhaps this is due in part to the low specific content of water in lipid membranes (phospholipids) and lipid droplets (triglycerides). They can be hydrolyzed by

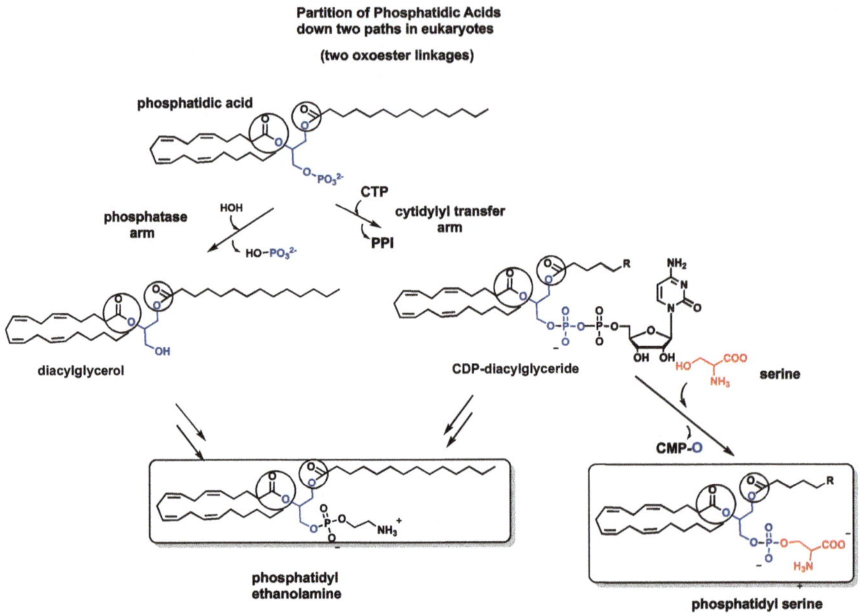

Figure 3.26 Partition of phosphatidic acids *via* diacylglycerols or CDP-diacylglycerides on the way to either phosphatidyl ethanolamine or phosphatidyl serine.

the (hydrolytic) esterase action of lipases during local membrane component reorganizations, but otherwise the oxoester linkages are indefinitely stable. These acyl oxoester linkages are used as permanent chemical bonds, unlike the balancing act of formation and subsequent transfer in both acetycholine and also in the aminoacyl oxoester bridges of aminoacyl tRNAs in ribosomal catalyzed protein biosynthesis.

3.10 Diverse Routes to Lactones (Cyclic Oxoesters)

Lactone functional groups are oxoesters in cyclic scaffolds as opposed to the acyclic frameworks of oxoester metabolites. Three routes to biologic lactones illustrate distinct strategies for their formation. We observed one route in Section 3.4.1 that the predominant hemiacetal in glucose can be oxidized directly by two electrons in NAD/NADP$^+$-dependent hydride transfers to the cyclic 6-P-gluconolactone (Figure 3.18). That is aldehyde to acid oxidation logic *via* hydride transfer to NADP$^+$ but occurs in a cyclic framework of cyclic hemiacetal substrate to lactone product.

A second route involves the intramolecular cyclization of a carboxylate oxygen as nucleophile onto C_4 of the methionyl moiety of *N*-acylated *S*-adenosylmethionine derivatives. The product is the *N*-acylated homoserine lactone and the methylthioadenosine fragment (Figure 3.27). The *N*-acylhomoserine lactones are quorum signaling molecules secreted by bacteria when cultures have reached sufficient density that starvation

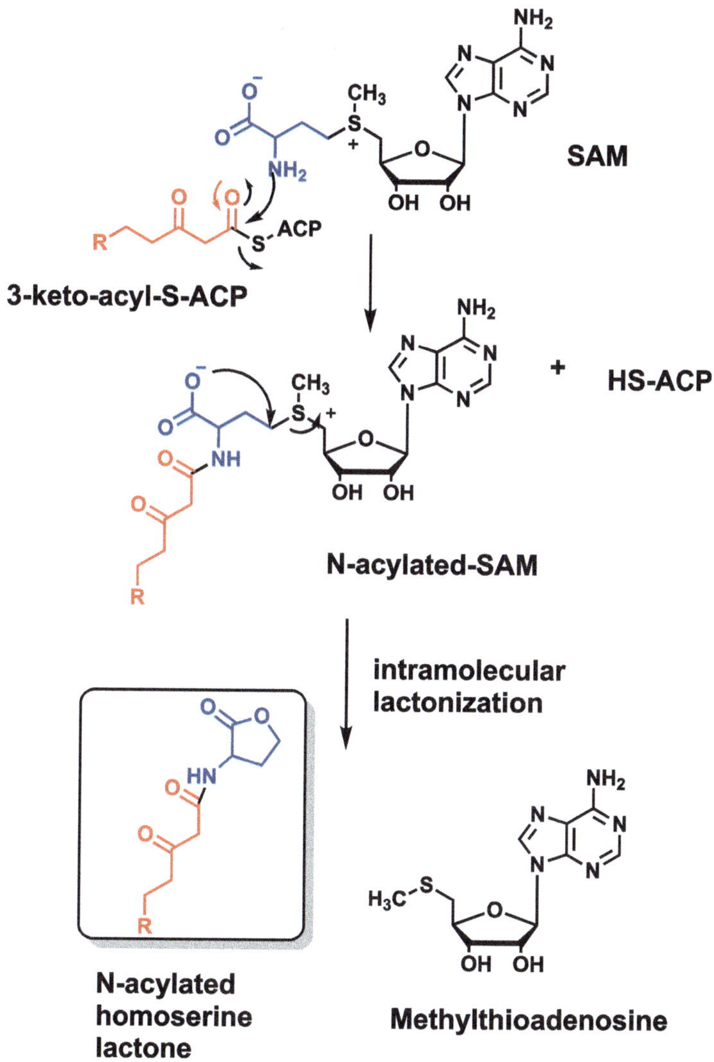

Figure 3.27 *N*-Acylhomoserine lactone formation from *N*-acylated *S*-adenosylmethionine.

conditions are about to set in.[30] These lipidated lactones pass into neighboring bacterial cells in the population and act as ligands for specific transcription factors to turn on or off genes that encode proteins to alter metabolic pathways in the recipient bacteria.

The reaction logic depends on two different reactivity aspects of *S*-adenosylmethionine. First is the reaction of the methionyl deprotonated amine group as nitrogen nucleophile on the thioester carbonyl of *S*-acylated-acyl carrier proteins (components of the fatty acid synthase machinery) to form amide bonds in the *N*-acylated-*S*-adenosylmethionine products. Among the intermediates in fatty acid biosynthesis, the 3-keto-acyl-*S*-ACPs serve as preferred cosubstrates. The second reactivity feature is the cationic sulfonium center in the *N*-acylated *S*-adenosylmethionine intermediates that labilizes the three attached carbons to function as electrophiles. In the active site of *N*-acylhomoserine lactone synthases, the carboxylate group acts as an oxyanion and attacks the C_4 carbon of that same methionyl chain as internal electrophilic partner. In the homoserine lactone product both oxygens in the lactone are from the carboxylate group. There was no alcoholic group in the starting substrate: a very different chemical strategy to form a lactone bond.

A third unique strategy to generate lactones starts from cyclic ketones and involves oxygen insertion to form the ring expanded lactone[31] (Figure 3.28).

This route is catalyzed by a subset of flavin-dependent monooxygenases that catalyze Baeyer–Villiger type insertions of an oxygen into a cyclic ketone as exemplified for the conversion of the five member cyclopentanone to the six member cyclic cyclohexyl lactone (Figure 3.29).

Three examples of ring expansion by oxygen insertion into cyclic ketones show some of the structural diversity of lactones constructed by this oxygenative strategy (Figure 3.28). The first is conversion of the cyclopentanone to the fused six member lactone ring in the tetracyclic sterol framework of testololactone. The second is in microbial metabolism of the monoterpene metabolite camphor. Enzymatic ring expansion of 2,6-diketocamphane creates the lactone that undergoes spontaneous hydrolysis: a carba-bicyclic alkane is thereby fragmented to a monocyclic pentene carboxylate. The third case is from plant biosynthetic metabolism in cytochalasin: maturation of cytochalasin E1, an anti-actin metabolite. Tandem insertion of one oxygen twice (*via* two catalytic cycles) into the large ring ketone generates the unusual vinyl carbonate ester group in cytochalasin E1. The mechanism for oxygen insertion by many Baeyer Villigerases is thought to be reaction of a flavin-hydroperoxy anion as oxygen nucleophile to attack the electrophilic carbonyl carbon of the ketone cosubstrate (Figure 3.29). Decomposition of the tetrahedral adduct occurs with migration of a C–C bond to expel flavin-OH as the expanded lactone ring forms.

3.11 Epoxides and Larger Ethers

Acyclic ethers, C–O–C functional groups, are relatively rare in metabolism, although methyoxyphenyl groups (aryl methyl ethers) are common ways to deactivate phenol –OH groups in conditional metabolic pathways in neurotransmitter signaling and in plant and in microbial secondary metabolism (Figure 3.30).

The cyclic ethers, starting with the smallest strained three member ring epoxides, deserve attention in chemical biology reactions. The epoxides are themselves both products of olefin oxygenations (noted in Chapter 2) and, once formed, can serve as reactive intermediates that can function as carbon electrophiles for capture by nucleophilic cosubstrates. The next higher cyclic ether is the four membered oxetane ring, then five member

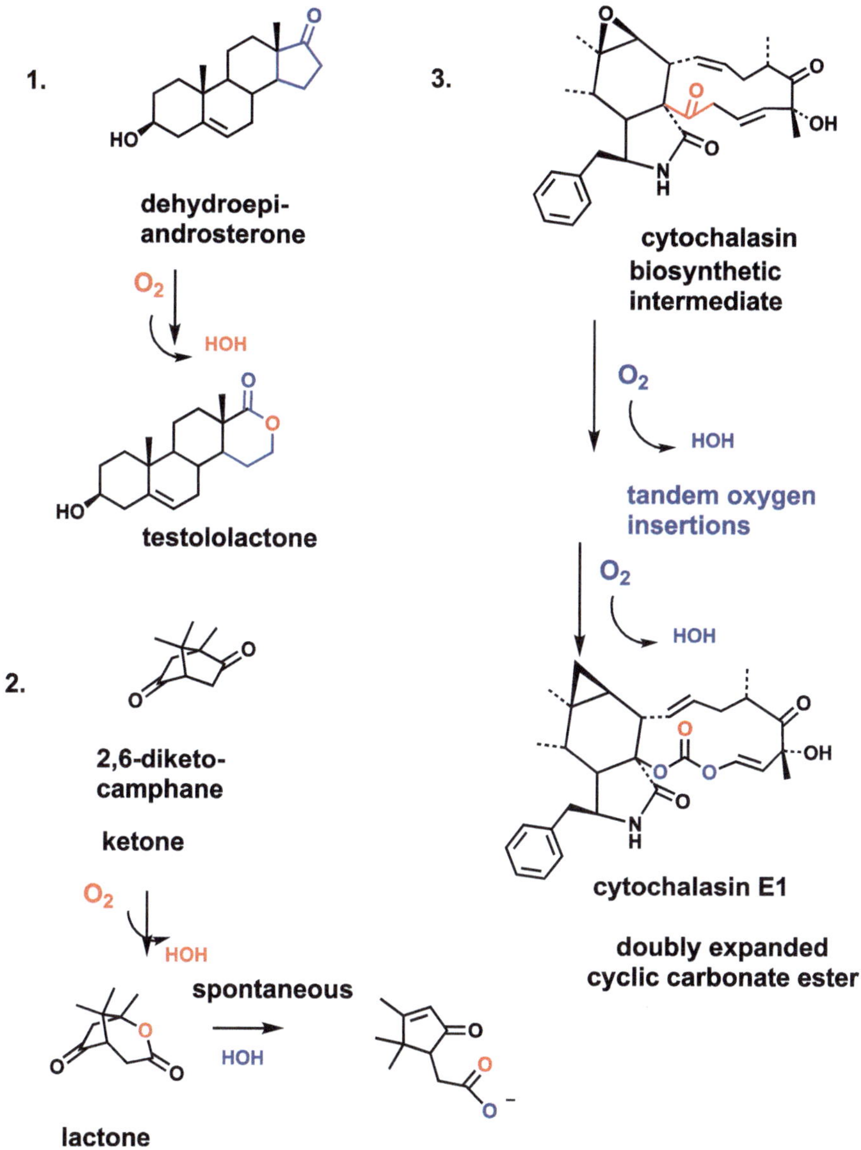

Figure 3.28 Three examples of lactone metabolites: testololactone, a camphane lactone, lactone to cyclic carbonate ester in cytochalasin E1.

tetrahydrofurans, and six member tetrahydropyran ring systems (see Figure 3.30, and also see Figure 7.1).

Enzymatic conversion of olefins to epoxides is the most common (but not the only) route to these strained three membered rings systems as noted in the preceding chapter. They can be formed in one electron pathways by iron-based epoxidases, from oxoiron species that deliver [OH•] equivalents in the

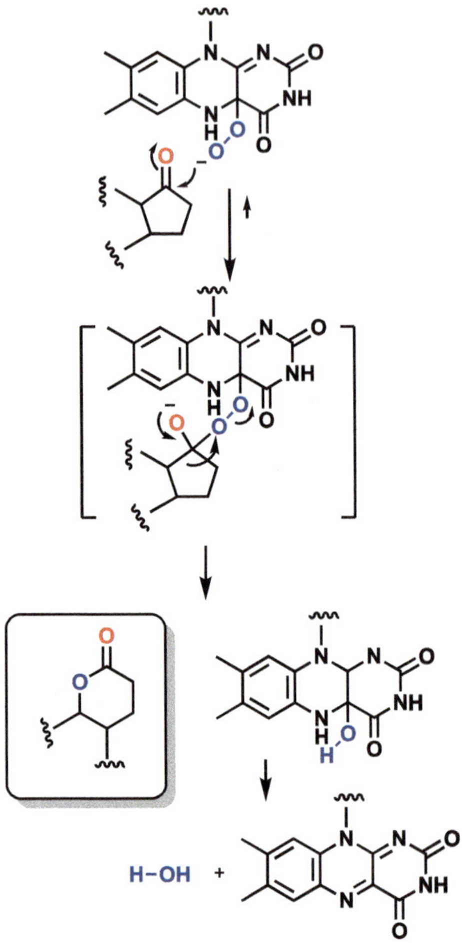

Figure 3.29 Ring expansion of cyclopentanone to the six-member lactone is catalyzed by a flavoenzyme cyclic ketone oxygenase.

iron-containing oxygenase active sites. Alternatively, as noted for squalene epoxidase in Chapter 2, the single oxygen atom can be delivered as an [OH^+] equivalent by flavin-containing epoxidases (Figure 2.19).

While the iron-containing epoxidases may be more widespread, the flavoenzymes can create epoxy scaffolds that go on to be captured by cosubstrate nucleophiles to form five- and six-ring cyclic ethers-tetrahydrofurans and tetrahydropyrans. One example is in biosynthesis of the potassium ionophore lasalocid and congeners[32] (Figures 3.28 and 3.31).

While still tethered as a covalent thioester to a polyketide synthase complex, the 1,4-diene in the full-length scaffold is epoxidized twice by the flavoenzyme lsd 18, consuming a molecule of O_2 in each catalytic cycle. The proximal oxygen transfer agent is thought to be the flavin 4a-OOH adduct. Then the next enzyme lsd19 catalyzes intramolecular ring opening of the

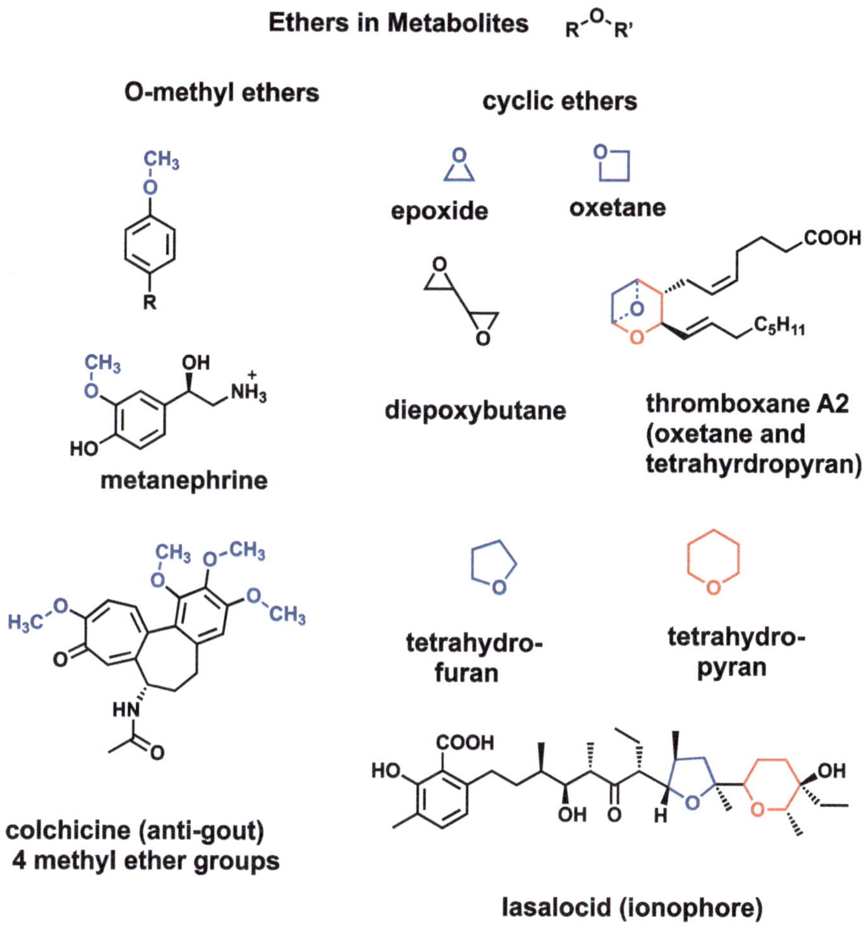

Figure 3.30 Examples of metabolic ethers, including *O*-methyl ethers, and cyclic ethers from epoxides to oxetanes to tetrahydrofurans and tetrahydropyrans.

epoxides by an internal OH in a cascade that forms first the tetrahydrofuran and then the pyran rings, and five membered and six membered cyclic ethers, respectively (Figure 3.30). In natural product biosynthesis there are many examples of disappearing epoxides as strained ring electrophiles to increase scaffold complexity amid C–O bond formations.

A three membered epoxide is also progenitor to a rare four membered cyclic ether an oxetane, that is part structure of the antitumor natural product taxol.[33] As depicted in Figure 3.32 the process again involves olefin conversion to epoxide and then a rearrangement to the ring-expanded four membered oxetane cyclic ether.[34] The exocyclic C=C double bond in the maturing framework of taxol is epoxidized, then attacked by the neighboring acetoxy oxygen to open the epoxide. The incipient alkoxide ion then forms the oxetane with migration of the acetoxy substituent to give mature taxol.

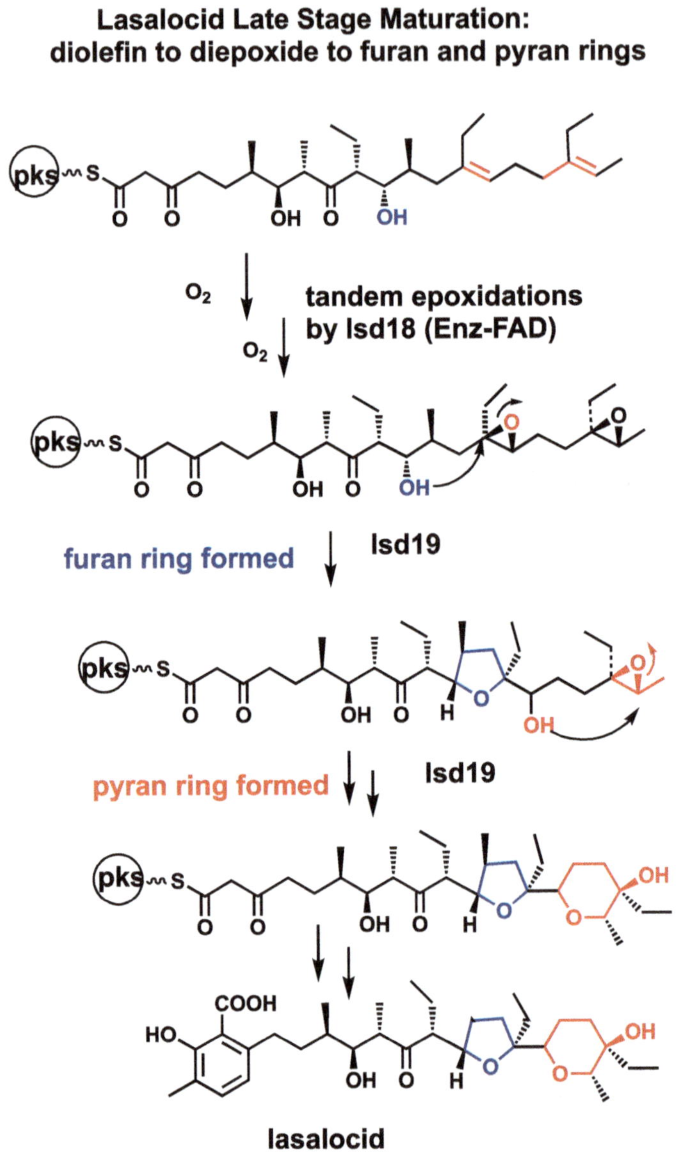

Figure 3.31 Lascalocid maturation. Enzymatic conversion of two olefins to two epoxides and intramolecular ring expansions to a tetrahydrofuran and tetrhydropyran ring.

A number of other, rare C–O containing functional groups are seen in natural products, including β-lactones and β-lactams (Chapter 7). Some of these are noted in Chapters 8 and 9 on natural heterocyclic metabolites and in Chapter 12 for isopenicillin N synthase.

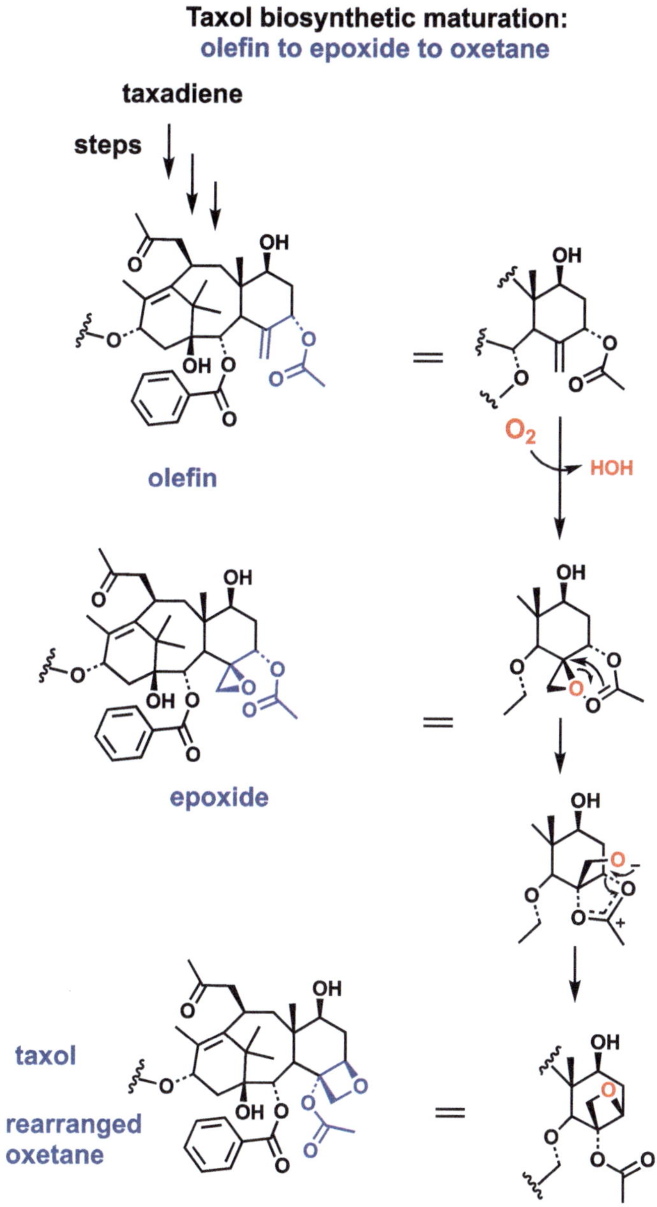

Figure 3.32 Biosynthetic maturation of taxol: olefin to epoxide to oxetane ring.

3.12 Decarboxylases and Carboxylases: The Chemical Biology of CO_2

Carbon dioxide, a central biologically active one carbon molecule, is the last of the carbon–oxygen functional groups that we take up in this chapter.[35]

Figure 1.12 noted five one-carbon compounds from the most reduced, methane, to the most oxidized, CO_2, separated by four two-electron traverses, through methanol, formaldehyde, and formic acid. Both methane and carbon dioxide are greenhouse gasses. Their prodigious emissions have been widely identified in recent decades with global warming.

Carbon dioxide is the final oxidized metabolic endpoint from reduced carbon compounds, such as dietary fatty acids and glucose from polysaccharides ingested from diet and oxidized to remove electrons for energy metabolism (see Chapter 14). Each adult human exhales about 500 L of CO_2 (~1 kg, ~22 mol) per day. The major metabolic sources of CO_2 are decarboxylation of carboxylate-containing metabolites. The variety of additional functional groups in such carboxylate metabolites determine the enzymatic routes for decarboxylations and the need for any cofactors.

Figure 3.33, left hand column, shows a generic two electron path for enzymatic decarboxylations. The predominant mode of decarboxylation involves heterolytic cleavage of a C–COO$^-$ bond. The departing CO_2, as a one carbon electrophile, leaves the electron pair on that adjacent carbon. (There are a few examples of homolytic/radical routes for decarboxylases, see Chapter 16.) Consequently, the energy barrier for decarboxylations

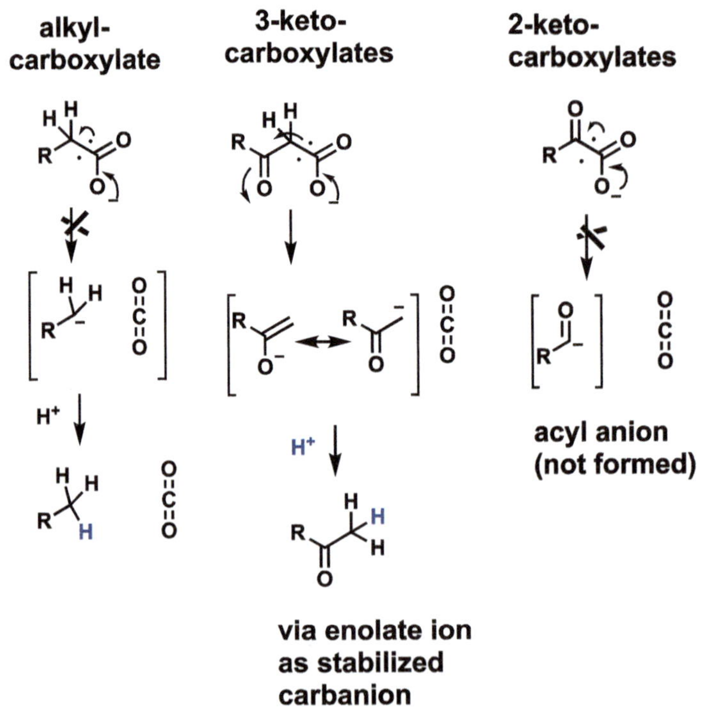

Figure 3.33 Barriers confronting decarboxylation of carboxy metabolites: alkyl carboxylates, *vs.* 2-keto-carboxylates *vs.* 3-keto-carboxylates.

can depend on ways in which that incipient product carbanions can be stabilized, lowered in energy. Resonance-mediated delocalization at that carbon is the dominant solution to create low energy paths for high flux metabolic decarboxylation of carboxylic acid anions. In the example shown in Figure 3.33 a primary carbanion is too high in energy to be formed in enzyme active sites and the $-CH_2-COO^-$ groups in metabolites are not decarboxylated without some additional electron sink.

In passing, while there is no requirement for an external reductant or oxidant as cosubstrate, the proposed products have undergone an internal redox change. The released CO_2 is two electrons oxidized up from the carboxylate group. The coproduct carbon has been converted from an internal CH_2 group to a terminal CH_3 group. The gain of one hydrogen is a formal two electron reduction. Finally, in this introduction to decarboxylations and product CO_2, the central carbon in carbon dioxide is electrophilic, as evident from considering the back reaction where a substrate carbanion equivalent would attack the electrophilic carbon in CO_2.

More generally, metabolically generated CO_2 is attacked by a water oxygen as nucleophile to undergo net hydration to HCO_3^- (bicarbonate anion) (Figure 3.34).

A family of zinc-dependent carbonic anhydrases catalyze the hydration/dehydration process some 10 000 to 1 million times per second, often as catalysts approaching perfection at the diffusion limit.[36] This rapid hydration not only allows the carbon dioxide/bicarbonate system to act as an important physiological buffer for blood pH but also interconverts an electrophile at carbon (CO_2) with a nucleophilic at oxygen (HCO_3^{2-}). These two different reactivities also condition how carboxylase enzymes, running in the C–COO^- bond-forming direction, choose distinct mechanistic strategies, as noted below.

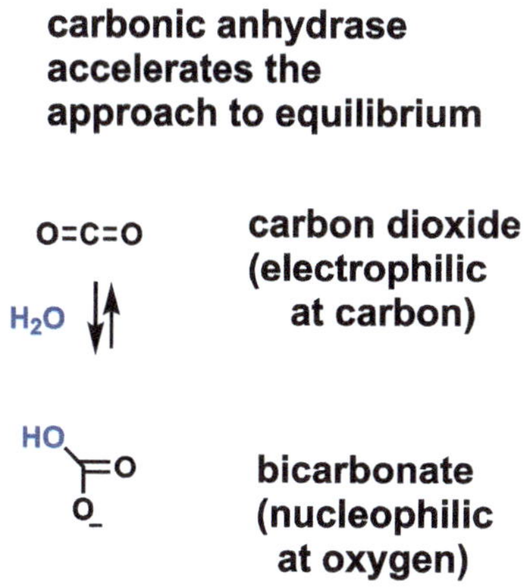

Figure 3.34 Carbonic anhydrase dramatically accelerates the approach to equilibrium between CO_2 and bicarbonate anions.

3.12.1 Strategies for Lowering Energy Barriers for Enzymatic Decarboxylations

From the above consideration, structural and electronic features that can stabilize the incipient decarboxyated carbanion will lower energy barriers and accelerate rates.[35] The classic case is the facile decarboxylation of β-keto acids (3-keto acids), *e.g.* acetoacetate to acetone and CO_2, that utilize an enolate anion/resonance-stabilized carbanion path to lower the energy barrier to carbanion formation, as shown in the second column of Figure 3.33.

The corresponding 2-keto acids are even more common primary metabolites, such as pyruvate, 2-oxalosuccinate in the citric acid cycle, and phenylpyruvate before its biosynthetic reductive amination to phenylalanine (see Chapter 5). There is no corresponding internal stabilization of the incipient carbanions (column 3 of Figure 3.33) that would have the characteristics of acyl anions, much too high in energy to form. To that end nature has invented thiamin-PP, the coenzyme form of vitamin B1 (see Chapter 9) which brings a thiazolium heterocyclic ring to form a β-iminium adduct with 2-keto acid substrates to provide an imine/enamine variation of the enolate/carbanion route to lower energy for decarboxylation (Figure 3.35). Otherwise, there is no low energy path for 2-keto acid decarboxylations.

In Chapter 5 we will note that many amino acids are decarboxylated to primary amines. A strategy comparable in principle to thiamin-PP setting up a β-iminium electron sink in the adduct between coenzyme-2-keto acid substrate are iminium adducts between the vitamin B6 coenzyme pyridoxal-P and the amino acid. Decarboxylation generates an equivalent low energy enamine/carbanion that protonates to the product amine (Figure 3.36) (see Chapters 5 and 9).

3.12.2 Carboxylases Fix CO_2/HCO_3^- as Carboxylate Groups

While most categories of decarboxylases run irreversibly to liberate CO_2 from carboxylate-containing substrates that can stabilize accessible nascent product carbanions, other enzyme families can act as carboxylases.[35] Such carboxylases have the option of utilizing CO_2 directly as an electrophilic partner in the carboxylation reaction or of using bicarbonate (HCO_3^-) ion as proximal reactant. The latter is an oxygen nucleophile, not a carbon electrophile and requires a different fixation strategy.

A small set of metabolically central carboxylases that use bicarbonate as nucleophilic substrate require vitamin B7 biotin and ATP to ultimately convert the one carbon fragment bicarbonate from a nucleophilic anion to an *N*-carboxy ureido derivative (see Chapter 9) that can be transferred as an electrophile at carbon. Specific cosubstrates are pyruvate or acetyl CoA where eneolate or thioester enolate serve as the accessible carbanion equivalents for C–C bond formation (Figure 3.37).

An alternate strategy is required by carboxylases that fix CO_2 directly, not bicarbonate ion. By far the most abundant protein on the planet is the plant-based

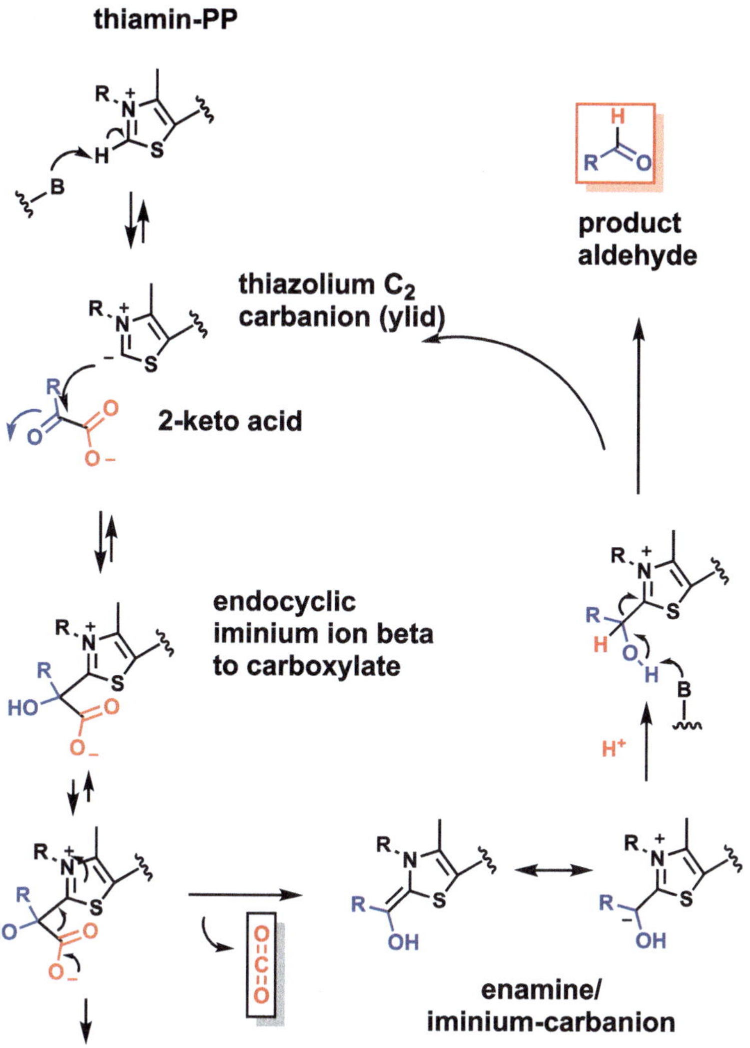

Figure 3.35 The quaternized thiazolium ring in the coenzyme form of vitamin B1, thiamin, can readily form a C_2 carbanion (an ylid) to attack 2-keto acids to form covalent adducts with an iminium ion beta to the carboxylate to act as an electron sink to stabilize incipient carbanion density on decarboxylation.

carboxylase, ribulose-1,5 bis-phosphate carboxylase.[37] It fixes an estimated 250 billion tonnes of CO_2 globally each year, dwarfing any other biological carbon fixation route.[38] The cosubstrate ribulose-bis-phosphate is ideal for generating a kinetically accessible low energy carbanion equivalent in the enzyme active site and using that to attack the electrophilic carbon in CO_2.

The initially accessible carbanion from the ribulose framework is at C_3 by enzymatic abstraction of the C_3–H as a proton to give the C_3-carbanion that

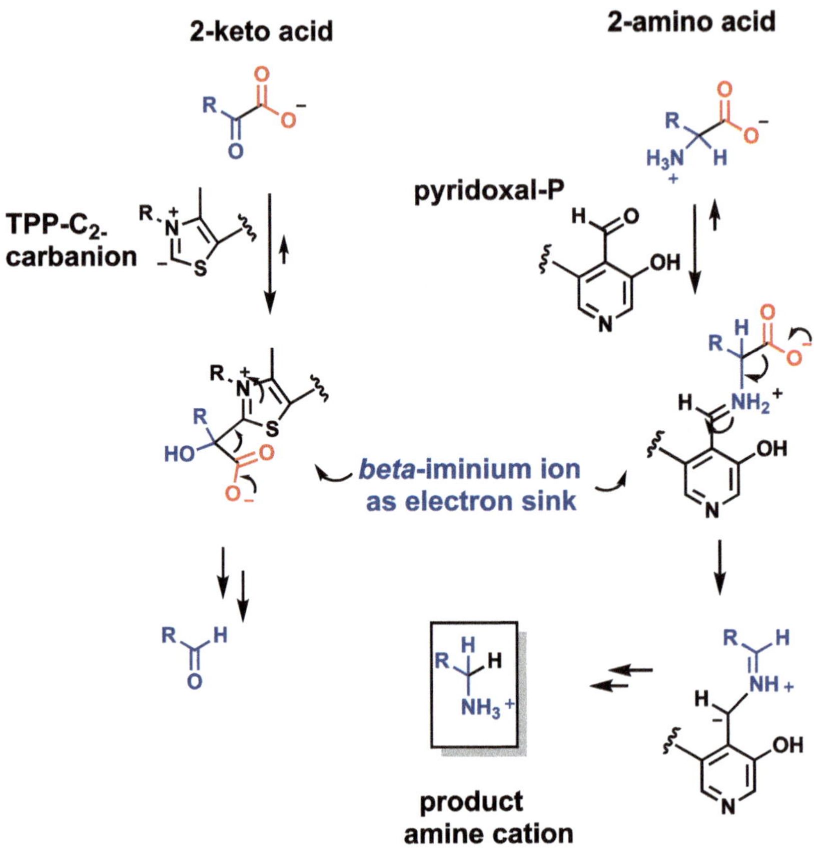

Figure 3.36 Analogy between thiazolium ring of B1 and pyridoxal ring of B6 as electron sinks to lower energy barriers for decarboxylation by delocalization of the resulting incipient carbanions.

is stabilized by delocalization of the negative charge onto oxygen. This resonance contributor is an enediolate anion. It is termed enediolate isomer 1 in Figure 3.38. Enzyme-catalyzed isomerization by differential protonation of the two oxygens in enediolate 1 leads to the isomeric enediolate 2. This is a resonance form of the C_2 carbanion of ribulose-bis-phosphate.

It is this regiospecific, isomerized C_2 carbanion that is utilized as the nucleophile to attack bound CO_2. This is the C–C bond-forming step, incorporating the carbon dioxide moiety as a carboxylate group in a branched six carbon scaffold. In particular, that nascent product has a ketone at C_3, isomerized from its initial C_2 locus during the enediolate isomer interconversions. Water addition into that ketone creates a transient ketone hydrate (Figure 3.35). That adduct is proposed to fragment to the two observed identical product molecules of 3-phophoglycerate, a readily processed intermediate in the glycolysis pathway of plants.

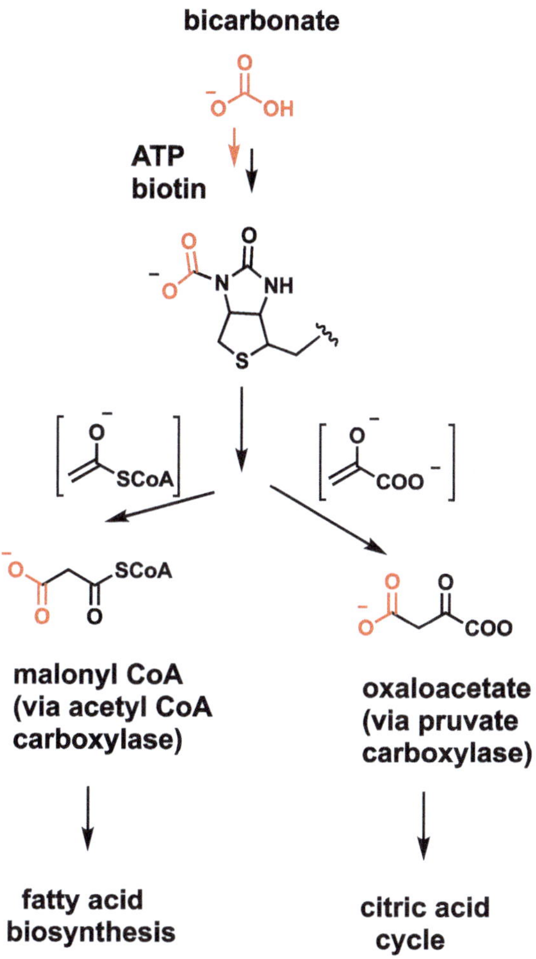

Figure 3.37 Biotin carboxylases use bicarbonate and ATP to generate an electrophilic form of carbon dioxide, capturable by cosubstrate carbanion equivalents in pyruvate carboxylase and acetyl CoA carboxylase.

The chemical wrinkle in RUBISCO catalytic cycles is that the enzyme generates an accessible C_3 carbanion equivalent, isomerizes it *via* enediolate contribution to the C_2 carbanion and uses only that one as the carbon nucleophile for CO_2 fixation. The presumption is that this enzyme catalyst, responsible for fixing 500 trillion grams of CO_2 in global plant metabolism evolved early in the chemical biology of C–O functional group transformations.

Back in Chapter 2, we noted that the NADH-dependent crotonyl CoA carboxylase (Figure 2.9) that fixes CO_2 gives ethylmalonyl CoA as product. The requisite carbon nucleophile to capture CO_2 is generated in the enzyme active site by prior addition of a hydride from NADH into the C_3-olefinic terminus of the conjugated C=C double bond of crotonyl CoA.

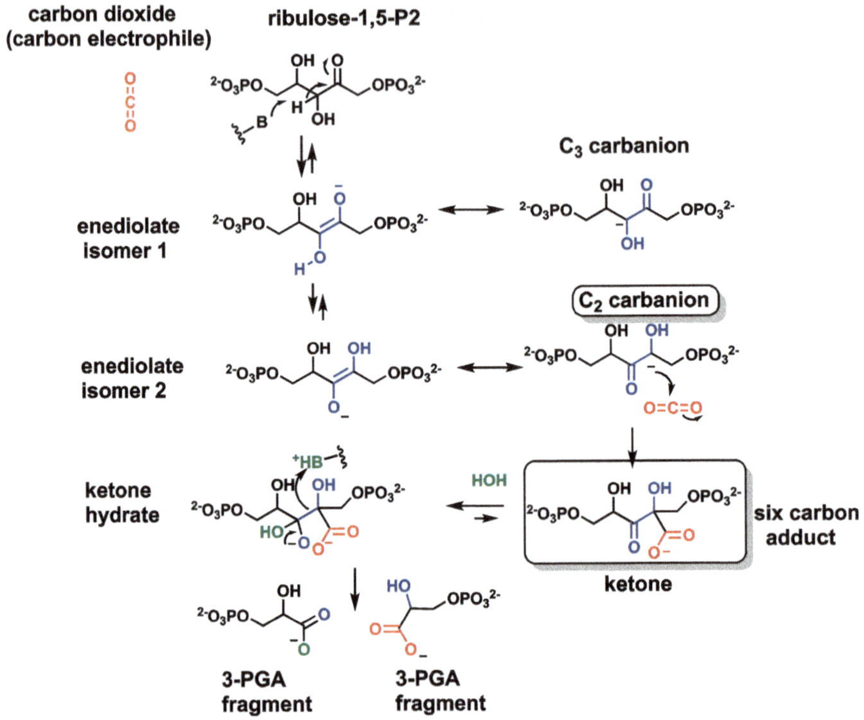

Figure 3.38 Ribulose-bis-phosphate carboxylase (RUBISCO) fixes CO_2 into the ribulose-1,5-P_2 cosubsrate *via* reversible formation of C_3 and C_2 carbanion equivalents.

3.13 Overview of C–O Functional Group Chemical Biology

The three most common C–O functional groups in metabolism are the sp^3-hybridized alcohols, the sp^2-hybridized aldehydes and ketones, and the sp^2-hybridized carboxylates, three oxidation states separated from each other by two electrons. The alcohol oxygens are nucleophilic and react with electrophilic groups, including phosphate and acyl derivatives as well as the sulfur and amine nucleophiles discussed in the next two chapters. In addition to aliphatic alcohol groups, phenolic oxygens are common, from the *para*-OH group in tyrosine, the *para*-OH in coumaryl CoA as building block for phenylpropanoids, and as precursor to the catecholic 1,2-diphenolic moieties, *e.g.* in the dopamine neurotransmitter. We note the radical-based oxygenation of L-phenylalanine to L-tyrosine in Chapter 12. Many of metabolite phenolic-OH groups arise by oxygenase action (Chapter 12).

Conversely, the aldehydes and ketone carbonyl carbons behave as the most abundant carbon electrophiles in cells. The third group, the carboxylates, when ionized (>99%) at physiologic pH, will react as nucleophiles in enzyme active sites. When acyl groups are converted to acyl phosphoric anhydrides or acyl thioesters, they instead behave as carbon electrophiles. We have noted the enzymatic formation and reactivity of acyl oxoesters, partners between alcohols and esters. In some cases oxoesters are intermediates between acyl thioesters and the alcohol and acid hydrolysis products of those oxoesters. In protein synthesis the acyl oxoesters are sufficiently activated to react with amines to form the more stable acyl amides. Lactones are cyclic versions of oxoesters, at the same oxidation state and with similar hydrolytic lability.

Carbon dioxide is the most oxidized form of carbon–oxygen molecules. It is the major end product of aerobic animal metabolism On average one kilogram per day per every adult human on the planet of carbon dioxide arises by action of various types of decarboxylases on carboxylate-containing substrates. The citric acid cycle releases two molecules of CO_2 in every turn of the cycle, one by β-keto acid decarboxylation, the second by α-keto acid decarboxylation. Conversely, plant-based ribulose bis phosphate carboxylase, annually fixing 500 trillion grams of CO_2 globally, operates at the other end of carbon cycles from decarboxylases.

Many of the low molecular weight metabolites are comprised of organic scaffolds with mixed C–O functional groups – alcohols, aldehydes and ketone and various acyl groups. Thus, polyhydroxy aldehydes comprise sugar frameworks. Hydroxy and keto acids are involved in the citric acid cycle as are di- and tricarboxylates. We have commented on the chemistry available to 1,2- and 1,3-hydroxycarbonyl systems as central to the first steps in glucose catabolic metabolism.

The inventory of carbon–oxygen functional groups, when joined by and expanded with the C–S and C–N chemical biology of Chapters 4 and 5, accounts for essentially all the carbon-based functional groups that nature uses to drive the multitude of transformations and interconversions that comprise the organic chemistry of life.

References

1. M. Engleder and H. Pichler, On the current role of hydratases in biocatalysis, *Appl. Microbiol. Biotechnol.*, 2018, **102**, 5841–5858.
2. E. Busto, V. Gotor-Fernández and V. Gotor, Hydrolases: catalytically promiscuouc enzymes for non-conventiona reactions in organic synthesis, *Chem. Soc. Rev.*, 2010, **39**, 4504–4523.
3. Western Oregon University C, Chapter 9 Organic Compounds of Oxygen, retrieved March 6, 2022.
4. D. V. Banthorpe and M. J. Osborne, Terpene epxoidases and epoxide hydrolases from cultures of Jasminum officiale, *Phytochemistry*, 1984, **23**, 905–907.
5. B. Persson, J. Hedlund and H. Jörnvall, Medium- and short-chain dhydrogenase/reductase gene and protein families: the MDR suprefamily, *Cell. Mol. Life Sci.*, 2008, **65**, 3879–3894.

6. C. J. Walsh and Y. Tang, *The Chemical Biology of Human Vitamins*, Royal Society of Chemistry, London, 2018.
7. A. Klungland and A. B. Robertson, Oxidized C5 methyl cytosine bases in DNA: 5-hydroxymethylcytosine; 5-formylcytosine, and 5-carboxycytosine, *Free Radical Biol. Med.*, 2017, **107**, 62–68.
8. R. S. Cahn, C. Ingold and V. Prelog, Specification of molecular chorality, *Angew. Chem., Int. Ed. Engl.*, 1966, **5**, 385–415.
9. A. Glasfeld, G. F. Leanz and S. A. Benner, The stereospecificities of seven dehydrogenases from Acholeplasma laidlawii, *J. Biol. Chem.*, 1990, **265**, 11692–11699.
10. M. Aoshima and Y. Igarashi, Nondecarboxylating and decarboxylating isocitrate dehydrogenases: oxalosuccinate reductase as an ancestral form of isocitrate dehydrogenase, *J. Bacteriol.*, 2008, **190**, 2050–2055.
11. D. A. Dougherty and E. V. Anslyn, *Modern Physical Organic Chemistry*, Univerist Science Books, 2006.
12. M. A. Fox and J. K. Whitesell, *Organic Chemistry*, Joned & Bratlett Learning, 2004.
13. W. P. Jencks, *Catalysis in Chemistry and Enzymology*, McGraw Hill, 1969.
14. P. A. Frey and A. D. Hegeman, *Enzymatic Reaction Mechanisms*, Oxford University Press, New York, 2007.
15. M. Garcia-Viloca, C. Alhambra, D. G. Trulahr and J. Gao, Hydride transfer catalyzed by xylose isomerase: mechanisms and quantum effects, *J. Comput. Chem.*, 2003, **24**, 177–190.
16. E. Lorentzen, B. Siebers, R. Hensel and E. Pohl, Mechanism of the schiff base forming fructose-1,6-bis phosphate aldolase: structural analysis of reaction intermediates, *Bicohemistry*, 2005, **44**, 4222–4229.
17. K. Shortall, A. Djeghader, E. Magner and T. Soulimane, Insights into aldehyde dehydrogenase enzymes: a structural perspective, *Front. Mol. Biosci.*, 2021, 659550.
18. C. Tristan, N. Shahani, T. W. Sedlak and A. Sawa, The diverse functions of GAPDH: views from different subcellular compartrments, *Cell. Signalling*, 2011, **23**, 317–323.
19. M. S. Cosgrove, C. Naylor, S. Paludan, M. J. Adams and H. R. Levy, On the mechanism of the reaction catalyzed by glucose-6-phosphate dehydrogenase, *Biochemsitry*, 1998, **37**, 2759–2767.
20. E. Miclet, V. Stoven, P. A. M. Michels, F. R. Opperdoes, J. Y. Lallemand and F. Duffieux, NMR spectroscopic analysis of the first two steps of the pentosde phosphate pathway addresses the role of 6-phosphogluconate lactonase, *J. Biol. Chem.*, 2001, **276**, 34840–34846.
21. C. T. Walsh, P. T. Benjamin and Y. Tang, Eight kinetically stable but thermodynamically activated molecules that power cel metabolism, *Chem. Rev.*, 2018, **118**, 1460–1494.
22. B. E. Haines, O. Wiest and C. V. Stauffacher, The increasingly complex mechanism of HMG-CoA reductase, *Acc. Chem. Res.*, 2013, **46**, 2416–2426.
23. H. M. Miziorko, Enzymes of the mevalonate pathway of soprenoid biosynthesis, *Arch. Biochem. Biophys.*, 2011, **505**, 131–143.

24. W. Yang and D. G. Drueckhammer, Understanding the relative acyl-transfer eractivity of oxoesters and thioesters: computational analysis of transition state delocalization effects, *J. Am. Chem. Soc.*, 2001, **123**, 11004–11009.
25. D. Nachmansohn and A. L. Machado, The formation of acetylcholine. A new enzyme. Choline acetylase, *J. Neurophysiol.*, 1943, **6**, 397–403.
26. D. M. Quinn, Acetylcholinesterase: enzyme structure, reaction dynamics, and virtual transition states, *Chem. Rev.*, 1987, **87**, 955–979.
27. S.-F. Yet, Y. K. Moon and H. S. Sul, Glycerol-phosphate acyltransferases, *Biochemistry*, 1995, **34**, 7303–7310.
28. R. M. Bradley and R. E. Duncan, The lysophsophatidic acid acyltransferases: one reaction, five enzymes, many roles, *Curr. Opin. Lipidol.*, 2018, **29**, 110–115.
29. P. Fagone and S. Jackowski, Membrane phospholipid synthesis and endoplasmic reticulum function, *J. Lipid. Res.*, 2009, **50**, S311–S316.
30. Y. Jiang, M. Camara, S. R. Chhabra, K. R. Hardie, B. W. Bycroft, A. Lazdunski, G. P. C. Salmond, G. S. A. B. Stewart and P. Williams, *In vitro* biosynthesis of the Pseudomonas aeruginosa quorum-sensing signal molecule *N*-butanoyl-L-homoserine lactone, *Mol. Microbiol.*, 2002, **28**, 193–203.
31. C. T. Walsh and Y. Tang, *Natural Product Biosynthesis: Chemical Logic and Enzymatic Mechanisms*, Royal Society of Chemistry, London, 2nd edn, 2022.
32. A. Minami, H. Oguri, K. Watanabe and H. Oikawa, Biosynthetic machinery of ionophore polyether lasalocid: enzymatic construction of polyether skeleton, *Curr. Opin. Chem. Biol.*, 2013, **17**, 555–561.
33. D. Willenbring and D. J. Tantillo, Mechanistic possibilities for oxetane formation in the biosynthesis of Taxol's D ring, *Russ. J. Gen. Chem.*, 2008, **78**, 723–731.
34. J. A. Bull, R. A. Croft, O. A. Davis, R. Doran and K. F. Morgan, Oxetanes: recent advances in synthesis, reactivity and medicianl chemistry, *Chem. Rev.*, 2016, **116**, 12150–12233.
35. C. T Walsh, Biologically generated carbon dioxide: nature's versatile chemical strategies for carboxy lyases, *Nat. Prod. Rep.*, 2030, **37**, 100–135.
36. T. Maren, Carbonic anhyydrase: chemistry, physiology and inhibition, *Physiol. Rev.*, 1967, **47**, 595–781.
37. I. Anderson, Catalysis and regulation in Rubisco, *J. Experimetal. Bot.*, 2008, **59**, 1555–1568.
38. I. C. Tommasi, The mechanism fo Rubsico catalyed carboxylation reaction: chemical aspects involving acid-base chemistry and functioning of the molecular machine, *Catalysis*, 2021, **11**, 813.

CHAPTER 4

Carbon–Sulfur Bonds: Adding to Functional Group Inventory

4.1 Inorganic Sulfur and Organosulfur Metabolites in Biology

Sulfur sits above oxygen in the periodic table, having 16 electrons,[1] with electronic configuration where the six valence electrons fill the 3s and 3p orbitals as noted: $3s^23p^4$. That is the ground state electronic configuration in divalent compounds such as the inorganic hydrogen sulfide and the organic thiol in the amino acid L-cysteine (Figure 4.1). Orbital hybridization to six equivalent sp^3d^2 molecular orbitals allows sulfur to form compounds with six valence electrons as in sulfuric acid. At neutral pH, dissolved sulfate is present as the dianionic sulfate ion (Figure 4.1). That sulfate dianion (SO_4^{2-}) is the most abundant form of sulfur available to organisms, at concentrations up to 28 mM in global oceans.[2] Sulfur in sulfate has a formal oxidation state of +6, its most electropositive electronic state. In H_2S gas, sulfur is electronegative at an oxidation state of −2, a dramatic distinction not available to its resolutely divalent lower homolog oxygen. The pK_a for H_2S is 7, so it is 50% dissociated to anionic, nucleophilic hydrosulfide anion, HS^-, in aqueous solutions at neutral pH.

During evolution anaerobic microbes evolved to use those most abundant sulfate dianions as terminal electron acceptor. The eight electron traverse of sulfur from +6 to −2 is typically initiated by an AMP-sulfate reductase and then a six-electron sulfite reductase to release the HS^- anion. Later, as the planet oxygenated sulfide oxidizing microbes could run the eight-electron redox span in the other direction, growing on inorganic sulfide as energy source and dumping the removed electrons into O_2 by normal aerobic membrane respiratory chain electron transfer[3] (Figure 4.2).

Chemical Biology No. 21
The Chemical Biology of Carbon
By Christopher T. Walsh

Published by the Royal Society of Chemistry, www.rsc.org

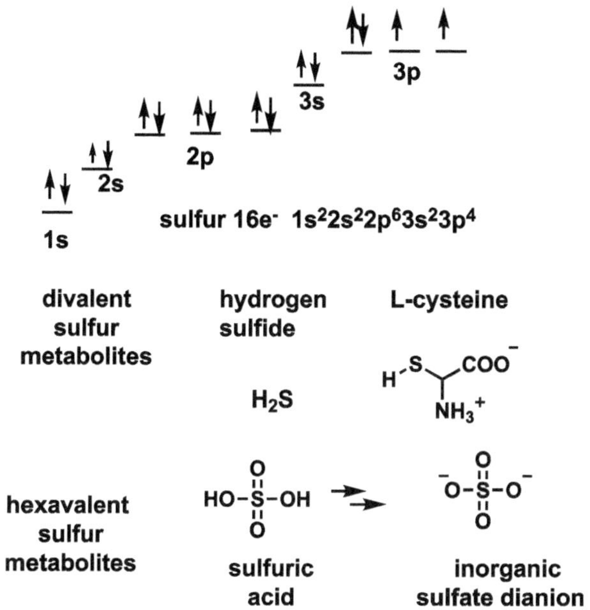

Figure 4.1 Orbital filling diagram for sulfur. Sulfur forms divalent to hexavalent inorganic compounds. Divalent, trivalent, tetravalent (sulfinates) and hexavalent (sulfate) oxidation states of sulfur are found in organosulfur metabolites.

Sulfur, as one of the six essential macro elements in humans (0.2–0.3% body weight, ~150–200 g), offers different chemistry at different sulfur oxidation states. Among the most common forms are organosulfur metabolites at the (S = − 2) oxidation state in thiol groups as in the side chain of cysteine[4] (Figures 4.1 and 4.3). Alkylation of the thiol in homocysteine leads to the thioether linkage in methionine, the second of the two sulfur-containing twenty proteinogenic amino acids (Figure 4.3). A further enzymatic alkylation of the methionine thioether side chain *via* the 5′-deoxyadenosyl moiety of ATP generates the third sulfur functional group, a trialkyl sulfonium cation with permanent charge on sulfur in *S*-adenosylmethionine[5] (Figure 4.3).

Additional carbon sulfur groups are disulfide bonds that arise from autoxidation of thiols. These often arise *in vivo* from conversion of an autoxidizable organic thiol group to a monooxygenated sulfenic acid. The sulfur has now gone from nucleophile to electrophile and is capturable by a second molecule of thiol (Figure 4.3). Sulfur in monooxygenated sulfenates (–C–SOH) can be further oxidized to a dioxy sulfur group known as a sulfinate (–C–SO_2H) or a trioxy group termed a sulfonate (–C–SO_3H). A structural analog of the disulfide group (RS-SR) at an equivalent oxidation state of the thiol sulfur is the persulfide (RSSH) where the distal sulfur can be nucleophilic as RSS^- or electrophilic as RSSH. These persulfides are key carriers of sulfur atoms for building the redox active protein bound Fe_2S_2, Fe_4S_4, and higher order Fe/S clusters[5] (Figure 4.3).

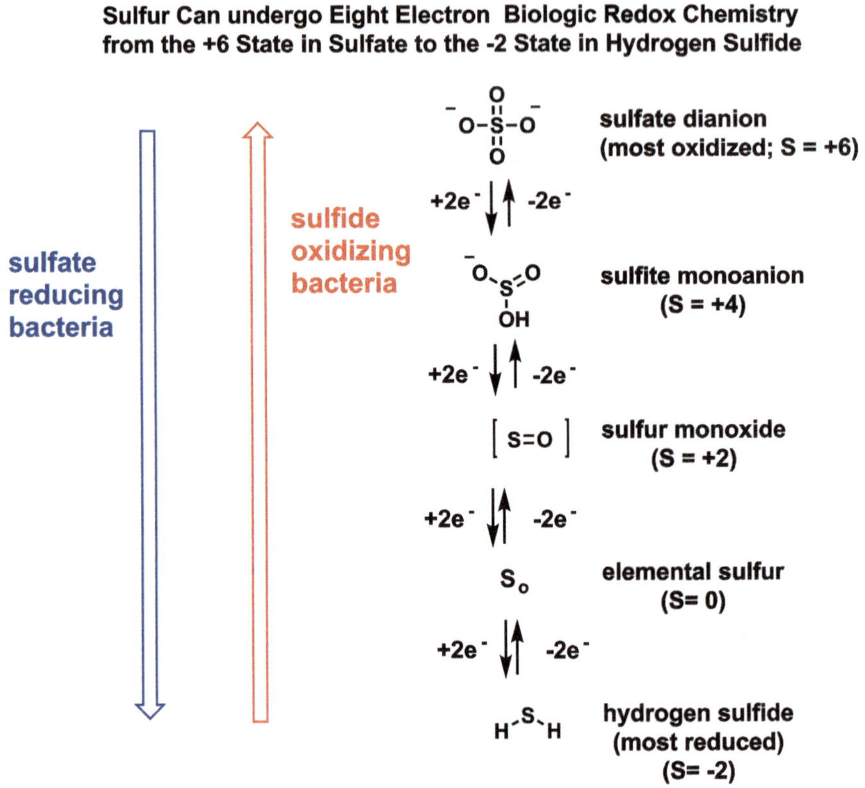

Figure 4.2 Sulfate-reducing and sulfate-oxidizing bacteria can catalyze eight-electron redox traverses between inorganic sulfate ions and hydrogen sulfide.

Central in terms of scope of its chemical biology, the organic thiolate side chain can act as nucleophile toward activated electrophilic acyl groups to form thermodynamically activated, kinetically stable acyl thioesters (Figure 4.4).

As an additional comparison of oxygen and its periodic second row relative sulfur, oxygen in biology is resolutely divalent in its stable chemistry. While we note briefly inorganic sulfate and sulfite biology in Figure 4.1, the comparison of divalent sulfur to divalent oxygen is displayed in functional group comparisons in Figure 4.5. Both alcohol and thiol oxidation states are central as reactive chemical handles in carbon chemical biology. While thioaldehydes are operant in a few enzymatic catalytic cycles, including a key decarboxylation step in coenzyme A biosynthesis,[6] thioaldehyde chemical biology is much less widely practiced than aldehyde and ketone carbonyl chemistry in nature. The thioaldehydes and thioketones can be unstable to polymerization. Also, the C=S bond is weaker than the C=O bond, such that the enethiol is a more prevalent component at equilibrium, consistent with its participation in the pantetheinyl cysteine decarboxylase reaction. The same holds for relative importance/abundance of amide over thioamide and carboxylate over thiocarboxylate in organisms.

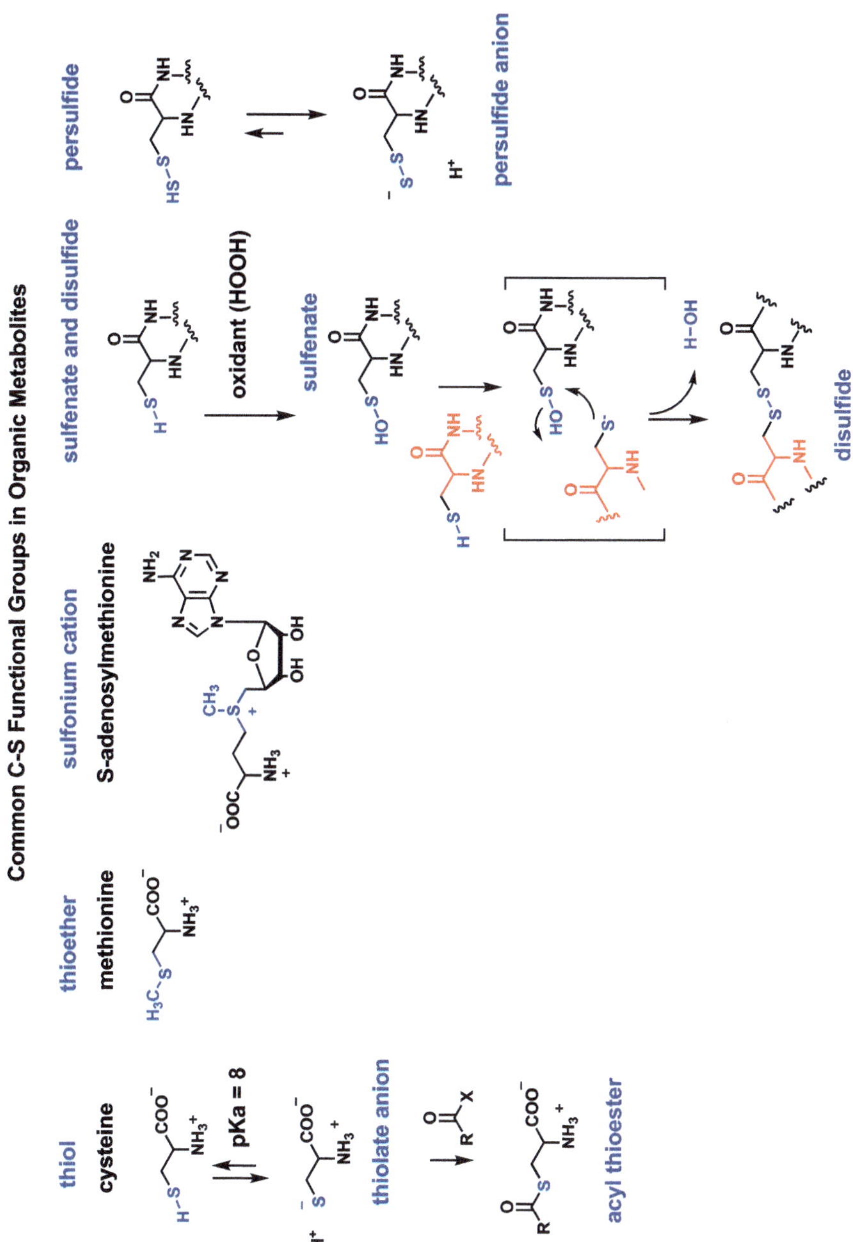

Figure 4.3 Common carbon–sulfur functional groups in metabolites.

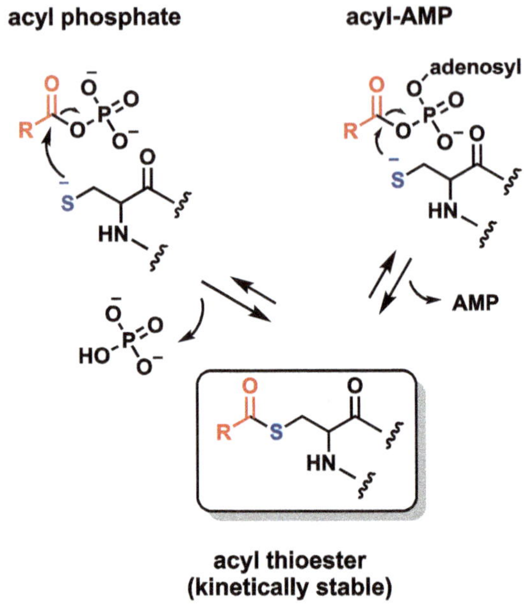

Figure 4.4 Both forms of acyl-phosphoric mixed anhydrides, acyl phosphates and acyl AMPs, can transfer acyl groups to thiols to create thermodynamically activated but kinetically stable acyl thioesters.

On the other hand, acyl thioesters (Figure 4.4) are more widely used as activated metabolites[7] than oxoesters, with the exceptions we noted in Chapter 3 that include protein synthesis at the ribosome peptidyl transferase center, glycerol lipid acyl esters, and formation of acetylcholine as a neurotransmitter. Likewise, the methionine thioether is a key group in the side chain of one of the twenty proteinogenic amino acids. A cyclic thiane (thioether ring) is found in biotin (see Chapter 9). The trialkyl sulfonium cation in *S*-adenosylmethionine is second only to ATP in its broad reach as activating group for two electron chemistry and outshines adenosyl coenzyme B12 in one electron reaction manifolds (Chapter 12). There are no comparable roles for reactive trialkyl oxonium compounds in chemical biology.

At the oxidation state of alkyl persulfides that are the key precursors to the sulfur atoms in iron sulfur clusters, the only comparable role for alkylperoxides is found in the flavin-4a-hydroperoxide intermediate in flavoenzyme oxidases and oxygenases. Likewise, the dialkyl disulfides (*e.g.* oxidized glutathione, disulfide bonds in extracellular proteins, Figure 4.3) have much more useful stability characteristics than dialkyl peroxides, yet are readily reducible back to a pair of dithiols. This reduction is exquisitely accomplished for oxidized glutathione by tandem employment of NADPH and FAD coenzymes in the active site of glutathione reductase (as detailed later in this chapter).

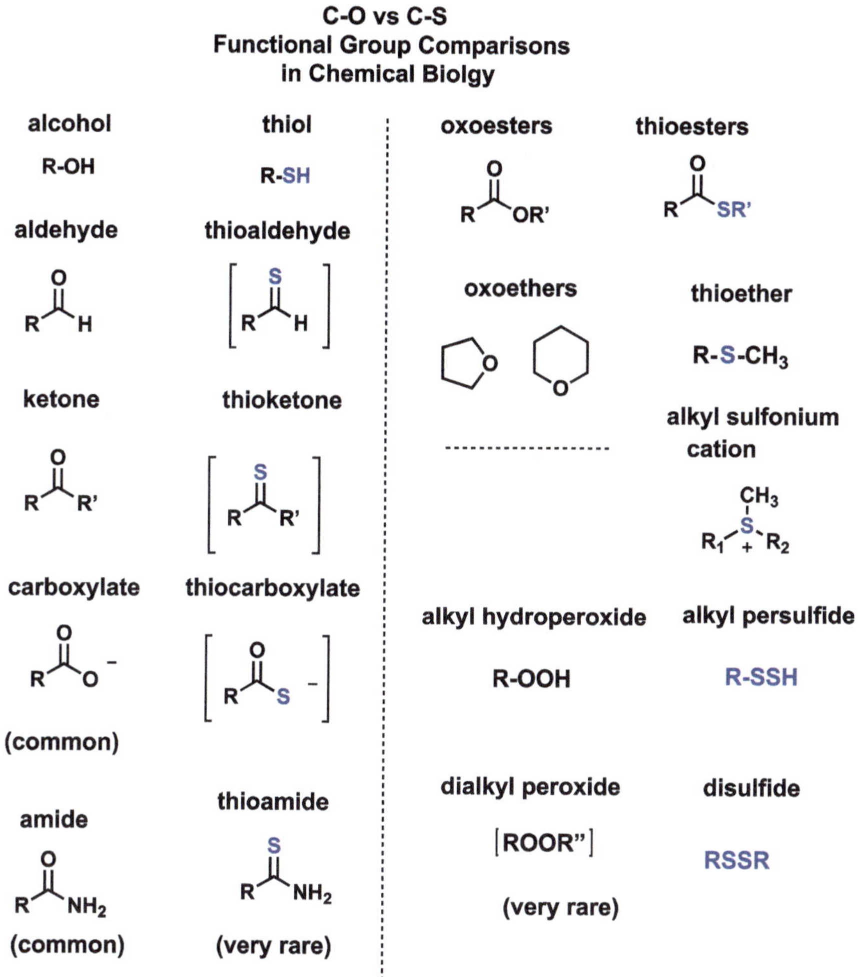

Figure 4.5 Carbon–oxygen *vs.* carbon–sulfur functional groups in carbon chemical biology.

4.2 From H_2S to the Thiol Group of Cysteine, Glutathione, and Coenzyme A

The single crossover point from inorganic sulfur chemistry to organic thiol chemistry in microbes and plants occurs in the enzymatic fixation of the hydrosulfide anion HS^- into the scaffold of *O*-acetyl serine to give the amino acid L-cysteine. Although we focus on the chemistry of cysteine's thiol group in this chapter, we note in passing that all three carbons of cysteine are functionalized with different groups for distinct reactivities thiol, amine, carboxylate, suggesting a plethora of potential *in vivo* reactivities (Figure 4.6). Only at that most reduced oxidation state is sulfur

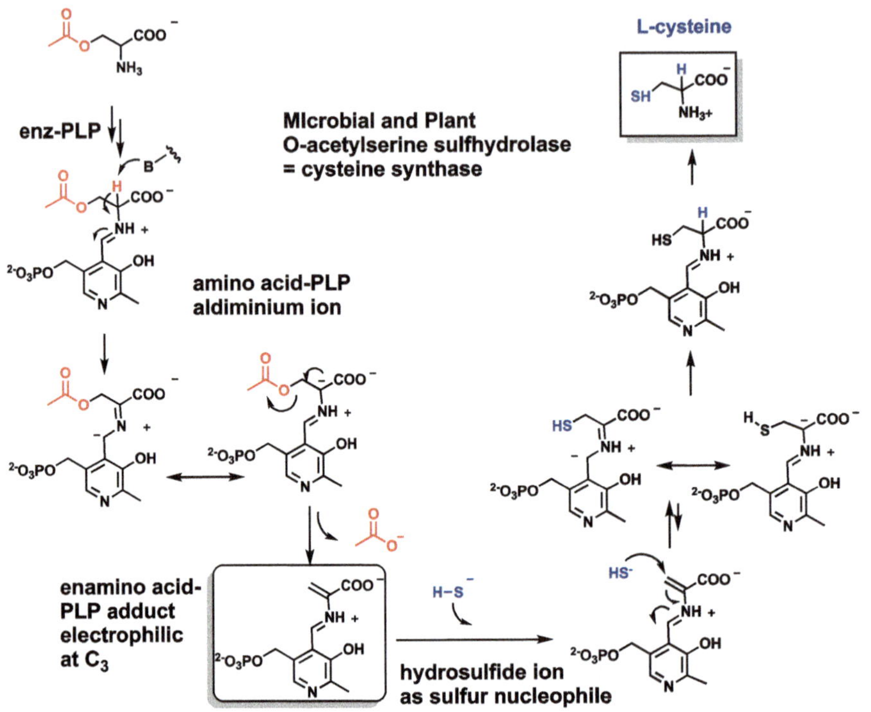

Figure 4.6 *O*-acetylserine sulfhydrylase = cysteine synthase catalyzes the crossover from inorganic H_2S to the organothiol linkage in the amino acid cysteine. The enzyme is a PLP-dependent beta-replacement enzyme.

nucleophilic (Figure 4.1). The enzyme *O*-acetyl serine sulfhydrylase, also known as cysteine synthase,[8] requires tightly bound pyridoxal-P[9] (vitamin B6, Chapter 9) to carry out the β-replacement chemistry (–SH for –*O*-acetyl) at C_3 of the serine scaffold.

The *O*-acetyl serine amino nitrogen forms an imine to the aldehydic carbon of PLP, then undergoes enzyme-catalyzed deprotonation at C_2 to give the accessible, delocalized amino acid-PLP C_2 carbanion. The electron pair at C_2 can then be used as a neighboring group nucleophile to cleave the C_3–*O*-acetyl C–O bond, generating acetate and the tightly held enamino acid-PLP adduct (Figure 4.7). As product acetate departs in the first half reaction, taking with it the OH group originally attached to C_3 of serine, the incoming nucleophilic hydrosulfide anion (HS^-) forms a HS–C_3 bond by attacking the polarized terminal carbon of the enamino-PLP adduct. Regisospecific reprotonation of the resultant PLP-amino acid adduct's carbanionic intermediate at C_2, followed by product release by hydrolysis of the iminium ion linkage, yields L-cysteine.[10]

This is carbanion chemistry, in the guise of iminium/enamine functional groups. The net replacement of the *O*-acetyl group by the incoming HS anion is indirect. The enamino acid-PLP adduct offers an electrophilic conjugated

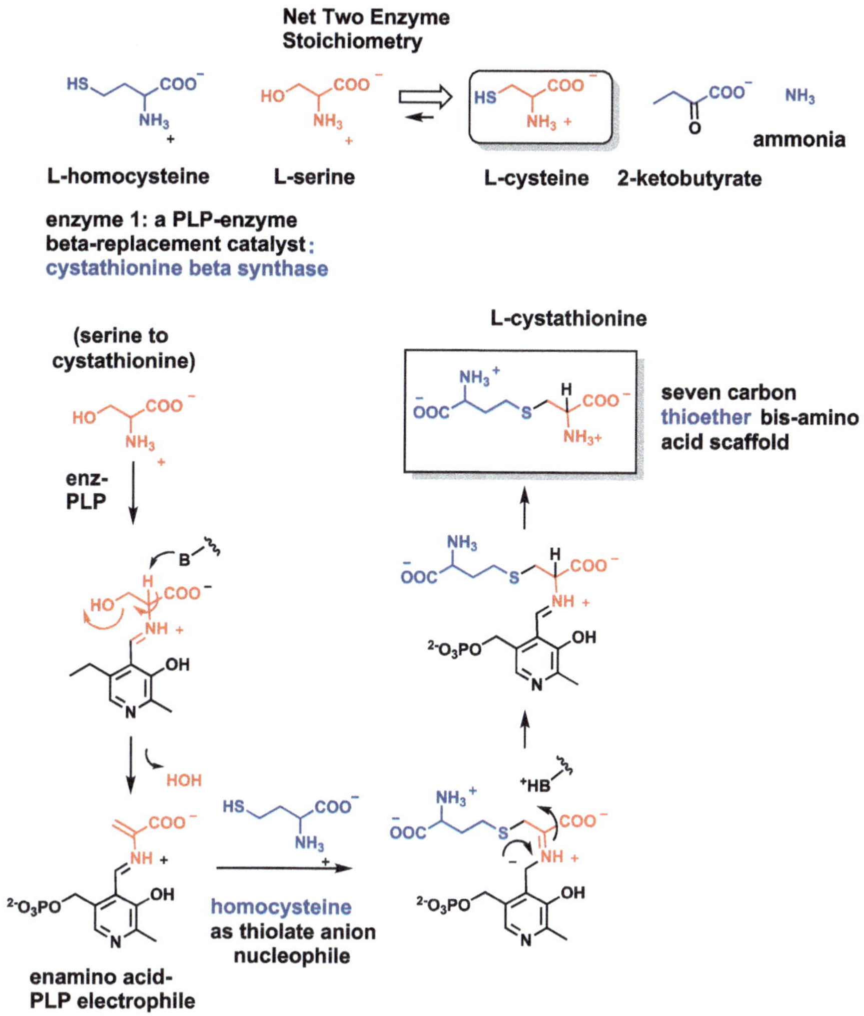

Figure 4.7 Animals, lacking cysteine synthase, use a two enzyme strategy to move the terminal thiol group from a C_4 scaffold (homocysteine) to a C_3 scaffold (cysteine). The first enzyme is cystathionine-beta-synthase, using the thiolate anion of homocysteine as a sulfur nucleophile to generate the seven carbon thioether cystathionine as a diffusible metabolite.

C=C double bond for trapping of the HS^- nucleophile, logically analogous to the hydride ion attack on conjugated enoyl thioesters discussed in Chapter 2.

From this entry point the inorganic sulfur that has now been installed as a cysteine-SH group can be converted enzymatically to all other C–S functional groups. Note that, although sulfate dianion is by far the more abundant source of available sulfur to organisms, only in the $S = -2$ oxidation state of HS^- is sulfur nucleophilic as required for its cross-over to covalent C–S bond formation, net organification of sulfur.

Animals, including humans, do not encode an *O*-acetylserine lyase counterpart in their genomes. Instead, they start with the four carbon L-homocysteine (itself derived from methionine) and effect a net transfer of the –SH group onto a three carbon amino acid scaffold, termed a transsulfuration process.[11] This requires two tandemly acting PLP-dependent enzymes (Figures 4.7 and 4.8). The net two enzyme reaction stoichiometry is conversion of a two amino acid substrate, L-homocysteine (C_4 scaffold) plus serine (C_3 scaffold), to give one three carbon amino acid cysteine (C_3 scaffold) and the four carbon 2-ketobutyrate plus free ammonia. This net accumulation of cysteine involves formation and then deconstruction of a (disappearing) thioether bridge in the seven carbon diamino acid intermediate cystathionine.

The two enzymes are named cystathionine β-synthase and cystathionine γ-lyase, respectively. The β(beta) and γ(gamma) descriptors are key to where the chemistry occurs, first on homocysteine (Figure 4.7) and then on

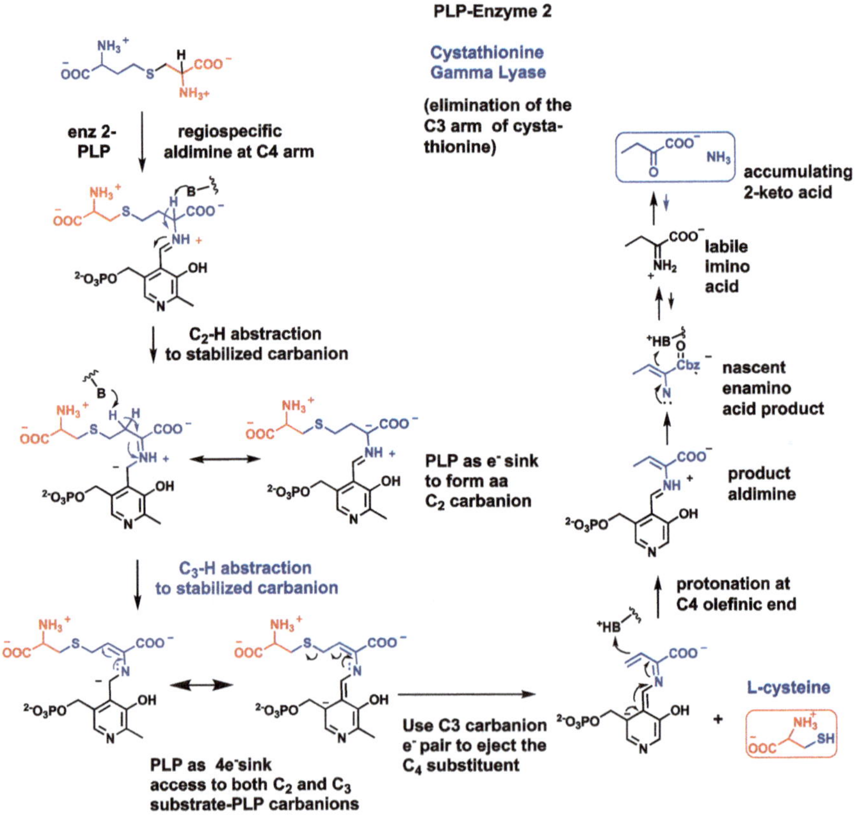

Figure 4.8 The second enzyme in the two enzyme transsulfuration pair is cystathionine-γ-lyase. To eliminate the cysteine arm at C_4 of cystathionine, the PLP-enzyme must sequentially generate C_2 and then C_3 carbanion equivalents of the cystathionine framework in the PLP adducts.

cystathionine (Figure 4.8). Finally, this is carbanion chemistry practiced on amino acids for transsulfuration. The focus on C–S chemistry also prefigures C–N amino acid chemistry enabled again by the pyridoxal-P coenzyme form of vitamin B6.

In the cystathionine β-synthase active site (Figure 4.7), the thiolate anion on the four carbon free homocysteine substrate attacks a C_3 enamino acid PLP adduct, derived from cosubstrate L-serine.[12] That enamino acid arises by the same chemical logic just seen with the microbial enzyme *O*-acetylserine sulfhydrylase (Figure 4.5). Capture of the C_3-enaminoacid-PLP adduct generates a thioether bond embedded in a nascent seven carbon amino acid scaffold. That new thioether functional group bridges the three carbon and four carbon amino acid arms. This seven carbon double-headed amino acid is cystathionine.

Then, this thioether is fragmented by the next enzyme cystathionine γ-lyase (Figure 4.8). The regiochemistry of iminium ion formation with PLP could be either at the three carbon arm, as was the case for the first enzyme cystathionine β-synthase. Or, as it happens, the PLP-cystathionine aldimine could be directed to form on the four carbon arm with opposite regiochemistry. This will require some additional carbanion chemistry in the cystathionine γ-lyase to get to an intermediate that can cleave that newly formed thioether, this time ejecting cysteine. The resultant four carbon amino acid-PLP adduct is unraveled to an oxidized 2-keto-butyrate: the original homocysteine arm has suffered a net oxidation at carbon 2 to the keto acid and ammonia.

Mechanistic analysis of the two kinds of amino acid-PLP carbanion equivalents that need to be generated by the tandem action of the β-synthase and the γ-lyase reveal the extra features of the second reaction.[13] Thioether formation by the first enzyme requires the PLP coenzyme to stabilize a C_2-aminoacid-PLP carbanion equivalent to achieve C_3–OR bond cleavage (Figure 4.7). In contrast, the cystathionine-γ-lyase requires transient stabilization of both a C_2 and C_3 substrate-PLP carbanionic transition state to effect the C_4–SR bond cleavage (Figure 4.8). One can envision the PLP coenzyme as a two electron sink in the β-synthase reaction cycle, but a net four electron sink in the γ-lyase reaction, stabilizing both the initial C_2-derived carbanion and also the transient C_3 carbanion needed to eject the cysteine arm.

This two-PLP enzyme pair that tandemly carry out a net transsulfuration of a C_4 thiol amino acid to the C_3 thiol-amino acid cysteine reveal an unusual chemical strategy as a C_4 thiol goes to a C_4–C_3 thioether and then to the C_3 thiol. This transformation reveals both the nucleophilicity of thiolate anions, and a transient alkykation/dealkylation sequence in the formation, and then cleavage of the thioether group in cystathionine. Also, execution of this sulfur chemistry depends crucially on the ability of PLP adducts to serve as electron sinks not only for two electrons stored transiently from an amino acid substrate, but also to store a second pair, effectively generating *ad seriatim* a C_2 and then a C_3 carbanion equivalent to get to a low energy path for cleavage of the C_4–SR bond.

The two most abundant thiol-containing low molecular weight metabolites in cells are the tripeptide glutathione[14] and coenzyme A[13] (Figure 4.9). Glutathione in the reduced, thiol form is often abbreviated as GSH and coenzyme A as CoASH to emphasize the importance of the cysteine-derived thiol in each of these two central molecules. Glutathione has the composition γ-glutamyl-cysteinyl-glycine. The thiol group and the deprotonated thiolate anion dominate its biology as we shall explore below. Coenzyme A has an AMP moiety, presumed to provide binding energy to partner enzymes and a cysteamine terminus, derived during biosynthesis from enzymatic decarboxylation at the stage of the intermediate pantetheinyl cysteine. Adenylylation and 2′-*O*-phosphorylation complete the biosynthesis of coenzyme A. Its major role is to serve as a thiolate nucleophile to capture electrophilic acyl group fragments, both short chain acyl groups, to form acetyl CoA, malonyl CoA, succinyl CoA, 3-hydroxybutyryl CoA, enoyl CoAs, and long chain fatty acyl CoAs myristoyl (C_{14}) and palmityl (C_{16}) (Figure 4.10).

The metabolic roles of these two abundant thiol molecules are largely nonoverlapping. Glutathione, at cellular concentrations of 1–10 mM is a key scavenging molecule for potentially toxic electrophiles. Coenzyme A is utilized to acylate electrophilic acyl phosphates and acyl AMPs to create kinetically stable, thermodynamically activated acyl-*S*-CoA thioesters (Figure 4.10) that serve as diffusible acylation currencies in cell metabolism.

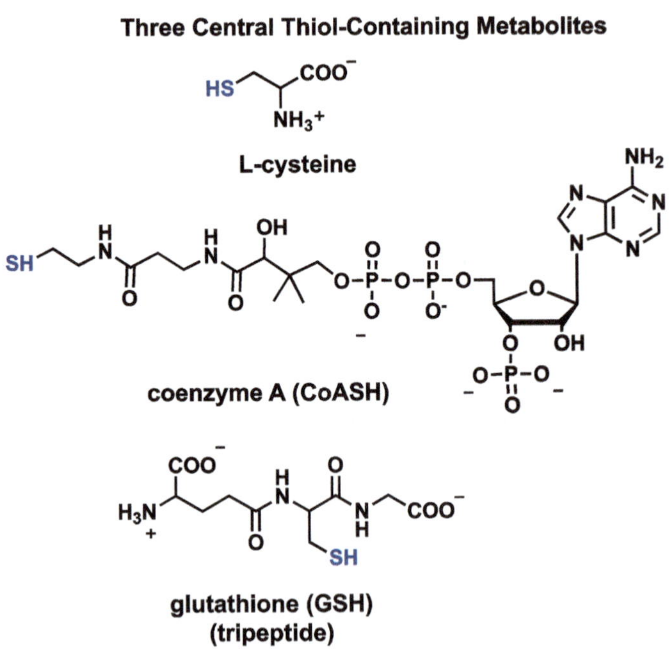

Figure 4.9 Three thiol group-containing central metabolites: cysteine, coenzyme A, glutathione.

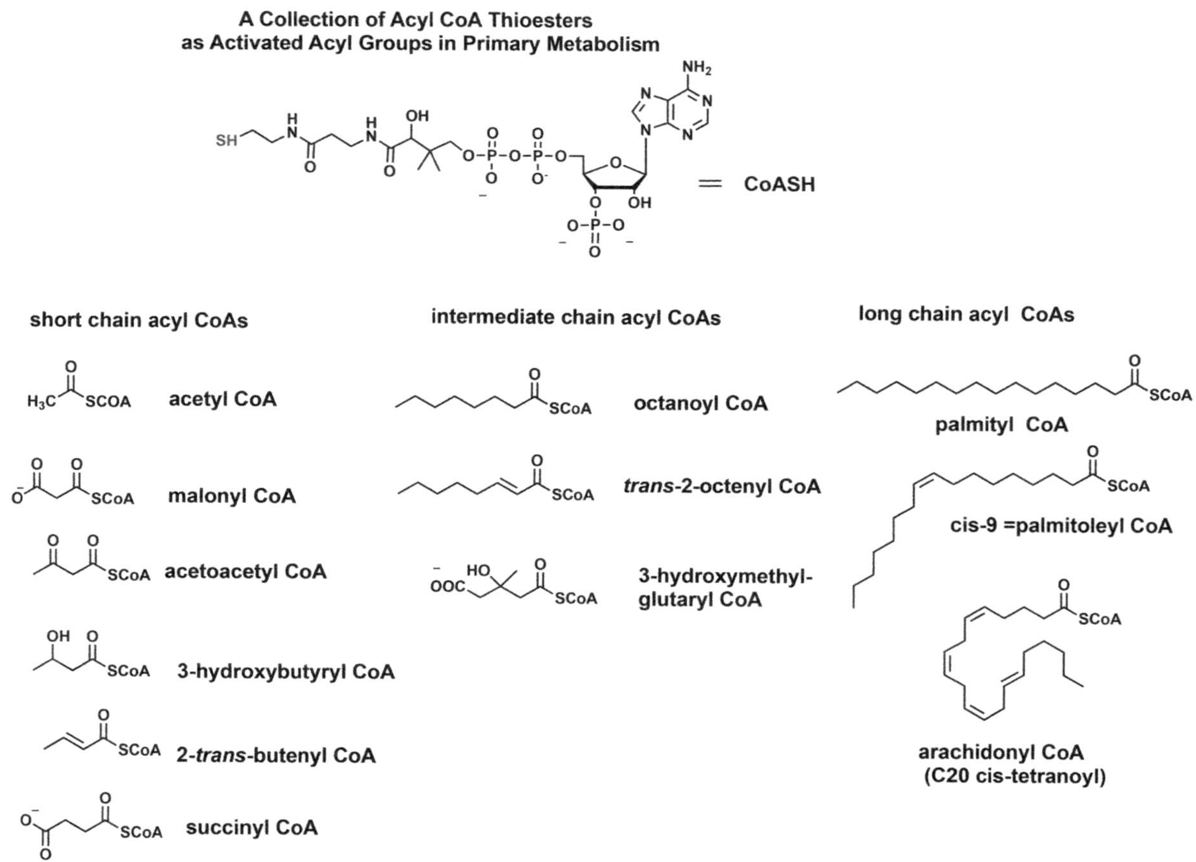

Figure 4.10 A collection of some of the acyl CoA thioesters prevalent in metabolism.

4.3 The Power of Thiolate Anions as Cellular Nucleophiles

L-Cysteine is not only the starting point for entry of HS^- into organic metabolite frameworks. With a pK_a of 8.18 for dissociation of the side chain S–H to the thiolate anion and a proton, the thiolate anion is a strong, accessible biologic nucleophile,[15] a million fold more accessible in solution than the serine alkoxide anion ($pK_a \approx 14$). Free cysteine is actually toxic to cells when intracellular concentrations are allowed to rise because of that thiolate anion reactivity. Animal cells instead insert the cysteinyl moiety between glutamyl and glycyl residues in the tripeptide glutathione (Figure 4.9) for it to serve as the main low molecular weight intracellular thiol/thiolate nucleophile.

4.4 Glutathione Nucleophilic Biology

4.4.1 Nucleophilic Thiolate Addition Toward Organic Cosubstrates

In principle, glutathione could have substituted chemically for CoASH as the general carrier of activated acyl groups in metabolic pathways (*e.g.* as *S*-acylated glutathiones). Instead, this tripeptide plays a major surveillance role in cells, using the side chain thiolate anion as a nucleophile to scavenge both inorganic and organic electrophiles and radicals that get loose inside cells.

Glutathione plays a key biosynthetic role in maturation of the lipid signaling molecules active in white cell networks. Illustrative capture of organic electrophiles occurs in the opening of the epoxide ring in leukotriene A4 by the glutathione thiolate anion to yield leukotriene C4.[16] This reaction highlights both the electrophilicity of the strained epoxide (cyclic ether) and the nucleophilicity of the glutathione thiolate anion to form the branched chain thioether linkage (Figure 4.11). The origin of leukotriene A4 was arachidonate, released by phospholipase A2 action on a membrane phospholipid. This phospholipase falls in the oxoesterase category noted back in Chapter 3. The epoxide group at $C_{5,6}$ of leukotriene A4 arose by an unusual doubly vinylogous participation of the 6,8 diolefinic moiety on the 5-hydroperoxy adduct (diene as nucleophile toward the distal oxygen in the O–OH bond).

Once the glutamyl thioether linkage has been formed in leukotriene C, a pair of hydrolases trim away the glutathionyl side chain. A hydrolytic γ-glutamyl transferase removes the glutamyl moiety from the glutathionyl arm of LTC_4 to give LTD_4, followed by dipeptidase action to cleave the remaining cys-gly peptide bond as LTE_4 is generated.

Leukotriene maturation is thereby seen as a three molecule, nested set of cysteinyl leukotrienes C-E, each with a thioether linkage, *trans* to an alcohol group. The set of leukotrienes act as lipid mediators in both acute and chronic inflammation by signaling through membrane protein receptors in target cells for bronchoconstriction and airway mucous secretion.

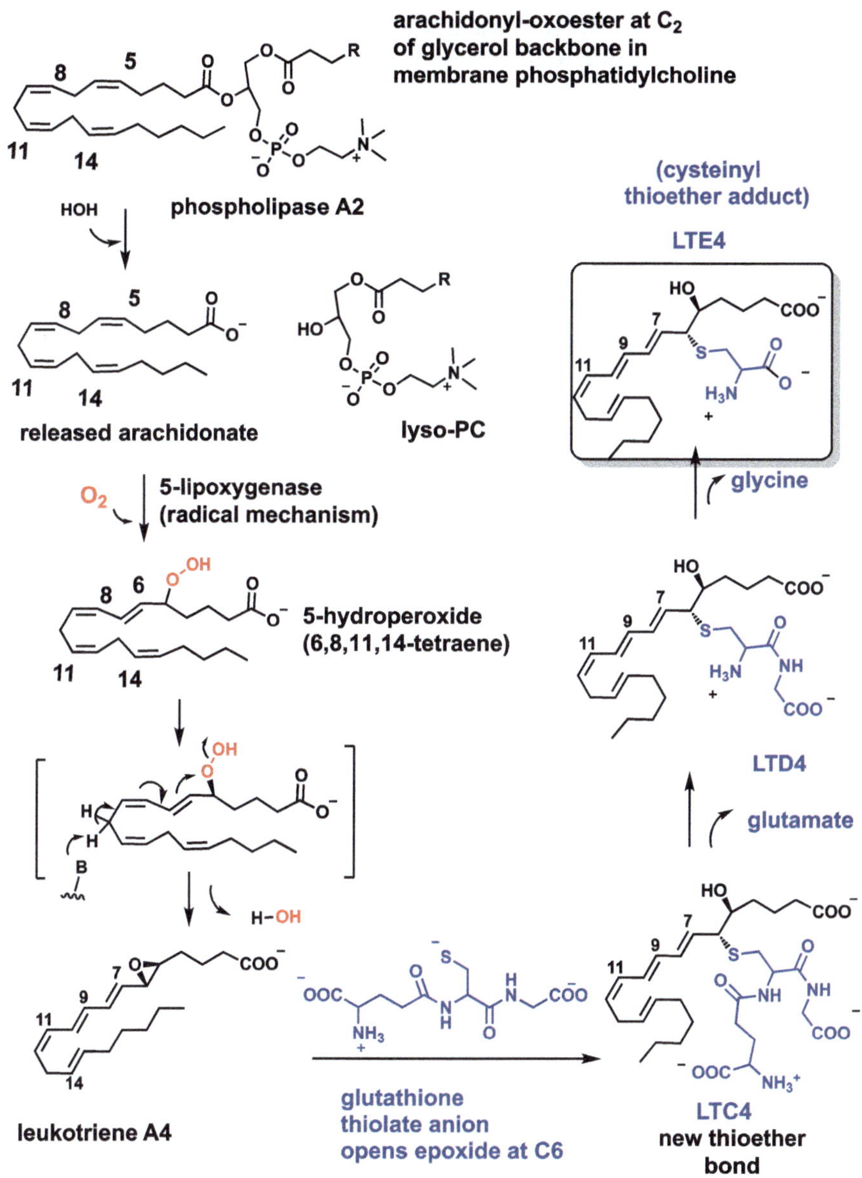

Figure 4.11 Glutathione as a thiolate nucleophile opening the epoxide ring in leukotriene A4 to create LTC4 on the way to LTD4 and LTE4, a family of leukotriene signaling molecules. The generation of LTA4 involves both radical-based oxygenation and abstraction of a C10–H and the use of the 6,8-diene as doubly vinylogous carbon nucleophile to cleave the O–OH bond and form the epoxide.

4.4.2 Surveillance Role Towards Inorganic Hydrogen Peroxide: Disulfide Formation

Complementary to the role of glutathione as a thiolate nucleophile towards organic cosubstrates, a surveillance role for protection by glutathione against the inorganic hydrogen peroxide molecule, HOOH, has been well studied. A number of cellular oxidase enzymes reduce O_2 to HOOH while oxidizing cosubstrates, *e.g.* hydroxy acids to ketoacids (see Chapter 3). While trace amounts of HOOH may function in oxidant signaling roles, HOOH buildup can be chemically promiscuous and thereby pathogenic.

Glutathione (GSH) is spent as a cellular defense response, by the enzyme glutathione peroxidase,[17] removing the HOOH potential oxidant by cleavage of the weak O–O peroxide bond. The products are a molecule of water and the *S*-hydroxylated form of glutathione, glutathione sulfenic (GS–OH) acid (Figure 4.12). This sulfenate group is shown as an intermediate on the way to disulfides back in Figure 4.3.

In the glutathione peroxidase active site hydrogen peroxide has been detoxified by two-electron reduction to two water molecules. Concomitantly, the thiol in glutathione has been oxidized by two electrons to the sulfenate.[18] In this oxidation state the glutathione sulfur is much less nucleophilic, not usable for molecular defenses and surveillance functions.

Indeed, the major cellular fate of glutathione sulfenate is to function as an electrophilic sulfur partner when the thiolate anion of a second glutathione molecule attacks to form an S–S bond, ejecting the OH as a second product water molecule (Figure 4.12). The two step reductive detoxification of hydrogen peroxide leads to release of two water molecules and an accumulating intermolecular oxidized disulfide form of glutathione (GS–SG). This enzyme glutathione peroxidase is a major scavenging enzyme in cells as it transiently forms the sulfenic acid and then holds it for dimerization to the disulfide form of glutathione. The sulfenate and disulfide are at the same oxidation states.

4.4.3 Glutathione Reductase Spends an NADPH Molecule to Convert the Disulfide Form of Glutathione Back to a Pair of Nucleophilic Glutathione Thiols (GSH)

The oxidized glutathione disulfide (GS–SG) is no longer useful as a scavenging, protectant nucleophile. The disulfide sulfur atoms are in fact electrophilic, not nucleophilic. If all the reduced glutathione is used in such protective scavenging against xenobiotic inorganic and organic electrophiles, the ability of cells to avoid damage to key macromolecular targets such as DNA and proteins falls rapidly.

Therefore, cells spend NADPH electron currency to reduce the intermolecular GS–SG disulfide back to the pair of nucleophilic GSH dithiols. That resuscitative enzyme is glutathione reductase, an FAD-containing enzyme, which also has an unusual active site cystine disulfide bridge between two

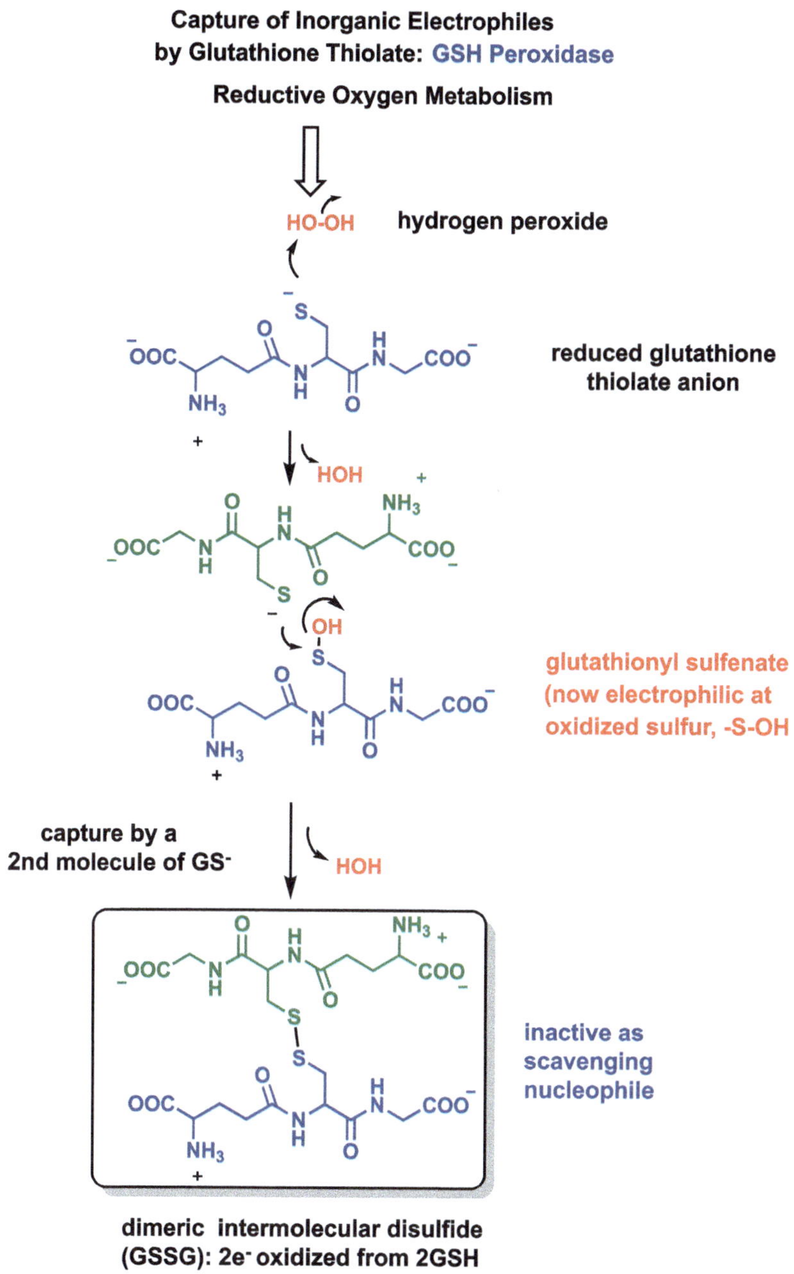

Figure 4.12 Capture of the inorganic electrophile hydrogen peroxide by reduced glutathione in the active site of glutathione peroxidase.

cysteine residues.[19] This enzyme therefore has two redox active groups. At the start of each catalytic cycle the FAD is oxidized and so is the pair of active site cysteine residues as the cystine disulfide linkage. The active site architecture during a catalytic cycle can be viewed as four layers of redox active species: NADPH on top of the FAD on top of the active site cystine disulfide on top of the bound glutathione disulfide, reducing electron flow in that direction from NADPH ultimately to the bound glutathione disulfide cosubstrate. The actual reduction step for bound GS–SG is by thiol disulfide interchange between reduced active sites cysteine pair and the bound glutathione.

Figure 4.13 shows the first half reaction where NADPH delivers its hydride ion to FAD, generating $FADH_2$ and $NADP^+$. This is the first of two forms of dihydro enzyme, EH_2 form 1. It undergoes conversion to EH_2 form 2 as an electron pair flows from the bottom face of $FADH_2$ to reduce the active site cys_{57}-*S*-*S*-cys_{62} disulfide to the dithiol. The mechanism for two-electron flow is proposed to be through the flavin-4a-thioether adduct shown in Figure 4.13. This EH_2 form 2 now has the right constellation of redox components, FAD_{ox}, active site dithiol, to carry out the second half reaction.

The half reaction shown in Figure 4.14 comprises attack of one of the active site cysteine thiolate side chains in bound GS–SG. This is an energy neutral thiol-disulfide exchange reaction. The result is formation of a mixed GS–*S*-Cys_{57} disulfide-enzyme adduct as the first GSH product molecule is released. A second thiol-disulfide exchange then occurs as the thiolate of cys_{62} attacks the mixed disulfide regioselectively. The second product molecule of reduced glutathione is released as the active site cys_{57}-*S*-*S*-cys_{62} disulfide reforms. The enzyme is back to its original state with both FAD and the active site pair of cysteines oxidized as the cystine disulfide. The net reaction of spending a molecule of the NADPH hydride donor to reduce the dimeric oxidized GS–SG has been executed but the reductive mechanism is indirect, using both FAD and an enzyme disulfide as redox intermediaries. Given that cellular GSH levels can be as high as 1–10 MM for cell protection, this is a high flux self-defense mechanism that consumes a large fraction of cellular NADPH.

Cells use the tandem action of glutathione peroxidase and glutathione reductase to lower the concentration of metabolically derived H_2O_2 and thereby lower the propensity for off target oxidative chemistry of hydrogen peroxide. The two enzyme pair deals with different chemical states of organosulfur metabolites, with both sulfenate and disulfide forms of C–S bonds. This two enzyme sequence also illustrates both the effective nucleophilicity of the thiolate anion and the electrophilicity of the disulfide sulfurs.

4.4.4 Protein Disulfides

The formation of protein disulfides crosslinking two cysteine residues similarly occurs in aerobic environments by such thiolate capture of autoxidative sulfenate intermediates.[20] Within the cytoplasm of cells, GSH concentrations are high enough to provide a sufficiently reducing environment that most proteins have their cysteine residues in the reduced thiol forms (glutathione reductase is an outlier but uses the active site disulfide for catalytic purposes). As those proteins move to the endoplasmic reticulum

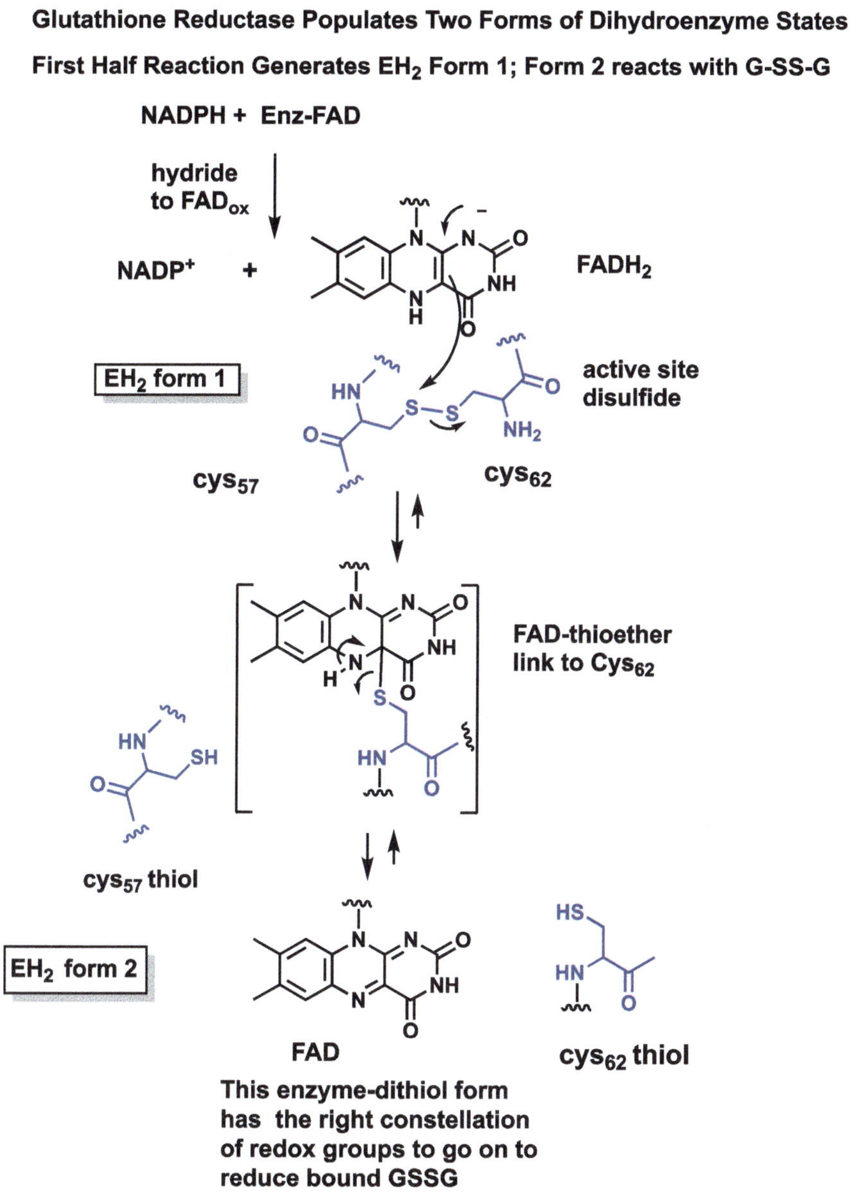

Figure 4.13 Glutathione reductase, a flavoenzyme that reduced the glutathione disulfide back to two reduced glutathione molecules with NADPH as external reductant. Hydride transfer from NADPH to Enz-FAD generates the first of two EH_2 states with $FADH_2$ and active site disulfide. That equilibrates with a second EH_2 containing oxidized FAD and active site cysteine dithiols. Then thiol-disulfide interchange with bound GS–SG completes the reduction of oxidized glutathione. This figure shows the first half reaction to produce an EH_2.

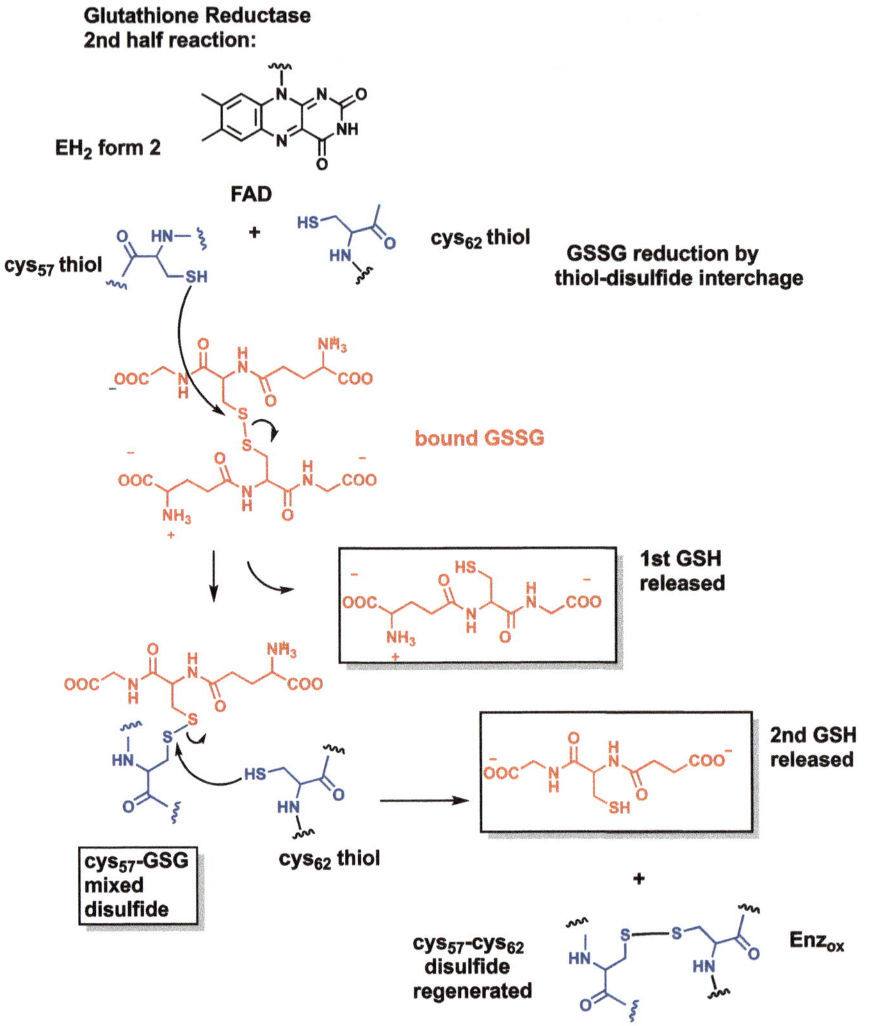

Figure 4.14 This figure shows the second half reaction for glutathione reductase, involving FAD-4a-thoether adduct on the way to the second form of EH_2.

and Golgi complex subcellular compartments of the secretory pathway, often all the way to extracellular spaces, reduced glutathione concentrations become low to negligible. Cysteine disulfides form and persist as proteins fold into native conformations. That persistence can serve as major conformation-stabilizing covalent crosslinks in the three-dimensional folded architectures of many/most of proteins at cell membranes and in the proteins that comprise the secretome. The covalent disulfides are mechanically and chemically stable, except to reduction and reductants are not available in the extracellular milieus.

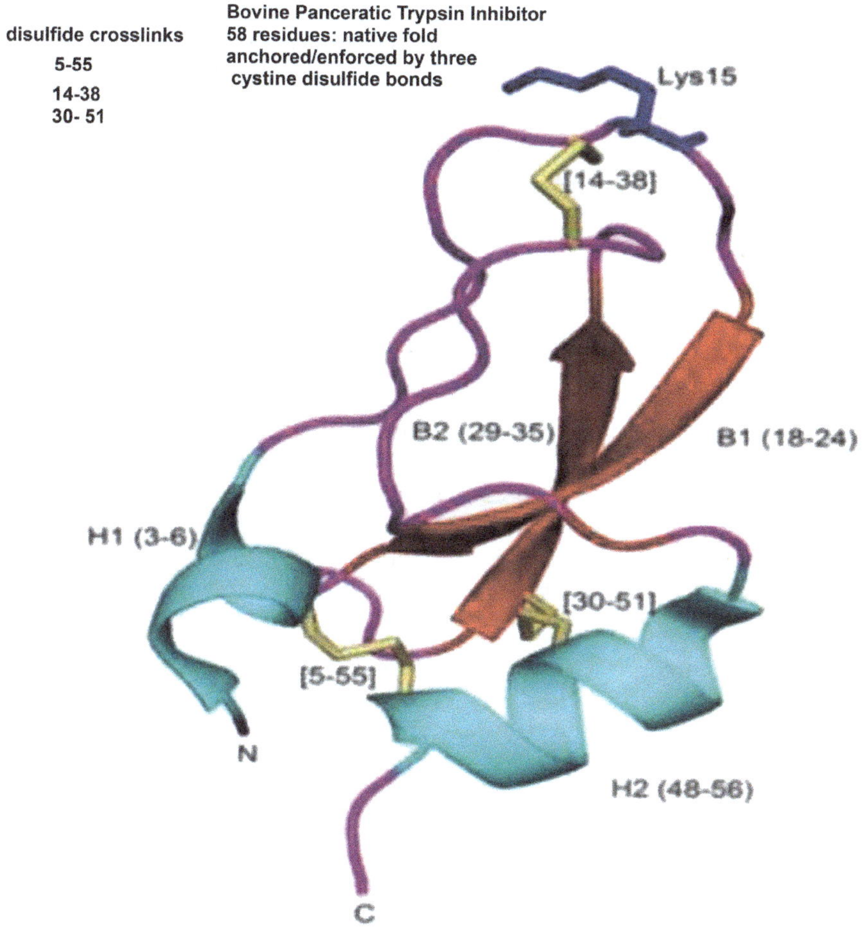

Figure 4.15 Persistence of three cystine disulfide bonds in the secreted protein bovine pancreatic trypsin inhibitor. In an aerobic extracellular micro-environment the disulfide bonds are stable and key structural elements in maintenance of an active 3D conformation.

The conformational stabilizing effect of cystine disulfide bonds are displayed in Figure 4.15. The 58 residue bovine pancreatic trypsin inhibitor is a secreted protein with three disulfides linking cysteines 5–55, 14–38, and 30–51, respectively that serve as key structural anchoring elements in the extracellular oxidizing environment.

4.5 Acyl Thioesters: The Preferred Cellular Acylation Reagents

The enhanced nucleophilicity of thiols over alcohols is evidenced in the preferred use of acyl thioesters over acyl oxoesters as activated acyl groups in chemical biologic systems.[7] The thiols are lower energy leaving groups than the alcohols would be from acyl oxoesters. By pK_a differences for thiol *vs.* alcohol dissociation to the corresponding (nucleophilic) anions, the energy barriers for acyl transfer from a thioester could be up to 8.4 kcal mol^{-1} lower than the barrier from an equivalent oxoester.[15] If all this energy difference were converted to rate effects, acylation of a cosubstrate nucleophile from an acyl thioester over an acyl oxoester could be faster by up to 10^6 fold. It is no chemical accident that organisms use acyl thioesters rather than acyl esters as their acylation currency (acyl amides are even more stable than oxoesters and are inert in biologic acylation chemistry, Figure 3.16).

A second chemical attribute in acyl thioesters over acyl oxoesters arises from the lower orbital overlap of the α-carbon to the large sulfur atom at further distance compared to oxygen. Thioester carbonyls are sometimes described as more ketone like. This effect lowers the energy barrier for adjacent C–H bond cleavage to give as delocalized, stabilized carbanion, the thioester enolate (Figure 4.16) compared to the ester enolate.

A practical outcome is that acyl CoAs display electrophilic acyl groups at C_1 of the acyl group coupled to the ability of C_2 of those acyl groups to act as accessible low energy carbanions. The thiolase family of enzymes can

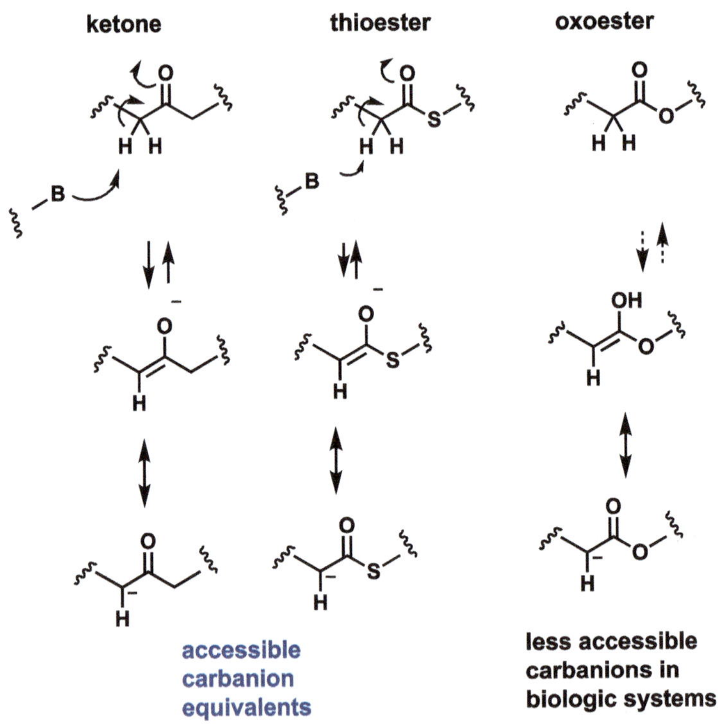

Figure 4.16 Enolate anions from ketones and thioester enolates from thioesters are used as accessible carbanion equivalents in metabolism. The corresponding oxoesters are rarely used as carbanion equivalents.

generate a C_2 carbon nucleophile on one acetyl CoA molecule to react with the electrophilic acyl C_1 carbon of a second acetyl CoA molecule.[7] The C–C bond-forming thio-Claisen chemistry generates a four carbon linear scaffold. Loss of one CoASH produces the 3-keto-butyryl CoA, the first committed metabolite in the isoprenoid/steroid biosynthetic pathway (see Chapter 2, Figure 2.3)

While coenzyme A thioesters predominate as diffusible metabolites, fatty acid synthases and polyketide synthase enzyme assemblies tether growing acyl chain thioesters to covalently attached phosphopantetheinyl prosthetic groups on acyl carrier protein domains.[21] Coenzyme A serves as substrate for post-translational modification of a seryl residue side chain in the 8–10 kDa acyl carrier protein (Figure 4.17). The phosphopantetheinylated carrier protein has the cysteamine terminal thiol characteristic of CoASH and functions equivalently as a nucleophile for attachment of growing fatty acyl chains (Figure 4.17). The acyl thioester activation logic is parallel but the

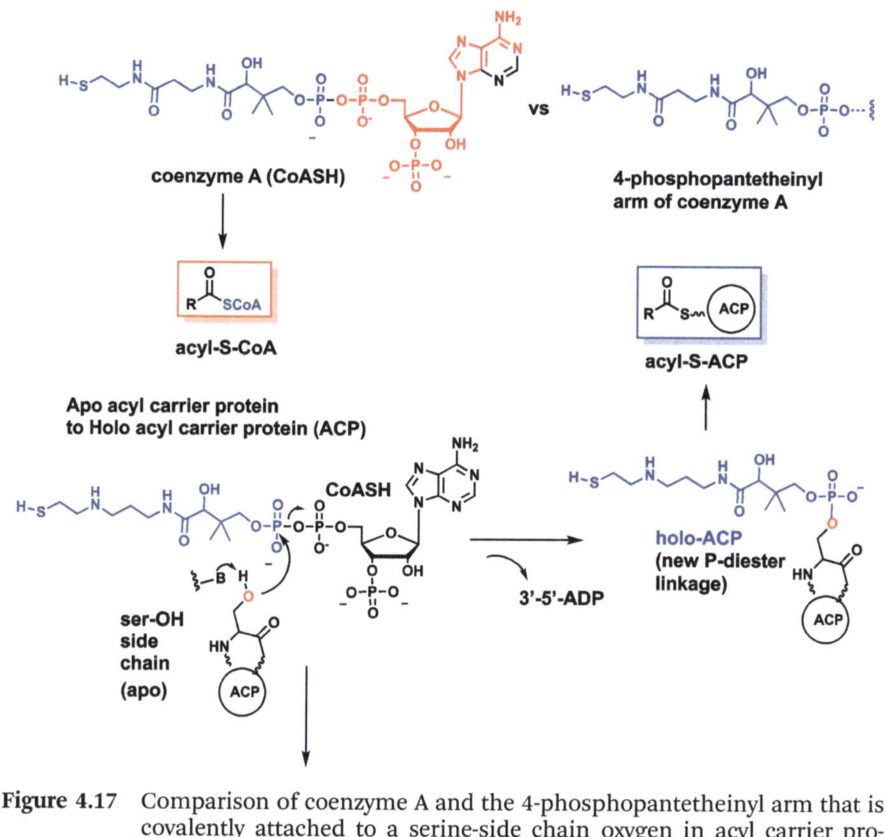

Figure 4.17 Comparison of coenzyme A and the 4-phosphopantetheinyl arm that is covalently attached to a serine-side chain oxygen in acyl carrier proteins. The acyl CoAs are used as diffusible sources of electrophilic acyl groups. The HS-ACPs are the site where acyl chains are elongated as covalent acyl thioester adducts during action of fatty acid synthases.

sequestration of the thermodynamically activated acyl groups away from solvent water differs, perhaps to ensure that none of the growing fatty acyl chains are adventitiously hydrolyzed to release partially elongated acyl chains as free carboxylates. Intact fatty acyl synthases typically only release C_{14}–C_{18} acyl CoAs by transfer of the full length acyl groups from one thioester linkage to another, an energy neutral acyl transfer from one thiol nucleophile (the phosphopantetheinyl arm on ACP) to the other (the phosphopantetheinyl arm of coenzyme A).

4.6 Alkylation of Homocysteine Thiol to Form the Thioether of Methionine

The second of the two proteinogenic amino acids that has a carbon–sulfur functional group in its side chain is L-methionine[5] (Figure 4.3). Compared to cysteine's dissociable thiol side chain to the strongly nucleophilic thiolate anion, the thioether linkage in methionine is more restrained in its nucleophilicity. Indeed, cells use the potentially nucleophilic divalent sulfide (thioether) in methionine for one remarkable enzyme-catalyzed formation of the trialkyl sulfonium linkage in *S*-adenosylmethionine described below. First, though, we examine the formation of methionine from homocysteine.

The chemical logic for generating the methyl thioether linkage in methionine is to alkylate the thiol/thiolate anion of the four carbon L-homocysteine with an electrophilic methyl group equivalent (Figure 4.18). The canonical choice of such a methyl group donor in cells would be *S*-adenosylmethionine (SAM) but until methionine is formed as the core scaffold, the SAM molecule will not be available. Instead, cells turn to N^5-methyl tetrahydrofolate as methyl donor.

In Chapter 9 on vitamin heterocycles we note that the tetrahydrofolate scaffold can carry one-carbon units at three different oxidation states. N^5-CH_3-THF has the one-carbon unit at the oxidation state of methanol. N^5,N^{10}-CH_2-THF has the one-carbon unit at the oxidation state of formaldehyde. N^{10}-CHO-THF has the one-carbon at the oxidation state of formic acid (see Chapter 9, Figure 9.23). The CH_3 group covalently attached to N^5 of the tetrahydropyrazine ring of the folate framework serves as the electrophilic partner, the thiolate anion of homocysteine acts as the sulfur nucleophile to form the thioether link in methionine under direction of the enzyme methionine synthase.

The alkylation can be a direct S_N2 type sulfur alkylation catalyzed by one form of methionine synthases[22] (Figure 4.18). Humans have a different isoform of the enzyme that contains tightly bound vitamin B12.[23] The B12 in the resting state has the cobalt ion embedded in the equatorial plane of the tetrapyrrolic corrin macrocycle in the air stable Co^{III} oxidation state (Figure 4.18). The chemistry of methyl B12 and adenosyl B12, two organocobalt metallo-organic complexes, is discussed in Chapters 9 and 12.

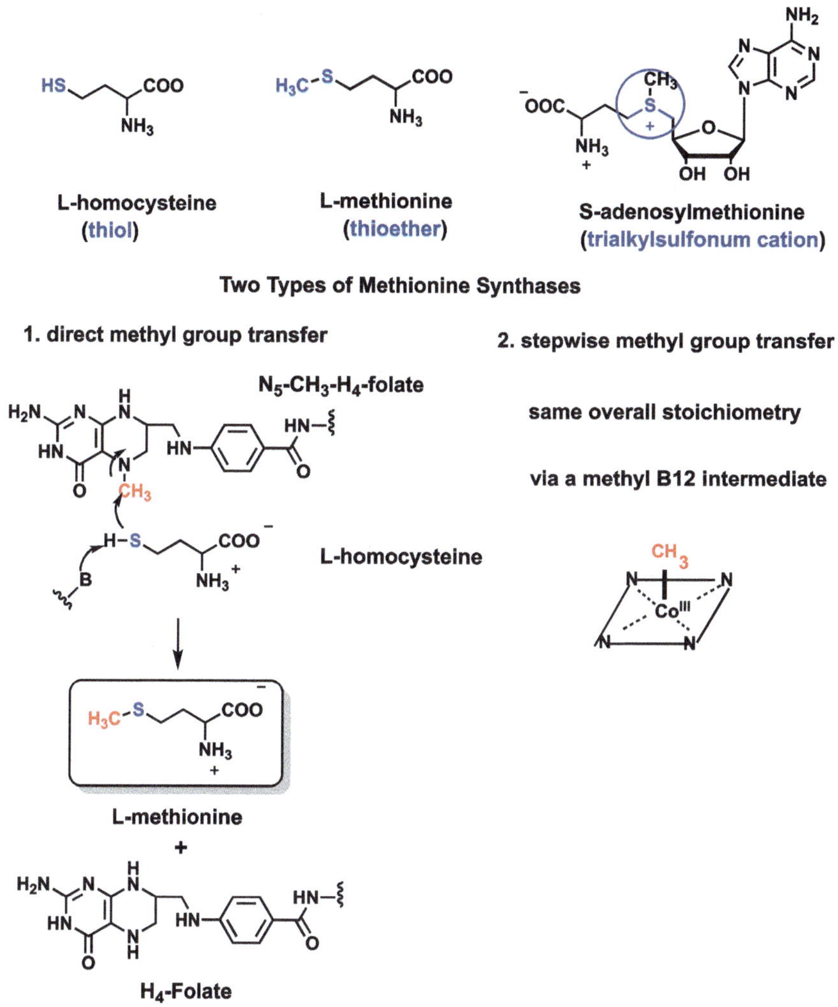

Figure 4.18 Homocysteine to methionine to *S*-adenosyl methionine. Two types of methionine synthase.

4.7 Alkylation of the Methionine Thioether by the 5′-Deoxyadenosyl Moiety of ATP: The Sulfonium Cation of *S*-Adenosylmethionine

The utility of the alkylation of the thiol of homocysteine to the dialkyl methyl thioether (sulfide group) of methionine is fully revealed in the subsequent enzymatic alkylation of that thioether sulfur to a trialkyl sulfonium cation in *S*-adenosylmethionine (Figure 4.19). The nucleophilicity of the thioether sulfur is utilized in the active site of *S*-adenosylmethionine synthetase to attack cosubstrate ATP at an unlikely position, the $C_{5'}$ carbon of the ribosyl

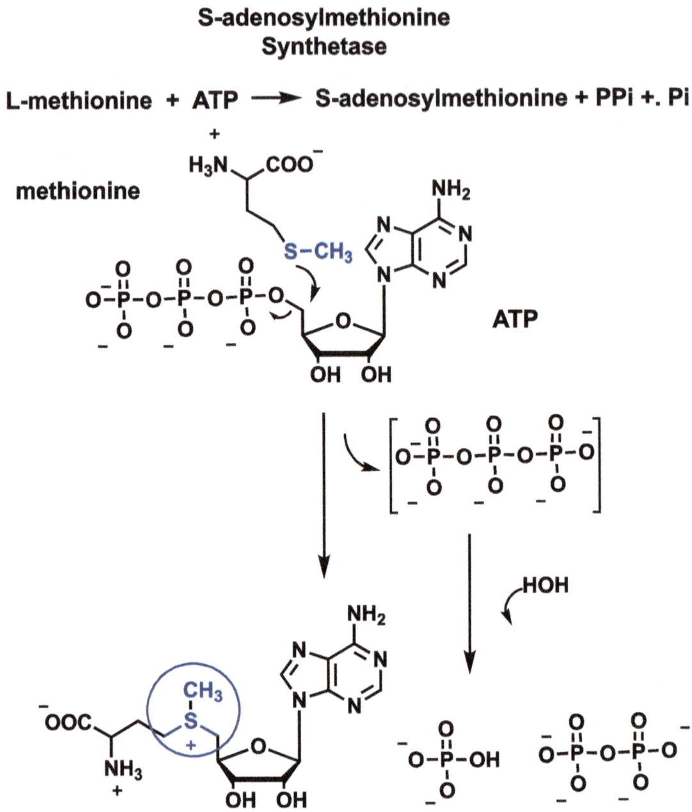

Figure 4.19 *S*-Adenosylmethionine synthase reaction.

ring rather than any of the three electrophilic phosphorus atoms in the two phosphoric anhydride linkages (Figure 4.19).

The result is an alkylation of the thioether sulfur by the 5′-deoxyadenosyl moiety of ATP and fragmentation of that ATP to nascent inorganic triphosphate. The methionine sulfur has now become a trialkyl sulfonium cation, permanently charged. This unique trialkyl sulfonium center activates all three of its alkyl substituents as electrophilic fragments for subsequent capture by nucleophiles by a variety of SAM-cleaving enzymes (Figure 4.20). There is no comparable trialkyl oxonium metabolic equivalent of SAM. The oxonium ion would not be stable under physiologic conditions. *S*-Adenosylmethionine has the twin properties of thermodynamic activation for alkyl transfers coupled to sufficient chemical (and stereochemical stability at sulfur) stability to be a diffusible reagent in cells as well as displaying orthogonal reactivity of homolytic cleavage to methionine and the 5′-deoxyadenosyl carbon radical when bound to a large family of enzymes known as radical SAM enzymes[24] (addressed in Chapter 12 on radical pathways).

While any of the three alkyl substituents attached to the sulfonium cation of SAM could be transferred to cosubstrate nucleophiles, methyl transfer in

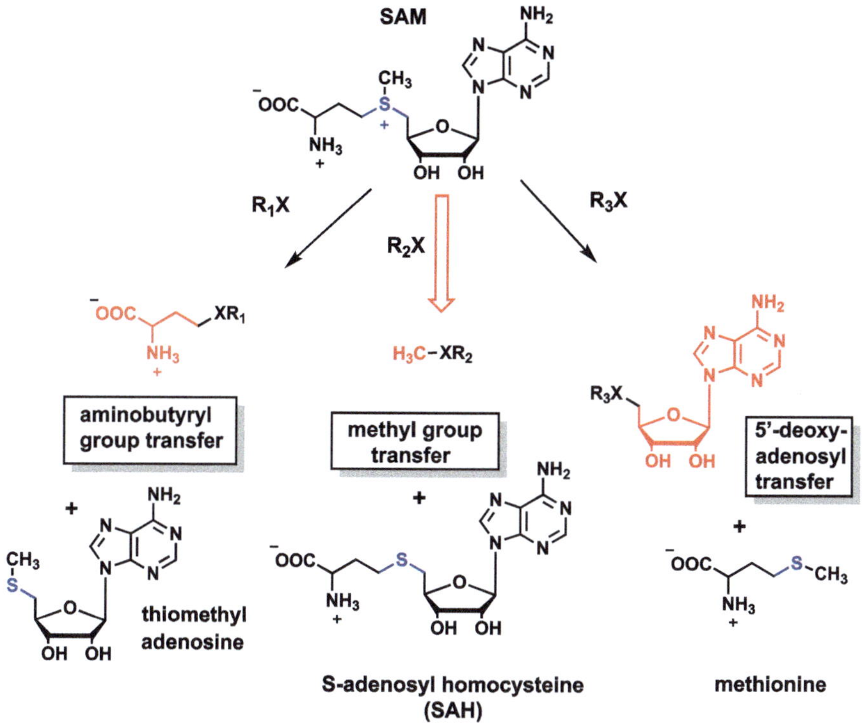

Figure 4.20 Enzymatic transfer of any of the three carbon substituents attached to the sulfonium cation in *S*-adenosylmethionine as electrophilic fragments to cosubstrate nucleophiles.

S_N2 transition states to nucleophilic oxygen, nitrogen, sulfur, or carbon atoms in cosubstrates is most pervasive (Figure 4.21). There are a few well characterized cases of enzymatic cleavage of the aminobutyryl (alkyl) C4–S^+ bond and almost no enzyme-catalyzed transfers of the deoxyadenosyl moiety by two-electron reaction paths. One-electron paths dominate in SAM homolysis at the sulfur–deoxyadenosyl bond (Chapter 12).

4.8 Enzyme Persulfides

An additional organosulfur functional group in chemical biology, the cysteine persulfide (Figure 4.3) is called into play to provide the inorganic sulfide ions that populate Fe_2S_2, Fe_4S_4, and higher order iron–sulfur redox clusters (Figure 4.22) in hundreds of proteins involved in one-electron transfer paths.[25] The Fe_2S_2 clusters with four thiol ligands from protein cysteine residues form initially and then by a still mysterious process are matured into Fe_4S_4 distorted cubane architectures. Four thiol ligands from cysteine residues on a target protein fill out the iron coordination shells.

While the redox cluster-containing sulfide ions might *a priori* have come from inorganic HS^- ions, the concentration of free hydrosulfide ions in cells

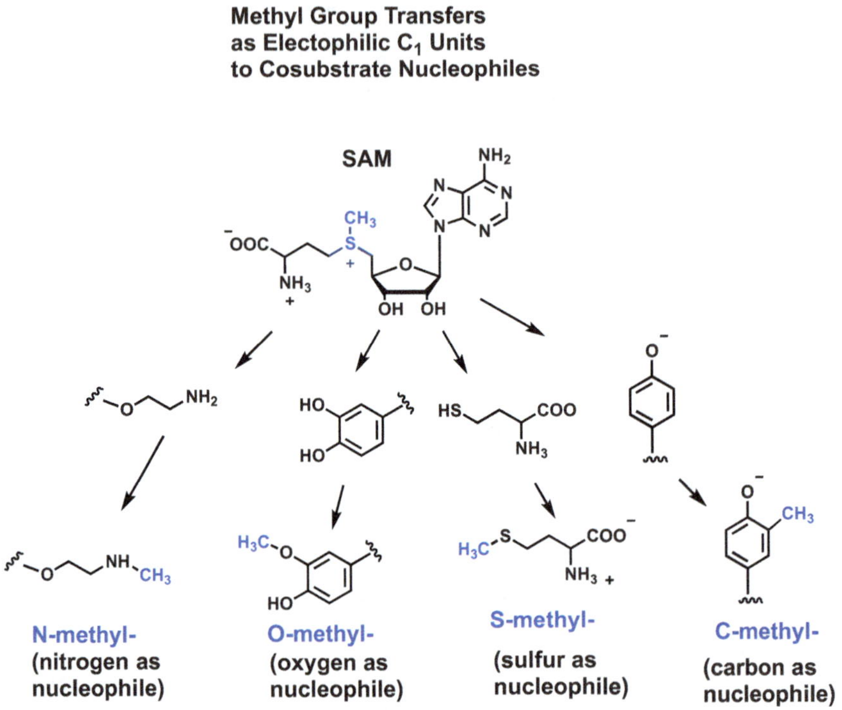

Figure 4.21 Methyl group transfers from SAM to a range of cosubstrate nucleophiles. S_N2 type transfers of the methyl group as a net electrophilic fragment.

is kept low to avoid unwanted nucleophilic chemistry. Instead, cysteine is the source of the inorganic sulfide atoms (Figure 4.22) that are ferried from carrier protein to the site of Fe/S cluster formation in dozens of target proteins. In turn, that requires some unusual C–S bond cleavage route to transfer the sulfur as an electron deficient sulfane atom rather than the usual hydrosulfide anion HS^-.

This unusual, electrophilic, six-electron form of sulfur, thiane sulfur, is generated *in situ* in PLP-containing enzyme active sites and captured by a cysteinyl side chain in those enzymes to create the unusual covalent Cys-*S*-SH residue, a persulfide functional group (Figure 4.23). Until a partner protein subunit can remove that persulfido sulfur the cysteine lyase has transiently committed suicide and then resuscitated in each turnover.[26]

4.8.1 Comparison of PLP-β-Elimination *vs.* β-Replacement to Understand Single Turnover Production of an Enzyme Covalent Persulfide Product Complex

To analyze two distinct modes of C–S bond cleavage in cysteine, we turn to representatives of two PLP-containing enzyme types that transfer either HS^- from C_3 or a six-electron form of sulfur that is an $[S^+]$ equivalent in flight.

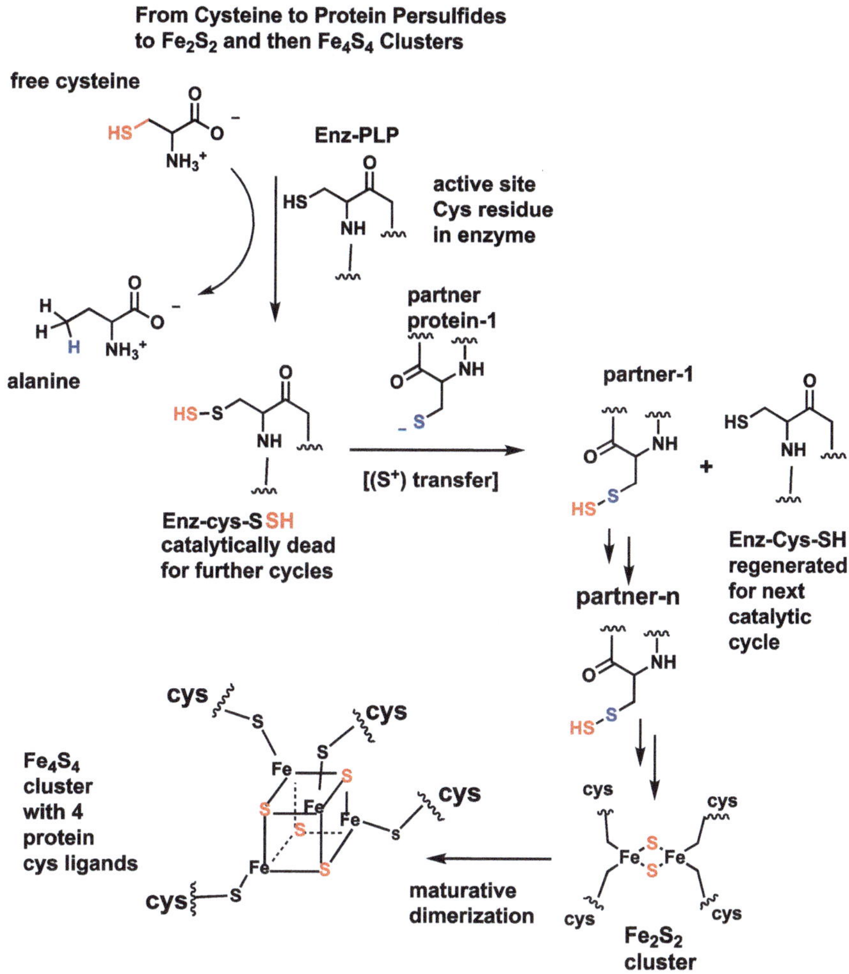

Figure 4.22 Transfer of a sulfane sulfur atom from cysteine to protein persulfides (–Cys-*S*-SH) on the way to Fe_2S_2 and Fe_4S_4 iron–sulfur clusters.

Cysteine lyases perform the first conventional elimination of the β-thiol substituent as HS^- leaving group, generating pyruvate and ammonia as coproducts, a classic β-elimination stoichiometry (Figure 4.23).

The companion persulfide-generating enzyme converts L-cysteine to L-alanine, a net β-replacement of the C_3–SH group by a proton (–CH_2SH to –CH_3) as L-alanine is released as product (Figure 4.24). This is a net two electron reduction at C_3 of cysteine to alanine. The reaction stoichiometry is balanced as the itinerant sulfur ends up oxidized, in a new S–SH bond on what started as a cysteine residue in the enzyme's active site. That cysteine residue has become oxidized by two electrons as the persulfide forms. In each turnover this β-replacement enzyme self modifies on an active site cysteine residue. That covalent modification has inactivated the enzyme.

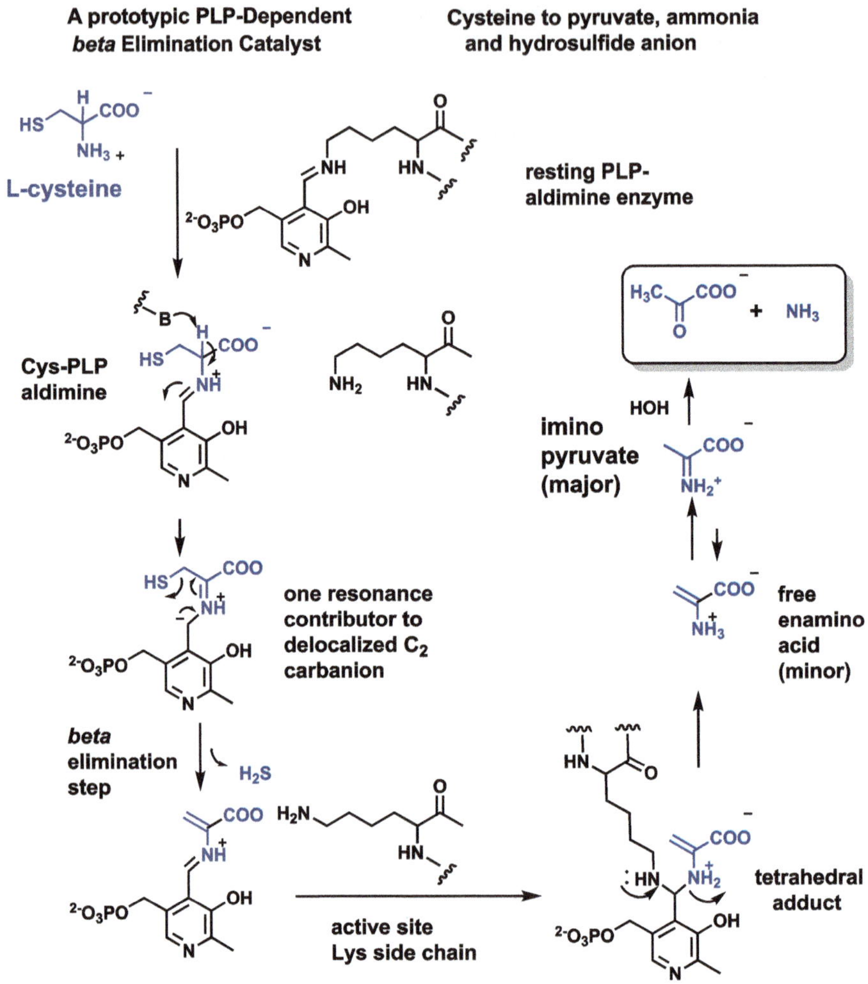

Figure 4.23 Prototypic enzyme-PLP beta-eliminases expel HS^- from cysteine-PLP adducts with the electrons that had been in the C–SH bond. The three carbon substrate scaffold ends up as pyruvate and ammonia products.

To be reactivated the enzyme must be able to transfer that same electrophilic sulfane sulfur atom to a partner protein with its own sacrificial cysteine residue thiolate side chain.

Each subsequent catalytic cycle requires a companion protein subunit to use one of its own cysteine thiolate side chains as nucleophile to move the persulfido sulfur from the catalytic subunit (Figure 4.22). A chain of cysteinyl thiol proteins act to move the terminal persulfido sulfur as an electrophilic thiane group until it is transferred to a protein where an Fe_2S_2 cluster self assembles[27] (requiring at least two such thiane sulfurs).

The mechanism of the C–S bond cleavage in the classic β-elimination enzymes is straightforward (Figure 4.23). The initial cysteinyl-PLP aldimine adduct forms in the active site and undergoes C_2–H bond cleavage by proton abstraction by an enzymatic basic group. The stabilized C_2 carbanion-PLP adduct uses the pair of electrons that constitute that negative charge as a neighboring group to form an olefin between C_2 and C_3 as the C_3–SH bond breaks. The two electrons in that bond depart with the SH as the hydrosulfide anion (viewed in the back direction this would be the key half reaction in cysteine synthases noted in Figure 4.6). The initial enamino acid-PLP adduct can isomerize to the more stable pyruvyl ketimine-PLP adduct, preparatory to release of iminopyruvate which is rapidly hydrolyzed nonenzymatically to the observed product pair, pyruvate and ammonia.

The β-replacement enzyme (cysteine to alanine) starts out the same way and arrives at the same cysteinyl C_2-carbanion-PLP complex. One resonance contributor in a fully planar system would be the *para*-quinonoid contributor with the electron pair stored all the way down on the pyridine nitrogen (Figure 4.24). That resonance form will have an iminium double bond between the cysteinyl amino group on the PLP aldehyde group. That is the resonance form proposed to be attacked by an active site cysteinyl residue thiolate group on the C_3–SH of the amino acid-PLP quinonoid adduct. The C_3–SH bond of the cysteinyl substrate is cleaved but now with opposite polarity from the β-elimination enzyme.[5] The sulfur is in flight as a formal transient $[HS^+]$ equivalent rather than the anticipated $[HS^-]$ species. The electron pair left at C_3 of the amino acid-PLP quinonoid adduct can be delocalized by a resonance contributor as shown (Figure 4.24). Thus, the PLP coenzyme is functioning transiently as a four-electron sink. Two subsequent sequential protonation events, first at C_3 (two electrons out), then at C_2 (two more electrons out of the PLP coenzyme), yield the alanyl-PLP aldimine nascent product, just prior to release of free L-alanine. (We have seen that PLP is likewise pressed into service as a four e^- sink during cystathionine γ lyase catalytic cycles, Figure 4.8.) We noted above that this reaction path constitutes a single turnover active site inactivation. Until the persulfide terminal sulfur atom is transferred onto (a chain of) partner proteins this PLP cysteine lyase can only carry out that one turnover.

The enzyme persulfide adduct is an intriguing carbon-based functional group in chemical biology. Clearly the distal SH of the alkyl persulfide behaves as an electrophile, in its initial formation and subsequent transfers to cysteine thiolate side chains in partner proteins (Figure 4.22). However, the pK_a of persulfides for dissociation to a persulfide anions is 6.3. Thus, the persulfide group is 100 fold more acidic than the cysteine thiolate ($pK_a = 8.2$) itself. The distal anionic sulfur in such persulfide anions ($RS–S^-$) act as expected as nucleophiles. Finally, the proximal cysteine in a persulfide, like the cystine sulfurs in disulfide links are electrophilic, because it is two electrons more oxidized than the thiol forms. The analogous alkyl peroxides (R–OOH) are known in biologic systems but typically as potentially harmful oxidants to be captured reductively (see the glutathione peroxidase section above). There are no comparable biosynthetic metabolic roles for alkyl peroxides as $[OH^+]$ donors, compared to these enzyme-bound alkyl persulfides.

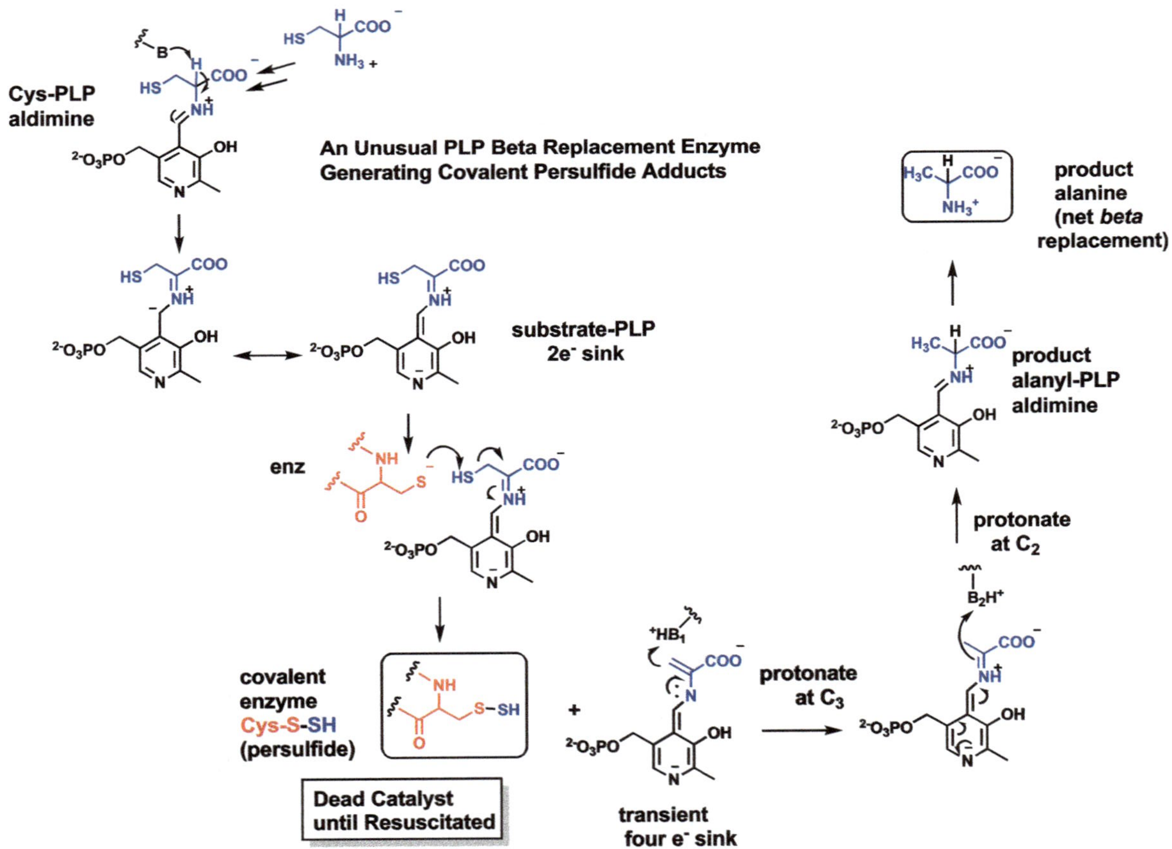

Figure 4.24 The cysteine lyases that generate protein persulfide adducts are beta-replacement PLP-enzymes. The cysteine C_3–SH bond cleaves with the electron pair remaining on carbon, not departing with sulfur.

(One could, however, consider flavin -OOH as adducts as alkyl hydroperoxides that deliver [OH^+] equivalents.)

It may be worth comparison of these protein bound cysteinyl persulfide residues (Cys-*S*-SH) with the protein cys–cys disulfide residues (Cys-*S*-*S*-Cys) noted above (Figures 4.14 and 4.15). Both are two-electron oxidized, compared to cysteine thiols or H_2S molecules. In both persulfides and disulfides, the sulfur atoms behave as electrophiles, not nucleophiles. The persulfide group but not the disulfide group can ionize in analogy to thiol to thiolate anion. Both the thiolate anion and the distal sulfur in the persulfide anion are nucleophilic. No such deprotonation is available to protein or glutathione disulfides so they are not capable of nucleophilic behavior. Disulfides such as oxidized glutathione (GS–SG) are reduced by hydride ions from NADPH. No such reductive cleavage enzymes are known for persulfides. The enzyme-persulfide functional group has been uniquely tuned to display malleable redox properties for Fe–S cluster assemblies.

4.9 Thiol Oxidations: Sulfenates, Sulfinates, Sulfonates

Continuing the theme of oxidation of cysteine-based thiol groups, in addition to disulfides and persulfides, we have noted that nonenzymatic autoxidation of the electron rich thiolate anion could result in net monooxygenation to the cysteine sulfenic acid, both free and in the tripeptide backbone of (reduced) glutathione. Cellular glutathione levels may be 100 fold higher than circulating cysteine levels so most of the cys-*S*-OH groups that form are embedded in the glutathione scaffolds. Capture by a thiol/thiolate anion in a second glutathione leads to the oxidized intramolecular disulfide (GS–SG) discussed in Sections 4.3 and 4.4 above.

Oxidation of free cysteine to the sulfenate can go on to give comparable disulfides that on reduction recycle back to cysteine in a cycle that consumes cellular NADPH. One of the strategies humans employ to minimize promiscuous nucleophilicity of the free cysteine sulfur group is to use a dioxygenase to incorporate both oxygen atoms from a molecule of O_2 into the doubly oxygenated product cysteine sulfinate[28] (Figure 4.25) In general, the sulfinic acid group formation is irreversible in biologic contexts (but see ref. 5). The oxidized sulfur in the cysteine sulfinate is much less nucleophilic than in free cysteine and the dioxygenase catalytic cycle avoids free cysyteine sulfenate.

The sulfinate is a metabolic intermediate to a triply oxygenated sulfur metabolite known as a sulfonate. Enzymatic decarboxylation of the sulfinate by a typical PLP-dependent decarboxylase yields hypotaurine, still at the sulfinate oxidation state. Recent efforts have identified a flavin-dependent monooxygenase as the third enzyme in the pathway, inserting one oxygen atom into the hypotaurine sulfinate group to produce the end product metabolite taurine as the trioxy-sulfonate group[29] (Figure 4.25). Taurine is used to form amides to bile acids, forming conjugates that enable increased enterohepatic circulation (Figure 4.25). The sulfonic acid ($pK_a = 1$) is a much

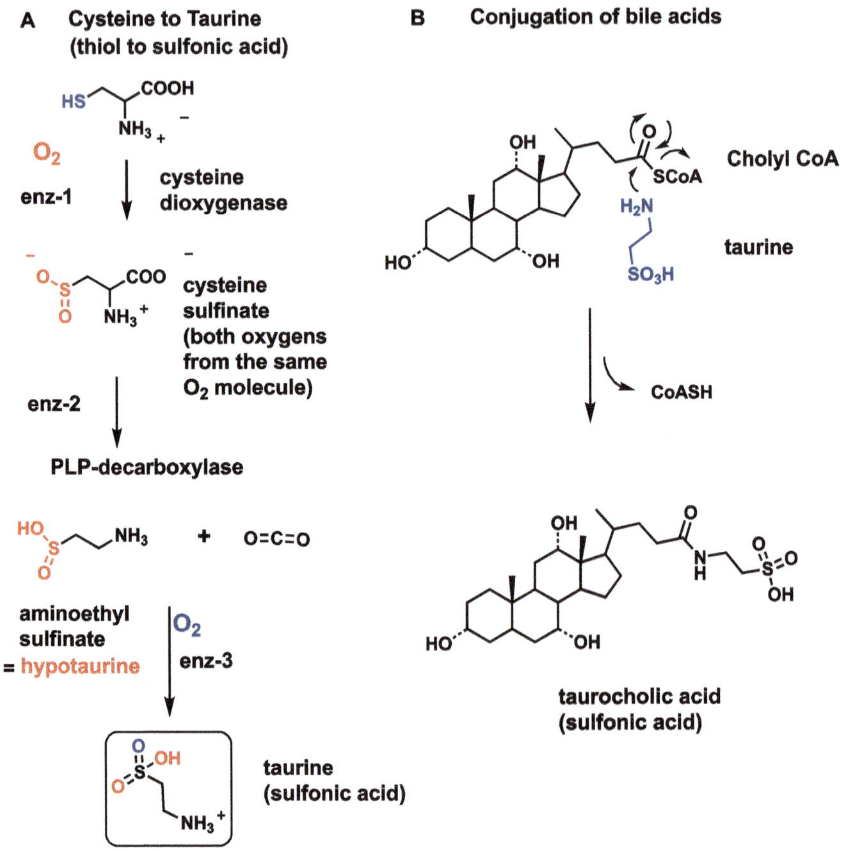

Figure 4.25 Cysteine triple oxidation at sulfur: thiol to sulfenate to sulfinate to sulfonate in taurine. Taurocholic acid is formed from cholyl CoA and the amine of taurine.

stronger acid than carboxylic acids, ($pK_a = 4–5$), so remains unchanged at all physiologic pH values allowing passage of the neutral conjugate through biologic membranes.

4.10 Thioaldehydes, Thiohemiaminals, Thioamides, Thiazolines, Thiazoles

Less common C–S functional groups in carbon sulfur chemical biology are found in secondary metabolite pathways. They include a thioaldehyde intermediate that occurs reversibly in pantetheinyl cysteine decarboxylase catalytic cycles during the biosynthesis of coenzyme A, as well as in some sulfatase catalytic cycles.[6] We have noted the thiohemicetal intermediate during catalytic cycles of aldehyde dehydrogenases and glyceraldehyde-3-phosphate dehydrogenase is a tetrahedral adduct arising from attack of a cysteinyl residue thiol side chain on the aldehyde of the bound substrate back in Chapter 3.

At the oxidation state of acyl groups, there are some nonribosomally generated peptide microbial natural products with multiple thioamide groups where the sulfur arises from inorganic hydrosulfide ions (HS^-).[30]

Also, in some peptides, either ribosomal or nonribosomal in origin, there are thiazoline and thiazole heterocyclic ring systems altering the typical peptide backbones. The thiazolines are heteroaromatic rings and serve as precursors to the thiazoles noted in Chapter 8.

4.11 Overview of C–S Chemistry *In Vivo*

Although sulfur weighs in at only 0.2–0.3% of human body weight, the C–S functional groups that are formed as inorganic hydrosulfide is converted to cysteine and are central to many of the key reaction types in carbon chemical biology. There are both clear parallels and distinctions between sulfur and oxygen that are set forth explicitly in this chapter. In analogy to functional groups with C–O bonds, C–S bonds are found in C–SH (thiols), CH=S (thioaldehydes), COSH (thioacids), and CSNHR (thioamides). The thiols predominate *via* cysteine-based thiol and thiolate anion side chains. Thioaldehydes are rare, albeit one is found as an enzyme intermediate in CoASH biosynthesis. Thioacids and thioamides are even rarer functional groups, found in a few microbial metabolites.

Thiols are stronger nucleophiles than alcohols and have lower pK_a values, so thiolate anions are central nucleophiles under physiologic conditions. Indeed, thiolates are among the strongest nucleophiles in chemical biology (the rare and restricted Co^I form of vitamin B12 is stronger).

Among the many reactivity differences between thiols (and accessible thiolate anions) *vs.* alcohols (and HOH), phosphate oxoesters are stable entities both in low molecular weight metabolites and in post-translationally modified proteins. The corresponding thiophosphates are restricted to transient thiophosphoryl enzyme intermediates in protein tyrosine phosphatases. In turn, acyl thioesters are more activated thermodynamically than corresponding acyl oxoesters. The predominant acetyl transfer currency in cells is the acetyl CoA thioester, albeit more stable oxoesters have their place as detailed for neutrotransmitter acetylcholine in Chapter 3.

Methyl ethers are used as largely dead end protecting groups in various microbial biosynthetic pathways. There is no comparable methoxy side chain in proteinogenic amino acids to compare with the metabolic centrality of the methyl thioether group in methionine. Even more singular to sulfur chemical biology is the trivalent sulfonium ion in *S*-adenosylmethionine (SAM). That cationic sulfonium atom in SAM activates the adjacent C–S bonds for both two electron and one electron transfer (see Chapter 12) reactions. In the two electron pathways the transfer of an electrophilic methyl group to cellular nucleophiles predominates. In one electron mode homolysis of the S–C_5-adenosyl bond creates the 5′-deoxyadenosyl radical as initiator of a variety of cosubstrate radical pathways. There is no comparable oxonium ion metabolite that mimics SAM reactivity and stability.

The enzymatic oxidation of thiols and the subsequent chemical stability of oxidized forms of organic sulfur atoms opens up a realm of chemical biology not utilized by oxygen. Most notably, persulfides (–CS–SH) and disulfides (–CS–SC–) are central redox states in sulfur proteomes. The corresponding alkyl hydroperoxides (ROOH) and dialkyl peroxides (CO–OC) have much weaker O–O bonds and are more in the category of reactive oxygen species than stable low molecular weight metabolites or post-translational protein modifications. Finally, the ability of sulfur to be oxidized leads from thiols to sulfenates, sulfinates, and sulfonates, with one, two, or three S–O bonds, respectively. No comparable metabolic oxygenations of alcohols on the divalent oxygen atom are observed.

Several of the enzymatic transformations of metabolites with sulfur functional groups occur in amino acid scaffolds as noted at several points in this chapter. These reactions have introduced and emphasized the functions of the coenzyme form of vitamin B6, pyridoxal-P, as both two-electron sink and occasionally as four-electron sink. In turn, these reactions center on imine and enamine chemistry and serve as a transition to Chapter 5 dealing with the major types of C–N bonds in organic metabolites.

References

1. B. Meyer, Elemenatal Sulfur, *Chem. Rev.*, 1976, **76**, 367–390.
2. P. Brimblecombe, The global sulfur cycle, in *Encyclopedia of atmospheric sciences*, 2003, pp. 649–679.
3. H. Cypionka, Oxygen respiration by Desulfovibrio species, *Annu. Rev. Microbiol.*, 2000, 827–848.
4. J. F. Honek, Biolgoical chemistry of the carbon sulfur bond, *Can. J. Chem.*, 2015, **93**, 1051–1060.
5. C. T. Walsh, *The Chemical Biology of Sulfur*, Royal Society of Chemistry, London, 2020.
6. E. Strauss and T. P. Begley, Mechanistic studies on phosphopantetheine decarboxylase, *J. Am. Chem. Soc.*, 2001, **123**, 6449–6450.
7. C. T. Walsh, B. P. Tu and Y. Tang, Eight kinetically stable but thermodynamically activated molecules that power cel metabolism, *Chem. Rev.*, 2018, **118**, 1460–1494.
8. B. Huang, M. W. Vetting and S. L. Roderick, The active site of O-acetylserine sulhyfrylase is the anchor point for bienzyme complex with serine acetylase, *J. Bacteriol.*, 2005, **187**, 3201–3205.
9. E. E. Snell, From bacterial nutrition and o enzyme structure in pyridoxal phosphate enzymes, *Annu. Rev. Biochem.*, 1993, **62**, 1–27.
10. W. Rabeh and P. Cook, Structure and mechanism of O-acetylserine sulfhydrylase, *J. Biol. Chem.*, 2004, **279**, 268–306.
11. G. J. McBean, The transsulfuration pathway, a source of cysteine for glutathione in astrocyctes, *Amino Acids*, 2012, **42**, 199–205.
12. R. Bannerjee and C. G. Zou, Redox regulation and reaction mechanism of human cystathionine-beta-synthase: a PLP-dependent heme protein snenor, *Arch. Biochem. Biophys.*, 2005, **433**, 144–156.

13. C. T. Walsh and Y. Tang, *The Chemical Biology of Human Vitamins*, Royal Society of Chemistry, London, 2018.
14. A. Meister, Glutathione metabolism ansd its selective modification, *J. Biol. Chem.*, 1988, **263**, 17205–17208.
15. W. P. Jencks, *Catalysis in Chemistry and Enzymology*, McGraw Hill, 1969.
16. D. Niegowski, T. Kleinschmidt, U. Olsson, S. Ahmad, A. Rinaldo-Mathis and J. Z. Haggestrom, Crystal structures of leukotriene C4 synthase in complex with product analogs: implications for enzyme mechanism, *J. Biol. Chem.*, 2014, **289**, 5199–5207.
17. R. Brigelius-Flohe and M. M. Maiorino, Glutathione peroxidases, *Biochim. Biophys. Acta*, 2013, **1805**, 3289–3303.
18. V. Gupta and K. S. Carroll, Sulfenic acid chemistry: detecrion and cellular lifetrime, *Biochim. Biophys. Acta*, 2014, **1840**, 847–875.
19. E. Pai and G. Sschulz, The catalytic mechanism of glutathione reductase as derived from X-ray diffraction analyses of reaction intermediates, *J. Biol. Chem.*, 1983, **258**, 1752–1757.
20. D. Fass and C. Thorpe, Chemistry and enzymology of disulfide cross-linking in proteins, *Chem. Rev.*, 2018, **118**, 1169–1198.
21. R. H. Lambalot, A. Gehring, R. S. Flugel, P. Zuber, M. Lacelle, M. A. Marahiel, R. Resi, C. Khosla and C. T. Walsh, A new enzyme superfamily: the phosphopantetheinyl transferases, *Cell Chem. Biol.*, 1996, **3**, 923–936.
22. S. Ravanel, B. Gakière, D. Job and R. Douce, The sepcific features of methionine biosynthesis and metabolism in plants, *Proc. Natl. Acad. Sci. U. S. A.*, 1998, **95**, 7805–7812.
23. R. V. Bannerjee and R. G. Matthews, Cobalamin dependent methionine synthase, *FASEB J.*, 1990, **4**, 1450–1459.
24. M. Fontecave, M. Atta and E. Mulliez, *S*-Adenosylemthionine: nothing goes to waste, *Trends Biochem. Sci.*, 2004, **29**, 243–249.
25. C.-M. Park, L. Weerasinghe, J. J. Day, J. M. Fukuto and M. Xian, Persulfides: current knowledge and challenges in chemisytry and chemical biology, *Mol. BioSyst.*, 2015, **11**, 1775–1785.
26. E. Behshad, S. E. Parkin and J. M. Bollinger, Mechaism of cysteine desulfurylase Slr0387 from Synechocystis 6803; kinteic analysis of cleavage of the persulfied intermediate by chemical reductants, *Biochemistry*, 2004, **43**, 12220–12226.
27. H. D. Urbina, J. Silberg, K. G. Hoff and L. E. Vickery, Transfer of sulfur from IcsS to IcsU during Fe/S cluster assembly, *J. Biol. Chem.*, 2001, **276**, 4521–44526.
28. C. A. Joseph and M. Maroney, Cysteine dioxygenase: structure and mechansim, *Chem. Commun.*, 2007, 3338–3349.
29. S. Veeravalli, I. R. Phillips, R. T. Freire, D. Varshavi, J. R. Everett and E. A. Shephard, Flavin-containing monooxygenase 1 catalyzes the production of taurin from hypotaurine, *Drug Metab. Dispos.*, 2020, **48**, 378–385.
30. N. Mahanta, D. Miklos Szantai-Kis, E. James Peterson and D. A. Mitchell, Thioamides: biosynthesis of natural compounds and chemical applications, *ACS Chem. Biol.*, 2019, **14**, 142–163.

CHAPTER 5

Carbon–Nitrogen Functional Groups

Nitrogen makes up 3% of the body mass in humans.[1] For a 150 pound adult male human that corresponds to about 4.5 pounds of nitrogen tied up in organic molecules. The bulk is in amino acid building blocks in the proteins that comprise up to 20% of body mass. Humans consist of $\sim 4 \times 10^{13}$ cells, all but germ cells with diploid genome content, corresponding to 6 billion nucleotide building blocks in DNA in each cell. The four nitrogen heterocyclic nucleotide bases in DNA sum to about 1% of total body mass. Carbon nitrogen bonds are critical in the major bioinformational macromolecules of DNA, RNA, and proteins, and their building blocks.

5.1 Common Types of C–N Functional Groups in Metabolites

C–N bonds comprise the third set of carbon–heteroatom organic functional groups, to go along with the C–O and C–S bonds in the previous two chapters. In aggregate the three sets of functional groups enable and account for most of carbon chemical biology reactions in primary metabolism (metabolism in all cells of an organism). In this chapter we summarize the major types of C–N functional groups that parallel and complement comparable C–O and C–S functional groups (Figure 5.1). Some aspects of amide and amino acid reactivity have come up tangentially in the C–O and C–S functional group foci of Chapters 3 and 4.

This examination includes the crossover step for enzyme-catalyzed incorporation of inorganic NH_3 into organic molecular scaffolds. In both Chapters 6 and 8 we specifically examine the essential role of chemistry available to nitrogen-containing heterocyclic rings in metabolism.

Chemical Biology No. 21
The Chemical Biology of Carbon
By Christopher T. Walsh

Published by the Royal Society of Chemistry, www.rsc.org

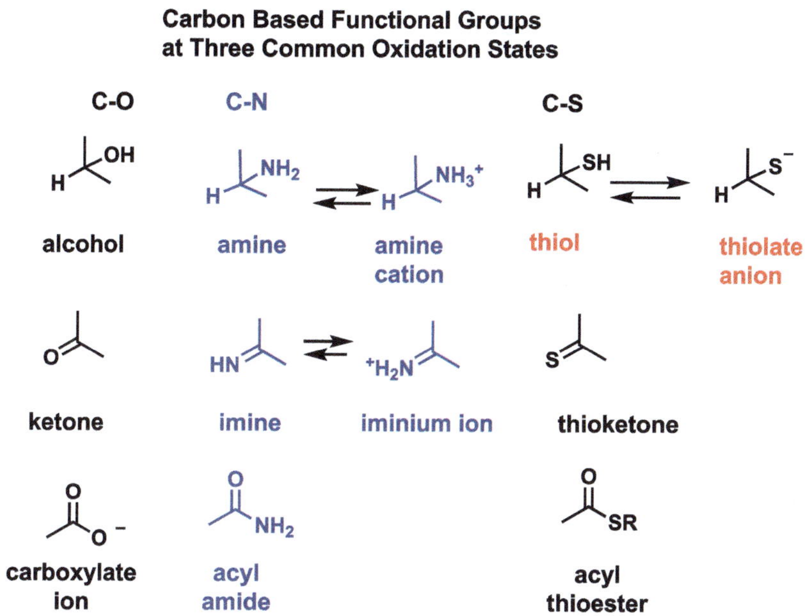

Figure 5.1 Carbon–oxygen, carbon–sulfur, and carbon–nitrogen functional groups at three comparable oxidation states.

In parallel to the three C–O oxidation states of alcohols, aldehydes/ketones and carboxylic acids and the three C–S oxidation states of thiol, thioaldehyde/thioketone (rare), and thio acids (rare), Figure 5.1 notes amines, imines (aldimines/ketimines) and amides. In addition to the imines we will note ureas, amidine linkages, guanidine linkages and their accessible cationic forms in nitrogenous metabolites (Figure 5.2).

Another nitrogen functional group two electrons more oxidized than imines are nitriles with a C–N triple bond (Figure 5.3). Naturally occurring nitriles have the fourth substituent to carbon as a carbon atom (–C–CN). By contrast, when that fourth substituent is hydrogen, the substance is hydrogen cyanide (H–CN). The pK_a of HCN is ∼9.2, such that cyanide ion, a strong carbon nucleophile, will be present at about 1% near neutral pH with dissolved hydrogen cyanide (Figure 10.4). The biosynthesis of organonitriles by plants starts with the amine of amino acids and involves tandem *N*-hydroxylation events.[2] Even less common are isonitrile functional groups in a few known metabolites.

The troika of amines, imines, and amides (Figure 5.1) are the most widespread of the carbon–nitrogen functional groups in chemical biology. We examine the organismal strategies for formation of these three nitrogen functionalities and the less common ureas, amidines, and guanidines in this chapter. In the next chapter we take up heterocycles and discuss the chemical properties of a core set of nitrogen-containing heterocycles, including the purine and pyrimidine building blocks for nucleic acids. The urea, amidine, and guanidine functional groups are information-rich in base pairing of DNA double helices.

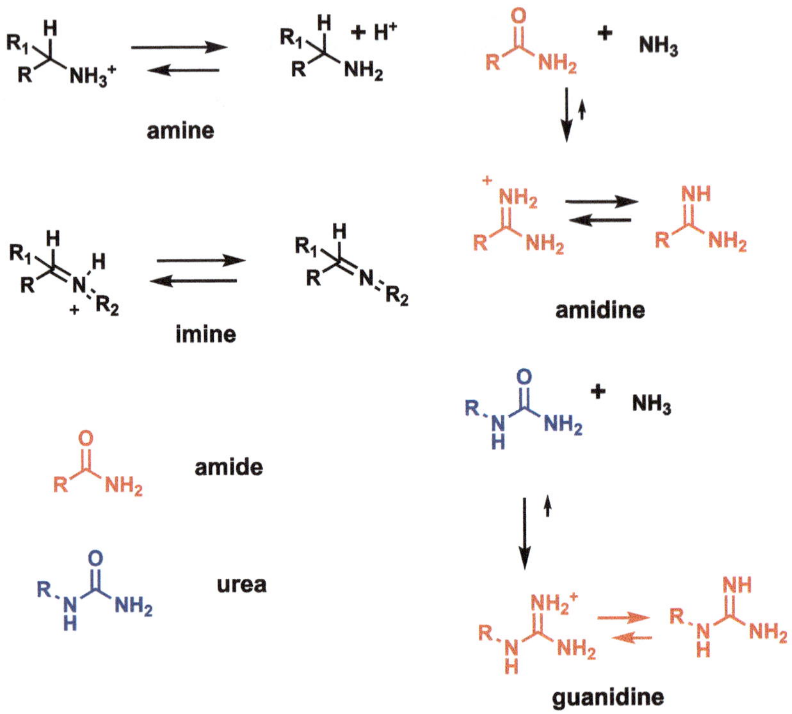

Figure 5.2 Six common C–N functional groups in biology: amines, imines, amides, amidines, ureas, guanidines.

The basicity of nitrogen in amines and imines is stronger than the basicity of corresponding oxygens in alcohols and ketones.[3] The pK_a values of amines and imines can be in the physiologic pH ranges and thus substantial accumulation of both protonated and protonated cationic forms of amines and imines are possible. In the deprotonated, but not the protonated forms, amines are nucleophilic. Imines can be protonated on nitrogen to iminium ions under physiologic conditions where ketones would never be protonated. The cationic iminium ions function as more powerful electron sinks than the neutral ketone carbonyl groups. Analogously, the positively charged amidinium and guanidinium ions are often the accumulating forms of the amidine and guanidine groups in biological molecules.

5.2 Incorporation of NH_3 into Amine and Amide Groups

5.2.1 Imine Formation and Reductive Amination

Nitrogen incorporation into carbon scaffolds occurs from NH_3, ultimately generated by microbial reduction of N_2 and/or plant-mediated reduction of nitrate ions (NO_3^-).[4] The pK_a for protonation of ammonia is 9.3. At pH 7.3, 99%

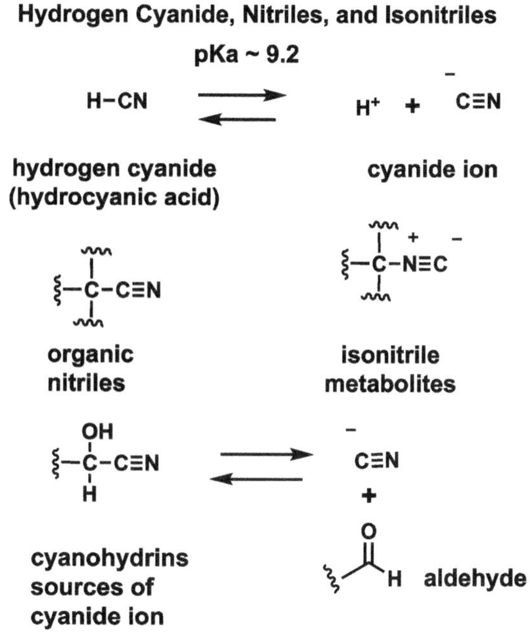

Figure 5.3 Rare C–N functional groups: HCN, organonitriles, isonitriles, cyanohydrins.

will be present as nonnucleophilic ammonium ion and 1% as the nucleophilic NH_3 (Figure 5.4). This minor nucleophilic NH_3 species is the one that is utilized by enzymes to make two types of organic C–N functional groups: amines and amides.

Two types of enzymes incorporate NH_3, one as an imine that is reductively trapped to an amine. The second converts ammonia to an amide linkage by reaction with an activated acyl group of a mixed acyl phosphoric anhydride (see Chapter 3) (Figure 5.4). The first strategy requires a cosubstrate with an electrophilic carbonyl group to engage in imine formation (Figure 5.4). The second strategy requires a cosubstrate with a carboxylate and also a cosubstrate molecule of ATP. The organic framework turns out to be 2-ketoglutarate for the first enzyme and L-glutamate for the second enzyme.

This first strategy for ammonia incorporation depends on a ubiquitously distributed enzyme that can operate in either direction, as a glutamate dehydrogenase[5] or as an iminoglutarate reductase (Figure 5.5). In the ammonia fixation direction, the minor form of nucleophilic NH_3 in a sea of protonated NH_4^+ ions attacks the electrophilic ketone carbonyl of bound ketoglutarate, first to generate a tetrahedral adduct, a carbinolamine with both a C–O and a C–N bond. These sp^3 tetrahedral adducts are unstable with regard to the planar sp^2 carbonyl compounds. While the carbinolamine can revert back to ketone and ammonia, it can also lose water and in the front direction proceed to the imine. Because nitrogen is more basic than oxygen, the ketimines are often protonated at nitrogen under physiologic conditions

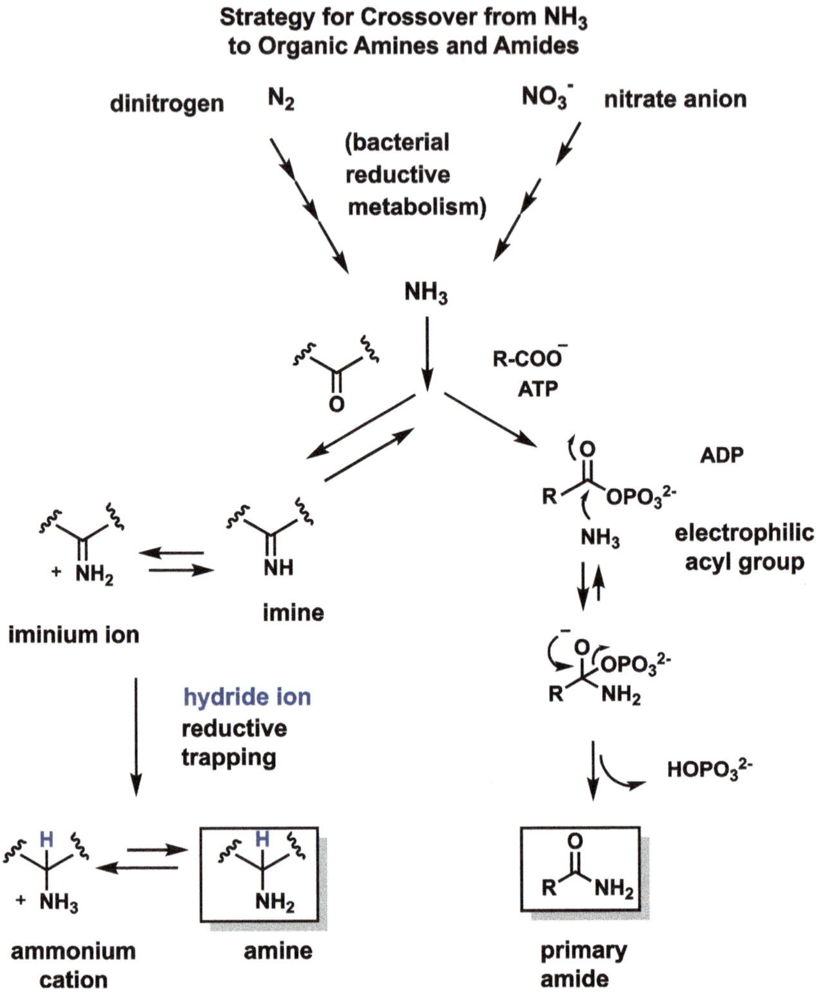

Figure 5.4 Strategy for crossover of inorganic NH_3 to organic amines and amides.

to the cationic nitrogen iminium ion (Figure 5.5). The condensation product from ammonia and α-ketoglutarate is the iminium form of iminoglutarate, held transiently in the enzyme's active site.

In a second catalytic step, the accumulating iminium ion in the glutamate dehydrogenase active site is then captured reductively by hydride transfer from cosubstrate NAD(P)H, reducing the imine with chiral control to 2*S* (L-)-glutamate (Figure 5.5).[6] The net result for glutamate dehydrogenase acting as an iminoglutarate reductase is initial addition of NH_3 to a ketone to yield the imine that is then reduced to the amine. NH_3 has been covalently captured in a stable amine linkage. The imine is a hydrolytically labile, reversible adduct of a ketone and ammonia and would not represent a kinetically stable fixation of nitrogen into a carbon scaffold, absent

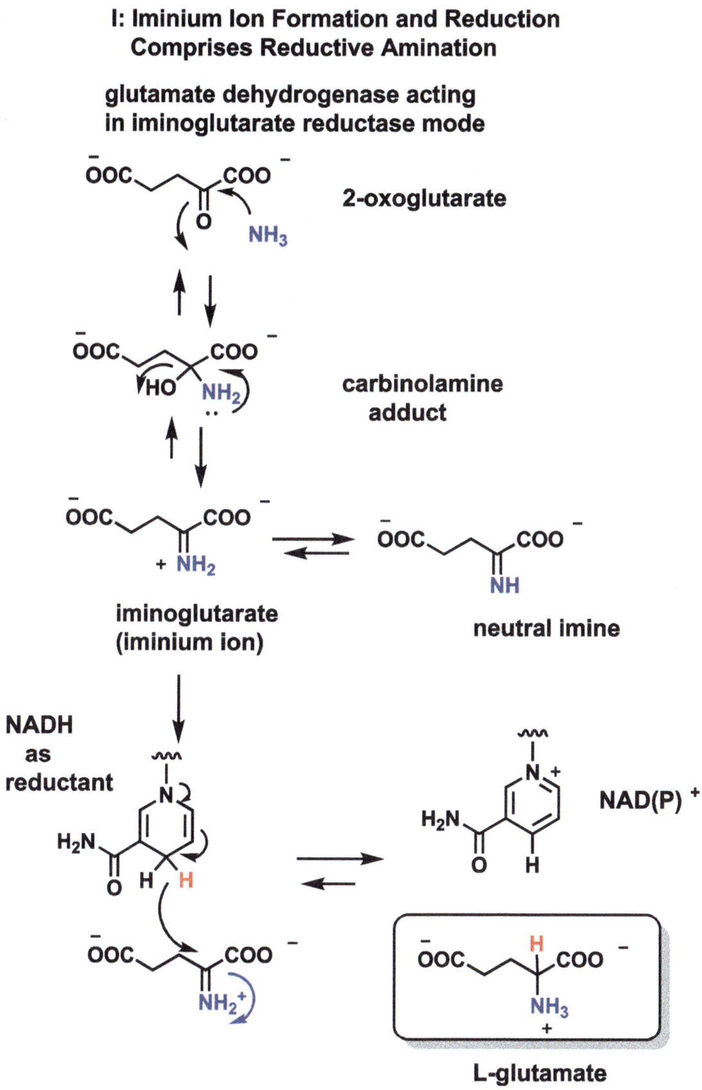

Figure 5.5 Covalent incorporation of NH_3 into organic frameworks by reaction with a ketone carbonyl, dehydration to an imine, and hydride ion reduction from NADH: glutamate dehydrogenase acting in the iminoglutarate reductase direction.

subsequent reduction (Figure 5.5). The iminoglutarate reductase reaction thus generates two types of C–N functional groups, the hydrolytically unstable imine/iminium intermediate and then hydrolytically stable amine group.

With a single C–H bond, the C_2-amino group of glutamate is a secondary amine, as are 19 of the 20 proteinogenic amino acids (only the unsubstituted

C_2 of glycine has a primary amine group). The glutamate dehydrogenase enzyme can function in the other direction. For instance, as a consequence of a high protein meal liberating an acute load of dietary amino acids, glutamate dehydrogenase molecules function in the oxidative direction to release ammonia. In this direction the enzyme is an amine to imine oxidation catalyst and the released iminoglutarate will rapidly decompose hydrolytically to free ammonia/ammonium ions and 2-ketoglutarate. Microbes also encode imino acid reductases to fix nascent NH_3 into 2-keto-isocaproate and into phenylpyruvate to generate L-leucine and L-phenylalanine, respectively.

More generally, organic amine addition to carbonyl groups to generate imine/iminium ions is central to the logic of amino acid metabolism mediated by the coenzyme form of vitamin B6, pyridoxal phosphate,[7] discussed below.[8] In those cases the organic amine is from the amino acid, the aldehyde from the pyridoxal coenzyme. The amino acid-PLP aldiminium adducts are formed in the enzyme sites of >100 amino acid metabolizing enzymes, including several described back in Chapter 4. Typically, the PLP coenzyme itself is tethered in aldimine linkage to the sidechain N_6 amine of an active site lysine residue in essentially all PLP enzymes at rest.[9] All PLP-dependent catalysts therefore start by switching from the resting enzyme=PLP aldimine adduct to the amino acid substrate=PLP aldimine adduct, a process sometimes called transaldimination (Figure 5.6). Ketones and aldehydes react with amines to form aldimines or ketimines reversibly and bring C=N double bonds into play from starting C=O bonds.

5.2.2 Amide Formation *Via* Acyl-phosphoric Anhydride Strategy

L-Glutamate, once formed, is then a substrate for the second class of enzymatic ammonia incorporation reactions: C_5 carboxamide formation (Figure 5.7). This reaction is catalyzed by glutamine synthetase, converting the C_5 carboxylate of L-glutamate to the carboxamide in product L-glutamine.[10] From a functional group chemistry perspective this is a distinct type of enzymatic transformation: a carboxylate oxygen is replaced by the incoming nucleophilic ammonia to create the carboxamide and expel one of the carboxylate oxygens. To achieve this fundamental carboxylate to carboxamide transformation, NH_3 is appropriately nucleophilic but both carboxyl groups of glutamate are negatively charged. At physiologic pH with pK_a values for each of the C_1 and C_5 carboxylates between 4 and 5, more than 99% of glutamate molecules in cells are present as the dianion. In particular, the C_5 carboxylate oxyanion, itself a potential nucleophile, is not attackable by electron rich ammonia.

Glutamine synthetase solves this problem, generalizable to many other amide-forming enzymes throughout chemical biology,[11] by utilizing the nucleophilicity of the glutamate C_5 carboxylate oxyanion first to attack the electrophilic γ-phosphoryl group of cosubstrate ATP. The result is formation of the mixed glutamic-phosphoric anhydride, glutamyl-5-phosphate, along with ADP (Figure 5.6). In this mixed anhydride both the phosphorus atom

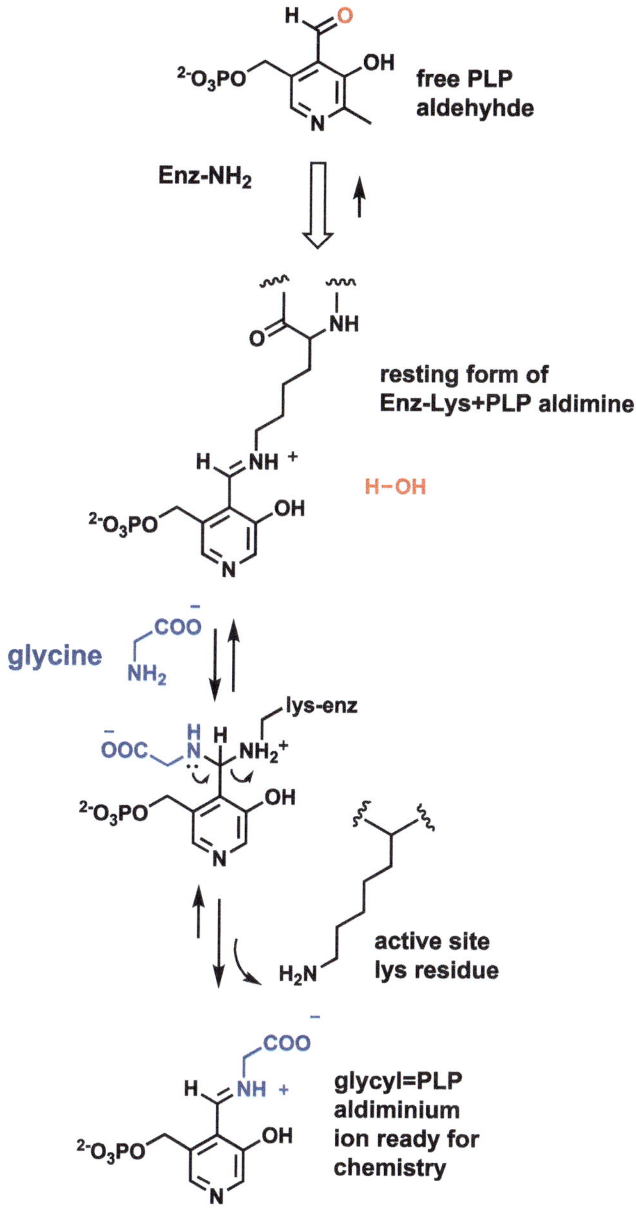

Figure 5.6 PLP-enzymes. The resting enzyme has PLP attached in aldimine linkage to side chain $-NH_2$ of a lysine residue. Binding of an amino acid substrate induces a net transaldimination to release the lys-NH_2 and form an amino acid=PLP aldimine ready for carbanionic chemistry.

and the C_5 carbonyl are electrophilic and thermodynamically activated, for either phosphoryl transfer or acyl transfer to a cosubstrate nucleophile. In glutamine synthetase active sites (the bacterial enzyme is a highly

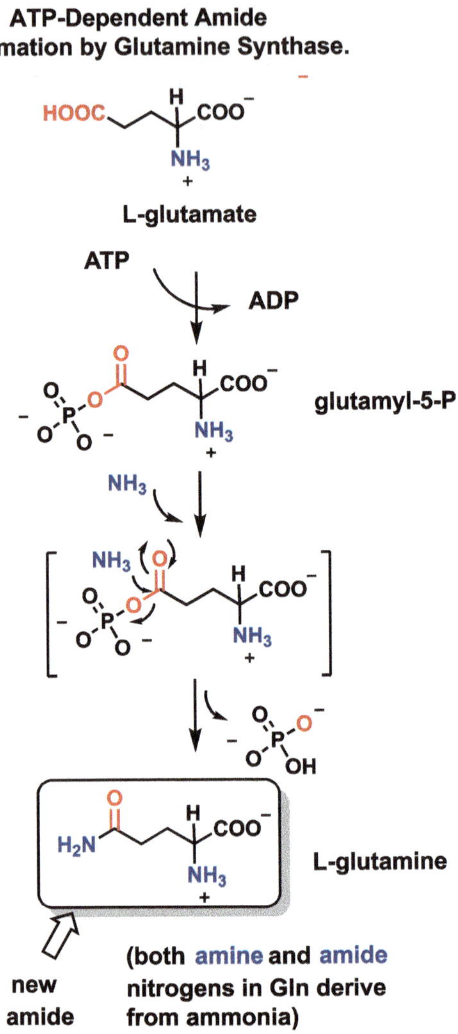

Figure 5.7 Glutamine synthase: ATP dependent formation of an amide from a carboxylate *via* a strategy to form a mixed acyl-phosphoric anhydride (acyl phosphate) that is attacked by NH_3.

regulated dodecamer),[10] NH_3 is the cosubstrate, regiospecifically directed to attack the carbonyl group of glutamyl-5-P. That attack first yields the indicated tetrahedral adduct; breakdown in the forward direction, with expulsion of inorganic phosphate dianion, produces L-glutamine.

Three observations: (1) first, three types of acyl group carbonyls are involved: starting with carboxylate, mixed acyl-phosphoric anhydride, and product carboxamide with relative energies as noted back in Figure 3.16. Provided the activation energy barrier for each step can be surmounted, the aminoacyl-phosphate can proceed downhill to the product amide.

The intermediate mixed anhydride (acyl phosphate) is the least stable, consistent with its function as a high energy reactive intermediate.

ATP cleavage is utilized mechanistically to activate the C_5 carboxylate from a nucleophilic resonance-stabilized oxy anion to the electrophilic mixed anhydride.[11] It is used thermodynamically to drive the equilibrium from carboxylate to the uphill acyl phosphate. The acyl phosphate mixed anhydride grouping is one of the core strategies that enable electrophilic acyl group transfers.

(2) Second, the C–O bond cleavage at C_5 of the glutamate carboxylate involves departure of that oxygen as part of the resonance-stabilized inorganic phosphate. The pK_{a2} value of inorganic phosphate = 6 (pH 6 = 50% monoanion, 50% dianion). If the reaction path had instead somehow proceeded by direct attack of NH_3 on the glutamate carboxylate, the C–O bond cleavage would have released OH^- (water $pK_a = 14–16$). By (overly) simple pK_a comparison of energetics of the C–O cleavage step, the intermediacy of glutamyl-5-P lowers the energy barrier by up to 10 orders of magnitude (pK_a scale is logarithmic).

(3) Third, the cleavage of a phosphoric anhydride side chain linkage of ATP is the generalized route nature has evolved for conversion of carboxylates and amines to form amide/peptide bonds, driving the equilibrium to accumulate the carboxamides.[11] There is a mechanistic dichotomy between a given enzyme directing substrate carboxylate oxygen to attack either the γ-PO^{2-} of ATP or the α-PO_3^{2-} of ATP. As shown in Figure 5.8 asparagine synthetase uses the parallel activation strategy. It converts L-aspartate and nascent ammonia to aspartyl-AMP as the mixed acyl-phosphoric anhydride intermediate. Subsequent attack by NH_3 and forward deconvolution of the tetrahedral adduct releases AMP and L-asparagine.[12]

5.2.2.1 *Phosphoryl* vs. *Nucleotidyl Transfers in Amide Bond Formations*

The asparagine synthetase uses a nucleotidyl rather than phosphoryl transfer.[12] The two examples of glutamine synthetase and asparagine synthetase illustrate electrophilic aminoacyl group transfers by either a mixed aminoacyl-OPO_3^{2-} or aminoacyl-*O*-AMP^{3-} intermediate strategy. These variants reflect the general logic for amide bond formations.

Nucleotidyl transfer logic and machinery has won out in all peptide/amide bonds formed during protein synthesis on ribosomes and in every chain elongation step in both RNA and DNA chain elongations.

The aminoacyl-AMP route is the one used to activate aminoacyl groups in the first half of aminoacyl tRNA synthetase catalytic cycles as noted back in Chapter 3. In the second half reaction the aminoacyl groups are transferred to the 3′-OH of a ribose group at the 3′ end of a cognate transfer RNA. This generates aminoacyl-oxoester linkages to the tRNA (Figure 5.9). Finally, at the ribosome, aminoacyl tRNAs act as amine nucleophiles towards the oxoester carbonyl group of a peptidyl tRNA for chain elongation by amide (=peptide) bond formation (Figure 5.9). This strategy elegantly takes an

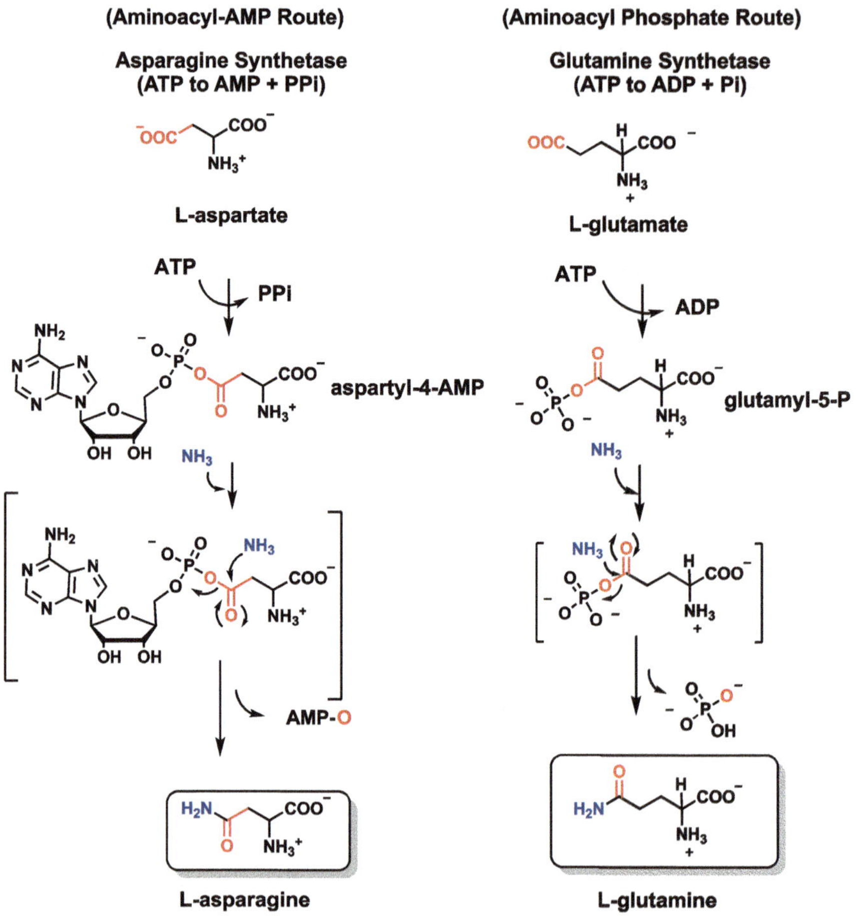

Figure 5.8 Asparagine synthetase *vs.* glutamine synthetase: side by side comparison of two kinds of acyl-phosphoric anhydrides: acyl phosphates *vs.* acyl AMPs.

amino acid building block from carboxylate to mixed acyl-AMP anhydride, to acyl oxoester, and finally to the thermodynamically stable amide.

For comparison, the two amide bonds in tripeptide glutathione are made by way of two aminoacyl phosphates (tandem phosphoryl transfers). The first is glutamyl-5-P, just like the glutamine synthetase logic above. It is captured by the amino group of cysteine. This is not a regular peptide bond, but instead is a type of amide termed an isopeptide bond (Figure 5.10). Then the next enzyme analogously makes the dipeptidyl-phosphoric anhydride, spending an ATP. The γ-glutamyl-cysteinyl-phosphate bound intermediate is capturable by the amino group of glycine to yield glutathione with its reactive thiol side chain[13] (Figure 5.10). This is a standard peptide bond between amine of one residue and the C_1 carboxylate of the next residue.

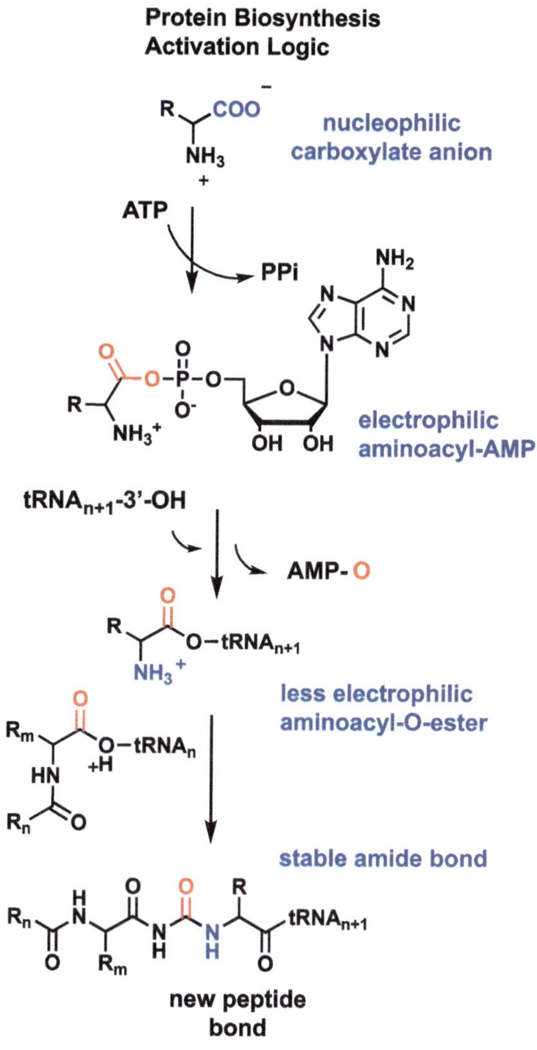

Figure 5.9 Protein biosynthesis activation logic: amino acid carboxylate to aminoacyl-AMP to aminoacyl-oxoester to aminoacyl amide (peptide bond).

Amide bonds are always formed by electrophilic acyl transfers to nucleophilic amine nitrogens.

5.3 Glutamine as a Latent Carrier of Nascent Ammonia for Aminations

Glutamate and its carboxamide derivative glutamine are the first two crossover metabolites that have incorporated inorganic NH_3 into organic

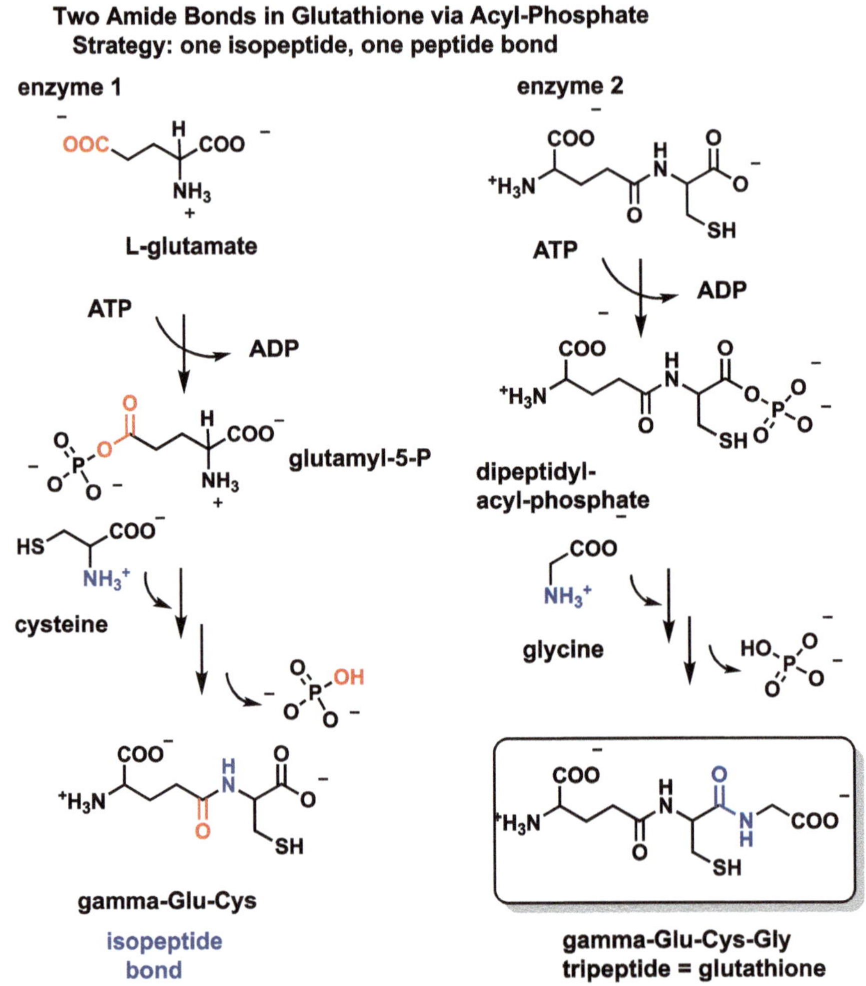

Figure 5.10 Two amide binds in tripeptide glutathione *via* acyl phosphate intermediates: one isopeptide, one peptide bond.

scaffolds: the amine group in glutamate and the amide group in glutamine. Although the amide group in glutamine is chemically stable, there is a family of some 20 enzymes that have glutaminase domains coupled to a second domain that carries out a "nascent ammonia"-dependent amination. They have been termed glutamine amidotransferases to indicate the amino group comes from the amide of glutamine.[14] However, these are not actually amido group transfers. They are net amino group transferases, despite the historical nomenclature. These enzymes involve initial hydrolysis of glutamine to nascent ammonia and glutamate. The ammonia diffuses down an internal protein channel to reach the active site of the aminase domain where it functions as a nucleophile[15] (Figure 5.11). The sequestration of nascent

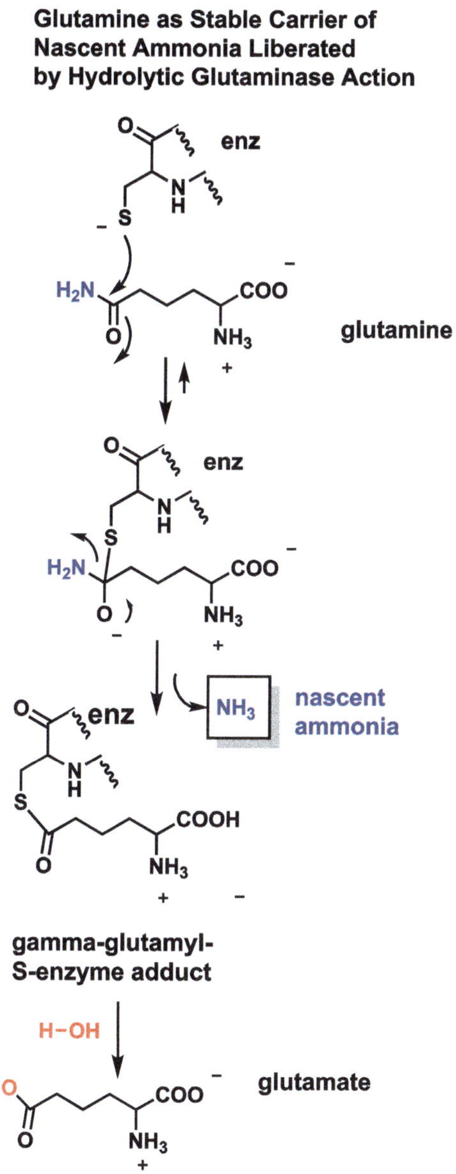

Figure 5.11 Glutamine as a stable carrier of NH_3 releasable by a glutaminase action at a target site.

ammonia within the double headed enzymes active sites, lowers any promiscuous side reaction nucleophilicity that would be exhibited by free NH_3.

Glucosamine-6-P synthase has both a glutaminase and synthase (aminase) domain[16] (Figure 5.12). The glutaminase half reaction proceeds *via* an initial attack of the thiolate side chain of an active site cysteine to generate a covalent glutamyl-*S*-thioester enzyme intermediate and the nascent NH_3.

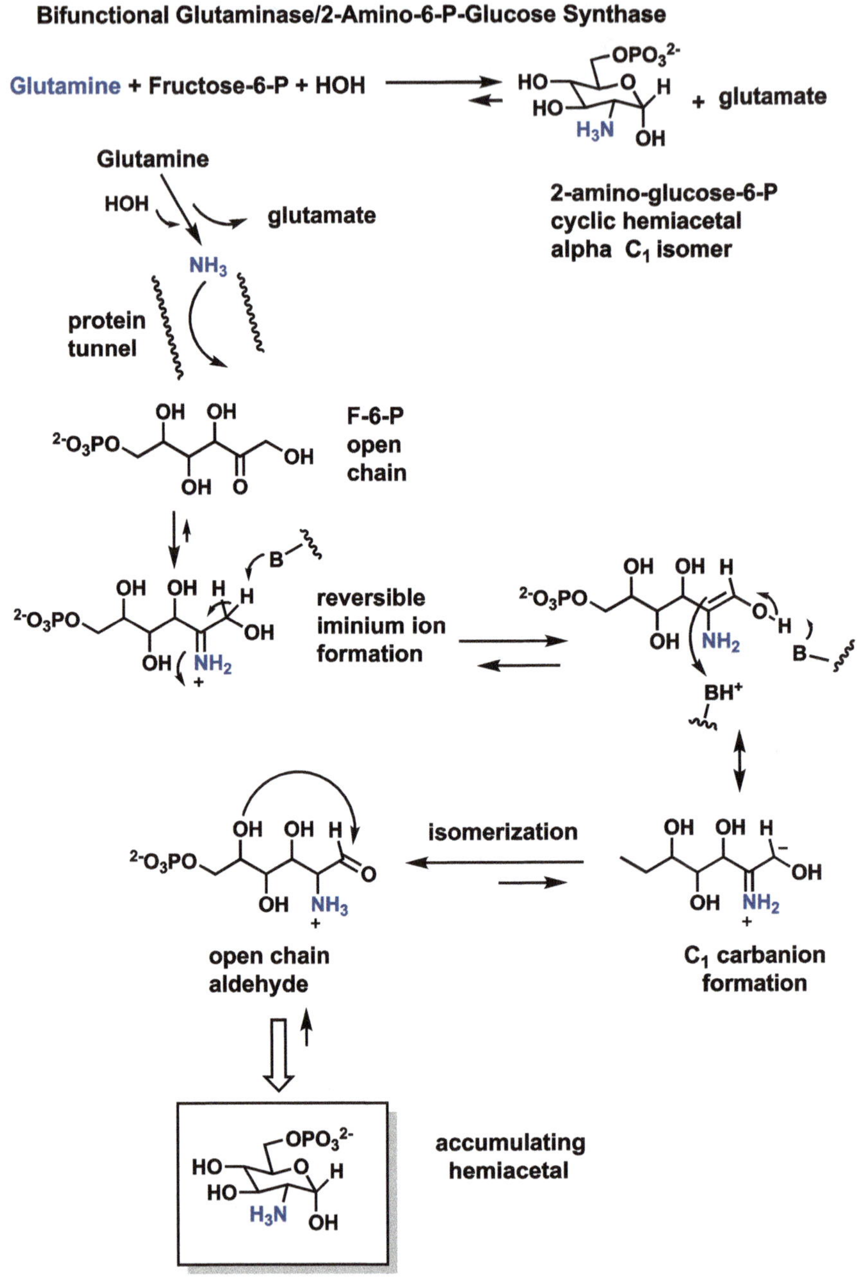

Figure 5.12 Bifunctional glutaminase/2-amino glucose-6-phosphate synthase.

That glutamyl-*S*-enzyme is then subjected to thermodynamically favorable hydrolysis to liberate the glutamyl moiety and set up the glutaminase domain for the next catalytic cycle.

The nascent NH_3, meanwhile, diffuses internally down a channel to the sugar aminase site where fructose-6-P is bound and has been opened from its cyclic hemiketal form in solution to its acyclic keto form. Imine formation between nascent NH_3 and the open chain fructose-6-P ketone generates the imine reversibly. Before release, that F-6-P-2-iminium ion undergoes C_1–H proton abstraction to the stabilized C_1 carbanion that has a resonance contributor that is the stable enamino-alcohol. Isomerization from the enamino alcohol to the glucose framework with internal oxidation at C_1 and reduction at C_2 yields the open chain form of the glucosamine product. That acyclic aldehyde cyclizes intramolecularly to the alpha anomer of the cyclic hemiacetal form of 2-amino-6-P-glucose, the dominant product form.

When the dust has settled, the amine group in the glucosamine scaffold has derived from *in situ* hydrolysis of the glutamine amide group. The original 6-P-fructose-2-imine is isomerized *via* iminum ion-enamine isomerization. The glucosamine product scaffold is trapped out as the cyclic hemiacetal.

Carbamyl phosphate synthetase, noted below in discussion of carbamates and ureas, also uses glutamine in this fashion as a carrier of a latent NH_3 equivalent, releasable only in a target enzyme active site[17] (Figure 5.12). The nascent ammonia that has diffused internally down to the carbamic acid forming active site is directed to attack the mixed anhydride carboxy phosphate to produce the covalent N–C bond in carbamic acid. Further expenditure of a second ATP yields thermodynamically activated carbamyl P as the released acyl phosphate product.

5.4 Transaminases Shuffle Amino Groups Between Keto Acid Scaffolds

While the aminase/glutaminase enzyme family uses nascent NH_3 as direct aminating reagent, a larger set of enzymes known as transaminases move the amino group of glutamate to distribute it among the framework of a variety of available α-keto acid metabolites. At no point is free NH_3 generated during transaminase catalytic cycles. All transaminases are PLP-containing enzymes that start with amino acid=PLP aldimine adducts.[9] As depicted in Figure 5.13, the itinerant amino group is covalently held by the B6 coenzyme while partner keto acid substrates go and come.[7]

The logic of transaminases is to carry out two half reactions.[8] The first half reaction takes the amino acid-1, which will be the amine group donor, from aldimine-PLP to ketimine-PNP (PNP = pyridoxamine-P). The amino acid framework has been oxidized by two electrons, the B6 coenzyme reduced by two electrons. C_2 of amino acid-1 has gone from amine to ketimine. The aldehyde carbon of PLP has gone from aldehyde to amine oxidation state. At this juncture water can hydrolyze the imine linkage, releasing the keto acid-1 and leaving the coenzyme in the two electron reduced pyridoxamine-P form, clear evidence that transaminases are reversible oxidoreductases.

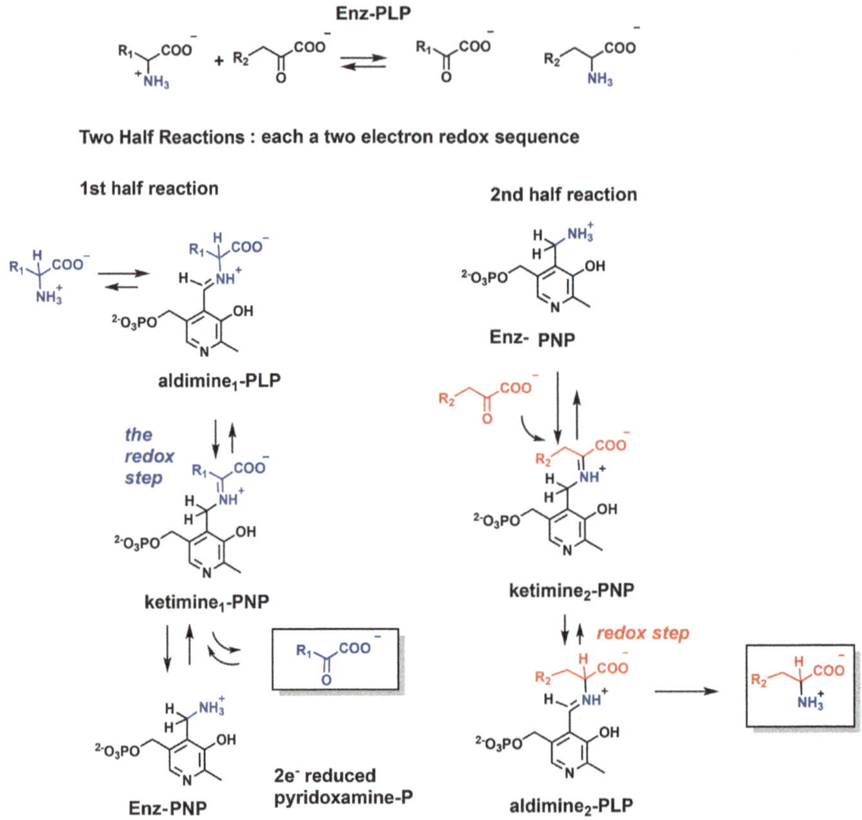

Figure 5.13 Transaminases: PLP-enzymes that move amino groups between amino acid and keto acid scaffolds.

The second half reaction then begins with the enzyme in this two electron reduced PNP complex. Although keto acid-1 could rebind and send the flux back to amino acid-1, if keto acid-2 binds it can form the ketimine-2-PNP iminium adduct (Figure 5.13). Now the redox step can occur in the opposite direction. The ketimine-2-PNP complex is converted to the amino acid-2-PLP complex. The ketone of substrate-2 has been reduced to the amine as the amino group of PNP has been oxidized back to the aldehyde. Hydrolysis of that product aldimine releases amino acid-2.

The tandem action of two oxidoreductase half reactions is a net movement of the amine group from amino acid-1 to keto acid-2, forming keto acid-1 and amino acid-2; an elegant solution for reductive amination of commonly available keto acid metabolites to amino acids. Dozens of transaminases with different specificity for amino acid and keto acid partners allow an amine group diaspora from the initial glutamate and glutamine to form hundreds of other amino acids. Note that this spreading of amino groups never involves free ammonia.

The mechanisms of the redox step in each half reaction are mirror images and rely on the ability of amino acid-PLP aldiminium ions and ketoacid-PNP ketiminium ions to act as electron sinks to delocalize carbanions to make the energy barriers low enough to be kinetically accessible. In the front direction of Figure 5.13, abstraction of the substrate C_2–H as a proton generates the delocalized carbanion (as noted in several examples back in Chapter 4). The specific wrinkle with transaminase active sites is the presence of a BH^+ group placed close to the benzylic carbon of the PLP (Figure 5.14). Delivery of a proton to that resonance contributor quenches the delocalized carbanion. This can be viewed as a net 1,3-proton transfer. It is also the redox step, reducing PLP to PNP as the substrate is oxidized from amino acid to keto acid, all while in iminium linkage. That iminium description changes from aldiminium to ketiminium linkage with the net 1,3-proton transfer.

Precisely the reverse happens in the redox step of the second half reaction of transaminases. The C–H bond targeted is the benzylic carbon of the PNP component of the ketimine-2-PNP complex. Abstraction of one of the two prochiral hydrogens leads to the same type of delocalized carbanion that was an intermediate in the first half reaction. Protonation at C_4 of the PNP returns the carbanion to starting material. Protonation at C_2 of the ketimine moiety constitutes the 1,3-net proton transfer, the redox step, and the conversion of PNP-ketimine to PLP-aldimine. We show a transaminase active site in Figure 5.14 as needing two side chains capable of proton transfers, a B_1 and B_2H in the first half reaction and a B_1H, B_2 protonation state in the second half reaction.[9] Depending on the stereochemistry of the proton abstraction and proton donation across the developing extended planar carbanion, one base could act in a suprafacial 1,3-proton transfer (not shown).

The two families of glutamine amidotransferases and transaminases account for the great bulk of movement of the amino group originally in glutamate to other amine containing metabolites. Some of the many amino acids created by transaminases are then substrates for decarboxylation to amines, as noted in Section 5.5 below. The glutaminase-aminase didomain proteins generate nascent ammonia *in situ*, and use it as a nucleophile. Those enzymes use the physical sequestering strategy to avoid having nascent NH_3 get free in solution. In contrast, the large transaminase family never generates the potential toxic free NH_3 as a reaction intermediate.

5.4.1 Peptide Bonds *vs.* Isopeptide Bonds

Peptide bonds are a subset of amide bonds where the two partner substrates are amino acids and the amide bond is between the α-amine of one amino acid and the α-carboxylate of the second amino acid. Isopeptide bonds are still amide linkages but with distinct regiochemistries of amine and carboxyl group coupling, often in some dibasic or diacidic amino acids, like lysine, aspartate or glutamate. The tripeptide scavenging metabolite glutathione is γ-Glu-Cys-Gly: one isopeptide bond then one peptide bond: both are amide bonds, noted in Figure 5.10.

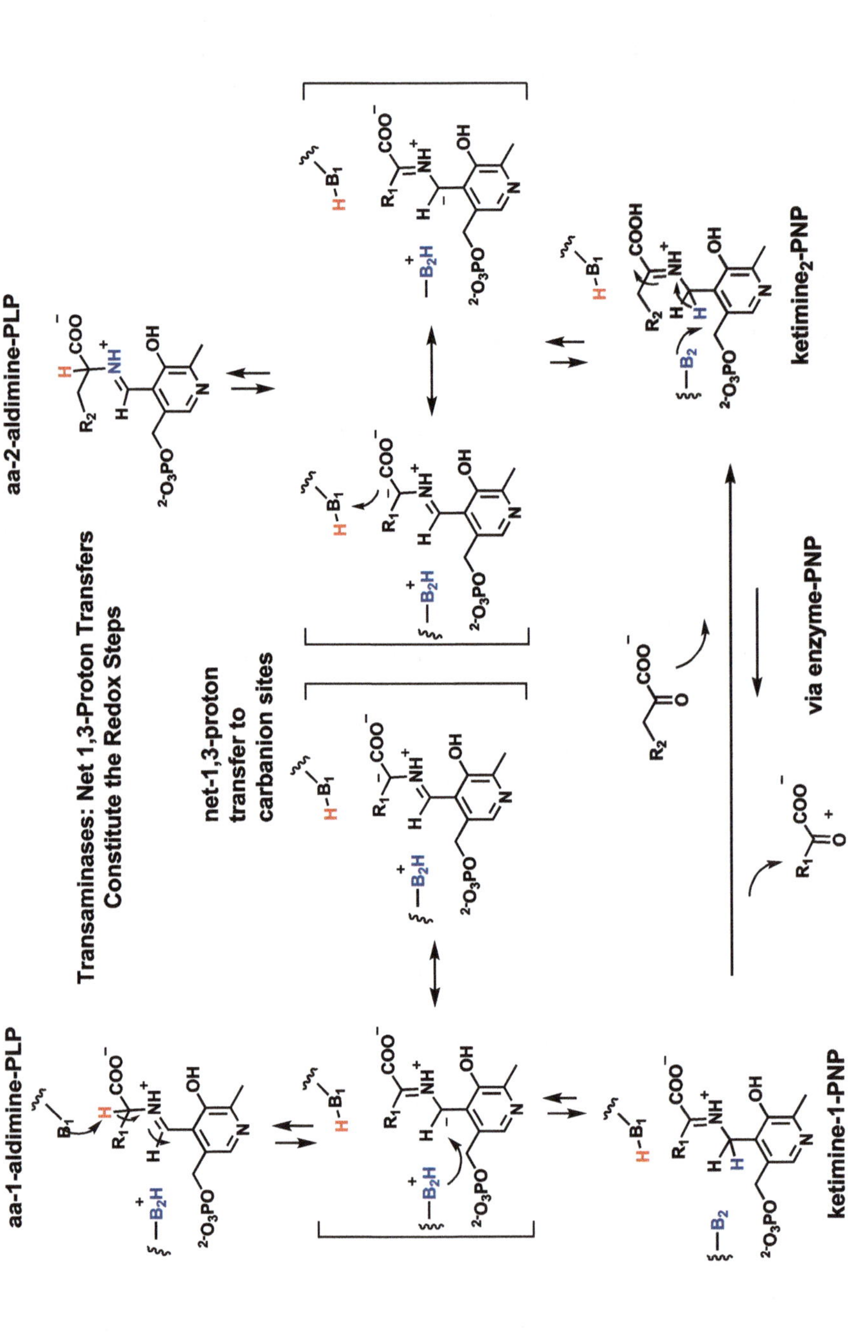

Figure 5.14 Transaminases: 1,3-proton transfers constitute redox steps.

5.4.2 Transamidases and Transpeptidases

Two more specialized versions of nitrogen transfers occur in protein and peptide contexts, catalyzed by transamidases and transpeptidases, respectively. Both types of enzymes take an amide and a carboxylate anion substrate, break an existing amide bond in a substrate, and form a new one before release of products (Figures 5.15 and 5.16). The net reaction, the sum of the amide/peptide bonds remains constant (isoenergetic), but the switch results in conversion of an intrachain amide bond in one substrate to a crosslinking interchain amide bond between the two products. Each of the enzymes carry out indirect acyl transfers, first forming covalent acyl-enzyme adducts, acyl-*S*-thioester in transamidases *vs.* acyl-*O*-thioester enzyme adducts in transpeptidases. The incoming amine nucleophiles on acceptor chains capture the acyl enzyme intermediates for the net interchain crosslinks.

Transamidases use a glutamine side chain on one protein as the electrophile and the side chain amine of a lysine residue on a partner protein as the nucleophile[18] (Figure 5.15). The strategy for the first half reaction mirrors that of the glutaminases noted in Section 5.3 above, formation of a 5-glutamyl-*S*-enzyme where the glutamyl C_5 carbonyl now has heightened

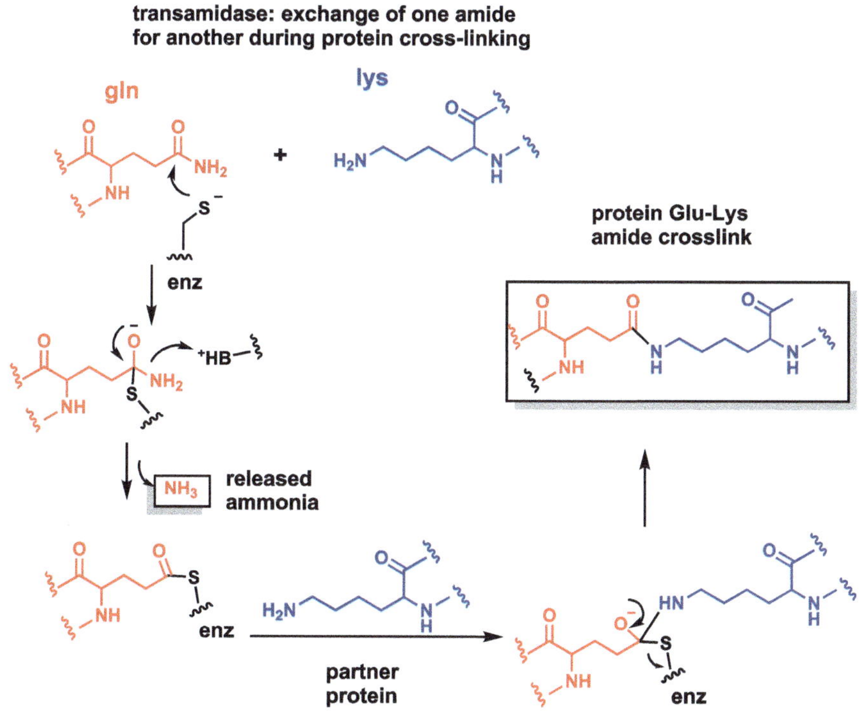

Figure 5.15 Transamidase enzymes exchange on an amide bond by another during protein cross linking.

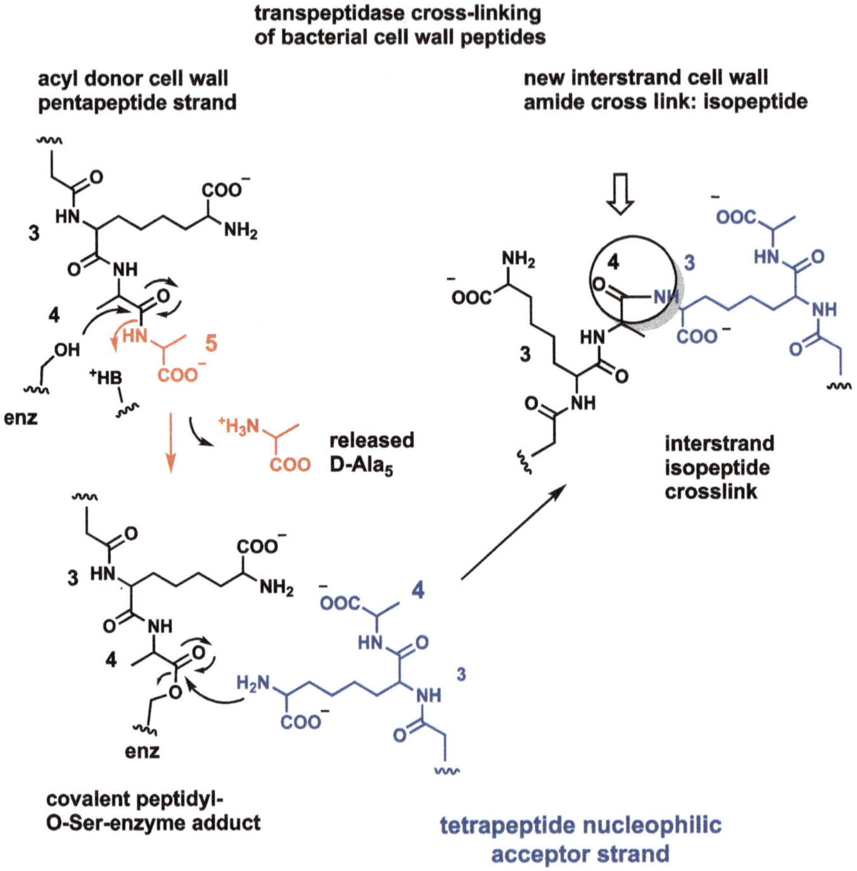

Figure 5.16 Bacterial transpeptidases generate interstrand peptidoglycan crosslinks during cell wall biosynthesis.

reactivity as an electrophile along with generation of nascent ammonia. In this case, however, the released ammonia is "thrown away", while the glutamyl-*S*-enzyme covalent adduct is then captured by the lysine side chain amine from the partner protein. The new crosslinking γ-glutamyl-lysine amide linkage is an isopeptide bond that connects the two proteins.

Transpeptidases that generate mechanical-strengthening cross links in the peptide chains of bacterial cell wall muropeptides[19] employ a related catalytic strategy of creating a peptidyl-enzyme covalent adduct as electrophile. The cosubstrate nucleophile in Figure 5.16 is the side chain amine of a *meso*-diaminopimelate residue on an adjacent muropeptide. As denoted in the figure, the transpeptidases works on the terminal D-Ala$_4$-D-Ala$_5$ terminus of a cell wall pentapeptide. The enzyme active site catalytic nucleophile in this case is a serine-OH rather than a cysteine-SH in the transamidases. D-Ala$_5$ is released as a free amino acid as the tetrapeptidyl-*O*-ser-enzyme adduct forms. This is intercepted as an electrophilic fragment by the amine of meso-DAP_3 on an adjacent tetrapeptide strand. The result is a new interstrand isopeptide bond between meso-DAP_3 on the attacking strand to D-Ala_4 on the receiving strand. Both the purposes and the strategy of transpeptidases in bacteria and transamidases in animals operate in parallel to provide covalent bonds to achieve mechanical strength.

5.5 PLP-containing Decarboxylases

A second major category of PLP-containing enzymes[20] that work on amino acid substrates are amino acid decarboxylases to provide many of the amino acid-derived metabolic amines such as serotonin, γ-aminobutyrate (GABA), and dopamine, among many others ( ). We have prefigured some of the following analysis in considering decarboxylases and CO_2 metabolism at the end of Chapter 3.[21] The decarboxylases utilize a similar initial chemical strategy as transaminases in starting catalytic cycles *via* formation of amino acid-PLP aldiminium ions. Then, the decarboxylases and transaminases differ.

Transaminases start catalytic cycles by removal of the amino acyl C_2–H as a proton to get to a delocalized, extended amino acid-PLP carbanion. Instead, PLP-containing decarboxylases orient the carboxylate anion of the amino acyl moieties orthogonal to the plane of the PLP coenzyme.[22] As the C_2–C_1OO^- bond breaks the incipient electron pair that stays behind at C_2 has maximal overlap with the π electron cloud of the pyridine aldehyde system, thus lowering the energy for carbanion formation during that bond cleavage specifically (Figure 5.18). The transaminases and decarboxylases get to stabilized substrate-PLP aldiminium carbanion adducts by different routes. Transaminases cleave the substrate C_2–H bond. Decarboxylases cleave the C–COO^- bond as CO_2 departs as an electrophile.

The extended carbanions are quenched by protonation by the different subfamilies of PLP enzymes. While transaminases can quench the extended substrate-PLP carbanion at the benzylic carbon of the cofactor, the decarboxylases appear only to be able to quench the carbanion by protonation at what had been C_2 of the substrate. The resultant product amine-PLP aldimines then release the primary amine coproduct on imine hydrolysis (Figure 5.17).

5.6 Amine Nucleophiles for a Range of Electrophilic Group Transfers

The many amine metabolites, typically derived from such enzymatic decarboxylations, are major cellular nucleophiles for capture of the core set of thermodynamically activated, kinetically stable electrophilic groups in partner molecules. The deprotonated amines undergo *N*-acylations from acyl CoAs, *N*-alkylations from *S*-adenosylmethionine, *N*-glycosylations from UDP-glucose, *N*-phosphorylations from ATP, and *N*-sulfurylations from 3′-phospho-5′-adenosine-phosphosulfate in a variety of biologic contexts (Figure 5.19). Two specific metabolic examples are the acetyl CoA-derived

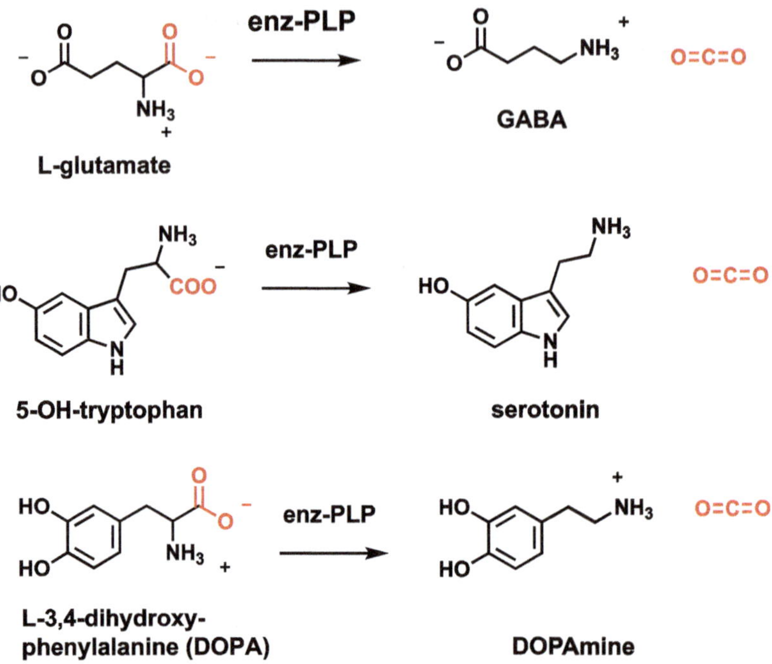

Figure 5.17 PLP-dependent decarboxylases convert amino acids into neurotransmitter amines. L-glutamate to GABA; 5-hydroxy-L-tryptophan to serotonin; L-3,4-dihydroxyphenyalanine (DOPA) to DOPAmine.

N-acetyl group in 2-*N*-acetyl-glucosamine-6-P and the *N*-methyl group derived from *S*-adenosylmethionine in the hormone adrenaline.

Once NH_3 has been captured either as imine that undergoes reduction to the stable amine or as an amide linkage, all three C–N functional groups (imine, amine, amide) are central to nitrogen metabolism and enable the diversity of those metabolic universes. Amines are the primary nucleophiles for capture of electrophilic groups. Imines are the direct connectors between carbonyl group and amine chemistry. Amides are the stable derivatives of carboxylic acids that are the backbone covalent bonds in all proteins.

5.7 Imines, Amidines, Ureas, and Guanidines

Beyond the central troika of imines, amines, and amides, three additional organic nitrogen functional groups occur in primary metabolic pathways (Figure 5.2). The amidine group occurs in the pyrimidine nucleotide cytidine triphosphate, a nucleic acid building block. We defer discussion of amidine group biogenesis until Chapter 6 which examines the pyrimidine and purine nitrogen heterocycles where we observe that an amidine group is effectively the imine derivative of an amide (Figure 5.2).

Parallel to the amidine group as the imine form of an amide precursor, the guanidine group can be viewed as the imine of a urea (Figure 5.2).

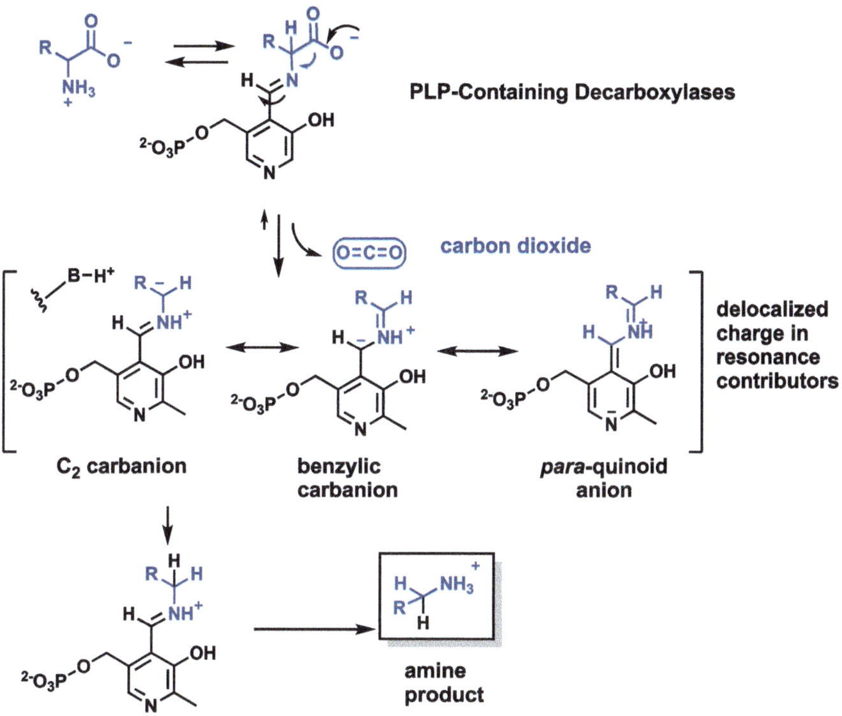

Figure 5.18 PLP-containing amino acid decarboxylases accelerate CO_2 release by stabilization of the incipient C_1 carbanion in amine-PLP delocalized adducts.

The metabolic cycle for generation of urea involves both ureido and guanidine C–N functional groups (Figure 5.20). The purpose of the urea cycle in humans is excretion of ammonia in a nontoxic, nonnucleophilic highly soluble form as free urea.[23]

As shown in Figure 5.20 the metabolic logic is to convert the carbamyl group of carbamyl-P first to a ureido group, then to a substituted guanidinium array. Elimination of fumarate yields the cationic guanidinium side chain ($pK_a = 13.8$) of arginine, one of the 20 proteinogenic amino acids. Hydrolysis of that guanidinium group by the enzyme arginase liberates urea and recycles the pathway intermediate citrulline (the ureido metabolite).

The starting metabolite for the nitrogen and carbon flux into the urea cycle is the one carbon scaffold of carbamyl phosphate, the mixed anhydride between carbamic acid and inorganic phosphate. Carbamyl phosphate synthetase enzymes are members of the glutamine amidotransferase family noted in Section 5.3. The NH_3 for carbamic acid comes from glutaminase action and diffusion of nascent ammonia within the double headed protein to the aminase active site.[17] The one carbon fragment that acts as cosubstrate is the mixed anhydride of carbonic acid and inorganic phosphate,

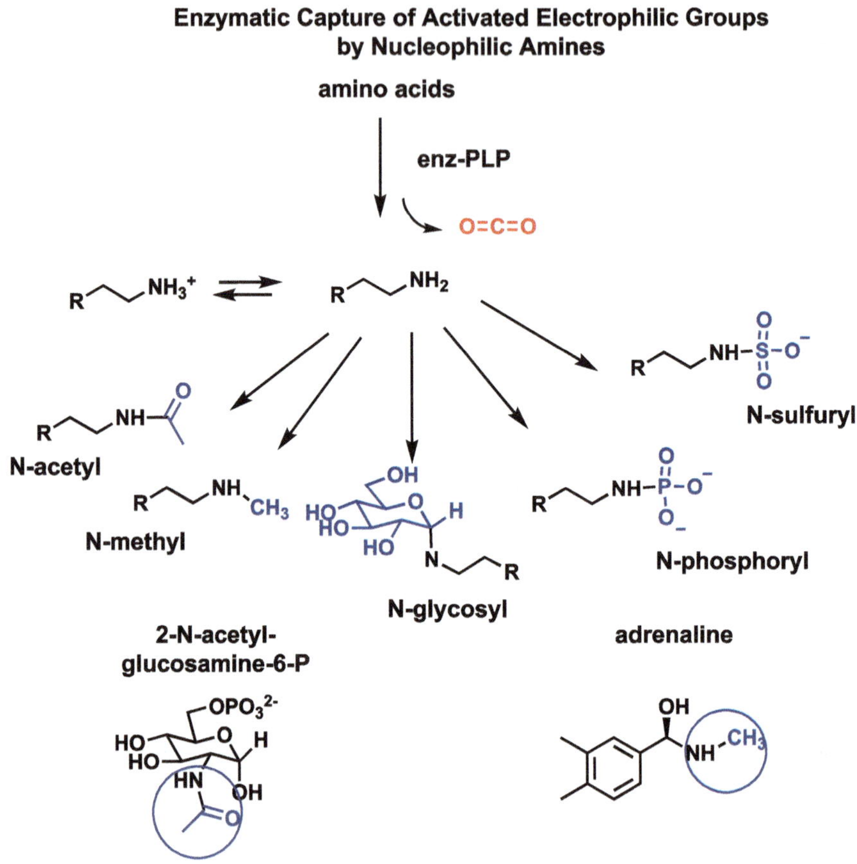

Figure 5.19 Cellular amine groups act as nucleophiles to capture a variety of electrophilic organic and inorganic groups.

termed carboxy phosphate. In turn, that can arise in the synthase active site by attack of a bicarbonate oxygen on the outer phosphoryl group of ATP (Figure 5.21). Next, an attack of nascent NH_3 on the carbon of the carboxy phosphate mixed anhydride releases inorganic phosphate as the N–C bond in carbamate is generated. Before release, one of the carboxylate oxygens of that carbamate attacks a second molecule of ATP to produce another mixed acyl-phosphoric anhydride carbamyl-P that is stable enough to be the released product.

The specifics of the urea cycle are as follows. One of two central metabolic fates of carbamyl-P is to be captured by the amine of the amino acid L-ornithine (first discovered in birds) in the active site of ornithine transcarbamylase.[24] (The other major metabolic fate of carbamyl-P is the first step in pyrimidine biosynthesis, taken up in the next chapter). The deprotonated amine from of the C_5 amino group of ornithine is the nitrogen nucleophile that attacks the electrophilic carbonyl group of carbamyl phosphate.

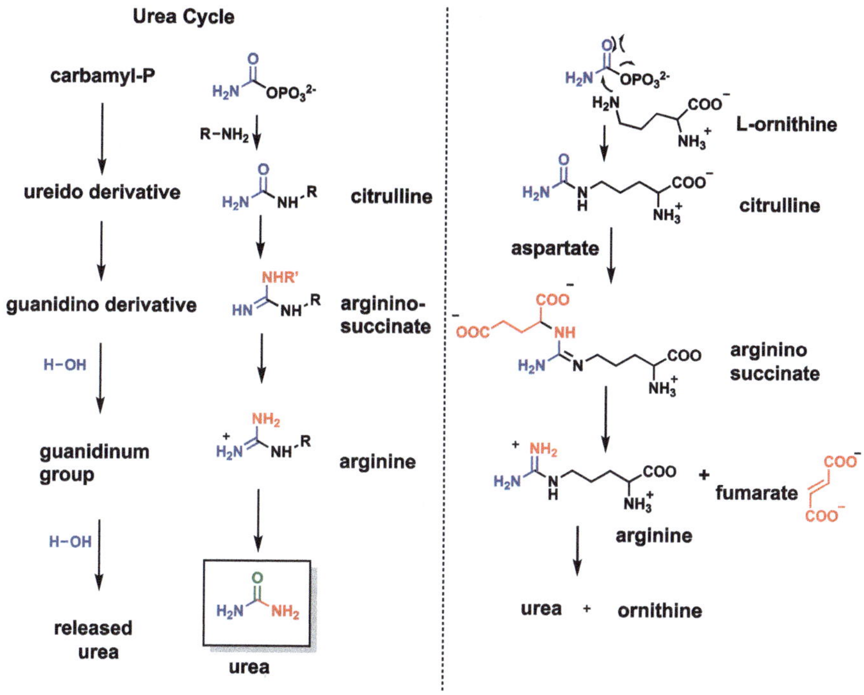

Figure 5.20 The urea cycle involves a series of C–N functional groups from carbamyl to ureido to guanidino on the way to release of free urea in which both nitrogen atoms are nonnucleophilic and nontoxic.

After decomposition of the initial tetrahedral adduct in the forward direction, the released products are inorganic phosphate and the amino acid known as citrulline (Figure 5.20). The transfer of the carbamyl group to the amine of ornithine has generated a substituted urea in citrulline (a ureido group).

To go from the ureido group to the imine derivative known as a guanidino group requires an enzyme that can add a nucleophilic nitrogen to the ureido carbonyl group. In the event animals/humans use a third source of an "NH_3" equivalent: the C_2-amine of L-aspartate. One can notionally trace the source of the aspartate amine group back through transaminase action to glutamate and from there to free NH_3 *via* glutamate dehydrogenase acting as an iminoglutarate reductase, described earlier in this chapter.

The requisite enzyme is named argininosuccinate synthase,[25] for the name of the product arising from attack of the aspartate amine group on the ureido carbonyl of citrulline (Figure 5.22). The amine has been installed but it is still attached to the four carbon succinyl scaffold derived from aspartate. A tandemly acting enzyme, argininosuccinase, removes that four carbon scaffold as the *trans*-olefin fumarate (Figure 5.22) (this olefin-forming elimination might also fit in Chapter 2 on olefin chemical biology).[26] The coproduct is arginine, bearing the guanidino side chain, the imine of the starting ureido group in citrulline. The pK_a of the guanidinium group is

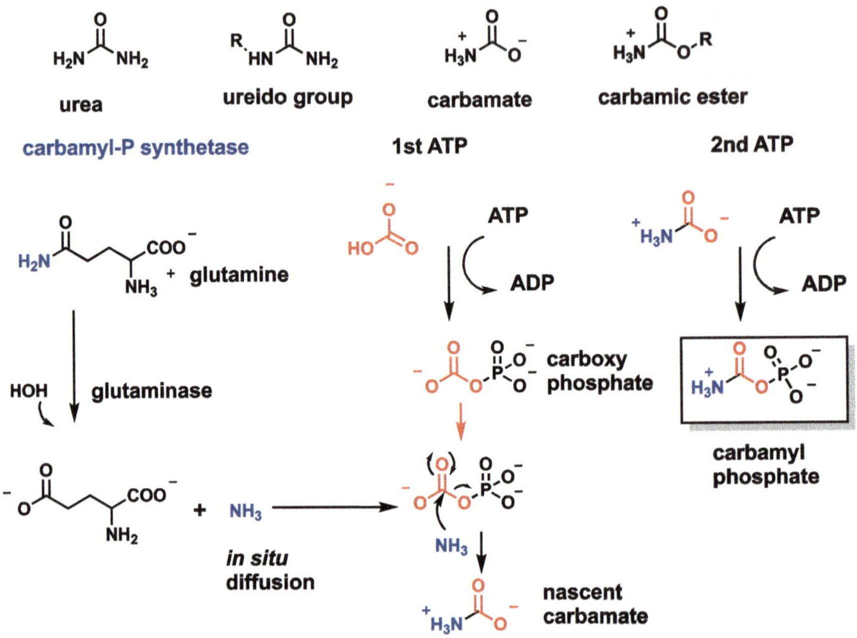

Figure 5.21 Carbamyl phosphate, a thermodynamically activated but kinetically stable mixed anhydride of carbamic and phosphoric acids, is generated from nascent NH_3 derived by the *in situ* glutaminase action of carbamyl-P synthase on carboxy-P.

around pH 13.8, so this nitrogen functional group bears a positive charge as a guanidinium group, 99.9999% of the time in physiologic fluids (Figure 5.2).

Figure 5.22 delineates the two amino acids, glutamine through its amido nitrogen, and aspartate through its α-amino group that contribute each of the two nitrogens in arginine's new guanidino group and from there by hydrolysis to the two nonnucleophilic nitrogens in urea. Together, glutamine and aspartate, *via* a host of transaminases, can serve as removal mechanisms of amino groups from any excess cellular amino acid and dump the ammonia equivalent into urea for excretion. While glutamine releases free NH_3, restrained physically within the generating enzyme structure, the aspartate amino group is never a free amino group.

To complete the cycle of converting nucleophilic ammonia to nontoxic nonnucleophilic free urea that can be excreted on multigram quantities per day by the kidneys, the enzyme arginase adds water into the cationic guanidinium group,[27] releasing urea while regenerating ornithine for the next round of carbamylation by carbamyl-P.

We will note the presence of both amidino groups and guanidino groups in the cytidine and guanine nucleotides, respectively, in the next chapter on nitrogen heterocycles.

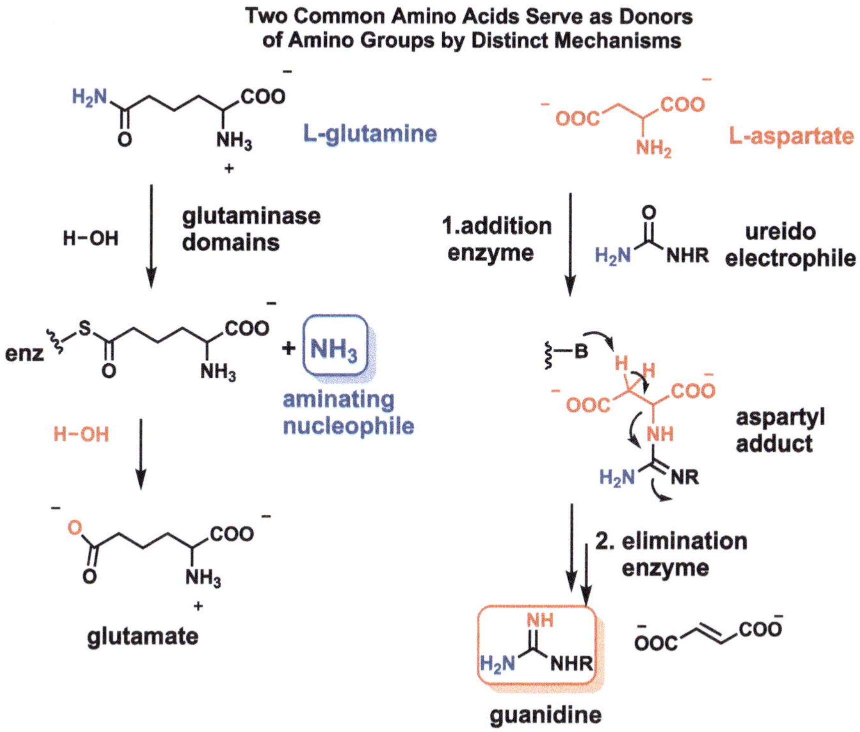

Figure 5.22 Glutamine and aspartate serve as amino donors by distinct chemical strategies: amide hydrolysis controlled in space ad time *vs.* addition/elimination of the aspartate C_2 amine as nucleophile.

5.8 Other Organonitrogen Functional Groups

There are several additional types of organonitrogen functional groups that are found in more specialized metabolites. These include *N*-hydroxy groups, often as part of hydroxamic acids (*N*-hydroxyamides), oximes, arising from tandem hydroxylation of amines, nitriles (triply bonded C–N groups, and even isonitriles (Figure 5.3). Interested readers can find entry into the literature in the accompanying volume The chemical biology of nitrogen.[1]

5.9 Overview of C–N Bonds in Chemical Biology

The crossover of inorganic NH_3 to become organic nitrogen compounds was clearly essential for life. In turn, so was the enzymatic six electron reduction of ubiquitous but unusable N_2 to NH_3. The most consequential port of entry into organic frameworks was fixation of NH_3 into the scaffold of glutamate as an amine group and into glutamine as an amide group. The chemical troika of amine, imine, and amide groups dominate much of carbon–nitrogen chemical biology. Ammonia is a sufficiently strong and promiscuous nucleophile that its intracellular concentrations are kept very low, and excretion of ammonia equivalents by humans is in the form of highly soluble, nonnucleophilic urea. To avoid high circulating levels of intracellular NH_3, cells have generalized the use of glutamine as a latent carrier of NH_3 in the C_5 nonnucleophilic carboxamide group, hydrolyzable back to glutamate and nascent ammonia in the active site of enzymes that aminate cosubstrates.

Among the trio of amines, imine and amide groups, amines are nucleophilic, while imines, especially as iminium ions, are excellent electron sinks (electrophiles) at the carbon atom. On adjacent C–H bond cleavage the resultant enamines are stabilized carbanion equivalents. The amide bonds are chemically stable, the nitrogens delocalized by isoamide resonance contributors, while planar, are nonnucleophilic. Amides form the stable backbone linkages of all proteins in the specific form of peptide bonds. The 20 proteinogenic amino acids are the building blocks for all those peptide bonds by the coupling of aminoacyl-tRNAs, oxoester linkages, in the ribosome's peptidyl transferase center. Meanwhile, nitrogen heterocyclic building blocks for RNA and DNA are the separate subject of the next chapter because of their chemical and metabolic centrality.

A second set of C–N functional groups function both in control of toxic ammonia levels and are central in nucleic acid information storage and transfer. They comprise a set of four functional groups – amides, amidines, ureas, and guanidines. Only the amides are a crossover from the set of three C–N functional groups above. Both biosynthetically and functionally, it is useful to consider amidines as the imine forms of amides and likewise to consider guanidines as the imine forms of urea (ureido) groups.

5.9.1 Summary of Section II

To this point, in Chapters 2–5, we have examined the four major types of carbon-based functional groups in the primary organic molecules of life: C=C, C–O, C–S, and C–N functional groups. These include mixed functional groups such as the *O,S*-containing acyl thioesters and the *O,N*-containing acyl amides. The chemistry open to this organic functional group ensemble, accelerated in rates by enzyme catalysts, comprises all the major reactions in primary metabolic pathways. Each of Chapters 2–5 contains numerous specific examples of organic metabolite transformations that emphasize how the rules of organic chemistry play out in actual metabolic transformations.

In the next section, Section III, we turn to consideration of how many of these functional groups serve as substituents in carbacyclic and heterocyclic ring frameworks. The carbacyclic ring systems include both saturated, unsaturated, and aromatic rings. The oxygen- and nitrogen-containing heterocycles impart architectural and reactivity options and constraints to the functional group ensembles covered to date. We will see that many of the nitrogen heterocycles, in particular, bring novel and essential accessible chemistry to the organic chemistry of life.

References

1. C. T. Walsh, *The Chemical Biology of Nitrogen*, Royal Society of Chemistry, London, 2021.
2. R. A. Kahn, T. Fahrendorf, B. Halkier and B. L. Moller, Substrate specificity of the cytochrome P450 enzymes CYP79A1 and CYP79E1 invoved in the biosynthesis of the cyanogenic glycoside dhurrin in Sorgum bicolor (L.) monarch, *Arch. Biochem. Biophys.*, 1999, **363**, 9–18.
3. https://chem.libretexts.org. Basicity of amino acids, 2022.
4. D. Fowler, M. Coyle, U. Skiba, M. A. Sutton, J. N. Cape, R. Reis, L. J. Shepard, A. Jenkins, B. Grizzetti, J. N. Galloway, P. Vitousek, A. Leach, A. F. Bouwman, K. Butterbach-Bahl, F. Dentener, D. Stevenson, V. Amann and M. Voss, The global nitogen cycle in the twenty-first century, *Philos. Trans. R. Soc., B*, 2013, **368**, 20130–20164.
5. C. Frieden, L-Glutamate dehydrogenase, in *The Enzymes VII*, ed. P.D. Boyer, H. Lardy and K. Myrback, Academic Press, New York, 1963, pp. 3–24.
6. A. Hallen, J. F. Jamie and A. J. L. Cooper, Imine reductases: a compariuson of glutamate dehydrogenase and ketimine reductases in the brain, *Neurochem. Res.*, 2014, **39**, 527–542.
7. E. E. Snell, From bacterial nutrition and o enzyme structure in pyridoxal phosphate enzymes, *Annu. Rev. Biochem.*, 1993, **62**, 1–27.
8. C. T. Walsh and Y. Tang, *The Chemical Biology of Human Vitamins*, Royal Society of Chemistry, London, 2018.
9. P. K. Mehta and P. Christen, The molecular evolution of pyridoxal-P dependent enzymes, in *Advances in Enzymology and Related Areas of Molecular Biology*, 1999, vol. 74, pp. 129–184.
10. D. Eisenberg, R. J. Almassy, C. A. Janson, M. S. Chaopman, S. W. Suh, D. Cascio and W. W. Smith, Some evolutionary relationships of the primary biological catalysts glutamine synthetase and RuBisCO, *Cold Spring Harbor Symp. Quant. Biol.*, 1987, **52**, 483–490.
11. C. T. Walsh, B. P. Tu and Y. Tang, Eight kinetically stable but thermodynamically activated molecules that power cel metabolism, *Chem. Rev.*, 2018, **118**, 1460–1494.
12. C. L. Lomelino, J. T. Andring, R. McKenna and M. S. Kilberg, Aspagine synthetase: function, structure and role in disease, *J. Biol. Chem.*, 2017, **292**, 19552–19558.
13. M. V. Fawaz, M. E. Topper and S. M. Firestine, The ATP grasp enzymes, *Bioorg. Chem.*, 2011, **39**, 185–191.
14. F. M. Raushel, J. Thoden and H. Holden, The amidotransferase family of enzymes; molecular machines for the production and deliveryb of ammonia, *Biochemistry*, 1999, **38**, 7891–7899.
15. F. M. Raushel, J. Thoden and H. Holden, Enzymes with molecular tunels, *Acc. Chem. Res.*, 2003, **36**, 539–548.
16. S. MIlewski, Glucosamine-6-phosphate synthase: the multi-facets enzyme, *Biochim. Biophys. Acta*, 2002, **1597**, 173–192.

17. H. M. Holden, J. Thoden and F. M. Raushel, Carbamoyl phosphate synthetase: an amazing biochemical odyssey from substrate to product, *Cell. Mol. Life Sci.*, 1999, **56**, 507–522.
18. M. Griffin, R. Casadio and C. M. Bergamini, Transaminases: nature's biological glues, *Biochem. J.*, 2002, **368**, 377–396.
19. W. Lee, M. A. McDonough, L. P. Kotra and S. Mobashery, A 1.2 angstrom snapshot of the final step of bacterial cell wall biosynthesis, *Proc. Natl. Acad. Sci. U. S. A.*, 2001, **98**, 1427–1431.
20. J. Liang, Q. Han, Y. Tan, H. Ding and J. Li, Current advances on structure-function relationships of pyridoxal-phosphate dependent enzymes, *Front. Mol. Biosci.*, 2019, **6**, 4.
21. C. T Walsh, Biologically generated carbon dioxide: nature's versatile chemical strategies for carboxy lyases, *Nat. Prod. Rep.*, 2020, **37**, 100–135.
22. H. C. Dunathan, Conformation and reaction specificity of pyridoxal phosphate enzyme, *Proc. Natl. Acad. Sci. U. S. A.*, 1966, **55**, 712–716.
23. D. Atkinson, *Functional roles of urea in vertebrates*, U of Chicago Press, Chicago, 1991.
24. C. J. Lusty, R. L. Jilka and E. H. Nietsch, Ornitihine transccarbamylase from rat liver. Kinetic, physical, and chemical propertiers, *J. Biol. Chem.*, 1979, **254**, 10030–10036.
25. C. Ghose and F. M. Raushel, Determination of the mechanism of arginionsuccinate synthetase by dynamic and static quench experiments, *Biochemistry*, 1985, **24**, 5894–5898.
26. S. M. Morris, Enzymes of arginine metabolism, *J. Nutr.*, 2004, **134**, 2743S–2747S.
27. D. E. Ash, Structure and function of arginases, *J. Nutr.*, 2004, **134**, 2760S–2764S.

Section III: The Central Metabolic Roles of Carbacycles and Heterocycles

Chapters 6–9 address common types of carbacycles and heterocycles and their central roles in chemical transformations in carbon chemical biology. Heterocycles typically do not receive mainstream attention in the early phases of syllabi in abiotic organic chemistry courses. However, the chapters in this section illustrate the crucial roles carbacyclic and heterocyclic molecular frameworks play in primary metabolic chemistry.

Chapter 6 is in a sense a chapter bridging the C–N bond focus of Chapter 5 and the nitrogen heterocycles in Chapters 8 and 9. The monocyclic pyrimidine ring and bicyclic purine (imidazolo-pyrimidine) ring systems are so central to life that they are broken out as a separate chapter. The four C–N functional groups noted in Chapter 5, ureas and guanidines, amides, and amidines, turn out to be the crucial hydrogen bond donors and acceptors that determine the stability of G–C and A–T base pairs. These are the famous Watson–Crick base pairs of double helical DNA and the main determinants of base-pairing rules. They lie at the heart of the logic of double helical DNA and the many kinds of RNAs that are information-carrying biopolymers. It is likely that both pyrimidine and purines are among the most ancient metabolites. Indeed, the chemical logic for their *de novo* ring constructions shows assembly from extremely simple (ancient?) metabolic building blocks and core reactions of amines, amides, dehydrative cyclizations, mixed acyl phosphoric anhydride formation, and capture of the ribose rings by amine nucleophiles by pyrimidine N_3 or the amine that becomes purine N_9.

Chapter 7 summarizes some of the biosynthetic routes to common carbacyclic ring structures, molecules that may bear several kinds of functional groups but do not have an oxygen, nitrogen, or sulfur as one of the constituent atoms that make up the carbacyclic ring. Four types of enzyme-catalyzed cyclizations of linear substrates to carbacyclic products occur. Three such cyclizations lead to aromatizing polyketide pathways, the shikimate pathway from glucose to phenylalanine, and some radical paths in phenylpropanoid metabolic pathways in plants.

The fourth pathway includes classic volatile fragrance molecules that are cyclic C_{10} monoterpenes generated from the acyclic geranyl-PP precursor by internal capture of allylic carbocations at C_1 by the internal double bond at C_7. Analogously, there are many C_{15} and C_{20} cyclic isoprenoids from cyclases acting on the corresponding linear sesquiterpene and diterpene isoprenyl-PP substrates.

A large variety of aromatic phenylpropanoid metabolites bearing a single phenyl ring derived from phenylalanine involve either carbon radical chemistry or thioester carbonyl intermediates for chain elongations in plant chalcone and flavonoid metabolites. The route from glucose to the phenyl ring in phenylalanine is an edifying *in vivo* lesson in both aldol cyclizations and successive dehydrations to achieve aromaticity.

Chapters 8 and 9 take two additional looks at the scope and utility of heterocyclic scaffolds in carbon chemical biology. Chapter 8 starts with consideration of routes to, and reactivity of, cyclic ethers, starting with electrophilic, strained epoxides. The cyclic hemiacetal forms of common hexoses (glucose) and pentoses (ribose) figure prominently in cellular metabolic routes. Large ring lactones and cyclic ethers are products from some types of polyketide assembly lines.

Common nitrogen heterocycles, beyond the nucleic acid building block purines and pyrimidines of Chapter 6, include pyrrolidine rings and the fully heteroaromatic pyrrole ring. The chemical versatility of 2-aminomethylpyrroles deserves its own chapter, Chapter 16, to enable biogenesis of tetrapyrrole macrocyclic pigments of life. The bicyclic indole ring system has a benzene ring fused to the pyrrole and engages in enamine-based carbon nucleophilic chemistry at the β-carbon of the pyrrole ring. The two nitrogens, in 1,3-relationship, in the imidazole heterocycle generate a pK_a of 6 for the first nitrogen, making it the strongest nitrogen base in metabolism.

Chapter 9 concludes this section with specific focus on the heterocyclic ring systems in 10 of the 13 human vitamins. Seven of those vitamins use their natural nitrogen heterocyclic ring systems as coenzymes for some central reactions in primary metabolism. For example, the coenzyme forms of vitamin B1 (thiamin-PP), B3 nicotinamide adenine dinucleotides-NAD(P)$^+$, and B6 (pyridoxal-P) present a thiazolium ring, pyridinium ring, and pyridine aldehyde, respectively. These nitrogen heterocycles enable generation of substrate-coenzyme β-iminium ion electron sinks (B1), the cationic pyridinium ring to function as hydride ion acceptor (B3), and heteroaryl aldehyde as electron sink for amino acid carbanion metabolism (B6). Likewise, the tetrahydrofolate form of vitamin B9 and the bicyclic biotin ring of biotin (vitamin B7) utilize heterobicyclic rings as reactive loci. Finally, the tricyclic isoalloxazine ring, containing fused benzene-pyrazine-pyrimidine rings, of riboflavin-based coenzymes, can access both two electron and one electron redox chemistry, including service as a $2e^-/1e^-$ redox bridge between NADH and O_2.

An appreciation of the reactivity of carbacyclic and heterocyclic substrates, products, enzyme side chains, and coenzymes is essential to deciphering the chemical logic of the chemistry of organic molecules in biologic systems.

CHAPTER 6

Purines and Pyrimidines: Essential Nitrogen Heterocycles

The four chapters in this section address the importance and variety of carbacyclic and heterocyclic ring systems in carbon chemical biology. Carbon ring systems with one or more oxygen, nitrogen, or sulfur atoms in a ring are defined as heterocycles. The rings impart steric and electronic constraints and opportunities to a large swath of organic metabolites. The underlying logic of metabolic transformations is incomplete without explicit attention to the types of chemistry available to carbacycles and heterocycles *in vivo*.

The case for central chemical roles of heterocycles is perhaps most clear in the pyrimidine and purine building blocks of RNA and DNA. They are the subject matter of this chapter, continuing a focus on C–N bonds that started in Chapter 5.

Chapter 7, in addressing biologic carbacycles, is in part a detour away from the subsequent C–N heterocycle focus of Chapters 8 and 9, but introduces the key concept of cyclic structures for organic metabolites beyond the isolated functional groups covered in the previous section. Most notable is the main strategy by which organisms ultimately convert glucose into the aromatic amino acid phenylalanine.

Chapter 8 returns to nitrogen heterocyclic ring systems in both low molecular weight substrates and in the imidazole and indole side chains of proteins. Indole chemistry in the framework of tryptophan and tryptamine is a staple of plant secondary metabolism in the vast family of alkaloid metabolites. Chapter 9 examines the distinct and distinctive nitrogen heterocycles that are the central functional groups in the action of nine of the 13 human vitamins. Seven of those vitamins are nitrogen heterocyclic coenzymes that lower energy barriers for hundreds to thousands of reactions of organic substrates in cell metabolism.

Chemical Biology No. 21
The Chemical Biology of Carbon
By Christopher T. Walsh

Published by the Royal Society of Chemistry, www.rsc.org

6.1 The Four C–N Bond Types in Pyrimidine and Purine Heterocycles Dictate Hydrogen Bonding and Base Pairing Strategies

There are many ways to parse the chemical logic and enzymatic machinery utilized at different stages of RNA and DNA biosynthesis and the underlying roles of each of the four nucleotide purine[1] and pyrimidine[2] heterocycles. One is to consider the assembly routes to the two RNA pyrimidine heterocycles uridine triphosphate and cytidine triphosphate and how the 5-methyl-2′-deoxy-uridine diphosphate (thymidine) is then built for substitution of uracil building blocks in DNA assembly. Analogously, the bicyclic purines, adenosine- and guanosine 5′phosphates, have a more complex route to build the 5,6-fused bis-heterocyclic structure. Both prefigure the utility of other C–N and C–O containing heterocycles in chemical biology elaborated in Chapter 7.

Figure 6.1 reminds us that the canonical adenine–thymine (A–T) and guanine–cytosine (G–C) base pairs specify information in DNA double helices.[3] The ensemble of heterocyclic base pairs employ the tetrad of C–N

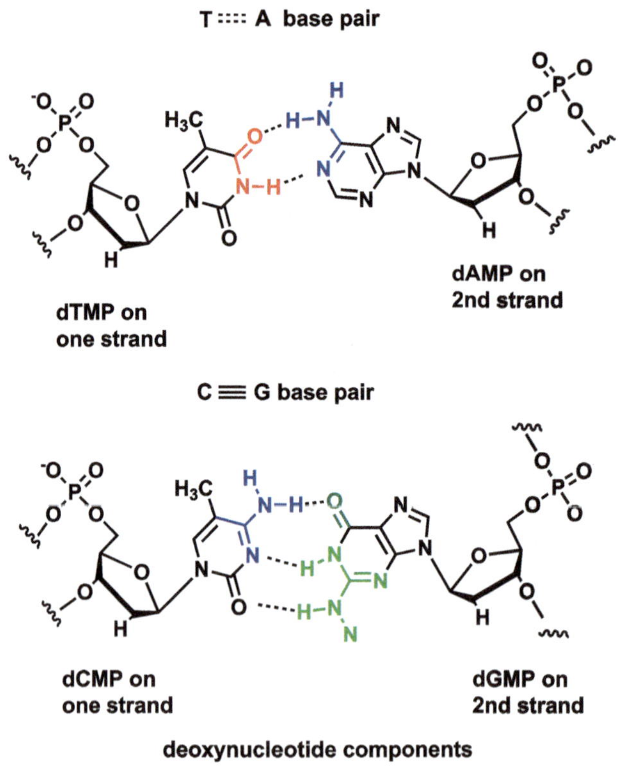

Figure 6.1 Structure of A=T and G–C base pairs mediated by internucleotide hydrogen bonds.

functional groups enumerated in Chapter 5 for instructional hydrogen bonding: amides, amidines, ureas, and guanidines.[4]

In the preceding Chapters 2–5 on types of carbon-based functional groups, we have noted the participation of several heterocyclic cofactors, focusing on how they serve their constituent enzymes, more than on the chemistry available because they are heterocycles. We devote this chapter entirely to the pyrimidine and purine heterocycles, how they are assembled and how they function for information in the two strands of DNA double helices. In DNA biopolymers the complementary hydrogen bonding capacity of the adenine–thymine (A–T) and guanine–cytosine (G–C) pairs in constrained ring scaffolds are the basis of base-pairing in the antiparallel double helical DNA strands (Figure 6.1).

A starting point is to note the distinct kinds of carbon–nitrogen functional groups in the uracil/thymine and cytosine pyrimidine monocyclic frameworks and in the comparable bicyclic scaffolds of the purines adenosine and guanosine (Figure 6.1). By simple counting, there are two nitrogens in uracil and three (one exocyclic) in cytosine. The bicyclic adenosine has four endocyclic nitrogens and one exocyclic for a total of five. Likewise, guanine has four endocyclic and one exocyclic nitrogen but with distinct regiochemistry that leads to the unique guanidine functional group.

The simplest of the four building blocks in terms of functional group is the monocyclic uracil. With an array of alternating NH and CO groups at positions 1–4, there are elements of a ureido group (residues 1–3) and an amide (3,4) mixed in (Figure 6.2). Thymine is 5-methyluracil with a new C–CH_3 bond, while positions 1–4, containing the embedded nitrogens, are unaltered. Cytosine is uracil with an exocyclic amino group. That added amino group creates an amidine functional group part structure[5] (Figure 6.1). Amidines can be thought of as imine derivatives of amides and can populate either of two tautomeric forms, although the major form of cytosine is with the exocyclic NH_2 rather than C=NH (Figure 6.2).

In the bicyclic purine ring building blocks one can parse a 3,5 bis-amide part structure in the inosine ring of inosine monophosphate (IMP), the immediate biosynthetic precursor of AMP and GMP (see below). That amide part structure persists in guanine. However, guanine has gained an exocyclic amine at C_2 that has now become a guanidine part structure[6] (Figure 6.3). The exocyclic amine at C_6 of the adenine scaffold is now part of an amidine grouping. Thus, inosine's enzymatic conversion on to AMP or GMP has created two new C–N functional groups. Just as one can consider an amidine the imine of an amide, one can consider the guanidine group the imine of a urea (Figure 6.2). In sum, these pairs of pyrimidine and purine heterocycles used for nucleic acid construction have among them a wide array of related but distinct C–N functional groups as part structures.

While the electronic demands of the full heterocyclic ring structures will condition the behavior of those C–N functional group part structures it is instructive to look at the canonical C–G and A–T base pairs that are the key informational constraints that stabilize double helical forms of DNA

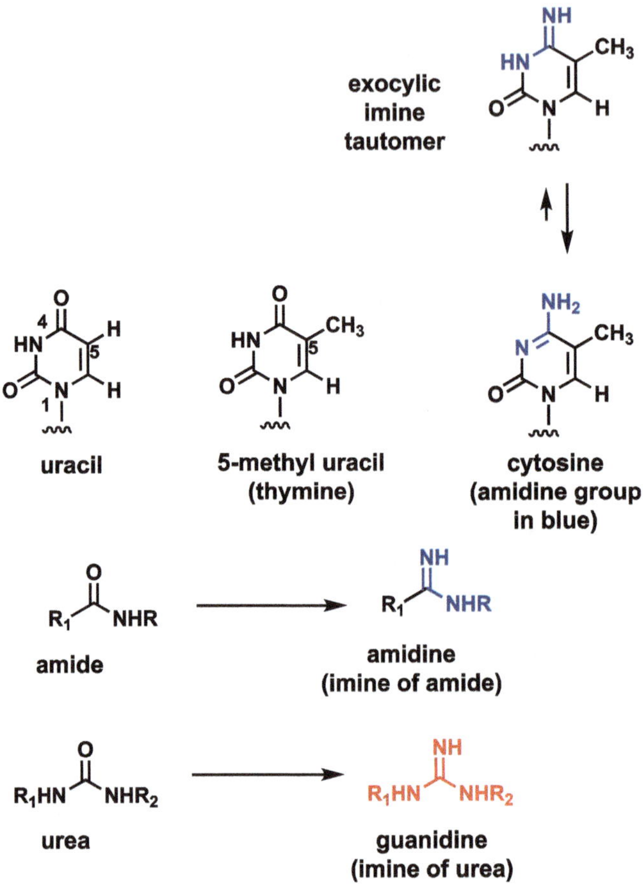

Figure 6.2 Amide, amidine, urea, and guanidine functional groups in nucleotides.

(and RNA) (Figure 6.1). The C–N part structures dictate the selectivity and the polarity of the three hydrogen bonds in C–G/G–C base pairs and the two H bonds in the weaker A–T/T–A base pairs.

As a reminder, the four C–N functional groups are distinct. The amidine part structures in cytosine and adenine are drawn in blue in Figure 6.3, the guanidine part structure in iridescent green, the ureido part structure of uracil, thymine, and cytosine in red, the amide moiety of guanine in forest green.

A–T base pairs share two stabilizing hydrogen bonds. The amidine moiety (blue) of adenine is involved in both, once as a hydrogen donor (exocyclic NH_2), once as a hydrogen acceptor (endocyclic N_3). Correspondingly, the thymine partner uses its amide part structure as H acceptor and H donor respectively. This is amidine-amide H bonding as information transfer in chemical biology.

G–C base pairs have three hydrogen bonds that stabilize and select for this pairing in nucleic acid double strands. The cytosine partner is the H donor at

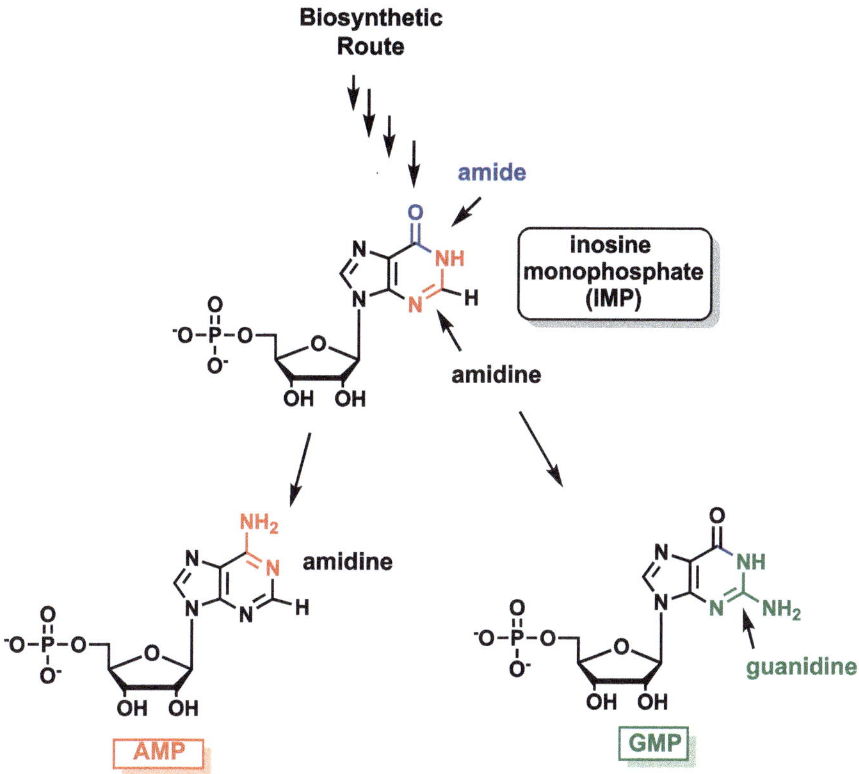

Figure 6.3 Purine biosynthetic route leads to inosine monophosphate (IMP). The purine ring is enzymatically modified to an amidine (AMP) or a guanidine (GMP) for the two building blocks of RNA and DNA.

the exocyclic NH_2 of the amidine part structure and is the H-bond acceptor at the endocyclic N_3 of the amidine moiety. This is the same usage of the bifunctional amidine as seen in A–T base pairs. The third H bond on the cytosine side is to the C_2 carbonyl oxygen of partner G as acceptor. Thus, the cytosine partner in a C–G base pair donates one H bond and accepts two. On the guanine side the unique guanidine part structure is the donor of those two hydrogen bonds. The exocyclic amide carbonyl is the acceptor of the other H bond.

These simple four C–N functional groups, amides, ureas, amidines, and guanidines[4] (Figures 6.1 and 6.3) provide the essence of information transfer instructions as DNA is replicated from a template strand and as RNA is transcribed from DNA into the various functional subsets of RNAs (ribosomal rRNAS, transfer tRNAs, messenger mRNAs, noncoding ncRNAs). We commented back in Chapter 5 on glutamine synthetase as a paradigm for the chemical logic of amide bond biosynthesis. We turn now to examine how the ureido, amidino, and guanidino functional groups are constructed in the context of the pyrimidine and purine heterocyclic cores.

It is likely that availability of these heterocycles was a precondition for life; their orderly supply is a continuing need for all contemporary organisms. In turn, the biosynthetic pathways rely on simple chemical reactivities that were likely available to organisms early in evolution. The chemistry of assembly is so straightforward as likely to have been available primordially.

6.2 Pyrimidine Ring Construction Biosynthetic Logic

The simplest nucleoside, the monocyclic nucleotide uridine monophosphate, is assembled by a short efficient pathway from two metabolically central substrates. One is the amino acid L-aspartate. The second is the mixed acyl phosphoric anhydride carbamyl phosphate, noted back in Chapter 5. There, we discussed the capture of the carbonyl of carbamyl-P by the deprotonated amine group at C_5 of ornithine in the active site of ornithine transcarbamyalse. Here, we note the parallel action of aspartate transcarbamylase[7] using instead the deprotonated, nucleophilic amine of L-aspartate as the initial step in the biosynthetic pathway.

The enzyme aspartate transcarbamylase catalyzes the carbamyl transfer to the amino nitrogen to generate the *N*-carbamyl-L-aspartate product with a newly formed ureido group (N_1–C_2–N_3) (Figure 6.4). Indeed, enzymatic

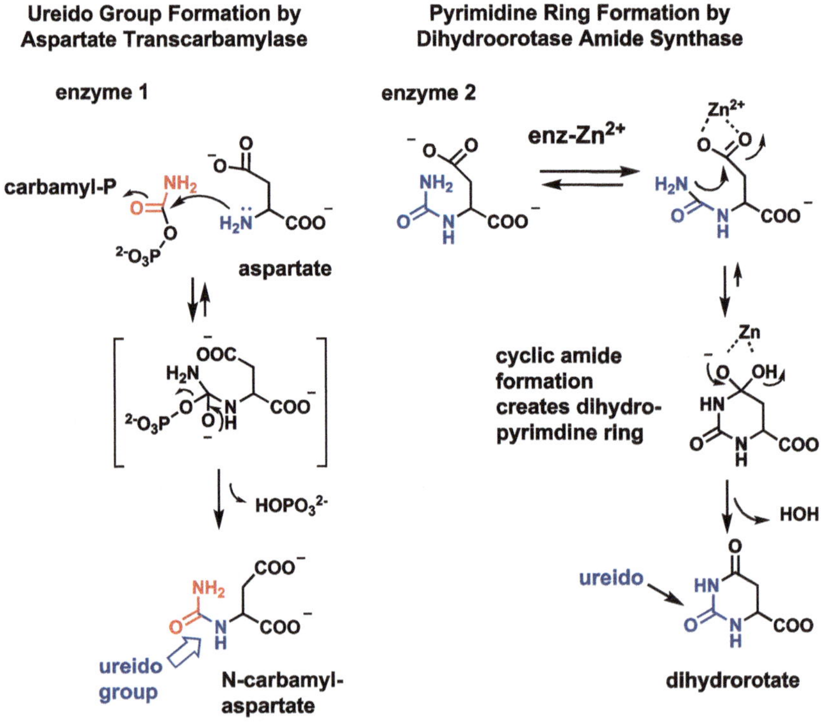

Figure 6.4 The first two enzymes in the pyrimidine biosynthetic pathway. Ureido group formation by aspartate transcarbamylase; pyrimidine ring formation by dihydroorotase.

carbamylation of amine nucleophiles is the general route to urea/ureido function groups in chemical biology.

The *N*-carbamyl-L-aspartate is only one enzymatic step away from the pyrimidine core. The zinc-containing enzyme dihydroorotase catalyzes the ring closing dehydration between the ureido N_2 nitrogen and the C_4 carboxylate of the aspartate moiety, creating the heterocyclic scaffold of dihydroorotate[8] (Figure 6.4). This is dehydrative amide bond formation between a weak nitrogen nucleophile and a poorly activated carboxylate group (maybe it is protonated in the enzyme active site).

At this point the fundamental connectivity of pyrimidines at N_1–C_4 has been established. What remains is desaturation of the C_5 and C_6 adjacent methylene groups (alkanes) to the corresponding alkene. By olefin formation precedent described back in Chapter 2 (*e.g.* fatty acid desaturase, succinate dehydrogenase), one anticipates a proton hydride mechanism for desaturation with hydride transfer to enzyme-bound flavin coenzyme. That is the strategy of the flavoenzyme dihydroorotate dehydrogenase[9] to yield orotate (Figure 6.5). Reoxidation of the reduced flavin (FlH_2) is by one-electron steps to iron-containing acceptor proteins.

Only two additional steps remain to reach UMP, formation of an N_1–carbon ribosyl bond to ribosyl-5-phosphate and decarboxylation of the original aspartate carboxylate moiety (now lodged as a C_6 substituent on orotidine monophosphate). The phosphoribsylation step is accelerated by an enzyme[10] that uses N_1 of the orotate as nucleophile to capture the incipient C_1–O_5 oxocarbenium ion from cosubstrate 5-P-ribose-1-PP as the C_1–OPP bond cleaves early along the reaction coordinate (Figure 6.5). OMP (orotidine monophosphate) is the product. In humans the dihydroorotate phosphoribosyltransferase and the tandemly acting orotate decarboxylase are two domains of a single protein UMP synthase.[10] Its subsequent enzymatic processing is by decarboxylation. Despite its metabolic centrality and decades of mechanistic analysis, the mechanism for lowering the energy for decarboxylation at the sp^2 hybridized carbon 6 of OMP is not well understood.[11] UMP is the product from action of OMP decarboxylase (Figure 6.5). UMP undergoes subsequent sequential enzymatic phosphorylation from 2 ATPs, up to the UDP and then UTP levels before CTP synthetase installs the amidino group at C_4 of the uracil ring of UTP (Figure 6.5).

CTP synthetase strategy solves two chemical issues. One is to avoid toxic levels of free NH_3 by action of a glutaminase domain, utilizing glutamine as a latent carrier of nascent NH_3 on a "just in time" inventory control manner (see Chapter 5). The second issue is to create a form of C_4 on the uracil ring that can be captured by nascent NH_3. This challenge is solved by spending an ATP molecule to derivatize the C_4 carbonyl of UTP as an isoamide-*O*-AMP, still electrophilic at that imino carbon, set up for capture by the nascent ammonia in the amination active site (Figure 6.6). The isoamide, nucleophilic at oxygen, is much higher in energy than the amide contributor and unlikely to be available at useful concentrations in free solution. That low proportion may be elevated in the active site of CTP synthetase.[12]

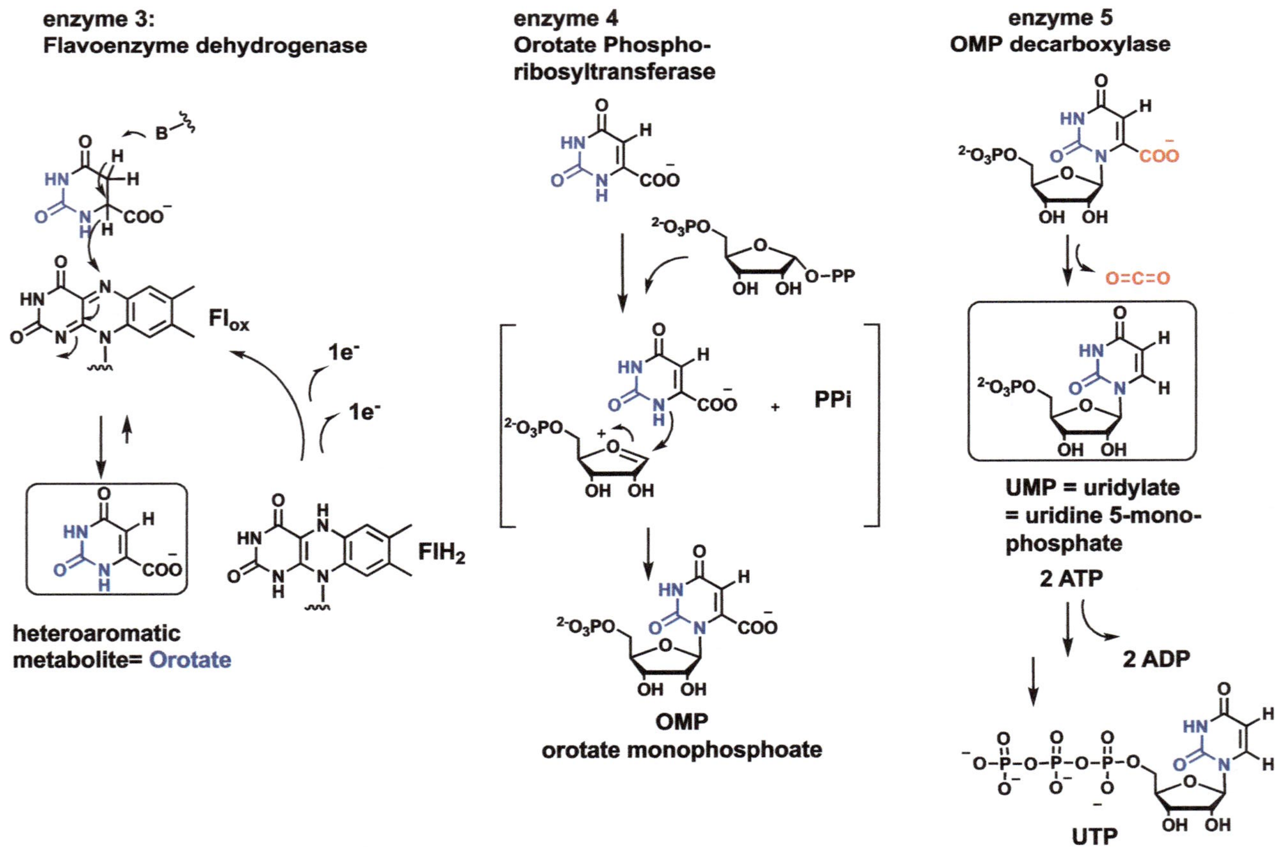

Figure 6.5 Enzymes 3–5 in pyrimidine biosynthetic pathways: dehydrogenation of dihdroorotate to orotate; phosphoribosylation of orotate by attack of N_1 on the 5-P-ribosyl C_1 carbenium ion from PRPP; decarboxylation of OMP to UMP.

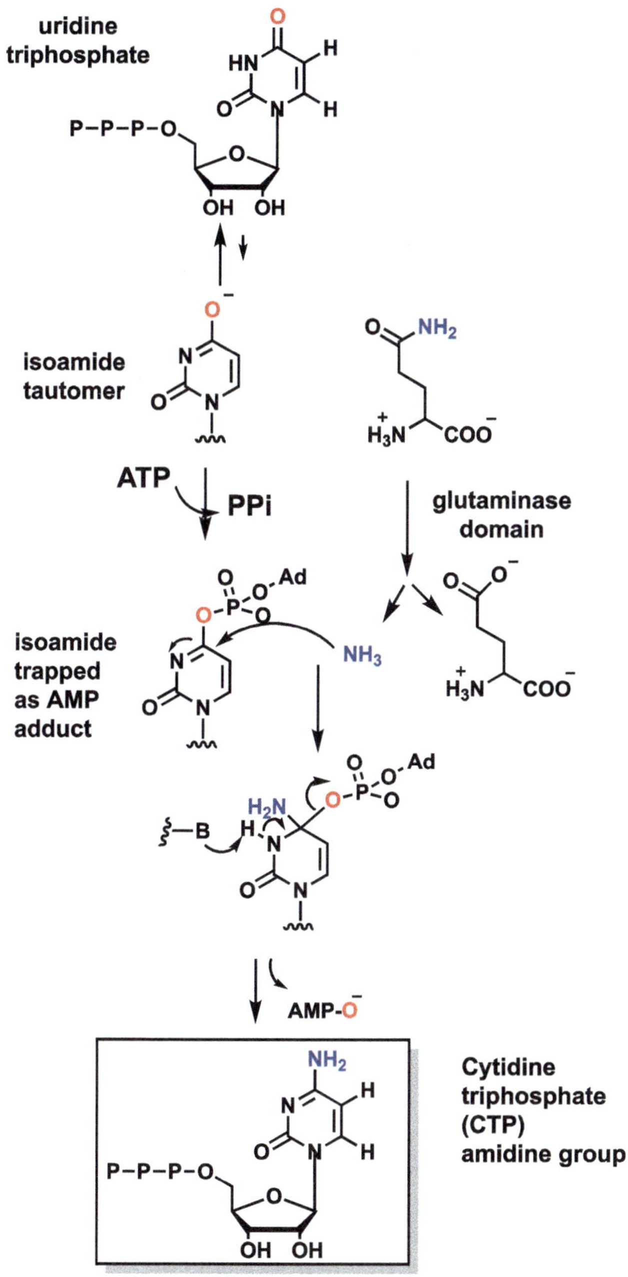

Figure 6.6 CTP synthetase uses ATP to generate a phosphorylated isoamide form of UTP to be captured by nascent NH_3.

The result of this two pronged strategy is to convert UTP to the second pyrimidine triphosphate, CTP with that newly formed amidino group. The almost identical chemical strategy will be used in amination of IMP at C_6 to create AMP and at C_2 to create GMP as noted in a section hence.

6.3 *De Novo* Biosynthesis of the Heterocyclic Purine Ring System: Imidazolo-pyrimidine Bicyclic Scaffold

Construction of the bicyclic purine heterocyclic ring system[13] (= imidazole-pyrimidine) uses a distinct strategy for assembly of the 5 and 6 member fused rings. Yet, simple, likely ancient, chemistry is in play by the sequential action of 10 enzymes. Inspection of the source of the five carbons and four nitrogens that form a purine ring biosynthetically are consistent with an ancient, even prebiotic route, using the simplest of metabolic building blocks. The first fused 5,6-imidazolopyrmidine (purine) ring metabolite that emerges from the purine biosynthetic pathway is inosine monophosphate (IMP) (Figure 6.3).

Of the five carbons in IMP, three are installed as one carbon units, two at the level of formaldehyde and the third as free carbon dioxide. The other two carbons are provided by two carbons of glycine, itself the simplest amino acid. Of the four nitrogens in IMP, one comes in with the glycine framework, and two from nascent ammonia released *in situ* by glutaminase action. The fourth nitrogen atom is extracted from aspartate in a two enzyme addition/elimination sequence that ejects the four carbons as fumarate while inserting the amine of aspartate.

This is the same strategy used by the enzyme pair (argininosuccinate synthetase/arginase) in the urea cycle to install only the aspartate α-amino group in arginine. (see Chapter 5, Figure 5.22). Nature uses a few reaction strategies over and over again.

Ten enzymes are committed to this pathway. The first five complete the formation of the imidazole heterocyclic ring in aminoimidazole ribotide (AIR) (Figure 6.7). The next five enzymes build the pyrimidine ring fused to the imidazole, creating the bicyclic purine ring system of IMP (Figure 6.8). Five of the enzymes split ATP to ADP and Pi during creation of five transient electrophilic groups in the growing purine ribotide scaffold. Three of those five ATP cleavages generate acyl phosphate intermediates[14] to give acyl groups that get captured by an amine. The other two enzymes create phosphorylated isoamides as electrophiles[4] (analogous to the CTP synthetase logic) for capture by nitrogen nucleophiles. These are recurring strategies for building C–O and C–N bonds in biologic systems, as elaborated in Chapters 3 and 5.

The unusual ATP-dependent carboxylation of AIR generates carboxyphosphate as a one carbon electrophile while using C_2 of the aminoimidazole as an enamine carbon nucleophile to construct the C–C bond in CAIR

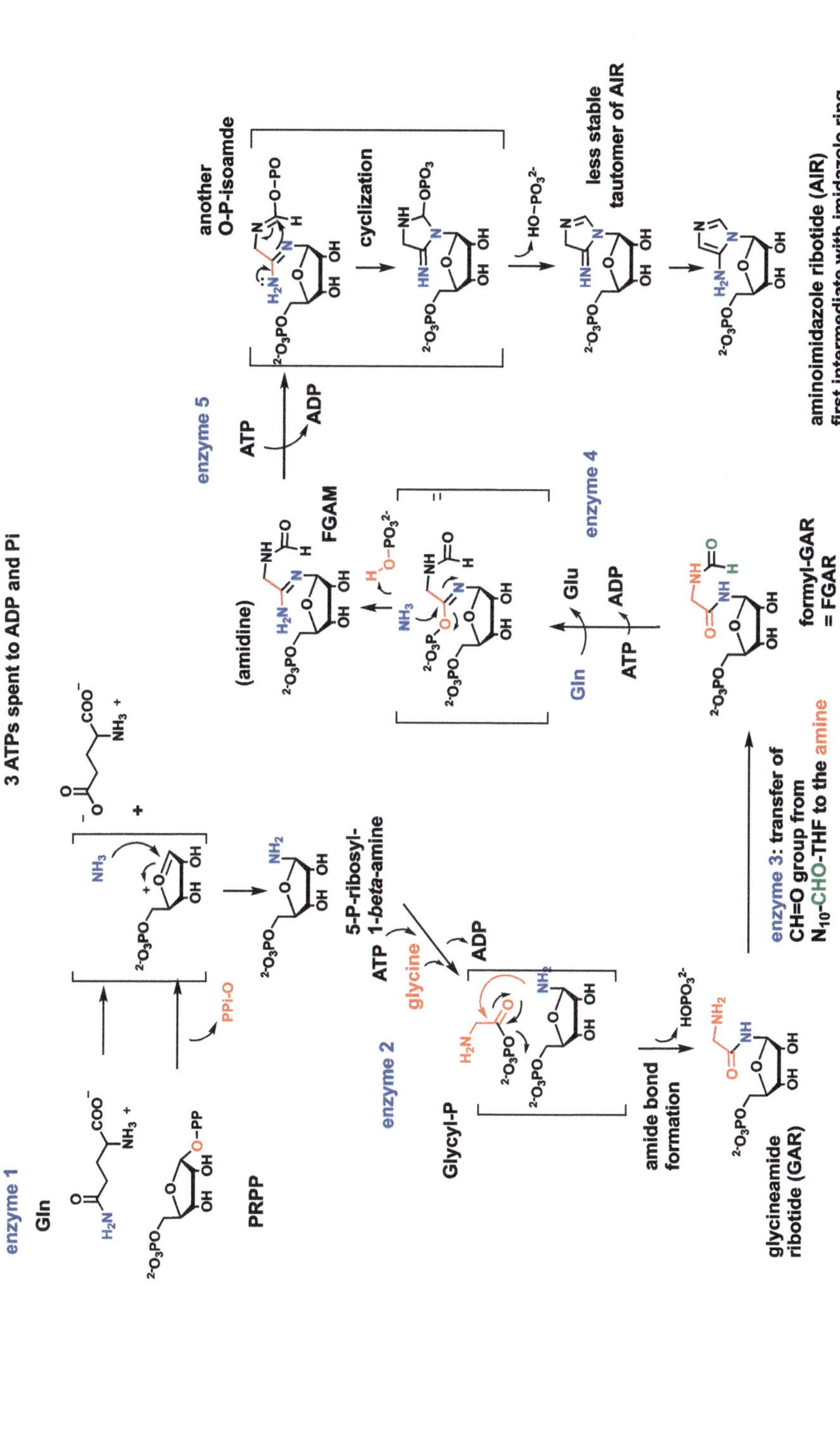

Figure 6.7 Purine ring biosynthesis. The first five enzymes assemble the aminoimidazole ring of aminoimidazole ribotide (AIR).

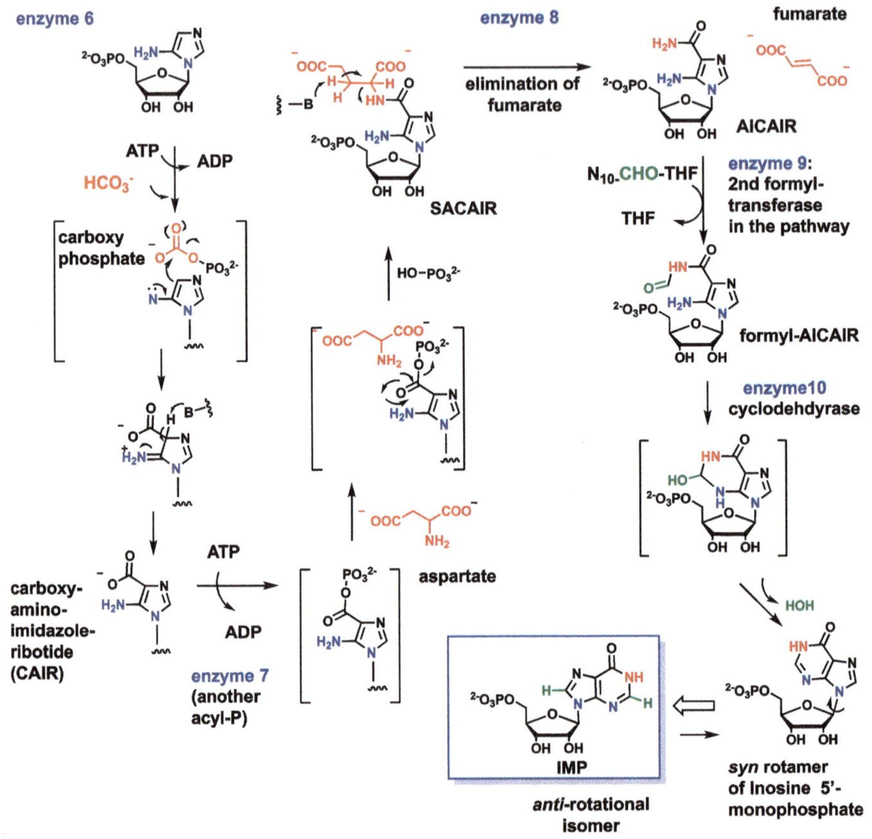

Figure 6.8 The second set of five enzymes elaborate the fused pyrimidine ring onto the preexisting imidazole ring to generate IMP as the first purine metabolite.

(carboxy-aminoimidazole-ribotide-P) (Figure 6.8). Transient amide linkages, four of them, in the growing imidazole and pyrimidine rings undergo enzyme-catalyzed dehydrations to create both imidazole and amidine part structures along the way. The transiently detectable amides are created by nucleophilic amines attacking electrophilic carbonyl groups as the linear, then monocyclic, framework takes shape as an amino-carboxy-imidazole ring.

The operant strategy for creating the purine fused bicyclic ring scaffold is to first build the five membered imidazole heterocycle up from an amine in a series of amide bond formations and subsequent dehydrations to imines to get to a 2-aminoimidazole-1-ribose-5′-P (Figure 6.7). Then, the six membered pyrimidine is elaborated from that exocyclic amine of the imidazole (Figure 6.8)

The pathway starts by releasing nascent ammonia (again from glutaminase action as noted above for CTP synthetase) to attack C_1 of 5-P-ribose-1-PP (in mechanistic analogy to the just noted amine phosphoribosylation of orotate in pyrimidine assembly). This releases inorganic pyrophosphate and yields the β-isomer of 1-aminoribose-5-P (Figure 6.7).

As discussed earlier in Chapters 3 and 5, amide bond formations require the carboxylate component to be activated as an electrophilic functional group, typically as the mixed acyl-phosphoric anhydride. That is the strategy in constructing the aminoimidazole ring. The first step is amide bond formation between the carboxylate of glycine and the 1-amino ribose-5-P as nucleophile. To activate the glycine carboxylate the glycinamide ribotide (GAR) synthase spends an ATP to make the glycyl phosphate that is then the electrophile for capture by the nucleophilic amine (Figure 6.7). The product glycinamide ribotide (GAR) has a free amine group as nucleophile for the next amide bond-forming enzyme. The carbon electrophile is a formaldehyde equivalent presented as a (tetrahydro)-folyl–N_{10}–CH=O adduct (see Chapter 9 for the use of vitamin B9 folate as a one carbon unit carrier at three different oxidation states). The product of this this enzyme is *N*-formyl-GAR.

A third amide bond then arises as nascent ammonia, *via* glutaminase action, and attacks the electrophilic carbonyl of the formyl group to yield the formamido adduct of GAR, producing FGAM. Now, facile dehydration leads to ring closure of a five membered ring with two endocyclic nitrogens and the exocyclic amine. Construction of this aminoimidazole (AIR) has been effected by simple amide bond chemistry from C_1 and C_2 equivalents with participation of three nucleophilic amines, two as nascent NH_3, the third as the amine of the glycyl moiety.

Elaboration of the AIR framework to the fused pyrimidine (Figure 6.8) then starts with ATP-dependent incorporation of bicarbonate/CO_2 to form carboxy-aminoimidazole ribotide (CAIR). Presumably this involves carboxy phosphate as electrophile and the vinylogous enamine of AIR as carbon nucleophile for an unusual C–C bond formation by net carboxylation.

Next is another ATP-dependent amide bond-forming step. The just added carboxylate in CAIR is activated as an electrophilic acyl phosphate for capture by an amine. In this step the amine is the amino group of L-aspartate, yielding succinyl CAIR. This is the third amide bond forming enzyme in the pathway. The next enzyme eliminates the four carbon framework originally in aspartate (the succinyl moiety) as the alkene fumarate. These two enzymes work in tandem in an addition/elimination sequence to install just the amino group of aspartate while throwing away the carbon framework as fumarate (fed into the citrate cycle: no waste). Two carbon-based functional groups have been generated: a new amide bond and a new alkene in fumarate.

At this juncture the AICAR scaffold is just two steps away from IMP (Figures 6.6 and 6.8). Next, another amide forming step, the fourth in the pathway, is essentially a repeat of the earlier transfer of the one carbon formyl group from N_{10} of tetrahydrofolate to the exocylic amine of the aminoimidazole moiety. This enzyme action yields aminoimidazole-carboxamidoribotide (AICAIR) as the penultimate metabolite in the purine ring-forming pathway. Internal amide formation with the carboxamide nitrogen acting as weak nitrogen nucleophile (effective due to intramolecular proximity?) on the just added formyl group carbonyl gives a tetrahedral adduct. On elimination of water that adduct closes the pyrimidine ring as the 5,6-fused bicyclic purine scaffold of IMP.

Although multiple amide bond formations between nucleophilic amines and carboxylates activated as acyl phosphates are the iterated functional group biosynthetic strategy, only one of the amide linkages persists in IMP. The others have been dehydrated along the way to expel the oxygen substituents en route to imidazole and pyrimidine ring formation/heteroaromatization. A review of the ten enzymatic steps reveals that each of the mechanisms has precedent in one or more of the enzymes we have analyzed to date. Each step is a simple chemical step, albeit the growing scaffold complexity can obscure the underlying simplicity of the chemical strategy. Note that both the imidazole ring (first) and then the pyrimidine ring moiety are closed by cyclizing dehydrations of amides, reaching the stable heteroaromatic ring systems.

6.4 Enzymatic Partition of IMP to AMP and GMP: Two Amination Strategies to Create Amidine *vs.* Guanidine Groups

IMP is not itself a building block for RNA or DNA. It is acted on enzymatically to form two distinct exocyclic amine derivatives AMP and GMP that are direct building blocks (Figure 6.3). IMP is the direct precursor to AMP. But IMP to GMP requires passing through XMP (xanthosine monophosphate). In that route, IMP is first oxidized to the C_2-ketone (actually becoming part of a ureido group from N_1 to N_3) in xanthosine monophosphate (XMP) by action of IMP dehydrogenase (Figure 6.9).

IMP dehydrogenase[15] has the redox task of transferring a hydride ion from C_2 of IMP to NAD^+. The catalytic strategy is the nitrogen analog of that used by aldehyde dehydrogenases (Chapter 3, Figure 3.14). Carbon 2 of IMP is an imine part structure with an electrophilic carbon that is capturable by the thiolate side chain of an active site cysteine residue (Figure 6.9). Oxidation now proceeds from the tetrahedral substrate-enzyme adduct (a hemithioaminal, with both C–S and C–N heteroatom bonds) by hydride ejection to cosubstrate NAD^+. This transfer constitutes oxidation of C_2 to a thioimidate adduct (new functional group). Attack of a water molecule on the electrophilic carbon of the thioimidate generates a new tetrahedral adduct, from which elimination of the initially added enzyme cysteine thiolate gives the iminol derivative of XMP. Tautomerization to the more stable ureido part structure gives the accumulating form of XMP. This is the first half reaction: an aldehyde oxidase variant. The second half reaction traps a minor isoamide anion form as a phosphate ester, then adds in nascent NH_3, the same strategy for creating the amidine linkage in CTP in Figure 6.6.

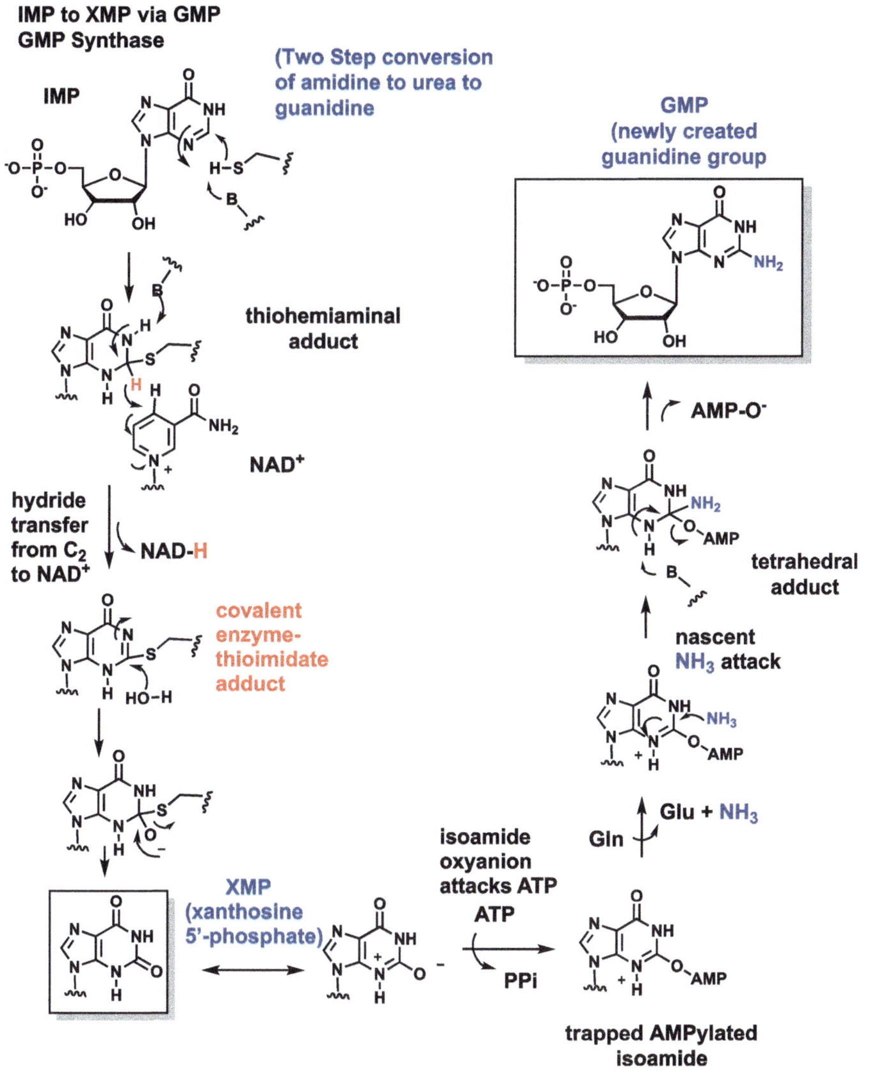

Figure 6.9 Two step conversion of IMP to GMP *via* xanthosine monophosphate by GMP synthase; a covalent thioimidate-enzyme adduct.

Either IMP or XMP are subjected to distinct regiospecific aminations at exocyclic carbonyl groups at C_6 and C_2 by two distinct enzymes, respectively (Figure 6.3). Both the C_2 and C_6 carbonyl oxygens require prior derivatization to be converted to low energy leaving groups. Organisms turn to the by now familiar chemical strategy of capturing the minor isoamide forms by using the now nucleophilic oxygen of those minor isoamide forms to attack a phosphoryl group in the side chain of a nucleotide triphosphate cosubstrate to form a stable derivative of the isoamide. GTP is the activating cosubstrate on the AMP arm and ATP on the GMP arm, perhaps for separate regulation and inventory control. Two different regiochemistries of attack occur on GTP (Pα) *vs.* ATP (Pγ). The *O*-phospho-IMP is formed in the arm leading to AMP, the *O*-AMP adduct in the GMP synthase arm (Figure 6.9). Now those adducts are capturable by nitrogen nucleophiles.

The GMP synthetase uses nascent NH_3 liberated by glutaminase domain activation[16] and is the more straightforward of the amination strategies (Figure 6.9). The immediate product is GMP: a ureido group has been converted to the guanidino group that is the crucial participant in two of the three hydrogen bonds in G–C base pairs in double helical DNA and corresponding double helical stretches of RNA.

However, the conversion of IMP to AMP by the AMP synthase has evolved to use aspartate instead of glutamine as amino donor. The two step process thus requires two enzymes, adenylosuccinate synthetase[17] and adenylosuccinate lyase[18] (Figure 6.10).

This option for the source of the amino group in the amidine linkage of AMP is in strict analogy to the tandem two enzyme action during introduction of the fourth nitrogen on the way to IMP above (Figure 6.9: SAICAIR to AICAIR). The aspartate amino group adds into the trapped *O*-phosphoryl isoamide at C_6 of IMP. The phosphate group departs as adenylosuccinate forms. Then, the lyase acts to eliminate the four carbon fumarate framework to trim away the aspartate-derived carbon framework, leaving only the amino group tethered as the exocyclic amine of the AMP product.

A series of phosphoryl group transfer enzymes take both AMP and GMP up to the diphosphate and then the triphosphate forms (Figure 6.11). The crossover to 2′-dexoynucleotide building blocks for DNA occurs at the diphosphate level by members of the ribonucleotide reductase family, all of which use some form of radical chemistry to convert the ribose-2′-OH to the two electron reduced 2′-H in product deoxyribonucleotide diphosphates.

6.5 Overview of the Pyrimidine and Purine Nitrogen Heterocycles

The construction of the pyrimidine and purine heterocycles as nucleic acid building blocks were/are absolutely essential for life. They are biosynthesized from three common proteinogenic amino acids, glycine, aspartate, glutamine, and one carbon unit (formyl) and one nitrogen unit (ammonia), simple metabolites. Yet, the four carbon–nitrogen functional groups – amides, amidines, ureas, guanidines – are the essential bits for information retention and information transfer in DNA and RNAs (Figure 6.1).

These central carbon–nitrogen groups are distributed differentially through other scaffolds in chemical biology. Amides are widely utilized functional groups throughout chemical biology, including the backbones

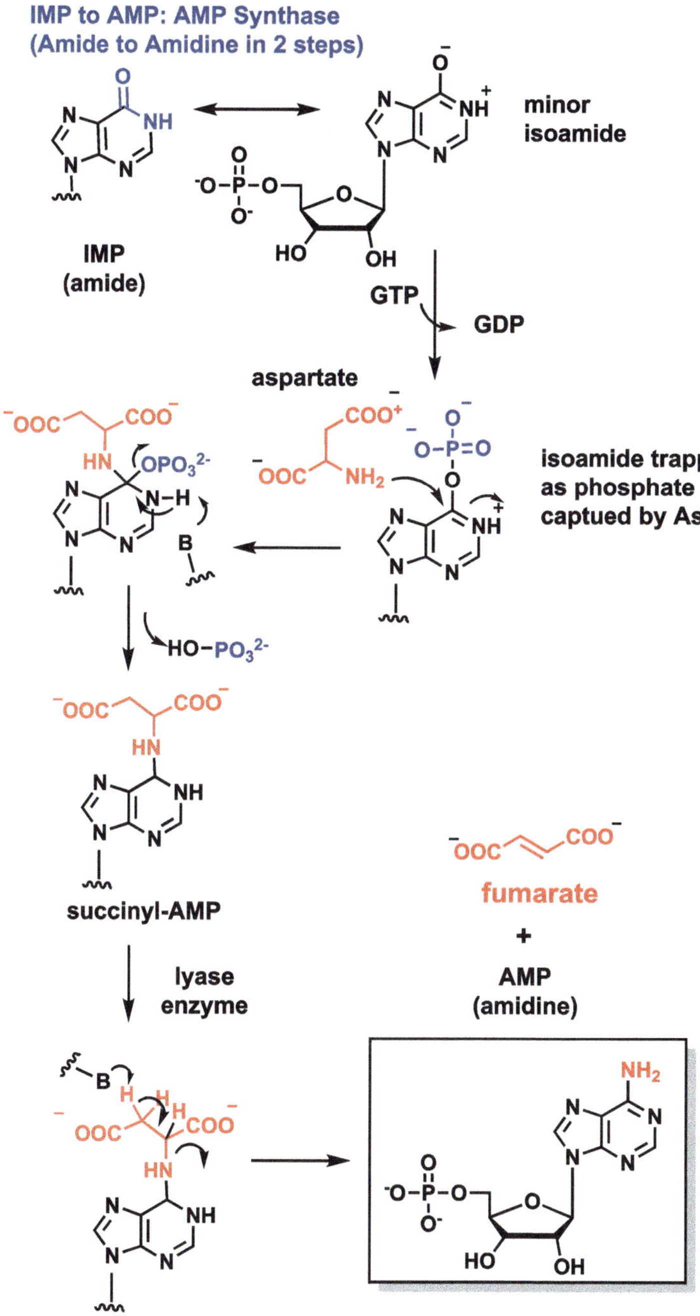

Figure 6.10 Conversion of IMP to AMP by AMP synthase; reprise of a trapped isoamide phosphate ester. The attacking amine comes from aspartate in a two step addition/fumarate elimination sequence.

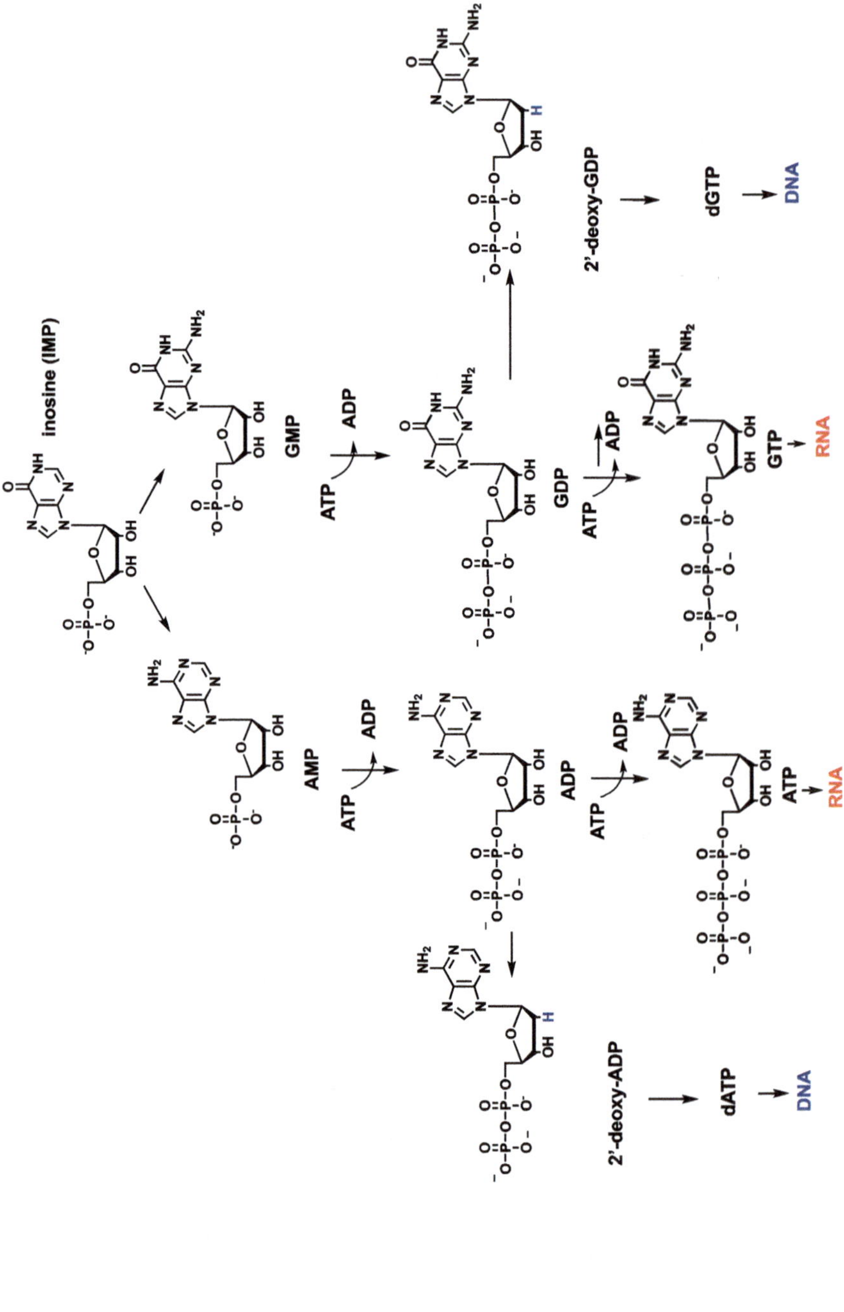

Figure 6.11 Overview of conversion of IMP to AMP and GMP and then to ATP and GTP for RNA or dATP and dGTP for DNA building blocks.

of all proteins and many examples of *N*-acetyl metabolites, including *N*-acetylated histone proteins. Urea is an excreted non-nucleophilic form of nitrogen to maintain nitrogen homeostasis in humans. The guanidine group of the proteinogenic amino acid arginine is built from converting the ureido part structure of citrulline to its imine. On the other hand, the guanidine group of creatine and creatine-P in muscle is created by an amidino group transfer enzyme. From a chemical dissection point of view, amidines are imine forms of amides while guanidines are imine forms of ureas, giving insight into how these C–N functional groups are interrelated.

The bifurcation of the initial bicyclic purine metabolite IMP to GMP and AMP leads to regiospecific installation of different exocyclic C–N functional groups in AMP and GMP. These are crucial specificity-determining hydrogen bonding elements in determining the identity of DNA base pairs. The bifurcating logic for enzymatic amination of IMP to the amidine group in AMP and of XMP to the guanidine group of GMP fix the key hydrogen bonding functional groups for A–T and G–C base pairs.

They also contrast the way organisms use the two amino acids glutamine and aspartate as latent carriers of free amine groups (recall Figure 5.22). The different strategies reflect the chemical constraints between hydrolysis of the C_5 amide of glutamine to liberate nascent NH_3 *vs.* the more difficult extraction of the amino group embedded within the aspartate carbon skeleton. Finally, aspartate delivers two of the five total nitrogen atoms in AMP by the two-step addition/elimination process, early in IMP assembly and then as the last step in AMP formation. By comparison, the complete aspartate scaffold, nitrogen and four carbons, are incorporated intact into uracil and cytosine.

At several steps in pyrimidine and purine ring maturations, amides are converted to amidines by intermediacy of minor isoamide contributors as oxygen nucleophiles to attack a phosphoric anhydride side chain of ATP. The resultant trapped isoamide phosphates are the electrophiles for capture by nascent NH_3 released *in situ* by glutaminase action. This strategy ties together amides, isoamides, phosphoric anhydrides, and nascent NH_3 as an iteratively used strategy to convert amides to their imine derivatives – the amidines.

The nitrogen heterocyclic nature of the monocyclic pyrimidines and the bicyclic purines are inextricably tied to their function. The equivalent size of the planar A–T and G–C heterocyclic base pairs fit within the antiparallel DNA double helices. The four C–N functional groups embedded as amide, urea, amidine, and guanidine validate the fundamental importance of nitrogen heterocyclic chemistry to life.

6.6 More Heterocycles to Come

We turn in the next three chapters to a broader look at carbacyclic and heterocyclic ring scaffolds containing different amounts of oxygen, nitrogen and/or sulfur, expanding the functional group chemistries available in chemical biology. The first of those chapters deals with strategies to assemble carbacycles of different size and functionality. The second chapter examines some additional key five and six membered heterocyclic metabolites found in nature. The third chapter focuses in on the fact that nine of the thirteen human vitamins contain heterocyclic ring structures, many of which enable central chemical transformations on the C–O, C–S, and C–N functional group inventories described thus far.

References

1. H. Rosenmeyer, The chemistry of purine as a constitent of natural products, *Chem. Biodiversity*, 2004, **1**, 361–401.
2. I. M. Lagoja, Pyrimidine as constituent of natural biolocially active compopunds, *Chem. Biodiversity*, 2005, **2**, 1–50.
3. M. Spencer, The stereochemistry of deoxyribonucleic acid II: hydrogen-bonded pairs of bases, *Acta Crystallogr.*, 1959, **12**, 66–71.
4. C. T. Walsh, *The Chemical Biology of Nitrogen*, Royal Society of Chemistry, London, 2021.
5. J. V. Greenhill and P. Lue, Amidines and Guanidines in Medicinal Chemistry, in *Progress in Medicinal Chemistry*, 1993, pp. 203–226.
6. L. B. Gavin, H.-B. Wang, A. L. Crowell, D. L. Begar, D. M. Martin, S. Sarkaen and N. G. Lewis, Stereoselective bimodular phenoxy radical coupling by an auxilairy (dirigent) protein without an active center, *Science*, 1997, **275**, 362–367.
7. E. Gouaux and W. N. Lipscomb, Three-imensional structure of carbamyl phopsphate and succinate bound to aspartate carbamoyltransferase, *Proc. Natl. Acad. Sci. U. S. A.*, 1988, **85**, 4205–4208.
8. J. P. Simmer, R. E. Kelly, A. G. Rinker, B. H. Zimmerman, J. L. Scully, H. Kim and D. R. Evans, Mammalian dihydrroorotase: nucleotide sequence, peptide sequences, and evolution of the dihydroorotase domains of the mutlifunctional protein CAD, *Proc. Natl. Acad. Sci. U. S. A.*, 1990, **87**, 174–178.
9. C. M. Alves, J. R. A. Silva and A. E. Roitberg, Insights into the mechsnism of oxidation of dihydroorotate to orotate catalyzed by the human cllass 2 dihydroorotate dehydrogenase: a QM/MM free energy study, *Phys. Chem. Chem. Phys.*, 2015, **17**, 17990–17996.
10. *Emery and Rimoin's Principles and Practice of Medical Gentics: Biochemical Features of UMP Synthase*, ed. D. Rimoin, R. Pyeritz and B. Korf, 2013.
11. B. G. Miller and R. Wolfenden, Catalytic proficiency: the unusual case of OMP decarboxylase, *Annu. Rev. Biochem.*, 2002, **71**, 847–885.
12. E. M Lynch, D. R Hicks, M. Shepherd, J. A. Endrizzi, A. Maker, J. M. Hansen, R. M. Barry, G. Zitai, E. P. Baldwin and J. M. Kollman, Human CTP synthetase filament structure reveals the active enzyme conformation, *Nat. Struct. Mol. Biol.*, 2017, **34**, 507–514.
13. R. H. Garrett and C. M. Grisham, *Biochemistry*, Boston MA, 6th edn, 2016.

14. C. T. Walsh, B. P. Tu and Y. Tang, Eight kinetically stable but thermodynamically activated molecules that power cel metabolism, *Chem. Rev.*, 2018, **118**, 1460–1494.
15. L. Hedstrom, IMP dehydrogenase: structure, mechanism, inhibition, *Chem. Rev.*, 2009, **109**, 2903–2928.
16. J. Nakamura, K. Straub, J. Wu and L. Lou, The glutamine hydrolysis function of human GMP synthetase, *J. Biol. Chem.*, 1995, **270**, 7347–7353.
17. B. W. Poland, M. M. Silva, M. A. Serra, Y. Cho, K. H. Kim, E. M. Harris and R. B. Honzatko, Crystal structure of adenylosuccinate synthetase from Eschericia coli. Evidence for convergent evolution of GTP-binding domains, *J. Biol. Chem.*, 1993, **268**, 25334–25342.
18. E. A. Toth and T. O. Yates, The structrure of adenylosuccinate lyase, an enzyme with dual activity in the de novo purine biosynthetic pathway, *Structure*, 2000, **8**, 163–174.

CHAPTER 7

Carbacyclic Metabolites: Alicyclic and Aromatic Rings in Chemical Biology

7.1 Common Cycloalkanes: From Cyclopropane to Cyclooctane

To this point, apart from the nitrogen heterocycles in Chapter 6, the discourses on C=C, C–O, C–S, and C–N functional groups have not directly addressed whether these reactive groups are embedded in linear scaffolds or are part of cyclic frameworks. As a prelude to a deeper dive into common heterocycles in chemical biology in Chapter 8, beyond the purines and pyrimidines of Chapter 6, we first take a closer look at carbacyclic systems[1] where there are no C–O, C–N or C–S bonds embedded within the ring structures.

Strictly speaking cycloalkanes abiotically have only C- and H-atoms. The common ring sizes range from the smallest ring, the three member cyclopropane, to the eight carbon cyclooctane metabolites. As shown in Figure 7.1 the typical planar view of these six rings can be augmented by three dimensional views of preferred conformers. Only the cyclopropane ring has all the cyclic carbons in the same plane. These other nonplanar conformers adopt these shapes to minimize angle strain from the preferred 109.5° angle found in acyclic methane. The most common ring sizes of carbacyclic organic metabolites are five member and six member rings – cyclopentanes and cyclohexanes. The lowest energy forms of cyclohexanes are a pair of chair forms, interconvertible by passage through a higher energy boat conformer (Figure 7.1).[2]

Chemical Biology No. 21
The Chemical Biology of Carbon
By Christopher T. Walsh

Published by the Royal Society of Chemistry, www.rsc.org

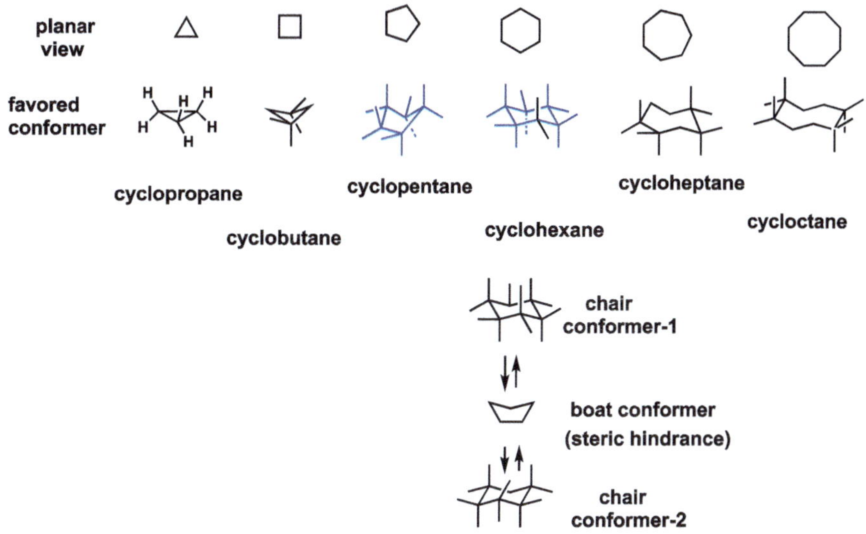

Figure 7.1 Cycloalkanes: cyclopropane to cycloctane; planar *vs.* conformatonal representations.

Among the most prevalent organic metabolites with fused cyclohexanes and cyclopentane rings are three of the cyclic triterpene metabolites noted in Chapters 2 and 15 (Figure 7.2). Hopene is pure cycloalkane (one olefinic link at C_{22}) assembled from the acyclic thirty carbon squalene by carbocation-driven cyclizations by squalene synthase. Hopene's fused pentacyclic scaffold (6-6-6-6-5) has four cyclohexanes and one cyclopentane ring with its isopropenyl substituent. Five of the six olefins in substrate squalene have been converted to new C–C single bonds to build the penta-carbacyclic product framework. The conformational perspective rendering of the hopene structure in Figure 7.2 shows all four cyclohexanes rings in the most stable chair forms. The lone fused cyclopentane ring is puckered.

β-Amyrin and cholesterol are tetracyclic and pentacyclic mono-alcohols, respectively, arising from action of the related family of oxidosqualene cyclases.[3] Amyrin, the major released sterol from oxidocyclases in many plants has a different regiochemistry of cyclization from hopene, with a four cyclohexane/one cyclohexene (6*) framework (6-6-6*-6-6), containing the one endocyclic double bond. That double bond reflects the position of quenching of the final carbocation by proton removal. The perspective drawing reminds us that the C-ring in amyrins is a cyclohexene not a cyclohexane and so has a boat shape rather than the adjacent cyclohexane chairs.

The major human sterol metabolite, C27-cholesterol, is downstream from the initial C30 lanosterol oxidocylase product. Both lanosterol and cholesterol have undergone yet a third cyclization regiochemistry, now to a tetracyclic (6-6-6-5) framework (with a linear eight carbon tail). In cholesterol, the cyclohexene is the B ring, imposing a distinct shape to the overall architecture of cholesterol from amyrin and hopene.

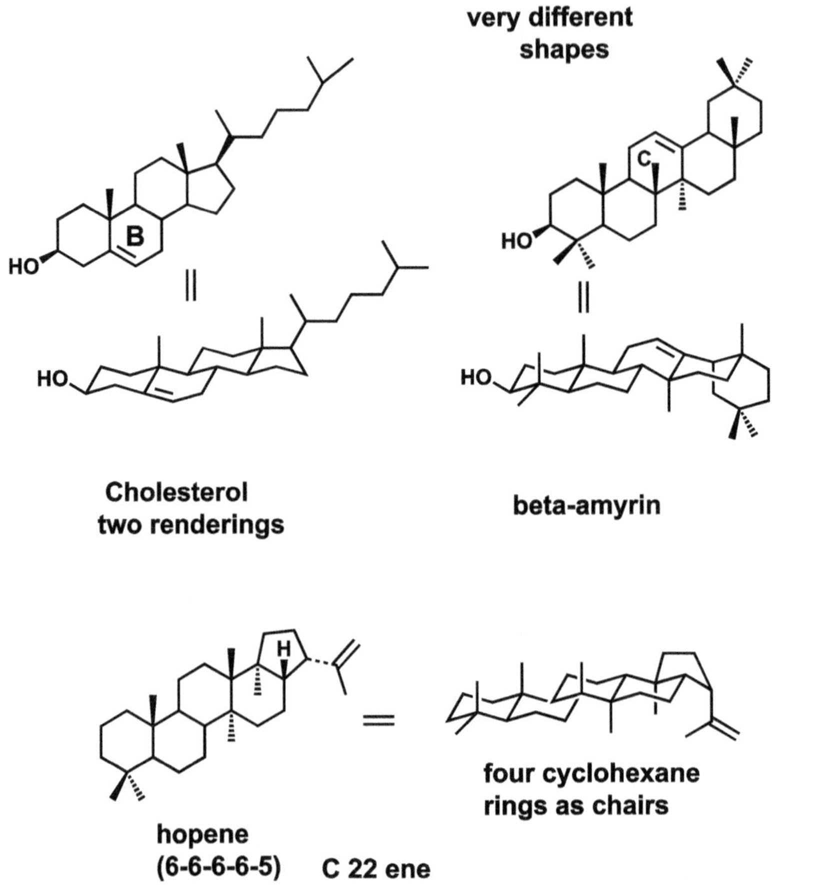

Figure 7.2 Different shapes of hopene, cholesterol, beta-amyrin.

Alicyclic carbon scaffolds in abiotic organic chemistry are *strictu sensu* restricted to organic molecules with carbon and hydrogen only. No oxygen, nitrogen, or sulfur atoms are present. This is the case for hopene but not for cholesterol or amyrin or hundreds of other oxygen- and even nitrogen-containing steroid scaffold alkaloids. Indeed, in this chapter we expand the working definition of alicyclic (carbacyclic) metabolites and prefer the term carbacyclic metabolites to include organic molecules in biologic systems that have O, N, and S atoms in functional groups where those three heteroatoms are not part of the carbacyclic rings. Thus, cyclic alcohols, ketones, carboxylates, amines, and amides qualify for discussion, but lactones, lactams, and pyrones do not. Those latter three functional groups are part of heterocyclic ring systems taken up in the next chapter.

7.2 Alicyclic *vs.* Aromatic Metabolites

Alicyclic is an adjective used interchangeably with cyclic hydrocarbon frameworks, also termed in this chapter as carbacycles. Alicyclic compounds can be fully saturated (cycloalkanes). Carbacyclics can also include C=C double bonds, as in cyclohexene and cyclohexadiene rings (Figure 7.3). As noted back in Chapter 2 on alkene redox interconversions, the traverse from cyclohexane to cyclohexene is a two electron redox step. Cyclohexadiene rings are four-electron oxidized from the cycloalkane. The predominant conformer shape of the carbacycle changes with oxidation state, from chair in cyclohexane, to twist-chair in cyclohexene, to essentially planar in cyclohexadiene. The cyclohexadiene ring is planar but the two sp^3-hybridized CH_2 carbons have hydrogens above and below the plane.

A further two-electron removal from cyclohexadiene formally generates the third C=C double bond. This product is the fully planar benzene ring, six electrons more oxidized than cyclohexane. Benzene is the most famous aromatic ring system.[2] Its structure is a hybrid between the two resonance forms shown in Figure 7.3. The ring is planar; the six hydrogen substituents are in plane. The six π-electrons are conjugated. Benzene is stabilized by the

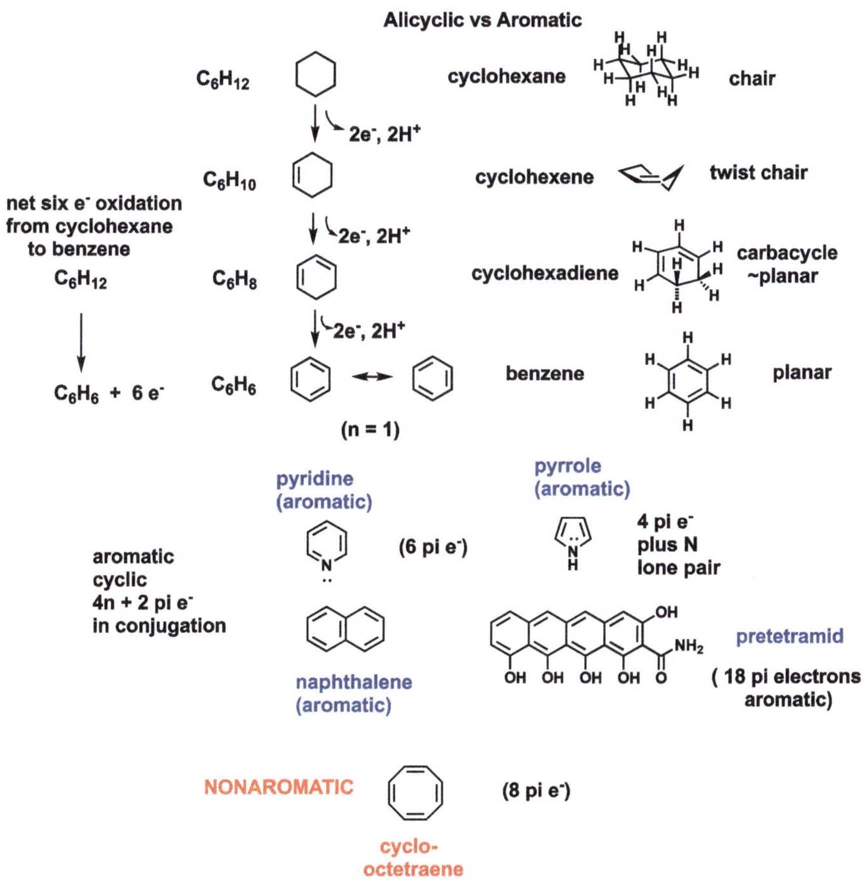

Figure 7.3 Alicyclic *vs.* aromatic metabolite examples: aromatic ($4n + 2\pi$ electrons) *vs.* nonaromatic ring structures.

energy gained by the π-electron delocalization *via* those resonance contributors. The planar geometry of the cyclic ring system allows maximal orbital overlap and the resonance delocalization of the conjugated π-electrons. That ring current is measurable indirectly by proton NMR and affects stability and reactivity of aromatic rings systems.

More generally, any cyclic planar system with $4n+2\pi$ electrons will be aromatic and be stabilized by the delocalization of those conjugate π-electrons and the consequent gain in resonance energy.[2] Figure 7.3 notes that while benzene is aromatic (six π-electrons), cycloctatetrane with eight π-electrons is nonaromatic. Naphthalene with 10 π-electrons is aromatic.

Two common nitrogen heterocycles that feature prominently in metabolic pathways are the pyridine ring (found in both vitamins B3 and B6, see Chapter 9) and the pyrrole ring. Both heterocycles, examined for their chemical roles in the organic chemistry of life in Chapters 8, 9, and 16, are heteroaromatic. The three double bonds in pyridine sum easily to six π-electrons. For the five ring pyrrole with two C=C bonds contributing four π-electrons, the lone pair on the nitrogen atom is available and participates in the resonance contributors, summing to six conjugated electrons.

Figure 7.3 also reminds us that enzymatic oxidation of the bicycloaromatic naphthalene to the 1,4-diketone creates the redox active naphthoquinone nucleus of the human vitamin K series. Two additional aromatic metabolites shown are the tetracyclic metabolite pretetramid in the bacterial pathway to the natural antibiotic tetracyline. Pretetramid's conjugated tetracyclic rings have 18 π-electrons, so it is an aromatic metabolite. The alkaloid cocaine contains multiple functional groups, among them an aromatic benzoyl ester group.

We take up a major route from carbohydrates to aromatic carbacycles three sections below, focusing on both the initial cyclization step and then each step of oxidation to cyclohexene and cyclohexadiene, before a third enzyme-catalyzed two-electron oxidation yields the aromatic phenyl ring.

7.3 A Gallery of Carbacyclic Metabolites

Carbacycles smaller than cyclohexanes have strained ring angles (Figure 7.1). Cyclopropane is the most strained at 60° angles, then cyclobutanes at 90° angles. Cyclopentane rings are less stained and more stable than three and four member rings. The chemical stability of the carbacycles follows ring strain. Cyclopropanes are more reactive than cyclobutanes to ring opening and relief of ring strain. Simple cyclopentanes and cyclohexanes are indefinitely stable under physiologic conditions in the presence of co-reactants but absence of specific enzyme catalysts.

The carbacyclic structures displayed in Figure 7.4 encompass some natural metabolites with cyclopropane, cyclobutane, and cyclopentane ring systems. They are only a small sample of representative metabolites, a subset of which are found in primary pathways in human metabolism. The majority of carbacycles are plant or microbial secondary metabolites.

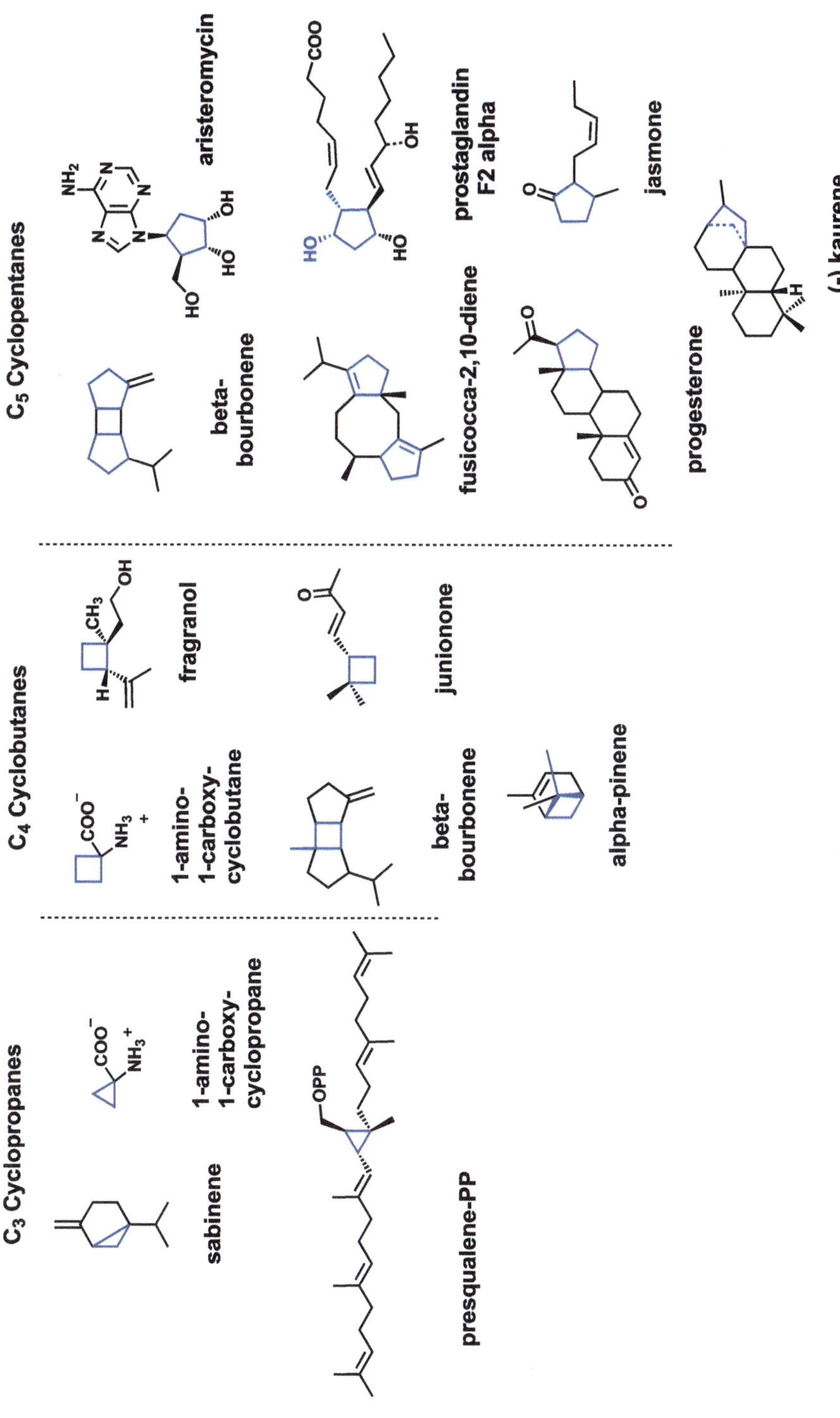

Figure 7.4 Metabolites with cyclopropane, cyclobutene or cyclopentane carbacyclic rings.

The C_3 strained cyclopropane rings are found in many biologic contexts, from terpenoids such as sabinene formed by cationic cyclizations, to 1-amino-1-carboxy-cyclopropane, generated by internal C_2 carbanion capture on C_4 of the aminobutyryl arm of *S*-adenosyl methionine. As discussed in Chapter 2 and also in Chapter 15, the mammalian pathway to steroids involves reversible cyclopropane ring formation and then ring-opening cleavage chemistry in presqualene-PP as a key intermediate on the way to the C_{30} hexaene squalene, the metabolic precursor of steroids. The strained cyclopropane rings, analogous to the three member strained epoxide rings in the next chapter, can serve as electrophiles for ring opening by a variety of biologic nucleophiles, depending on the context of the scaffolds in which the cyclopropane is a part (see next section).

The strained cyclobutane rings with 90° C–C bond angles are substantially rarer isoprenoid metabolites, five of which are depicted in Figure 7.3. The amino-carboxy-azetidine derives from *S*-adenosylmethionine but the other four are of terpenoid origin. The cyclobutane ring in the terpenoid hydrocarbon β-bourbonene is sandwiched between two cyclopentane rings while the cyclobutane ring in α-pinene is part of a bicyclic mono-olefinic framework.

Of the seven examples of cyclopentane rings in Figure 7.4, prostaglandin F2 and the D ring of steroids in the female sex hormone progesterone, occur in primary metabolic pathways. The other five cyclopentane entries are microbial or plant secondary metabolites, showing a range of functional groups and/or fused cyclic scaffolds. The carba-analog of ribose in aristeromycin suggests antimetabolite activity, while a cyclopentanone occurs in the volatile jasmone scent. The biosynthetic origin for the cyclopentane in aristeromycin is distinct from the radical cyclization of arachidonate to prostaglandins, different again from the carbocation cyclization chemistry in isoprenoid-derived scaffolds.

The six carbon cyclohexane rings are the most common cycloalkane scaffolds in chemical biology (Figure 7.5). A small subset of alicyclic (saturated to partially saturated) cycloalkanes are displayed on the left side and aromatic phenyl rings on the right side. They can arise by a number of routes, three of which are summarized in Section 7.5. The cyclohexane-containing metabolite frameworks in Figure 7.5, start from simple monocyclic examples and proceed up to pentacyclic frameworks in the steroid superfamily cycloalkane oleanane.

The first five entries are monocyclic monoterpene alkenes, alcohols, or ketones. Limonene, a classic fruit volatile odorant, has a cycloalkene ring while the related phellandrene has the two-electron more oxidized cyclohexadiene framework. The flavorant trio of menthol, piperitol, and carvone display a cyclohexanol, a cyclohexenol, and a cyclohexenone grouping, respectively; three related metabolic oxidation states. The next five entries continue the parade of cyclohexane frameworks but now in fused bicyclic contexts, as the 3-6 3-thujene and 3-carane, 4-6 (alpha-pinene), and bicycloheptane (carvone) isomers, all derived from an initial α-terpinyl cation (see Chapter 2, Figure 2.15).

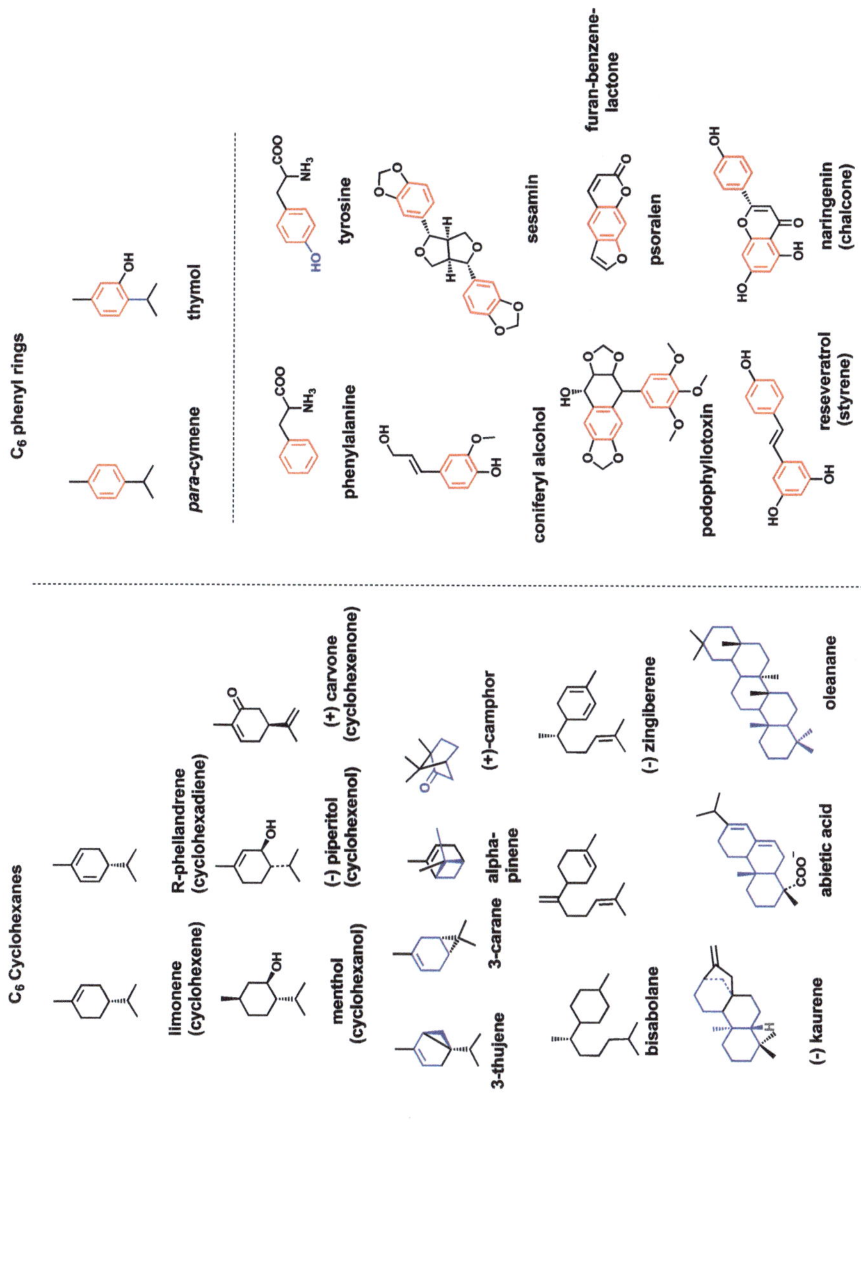

Figure 7.5 Metabolites with cyclohexane, cyclohexene, cyclohexadiene or phenyl rings.

From the C_{15} farnesyl-PP, hundreds of cyclic sesquiterpene plant metabolites are known (Chapter 2). The four metabolites shown contain a single cyclohexane in bisabolane, a decalin moiety in eudesmane, and then monocyclic cyclohexene and cyclohexadiene oxidation states in bisabolene and zingiberene. It would require one further two-electron oxidation (autoxidation?) of cyclohexadiene rings to generate the fully aromatic resonance-stabilized phenyl rings, exemplified by the two entries for *para*-cymeme and thymol, respectively (Figure 7.5).

This isoprenoid origin is one biosynthetic route from cyclohexyl scaffolds to aromatic phenyl (benzene) rings, but not operant in the route to phenylalanine and tyrosine discussed in the next section. The final three entries in the cyclohexane category of Figure 7.5 show three cyclohexanes and a fused cyclopentane in (−)-kaurene, and the tricyclic abietic acid. The abietic acid tricyclic framework has two fused cyclohexene rings to go along with the cyclohexane. Finally, the bacterial hydrocarbon oleanane is a fully reduced ring system with five fused cyclohexanes, showing that nature can fuse multiple sets of cyclohexyl rings from acyclic isoprenoid precursors (see Chapter 2).

The six-carbon aryl ring metabolites on the right side of Figure 7.5 are derivatives of the central proteinogenic aromatic amino acid L-phenylalanine. Phenylalanine in turn is assembled from sugar condensations and dehydrations outlined in the next section of this chapter. The next four metabolites with embedded six-carbon phenyl rings derive from oxygenations of phenylalanine-derived cinnamic acid that feature carbon radical pathways (Chapter 12). The coniferyl alcohol monomer is one of the building blocks of lignin, a major structural constituent in woody plants and trees. Coniferyl alcohol arises by oxygenase-mediated carbon radical chemistry, rather than the hallmark carbanion chemistry of polyketides or the alkyl cation-initiated chemistry of the isoprenoid pathways (see Chapter 12 in Section IV). Likewise, the sesamin dimer in sesame seed oil is a result of enzyme-initiated carbon–carbon radical coupling enzyme chemistry. Podophyllotoxin and psoralen show more complex fused aromatic-heteroaromatic scaffolds. Finally, the two metabolites resveratrol (a stilbene in red wine) and narigenin (a flavanone) are aromatized frameworks that arise by dehydrations of plant polyketide synthases that use typical enolate/carbanion chemistry for intramolecular cyclizations onto ketone groups.

Figure 7.6 gives two examples of rare cycloheptane and three examples of rare cyclooctane rings, all five as secondary metabolites. Colchicines from the autumn crocus have a long therapeutic history in humans. Notably, they have a pair of fused seven-member cycloalkane rings in a 6-7-7 tricyclic framework. We have noted taxadiene biosynthesis previously (Chapter 2) with cationic driven conversion of the acyclic diterpene metabolite geranylgeranyl diphosphate to the 6-8-6 tricyclic ring system. That plant hydrocarbon taxadiene subsequently undergoes eight oxygenation events around the periphery and three acylations that bring in three phenyl rings to create a hydrophobic/hydrophilic balance to this anticancer, tubulin-targeting natural drug. Nitidasin contains three types of cycloalkane rings – six, eight, and five.

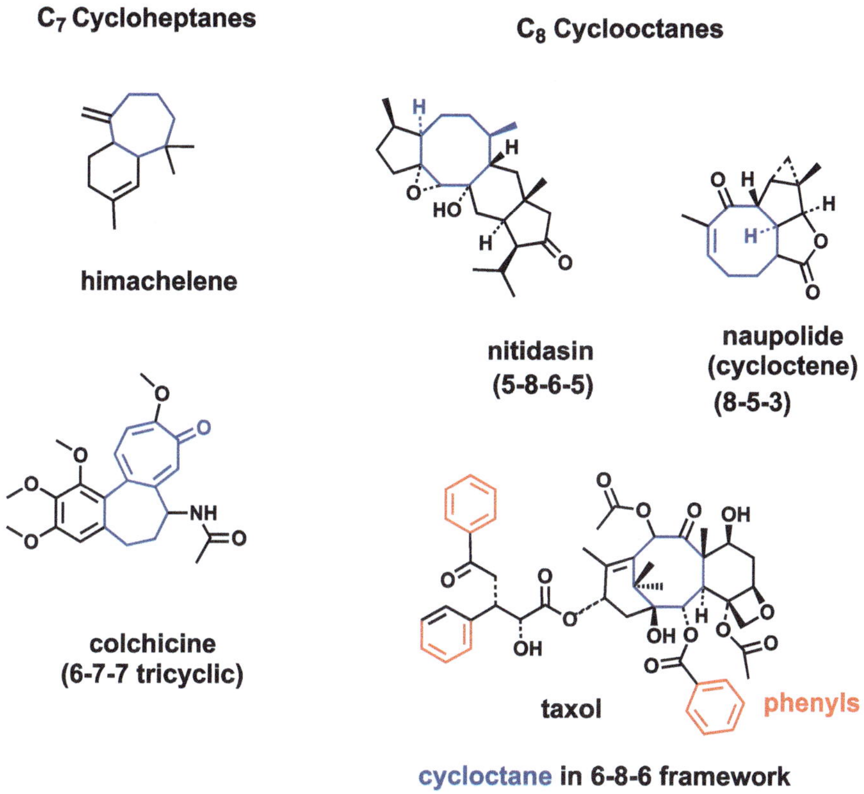

Figure 7.6 Metabolites with cycloheptane or cyclooctane carbacyclic rings.

In Section 7.5 we take a complementary look at polycyclic scaffolds in metabolites generated from polyketide synthase chemistry *vs.* those generated from prenyl transfer condensation chemistry. First, though, in the next section we look at the shikimate pathway logic that funnels the two primary metabolites, three-carbon phosphoenolpyruvate (PEP) and four-carbon erythrose-4-P, down the shikimate/chorismate pathway to *de novo* phenyl ring construction in phenylpyruvate (see Figure 7.9). Aromatic rings, both benzene and naphthalene, are built *de novo* in microbes and plants but not in humans. Thus, phenylalanine, the major source of phenyl ring metabolites, is a required dietary amino acid for humans while the naphthoquinone nucleus for electron transfer chemistry is provided by dietary vitamin K (Figure 7.3).

7.4 Cyclopropane Ring Enzymatic Formation and Breakdown Routes

The highly strained three carbon cyclopropane rings are perhaps the most reactive of the cycloalkanes. They can also be synthesized by multiple routes, including carbocation, carbanion and carbon radical routes,[4] illustrating three of the four major biologic routes to C–C bond formations (Chapters 10–12). Figure 7.7 shows four enzyme-catalyzed routes to cyclopropanes. Two of the four routes involve *S*-adenosylmethionine. The third arises by coupling of two 2-IPP molecules *via* carbocation chemistry to give the cyclopropane ring in chrysanthemyl-PP (This is also the route to the cyclopropane presqualene-PP (Chapters 2 and 15). In that case, a cyclopropane is opened as the electrophilic partner on addition of a hydride from NADPH).

The first two routes involve the methyl group of SAM as the donor of a "CH_2" group to an olefin. In column A1 of Figure 7.7, the olefin is proposed to act as transient carbon nucleophile on SAM, yielding a carbocationic transition state (see Chapter 2). Proton removal from the CH_3 group is proposed to be simultaneous with quenching the cation and forming the cyclopropane by a polar mechanism. This is contrasted by column A2 where the CH_3 group of SAM is proposed to act as a radical donor by prior formation of a SAM-$CH_2^{\bullet}$ species by a homolytic route in the active site of a radical SAM enzyme (see Chapter 12). A second one-electron input from an Fe_4S_4 cluster in the enzyme sets up a final two-electron cleavage step on the attached SAM fragment.

The fourth column, C, shows the amino acid-PLP carbanionic route in which intramolecular elimination of a 4-chloro-substituent by the C_2 carbanion of the amino acid-PLP imine yields the cyclopropyl amino acid product.[5] A variation of this amino acid-PLP stabilized C_2 carbanion route is shown in Figure 7.8 where a tandem pair of enzymes first create and then cleave 1-amino-1-carboxy-cyclopropane (ACPC).[6]

The first enzyme, ACPC synthase, also makes use of *S*-adenosylmethionine as substrate: now not as a donor of the one carbon fragment but rather of the four carbon aminobutyryl arm. The SAM-PLP C_2 carbanion attacks the electrophilic C_4 of its aminobutyryl arm. Cleavage of the C_4–S^+ bond quenches the positive charge on the sulfonium ion as thiomethyladenosine departs. Release of the ACPC amino acid provides the substrate in plants for the next enzyme ACPC oxidase.[7] This is an iron-based enzyme that catalyzes radical-based fragmentation of the cyclopropane to ethylene, cyanide ion and carbon dioxide as O_2 is reduced by four electrons to HOOH. Ethylene gas, deriving from carbons 3 and 4 of the methionyl arm is the major fruit-ripening hormone of plants.

7.5 Building Phenyl Rings by Cyclization and Successive Dehydration of Carbohydrates

Perhaps the most intriguing chemical logic for *in vivo* carbacycle assembly is the main route to the formation of aromatic rings, notably L-phenylalanine and L-tyrosine, carried out by microbes and plants.[8] Dietary phenylalanine is the major source of benzene ring scaffolds in humans. Microbes and plants are much more metabolically versatile, turning sugars into phenylpyruvate

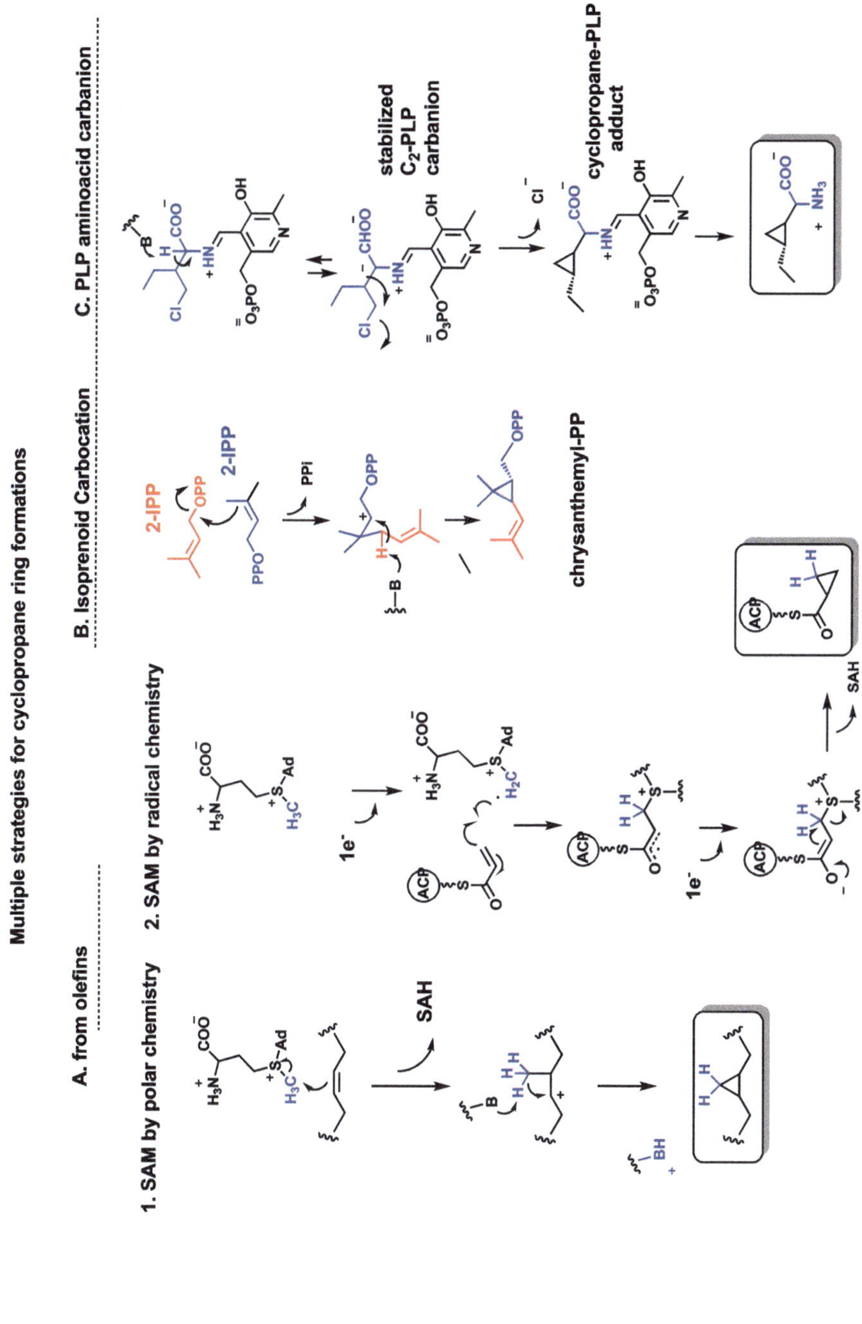

Figure 7.7 Multiple strategies for biosynthesis of cyclopropane rings.

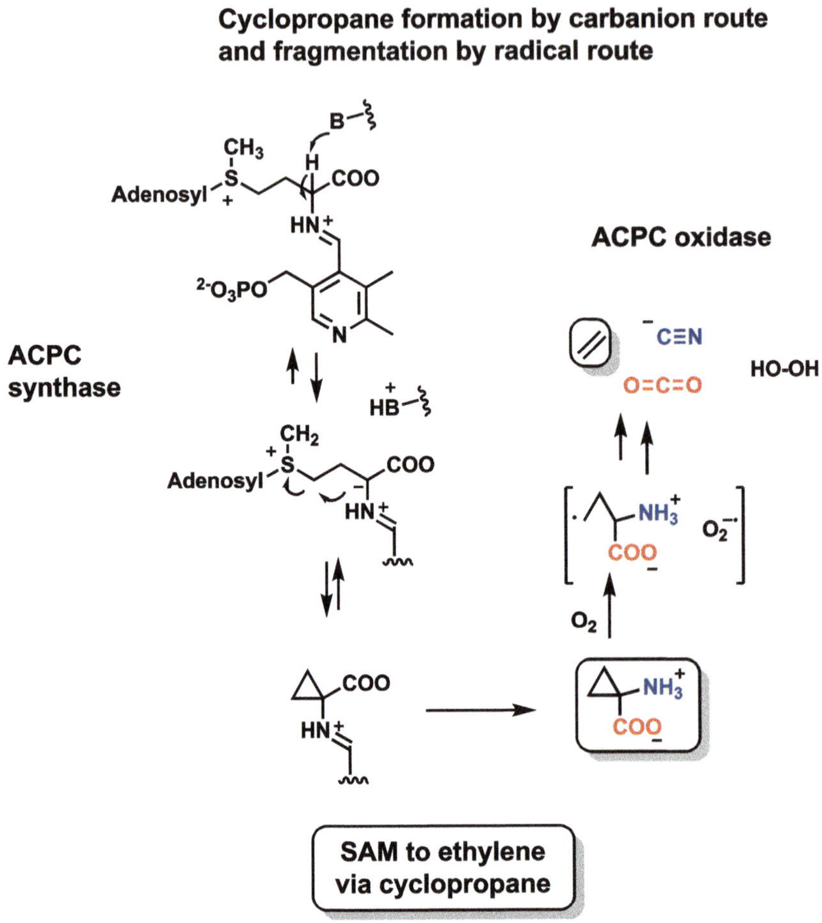

Figure 7.8 Cyclopropane ring formation by a carbanion route and then fragmentation by a radical route. ACPS synthase and ACPC oxidase. SAM to ethylene.

or 4-hydroxyphenylpyruvate by a series of controlled dehydrations of a seven-carbon sugar, 3-deoxy-D-arabinoheptulosonic acid-7-phosphate. That seven-carbon ketose is morphed into the trihydroxy cyclohexene carboxylate shikimate, the metabolite for which the first half of the pathway to phenylalanine is named (Figure 7.9).

This heptulosonic acid arises by enzymatic aldol condensation of phosphoenolpyruvate with erythrose-4-P (Figure 7.9). The C_1 aldehyde of the four carbon sugar is captured by the C_3 carbanion contributor to the enolate anion of pyruvate after P–O bond cleavage releases the trapped enolate to engage in an aldol condensation. Then, oxidation of the C_6–OH group by hydride transfer to NAD^+ sets up enolate/carbanion formation at adjacent C_7 and its use as carbon nucleophile in ring closure to the cyclohexane carboxylate framework in 3-dehydroquinate, a second (intramolecular) aldol

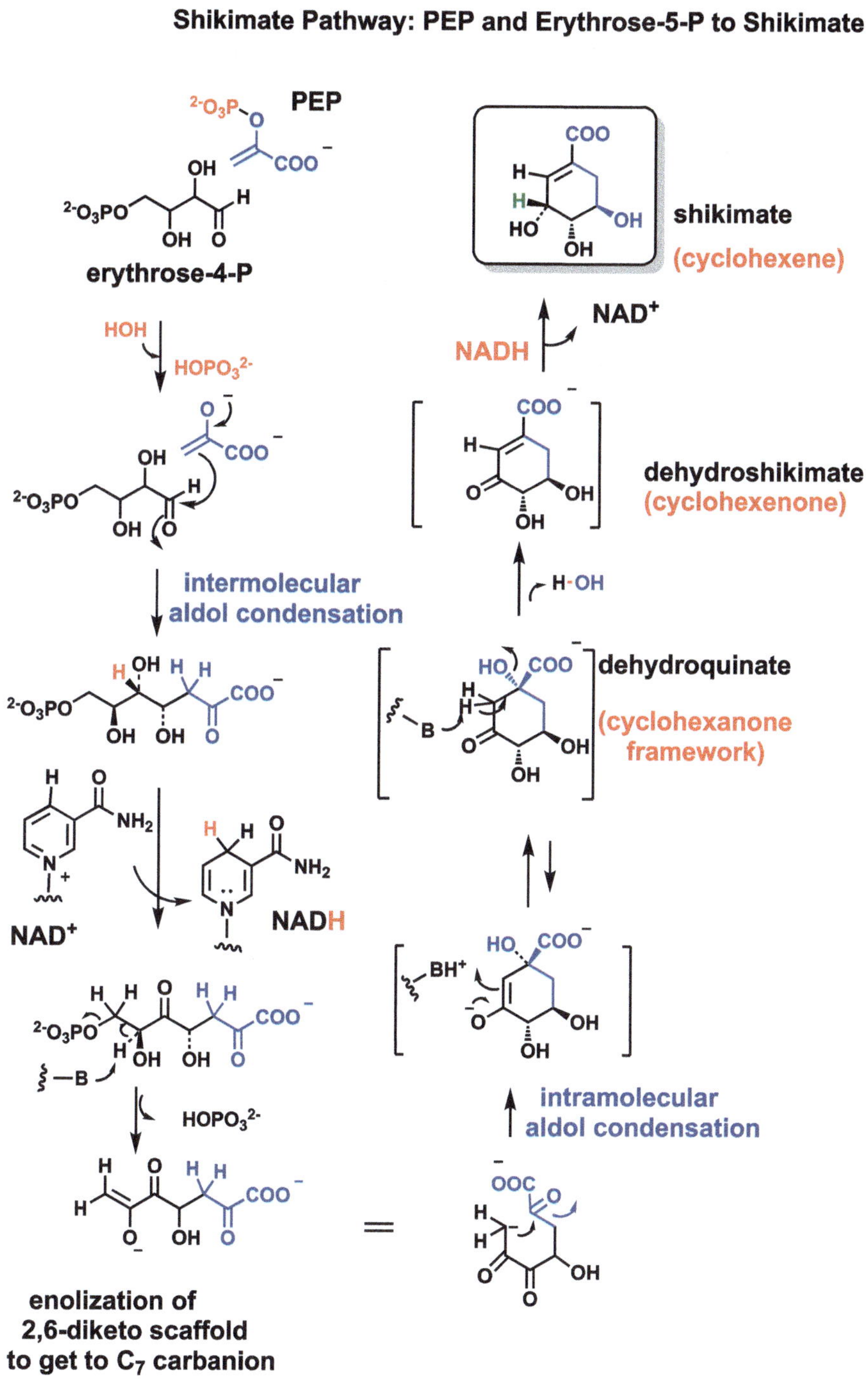

Figure 7.9 Shikimate pathway from initial aldol condensation of PEP and erythrose-4-P to shikimate cyclic scaffold includes cyclohexane and cyclohexene ring forming steps.

condensation. This aldol ring-forming reaction starts from the 2.6-diketo scaffold, enabling formation of the C_7 enolate carbanion to attack the electrophilic C_2 carbonyl group.[9] This is the cyclohexane ring-forming step. The immediate next step is a phosphatase elimination to give dehydroquinate, the first released carbacyclic intermediate in the pathway (Figure 7.9).

Loss of water across C_3–C_2 is mediated by an enolate type C_3 carbanion (Figure 7.9) to covert the cyclohexane-containing substrate to the cyclohexene-containing dehydroshikimate product. At this juncture the ketone has performed its two roles: to facilitate accessible enolate/carbanion formation for the cyclohexane ring-forming C–C bond formation and a subsequent dehydration.

The now superfluous ketone is next reduced to the alcohol oxidation state in shikimate by hydride transfer from NADH (Figure 7.9). To this point, three carbanionic species have been utilized, two to form C–C bonds, one to induce adjacent loss of OH, enabled by carbonyl and enolate, first intermolecular, then intramolecular chemistry. Shikimate is halfway to an aromatized phenyl ring end product. It still has to undergo a net four-electron oxidation, by way of a cyclohexadiene, to get to the aromatic phenyl ring product.

The strategy to get from a cyclohexene to a cyclohexadiene ring as an immediate precursor to aromatization starts with a phosphorylation, dephosphorylation sequence, interrupted by enolpyruvyl ether formation (Figure 7.10). Phosphorylation of the C_3–OH of shikimate by shikimate kinase (Figure 7.10) sets up a future elimination of phosphoric acid across C_2 and C_3 to get to a cyclohexadiene.

First, however, the shikimate-3-P metabolite is subjected to a remarkable enolpyruvyl ether formation at the C_5–OH by enzyme-mediated etherification chemistry. This step installs the enolpyruvyl ether that will be a critical component of the pericyclic reaction that tops off the pathway chemistry.

The formation of 5-enolpyruvyl-shikimate-3-P illuminates an unusual reactivity aspect of cosubstrate PEP. The reaction is featured as an addition elimination reaction on the PEP double bond with unusual regiospecificity.[10] As depicted in Figure 7.10, the shikimate-5 alcohol-oxygen acts as nucleophile at C_2 of the double bond of PEP. The terminal vinyl CH_2 is protonated to form a transient C_3 methyl group. The resultant tetrahedral adduct proceeds to the enol ether product by loss of the elements of inorganic phosphate across the C_2–C_3 bond of that adduct. This requires abstraction of one of the equivalent C_3-methyl hydrogens of the PEP scaffold as a proton to initiate $-OPO_3^{2-}$ loss from carbon 2. The enol ether group is key for the eventual 3,3-claisen rearrangement two steps hence.

The enolpyruvyl-shikimate-P product is then a substrate for the chorismate synthase enzyme.[8] It carries out a doubly vinylogous elimination of the elements of inorganic phosphate. Chorismate synthase abstracts one of the prochiral methylene hydrogens from C_6 to initiate double bond rearrangement and cleavage of the C–O bond at C_3 (the participation of the migrating double bond could be categorized as neighboring group nucleophilic participation). This sets up the cyclohexadiene functional group in

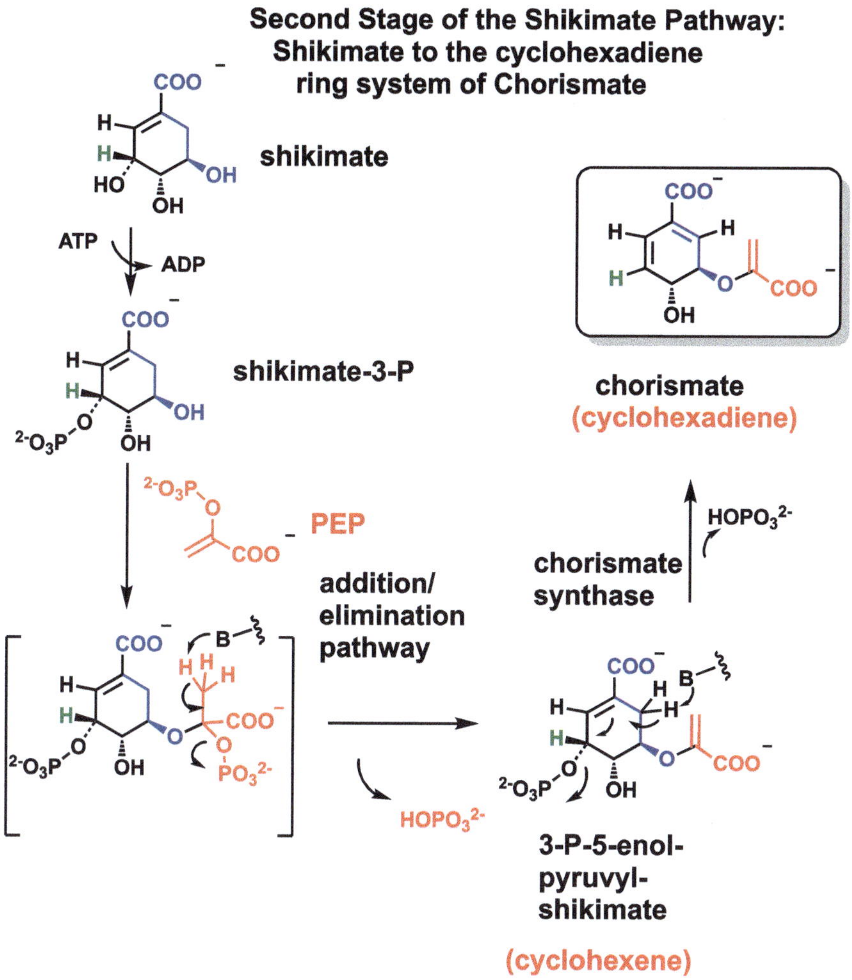

Figure 7.10 Second stage of the shikimate pathway: shikimate to the cyclohexadiene ring system of chorismate.

chorismate in a 2,6-diene relationship. The last two enzymes in the pathway, the chorismate mutase as a pericyclase and the two aromatizing decarboxylases that act on prephenate, are taken up in the next section.

From the perspective of carbacyclic ring formation, the cyclohexane ring is formed at the stage of dehydroquinate, by the intramolecular aldol condensation. The cyclohexane to cyclohexene net oxidation occurs in the dehydration of dehydroquinate to dehydroshikimate. The cyclohexene to cyclohexadiene occurs in the action of chorismate synthase, again through an elimination of the elements of phosphoric acid (this time doubly vinylogous). The cyclohexadiene functionality in chorismate is one oxidation step away from the aromatic phenyl ring.

In Figure 7.11 the two three-carbon frameworks from the two molecules of PEP added during the pathway are shown in blue and red, respectively, accounting for 6 of the 10 carbons in chorismate (the other four derive from the erythrose scaffold). *Inter alia*, the pathway to date reveals a route to C–C double bonds distinct from the main pathways noted back in Chapter 2. On subsequent conversion to the nine-carbon aromatic scaffold of phenylpyruvate, the original carboxylate from the first (blue) PEP molecule is lost as CO_2.

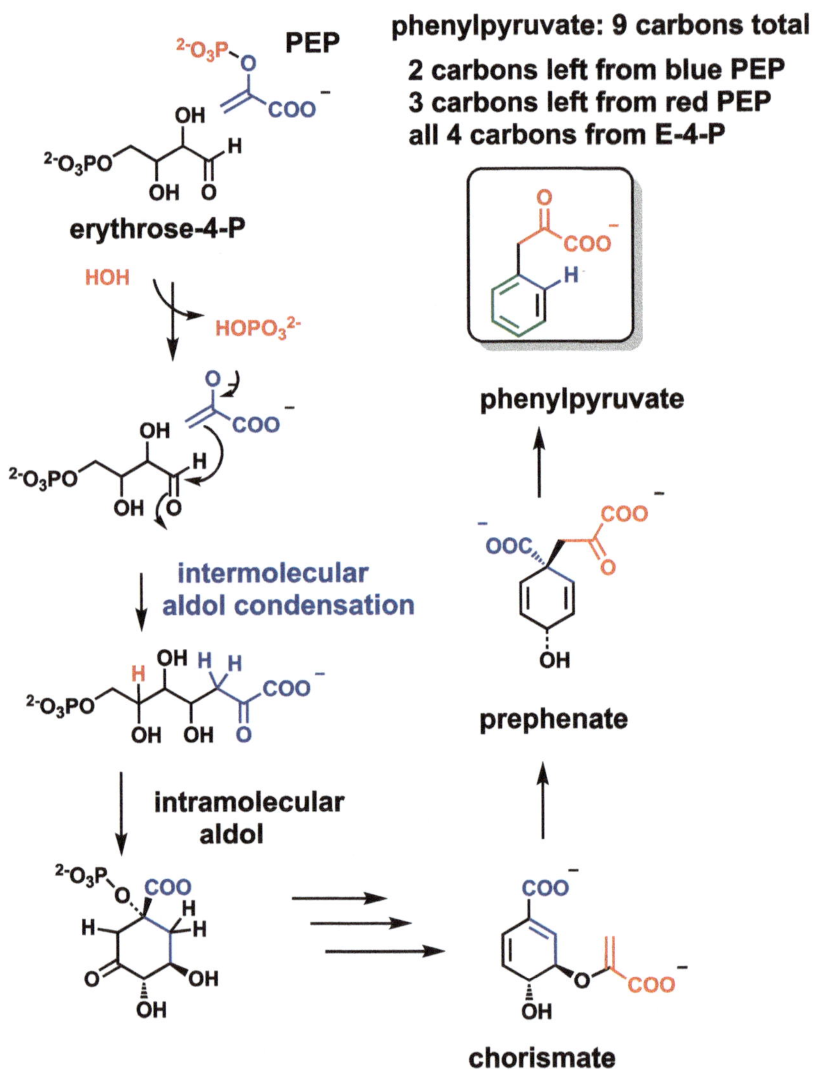

Figure 7.11 Source of the nine carbons in phenylpyruvate: two PEP, one erythrose-4-P.

7.6 Chorismate to Prephenate: Aromatizing Decarboxylation Outcomes

Chorismate is the substrate that undergoes a famous 3,3-pericyclic rearrangement (the Claisen rearrangement, see Chapter 13 for discussion of pericyclases) to prephenate[11] (Figure 7.12).

Finally, prephenate is subjected to enzymatic decarboxylation with two possible fates as the cyclohexadiene aromatizes. One decarboxylase uses the electrons left at the carbon from which CO_2 departs as a neighboring group to expel the *para*-OH substituent.[12] The driving force for loss of water is the gain in resonance energy as the cyclohexadiene ring aromatizes to the phenyl ring of phenylpyruvate (Figure 7.13).

The alternative decarboxylase also catalyzes aromatization but by ejection of the other substituent at the *para* position.[13] That is a hydrogen. Its ejection with the electron pair as that C–H bond breaks constitutes hydride ion ejection. The cosubstrate receiving the hydride ion in flight is NAD^+ to yield NADH and 4-hydroxyphenylpyruvate (Figure 7.13). Although both enzymes are decarboxylases, the second is termed a dehydrogenase[13] to emphasize the redox aspect of the aromatizing decarboxylation. In both cases a *para*-substituent is ejected with the electron pair that had been in the bond that is broken. The OH^- group can depart without a cosubstrate. The H^- needs NAD^+ as an acceptor.

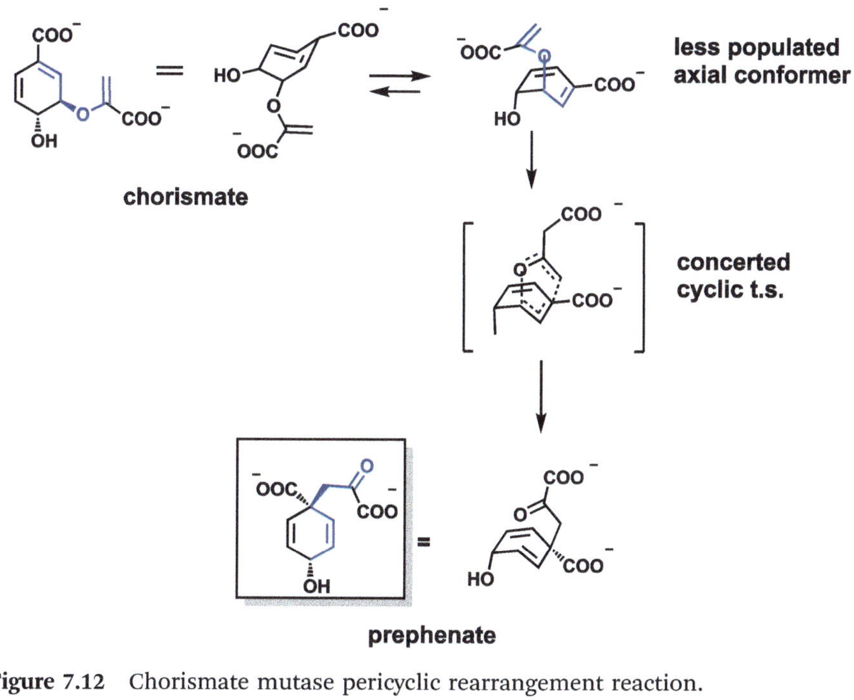

Figure 7.12 Chorismate mutase pericyclic rearrangement reaction.

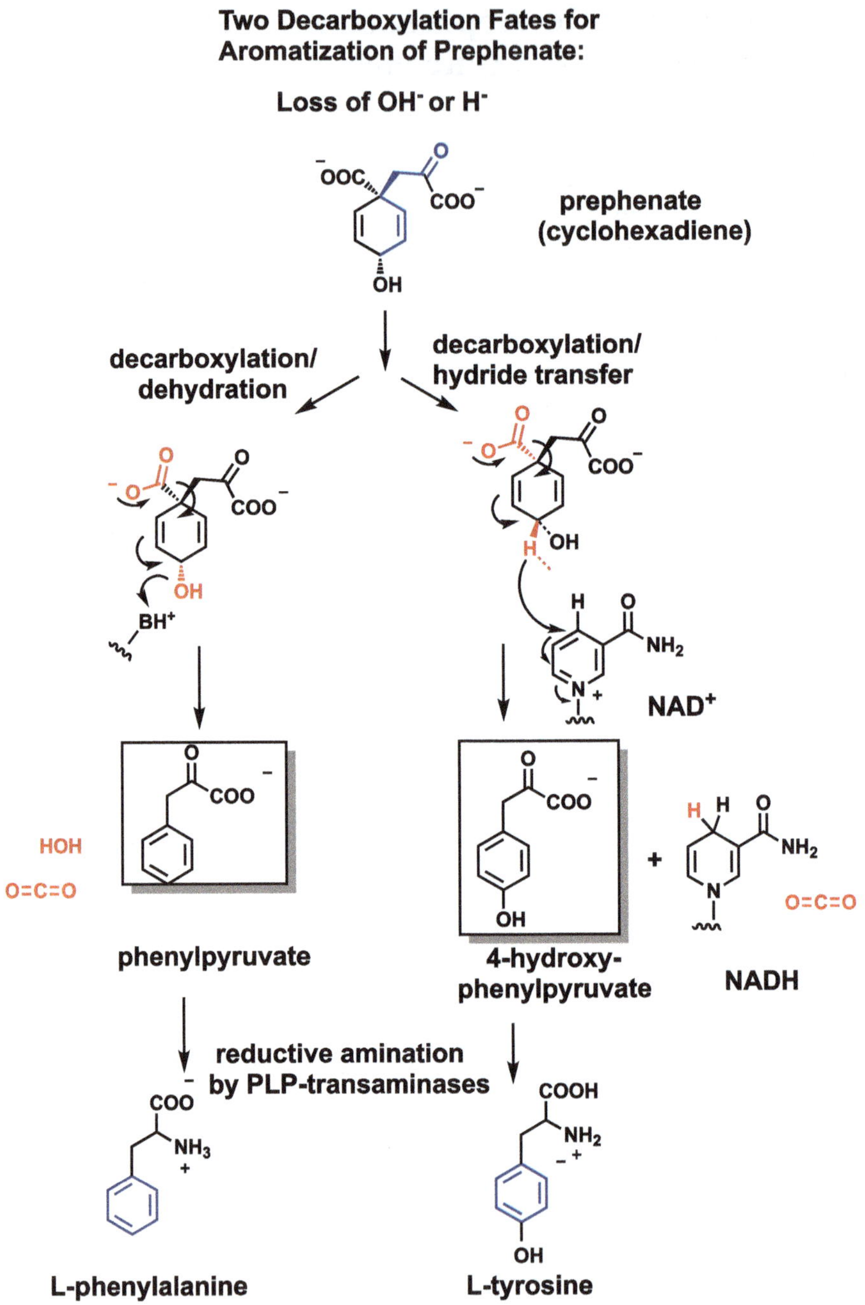

Figure 7.13 Two decarboxylation fates on enzymatic aromatization of prephenate.

A pair of reductive pyridoxal-P-dependent transaminases (Chapter 5) produce L-phenylalanine or L-tyrosine, respectively, creating the C–N bonds in the amino acids (Figure 7.13) phenylalanine and tyrosine, respectively by iminium ion reductions.

It is worth reconstructing the logic for conversion of two common C_3 and C_4 metabolites, PEP and erythrose-4-P, into the nine-carbon scaffold of phenylalanine and tyrosine. Two C_3 units (both C_3 units are PEP, one utilized as C_3 nucleophile, the second as carbon electrophile at C_2!!) plus erythrose make up the C_{10} scaffold in both chorismate and prephenate. It takes the 3,3-sigmatropic rearrangement step to convert the enol pyruvyl ether (O–C bond) to the C–C bond in the side chain of phenylpyruvate. Decarboxylation, driving aromatization of a cyclohexadiene ring system, converts the C_{10} framework to the C_9 framework.

The enzymatic strategy for cyclohexane ring formation utilizes two aldol condensations. The first is intermolecular, joining C_3 and C_4 fragments to form a C_7 product. The second aldol is an intramolecular condensation, creating the cycylohexane carboxylate framework. This enzyme-catalyzed cyclization event is followed by introduction of two double bonds, one by net dehydration, the second by doubly vinylogous loss of inorganic phosphate. The aromatizing step, introducing the third degree of unsaturation as the benzene ring forms, is coupled to decarboxylation by the prephenate dehydrogenase or prephenate decarboxylase, respectively. The latter half of this pathway is a lesson in C=C bond construction.

7.7 Shikimate to Cyclohexane Carboxylate: Olefin Reduction and Triple Dehydration During a Net Eight Electron Reduction

In the previous section we saw the strategy employed by plants and microbes to convert the keto-trihydroxycyclohexene metabolite dehydroquinate (DHQ) to the aromatic ring in phenylpyruvate and then phenylalanine. The cyclohexane ring of DHQ is converted enzymatically to the cyclohexene ring of dehydroshikimate by loss of water. Then, reduction of the keto creates the trihydroxycyclohexene ring of shikimate (Figure 7.9). The subsequent strategy leads to *ad seriatim* elimination of the three hydroxy substituents *via* a cyclohexadiene intermediate on the way to loss of the third and last hydroxyl group in the prephenate decarboxylase step (Figures 7.10 and 7.13) to create the aromatic ring. That phenyl group is eight electrons oxidized from the keto-trihydroxy cyclohexane ring of dehydroquinate.

Microbes can reroute the same trihydroxcyclohexene ring intermediate shikimate to reduce it by a net eight electrons to cyclohexanyl CoA for incorporation into more complex natural scaffolds[14] (Figure 7.14). Two-electron input is needed to reduce the single C=C double bond in the shikimate cyclohexene carbacyclic ring. Each of the steps eliminating the three hydroxy groups constitute two-electron reductions, for six more

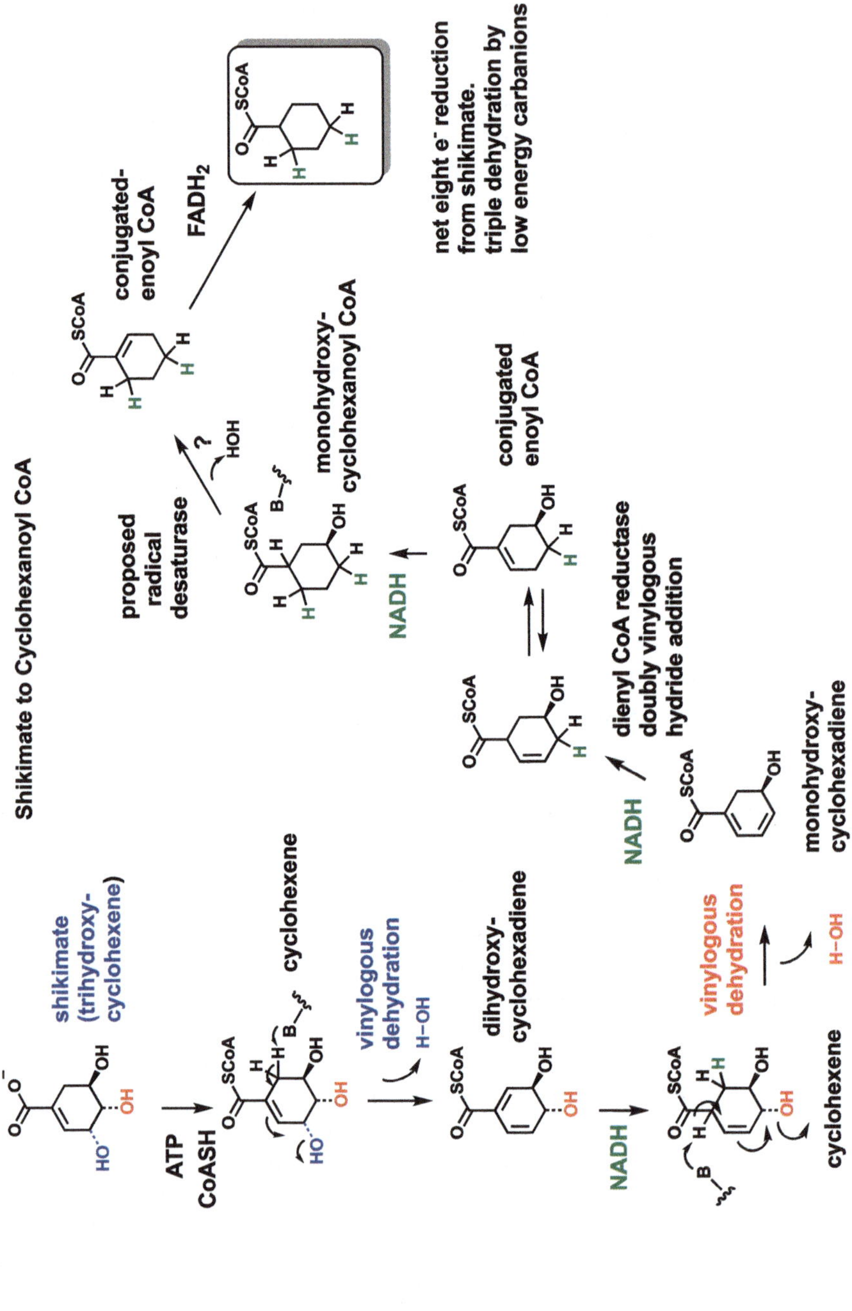

Figure 7.14 Shikimate reduction to cyclohexanoyl CoA.

electrons, leading to a total of eight-electron inputs to attain the cyclohexanoyl carboxy end metabolite. Three NADH and one $FADH_2$ equivalents power these reductive steps. The enzymatic strategies are a series of alternating dehydrations and then hydride reductions of conjugated enoyl thioesters, including one dienyl CoA reductase. That catalyst generates an unconjugated olefinic product that will require isomerase action to move the C=C bond back into conjugation.

Thus, shikimate, with its mono-olefinic-trihydroxy substituents on the carbacyclic ring, can be processed as a metabolic intermediate either for oxidation to the aromatic phenyl ring or for reduction to the cycloalkane redox state of cyclohexane carboxylic acid. In the oxidative aromatizing direction, a series of vinylogous dehydrations lead to the aromatic phenyl ring. In the reductive direction, the three dehydrative steps yield conjugated/conjugatable enoyl thioesters that can be reduced by hydride transfers from nature's two common hydride transfer coenzymes, NADH or $FADH_2$.

7.8 Chorismate to Isochorismate, 2-Aminobenzoate, and 4-Aminobenzoate

Chorismate is a remarkably versatile cyclohexadienyl metabolite, poised to be tipped over to different functionalized aromatic ring products. Due both to its cyclohexadienyl ring and the labile enolpyruvyl ether group, this substrate can be nudged down distinct pathways (Figure 7.15).

One such route is the enzymatic conversion of chorismate to isochorismate by the enzyme isochorismate synthase.[15] The reaction appears to be a regiospecific addition of a water-derived –OH group at C_6, setting off a doubly vinylogous elimination of the C_4–OH group. Both chorismate and isochorismate have two double bonds and an –OH group to go along with the enolpyruvyl substituent, but the relative functional group placements are distinct. The numbering of the ring substituents changes and isochorismate is now a 1-carboxy-2-hydroxy-3-enolpyruvyl-4,6-cyclohexadiene (Figure 7.15). The next enzyme in the bacterial pathway to the iron-chelating enterobactin hydrolyzes the enol ether to create the 1,2-dihydrodiol.[16] The third enzyme oxidizes one of the dihydrocatecholic alcohols in an NAD^+-dependent dehydrogenation to the cyclohexadienone framework.[17] That is the less stable tautomer of the aromatic catecholic 2,3-dihydroxybenzoate (Figure 7.15).

We note in Chapter 13 that the bacterial enzyme isochorismate lyase will catalyze the fragmentation of isochorismate to salicylate and pyruvate, an aromatizing elimination of an enol ether that has been proposed to be a concerted, pericyclic 1,5-sigmatropic rearrangement. Thus, isochorismate can be nudged down different pathways to aromatic rings. A difference between 2,3-dihydroxybenzoate and 2-hydroxybenzoate (salicylate) formation is the C_3–O bond cleavage or retention as the system aromatizes (Figure 7.15), by distinct routes.

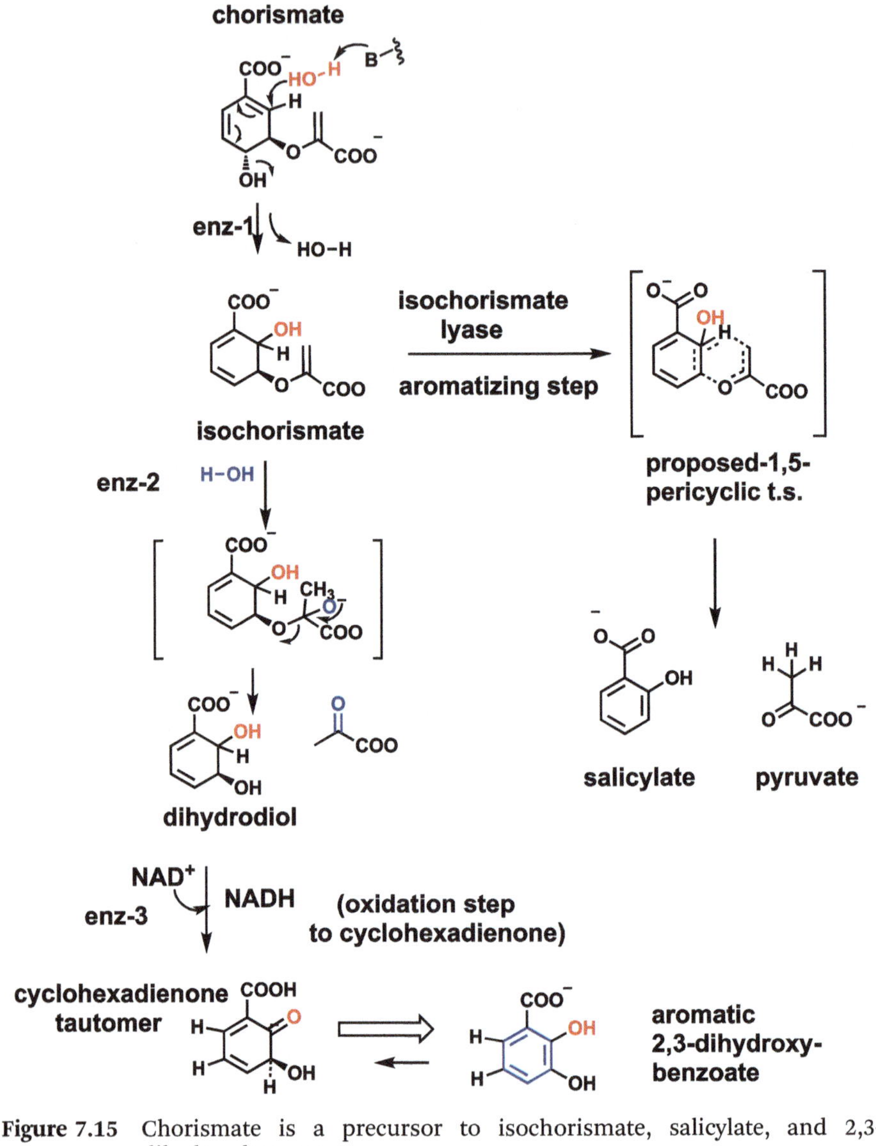

Figure 7.15 Chorismate is a precursor to isochorismate, salicylate, and 2,3-dihydroxybenzoate.

Chorismate, but not isochorismate, is also progenitor to 2-aminobenzoate (anthranilate) (Figure 7.16), an aromatic amine with a newly generated C–NH_2 bond. Instead of water addition to C_6 of chorismate, the two subunit enzyme anthranilate synthase adds nascent ammonia.[18] The NH_3 derives from *in situ* glutaminase activity to generate nucleophilic NH_3 with a "just in time" inventory control logic (see Chapter 5 for coupled glutaminase logic). The nascent NH_3 attack is concurrent with the doubly vinylogous elimination

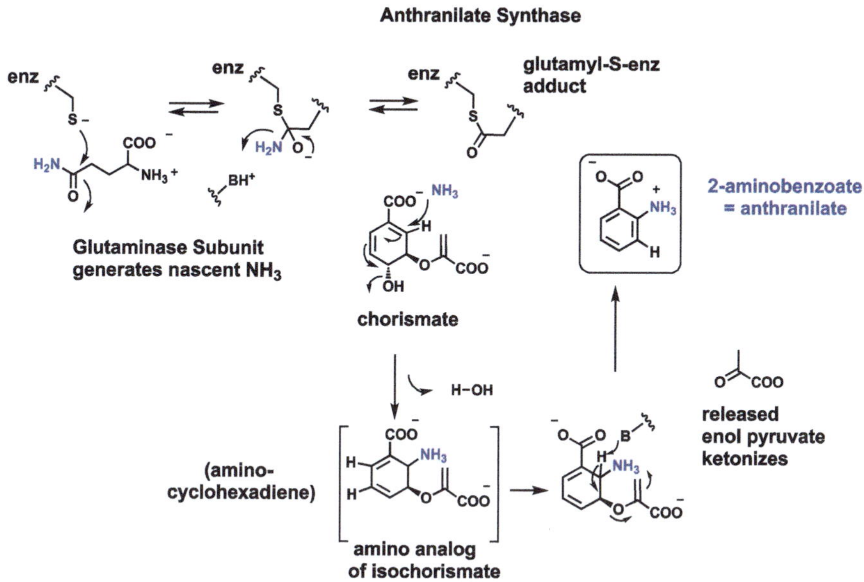

Figure 7.16 Anthranilate synthase aminates at C_2 and aromatizes chorismate.

of the C_4–OH. That action generates, with numbering change, a 1-carboxy-2-amino-3-enolpyruvyl, 4,6-cyclohexadiene. This can be aromatized by the second subunit of anthranilate synthase by abstraction of the C_2–H as a proton and use of the incipient carbanion as neighboring group electron pair to cleave the C_3–O bond. This eliminates pyruvate as the enolate and aromatizes the cyclohexadiene ring to the phenyl ring of 2-aminobenzoate=anthranilate. Note that the enzyme has directed nascent ammonia to attack a cyclohexadiene with -OH elimination rather than direct substitution on an aromatic ring. Anthranilate is the direct precursor of the indole ring of tryptophan (see Chapter 8).

Chorismate can also be converted instead to the 4-aminobenzoate regioisomer by a variant of the logic and enzymatic machinery just noted for 2-aminobenzoate synthase.[19] Instead of free ammonia acting as initial nucleophile at C_6 of bound chorismate, the PABA synthase is thought to catalyze addition of the side chain NH_2 of a lysine residue to create a covalent adduct that is the analog of the anthranilate synthase amino intermediate (Figure 7.17).

Then, the enzyme uses nascent ammonia to attack at C_4 and run the doubly vinylogous elimination in the reverse direction. That results in expulsion of the enzyme-Lys-NH_2 group and creates the 1-carboxy-3-enolpyruvyl-4-amino, 1,5-cyclohexadiene. Then, abstraction of the C_4–H as a proton initiates elimination of pyruvate as the enolate anion while aromatizing the benzoate ring in product 4-aminobenzoate. This *para*-regioisomer of aminobenzoate is a key building block in generation of the folate scaffold of vitamin B9 (see Chapter 9).

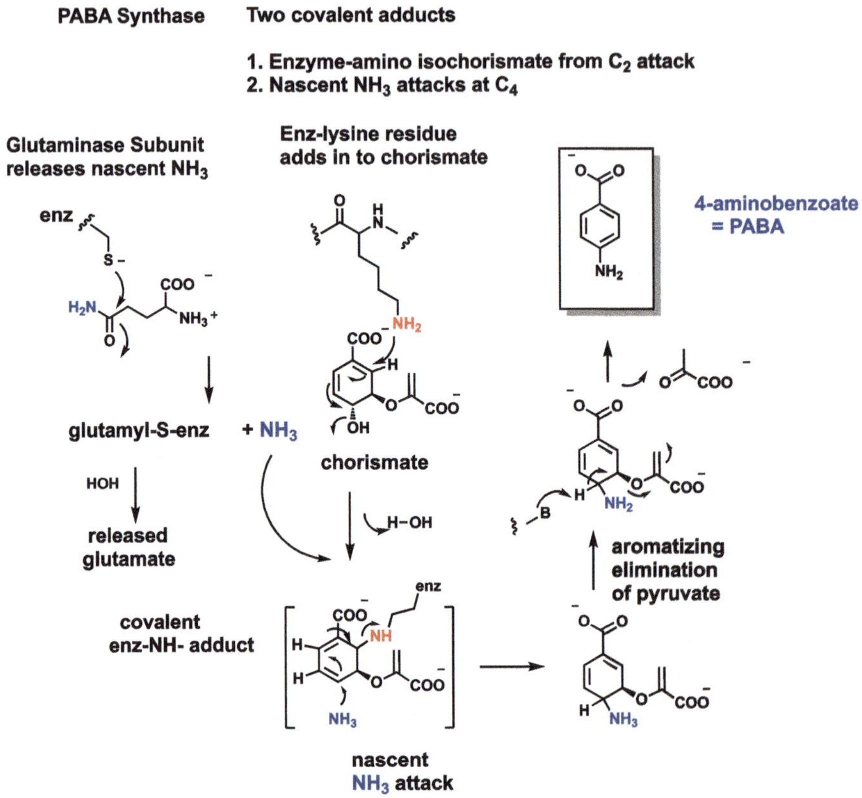

Figure 7.17 PABA synthase aminates at C4 and aromatizes chorismate by way of two proposed intermediates.

The above conversions show that the dihydro aromatic cyclohexadiene ring system of chorismate can not only undergo a 3,3-pericyclic rearrangement but also 1,4-aminating aromatizations. The corresponding 1,4-doubly vinylogous hydration to isochorismate sets up that isomer for either conversion to the 2,3-dihydroxybenzoate or the 2-hydroxybenzoate scaffold *via* a 1,5-enzyme catalyzed sigmatropic rearrangement of isochorismate. The timing and fate of additions, eliminations, and aromatizations of the cyclohexadiene group in chorismate or isochorismate by cognate enzymes can be exquisitely controlled. Figure 7.18 shows three aromatized end products from isochorismate: the iron-chelating siderophores enterobactin (2,3-dihydroxybenzoate) and mycobactin (*via* salicylate), as well as the chloroplast electron transfer agent phylloquinone. Analogously, chorismate yields tryptophan and folate as well as phenylalanine and tyrosine (Figure 7.18). Nature's aromatization logic and strategy from sugars *via* the shikimate/chorismate pathway has proven to be widespread and successful.

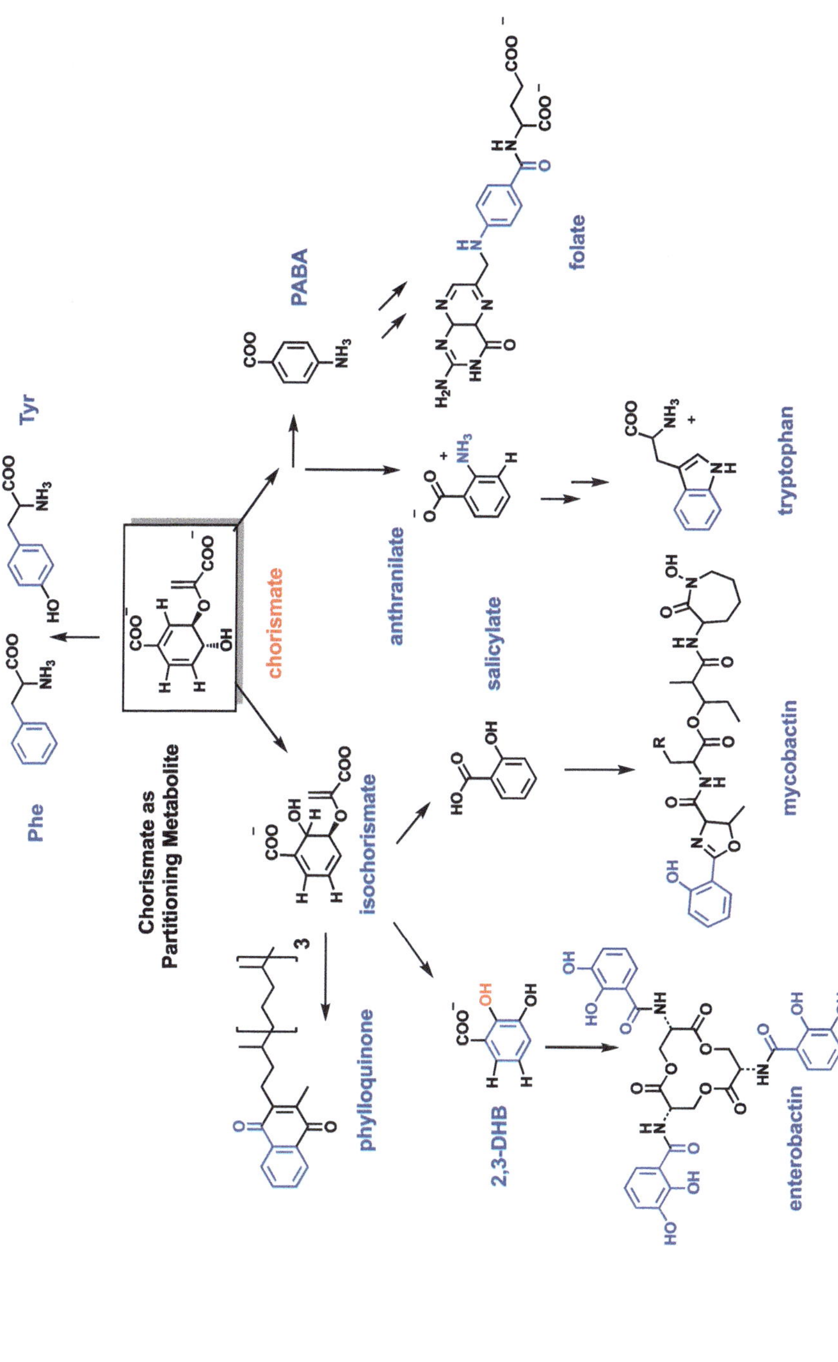

Figure 7.18 Chorismate as a partitioning metabolite.

7.9 Naphthoquinones

Plants synthesize a variety of 1,4-napthoquinones (including the phylloquinones of Figure 7.18), with major flux down the *ortho*-succinyl benzoate pathway to vitamin K and related metabolites.[20] One of the two benzene rings in the naphthalene scaffold comes from isochorismate, the second from the succinyl moiety of α-ketoglutarate[21] (Figure 7.19). Formation of the fused bicyclic naphthalene nucleus in menaquinones reflects not only aromatization of the cyclohexadiene moiety in isochorismate but also *de novo* construction and aromatization of the second, benzoquinone ring.

One key step is decarboxylation of the 2-ketoglutarate by a thiamin-PP-requiring decarboxylase (see detailed mechanistic role of vitamin B1 in Chapter 9). Then, decarboxylation, driving aromatization of a cyclohexadiene ring system, converts the C_{10} framework to the C_9 framework. This is an analogous strategy to the prephenate decarboxylations noted earlier. Capture of the diene of isochorismate at the C_1 double bond (with protonation at C_4) by the carbanionic succinyl-TPP adduct forms the new C–C bond in product 2-succinyl-5-enolpyruvyl-6-hydroxy-3-cyclohexene-1-carboxylate (SEPHCHC).

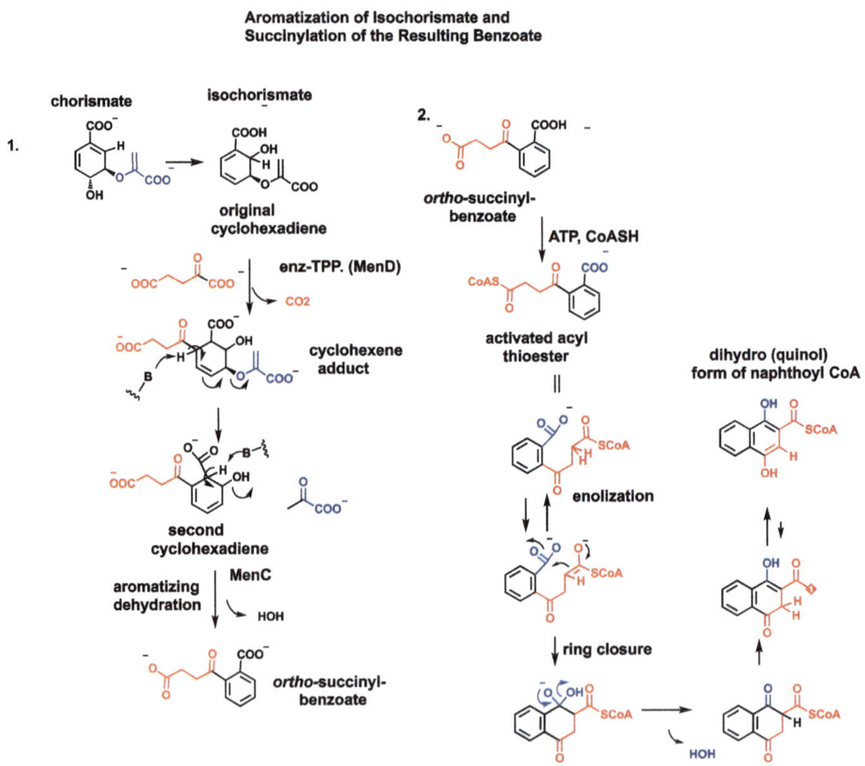

Figure 7.19 Aromatization of isochorismate and *ortho*-succinylation of the resulting benzoate on the way to the naphthoquinone ring.

That metabolite then undergoes a 2,5-elimination of the elements of pyruvic acid (Figure 7.19).

The net addition/elimination reaction transiently creates a trisubstituted cyclohexene from the starting cyclohexadiene. Upon subsequent elimination of the enol pyruvyl ether the product is (1*R*,6*R*)-2-succinyl-6-hydroxy-2,4-cyclohexadiene-1-carboxylate (SHCHC). The enol ether bond has been cleaved as C_2 has become succinylated *via* a new C–C bond. There is a cyclohexadiene functional group in the accumulating product, but it is isomeric to the diene in starting isochorismate. The ketone carbonyl of the succinyl side chain is the original C_2 of 2-ketoglutarate.

The next enzyme, MenC, then dehydrates (and aromatizes) the hydroxycyclohexadiene ring to the phenyl ring in 2-succinyl-benzoate (*ortho*-succinyl-benzoate). Aromatization has been achieved in the first six membered ring (Figure 7.19).

A survey of the strategies for progressive conversion of cyclohexanes to cyclohexenes to cyclohexadienes to aromatized phenyl rings in the chorismate-isochorismate diaspora involve: (1) elimination of water(dehydration), (2) elimination of inorganic phosphate (by 1,2 or 1,4 eliminations), and (3) elimination of the enolpyruvyl ether substituent.

Formation and aromatization of the second 1,4-hydroqunol ring *en route* to naphthalene metabolites requires two enzymes (Figure 7.19). Men E uses ATP and CoASH to convert the carboxylate of the succinyl side chain into a succinyl CoA thioester to set up the adjacent C_2 carbanion/thioester enolate. Men B forms the second, quinone ring using that accessible C_2 carbanion of the succinyl CoA side chain to attack the adjacent C_1-carboxylate, presumably as the protonated carboxylic acid in the MenB active site.[22] The tetrahedral adduct could break down in the forward direction with loss of water to form the ketone at C_1. This is a dihydro oxidation state of the naphthoquinone ring, tautomerizing to the naphthoquinol isomeric form of 1,4-dihydroxy-2-naphthoyl CoA.

The Men D enzyme catalyzes a thiamin-PP dependent decarboxylation of 2-ketoglutarate to create a TPP-succinyl carbanion equivalent (Figure 7.20). That TPP-succinyl carbanion equivalent is proposed to form the new C–C_1 bond *via* a doubly vinylogous displacement of water from C_4 of isochorismate. Ejection of TPP precedes loss of the enolpyruvyl side chain to produce the accumulating succinyl hydroxy-cyclohexadiene scaffold.

The path forward from the hydroquinone ring in naphthoyl CoA to vitamin K shows the versatility of hydroquinone/quinone ring systems to behave both as electrophiles (electron sink for decarboxylation) and then as carbon nucleophiles in the prenyl side chain attachment (Figure 7.21).

Hydrolysis of the CoA thioester precedes decarboxylation, reflecting the behavior of quinols as β-electron sinks to facilitate decarboxylation. In turn, that nascent carbanion is the nucleophile for capture of the C_1 allyl cation from cosubstrate octaprenyl diphosphate (see Chapter 2) to yield desmethyl-menaquinol K2. C-Methylation by methyl transfer from SAM to C_3, acting as nucleophile yields the final substituted menaquinone naphthyl framework.

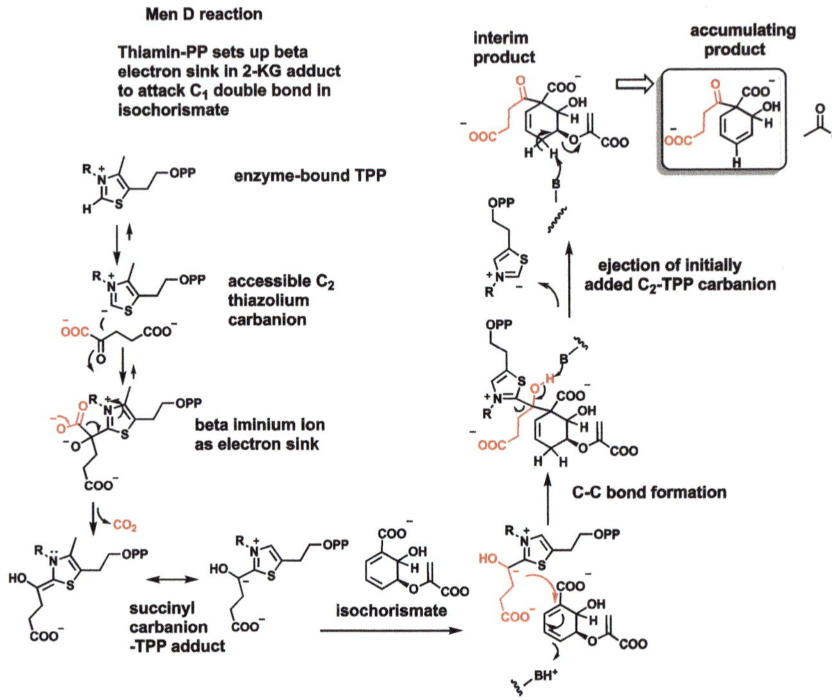

Figure 7.20 Mechanistic analysis of isochorismate conversion to naphthoyl CoA.

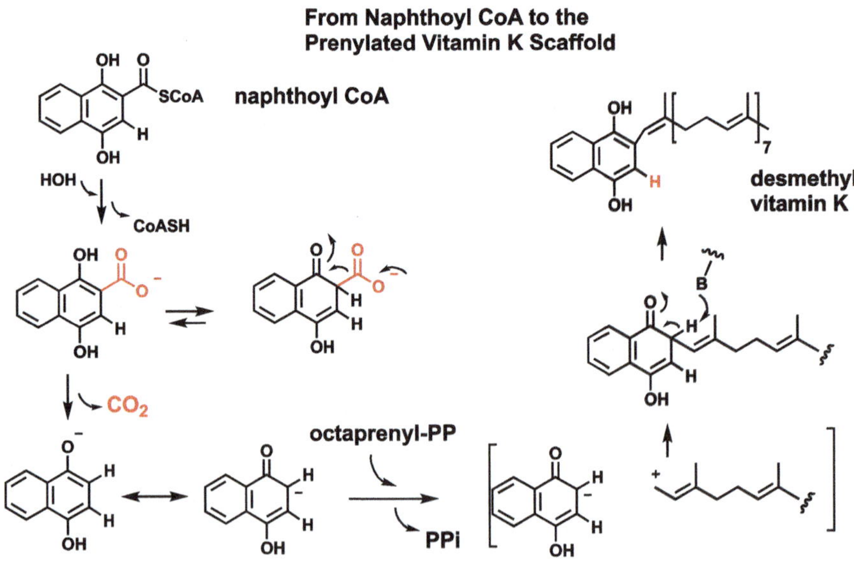

Figure 7.21 Napthoyl CoA to desmethyl vitamin K.

The last two C–C bond formations, prenylation at C_2 and methylation at C_3, are two examples of capture of two distinct forms of carbon electrophilic equivalents (prenyl/allyl cation and methyl cation equivalent) by the carbanion forms of C_2 and C_3 in the quinol oxidation state of the naphthoquinone nucleus. These two reactions show the ability of quinols to act as carbon nucleophiles while the quinone carbonyls act as electrophiles. Quinones can also react by one electron pathways as well as these two electron pathways, acting as remarkably versatile and useful redox stepdown agents located in membrane milieus.

In sum, chorismate and isochorismate are building blocks for both benzene and naphthalene aromatic ring frameworks in natural metabolites. The benzenoid scaffolds include the phenyl ring of all downstream phenylpropanoid metabolites (including phenylalanine), 2-hydroxy (salicylate), 2-amino (anthranilate), 2,3-dihydroxy (catecholic group in 2,3-DHB), 4-hydroxy (tyrosine), and 4-amino-phenyl metabolites (*para*-aminobenzoate). Chorismate *via* isochorismate *via* its annulation by the succinyl moiety of 2-ketoglutarate gives rise to the fused napthoquinone ring system with its characteristic C-prenyl and C-methyl substitution patterns (Figure 7.18).

7.10 Enzymatic Conversion of the Pyranose Ring of Glucose-6-P to the Hexahydroxy Cyclohexane Ring of Inositol-1-P

The generation of dehydroquinate from intramolecular aldol condensation of the transient 6-keto intermediate of 2-keto-3-deoxy-heptulosonic acid noted in Section 7.4 above (see Figure 7.8) is the key step in cyclohexane ring formation in all the shikimate-derived metabolites discussed above. A second example of an intramolecular aldol condensation from a sugar substrate to a polyhydroxy cyclohexane framework is catalyzed by inositol-P synthetase[23] (Figure 7.22). The starting substrate is the ubiquitous metabolite glucose-6-phosphate. It is the minor ring open form that provides the C_1 aldehyde as the electrophilic carbonyl group required to set up the internal aldol condensation.

To generate a low energy carbon nucleophile to attack that open chain C_1 aldehyde, inositol-P synthase transiently generates a second carbonyl, as the C_5-ketone of bound glucose-6-P. That internal C_5 ketone arises by transfer of the C_5 hydrogen as a hydride ion to tightly bound NAD (inositol-P synthase is one of the subset of oxidoreductases that bind NAD^+ so tightly it does not get released and so is cryptic in reaction stoichiometry by undergoing a return from a bound intermediate NADH oxidation state). Thus, an enzyme bound dicarbonyl version of substrate is generated reversibly.

Transient generation of 5-keto-glucose and NADH enables the subsequent low energy enolization to the C_6 carbanion. This is the carbon nucleophile that then attacks the open chain electrophilic C_1 aldehyde, forming a new

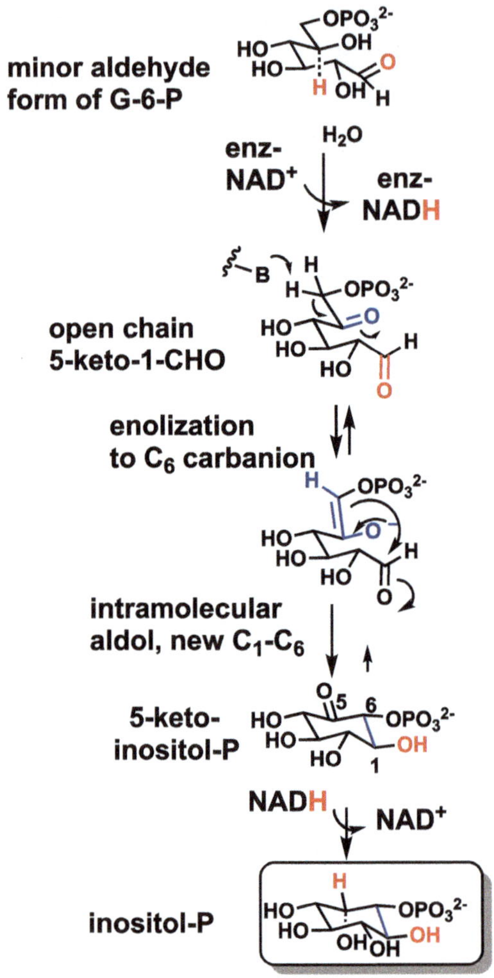

Figure 7.22 Enzymatic conversion of glucose-6-P to the cyclohexane scaffold of inositol-1-P.

C_1–C_6 C–C bond. This is the first cyclic intermediate, a 5-keto-6-phospho-inositol (still using glucose numbering). Return of the hydride from bound NADH to the C_5-ketone then renders the redox steps occult and generates inositol-P (inositol-1-P by inositol numbering).

This polyhydroxy cyclohexane is not as central a metabolite as chorismate but is incorporated into phospholipids (Figure 7.23). Phosphatidylinositols at different phosphorylation states, *e.g.* phosphatidylinositiol-4,5-P_2, are consequential lipid messengers in human signaling pathways[24] (Figure 7.22). On enzymatic hydrolysis inositol 1,4,5-triphosphate (IP_3) is also a signaling metabolite.

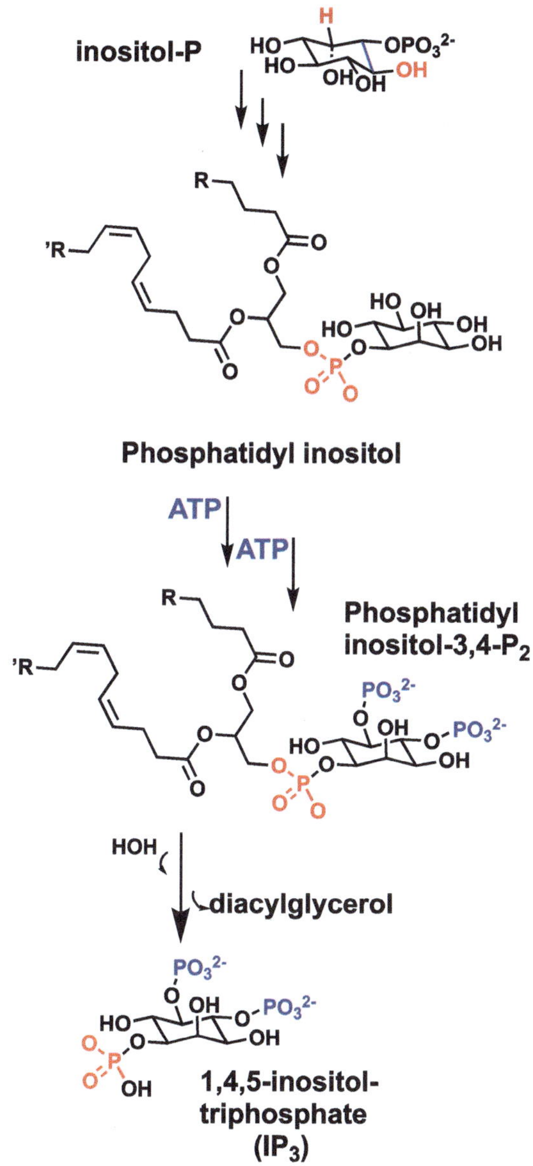

Figure 7.23 Inositol-based signaling molecules.

7.11 Three Additional Classes of Fused Carbacyclic Ring Systems in Biology

Three additional classes of carbacyclic metabolites in nature arise by routes distinct from chorismate pathway logic. Those routes build carbacyclic products from (1) polyketides, (2) isoprenoids, and (3) phenylpropanoid

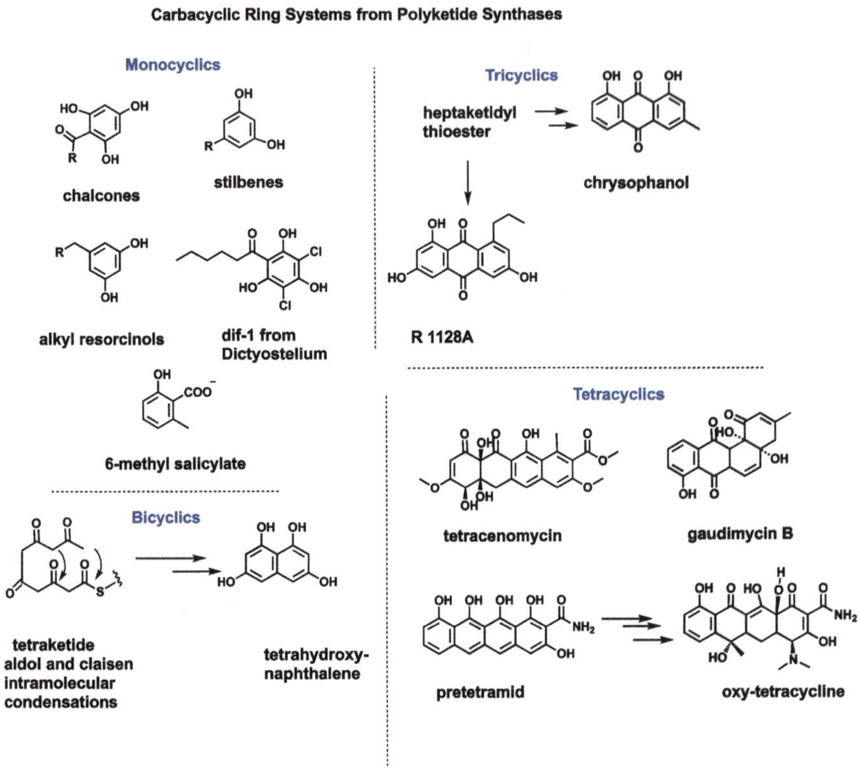

Figure 7.24 Carbocyclic ring systems form polyketide synthases: monocyclic, bicyclic, tricyclic, and tetracyclic representative metabolite scaffolds.

classes of secondary metabolites from plants, fungi, and bacteria.[25] Humans biosynthesize essentially one isoprenoid-based tetracyclic framework, cholesterol. The set of other human steroids are largely oxygenated downstream metabolites from cholesterol (see Chapter 15 for an overview).

This monograph is not a treatise on secondary metabolic pathways in plants and microbes, so we do not focus significantly on polyketides, isoprenoid subtypes, or the plant-based phenylpropanoids. For analysis of the biosynthetic logic for those three universes of natural products, interested readers are directed to the volume: Natural Product Biosynthesis: Chemical Logic and Enzymatic Machinery, 2nd edition, 2022.[25] However, there are a few trends, displayed in Figures 7.24 and 7.25 that give some sense of the range and complexity of carbacyclic frameworks from polyketides, isoprenoids and phenylpropanoids, respectively.

7.11.1 Arrays of Fused Aromatic Scaffolds from Polyketonic *S*-enzyme Species

As noted in Chapter 10, carbanion chemistry, accessed *via* enolate anions, constitutes the major C–C bond-forming strategy in polyketide metabolites.

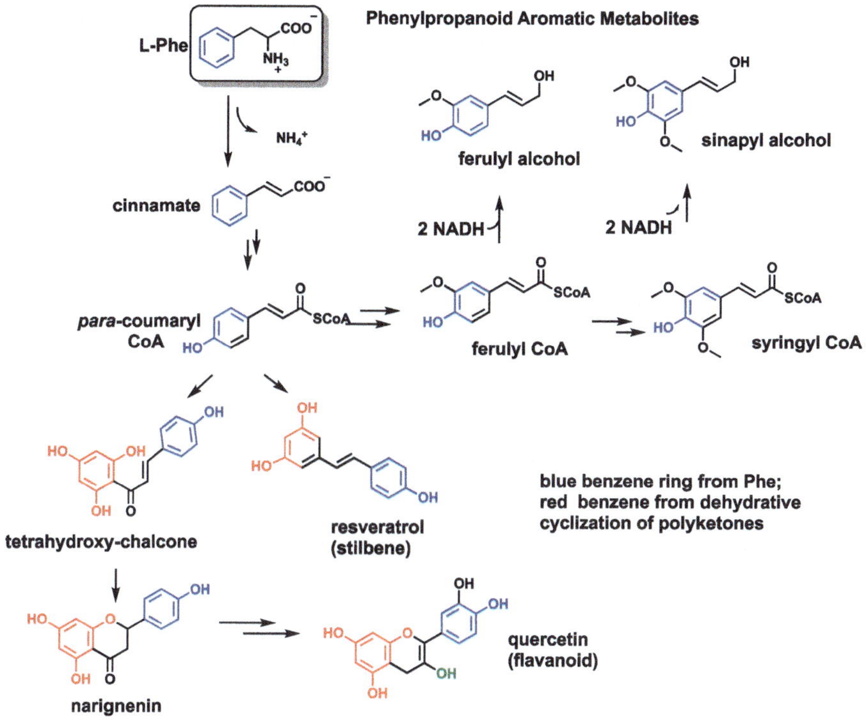

Figure 7.25 Some phenylpropanoid aromatic metabolites with ring systems from both phenylalanine and polyketide cyclases.

Subcategories of polyketide products include linear scaffolds (encompassing polyunsaturated chains), cyclic ethers, decalins and spirotetronates (from pericyclic reactions), macrolactones, and finally fused aromatic ring frameworks. It is that latter category that contain six-membered carbacyclic rings that have aromatized *via* successive dehydrations of initial cyclohexane and cyclohexenone precursors.

Figure 7.24 shows a few representative structures in the classes of monocyclic, bicyclic, tricyclic, and tetracyclic polyketides. The chalcones and stilbene monocycles are typically plant polyketides. Dif-1 is a differentiation signal for *Dictyostelium discoideum* under starvation conditions, while 6-methylsalicylate is among the simplest of the cyclic, aromatized polyketides. Its biosynthetic origin as a polyketide is clearly distinct from the origin of salicylate itself, covered in the preceding section as a chorismate pathway metabolite.

The bicyclic tetrahydroxynaphthalene, is likewise a polyketide end product, built by a distinct strategy from the naphthalene nucleus in menaquinones noted in Figure 7.17. A linear tetraketidyl-*S*-enzyme intermediate undergoes successive aldol and Claisen intramolecular condensations to build each of the two six-membered fused carbacyclic rings. Such iterative polyketidyl-*S*-enzymes are the progenitors to the whole class of fused aromatic polyketides. For example, a heptaketidyl-*S*-enzyme species can cyclize to the chrysophanol or R1128 tricyclic frameworks (Figure 7.22). Analogously, a terminally amidated linear nonaketidyl-*S*-enzyme is successively cyclized first to a tricyclic acyl-*S*-enzyme and then released as tetracyclic pretetramid (Figure 7.24) before a series of tailoring enzymes methylate and oxidize pretetramid to the antibiotic oxytetracycline. The register in which intramolecular cyclizations occur is a feature of how the acyclic polyketidyl chains are folded in their respective cyclase active sites. Such foldamers can yield linear tetracyclic products such as tetracenomycin or the angular angucyclines represented by gaudimycin B (Figure 7.24).

In contrast to the elaborate multistep logic and enzymatic machinery for building and then aromatizing chorismate and isochorismate to monocyclic and fused bicyclic aromatic scaffolds noted earlier in this chapter, the polyketonic-*S*-enzyme intermediates are reactive sets of enolyzable polyketones whose length and number of ketones determine the size of the ultimate fused carbacycle array in the aromatic polyketides. Typically, the polyketide-derived aromatic systems display many oxygen functional groups (many alcohols) while they are essentially devoid of nitrogen as the building blocks are all acyl CoAs.

7.11.2 Carbacyclic Scaffolds from Isoprenoid-based Carbocations

In Chapter 2 we noted that most isoprenoid cyclizations and/or chain extension enzymes employ carbocation chemistry at different stages to capture intramolecular or intermolecular oldefins in the buildup of more complex metabolite frameworks. The head to head condensation of two farnesyl-PPs to the C_{30} acyclic hexaene squalene (Figure 2.15) and then its oxygenative cyclization to the tetracyclic framework of lanosterol (see Chapter 15) represent cascades of carbocation reaction manifolds in enzyme active sites. Indeed, carbocations, whether acyclic allylic cations from Δ^2-prenyl diphosphates, cations on the way to and from the cyclopropane ring in presqualene-PP, or the cascade of cation-driven formation of tetracyclic and pentacyclic steroid carbacyclic scaffolds, are at work in all stages of isoprenoid chemical biology (Chapters 2 and 15). Many of the C_3, C_4, C_5, and C_6 carbacyclic structures in Figures 7.3 and 7.4 arise by such chemical logic. The aromatic A ring in estradiol and estrone arise by iterative oxygenative conversion of the angular methyl group to free formic acid as the cyclohexadiene A ring aromatizes to the phenol.

7.11.3 Phenylpropanoids: Downstream Metabolites from Phenylalanine *Via* Cinnamate

Phenylpropanoid metabolites (Figure 7.25) are largely of plant origin. As the name implies, the phenylpropane scaffolds derive from the nine carbons of phenylalanine as the ur phenyl metabolite. Phenylalanine is enzymatically deaminated to cinnamate, activated as the CoA thioester (Figure 7.25), and oxygenated to *para*-coumaryl CoA. That is the early common intermediate from which two main pathways diverge.

One route proceeds *via* action of two regiospecific iron-based oxygenases that act to hydroxylate the phenyl group, generating catechol or trihydroxyphenol functional groups, followed by mono- or dimethylation to the ferulyl and sinapyl CoA thioesters. Double reduction by two equivalents of NADH yields the ferulyl and sinapyl alcohols (Figure 7.25).

These alcohols can serve as building blocks for enzyme-controlled dimerization *via* radical pathways. Among the dimeric products is the molecule pinoresinol[26] (Figure 7.26). Pinoresinol is just one of several dimers that can be further oligomerized by radical mechanisms to the lignin polymers that provide mechanical strength to woody plant tissues. In turn, the monomethyl catechol groups on each aromatic ring of pinoresinol can be further oxygenated to the indicated hydroxymethyl groups and cyclized to the methylenedioxy bridges in sesamin in sesame seed oil[27] (Figure 7.26). This arm of phenylpropanoid metabolism is an example of radical oxygenation of (pre-existing) phenyl rings and the intervening carbon–oxygen radical coupling chemistry that increases scaffold complexity.

A separate universe of phenylpropanoids arises from coupling of *para*-coumaryl CoA with malonyl CoA units by polyketide synthase catalysts.[28] Figure 7.25 shows that *p*-coumaryl CoA carbons can be funneled either to the stilbene resveratrol (red wine) or to the tetrahydroxy chalcone. Conjugate addition of the OH group from the trihydroxyphenyl ring of that chalcone

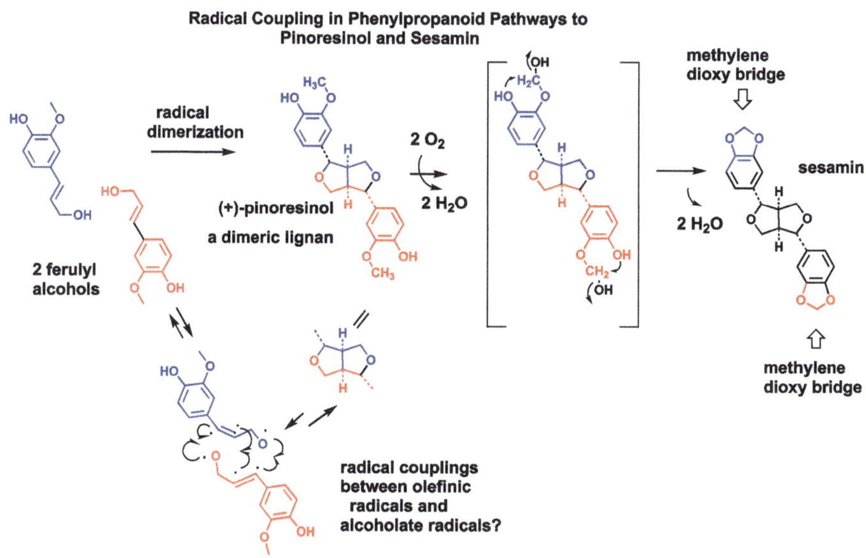

Figure 7.26 Radical couplings in phenylpropanoid pathways to pinoresinol and sesamin.

into the electrophilic enone group produces the flavanone naringenin, the predominant flavanone in grapefruit (Figure 7.25). Further oxygenations produce the flavanoid catecholic antioxidant metabolite quercetin. In Chapter 12 on radicals in chemical biology we note that the oxygenase converting the flavanone naringenin to the isoflavanone genistein involves carbon radicals and hydroxyphenyl group migration before oxygen transfer. In the action of polyketide synthases in the processing of *para*-coumaryl CoA molecules, the triketone moieties can and do cyclodehydrate to catechols and trihydroxyphenols by the kind of carbanion-based aldol chemistry, followed by dehydration, which is observed in the aromatic polyketides discussed earlier in the chapter. In Figure 7.25, those carbanion-based aromatic rings are shown in red while the phenylalanine-derived aromatic ring is shown in blue to differentiate the strategies employed in the biosynthetic logics of their construction.

It is an oversimplification to divide these three large families, polyketides, isoprenoids, and phenylpropanoid carbacycles, into three distinct mechanistic realms. Yet, there is some unifying value in focusing on thioester enolates as carbon nucleophiles for polyketide assembly, carbon radical routes for the lignol class of phenylpropanoids, and carbocations as the dominant motif in isoprenoid/steroid transformations. It is also worth analysis of the strategies for initial formation of cyclohexane rings and then the strategy for progressive oxidation to cyclohexene, cyclohexadiene, and aromatic benzene rings from the different monomers noted in this chapter. Finally, carbacycles are often fused with heterocyclic ring systems in metabolites, leading us to the next two chapters on nature's common heterocycles.

7.12 Overview of Carbacycles in Biologic Systems

The diversity of carbacycles in biologic systems is large. They go from simple 3-6 membered cycloalkanes, typically decorated with peripheral oxygen- or nitrogen-containing carbon-based functional groups to fused polycyclic scaffolds of three to six cyclohexane and/or cyclopentane rings. The carbacycles range from cycloalkanes to cycloalkenes to cyclodienes to aromatic rings systems. The most common aromatic ring systems include the simple phenyl group as part of benzoate, phenol, phenylamine groups. The bicyclic naphthalene nucleus is present in microbes, plants and animals, typically in the context of alkyl-1,4-naphthoquinones as redox carriers ($2e^-/1e^-$). The biosynthetic routes to the phenyl and naphthyl groups go through chorismate as a staging intermediate. In turn, chorismate arises by a tandem pair of intermolecular, then intramolecular aldolases.

Fused polycyclics arise by multiple routes but two common pathways use different chemical logic. The aromatic polyketide synthase machineries (tetracylines, anguicyclines) act *via* polyketonic intermediates complemented by the phenylpropanoid pathways that oxygenate cinnamoyl and coumaryl esters and then mediate radical couplings to dimeric lignols as advanced intermediates.

We turn next to cyclic metabolites where C–O, C–N, and/or C–S bonds are formed within the cyclic frameworks themselves. These are heteroatom metabolites expanding on the pyrimidine and purine heterocycles of Chapter 6. We will observe that the chemical machinery of nature substantially runs on heterocycles.

References

1. S. Samanta, A. Bhattacharyya, S. Das and M. Ghorai, Orrganocatalytic domino approaches for enantioselective formation of six membered carbacycles, *Curr. Org. Chem.*, 2014, **18**, 2842–2856.
2. E. Anslyn and D. A. Dougherty, *Modern Physical Organic Chemistry*, Univerist Science Books, 2006.
3. K. U. Wendt, Enzyme mechanisms for triterpene cyclization: new pieces of the puzzle, *Angew. Chem., Int. Ed.*, 2005, **44**, 3966–3971.
4. S. Ma, D. Mandalapu, S. Wang and Q. Zhang, Biosynthesis of cyclopropane in natural products, *Nat. Prod. Rep.*, 2022, **39**, 915–938.
5. F. H. Vaillancourt, E. Yeh, D. A. Vosburg, S. E. O'Connor and C. T. Walsh, Cryptic chlorination by a non-haem iron enzyme during cyclopropyl amino acid biosynthesis, *Nature*, 2005, **436**, 1191–1194.
6. S. Yang and N. E. Hoffman, Ethylene biosynthesis and its regulation in higher plants, *Annu. Rev. Plant Physiol.*, 1984, **35**, 155–189.
7. M. Houben and B. Van de Poel, 1-Aminocyclopropane-1-carboxylic acid oxidase: the enzyme that makes the plant hormone ethylene, *Front. Plant Sci.*, 2019, **10**, 695.
8. K. M. Herrmann and L. M. Weaver, The shikimate pathway, *Annu. Rev. Plant Physiol. Plant Mol. Biol.*, 1999, **50**, 473–503.
9. E. P. Carpenter, A. R. Hawkins, J. W. Frost and K. A. Brown, Structure of dehydroquinate synthase reveals an active site capable of multistep catalysis, *Nature*, 1998, **394**, 299–302.
10. K. S. Anderson, R. D. Sammons, G. C. Leo, J. A. Sikorski, A. J. Benesi and K. A. Johnson, Observation by carbon-13 NMR of the EPSP synthase tetrahedral intermdiate bound to the enzyme active site, *Biochemistry*, 1990, **29**, 1460–1465.
11. W. J. Guilford, S. D. Copley and J. R. Knowles, The mechanism of the chirismate mutase reaction, *J. Am. Chem. Soc.*, 1987, **109**, 5013–5019.
12. J. D. Hermes, P. A. Tipton, M. A. Fisher, M. K. O'Leary, J. F. Morrison and W. W. Cleland, Mechanisms of enzymatic and acid-catalyzed decraboxylation of prephenate, *Biochemistry*, 1984, **23**, 6263–6275.
13. W. Sun, S. Singh, R. Zhang, J. L. Turnbull and D. Christendat, Crystal structure of prephenate dehydrogenase from Aquifex aolicus: insight into the catalytic mechanism, *J. Biol. Chem.*, 2006, **281**, 12919–12928.
14. W. Skyrud, A. D. R. Flores and W. Zhang, Biosynthesis of cyclohexane carboxyl-CoA highlghts a promiscuous shikimoyl-CoA synthase and FAD-dependent dehydratase, *ACS Catal*, 2020, **10**, 3360–3364.

15. J. Liu, N. Quinn, G. A. Berchtold and C. T. Walsh, Expression, purification, characterization and substrate specificity of isochorismate synthase, *Biochemistry*, 1990, **29**, 1417–1425.
16. F. Rusnak, J. Liu, N. Quinn, G. A. Berchtold and C. T. Walsh, Subcloning of the entereobaction biosynthetic genene etbB: expression, purification, characterization and substrate specificity of isochorismatase, *Biochemistry*, 1990, **29**, 1425–1435.
17. M. Sakatani, F. Rusnak, N. R. Quinn, C. Tu, T. B. Frigo, G. A. Berchtolf and C. Walsh, Mechanistic studies on trans-2,3-dihydroxybenzoate dehydrogenase (EntA) in the biosynthesis of the iron chelator enterobactin, *Biochemistry*, 1990, **29**, 6789–6798.
18. H. Tamir and P. R. Srinivasan, Studies on the mechanism of anthranilate synthase reaction, *Proc. Natl. Acad. Sci. U. S. A.*, 1970, **66**, 547–551.
19. Q.-Z. Ye, J. Liu and C. T. Walsh, PABA Synthesis in Eschericia coli: purification and characterization of PabB as aminodeoxychorismate synthase and enzyme X as aminodeoxychorismate lyase, *Proc. Natl. Acad. Sci. U. S. A.*, 1990, **87**, 9391–9395.
20. J. R. Wildham and D. Rhodes, Biosynthesis and molecular actions of specialzed 1,4-napthoquinone natural products produced by horticutural plants, *Hortic. Res.*, 2016, **3**, 16046.
21. M. Jiang, Y. Cao, M. Chen, X. Chen and Z. Guo, Menaquinone biosynthesis in Eschericia coli: identification of 2-succinyl-5-enolpyruvyl-6-hydroxy-3-cyclohexene-1-carboxylate as novel intermediate and reevaluation of MenD activity, *Biochemistry*, 2007, **46**, 10979–10989.
22. J. J. Truglio, K. Theis, Y. Feng, R. Gadja, C. Machutta, P. J. Tonge and C. Kisker, Crystal structure of Mycobacterium tuberculosis Men B, a key enzyme in vitamin K2 biosynthesis, *J. Biol. Chem.*, 2003, **278**, 42352–42360.
23. A. J. Stein and J. H. Geiger, The crystal structure and mechaism of 1-L-myo-inositol-1-phosphate synthase, *J. Biol. Chem.*, 2002, **277**, 9484–9491.
24. X. Zhang and P. Majerus, Phosphatidylinositol signaling mechanisms, *Semin. Cell Dev. Biol.*, 1998, **9**, 153–160.
25. C. T. Walsh and Y. Tang, *Natural Product Biosynthesis: Chemical Logic and Enzymatic Mechanisms*, Royal Society of Chemistry, London, 2nd edn, 2022.
26. L. B. Gavin, H.-B. Wang, A. L. Crowell, D. L. Begar, D. M. Martin, S. Sarkaen and N. G. Lewis, Stereoselective bimodular phenoxy radical coupling by an auxilairy (dirigent) protein without an active center, *Science*, 1997, **275**, 362–367.
27. E. Ono, M. Nakai, Y. Fukui, N. Tomimori, M. Fukuchi-Mizutani, M. Saito, H. Satake, T. Tanaka, M. Kasuta, T. Umezawa and Y. Tanaka, Formation of two methylenedioxy bridges by a Sesamum CYP81Q protine yielding a furofuran lignan (+) sesamin, *Proc. Natl. Acad. Sci. U. S. A.*, 2006, **103**, 10116–10121.
28. D. Yu, F. Xu, J. Zeng and J. Zhan, Type III polyketide synthases in natural product biosynthesis, *IUBMB Life*, 2012, **64**, 285–295.

CHAPTER 8

Heterocycles: I

8.1 Introduction to Common Heterocycles in Metabolism

Along with the types of naturally occurring carbacyclic scaffolds just summarized in Chapter 7, nature uses a plethora of heterocycles to enact the chemistry of life. The six most common chemical elements in humans are C, H, N, O, P, and S so it is not surprising that natural heterocycles mix one or more oxygen, nitrogen and sulfur atoms with C–H and C–C bonds in heterocyclic metabolites. The various heterocycles can be free standing, fused to one another, or fused to carbacyclic rings. There are few if any direct C–P-containing biologic heterocycles identified, leaving C–O, C–N, and C–S heterocycles, including rings with more than one type of heteroatom. To emphasize the importance of heterocycles in carbon chemical biology, this is the first of two chapters on heterocycles. The second chapter focuses exclusively on vitamin heterocyclic frameworks and how they enable discrete functions in 9 of the 13 human vitamins.

8.1.1 Introduction to Oxygen Heterocycles in Metabolism

Depending on the starting substrates and the distinctive chemical logic of metabolic pathways, one can find products that have only oxygen-containing heterocycles.[1] Polyketides, isoprenoids and phenylpropanoids, surveyed for carbacycles in the last chapter, are essentially nitrogen-free metabolic zones; oxygen-containing heterocycles emanating from these pathways are typically nitrogen-free (Figure 8.1).

The simplest heterocycles have one heteroatom. The cyclic ethers fall into this class. For such oxygen heterocycles found in nature, ring sizes typically vary from three to six carbons, from epoxide to oxetane to tetrahydrofuran to

Chemical Biology No. 21
The Chemical Biology of Carbon
By Christopher T. Walsh

Published by the Royal Society of Chemistry, www.rsc.org

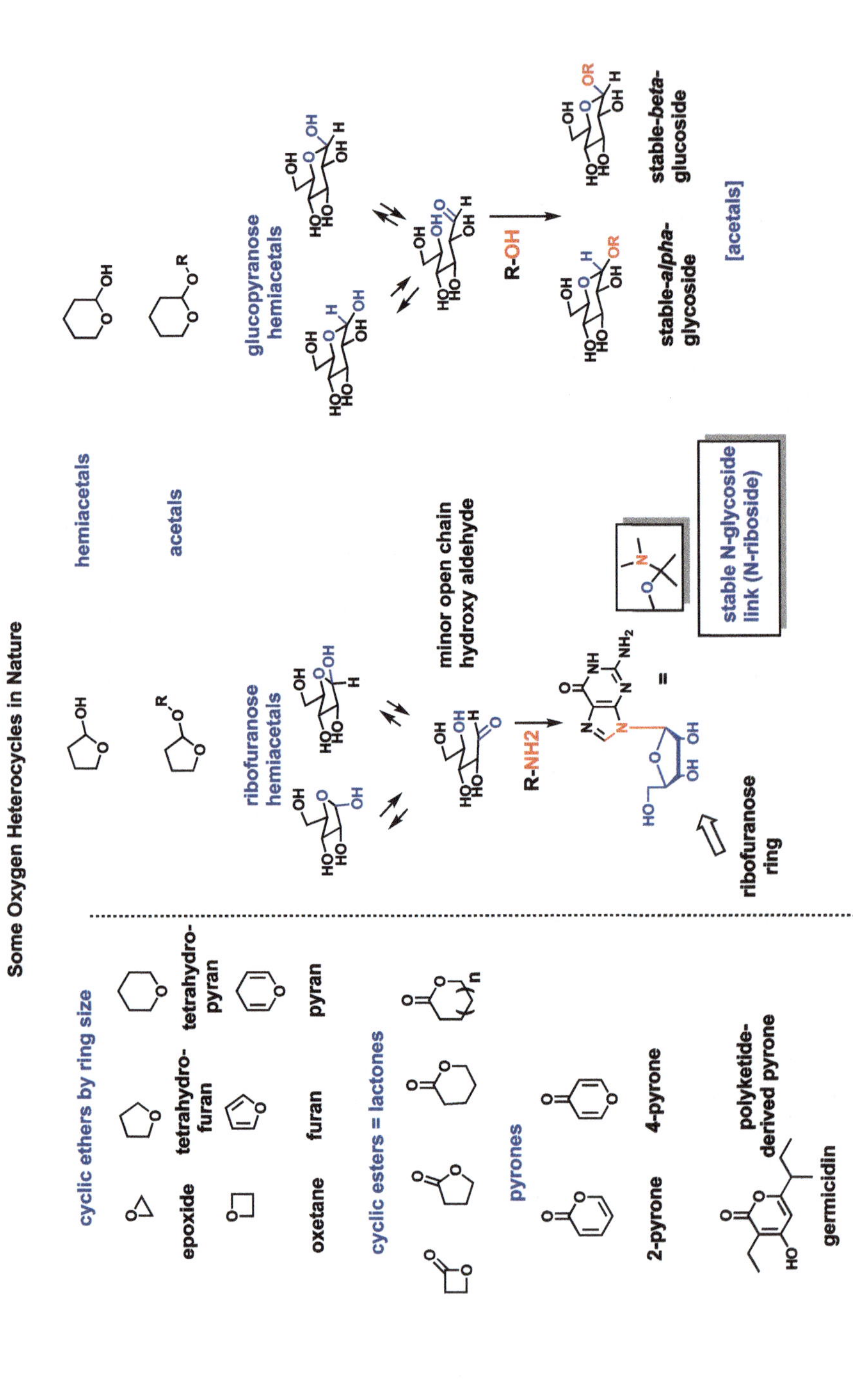

Figure 8.1 A set of some of the oxygen heterocycles in natural molecules: ethers, lactones, hemiacetals, and pyrones.

tetrahydropyran[2] (Figure 8.1). We have noted the existence of all four of these ring systems in previous chapters, notably Chapters 2 and 3. These are formally cyclic ethers. The oxidation state of the five- and six-ring ethers can vary by four electrons, for example from the saturated tetrahydro to dihydrofuran to heteroaromatic furan and pyran rings. In parallel with the small strained three member cyclopropanes and four member cyclobutanes of Chapter 7, epoxides and oxetanes have comparable ring strain, and reactivity, typically as electrophiles at carbon.

Also depicted in Figure 8.1 are lactones,[3] cyclic oxoesters, whose routes of biologic formation are summarized in Chapter 3. The third group of oxygen heterocycles in Figure 8.1 are hemiacetals[4] which represent protected cyclic forms of hydroxy-aldehydes, noted also in Chapter 3. A fourth class are the cyclic 2- and 4-pyrone isomers, cyclohexadienone-containing ethers/esters.

The hemiacetal forms of ribose and glucose may have the most central metabolic roles of the mono-oxygen-containing heterocyclic metabolites. Hexoses accumulate as reversible six member cyclic hemiacetals (pyranoses) while ribose presents as the five-ring hemiacetal isomers at C_1 (furanoses) (Figure 8.1). These oxygen functional groups are not simple cyclic ethers. These cyclic hemiacetal groups act as reversible tetrahedral protectants for the planar electrophilic carbonyl groups. When the –OH of the hemiacetal linkage is subsequently replaced by an alcohol group (ROH), *e.g.* in oligo- and polysaccharides, the cyclic pyranose and furanose structures become chemically stable acetal linkages[5] (Figure 8.1). When an amine ($R–NH_2$) instead acts to capture the C_1 hemiacetal carbon of the α-ribofuranose oxygen heterocycle, the stable tetrahedral product is an *N*-glycosidic linkage[6] (Figure 8.1). This is not a rare obscure linkage. As shown for guanosine in Figure 8.1, nature has chosen to employ these *N*-ribosides in the DNA genomes of cells.

Oxygen atoms are only rarely present adjacent to each other in a naturally occurring heterocycles. Three examples of metabolites with such cyclic endoperoxide links (–C–O–O–C–) are shown in Figure 8.2. In all cases the cyclic peroxide functionality derives from reductive metabolism of O_2, by one-electron routes (see Chapter 12).

The fungal metabolite verruculogen has five fused rings in a 6-5-6-6-5 array. The sixth ring is the eight member cyclic endoperoxide. Inspection of the scaffold indicates a tryptophan-prolyl dipeptide has undergone cyclization to the five-ring diketopiperazine heterocycle, followed by prenylation and finally endoperoxide formation. The plant antimalarial agent artemisinin has a 6-6-7 tricyclic core.[7] One of the rings is a six member lactone, another the seven member cyclic ether. Embedded across that seven member ether is a cyclic endoperoxide that creates two bridging six membered rings. Remarkably, artemisinin is stable enough chemically to be formulated as a human therapeutic antimalarial agent. The third molecule in Figure 8.2 is in the prostaglandin class of lipid hormones, arising by oxygenation of the 20 carbon eicosatetraenoate, identified as a polyolefin back in Chapter 2. Prostaglandin G2 has both a 9,11-cyclic endoperoxide and a 15-α-hydroperoxide group.

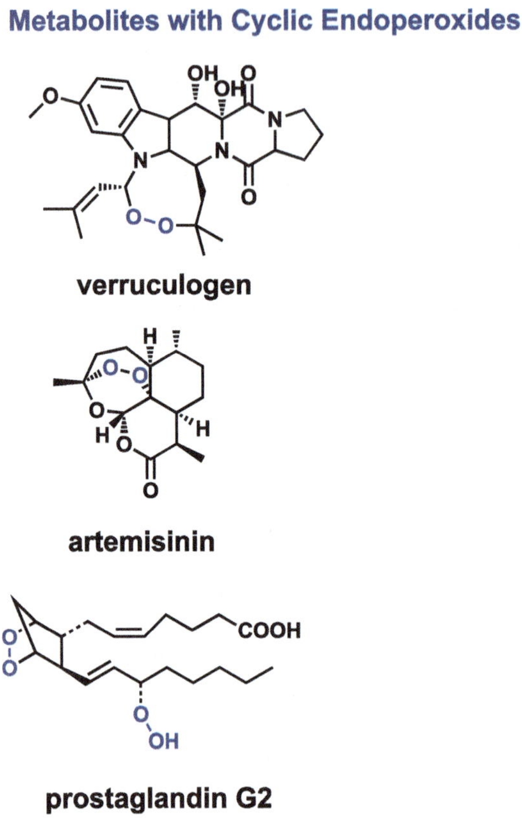

Figure 8.2 Metabolites with cyclic endoperoxides.

Oxygen heterocycles can be standalone ring systems or fused in bicyclic, tricyclic, and larger polycyclic arrays as shown both in Figure 8.2 and in Figure 8.3. The phenylpropanoid metabolite psoralen is a tricyclic metabolite with two of the three furan-benzene-lactone rings as different oxygen heterocycles. The furan is heteroaromatic, thanks to the electron lone pair on oxygen. Among plant natural products are coumarins, lactones, furans, and methylene dioxo ethers, highlighted in blue as oxygen heterocycles (Figure 8.3).

8.1.2 Common Nitrogen Heterocyclic Metabolites

Nature likewise builds nitrogen heterocycles.[8] Most central to life and already discussed in Chapter 6 are the pyrimidine and purine ring building blocks for both RNA and DNA synthesis (Chapter 6). Pyrimidines are monocyclic, while purines, formally imidazolo-pyrimidines are bicyclic, with a five member imidazole ring fused onto a pyrimidine core. We will not discuss the purine and pyrimidine heterocyclic building blocks further in this chapter.

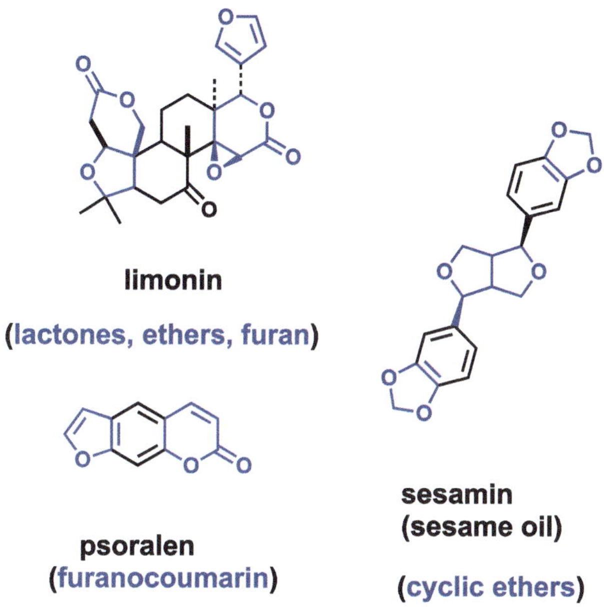

Figure 8.3 Complex oxygen heterocyclic scaffolds: limonin, psoralen, and sesamin.

Figure 8.4 displays some additional prototypic carbon–nitrogen-only ring systems in carbon chemical biology. They include the monocyclic amine analogs of the cyclic ethers of Figure 8.1, including highly reactive aziridine and azetidine, to the more stable pyrrolidine and piperidine cyclic secondary amines. Pyrrolidine rings are fully saturated and are represented by proline in proteins. The fully aromatic mono-nitrogen five membered ring is pyrrole and it is the core of the tetrapyrrole pigments of life (see Chapter 16).

Piperidine is the fully saturated six-ring system. The 1,4-dihydro analog, a cyclohexadienyl secondary amine also known as a dihydropyridine, is the group found in NAD(P)H coenzymes. The fully aromatized ring is a cationic pyridinium ring. *N*-Alkyl pyridines are cationic pyridinium ions that can function as electron sinks for electron transfer reactions (*e.g.* hydride ion transfers in $NAD(P)^+$).

The second set of nitrogen heterocycles in Figure 8.4 are a biologically relevant subset with two nitrogen atoms in a five or six member heterocyclic ring. These include pyrimidine, pyrazine and imidazole ring systems. The biologic role for pyrazine rings in vitamin B9 coenzyme forms are the dihydro and tetrahydro oxidation states. Folate vitamin tablets have the pyrazine ring in the air stable fully oxidized state. That ring must be reduced by four electrons *in vivo* to access the active tetrahydro oxidation state.[9]

Cyclic amides include the thermodynamically activated, strained β-lactams, the warhead in penicillin antibiotics, and the much more stable γ- and δ-lactam ring systems.

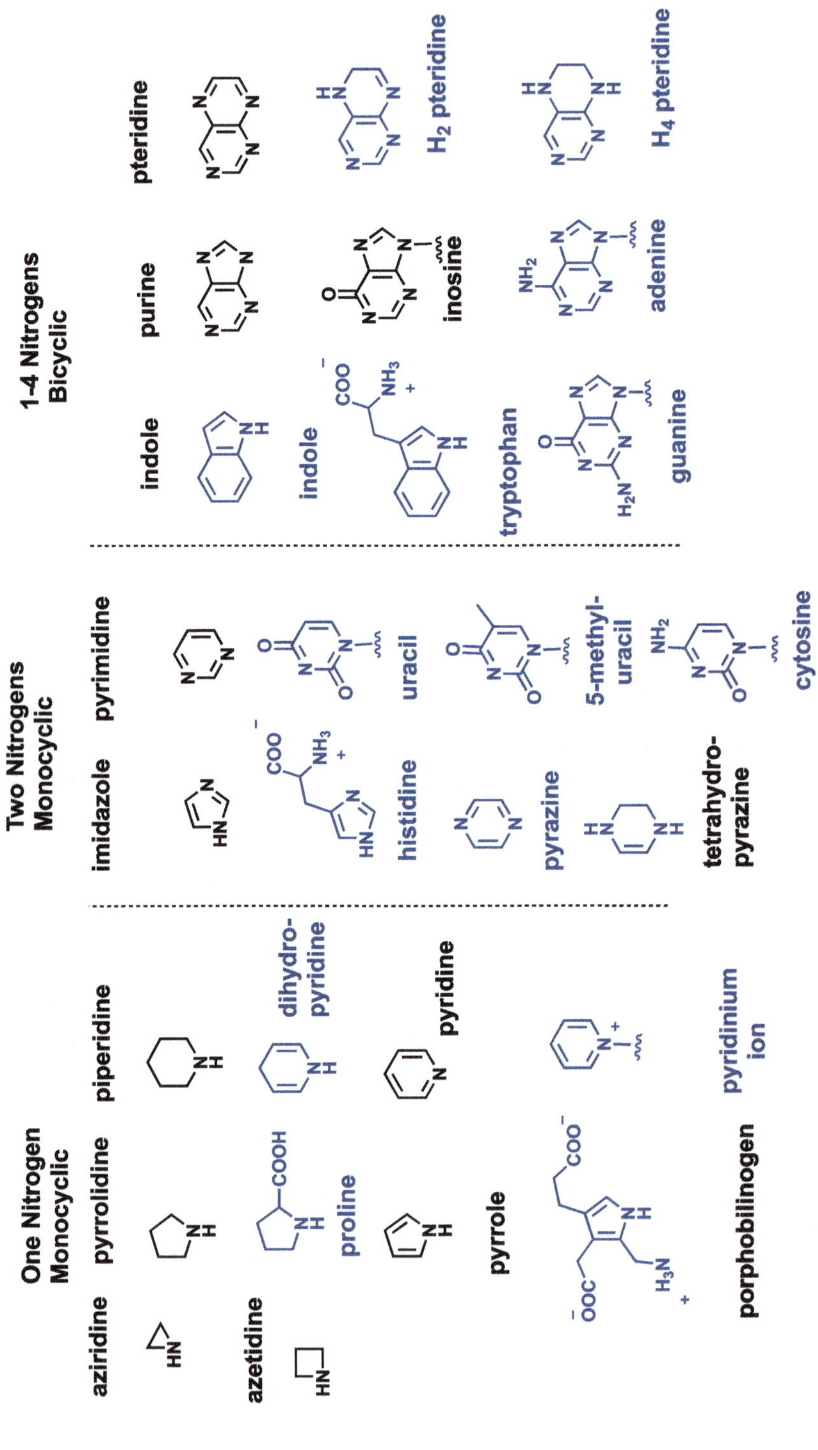

Figure 8.4 Representative nitrogen heterocycles in nature.

A somewhat more complex subset of biologic nitrogen heterocycle are bicyclic systems with anywhere from one to four nitrogen atoms (Figure 8.4). In this category are the indole, purine, and pteridine ring systems. We dealt with purine chemical biology in Chapter 6, note pterin chemical biology in the next chapter, and examine indole reactivity later in this chapter. Figure 8.4 is not of course an exhaustive list but gives some flavor to the widespread use of cyclic amines, as basic or nucleophilic groups, and their fully heteroaromatic versions across many metabolic pathways.

Biologic heterocycles with sulfur are more restricted.[10] The five member, saturated monosulfur ring system known as a thiane ring is part of the framework of biotin vitamin B7 (see Chapter 10) (Figure 8.5) and also penicillin antibiotics (see below). Thiazole rings are not found in the side chains of protein amino acids, but there is a remarkable tetra-heterocyclic core of a trithiazolylpyridine in about 100 thiazolyl peptide antibiotics (Figure 8.5). We will also examine the remarkable properties of a five member heterocycle (Figure 8.5) with both nitrogen and sulfur, a quaternized thiazolium ring, in vitamin B1 in the next chapter. Other heterocycles with combinations of two noncarbon atoms, such as nitrogen and oxygen, are also found as biologic metabolites. Among them are dihydroaromatic oxazolines and heteroaromatic oxazole rings. The N and O in heteroaromatic oxazoles are in a 1,3 relationship. The corresponding isoxazoline and isoxazole with 1,2 connectivity of the N and O atoms are rarer, found in isoxazolinone rings of D-cycloserine and in ibotenic acid, respectively (Figure 8.5).

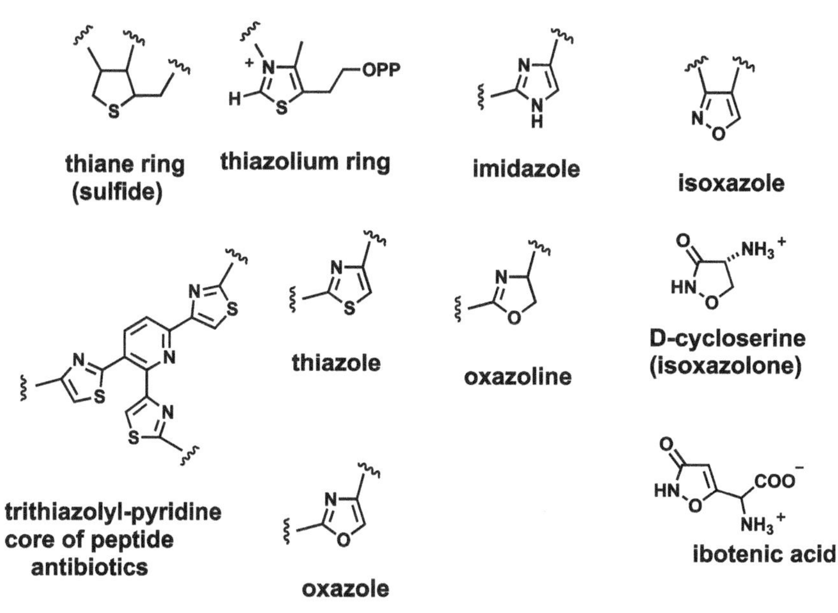

Figure 8.5 Additional five and six member nitrogen heterocyclic functional groups.

Thus, heterocyclic ring structures are pervasive in nucleic acids, in 3 of the 20 proteinogenic amino acids incorporated into proteins, and in sugar metabolism. In the rest of this chapter we examine some of the functional properties and consequences of these types of common oxygen and nitrogen heterocycles. We leave the workings of the nine different heterocycles in human vitamins to the next chapter, Chapter 9. We start with oxygen heterocycles and then turn to nitrogen heterocycles, including a brief overview of amino acids as building blocks for cyclic nitrogen scaffolds in alkaloids.

8.2 Cyclic Ethers and Lactones from Polyketide Assembly Lines

In addition to the subclass of aromatic carbacyclic polyketides exemplified in Chapter 7, two additional subclasses of polyketides lead to quite distinct types of nitrogen-free, oxygen heterocycles, the polyethers and the macrolactones. They arise by distinct strategies and yield very different product structural types. Figure 8.6 notes that three types of cyclic products are generated from polyketide synthase assembly lines.[11] One class contains carbacycle-fused linear polyaromatic scaffolds in metabolites such as oxytetracyline (Chapter 7). A second class has macrocyclic lactones in the erythromycin family. The third type contain oxo-ether heterocycles as the core of the polyether ionophores.

8.2.1 Cyclic Ether-containing Ionophores

The polyethers[12] such as monensin, lasalocid, and ionomycin have clusters of tetrahydrofuran and tetrahydropyran rings whose oxygen atoms serve as ligands for cations to be transported through membranes to act as ionophores (Figure 8.7). We note in Chapters 2 and 3 that the five member and six member cyclic ether functionalities typically arise by post-assembly line epoxidations and ring expansions from nascent polyene PKS assembly line products.

The polyenes emerging from the PKS enzymatic machinery (Chapter 2) are then subjected to tandem epoxidations by flavoenzyme monooxygenases (Chapter 3). In turn, the epoxides are then converted in cascades to tetrahydrofuran and tetrahydropyran cyclic ethers as depicted in Figure 3.28. In maturation of the nascent triene released from the monensin PKS enzyme machinery, triple epoxidation sets off an internal hydroxy group-initiated cascade (Figure 8.8) On opening, the epoxide oxygens participate as nucleophiles in formation of four cyclic ethers, three tetrahydrofurans and one tetrahydropyran. The resultant dehydroxymonensin is one P450-catalyzed hydroxylation (see Chapter 12) away from the monensin ionophore.

One extreme in the natural polyether marine metabolite series is brevetoxin B which is proposed to arise by a metabolic precursor with nine epoxides, presumed in turn to arise from nine intramolecular alkene

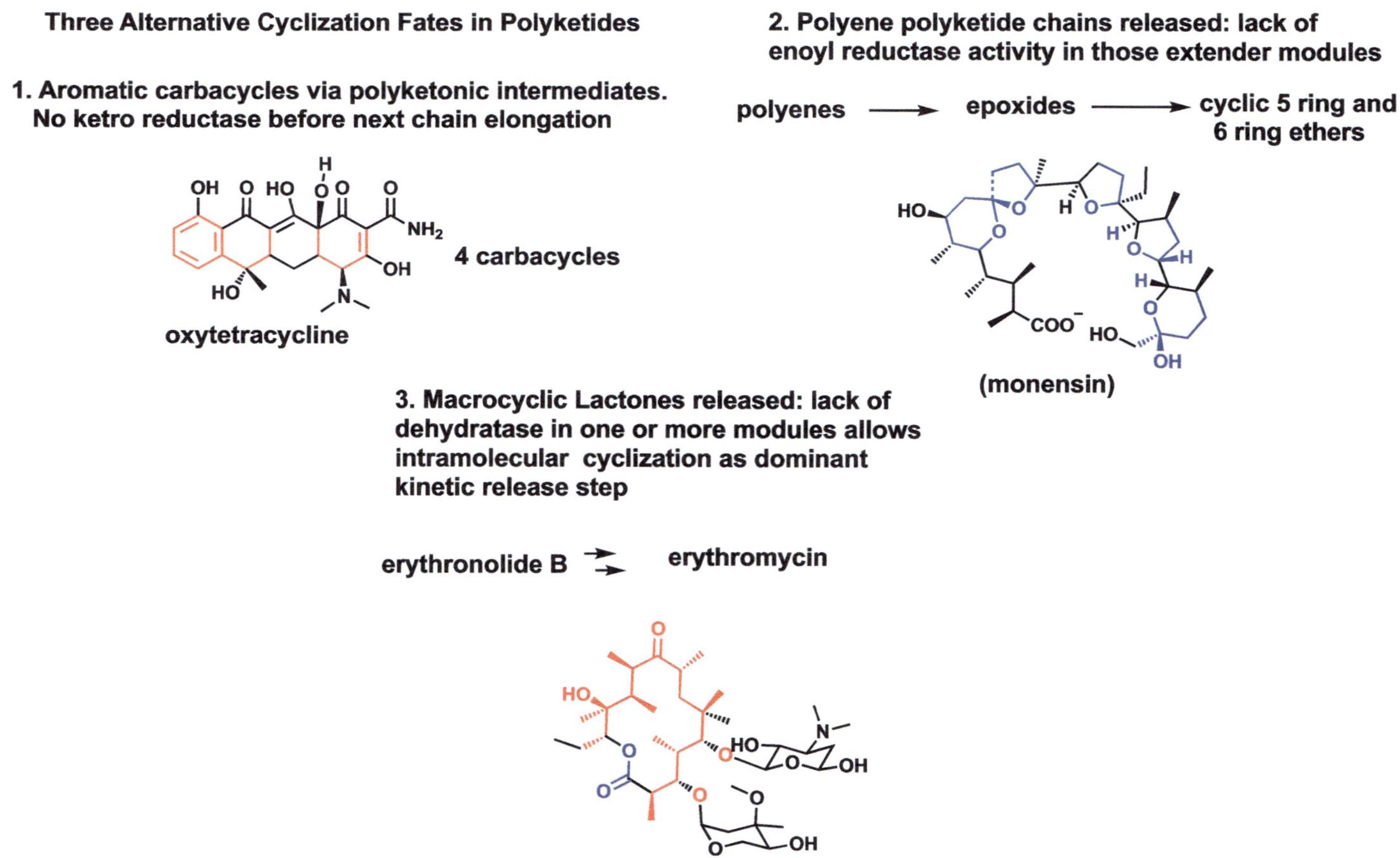

Figure 8.6 Three alternative cyclization fates from polyketide synthases: aromatic carbacycles; polyenes to epoxides to larger ring cyclic ethers; macrocyclic lactones.

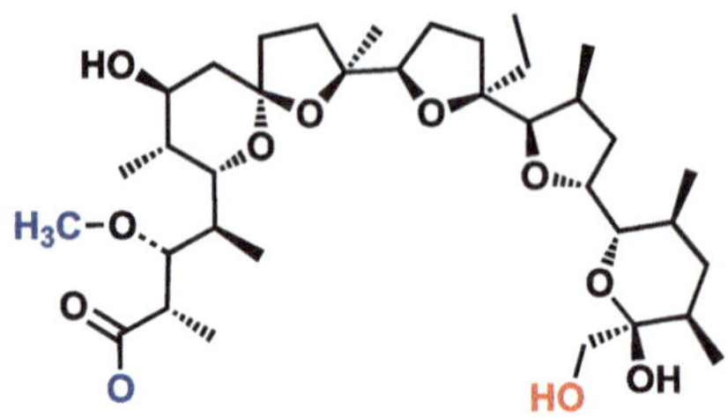

Monensin

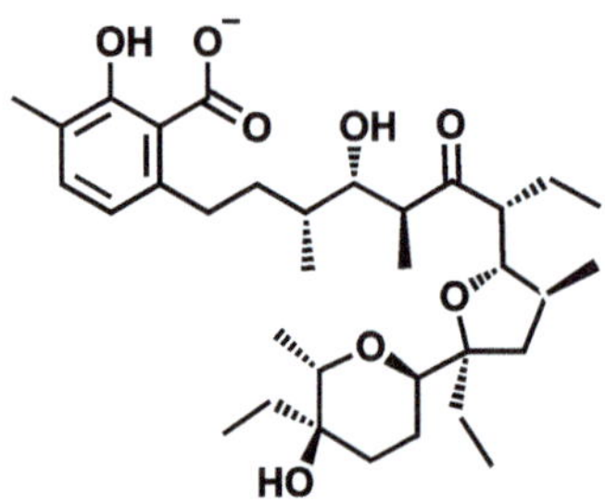

Lasalocid

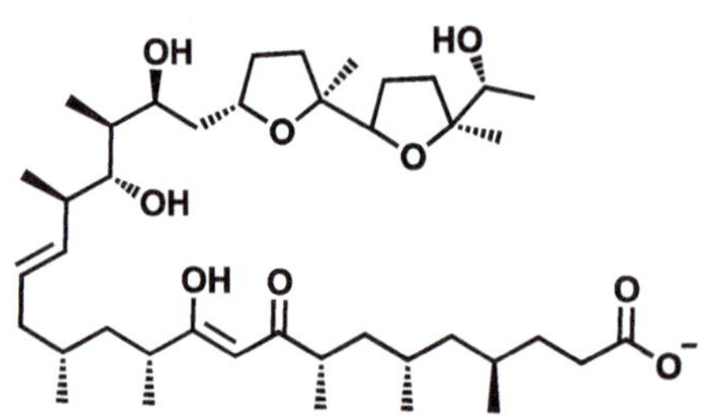

Ionomycin

Figure 8.7 Cyclic ether-containing ionophores: monensin, lasalocid, and ionomycin.

groups (Figure 8.9). The nine cyclic ethers in brevetoxin are: four tetrahydropyrans, one seven-ring, three eight-rings, one nine-ring ether.

8.2.2 Macrocyclic Polyketide Lactones

In contrast to type I PKS catalysts that build polyketonic chains on the way to fused polycyclic aromatic metabolites such as tetracyclines (see Chapter 7), type II bacterial polyketide synthases avoid the buildup of polyketonic intermediates in the growing chains by reducing the initial 3-ketoacyl chains sequentially to the set of 3-OH, 2-enoyl, or fully reduced methylene units (discussed in Chapters 2 and 3). Features distinguishing polyketide assembly lines from fatty acid synthases are the potentially incomplete reductive

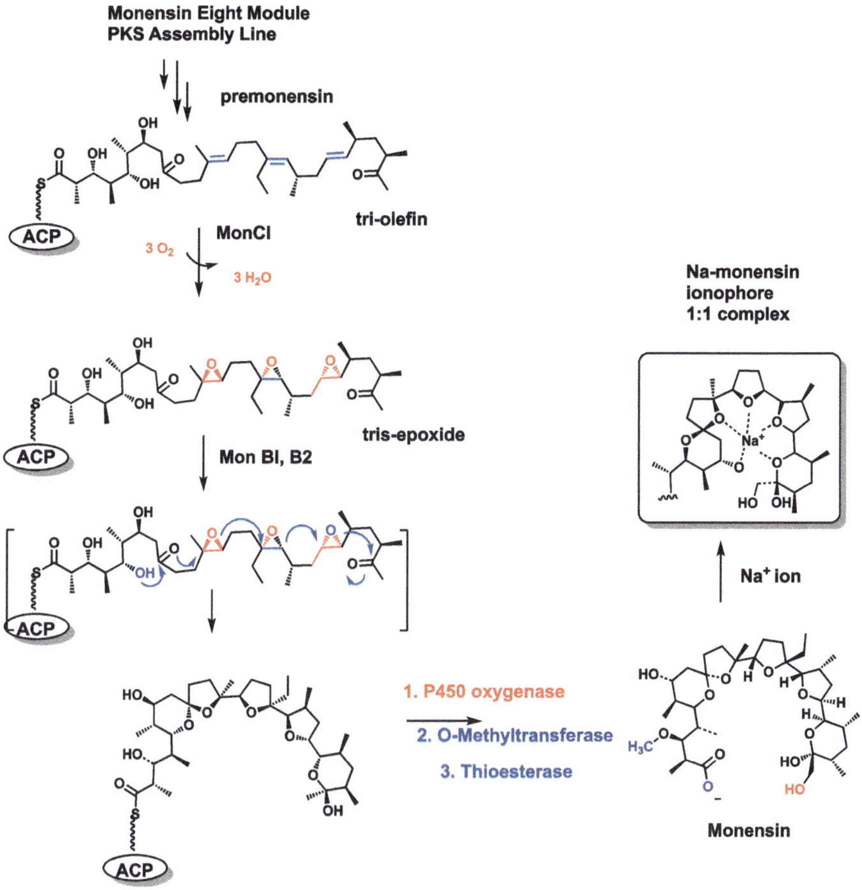

Figure 8.8 Late-stage steps in the biosynthesis of monensin.

processing of the 3-ketoacyl, the 3-hydroxyacyl, and the enoyl intermediates during chain elongations. Consequently, chain elongation cycles might result in persistence of any of those three functional groups, ketone, alcohol, or conjugated alkene. The full-length acyl chain that accumulates just prior to cyclization by deoxyerythronolide B synthase has four alcohol and two ketone groups embedded in the growing chain[13] (Figure 8.10). Only the most distal alcohol then serves as the internal oxygen nucleophile in cyclization of the covalent oxoester adduct in the thioesterase (TE) domain of the synthase to give the 14 member lactone deoxyerythronolide B.

Correspondingly, on the way to the 16 membered macrolactone released by tylactone synthase, the acyl chain contains two enoyls, one ketone, and one alcohol group (Figure 8.11). That lone alcohol at C_{15} attacks the oxoester C_1 carbonyl group to release tylactone.[14] In both cases there are post release enzymatic modifications, hydroxylation and glycosylation, to mature erythromycin A or tylosin, the bioactive forms of the two macrolactone antibiotics.

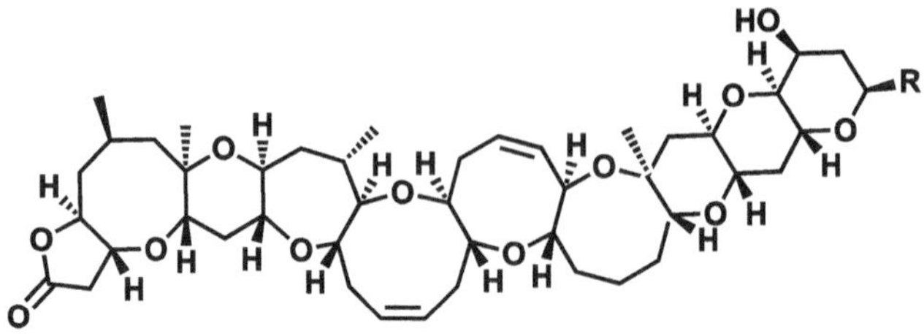

Figure 8.9 Brevetoxin: nine cyclic ethers and one lactone.

The persistence of ketones, alcohols, or alkenes in the acyl chains accumulating in the active sites of different subclasses of polyketide synthases may vary. In some instances, any of those three functional groups may persist in downstream released metabolites. In other cases, each of the three functional groups may enable chemical changes to product scaffolds. The conversion of alkenes to epoxides and on to tetrahydrofurans and tetrahydropyran ethers has been noted in several earlier contexts, including monensin maturation in Section 8.3. The persistence of ketones enable adjacent enolization to carbanions needed for the fused carbacycles in molecules such as pretetramide on the way to oxytetracycline in Chapter 7. Here, the third persistent functional group, an alcohol, can lead to its use as a regioselective oxygen nucleophile on the way to macrolactones of different ring sizes.

8.3 Baeyer–Villiger Route to (Cyclic) Lactones

An alternative, substantially restricted route to small and medium ring cyclic lactones was noted briefly in Chapter 3 in a subclass of flavin dependent monoxygenases. The flavin 4a-peroxyanion adducts act as oxygen nucleophile on cyclic ketones.[15] The resultant tetrahedral adduct breaks down in the forward direction with C–C single bond migration to assist in cleavage of the O–O peroxy bond (Figure 3.26). That is both the ring expansion step and the oxygen insertion step to create the ring expanded lactone.

8.4 Pyrones in Phenylpropanoids. Flavones in Foods and Spices

Dozens of fused phenol-pyrone bicyclic frameworks are found in polyhydroxy flavanoid scaffolds[16] where the phenolate and catecholate forms

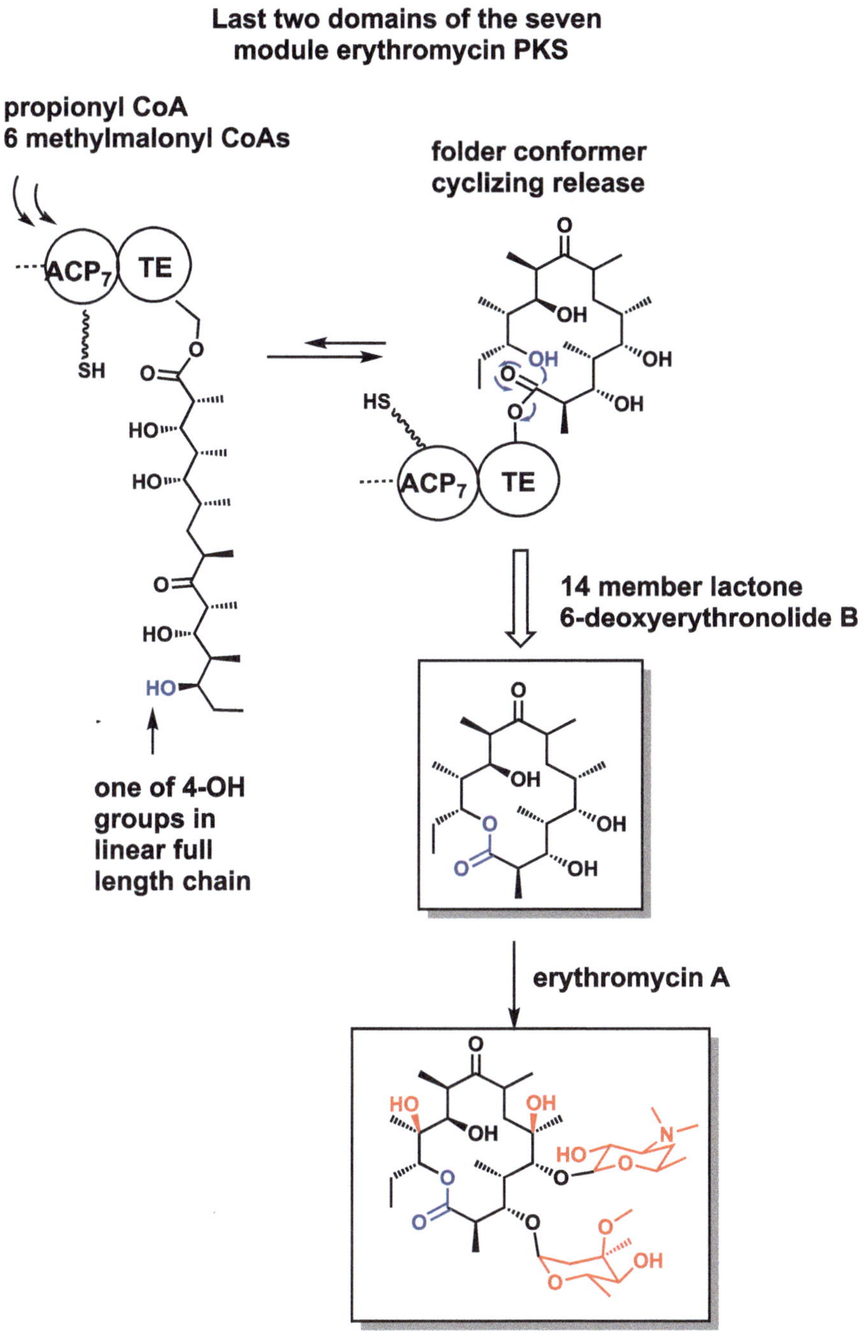

Figure 8.10 Last two domains of final module of deoxyerythronoactone B synthase releases the 14 member macrolactone.

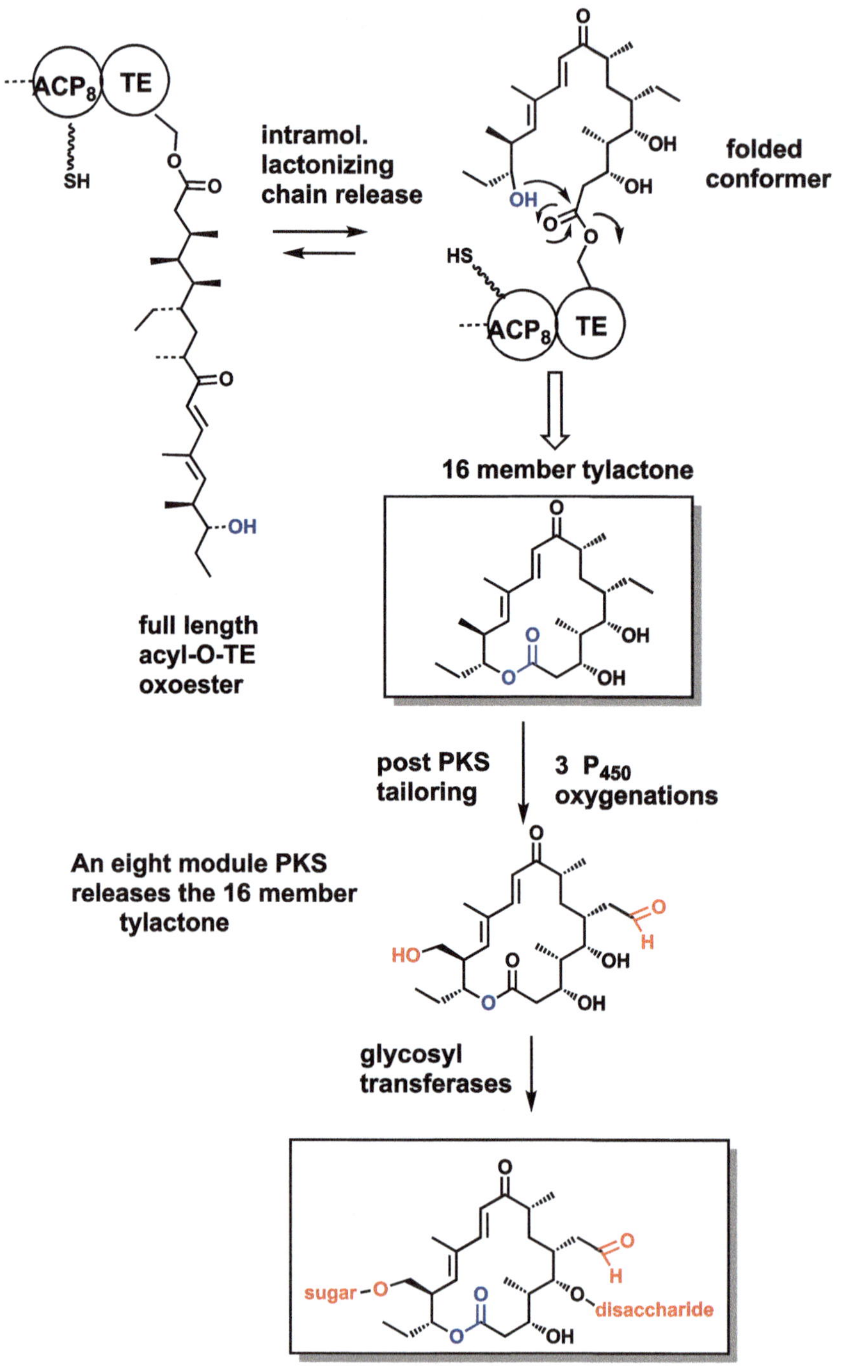

tylosin

Figure 8.11 An eight module tylosin polyketide synthase releases the 16 member macrocyclic tylactone.

are thought to be antioxidant metabolites. The classic flavanone naringenin arises from iterative addition of three malonyl CoA molecules to *para*-coumaryl CoA by polyketide synthase logic to create the indicated tetraketidyl thioester. The malonyl derived triketide moiety can cyclize by generation of the C_6 carbanion (C_5 enolate) onto the C_1 thioester carbonyl (Figure 8.12). Consecutive enolizations turn the triketo-cyclohexane ring into the aromatized trihydroxy phenol moiety. The product is naringenin chalcone. One of the phenolic-OH groups can then act as nucleophile in conjugate addition to the enone moiety to create the 4-pyrone ring fused to the diphenol ring in naringenin, a flavanone.[17] This pathway illustrates how each ring in a fused pyrone-benzene ring system is generated.

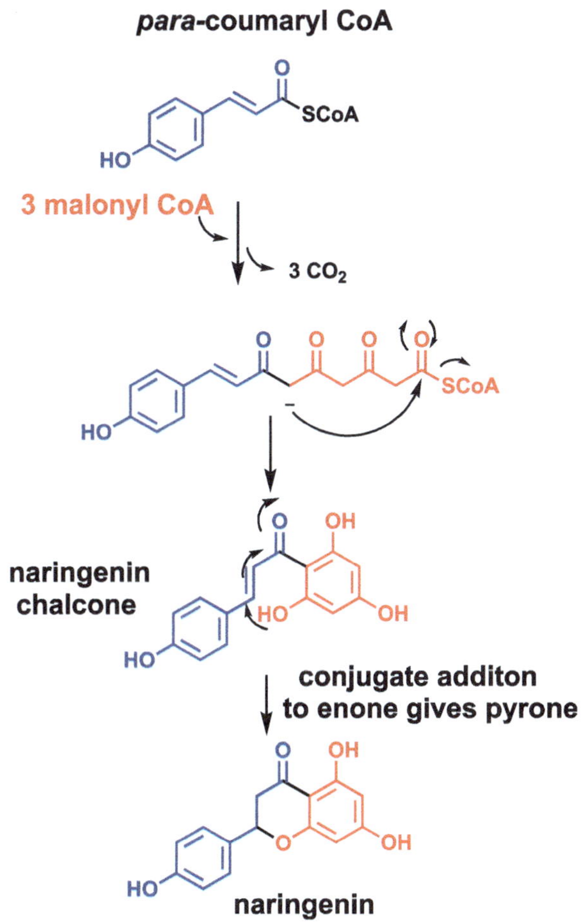

Figure 8.12 Chain extension of *para*-coumaryl CoA by three cycles leads to naringenin chalcone and then the pyrone ring in naringenin.

8.5 Nitrogen Heterocycles

While polyketides, isoprenoids, and phenylpropanoid-derived cyclic metabolites contain both carbocyclic and oxygen heterocyclic ring systems, they are largely devoid of nitrogen atoms. However, the vast class of alkaloid metabolites have been defined historically by a basic nitrogen atom that is protonatable under acid conditions but deprotonated under basic conditions, allowing partition in aqueous or organic phases at different stages of isolation. Not surprisingly, the building blocks for alkaloids[18] are several of the metabolically central amino acids.

Figure 8.13 shows several monocyclic, bicyclic, and tricyclic nitrogen heterocyclic ring systems that derive metabolically from the indicated starting amino acids. The first column shows saturated monocyclic rings, pyrrolidine, and piperidine as well as the 5,5-bicyclic pyrrolizidine and the 5,6-tropane ring system. The second column lists some bicyclic and tricyclic heteroaromatic rings systems.

Correspondingly, Figure 8.14 shows the structures of seven alkaloid natural products. Coniine, a hemlock poisonous metabolite, the first alkaloid to have its structure determined back in the 19th century, has only a piperidine ring system. The frog poison histrionicotoxin D has a core 5,5-pyrrolizidine ring system. Cocaine is a reminder of the pharmacologic effects of *N*-methyltropane ring systems. Nicotine has both a saturated pyrrolidine and an aromatized pyridine ring. Quinine has a bicyclic quinoline heteroaromatic core and an aminobicyclooctane moiety. Two metabolites derived from the indole ring of tryptophan are the tetracyclic lysergic acid and the unusual chromopyrrolic acid with a pyrrole ring wedged between two indole ring systems. These seven molecules, out of thousands of alkaloids, give a flavor of the diversity of nitrogen heterocycles built from amino acid starting units. Readers interested in the formation of these ring systems are directed to recent literature reviews on alkaloid biosynthesis.[11]

8.5.1 Pyrimidine and Purines

The two heterocycles most fundamental to life may be the (two) purine rings and (three) pyrimidine rings that are the building blocks for all RNA and DNA biopolymer chains (Figure 6.1). We devote Chapter 6 to purines and pyrimidines, outlining the simple chemical strategies that underlie the separate assembly of the pyrimidines and the imidazolyl-pyrimidines (the purines). Chapter 6 also focuses on the importance of the four kinds of nitrogen functional groups embedded in and attached to carbons in those heterocycles: amides, ureas, amidines, and guanidinium functional groups. These nitrogen functional groups and prototropic isomeric forms explain the Watson–Crick binding modes that govern A–T (A–U) and G–C base pairs, with their two pairs or three pairs of hydrogen bonds, respectively (Figure 6.1).

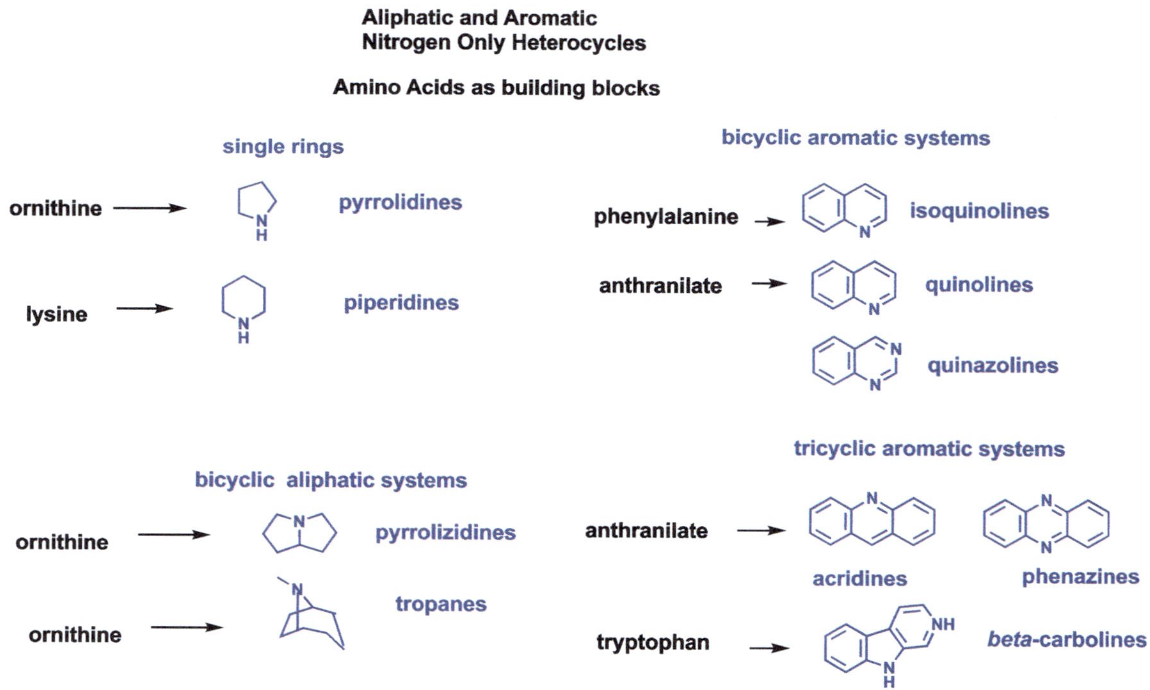

Figure 8.13 Aliphatic and aromatic naturally occurring nitrogen heterocycles in alkaloids, categorized by starting amino acids.

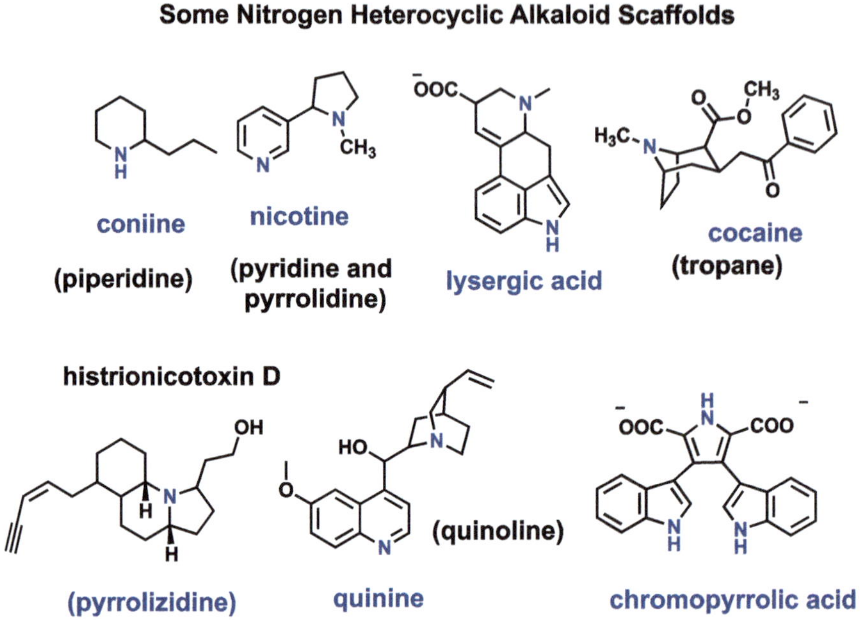

Figure 8.14 Seven heterocyclic nitrogen alkaloid scaffolds.

8.5.2 Pyrroles and Indoles

A comparable pair of heterocycles to the monocyclic pyrimidine and the bicyclic purine (imidazole-pyrimidine) are the standalone five membered pyrrole ring and the fused 5,6-pyrrolo-benzene ring system known as indole (Figure 8.4). Both are fully oxidized heteroaromatic systems.

The pyrrole ring is the building block for all the tetrapyrrolic pigments of life,[19] red heme, purple siroheme, green chlorophyll, salmon pink vitamin B12, and yellow nickel F430 (see Figure 16.1). Free pyrrole itself is not a common metabolic heterocycle. The key monomer unit for tetrapyrrole macrocycles is the 2-methyl-3-carboxymethyl-4-carboxypropyl pyrrole known as porphobilinogen, the precursor to bilanes and porphyrins (Figure 8.4).

The chemical logic for creating porphobilinogen from dimerization of an acyclic amino acid precursor, then the homologous chain extension and closure to tetrapyrrole macrocycles, is a fascinating progression in how nature has evolved catalysts and pyrrole-based metabolites to assemble a family of tetrapyrrole macrocycles (examined in more depth in Chapter 16). Here, we note only two features of the aminomethylpyrrole core of porphobilinogen that endow it with properties as carbon nucleophile and carbon electrophile at distinct positions. The pK_a of pyrrole is listed as 17.5, far outside the physiologic pH range. Thus, dissociation of the pyrrole-N–H to the N-anion does not happen in its chemical biology. However, the lone pair on the nitrogen atom is available and can be used to contribute to resonance structures. In the aminomethylpyrrole core of porphobilinogen one such

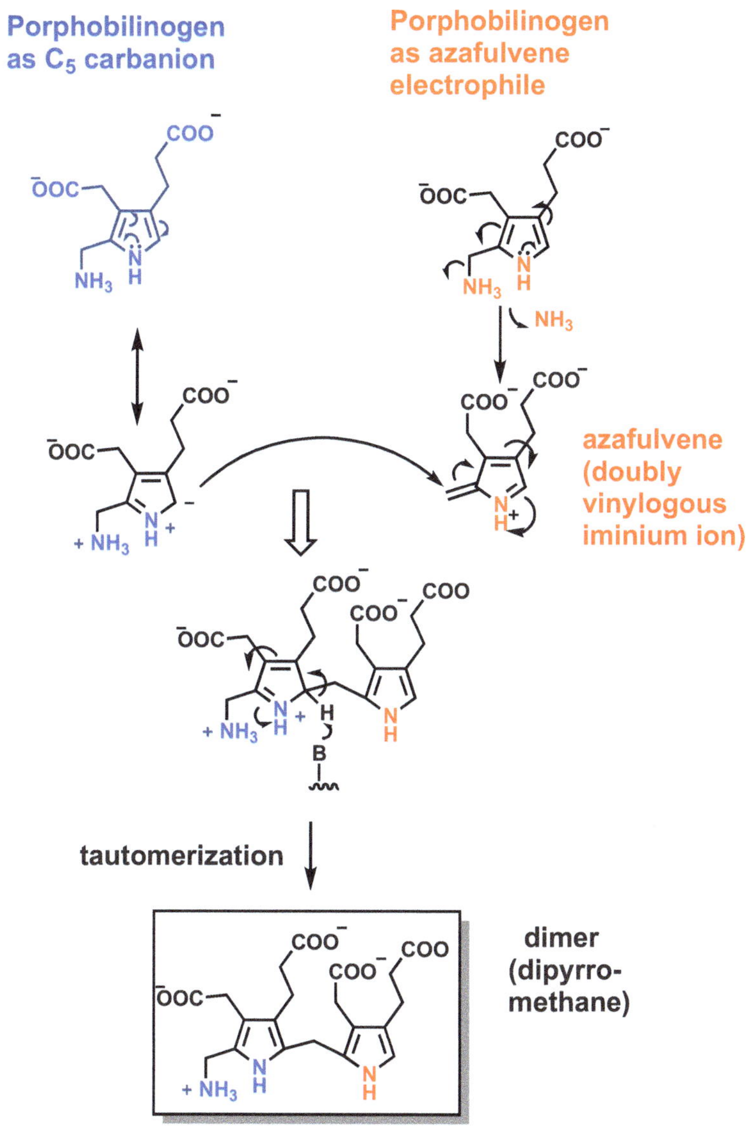

Figure 8.15 Coupling of porphobilinogen with its azafulvene elimination product creates the C–C bond in the iterative chain elongations to tetrapyrrole macrocycles.

resonance contributor would place carbanion density on carbon 5 (blue structure). This potential carbon nucleophilicity (Figure 8.15) is exactly the reactivity needed in chain elongations from the monopyrrole to the linear hexapyrrole intermediate state during tetrapyrrole biosynthesis.

A second potential use of the lone pair on the pyrrole ring nitrogen is to function as a neighboring functional group to expel the side chain amine as ammonia, while generating an exocyclic methylene species known as an azafulvene[20] (red structure, Figure 8.15). The methylene terminus of such azafulvene species is electrophilic as noted for addition of a nucleophile RX to restore the pyrrole ring to its heteroaromatic tautomer.

Putting these two opposite reactivity facets together, the carbanion character at C_5 and azafulvene form at the C_2 methylene group gives insight into how a bipyrrole adduct linked by the intervening methylene group can arise and be iterated for chain elongation (Figure 8.15) to a pyrrole dimer linked by the methylene bridge. Note the left-hand ring still has the aminomethyl substituent at C_2 and can function as electrophile in the next chain elongation cycle.

Although pyrroles show up elsewhere, in secondary metabolic pathways, including the chromopyrrolic acid noted in Figure 8.14 above, the flux to tetrapyrrolic pigments is by far the dominant role for pyrrole rings in chemical biology.[21] The central chemical logic for tetrapyrrole macrocycle assembly is examined more fully in Chapter 16.

8.5.3 An Enamine Part Structure in the Indole Ring System

In addition to the standalone pyrrole ring in the central metabolite porphobilinogen above, the second place one finds a pyrrole ring is in the indole side chain of the proteinogenic amino acid tryptophan and its decarboxylation product tryptamine (Figure 8.16). In addition to its role in protein

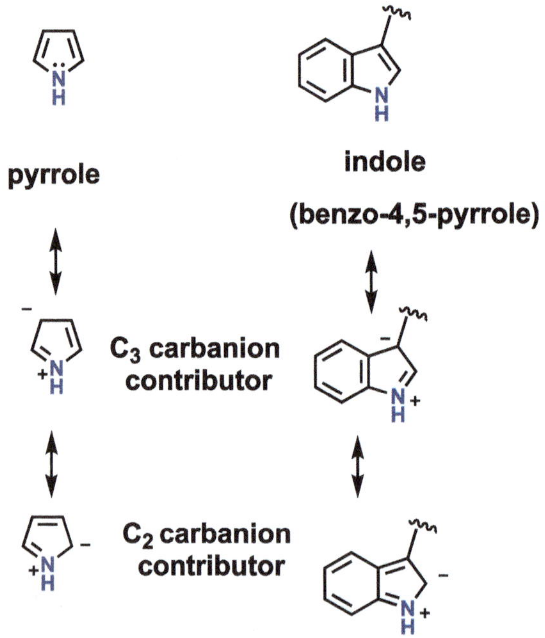

Figure 8.16 Enamine/carbanion characteristics of pyrrole and indole rings.

biosynthesis, free tryptophan and tryptamine are the starting materials for more than a thousand indole alkaloid metabolites. Many of those downstream metabolites arise from the indole ring acting as a C_3 carbanion equivalent Figure 8.16. The C_3 carbanion is a resonance contributor to the resting enamine structure[22] of the pyrrole ring of the bicyclic benzene-4,5-pyrrole framework that is the indole ring. The figure also shows that the standalone pyrrole ring in 2-substituted pyrroles such as porphobilinogen also has such a resonance contributor. However, the indole biosynthetic pathways more often use the C_3 carbanion, rather than the C_2 carbanionic contributor in the dipyrromethane reaction shown in Figure 8.14.

Only a small number of enzymes that act on the indole pyrrole ring portion of indole use C_2 as a carbanion equivalent. The vast majority of indole alkaloids arise by use of the enamine/C_3 carbanion site as carbon nucleophile. Two typical reactions are: (1) enzymatic Mannich condensations and (2) pyrroloindole ring formations (Figure 8.17). On one hand, the Mannich reactions[23] lead to 6,5,6 tricyclic β-carboline ring systems. On the other hand, pyrroloindole-6,5,5 tricyclic systems[11] can arise in tryptophan dipeptides by C_3-prenylation and then internal capture of the 2,3-iminium ion by the adjacent amide nitrogen.

Enzyme-catalyzed prenylations at the indole nitrogen and all the non-bridgehead carbons of indole rings are observed (Figure 8.18). Some amount of carbanion density by resonance contributors can make for carbon nucleophilic transition states (Figure 8.19), capturing allyl cations at either allyl-C_1 (normal regiochemistry) or allyl-C_3 (reverse regiochemistry of prenylations).[24] Surprisingly, the enamine/extended enamine chemistry available to the indole ring system, and evinced in so many alkyl indole scaffolds, is apparently suppressed in tryptophan residues incorporated into proteins.

8.5.3.1 De Novo *Assembly of the Indole Ring*

Indole ring biosynthesis on the way to its placement as the side chain of tryptophan starts with chorismate and glutamine in the active site of anthranilate synthase to yield anthranilate[25] (2-aminobenzoate). The additional five carbons needed to create both the fused pyrrole and the three-carbon side chain in the indoleglycerol-3-P precursor to tryptophan come from a molecule of ribose-5-P. To generate the requisite N–C bond between the amine of anthranilate and C_1 of ribose, cells turn to 5-P-ribose-1-PP as the precursor to the ribosyl C_1–O_5 oxocarbenium ion, capturable in an enzyme active site by anthranilate (Figure 8.19) to produce 5-phosphoribosyl anthranilate. In this product $C_{1'}$ of the ribose is a cyclic aminal. It can be converted to the ring opened iminium form and then enzymatically isomerized to the enamine (carboxyphenyl-amino-deoxyribulose-5-P) which undergoes one more (nonenzymatic) isomerization of the less stable 1,2-enol to the 2-keto open chain derivative of ribose. That metabolite is set up for an unusual enzymatic decarboxylation, ring closure and dehydration: an enzymatic cascade reaction, driven by the gain in aromatization resonance

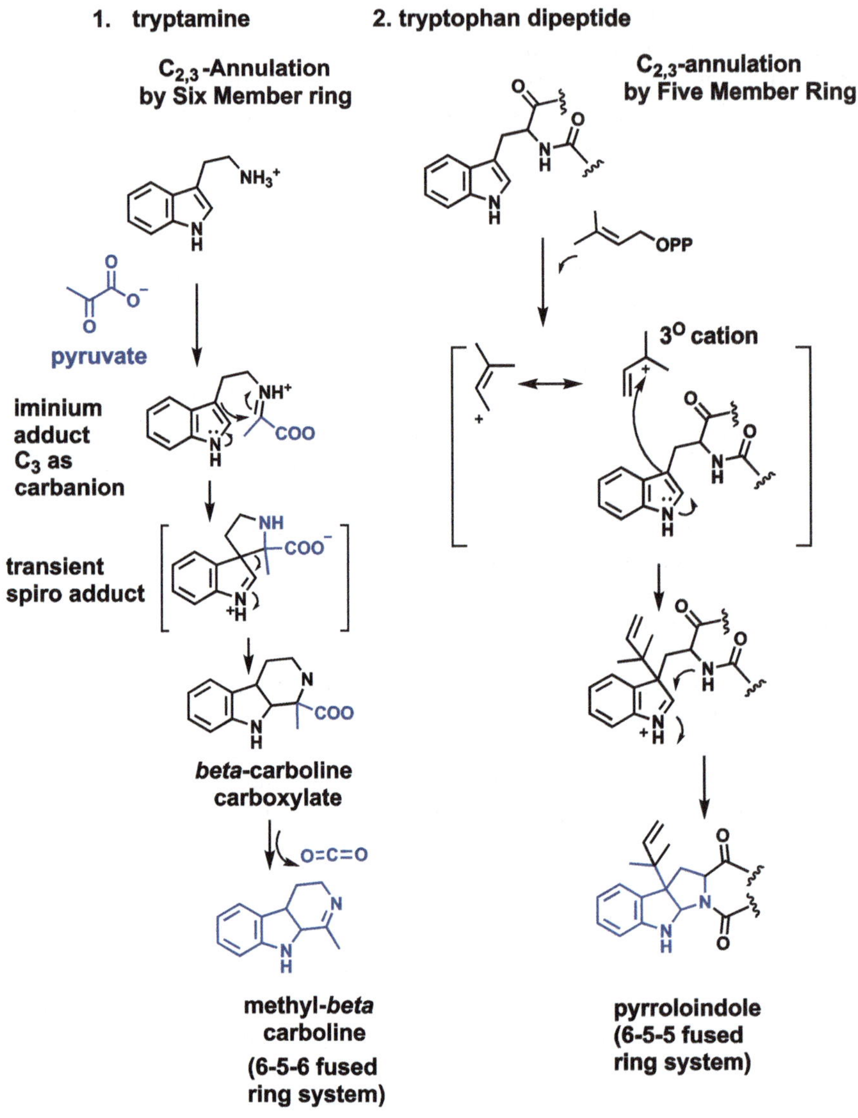

Figure 8.17 Two patterns of enzymatic annulation of the indole side chain of tryptamine or tryptophan diketopiperazines lead to β-carboline or pyrroloindole fused tricyclic systems.

energy as the pyrrole ring becomes fused to the preexisting phenyl ring in indole glycerol-P (IGP) (Figure 8.19). This is the step where the C–C bond is formed, creating the five membered ring fused to the benzene ring. Now, the scaffold is a dehydration across the 2,3-carbons away from introduction of the double bond of the heteroaromatic indole (benzene-5,4-pyrrole). The stable released product is indoleglycerol-3-P, the first metabolite with an indole ring framework.

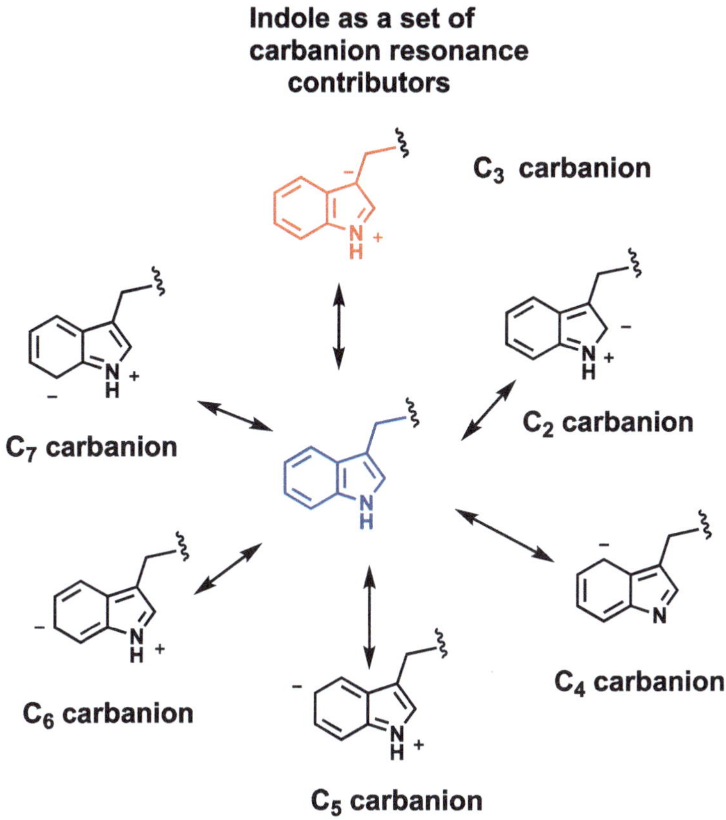

Figure 8.18 Resonance contributors to the structure of indole place negative charge density at carbons 2-7.

8.5.3.2 *Generation and Utilization of Free Indole* Via *Its C_3 Carbanion Chemistry*

The next two enzymes use both facets of the enamine moiety in the pyrrole ring of IGP. First is cleavage of the C–C bond between the indole and glycerol moiety to release free indole.[26] The subsequent enzyme tryptophan synthetase then uses the free indole as a carbon nucleophile at C_3 to form a new C_2–C_3 bond in the proteinogenic amino acid L-tryptophan.[27]

The first enzyme is an aldolase. Aldolase substrates typically have β-hydroxy carbonyl part structures (Figure 8.20) to create accessible low energy stabilized carbanions. To get to that energy-lowering functional array, this aldolase isomerizes the indole ring of bound substrate to the iminium ion tautomer (*via* protonation at the C_3 carbanion resonance contributor). That tautomer has a β-hydroxy iminium ion grouping with the frank positive charge on the indole imine as a powerful electron sink. Aldol cleavage severs the bond between C_3 of the indole ring and the three carbon sugar that ends up as the glycolytic metabolite glyceraldehyde-3-P. Free indole is the

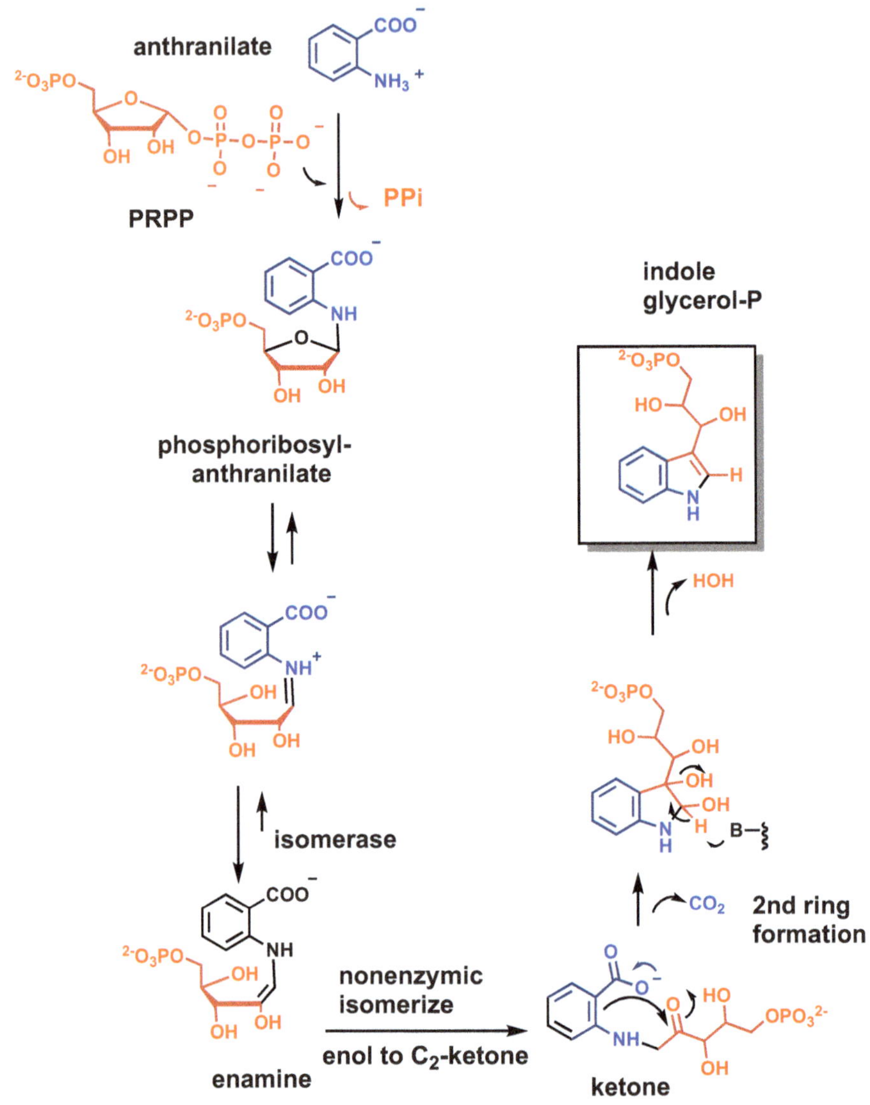

Figure 8.19 Anthranilate to indoleglycerol-3-P. Creation of the indole ring from 2-aminobenzoate and ribose carbons.

coproduct, a substrate for the next enzyme tryptophan synthase. This can be viewed as an internal redox reaction. One of the alcohol carbons of the glycerol-3-P moiety has been oxidized to the aldehyde product. Meanwhile C_3 of the substituted indole has undergone a formal reduction, as the indole product molecules have gained a hydrogen as a C–H bond at C_3.

Tryptophan synthase is one of the large family of PLP-dependent enzymes that use the aldehyde of tightly bound pyridoxal-P (see Chapters 3 and 5) to form initial aldimine with cosubstrate L-serine[28] (Figure 8.21).

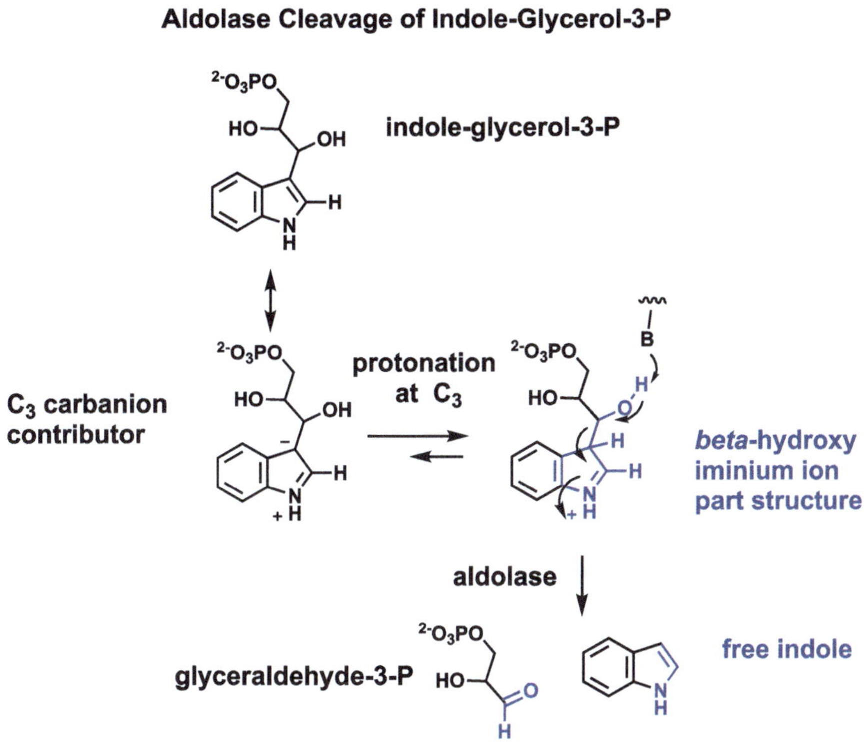

Figure 8.20 Aldolase cleavage of indolegycerol-3-P to glyceraldehyde-3-P and free indole relies on the C_3-enamine/iminium ion capacity of the indole ring.

The serine-PLP adduct is then subject to C_2 carbanion formation by abstraction of the C_2-hydrogen by an active site base. That stabilized carbanion can act in neighboring group mode to assist with expulsion of the C_3–OH group of the serine moiety. This net elimination of water generates the acrylyl-PLP aldimine adduct that is electrophilic at its olefinic terminus by virtue of conjugation to the PLP electron sink. This is the electrophile to be captured by free indole in C–C bond formation. Again, C_3 of the free indole cosubstrate must behave transiently as an enamine/carbanion equivalent, this time as the requisite carbon nucleophile. The resulting new C–C bond creates the scaffold that is two proton transfers away from product tryptophan-PLP aldimine (Figure 8.21). One such protonation occurs at C_2 of the tryptophanyl moiety. The second proton transfer is a deprotonation at C_3 of the indole iminium ion back to the starting enamine part structure in the pyrrole ring of the indole substituent.

Overall, the two enzyme traverse from indole glycerol-3-P plus serine to tryptophan and glyceraldehyde-3-P requires an aldol cleavage and then a net β-replacement of the OH group of serine by the indole group in tryptophan: a new C–C bond. Free indole is an intermediate. The indole C_3 carbanion resonance contributor is a key species at different stages of the two enzyme catalysts, as electron sink for aldolase and as a carbon nucleophile for C–C bond formation.

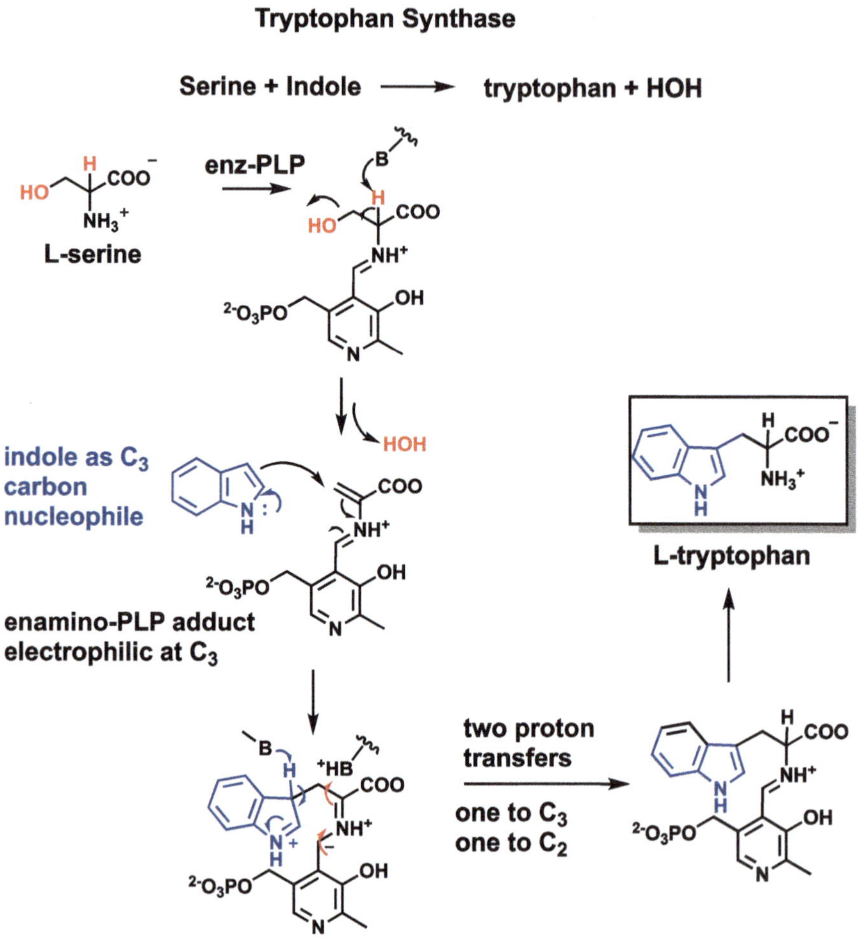

Figure 8.21 Tryptophan synthase is a PLP-containing beta replacement enzyme. Formal replacement of the –OH at C_3 of serine by the C_3–carbon of indole.

8.6 Proline

Two five-member nitrogen heterocycles, found in proteins, as well as in free standing molecules, in addition to the pyrrole ring, are the pyrrolidine ring in proline and the imidazole ring in histidine, both proteinogenic amino acids. Proline is the only secondary amino acid of the 20 building blocks for proteins. Its bulk when in peptide linkages can be a site of kinking in polypeptide chains and it can also populate both *cis* and *trans* peptide linkages appreciably. Otherwise, there is no significant chemical reactivity to the pyrrolidine ring of proline in enzyme catalysts, although it can undergo posttranslational modification (hydroxylation in proteins).

The biosynthetic origin of cyclic proline is typically the acyclic five carbon amino acid glutamate[25] (Figure 8.22). The initial enzymatic step activates the

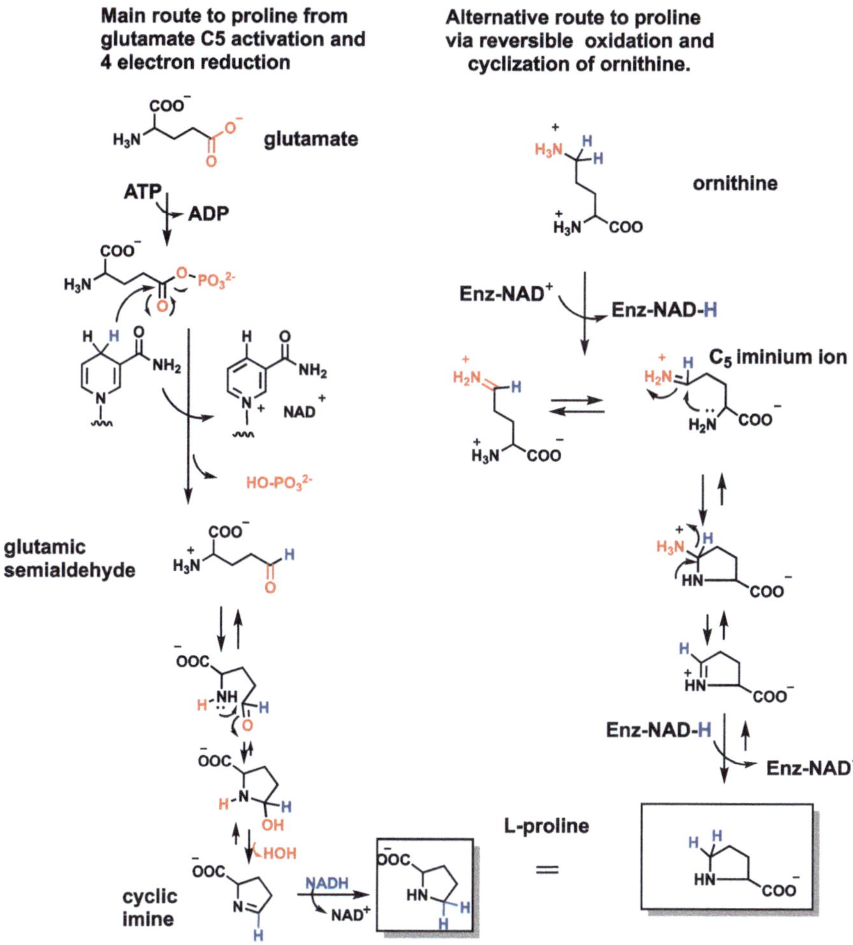

Figure 8.22 Two biosynthetic routes to proline. The common route is by two electron reduction of glutamate to the C_5 semialdehyde, cyclization to the imine and NADH-mediated reduction of the imine to L-proline. The alternate route involves two electron oxidation of L-ornithine to the imine, cyclization and reduction to L-proline. The first route consumes two NADH equivalents. The second route uses tightly bound NAD^+ for catalytic reduction and reoxidation, with no net redox change.

5-carboxylate of glutamate as the acyl-phosphate to increase the electrophilicity of that C_5 carbonyl group. This is exactly the strategy for the first half reaction of glutamine synthetase back in Chapter 5. In the glutamine synthetase active site the attacking cosubstrate nucleophile is ammonia. In this proline biosynthetic pathway, the nucleophile is instead a hydride ion from NADH, reducing the acyl phosphate to the aldehyde. The resultant glutamic-5-semiadehyde can undergo nonenzymatic internal cyclodehydration to the imine in pyrroline 5-carboxylate. This is the cyclization step. One more hydride reduction converts that imine to the secondary amine in L-proline. The consumption of two equivalents of NADH remind us that the C_5-amine linkage in proline is four electrons reduced from the C_5 carboxylate of starting L-glutamate.

In certain microbial contexts, a second source of proline is the non-proteinogenic linear amino acid metabolite L-ornithine. In this case, there is no net redox change from ornithine to proline but a mechanism needs to be found to convert the linear ornithine substrate to the cyclic proline product The solution to the challenge is transient oxidation of the C_5 amine to the imine (Figure 8.22) by ornithine cyclodeaminase.[29] This enzyme uses tightly bound NAD^+ catalytically. L-Ornithine is oxidized at C_2 to the imino acid and NADH, which both remain tightly bound in the active site. That imine is captured intramolecularly by the C_2–NH_2 as nucleophile. This constitutes the cyclization. The cyclic tetrahedral adduct can breakdown backwards to the ornithine C_5 imine or in the forward direction, ejecting NH_3 from what had been the ornithine C_5-amino group. That loss of NH_3 yields the cyclic imine, pyrroline-1-carboxylate, the distinct cyclic iminium regioisomer from the glutamate route. At this point chiral reduction of that cyclic imine from the bound NADH, produces L-proline and returns the nicotinamide coenzyme to the oxidized NAD^+ for the next catalytic cycle. NAD^+ is not consumed, does not show up in the reaction stoichiometry. The redox reaction is reversible and occult, but necessary chemically to provide a mechanistic route to cyclization.

The ornithine cyclodeaminase logic is predictive of the biosynthetic route for pipecolic acid, the next higher homolog of proline, using lysine rather than ornithine as starting substrate with occult redox chemistry through tightly bound NAD^+. Although not found in proteins, pipecolate is found in a number of nonribosomal peptide natural products, perhaps most notably as the sole amino acid in the scaffold of the immunosuppressant natural product drug rapamycin[30] (Figure 8.23).

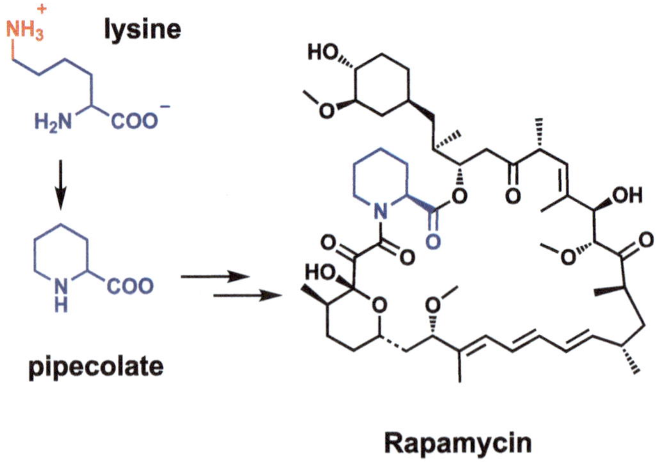

Figure 8.23 Rapamycin has a pipecolic acid moiety inserted between two parts of a polyketide chain.

8.7 Imidazole Ring of Histidine

In contrast to proline, the five membered imidazole ring, with its two nitrogens as the side chain of histidine, offers a dissociable NH group with a pK_a value of 6. Thus, at pH 7 in physiologic media the free base form of imidazole is 90% of the population (Figure 8.24). In that deprotonated form, the imidazole nitrogen, either N_1 or N_3 in specific enzyme active sites, can either act as a base towards protons or as a nucleophile toward electrophiles other than protons. Indeed, given that pK_a of 6, the side chain of histidine may exist as the most populous base/nucleophile for enzymatic catalyses (Figure 8.24).

In different enzymatic contexts, the imidazole side chain of histidine residues can act as a nucleophile toward P_γ or P_α of ATP, in histidine autokinases or DNA ligases respectively (Figure 8.24). Alternatively, in the superfamily of ~250 human protein serine hydrolases, the active site serine-OH acts as catalytic nucleophile while the imidazole side chain of a histidine residue acts as a catalytic base to remove the hydrogen from the ser-OH and generate the serine alkoxide equivalent in the transition state (Figure 8.25). Then, in the next step, the protonated imidazole group of that histidine residue delivers a proton, acts as an acid catalyst, as the leaving group departs during acyl-*O*-ser enzyme formation. Likewise, in the second half reaction, hydrolysis of the covalent acyl-*O*-ser-enzyme intermediate, the imidazole ring of that histidine again acts iteratively as a base and then an

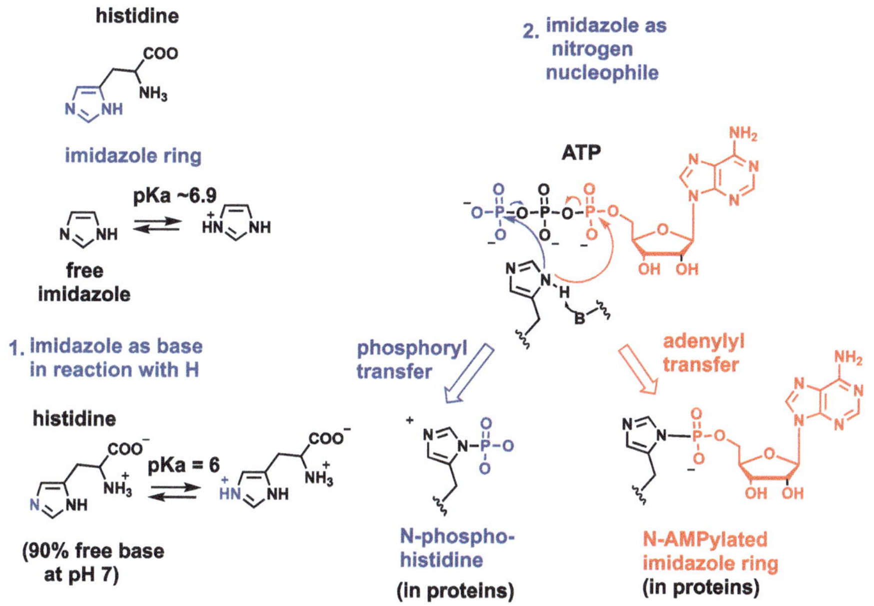

Figure 8.24 The pK_a of the imidazole side chain of histidine is 6.0 so 90% of histidine at pH 7 has the free base form of the heterocycle. The imidazole ring is a nucleophile undergoing both *N*-phosphorylation and *N*-AMPylation.

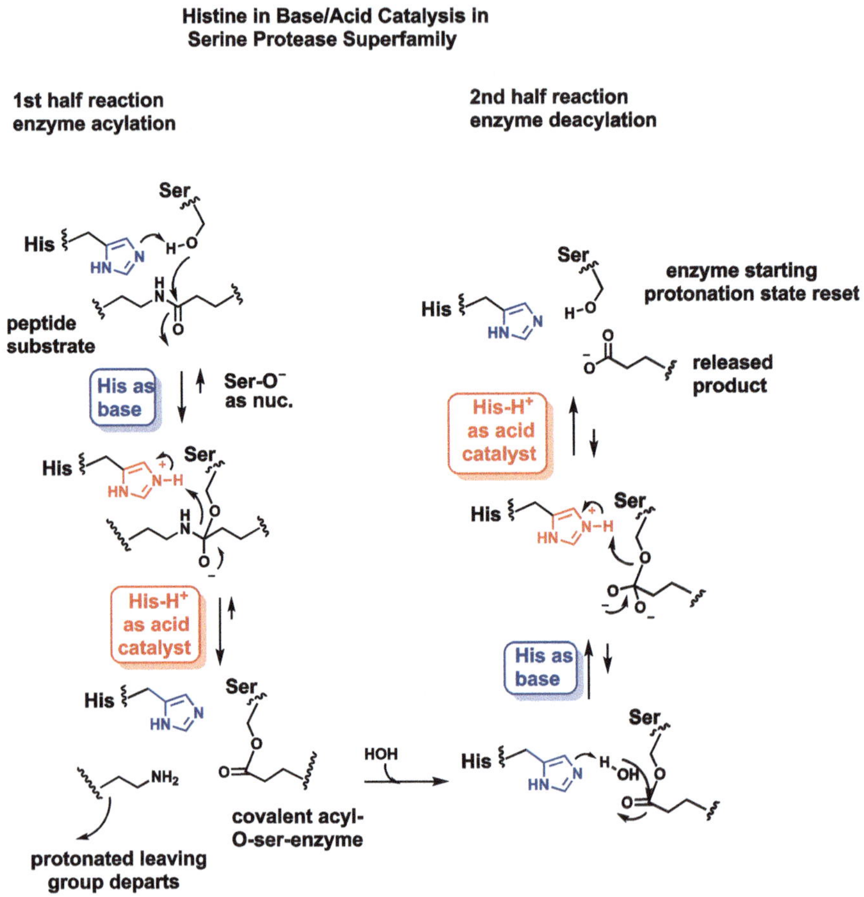

Figure 8.25 Histidine residues function as key base/acid catalysts reciprocally in the superfamily of "serine" hydrolases.

acid catalyst. This behavior emphasizes the energy-lowering function of acid-base catalysis in enzyme active sites and the special role of imidazole side chains that can exist in both protonated and deprotonated forms at physiologic pH.

8.7.1 Unobvious Biosynthetic Logic to the Imidazole Ring of Histidine

The biosynthesis of the imidazole side chain of histidine reveals distinct logic from *de novo* assembly of the imidazole ribotide in purine biosynthesis[25] (see Chapter 6). There, glycine was the core on which a one carbon formyl group, then nascent ammonia was added *via* *O*-phosphoryl isoamide strategies to close the incipient imidazole ring (Figure 6.8).

Instead, the bacterial and plant biosynthesis of the imidazole ring of histidine involves carving the framework of imidazole glycerol-3-P out of the

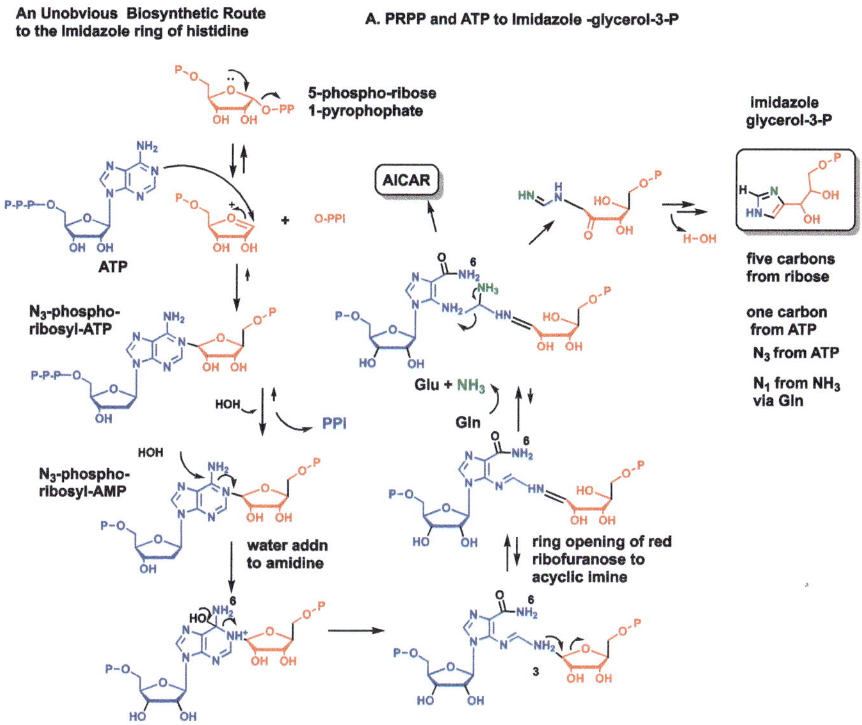

Figure 8.26 PRPP and ATP combine to provide an unobvious route to the imidazole ring of imidazole glycerol-3-P.

scaffold of ATP, starting with an unobvious attack of the N_3 endocyclic amidine nitrogen of ATP on the C_1–O_5 oxocarbenium ion transition state arising from 5-phospho-ribose-1-PP (Figure 8.26). The products are inorganic pyrophosphate and the unusual 5-P-ribosylated N_3-ATP.

All five carbons from the ribosyl moiety end up in the final histidine framework. The sixth carbon in histidine originates from C_2 of ATP Of the two nitrogens in the *de novo* imidazole ring of histidine, one is from nascent NH_3 released by glutaminase action. The second nitrogen is from N_3 of ATP. Thus, C_2 and adjacent N_3 from the pyrimidine ring moiety of ATP have been split out and fashioned into the imidazole ring. One challenge is the hydrolytic, regiospecific opening of that pyrimidine ring of N_3-ribosyated ATP. Histidine itself is three enzymatic steps away from the first imidazole-containing product imidazole glycerol-3-P (Figure 8.27).

Enzymatic hydrolysis of the ATP triphosphate side chain to the monophosphate side chain is prerequisite to enzyme-catalyzed hydrolytic ring opening of the adenine ring. That ring opening is directed water addition to the C_4 iminium carbon of the amidine group in N_3-ribosylated-adenine (Figure 8.26). This is the first of two ring-opening steps: the first opening the ATP pyrimidine ring, the next opening the ribose *via* isomerization of the aminal linkage to the acyclic imine.

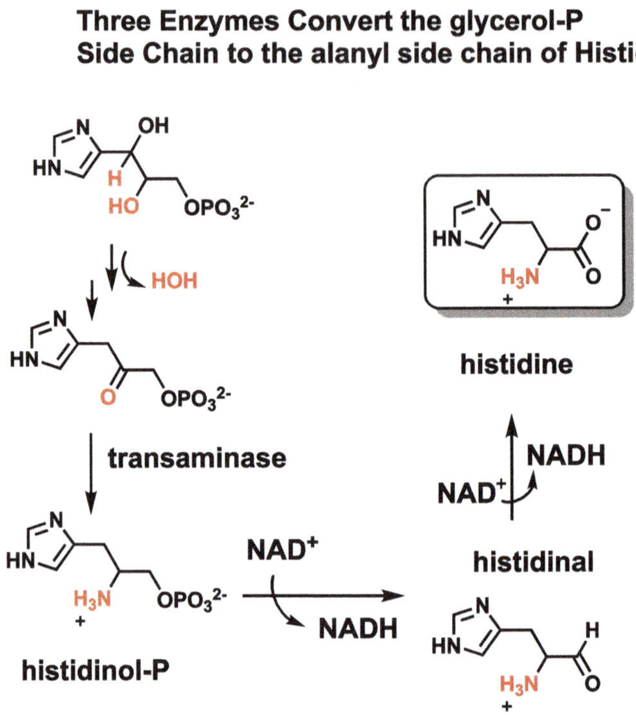

Figure 8.27 Three enzymatic steps convert the glycerol-P side chain in imidazole glycerol-3-P to the alanyl side chain in histidine.

At this juncture, nascent ammonia, hydrolyzed from glutamine on a "just in time" schedule, is proposed to attack the imine linkage encompassing N_6–C_4–N_3 of the original adenine ring to give a remarkable carbon with three nitrogen substituents. One of the ways it can decompose is fragmentation to liberate the metabolite AICAR (amino-imidazole-carboxamide-ribotide, an intermediate in the purine biosynthetic pathway, see Chapter 6). So that fragment is fed back into purine biosynthesis.

The other fragment is the acyclic iminoribulose derivative (Figure 8.26) with an *N*-formimido group. Isomerization to the enamine presumably precedes intramolecular attack on the 3-keto group. The result is a five membered bis-nitrogen ring that is precursor to the imidazole ring after dehydration (Figure 8.26) to stable imidazole glycerol-3-P.

The glycerol-3-P side chain attached to the imidazole heterocycle is the same side chain in the biosynthesis of indole glycerol-3-P on the way to tryptophan (Figure 8.20 above). However, they are dealt with in two different ways reflecting the electronic properties of indole *vs.* imidazole heterocycles. In the indole glycerol-3-P case an aldolase could fragment the indoline isomer by using its electron sink capacity on the way to free indole. Indole then has to be processed as C_3 carbon nucleophile by tryptophan synthetase (Figures 8.20 and 8.21).

The aldolase strategy does not apply to imidazole glycerol-3-P, due to lack of an internal electron sink. Instead, the three carbon glycerol-P side chain is morphed in three enzymatic steps to end product histidine (Figure 8.27). First, a regiospecific dehydratase creates the side chain enol that isomerizes to the more stable 2-ketone. Then a PLP-containing transaminase (Chapter 5) reductively aminates the ketone to histidinol-3-P. The alcohol oxidation state at C_3 is four electrons reduced compared to the carboxylate in histidine. Two equivalents of NADH are spent by the third enzyme to produce the aldehyde histidinal as intermediate on the way to the carboxylate of histidine. This is one of several cascades that interconnect alcohol, aldehyde, and acid oxidation states. The glutamate to proline four electron redox conversion (Figure 8.22) has underlying mechanistic analogy.

8.8 Metabolites with Both Oxygen and Nitrogen Heterocycles

Some alkaloids contain both nitrogen heterocyclic rings and oxygen heterocycles. Four such metabolites are shown in Figure 8.28. Morphine has a compact pentacyclic architecture. Three of the rings are six membered carbacycles (a phenol, a cyclohexane, and an allylic cyclohexanol). The one oxygen heterocycle is a five membered dihydrofuran cyclic ether. The nitrogen is a secondary amine embedded in a fused carbocyclic *N*-methylcyclohexylamine

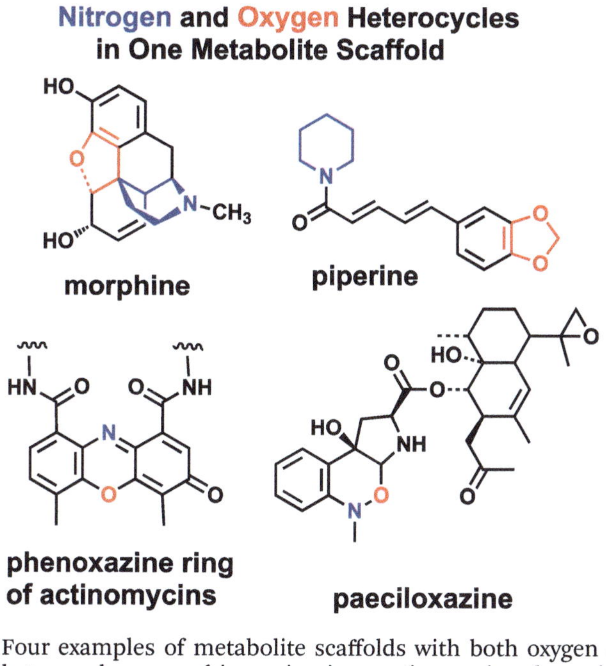

Figure 8.28 Four examples of metabolite scaffolds with both oxygen and nitrogen heterocycles: morphine, piperine, actinomycin phenazine ring, and paeciloxazine.

ring. The second metabolite, piperine, one of the major flavors of black pepper, has a simple structure: a piperidine in amide linkage to a dienoyl group with a catechol ring where the two catecholic oxygens are protected in a methylenedioxy five membered ring. Such cyclic diether ring systems typically arise by cytochrome P450 action on *O*-methylcatechols. The third example in Figure 8.28 is the tricyclic phenoxazine core of the actinomycin antibiotics. In this molecule the oxygen and the nitrogen are together in the same central ring system, termed a morpholino ring. The fourth example, paeciloxazine (31) has an unusual, direct 1,2-N–O bond in the six member oxazine ring,[31] compared to the 1,4-relationship in actinomycin. These two N–O six member rings give a glimpse into the less frequent heterocyclic rings of microbial secondary metabolic pathways.

8.8.1 Lactams (Cyclic Amides): Penicillins and Cephalosporins

Additional examples of heterocyclic rings with both nitrogen and sulfur or nitrogen and oxygen are lactam structures (cyclic amides). Most common are five member γ-lactams and six member δ-lactams (Figure 8.29), but the strained four member β-lactams have been famous therapeutic natural products. This rare but celebrated ring system in biology has been the backbone of the fungal-derived antibiotic class of penicillins and cephalosporins. While penams such as penicillin N were the first class of natural β-lactams discovered, the ring expanded cephems were then discovered, followed by clavams, carbapenems and monocyclic beta lactams[32] (Figure 8.29). In the clavams the sulfur of the thiane ring is replaced by oxygen, while in carbapenams that substituent is a carbon atom. The biosynthetic routes and logic have been exhaustively studied for these lactam variants. Perhaps most unexpected was the finding of free standing β-lactams without the fused five member heterocycle. In all five subclasses the chemical warhead is the strained β-lactam, with its strained 90° bonds in the activated β-lactam framework, opened by acylating agents.

The targets of the various lactam subclasses are typically multiple members of bacterial serine hydrolases involved in crosslinking and strengthening the peptidoglycan layer of bacterial cell walls. The β-lactams behave as pseudosubstrates, undergoing enzymatic processing as far as covalent acyl enzymes that then have long lifetimes, resistant to hydrolysis and thereby weakening the cell wall crosslinking process (Figure 8.30). The biosynthesis of the fused 4,5- and 4,6-ring systems involves radical catalysis as discussed in Chapter 12.

8.9 Xenoacremones: How to Build a Hexacyclic Scaffold with Five Ring-forming Strategies

Complex natural product scaffolds often have a mix of ring types, carbacyclic and heterocyclic, formed by distinct chemical strategies during biosynthesis.

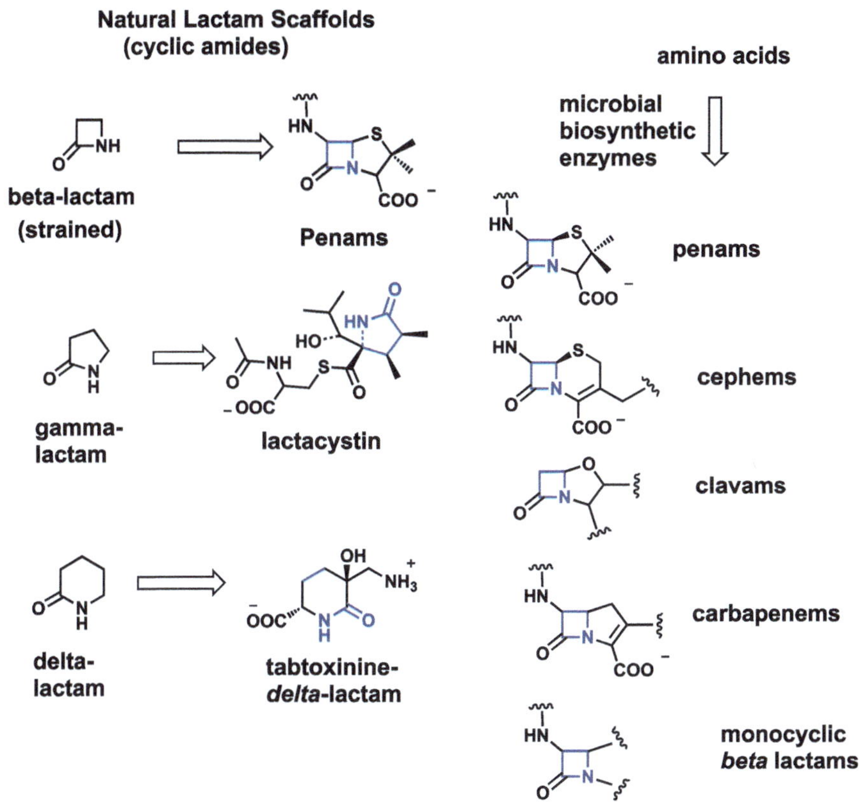

Figure 8.29 Naturally occurring β-, γ- and δ-lactam scaffolds. Five different members of the penam antibiotic family share the strained β-lactam warhead.

A recent report[33] on deciphering the biosynthesis of fungal xenoacremones gives insight into mixing and matching of routes to C–C bonds that illustrate many of the strategies discussed this far. The xenoacremones have a 13 membered *para*-cyclophane oxo-ether ring attached to a 6-5-6 tricyclic core (Figure 8.31). The cyclophane is attached at one end to a tetramate ring and at the other end to an oxygen from the cyclopentane core. There is also an epoxide ring, for a total of six rings, three fused carbacyles, and three heterocycles: a three member epoxide, a five member tetramate, and a 13 member *para*-cyclophane.

Figure 8.31 specifies that a hybrid polyketide synthase–nonribosomal peptide synthase assembly line releases a nascent metabolite with a hydroxyphenyl-substituted tetramate ring at one end and a triene at the other end. The framework is clearly that of a polyketide to which one amino acid, tyrosine, has been added in the last elongation step. Chain release has occurred *via* an intrachain (Dieckman type) cyclization to form the phenol-substituted heterocyclic tetramate ring, the first two of the five rings in the xenocremones.

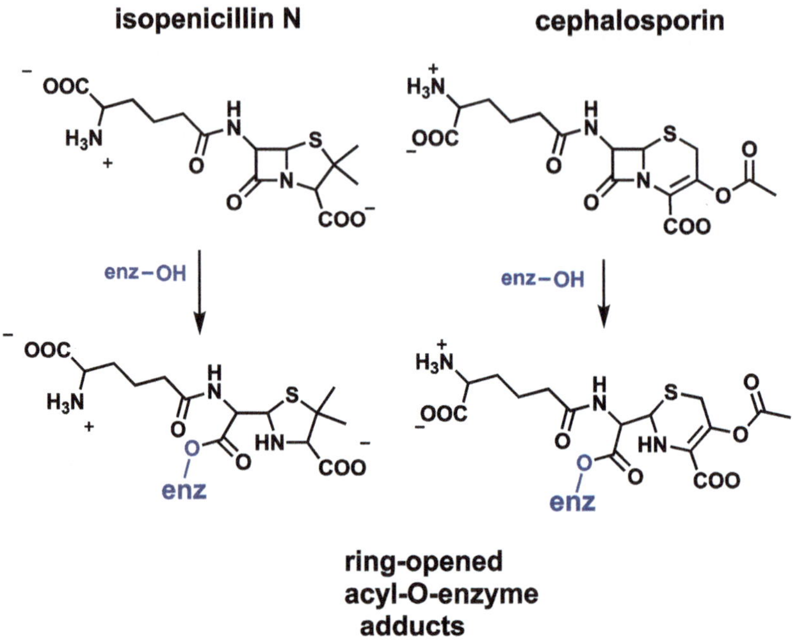

Figure 8.30 The strained beta lactam rings in penicillins and cephalosporins are susceptible to hydrolytic ring opening by the active site ser-OH side chain in bacterial cell wall target enzymes. The lifetime of the acyl-*O*-ester bond in the acyl enzyme adducts determines the lifetime of enzyme inactivation.

Then, the first cyclohexane ring forms by suggested (remarkable) participation of five of the olefins. The terminal three act as a triply vinylogous carbon nucleophile and the two internal olefins, four methylene groups further into the chain, act as an electrophile. C–C bond formation, resulting from those five double bond isomerizations, is the step that builds the dimethyl cyclohexane ring (blue). This proposed involvement of the π-electrons in the triene is reminiscent of the mechanism for LTA_4 conversion to LTB_4 (see Chapter 2). Vinylogous, (doubly, and triply vinylogous) olefin participations are an extension of the kind of C–C bond-forming logic used by the π-electrons of Δ^3-IPP as carbon nucleophile in all intermolecular prenyl transferase-catalyzed isoprenoid chain elongations.

Subsequent steps proposed on the way to xenoacremone are isomerization of the enol to the tautomeric keto tetramate preparatory to epoxidation of the endocyclic double bond in the tetramate ring by a monooxygenase (akin to squalene epoxidase action) (Figure 8.31).

At that point the 13 member *para*-cyclophane ring is posited to form as indicated, by connection of the phenoxide oxygen of the original tyrosyl residue to the olefin βγ- to the ketone in the product. Whether this is heterolytic (phenoxide anion as nucleophile on double bond isomerizing into conjugation), or homolytic (phenoxy radical adding to double bond and

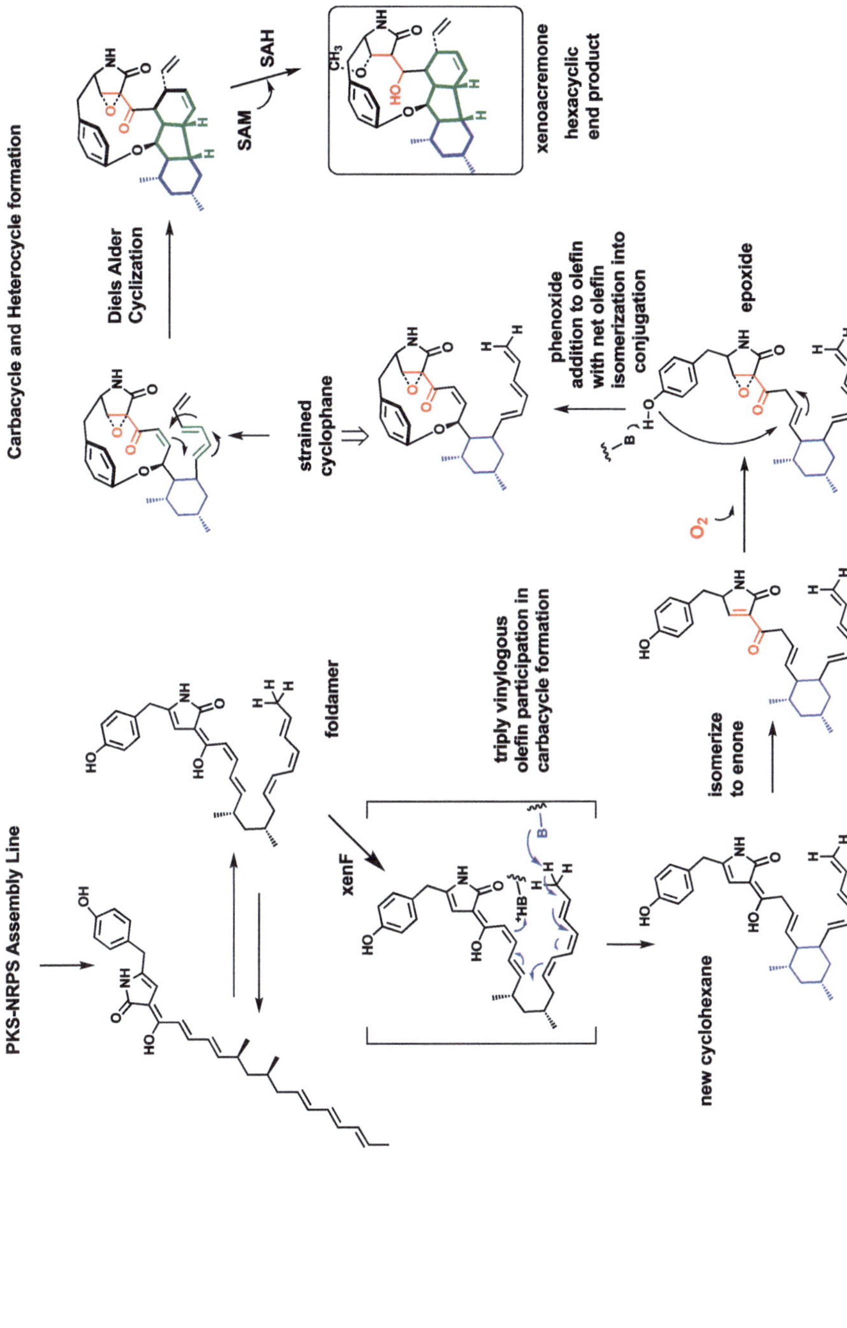

Figure 8.31 Xenoacremone formation illustrates underlying logic to build a fused pentacyclic metabolite scaffold.

subsequent quenching of cyclophane carbon radical) has not been determined. That 13 membered ring will impose rigidity and conformational bias on the rest of the scaffold. At this juncture, the investigators[33] propose a Diels Alder cyclization between the indicated diene and the isomerized dienophile double bond (Figure 8.30, also see Chapter 13). That concerted process forms the central five member cyclopentane and simultaneously the fused cyclohexene with the indicated stereochemistry. This new 5,6-bicyclic moiety is fused to the original cyclohexane ring to produce the fused 6-5-6 tricyclic core. In Chapter 13 on pericyclic reactions in biology, we note a variant of the 6-5-6 fused core in spinosyn where the 5,6-Diels Alder product forms first and then the adjacent carbons are molded into the fused cyclohexane ring.

There are many other heterocyclizing strategies, using any of the four reaction manifolds discussed in Chapters 10–13, carbanion, carbocation, carbon radical, and pericyclic, for assembling a cascade of complex multicyclic metabolite scaffolds.[34]

8.10 Overview

Heterocycles, with addition of one or more oxygen, nitrogen, or sulfur atoms to the carbon backbones, offer more constitutional combinations than carbacycles. They confer crucial properties to biomolecules small and large. We have not attempted a comprehensive list of natural heterocycles and their properties but in this chapter have noted some of the most central heterocyclic frameworks that are essential to life. Additional heterocyclic ring systems are typically restricted to subsets of natural product frameworks. Chapter 9 follows with an examination of the different heterocycles provided by the human vitamins and their chemical roles in key metabolic pathways.

Heterocycle chemical biology probably starts with bioassembly and then utilization of the monocyclic pyrimidine and bicyclic purine rings that are the four to five building blocks for RNA and DNA (Chapter 6). Embedded within those pyrimidines and purines are four nitrogen-containing functional groups, amides, amidine, ureas, and guanidines, that dictate the Watson–Crick base pairing in double helical DNA strands.

Centrally important to the redox reactions that sustain almost all independent life forms and the photosynthetic fixation of carbon by plants and microbes are the metalated tetrapyrroles. We note in this chapter both the nucleophilic and electrophilic capacities of the key 2-aminomethyl pyrrole porphobilinogen for building those tetrapyrrolic macrocyclic frameworks and examine the overall pathway logic in Chapter 16.

Additional five membered heterocycles of consequence in chemical biology are found in three proteinogenic amino acids. The imidazole ring of histidine (the purine ring of nucleic acids is an imidazolyl-pyrimidine), the pyrrole ring in the bicyclic indole ring side chain of tryptophan, and the pyrrolidine ring of proline are functional groups incorporated into the proteomes of all organisms.

We have included an initial discussion of the hemiacetal cyclic forms of the common sugars glucose, and ribose as oxygen heterocycles for two reasons. One is to emphasize these pyranose and furanose ring forms are the predominant bulk forms of these central sugar metabolites, This realization introduces the utility of these hemiacetals and hemiketals as protected but readily reversible forms of the open chain aldehyde and ketone forms of glucose and fructose, respectively. As the sugars interconvert between open chain aldehyde to cyclic hemiacetal isomers or from open chain ketone to cyclic hemiketal isomers, the chemistry at C_1 of glucose and C_2 of fructose in those forms change. The aldehyde and ketone carbonyl groups behave as electrophiles at carbon. The cyclic hemiacetal and hemiketal bulk forms behave as oxygen nucleophiles at the C_1–oxygen and C_2–oxygen, respectively. Second, these dichotomous properties of open chain carbon electrophiles or cyclic oxygen nucleophiles sets up the distinctive metabolic roles of glucose or fructose in their metabolic fluxes (Chapter 14). Finally, the hemiacetal and ketal ring structures molecules are converted to configurationally and chemically stable acetals, ketals, and substituted aminals. In hydrolysis reactions they behave as electrophiles at $C_{1'}$ whereas the hemiacetals behave as nucleophiles at that oxygen.

Vitamin-based heterocycles, required for health maintenance, interject an additional set of heterocyclic functional group scaffolds into chemical biology by enabling further heterocyclic chemical reactivity necessary for central metabolic transformations. They are the subject of the next chapter.

References

1. *Heterocycles Synthesis and Biologic Activities*, ed. B. P. Nandeshwarappa and S. O. Sadashiv, 2019.
2. Q. Lu, D. S. Harmalkar, Y. Choi and K. Lee, An overview of saturated cyclic ethers: bioloical profiles and synthetic stratgeies, *Molecules*, 2019, **35**, 3778.
3. S. K. Sartori, M. A. N. Diaz and G. Diaz-Munoz, Lactones: classification, synthesis, biologic activities, and industrial applications, *Tetrahedron*, 2021, **84**, 132001.
4. T. Soderberg, CLT, Hemiacetals, Hemiketals and Hydrates, 2021.
5. IUPAC, *Compendium of chemical terminology*, 2nd edn, 1997.
6. IUPAC, *Glycoisdes*, 2009.
7. S. R. Meshnick, T. E. Taylor and S. Kamchonwongpaisan, Artemisinin and the antimalarial endoperoxides: from herbal remedy to targeted chemotherapy, *Microbiol. Rev.*, 1996, **60**, 301–315.
8. C. Walsh, Nature loves nitrogen heterocycles, *Tetrahedron Lett.*, 2015, **56**, 3075–3081.
9. C. T. Walsh and Y. Tang, *The Chemical Biology of Human Vitamins*, Royal Society of Chemistry, London, 2018.
10. C. T. Walsh, *The Chemical Biology of Sulfur*, Royal Society of Chemistry, London, 2020.

11. C. T. Walsh and Y. Tang, *Natural Product Biosynthesis: Chemical Logic and Enzymatic Mechanisms*, Royal Society of Chemistry, London, 2nd edn, 2022.
12. J. W. Westley, *Polyether Antibiotics: Naturally Occurring Acid Ionophores*, Marcel Dekker, New York, 1982, p. 192.
13. C. Khosla, Y. Tang, A. Y. Chen, N. A. Schnarr and D. E. Cane, Structure and mechanism of the 6-deoxyerythronolide B synthase, *Annu. Rev. Biochem.*, 2007, **76**, 195–221.
14. A. Keating-Clay, A tylosin ketoreductase reveals how chirality is determined in polyketides, *Chem. Biol.*, 2007, **14**, 898–908.
15. C. T. Walsh and Y.-C. Jack Chen, Enzymatic baeyer-villiger oxidations by flavin-dependent monooxygenases, *Angew. Chem., Int. Ed.*, 1988, **27**, 333–343.
16. A. N. Panche, A. D. Diwan and S. R. Chandra, Flavanoids: an overview, *J. Nutr. Sci.*, 2016, **5**, e47.
17. Y. Zang, J. Zha, X. Wu, Z. Zheng, J. Ouyang and M. A. G. Koffas, *In vitro* naringenin biosynthesis from para-coumaric acid uisng recombinant enzymes, *J. Agric. Food Chem.*, 2019, **67**, 13430–13436.
18. R. G. Salomon, *Amino acids and alkaloids*, 2021.
19. A. R. Battersby, Tetrapyrroles: the pigments of life, *Nat. Prod. Rep.*, 2000, **17**, 507–526.
20. R. A. Jones and G. P. Bean, *The chemistry of pyrroles*, 1977.
21. C. T. Walsh, S. Garneau-Tsodikova and A. R. Howard-Jones, Biologic formation of pyrroles: nature's logic and enzymatic machinery, *Nat. Prod. Rep.*, 2006, **23**, 517–531.
22. W. I. Taylor, *Indole alkaloids, an introduction to the enamine chemistry of natural products*, Pergamion Press, New York, 1966.
23. F. F. Blicke, The Mannich Reaction, in *Organic Reactions*, 2011, vol. 1, pp. 303–341.
24. N. Steffan, A. Grundmann, W.-B. Yin, A. Kremer and S.-M. Li, Indole prenyltransferases from fungi: a new enzyme group with high poterntial for the production of prenylated indole derivatives, *Curr. Med. Chem.*, 2009, **16**, 218–231.
25. J. E. McMurry and T. P. Begley, *The organic chemistry of biological pathways*, MacMillan, New York, 2nd edn, 2015.
26. C. Yanofsky, The enzymatic conversion of anthranilic acif to indole, *J. Biol. Chem.*, 1956, **223**, 171–184.
27. S. Raboni, S. Bettati and A. Mozzarelli, Tryptophan synthase: a mine for enzymologists, *Cell. Mol. Life Sci.*, 2009, **66**, 2391–2403.
28. E. E. Snell, From bacterial nutrition and o enzyme structure in pyridoxal phosphate enzymes, *Annu. Rev. Biochem.*, 1993, **62**, 1–27.
29. J. L. Goodman, S. Wang, S. Alam, F. J. Ruzicka, P. A. Frey and J. E. Wedekind, Ornithine cyclodeaminase: structure, mechanism of action and implications for the mu-crystallin family, *Bicohemistry*, 2004, **43**, 13883–13891.
30. G. J. Gatto, M. T. Boyne, N. L. Kelleher and C. T. Walsh, Biosynthesis of pipecolic acid by RapL, a lysine cyclodeaminase encoded in the raoamycin gene cluster, *J. Am. Chem. Soc.*, 2006, **128**, 3838–3847.

31. D. Yan, K. Wang, S. Bai, B. Liu, J. Bai, X. Qi and Y. Hu, Flavin-dependent monooxygenase-mediated 1,2-oxazine construction via Meisenheimer rearrangement in the biosyntheris of paeciloxazine, *J. Am. Chem. Soc.*, 2022, **144**, 4269–4276.
32. K. Bush and P. A. Bradford, Beta-lactams and beta-lactamase inhibitors: an overview, *Cold Spring Harbor Perspect. Med.*, 2016, **6**, a025247.
33. Z. Liu, W. Li, P. Zhang, J. Fan, F. Zhang, C. Wang, S. Li, Y. Sun, S. Chen and W. Yin, Tricarbocyclic core formation of tyrosine-decahydrofluorenes imlies a three enzyme cascade with XenF-mediated sigmatropic rearrangement as a prerequisite, *Acta Pharm. Sin. B*, 2021, **11**, 3655–3664.
34. C. T. Walsh and B. S. Moore, Enzymatic cascade reactions in biosynthesis, *Angew. Chem., Int. Ed.*, 2019, **58**, 6846–6879.

CHAPTER 9

Heterocycles in Chemical Biology: II. Vitamins

9.1 Thirteen Human Vitamins

Vitamins by definition are organic metabolites required for healthy growth and nutrition for a particular organism (*e.g.* humans). Vitamins are required in small quantities (as micronutrients) in the diet because the organisms have lost the capacity to biosynthesize them. There can be vitamins for bacteria, yeast, animals, or humans. After more than a century of searching, 13 vitamins have been identified as essential for human health.[1]

Twelve of the thirteen human vitamins use carbacyclic or heterocyclic scaffolds for their life-sustaining functions (Figure 9.1). Seven of the eight B vitamins, plus vitamin C and vitamin E, operate through heterocyclic chemistry. Vitamins A, D, and K have carbocyclic rings in their scaffolds. Only the pantothenate structure of vitamin B5 is acyclic. In turn, it is four biosynthetic steps away from the active coenzyme form coenzyme A, which does harbor an adenine heterobicyclic moiety, admittedly far from the business end of the terminal thiol/thiolate anion.

Among the ways to characterize the 13 human vitamins are whether they are water soluble or lipid soluble.[2] In this categorization vitamins A, D, E, and K are the four lipid soluble molecules (Figure 9.1). Vitamins can also be classified according to physiologic function and what nutritional deficiencies they cure.[2] The B vitamins are the only class with more than one entry: eight molecules comprise the B vitamins. They differ dramatically in chemical structure and biologic function but are unified by the fact that they all function as coenzyme forms.[1] From a carbon chemical biology perspective, each enables central enzyme-catalyzed chemical transformations. Each brings a unique chemical functional group, heterocyclic for the seven

Chemical Biology No. 21
The Chemical Biology of Carbon
By Christopher T. Walsh

Published by the Royal Society of Chemistry, www.rsc.org

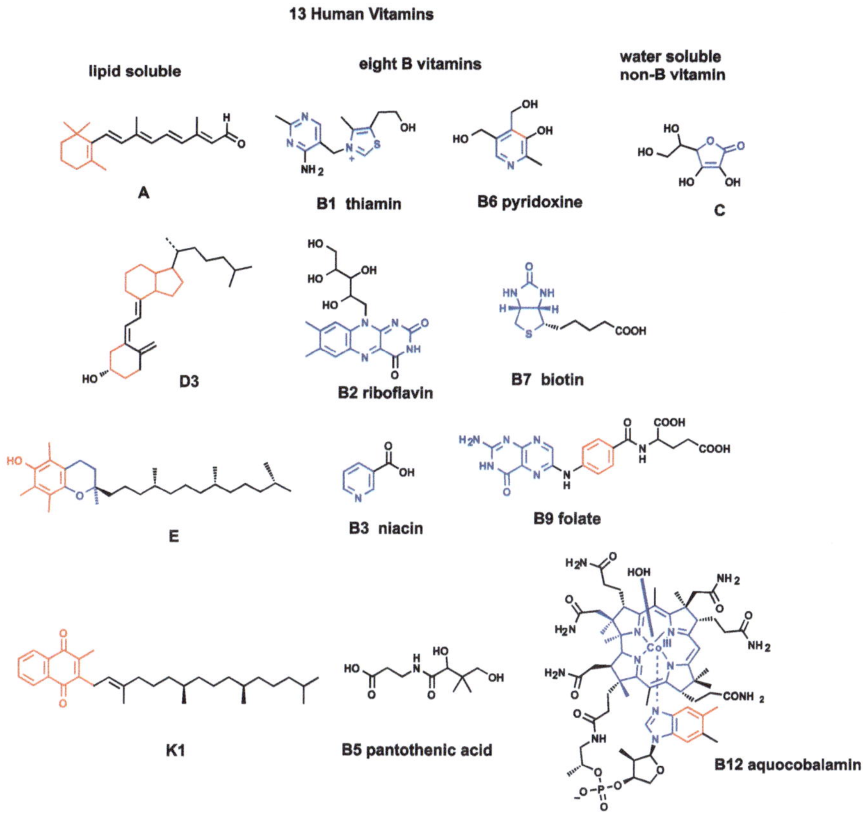

Figure 9.1 Thirteen human vitamins.

out of eight B vitamins that are absolutely necessary reagents for nature's catalytic inventory. From this perspective, vitamins C and K can also function as essential cofactors.

One more operational categorization, used below, is to parse the seven B vitamin coenzymes according to the number of heterocyclic rings that are the business ends of catalytic function. In this grouping, vitamins B1 (thiamin), B3 (niacin), and B6 (pyridoxine) provide mono-heterocyclic rings. Vitamin B7 (biotin) and B9 (folate) provide heterobicyclic ring systems as key reaction centers. Vitamin B2 (riboflavin) uses its tricyclic isoalloxazine ring as a versatile redox coenzyme. Vitamin B12 is by far the most complex structural scaffold, with its contracted tetrapyrrole ring macrocycle and its equatorially embedded cobalt^{III} ion. All these seven B vitamins contain nitrogen heterocyclic ring systems, and all of them are distinct, each enabling different chemistry.

Then, we turn briefly to vitamins C, K, and E in the continuing context of how specific heterocycles or carbacycles (*e.g.* vitamin K) condition essential vitamin functions to give a sense of how vitamin chemical biology is 85% heterocyclic chemistry. A quick note about vitamin B nomenclature: during the first four decades of the 20th century when vitamin discovery was at its peak, several candidate B vitamins were named (B-1 to B-20) but 12 of them ultimately failed the test of time, either because they were previously undiscovered endogenous metabolites or dispensable as dietary additives.[2] The survivors are B1, B2, B3, B5, B6, B7, B9, and B12 (Figure 9.1).

9.2 Monoheterocyclic B Vitamins

9.2.1 Vitamin B1 Thiamin

Vitamin B1 was the first human vitamin identified, as the deficiency factor in beri beri, predominant in nutritionally deprived residents of Java.[3] The discovery and characterization of this molecule gave rise to the whole class of vitamins as essential human nutritional factors, not made endogenously by humans and therefore required in microgram to milligram daily amounts in diet as micronutrients.

The structure of thiamin revealed a pyridine ring coupled to a positively charged thiazolium ring by a methylene bridge. The mechanistic role of the vitamin was unobvious until Breslow[4] in 1957 and 1958 showed that thiazolium compounds but not the neutral thiazole parent molecules exchanged the C_2–H with deuterons from solvent D_2O (Figure 9.2). This was *prima facie*

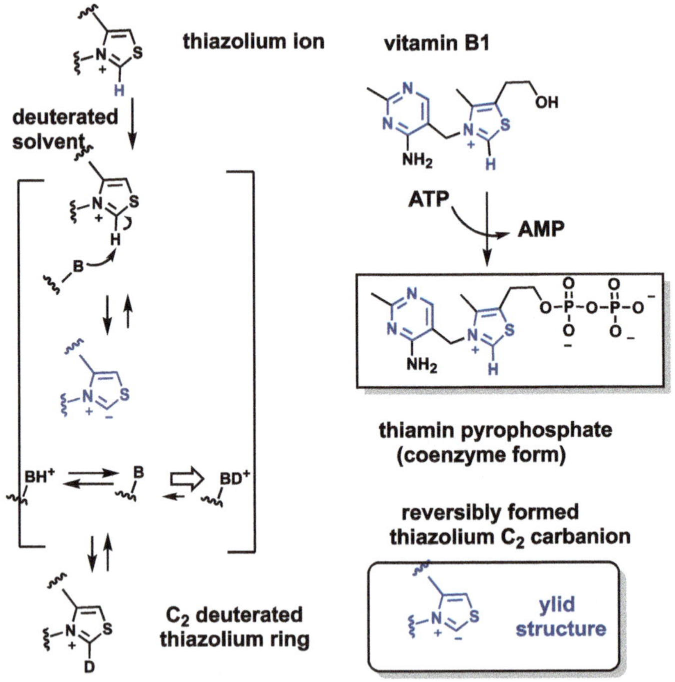

Figure 9.2 Coenzyme form of vitamin B1 thiamin. Deuterium exchange with solvent protons indicates reversible formation of a C_2 carbanion in the thiazolium ring.

evidence that the thiazolium ring was capable of low energy carbanion formation at C_2. That fact proved to be essential to the role of thiamin-PP (the active coenzyme form) as a key coenzyme that enabled pyruvate and α-ketoglutarate metabolic breakdown for energy metabolism.

The origin of the thiazolium C_2 acidity is thought to rest on that carbanion being part of an ylid (adjacent atoms with opposite charges). The permanent (positive) charge on the quaternized N_1 of the thiazolium ring is critical (Figure 9.2). A neutral thiazole would have a much lower C_2-acidity, a much higher energy barrier for generating the C_2 carbanion (this principle of a permanently quaternized cationic nitrogen acting as an electron sink is also utilized in nicotinamide coenzymes derived from vitamin B3). When thiamin-PP is bound in the active site of a partner enzyme, the thiazolium ring C_2–H exchange rate is faster by orders of magnitude such that C_2 carbanion formation is not rate-limiting in catalysis.

Thiamin-PP acts as an obligate coenzyme in two types of C–C bond forming enzymatic transformations. One set involves lower energy barriers for decarboxylation of α-keto acids, including pyruvate, α-ketoglutarate, and the branched chain α-keto acids arising from transamination of L-valine, L-leucine, and L-isoleucine amino acids[5] (Figure 9.3).

A second reaction type is practiced by transketolases in carbohydrate metabolism.[6] A typical reaction stoichiometry for a transketolase is to process fructose-6-P and glyceraldehyde-3-P by ketol transfer from C_1 and C_2 of fructose to the aldehyde C_1 carbon of glyceraldehyde-3-P (Figure 9.3) to form a four- and a five-carbon sugar phosphate. The net two-carbon hydroxy carbonyl transfer is termed a ketol transfer, hence the enzyme name. The key intermediate is a TPP-two carbon ketol adduct (blue in Figure 9.3).

In both cases the C–C bond cleavages are lowered in energy by reversible addition of the C_2 carbanion of the thiazolium ring of TPP into the substrate carbonyl (*e.g.* C_2 ketone of pyruvate or C_2 ketone of fructose-6-P). In those adducts the $C_2{=}N_1{}^+$ double bond of the thiazolium ring is now β to the substrate-adduct bond to be cleaved, creating the equivalent of a β-iminium electron sink.

Consider the case of the α-keto acids (*e.g.* pyruvate). Decarboxylation of the free acid anion would require a high energy acyl anion as CO_2 leaves without the electron pair in the pyruvate C_2–C_1OO^- bond. Such an acyl anion is too high in energy (unstable) to be accessed under biologic conditions. By contrast in the TPP–pyruvate adduct, there is now a polarized iminium electron sink β to the C_1OO^- group. C_2–C_1 bond cleavage generates CO_2 and the carbanion (now C_1) that is delocalized into and stabilized as the neutral enamine resonance form. This is the essence of the invention of the thiazolium ring in the coenzyme form of vitamin B1. Decarboxylation proceeds with low energy barrier, large rate acceleration, in the active site of those TPP-dependent decarboxylases.

The predominant contributor to the resultant hydroxyethyl-TPP carbanion is the neutral enamine,[7] but there is enough carbanion character to protonate at carbon to yield the hydroxyethyl-TPP adduct. That hydroxyethyl adduct can

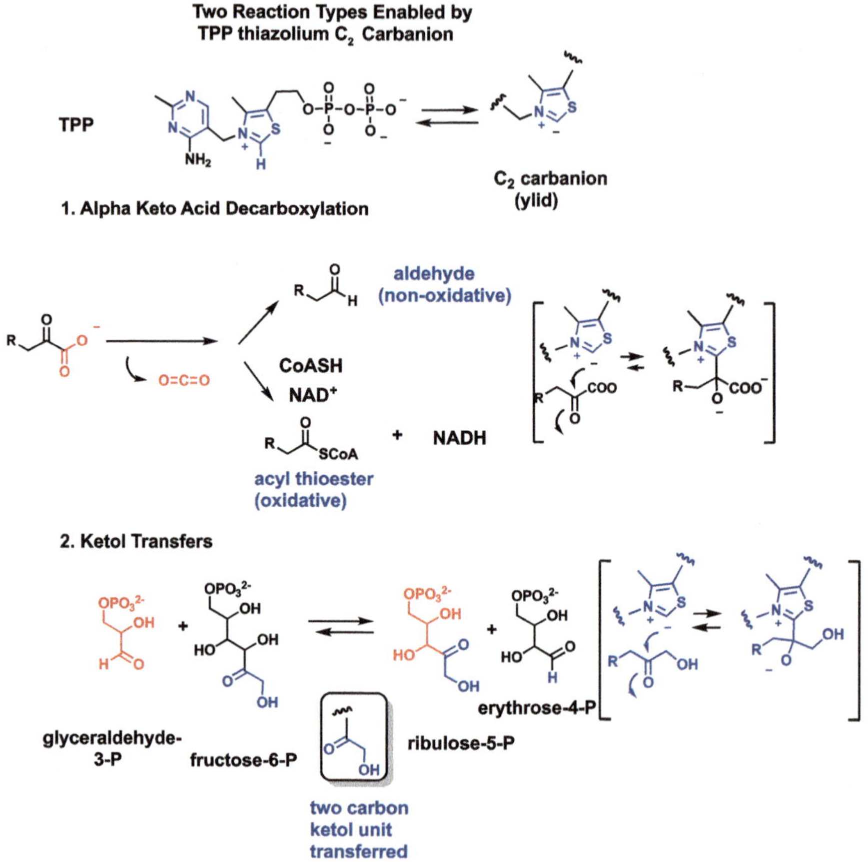

Figure 9.3 Thiazolium C_2 carbanion forms a covalent adduct with substrates in both α-keto acid decarboxylase and transketolase active sites.

partition down two routes. One is expulsion of the initial TPP thiazolium carbanion as acetaldehyde is released (Figure 9.4). This pathway constitutes a nonoxidative decarboxylation. The product C_1-aldehyde carbonyl is at the same oxidation state as the C_2 ketone of the starting α-keto acid.

The alternate partition, an oxidative outcome, occurs in the three protein α-keto acid dehydrogenase complexes[8] (Figure 9.5) where the hydroxyethyl group is transferred to the lipoamide disulfide of enzyme 2 of the three enzyme complex. The two carbon group in flight is initially an inter-enzyme two-carbon tether, spanning both the thiamin-PP of enzyme 1 and the lipoamide group of enzyme 2. As that adduct resolves in the forward direction it becomes clear that this is the oxidative step (Figure 9.5). The hydroxyethyl in flight is transferred into an acetyl group; the disulfide is opened to an acyl thioester and thiolate component (see Chapters 3 and 5).

In mechanistic parallel, transketolases utilize the readily accessible thiazolium C_2 carbanion of TPP for ketol transfer between sugar substrates, a

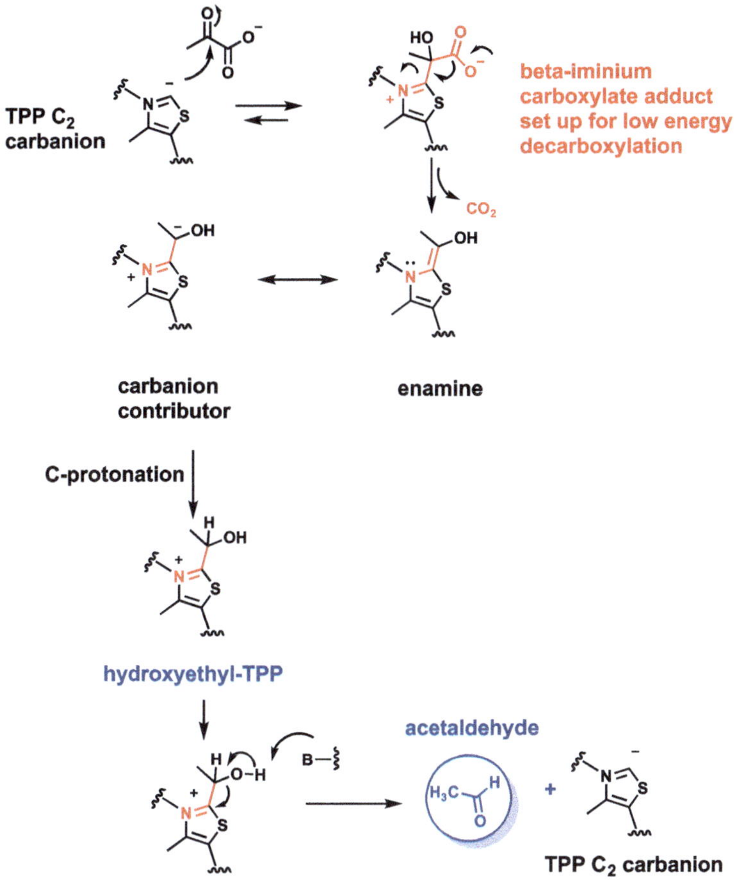

Figure 9.4 Mechanism for TPP-facilitated decarboxylation of 2-keto acids.

donor ketose, and an acceptor aldose (Figure 9.3). The two carbon ketol unit is transferred as a carbanion equivalent to attack the aldehyde carbonyl of aldose acceptors. The transfer occurs in two half reactions.

Figure 9.6 illustrates the first half reaction for the attack of that TPP C_2 thiazolium carbanion on the C_2 ketone of the (minor) open ring form of fructose-6-P. The resultant β iminium adduct is now set up for low energy cleavage between C_3 and C_2. This step releases the four carbon sugar erythrose-4-P, deriving from carbons 3-6 of fructose. C_3 has been transformed from an alcohol in starting substrate to aldehyde in the released four carbon aldose-4-P. Meanwhile, the two carbon scaffold from C_1 and C_2 of fructose-6-P is covalently tethered as a stabilized carbanion adduct to TPP, much like the hydroxyethyl carbanion adduct seen above in α-keto acid decarboxylases (Figures 9.4 and 9.5).

This hydroxycarbonyl carbanion adduct to TPP, stabilized as the exocyclic enamine, exists long enough in the transketolase active site for the erythrose-4-P

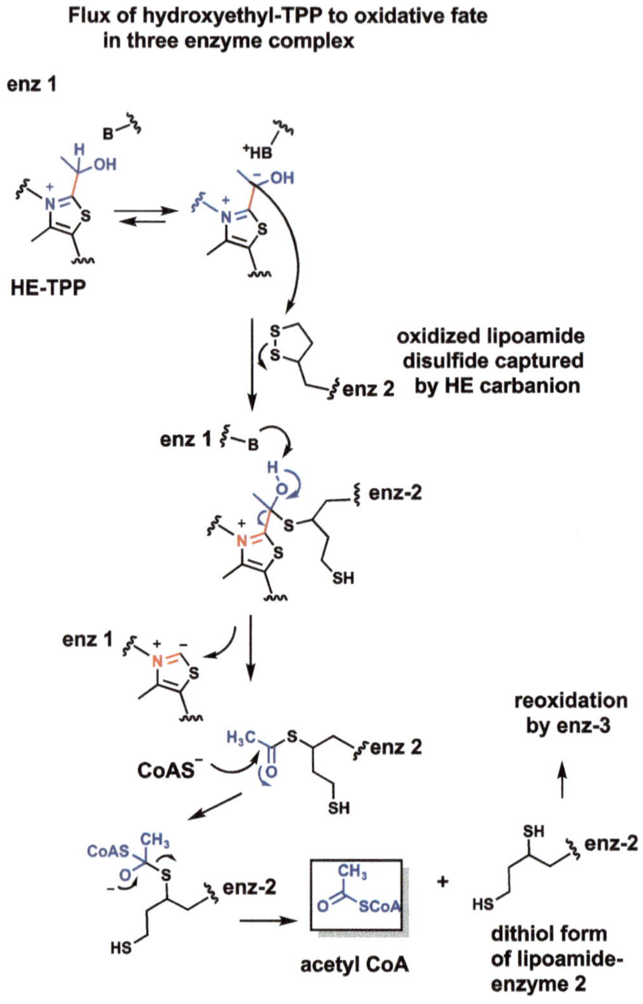

Figure 9.5 Flux of hydroxyethyl-PP to acetyl CoA and NADH in three enzyme pyruvate dehydrogenase complex.

product to dissociate and be replaced by another aldose. In the example show in Figure 9.7 for the second half reaction, glyceraldehdye-3-P is an abundant glycolytic aldehyde metabolite. C–C bond formation occurs by attack of the tethered two carbon ketol carbanion on that aldehyde carbonyl. This forms a new C–C bond (C_2–C_3) in a tethered ribulose-5-P adduct. Decomposition in the forward direction ketonizes C_2 while expelling the thiamin-PP thiazolium C_2 carbanion. The released five-carbon sugar is ribulose-5-P.

The enzymatic shuffling of ketol units between phosphorylated aldoses ensures a supply of the C_7, C_6, and C_5 phosphorylated ketoses, along with the phosphorylated C_3, C_4, and C_5 aldoses. In particular, the erythrose 4-P from TPP-transketolase fragmentation of fructose-6-P is the four-carbon aldose-P

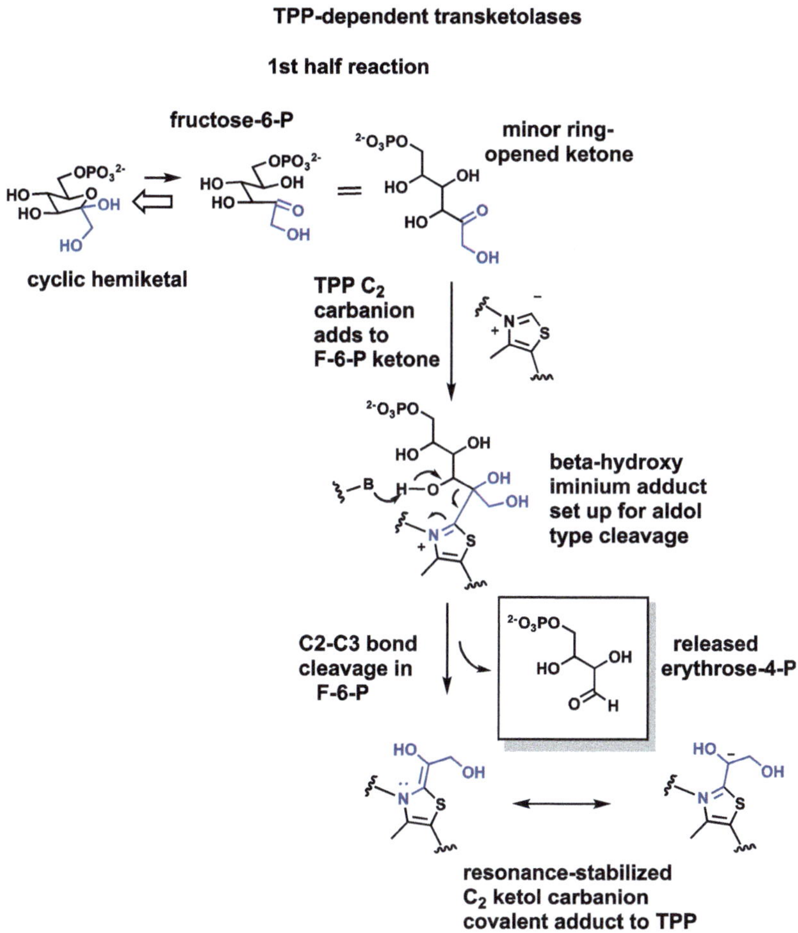

Figure 9.6 Transketolase first half reaction.

metabolite for aromatic ring biosynthesis *via* chorismate and prephenate (examined in Chapter 6 on aromatic carbacycles).

The permanently charged quaternized nitrogen in the thiazolium ring is the key to the chemical biology of the coenzyme form of vitamin B1. It can undergo low energy dissociation of the C_2–H as a proton to give the C_2–N_1 ylid. That is both stable enough to form and yet reactive enough as a carbanion to lower energy barriers for two key types of C–C bond cleaving/forming reactions in central metabolism. In facilitation of both α-keto acid decarboxylations and ketol transfers on hydroxy carbonyl sugars, the thiamin-substrate adducts create β-iminium ions as electron sinks to facilitate C–C bond cleavages with formation of resonance stabilized carbanions as carbon nucleophiles. Biosynthesis of the thiazolium ring in yeast and bacteria is complex and multistep but the resultant thiazolium ion functional group has been an enduring invention in chemical biology.

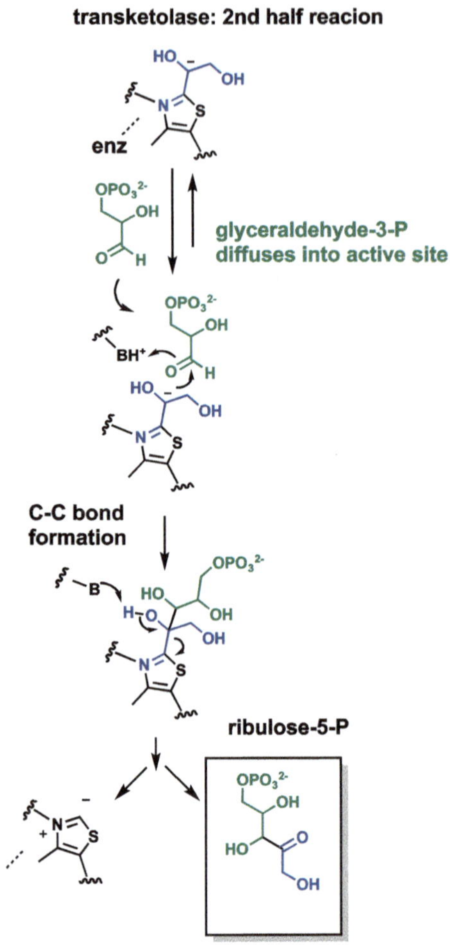

Figure 9.7 Transketolase second half reaction.

9.2.2 The Monocyclic Quaternized Nicotinamide Ring of Coenzyme Forms of Vitamin B3

The redox coenzyme functions of nicotinamide adenine dinucleotides, both the oxidized form of NAD^+ and $NADP^+$ and the reduced forms NADH and NADPH, respectively, have been exemplified many times in the preceding chapters. They serve as cosubstrates that undergo hydride transfers to and from one or more of specific cosubstrate carbonyl groups: alcohol and aldehyde dehydrogenases, glutamate dehydrogenase, inositol-P synthase, HMG CoA reductase, histidinol-1-P dehydrogenase, *etc.*

The human vitamin itself is nicotinic acid, 3 carboxypyridine[2] (Figure 9.1). Humans can elaborate the ingested nicotinate (also known as niacin) in several enzymatic steps[9] (Figure 9.8). First is enzymatic phosophoribosylation to nicotinate mononucleotide (NMN), when the pyridine nitrogen attacks the

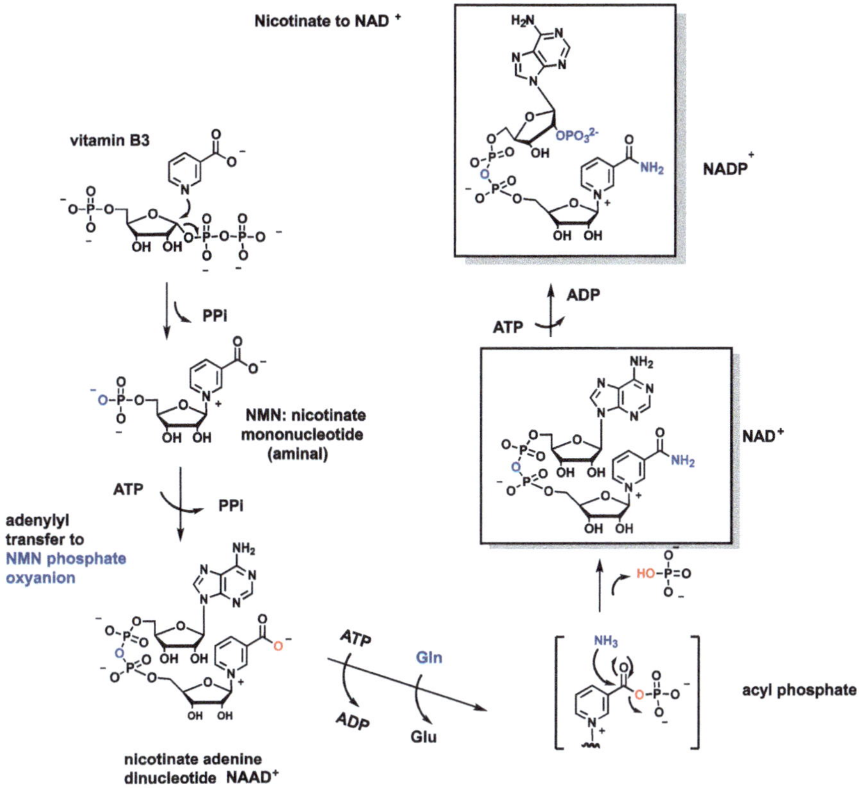

Figure 9.8 *In vivo* conversion of nicotinate, vitamin B3, to NAD.

electrophilic $C_{1'}$ of PRPP. Adenylyl transfer from ATP is the next enzymatic step, generating the nicotinic acid form of NAAD (nicotinic acid adenine dinucleotide). One enzymatic step remains, the conversion of the anionic 3-carboxylate substituent on the pyridinium ring to the neutral carboxamide. ATP is cleaved to make the acyl phosphate and glutaminase action releases nascent ammonia to attack the acyl phosphate carbonyl and form the amide of the NAD^+ form of the nicotinamide coenzymes[10] (Figure 9.8). The second nicotinamide coenzyme form at about 10% of the cellular NAD^+/NADH levels is $NADP^+$, arising from kinase mediated phosphoryl transfer to the 2′-OH of the ribose moiety attached to the nicotinamide ring[1] (Figure 9.8).

Nicotinamide coenzymes actually serve in two quite distinct cellular roles, redox *vs.* nonredox. In both contexts the permanently charged pyridinium ring nitrogen is a key electrophile, reminiscent of the permanently charged nitrogen in the thiazolium heterocycle of vitamin B1 above. The heterocyclic chemistry of the thiazolium ring in B1 and the pyridinium ring in B3-based nicotinamides is very different, but each is accessed because of the quaternized cationic nitrogen (Figure 9.9) and leads to electron sinks as targets for carbanions (B1) or hydride ins (B3).

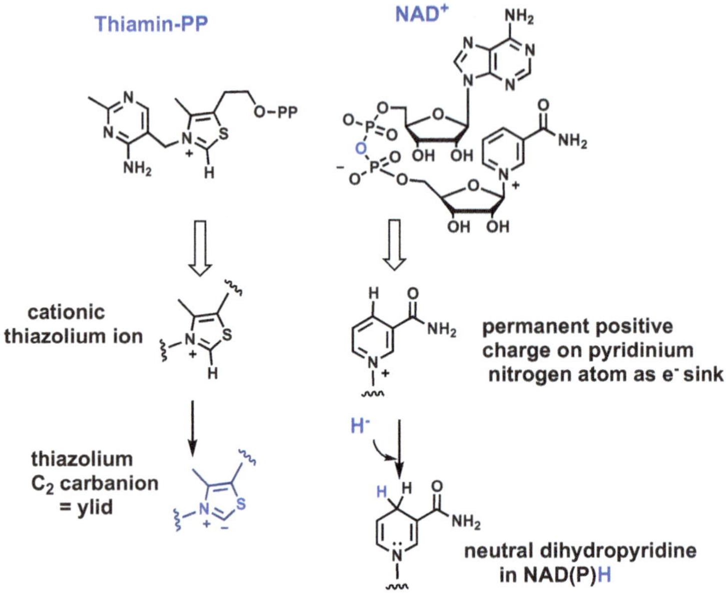

Figure 9.9 Comparison of the function of the two quaternized nitrogen heterocycles in thiamin-PP *vs.* NAD^+.

We note that pyridinamide/nicotinamide radical chemistry is out of the energy range available to organisms: the nicotinamides are instead obligate hydride transfer agents, acceptors in the oxidized form, donors in the reduced dihydro forms (NADH, NADPH). Two electron-only redox paths are incompatible with facile reductive chemistry with atmospheric O_2, itself reduced readily only by obligate one electron paths. The result is that NADH and NADPH, while thermodynamically unstable with regard to oxidizing re-aromatization, are kinetically stable in aerobic environs. This electronic mismatch with O_2 allows the reduced forms, NADH and NADPH, to diffuse through aerobic cells and serve as the major forms of cellular electron currencies. Indeed, 20 of 24 electrons removed from glucose as it is metabolized to six CO_2 product molecules accumulate as 10 molecules of NADH.

We noted back in Chapter 3 that hydride transfer into C_4 of the oxidized forms of the nicotinamides (NAD^+, $NADP^+$) creates a prochiral methylene carbon[11] at C_4 of the reduced forms. Incoming hydride ions are transferred to one face or the other of the planar pyridinium ring by families of nicotinamide-utilizing oxidoreductases. For example, alcohol dehydrogenases use the pro R (H_R) of NADH in the back direction to reduce acetaldehyde while glucose-6-P dehydrogenase carries out deuteride transfer from 6-P-glucose C_1 to yield only the 4S-deutero NADPH as product 6-P-gluconolactone is formed[12] (Chapters 3 and 6).

9.2.2.1 Nonredox Roles for NAD^+: Electrophilic Group Transfers

In the oxidized NAD^+, but not the two electron-reduced NADH oxidation state (where the nitrogen of the dihydropyridine is neutral/uncharged), the positive charge on NAD^+ at the pyridinium N_1-nitrogen labilizes that glycosidic link towards capture by cellular nucleophiles. In this mode NAD^+ acts as an electrophilic group (ADP-ribosyl) transfer reagent.[13] One such outcome is net hydrolysis to free nicotinamide and ADP ribose (Figure 9.10) where a water molecule acts as nucleophile at C_1' of the ribosyl moiety (fate 1).

Another is intramolecular capture by N_3 of the internal adenine ring to yield cyclic ADP ribose, an intracellular signaling molecule (fate 3). Intermolecular capture of the ADP ribosyl group of NAD^+ is mediated by a family of posttranslational modifying ADP ribosyltransferases, *e.g.* by an arginine residue side chain (fate 2).

A subset of histone deacetylases (HDACs) require NAD^+ as cosubstrate, cleaving it to free nicotinamide and ultimately 2′-*O*-acetyl-ADP ribose. A series of acetyl ribose intermediates have been proposed.[14] Since a family of histone deacetylases are known to hydrolyze the acetyl-Lys amide linkage directly, it has been suggested that the NAD^+-cleaving deacetylases may be monitoring cellular [NAD^+] levels as a surrogate for cellular energy supply.

A different mode of enzymatic cleavage occurs in microorganisms that use NAD^+ in place of ATP as consumable substrate for DNA gap repair by the enzyme DNA ligase[15] (Figure 9.11). Typically, ATP undergoes cleavage to release PPi and become adenylated by the side chain of an active site lysine, resulting in a covalent AMP-NH-lys (phosphoramidyl adduct to the enzyme). That AMP group is still sufficiently activated to be transferred on to one of

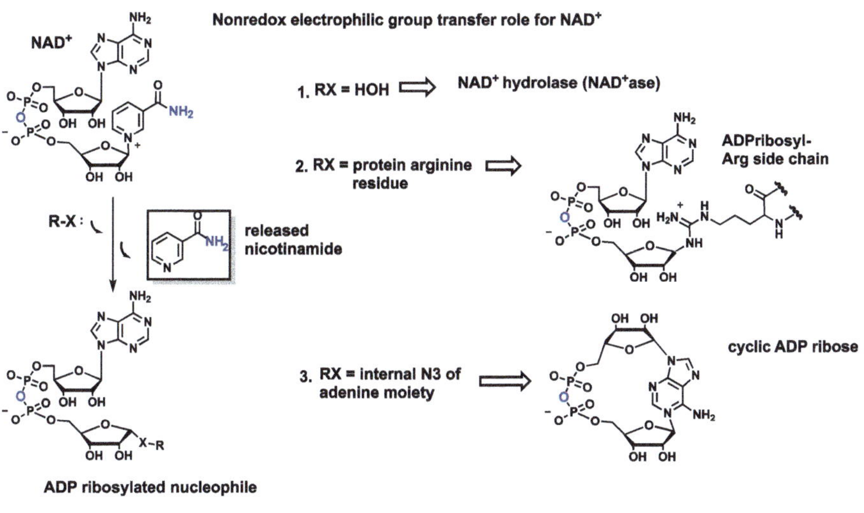

Figure 9.10 Nonredox electrophilic group transfer of the ADP-ribosyl moiety of NAD^+.

the 5′-phosphate oxyanions on the 5′ side of the gap. It is that 5′-ADP adduct that is repaired by nucleophilic attack of the 3′-OH of the deoxyribosyl unit on the 3′-side of the gap. Formation of the 3′-5′P-diester restores the integrity of the DNA strand and expels AMP.

When NAD^+ is used instead of ATP, the AMP moiety of NAD^+ is used identically in the downstream adducts for DNA strand repair (Figure 9.11). The initial enzyme-mediated fragmentation of NAD^+ by such DNA ligase action releases NMN as released fragment, as coproduct with AMP released at a later stage of the catalytic cycle.

9.2.3 The Coenzyme Form of Vitamin B6 Brings a Pyridine Aldehyde to Nature's Catalytic Inventory

The third of the B vitamins with a monoheterocyclic ring that enables hundreds of cellular transformations of amino acid is the coenzyme form of pyridoxine, vitamin B6[2] (Figure 9.12). The vitamin itself is a pyridine alcohol and must undergo *in vivo* enzymatic side chain alcohol phosphorylation and oxidation of the alcohol to the aldehyde (by a flavoenzyme oxidase).

The aldehyde group in the active coenzyme pyridoxal phosphate is the novel chemical group, not present in any of the twenty proteinogenic building blocks of proteins. Together with the conjugated pyridine heterocyle that provides a delocalized electron sink, the pair of functional groups enables accessible carbanion chemistry.[1] Adduct formation between that aldehyde group and amino groups of specific substrates (amino acids and amines) open up iminium ion chemistry by way of aldimines and ketimines, to set up the full range of carbonyl and ensuing stabilized carbanion chemistry.[16] In particular, the combination of aldehyde conjugated to the heteroaromatic pyridine ring creates a potent electron sink where substrate-derived electrons can be reversibly stored and delocalized during enzymatic catalytic cycles.

The full range of amino acid chemistry enabled by initial aldimine formation with pyridoxal-P at the active site of many enzymes makes this coenzyme synonymous with amino acid chemical biology. The scope of PLP-mediated enzymology encompasses essentially all types of metabolic transformations experienced by amino acids, accessed either by C_2–H cleavage or C_2–COO^- cleavage (Figure 9.13).

We have seen several variants of carbanion-enabled amino acid=PLP aldimine chemistry in Chapter 4 on C–S functional groups and in Chapter 5 on C-functional groups. The *O*-acetylserine sulfhydrolase/cysteine synthase enzyme[17] of microbes and plants that fixes inorganic hydrogen sulfide into the organic scaffold of cyctsine is a β-replacement enzyme (Figure 4.6). So is the cystathionine β-synthase where C_3 of seine is replaced by the thiol of cosubstrate homocysteine to produce the seven carbon thioether cystathionine as a metabolic intermediate[18] (Figure 4.8). In Chapter 8 on heterocycle chemical biology tryptophan synthase is also called out as a β-replacement

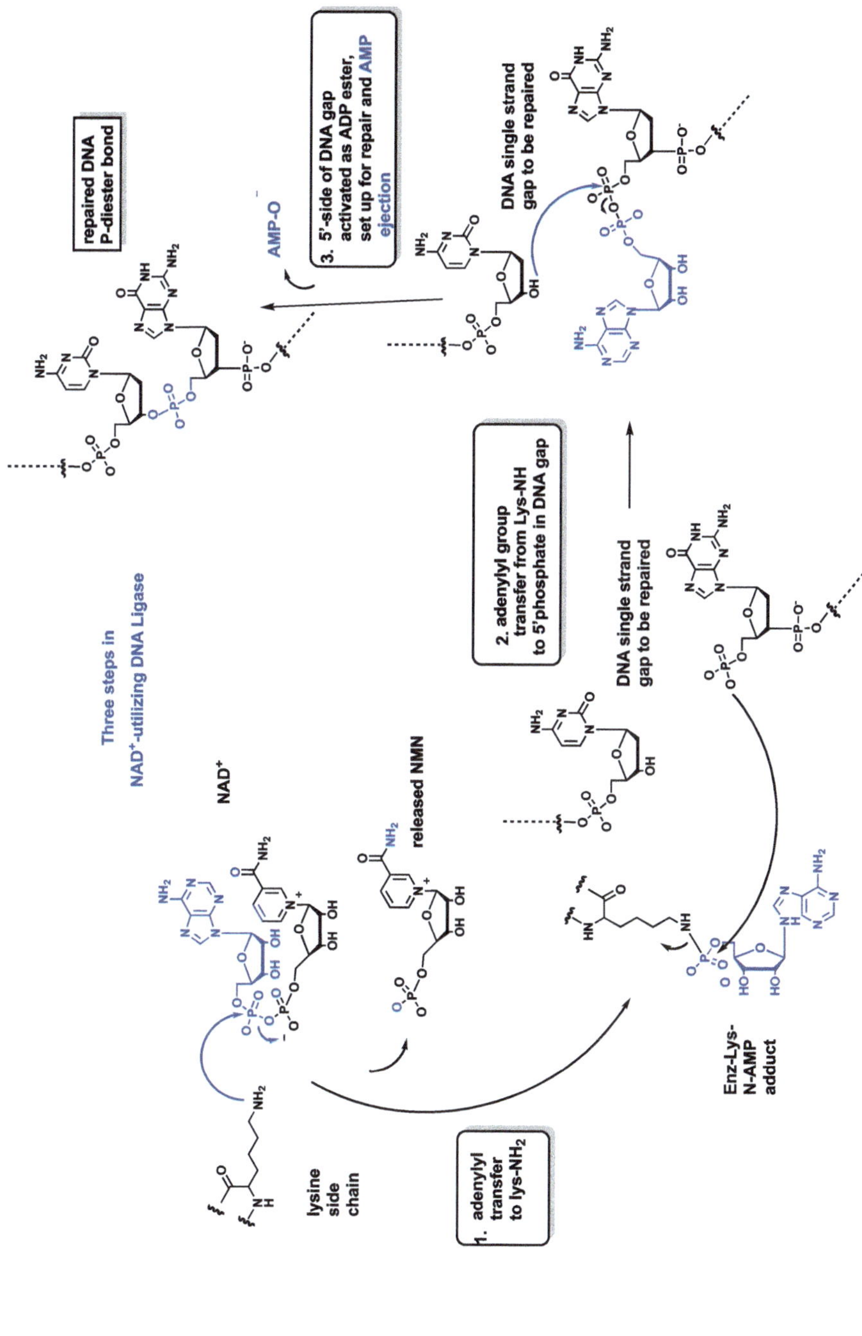

Figure 9.11 NAD$^+$-cleaving DNA ligase for DNA strand repair: three stages in the catalytic cycle.

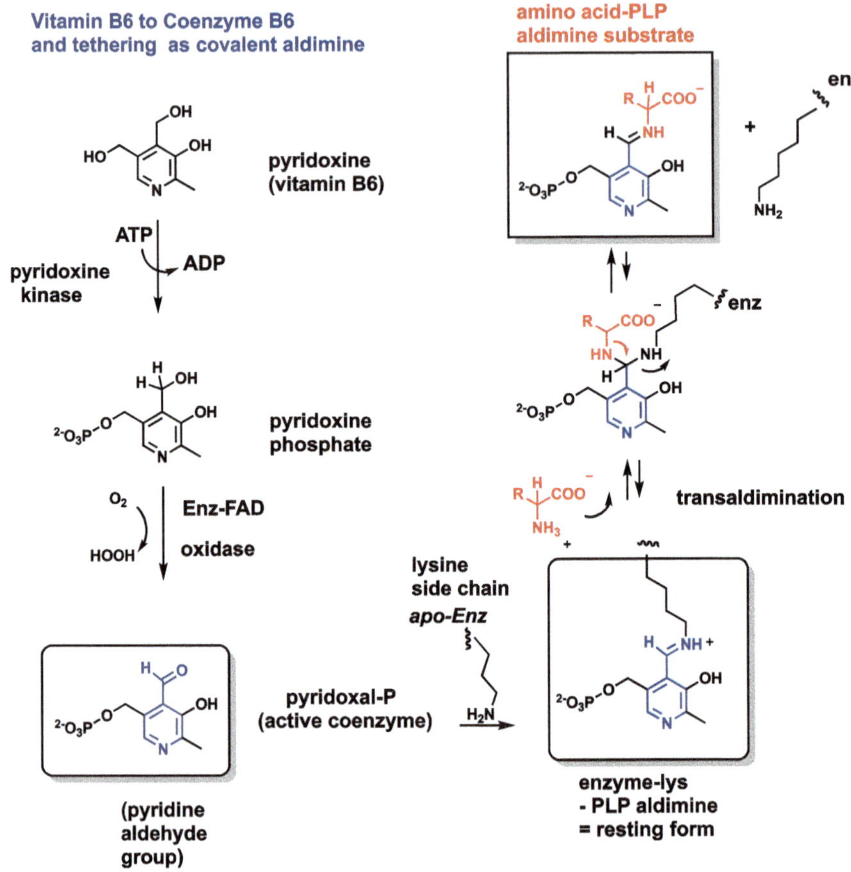

Figure 9.12 Conversion of vitamin B6 pyridoxine to coenzyme pyridoxal-P; covalent tethering of PLP *via* aldiminium ion linkage to an active site lysine side chain in PLP-enzymes.

enzyme (Figure 8.21). All of these catalysts convert initial serine=PLP aldimine complexes to stabilized C_2-PLP carbanion equivalents and then use that electron pair to eliminate the C_3–OH group. The resultant enamino=PLP is electrophilic at C_3, capturable by cosubstrate nucleophiles: by HS^- (cysteine synthase), homocysteine-S^- (cystathionine-β-synthase), or the enamino carbon 3 of indole (tryptophan synthase (Figure 9.14).

A subset of enzymes that expel a C_3-leaving group, such as threonine deaminase use the same catalytic mechanism as β-replacement PLP enzymes up to generate the common enamino acid=PLP adduct.[18] At that point the fates of that intermediate diverge (Figure 9.14). Whereas β-replacement enzymes use a cosubstrate nucleophile to capture the electrophilic C_3 terminus, β-elimination enzymes instead catalyze its hydrolysis to release the enamino acid and PLP (Figure 9.14). Such enamino acids tautomerize to the more stable imine and then undergo hydrolysis to the keto acid and NH_3.

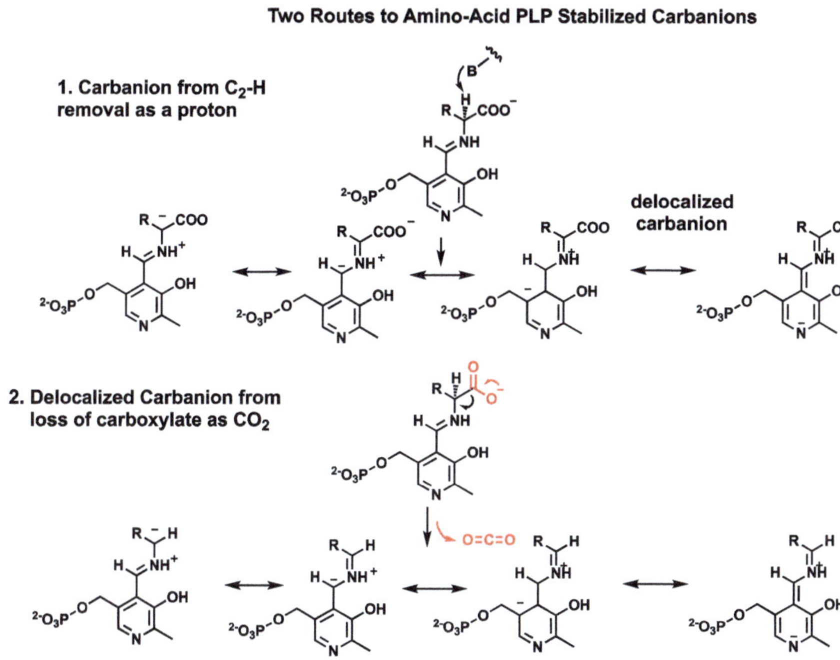

Figure 9.13 Two routes to amino acid–PLP stabilized carbanions. C_2H removal as a proton *vs.* decarboxylation.

For example, threonine deaminase releases water, ammonia and 2-ketobutyrate as a beta elimination enzyme.

In bacterial metabolism there are a number of PLP enzymes that racemize L-amino acids to D-amino acids, with a $K_{eq} = 1$. Those amino acid=PLP aldimines also start catalytic cycles by C_2–H cleavage by proton removal to get to the common stabilized carbanion. Reprotonation from the opposite face at C_2 leads to the enantiomeric amino acid=PLP aldimine before D-amino acid release[19] (Figure 9.15).

We also encountered several examples of amino acid decarboxylases in Chapter 5 that all orient the amino acid=PLP aldimine such that the carbanionic-PLP intermediate arises from C–COO^- bond cleavage. That carbanion is protonatable at C_2 to release the amine coproducts in the biogenesis of many neurotransmitter amines (Figure 5.18).

The above subcategories of PLP enzymes have resulted in cleavage of two of the four substituents at the C_2 carbon of alpha-amino acids, the C_2–H bond and the C_2–COO^- bond. There are also examples of PLP enzymes that lead to net cleavage of the remaining C_2 bond connections. One is the set of transaminases covered in Chapter 5 that oxidatively deaminate an amino acid substrate (net C_2–N bond cleavage) while transferring that amine for reductive amination of a cosubstrate 2-keto acid. In examination of the logic of transaminase catalysis, we observed that PLP can function as a formal

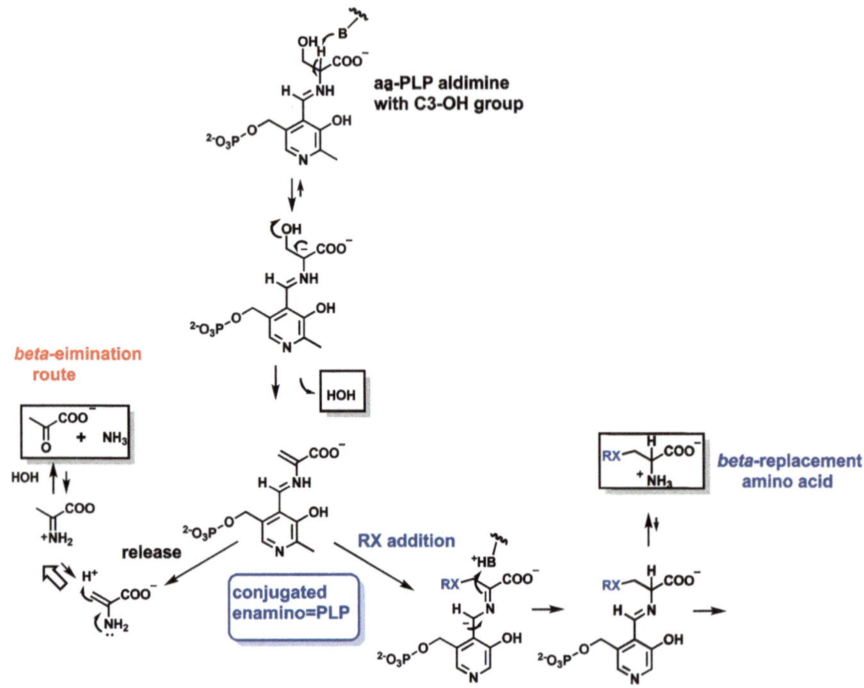

Figure 9.14 Mechanistic scheme for PLP-enzymes to act as β-elimination *vs.* β-replacement catalysts for amino acid substrates.

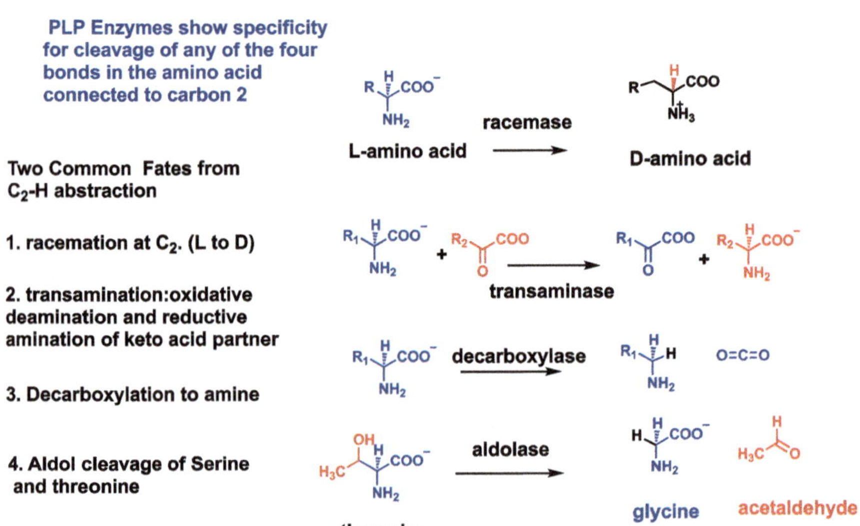

Figure 9.15 Cleavage of any one of the four bonds to C_2 of amino acids in amino acid=PLP aldimine adducts by subtypes of PLP-enzymes.

redox coenzyme, undergoing reversible reduction to pyridoxamine-P and then back to PLP[20] (Figure 5.13).

The fourth substituent at C_2 is C_3 and its substituents. That C_2–C_3 bond is cleaved by PLP-dependent aldolases that fragment 3-hydroxy-amino acids such as serine or threonine, generating glycine and either formaldehyde or acetaldehyde equivalents, respectively (Figure 9.16). Both threonine aldolases and serine aldolases have oxidized the C_3–OH to the corresponding aldehyde during fragmentation. On protonation of the intermediate C_2-carbanionic-PLP intermediate, the coproduct is glycine, formally reduced by two electrons at in its C_2 methylene group. We will note in Section 9.2.4.2 on the chemical biology of folate coenzymes that the serine aldolase, actually termed serine transhydroxymethylase, requires tetrahydrofolate for turnover to remove and derivatize potentially toxic formaldehyde product. Acetaldehyde does not cause equivalent problems of toxicity and is not subjected to comparable scavenging by tetrahydrofolate.

We have also previously discussed one additional feature of amino acid=PLP adldimine/aldiminum chemistry in Chapter 4 and in Chapter 5 where not just two substrate-derived electrons are stored (formally) in the PLP coenzyme as electron sink. In both cystathionine-γ-lyase (Figure 4.8) and

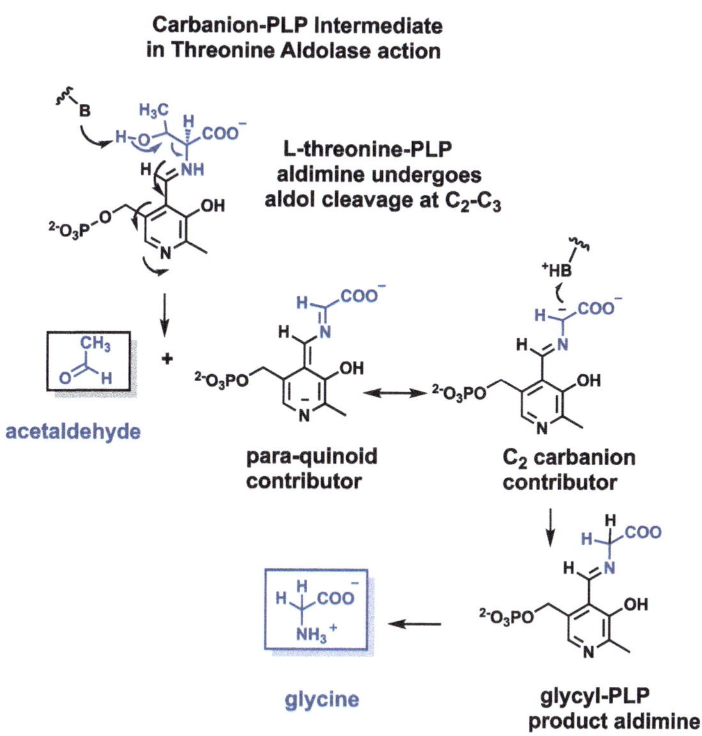

Figure 9.16 Nascent carbanionic product=PLP aldimine intermediate in the threonine aldolase reaction.

in the variants of cysteine β-lyases that release six-electron sulfane sulfur[21] (for FeS cluster assembly in dozens to hundreds of proteins) (Figure 4.24), an additional pair of two electrons are stored transiently in the amino acid–PLP adducts. That is a remarkable capacity to transiently store two electron pairs (two carbanion equivalents) to enable γ-lyase/γ-replacement enzyme chemistry or cleavage of the C_3–SH bond of cysteine where the electron pair in the C–S bond are retained on carbon rather than expelled with sulfur (as HS^-).

Tetrapyrrole macrocycle biosynthesis,[22] elaborated in Chapter 16, emphasizes two additional variants of PLP-enzyme chemistry early in the pathway.[23] One is 5-aminolevulinate synthase where a C–C bond is cleaved (decarboxylation of glycine) and a C–C bond is formed (between C_2 of glycine and C_1 of succinyl CoA) (Figure 9.17). The second intriguing, alternative PLP-enzyme in that pathway interconverts glutamic-1-semialdehyde and 5-aminolevulinate by net intramolecular isomerization of an amino-aldehyde to a keto-amine product by way of a transamination intermediate with a tightly bound ketoaldehyde (Figure 9.17). The first enzyme works in animals, the second in plants to create the same aminolevulinate product from distinct starting materials by different chemistry, both PLP-dependent.

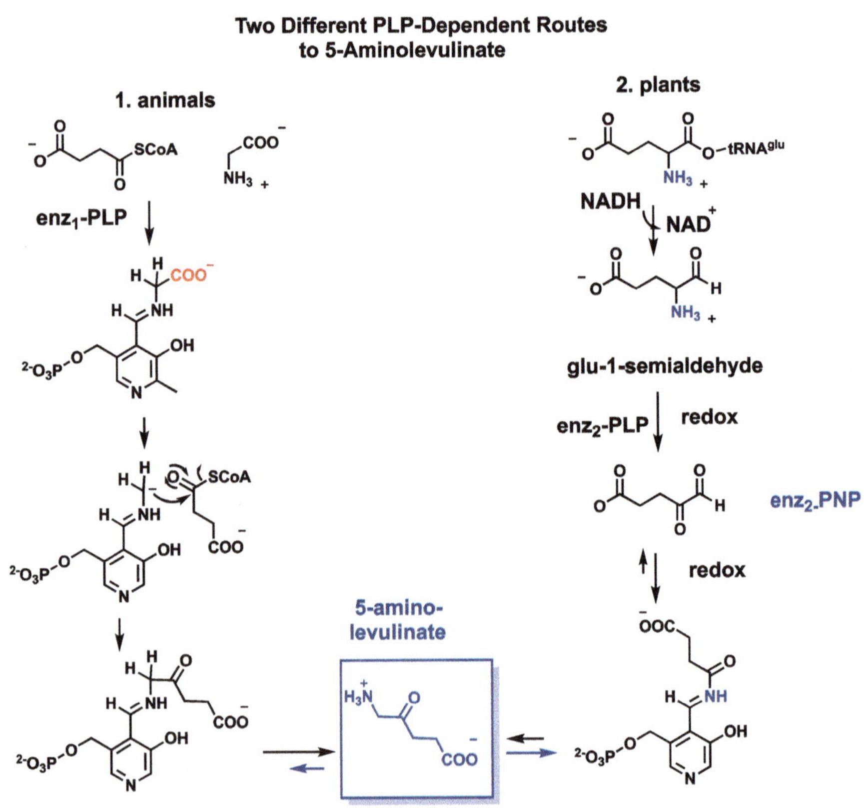

Figure 9.17 Two different PLP-dependent routes to δ-aminolevulinate.

Finally, although of relevance to microbial metabolism rather than human metabolism, the pyridine benzyl aldehyde group in PLP can also participate in one-electron redox reactions, providing stabilized benzyl radicals, both in radical SAM enzyme active sites and in O_2-reducing oxidase active sites. As we note in Chapter 12, those are the two enzyme types that generate carbon radical intermediates.

The coenzyme form of vitamin B6, pyridoxal-P with its bifunctional pyridine aldehyde core demonstrates the remarkable breadth of chemical transformations that nature can catalyze from amino acid=PLP aldimine-derived carbanion chemistry. A comparison of the role of the pyridine aldehyde in B6 coenzyme and the pyridinium/dihydropyridine redox traverse in the B3-based nicotinamide coenzymes further emphasizes the versatility of the pyridine ring for accessing much of the chemistry of life.

9.2.4 Bis-heterocyclic Coenzyme Forms of Vitamin B7 and B9

The two B vitamins with bis-heterocyclic reactive centers in the active coenzyme forms are biotin (B7) and folate (B9)[2] (Figures 9.1 and 9.18). The pair of fused heterocyclic rings are distinctive in each coenzyme, a ureido-thiane 5,5-ring system in biotin *vs.* a fused pteridine heterocycle framework in folate. Biotin is not redox active and only requires covalent tethering to a specific carrier protein region in its host carboxylases to convert inactive apo forms to covalently loaded holo forms (Figure 9.18).

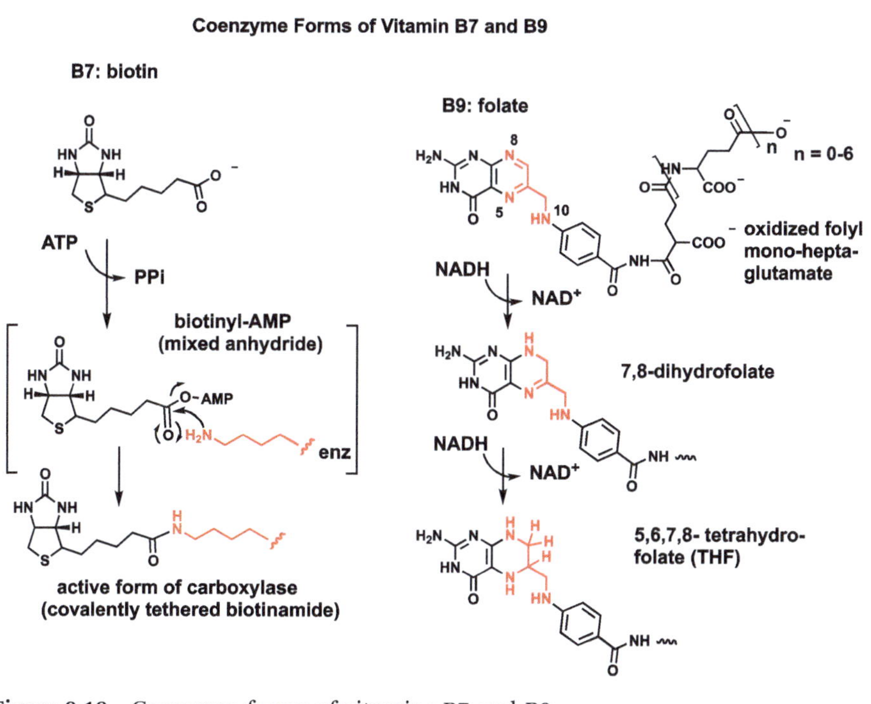

Figure 9.18 Coenzyme forms of vitamins B7 and B9.

By contrast, dietary folate needs to undergo successive two-electron reductions to the 7,8-dihydrofolate and then the 5,6,7,8-tetrahydrofolate oxidation state to get to the active coenzyme form (Figure 9.18). The tetrahydrofolates may have from one to seven oligoglutamates in the peptidic tail. Human folate-recognizing enzymes require the polyanionic oligoglutamate tails for activity.

Biotin and the tetrahydro form of folate together carry one-carbon units at four different oxidation states – from CO_2 (biotin) to formate, formaldehyde, and methanol (tetrahydrofolate) (Figure 9.19). There is no mammalian carrier for methane. The other four oxidation states of one carbon metabolites are carried around on transferable forms by the B7 and B9 coezyme forms, deliverable to specific enzyme active sites that contain cosubstrates with nucleophilic atoms to capture the differentially oxidized one carbon electrophilic groups.

9.2.4.1 Biotin

Biotin has a ureido ring fused to a five membered thiane ring.[2] The thiane has a sulfur in thioether linkage. That thiane ring is fused *cis* to the ureido ring, imparting a bent geometry to the bicyclic system that is important for function

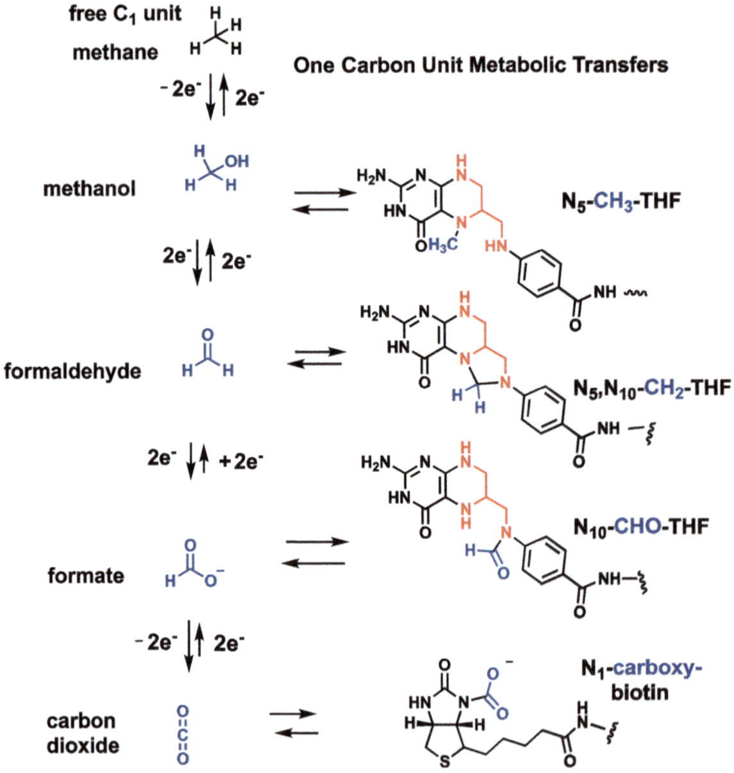

Figure 9.19 One-carbon unit metabolic transfers.

(Figures 9.1 and 9.18). Attached to the thiane ring is a valeric acid side chain that becomes covalently anchored to a lysine residue when the nascent apoenzymes are posttranslationally modified to generate the active holoenzyme. The ureido group nitrogen at N_1 is the reactive functional group that carries CO_2 between two subsites of the carboxylase family of enzymes. This is one of several cases in chemical biology where an amide nitrogen, a weak nucleophile at best, is utilized as the attacking species in N–C bond formations. (It may be that the ureido group is first enzymatically isomerized to the less stable isourea oxyanionic form to act as nitrogen nucleophile.)

Two metabolically central carboxylations are catalyzed by biotin-tethered carboxylases[24] (Figure 9.20): the carboxylation of acetyl CoA to malonyl CoA in the first committed step of fatty acid biosynthesis and in the action of pyruvate carboxylase to convert pyruvate to oxaloacetate. Each enzyme cleaves a molecule of ATP to ADP and Pi in each catalytic turnover. Carbon dioxide is in rapid hydration equilibrium with HCO_3^- by action of the highly efficient enzyme carbonic anhydrase and mechanistic analyses have shown the actual substrate is bicarbonate (HCO_3^-) not CO_2. One of the three oxygens from bicarbonate substrate ends up in the inorganic phosphate released from ATP.

To make a C–C bond (here a CH_2–COO^-) bond heterolytically requires a carbon nucleophile and carbon electrophile. The carbon nucleophile candidates are the substrates acetyl CoA and the enolate anion of pyruvate for the acetyl CoA carboxylase and pyruvate carboxylase, respectively. We will note in Chapter 10 that two of the most readily accessible carbanion equivalents in central metabolites are the thioester enolate of acetyl CoA (C_2 carbanion equivalent) and the enolate anion from pyruvate (C_3 carbanion equivalent) (Figure 9.20). Those are the accessible carbanion species in the second half reaction of these biotin-tethered carboxylases.

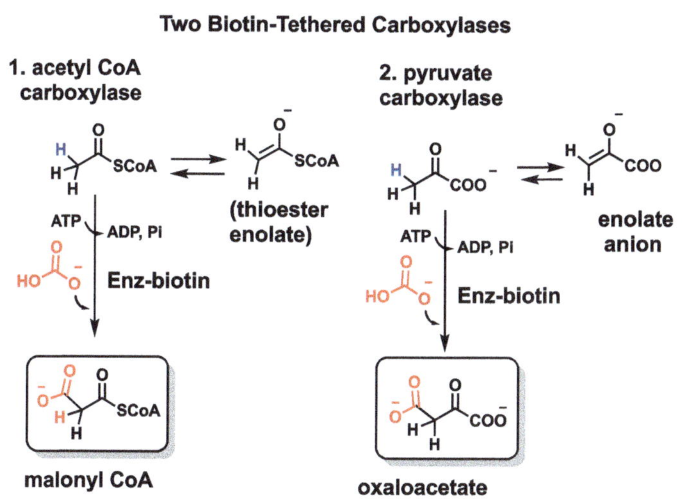

Figure 9.20 Two biotin-tethered carboxylases for acetyl CoA and pyruvate.

The problem that immediately arises is the abundant HCO_3^- is anionic (like a carboxylate anion) and non-electrophilic, unlike CO_2. Thus, the election of the biotin-tethered carboxylases to use HCO_3^-, instead of CO_2, requires them to convert bicarbonate into an electrophile. The general strategy used for turning carboxylate anions into electrophilic carbonyl groups is to turn them into acyl phosphoric anhydrides.[25] That is the strategy followed here as well. The oxyanion of bicarbonate attacks Pγ of ATP in the carboxylase subsite that binds ATP. The product ADP dissociates, leaving the mixed anhydride of carbonic acid and phosphoric acid, known as carboxy phosphate (see Chapter 5), still bound. Now the carbonyl group is electrophilic. The immediate attacking carbanion species is neither the thioester enolate of acetyl CoA nor the enolate of pyruvate. They bind in a distinct subsite (Figure 9.21) from ATP and bicarbonate.

Instead, the weak nitrogen nucleophile at N_1 of the ureido ring of the biotinamide group is the kinetically competent nucleophile attacking carboxy phosphate. The immediate result is formation of inorganic phosphate (net transfer of one oxygen from bicarbonate to inorganic phosphate) and N_1-carboxybiotinamide. This carbamate grouping is stable enough for the *N*-carboxy-biotinyl group on its long valery-lysine amide side chain to swing

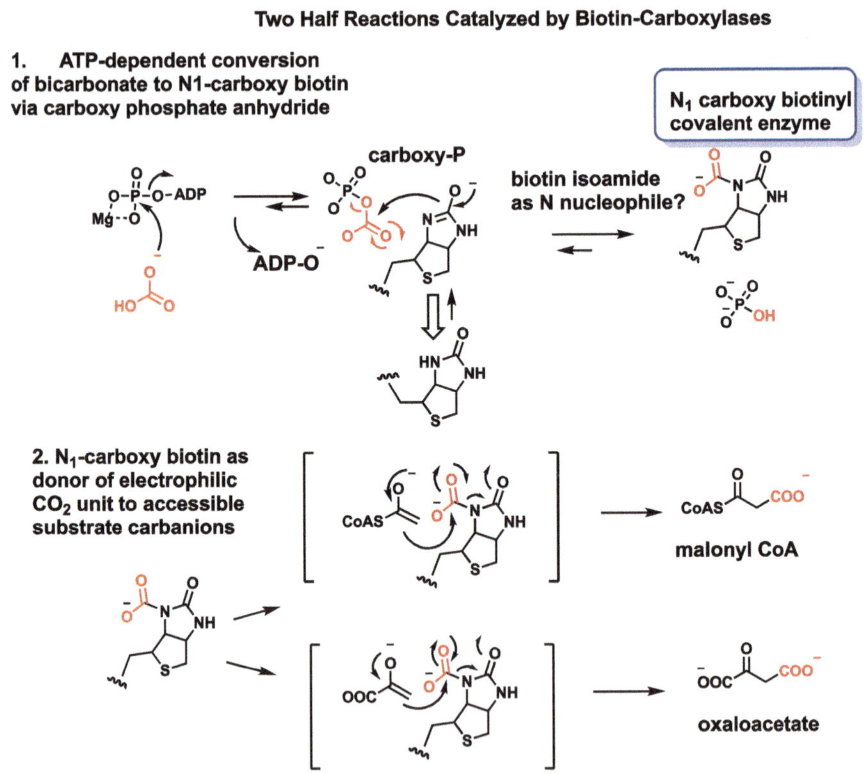

Figure 9.21 Two half reactions catalyzed by biotin carboxylases.

to the subsite where acetyl CoA or pryuvate is bound and enolized to the carbanionic transition state (Figure 9.21). CO_2 transfer from the ureido N_1 of the biotin cofactor to either the acetyl CoA C_2 carbanion or the enolpyruvate C_3 carbanion generates the CH_2–COO^- bonds and completes the carboxylase catalytic cycles.

It is worth contrasting the reaction strategies of the ATP-cleaving biotin-linked carboxylases with that of plant ribulose-1,5-diphosphate carboxylase which utilizes CO_2 rather than HCO_3^- and does not spend ATP molecules in a one for one stoichiometry of carboxylation (see Chapter 3). The ribulose-bis-phosphate carboxylase requires less energy expenditure, by developing a strategy for capturing the minor partner CO_2 in the presence of an excess of HCO_3^-.

9.2.4.2 *Folate (Vitamin B9)*

Folate was named for its high content in leafy (*folia*) vegetables.[2] Discovery and characterization was slowed in the first half of the 20th century by the existence of multiple forms of folates, which turned out to be a variable number of oligoglutamyl chains attached to the *para*-aminobenzoyl moiety of folates from different biological sources[1] (Figure 9.18). Because folate deficiency can figure into spina bifida pathologies in *in utero*-developing infants, USDA regulations include folate supplements in milk and bread. The business end of the folate coenzyme forms is the bicyclic pterin ring system comprised of a pyrazino[2,3-*d*] pyrimidine core. As shown in Figure 9.18, the pyrimidine ring is the left hand ring, fused to a six membered ring with two nitrogens (in a 5, 8 position). The parent right hand heterocycle is a pyrazine ring.

The active form of the B9 coenzyme is the tetrahydro oxidation state of the pyrazine ring, leaving the pyrimidine ring unaffected (Figure 9.18). In the tetrahydropyrazine ring system both N_5 and N_8 are basic secondary amines rather than the nonbasic aromatic nitrogen atoms of the conjugated, oxidized bicyclic pterin ring system. The basicity of N_5 is essential for coenzyme function of tetrahydrofolates. So is the basicity of N_{10} provided by the NH of the *para* aminobenzoyl moiety tethered by a methylene bridge to the bicyclic pterin ring system. The carboxyl group of the *para*-aminobenzoyl moiety is in amide linkage to the amino group of a glutamate. The simplest tetrahydrofolate is the monoglutamate (Figure 9.18). However, folates containing up to seven glutamates accumulate in different organisms and tissues. The additional six glutamates are also in amide linkage but *via* isopeptide rather than normal peptide bonds. The isopeptide bonds connect a glutamyl amine group to the C_5 carboxyl rather than the C_1 carboxyl of the next glutamate residue. The polyanionic oligoglutamate forms are the functional forms of tetrahydrofolate coenzymes in human metabolism.

The necessity for the tetrahydro oxidation state in the pyrazine ring of tetrahydrofolates is for N_5 and N_{10} to sponge up nascent formaldehyde before it can diffuse out of enzyme active sites and function nonselectively as a promiscuous crosslinking agent for pairs of amine groups. (That promiscuity is the basis for historical use of formaldehyde as embalming fluid *via* protein crosslinking.) The cellular source of formaldehyde is the PLP-linked enzyme serine transhydroxymethylase.[26] We note in Section 9.2.3. that some PLP enzymes will catalyze aldol cleavages on the β-hydroxy amino acids serine and threonine. Serine transhydroxymethylase is the major agent of one carbon flux in cells by cleaving L-serine to glycine and nascent formaldehyde (Figure 9.22). The cosubstrate for this enzyme is a tetrahydrofolyl-oligoglutamate. The enzyme does not turn over and release formaldehyde in the absence of THF to serve as a formaldehyde sponge.

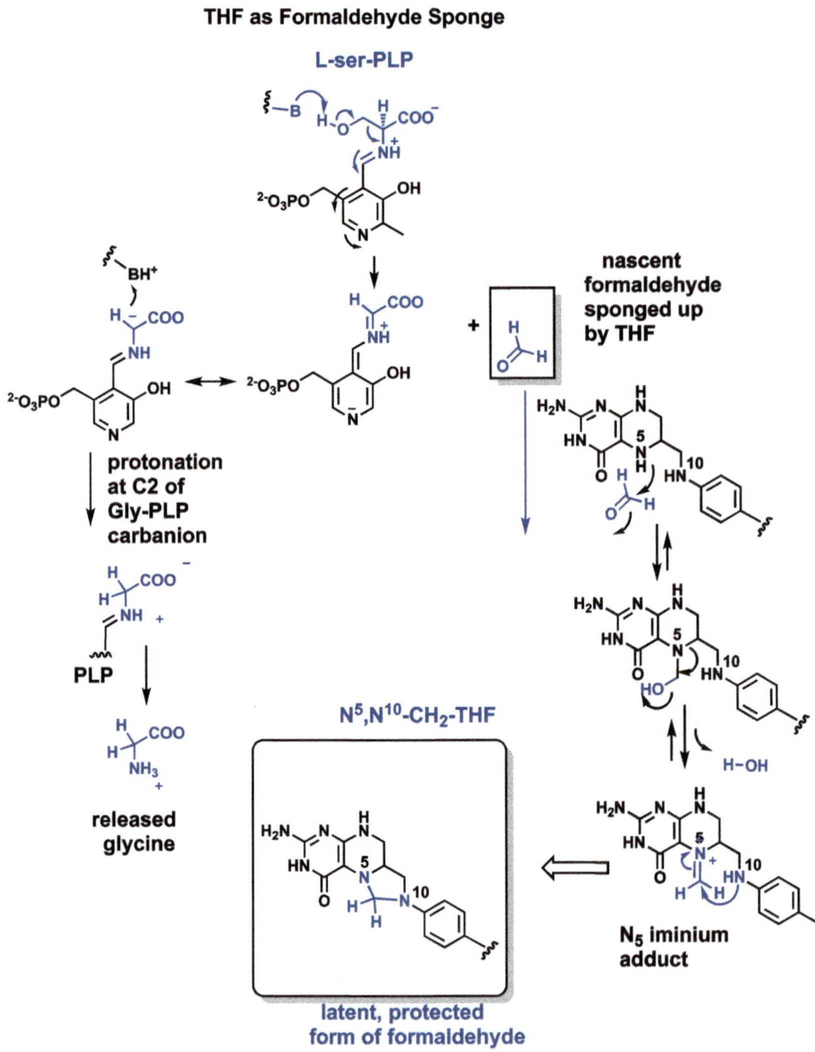

Figure 9.22 Tetrahydrofolate as a formaldehyde sponge in the active site of serine transhydroxymethylase.

The capture of nascent formaldehyde in the enzyme active site by the tetrahydropyrazine of H_4-folate is thought to be initiated by attack of N_5–H on the formaldehyde carbonyl. The immediate carbinolamine adduct can collapse to the iminium ion with net loss of water (Figure 9.22). That $N_5{}^+{=}CH_2$ iminium adduct can then be captured by N_{10} to form the cyclic N_5,N_{10}–methylene tetrahydrofolate adduct. This is a strain free, entropically favored five membered heterocycle that serves as a latent source of formaldehyde in a nontoxic form. The strategy of carrying around a latent nontoxic form of a metabolite is reminiscent of the sequestration of ammonia in the latent form of the carboxamide of glutamine, releasable on demand by glutaminase action (Chapter 5).

9.2.4.2.1 Tetrahydrofolate as Carrier and Donor of One-carbon Fragments at Three Distinct Oxidation States. The initial role of tetrahydrofolate is to sponge up metabolically derived formaldehyde generated by the aldolase action of serine transhydroxymethylase. The device of carrying latent formaldehyde equivalents has been extended by a pair of enzymes to allow THF to also carry around two electron-oxidized formate equivalents or two electron-reduced methanol equivalents, summarized in Figure 9.19. The details of the redox interconversions between the original latent heterocyclic formaldehyde adduct (N^5,N^{10}–CH_2-THF), latent methanol (N^5–CH_3-THF), and latent formate (N^{10}–CHO-THF) are sketched out in Figure 9.23.[27]

To go from the initial N_5,N_{10}–CH_2-THF to the N_{10}–CHO-THF requires an oxidation and then hydrolysis of the N_5–CH bond. The oxidoreductase uses $NADP^+$ as hydride acceptor from the formaldehyde-derived CH_2 group to form the initial cyclic N_5-imino adduct (N_5,N_{10}-methenyl-THF). Hydrolysis of that cyclic iminium, *via* the tetrahedral carbinolamine gives the ring-opened N_{10}-formyl-THF (Figure 9.23, right hand partition). This oxidoreductase is reversible.

The reductase operating in the other direction, from the tricyclic N_5,N_{10}–CH_2-THF also carries out a net two electron transfer, this time *via* a hydride ion transfer in from $FADH_2$ at the active site of methylenetetrahydrofolate reductase (Figure 9.23, left hand partition). In turn, the $FADH_2$ form of the reductase is generated by hydride transfer into the FAD from the external hydride donor NADPH. This enzyme is functionally irreversible. The directionality of the ring opening of the cyclic N_5,N_{10}–methylene ring is controlled in the reductase active site so that it is uniquely the CH_2–N_{10} bond that is cleaved as the N_5–CH_3-THF is generated.

Thus, the pair of oxidoreductases adjusting the oxidation state of the attached one carbon group ultimately use the major cellular electron currency in the form of $NADP^+$ as hydride acceptor to go to the N_{10}-formyl metabolite and in the form of NADPH as hydride donor to push the latent formaldehyde to become latent methanol (N_5–CH_3-THF). In the reduction direction it is the bound $FADH_2$, rather than the external NADPH that is the proximal hydride donor, reminding us that both B2 and B3 coenzymes can be hydride donors.

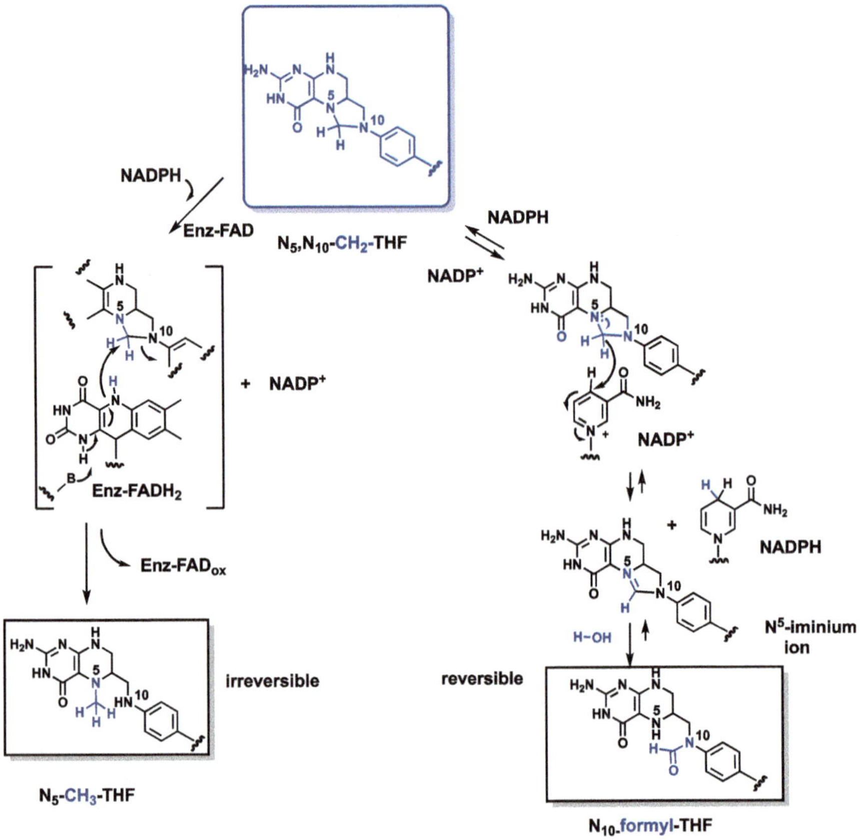

Figure 9.23 $N^{5,10}$–CH_2-THF. Reduction to N^5–CH_3-THF or oxidation to N^{10}–CHO-THF.

Given the distribution of the one carbon fragments covalently tethered to N_5 (N_5–CH_3–), N_{10} (N_{10}–CHO–), or both N_5 and N_{10} of tetrahydrofolate (N_5,N_{10}–CH_2), how do these three distinct oxidation states function as electrophilic groups for transfer to cosubstrate nucleophiles?

The simplest reactions may be the transfer of the N_{10} formyl group as an electrophilic equivalent of formic acid. For example, at two enzymatic steps in *de novo* purine biosynthesis (also see Chapter 6) the formyl group is transferred to a nucleophilic amine (Figure 9.24). The first case is at step 2 when the amine of glycinamide ribotide (GAR) attacks N_{10}-formyl-THF. Five steps later, with the imidazole ring constructed, AICAR (aminoimidazole carboxamide ribotide) uses the exocyclic amino group of the imidazole ring as nitrogen nucleophile to capture the formyl group to become formylated (N-formyl-AICAR). Thus carbons 2 and 8 of the imidazolopyrimidine (= purine) ring derive from an electrophilic form of the one-carbon formate (net transfers of the formyl group from N_{10} of THF to a basic nitrogen on the growing purine scaffold) (Figure 9.24).

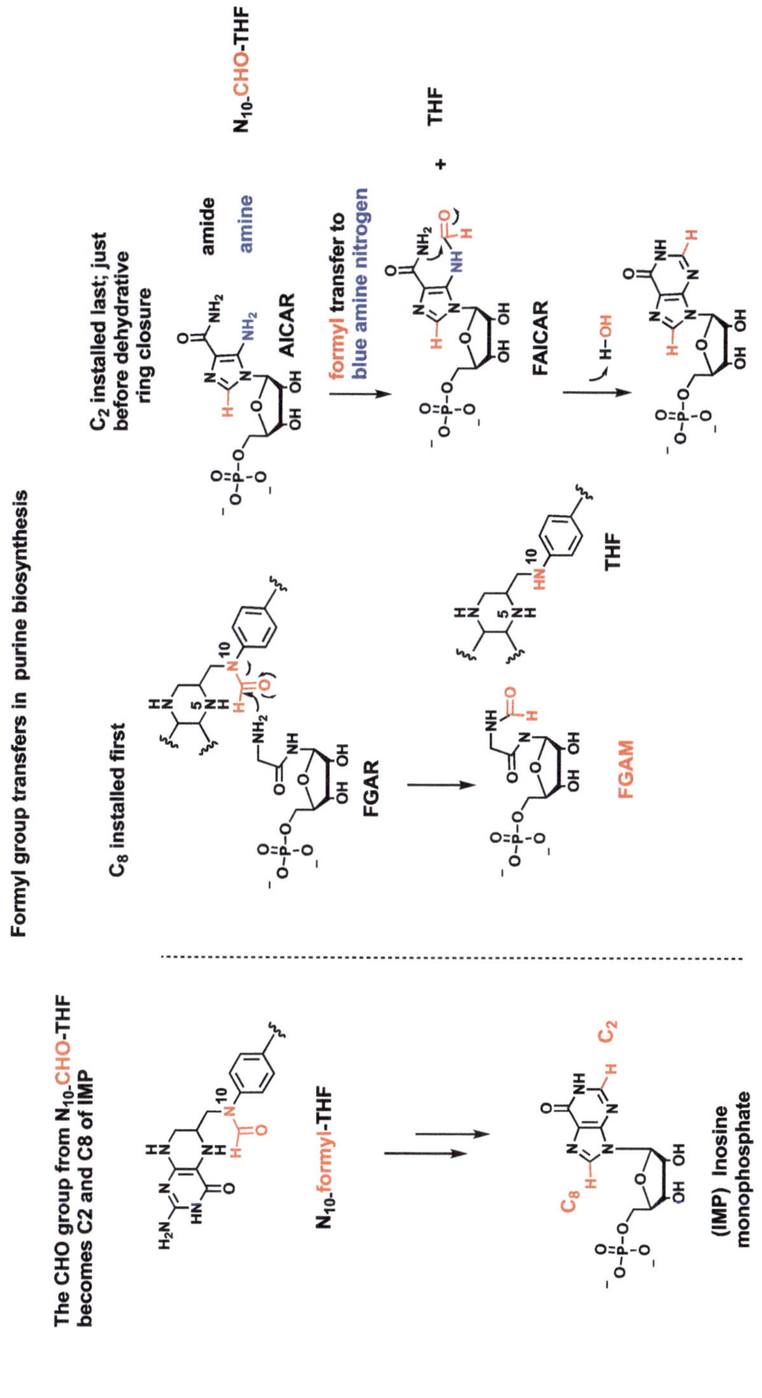

Figure 9.24 Two formyl transfers from N^{10}-CHO-THF at two stages of purine biosynthesis.

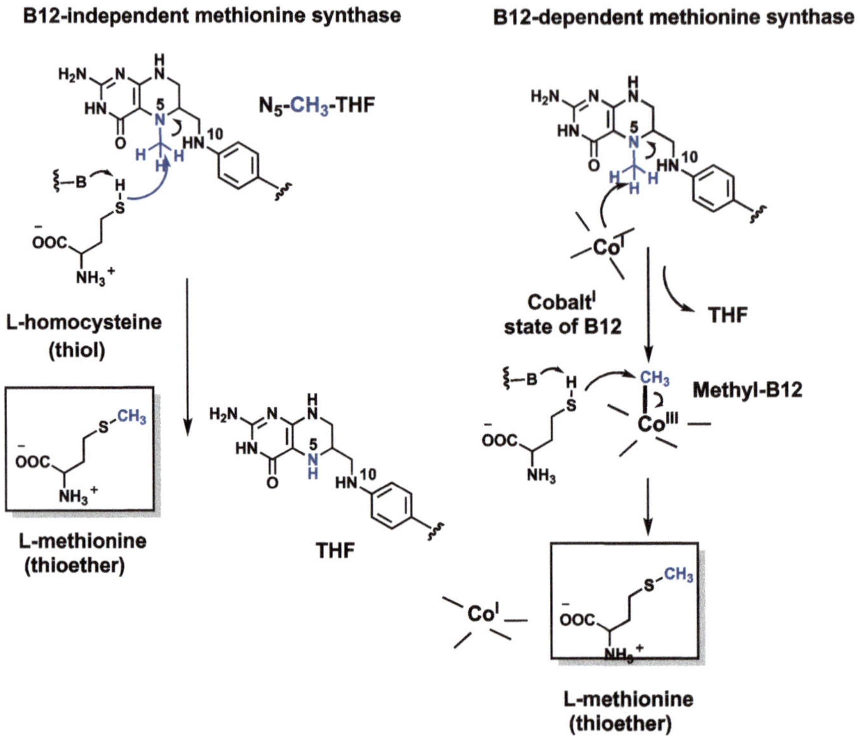

Figure 9.25 B12-independent *vs.* B12-dependent methionine synthases from N^5–CH_3-THF.

At the other extreme of the one carbon fragment oxidation state, four electrons more reduced, the N_5–CH_3 group of THF serves its only major cellular role as methyl donor to the thiolate anion of homocysteine in the active site of methionine synthase (Figure 9.25). We have noted in Chapter 4 that there are two versions of methionine synthases found in different organisms. One enzyme form catalyzes the direct transfer of the methyl group (with the polarity of a $[CH_3^+]$ group) in an S_N2 attack by the nucleophilic thiolate anion. The second version of methionine synthases, *e.g.* in humans, requires the Co^I form of coenzyme form of vitamin B12 (see Section 9.2.6) as an intermediate nucleophile.[28] The organometallic methyl-Co-B12 is an intermediate, accepting the methyl group from N_5–CH_3-THF and donating it to the thiolate anion of homocysteine.

9.2.4.2.2 Thymidylate Synthase for DNA Biosynthesis: Redox Chemistry During One-carbon Transfer. The tricyclic N_5,N_{10}–CH_2-THF, the immediate scavenger of nascent formaldehyde in the serine transhydroxymethylase active site, and thus the primary carrier of the tethered one carbon formaldehyde unit, is famously the source of the methyl group that ends up at C_5 of the pyrimidine ring of 2′deoxy-5-methyl-dUMP=thymidylate triphosphate.[29] dTTP is

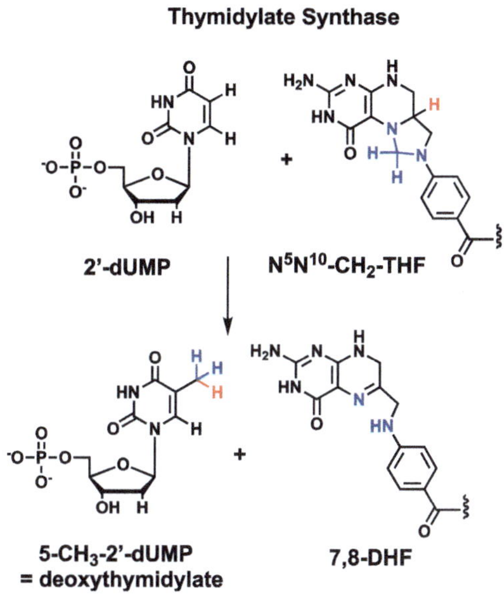

Figure 9.26 Stoichiometry of the thymidylate synthase reaction.

the unique building block found in DNA but not RNA (Figure 9.26). Close to a billion dTTP molecules must be made in each human cell cycle as DNA replicates. The enzyme thymidylate synthase acts on 2′-deoxyUMP to produce 2′-deoxy-5-methyl-UMP=dTMP. The cosubstrate is the N_5,N_{10}–CH_2-THF. However, there is an immediate redox mismatch. The dTMP product has a CH_3 with three hydrogens bonded to carbon while the transferring methylene group (CH_2) in N_5,N_{10}–CH_2-THF has only two C–H bonds. It is at the oxidation state of formaldehyde, while the C_5–CH_3 in dTMP is at the formal oxidation state of methanol. Examination of the reaction stoichiometry shows that while a tetrahydrofolyl scaffold substrate bears the latent formaldehyde, the product is the 7,8-dihydrofolate (Figure 9.26). The tetrahydropyrazine ring has been oxidized by two electrons to compensate/account for the two-electron reduction of the formaldehyde group at some point in the catalytic cycle.

In this redox sense thymidylate synthase is somewhat baroque as catalyst. One might have imagined using *S*-adenosylmethionine as a direct methyl donor to dUMP to form dTMP without any redox change required of the CH_3 group in flight. Indeed, we will compare thymidylate synthase with the cytidine 5-methyltransferases that do use SAM, momentarily.

First, though, back to thymidylate synthase. One of the chemical problems this enzyme must solve to accelerate reaction is to make C_5 of the pyrimidine ring of dUMP a transient carbon nucleophile to capture the latent electrophile formaldehyde nestled between the N_5 and N_{10} amine pair in N_5,N_{10}–CH_2-THF. At first glance this looks monumentally difficult given that C_5 and C_6 of the uracil ring are sp^2 carbons and C_5 is unreactive (C_6 in principle is

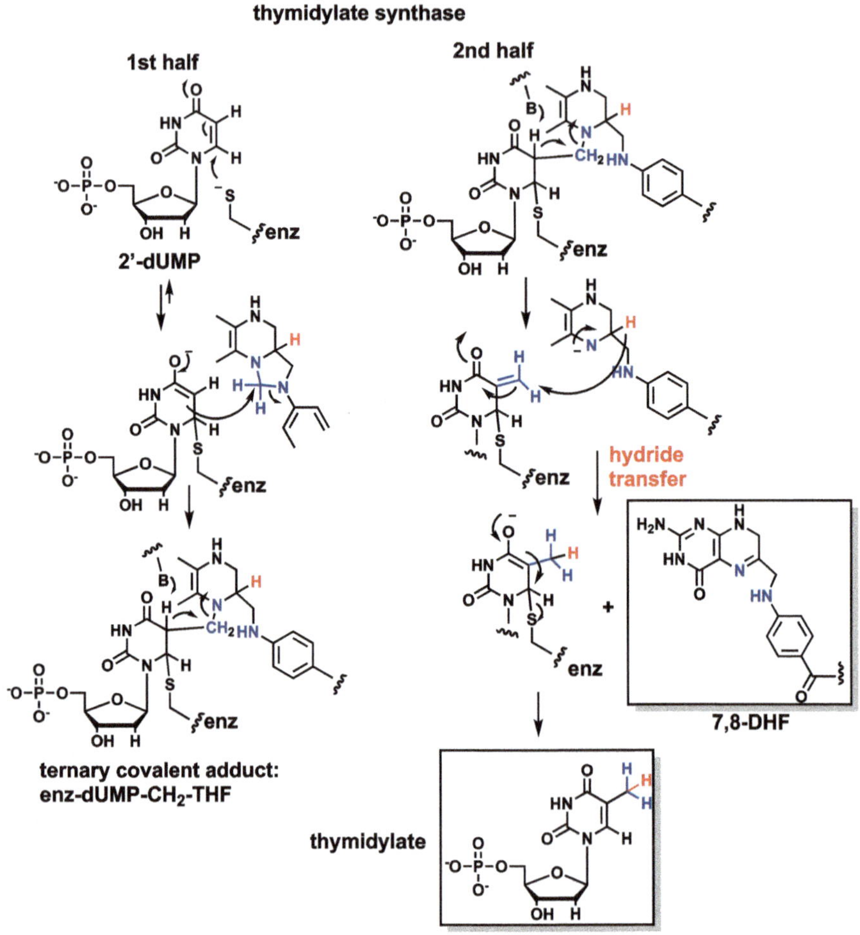

Figure 9.27 Thymidylate synthase reaction mechanism with covalent ternary adduct between enzyme-dUMP-THF.

part of an enone moiety and could be electrophilic). To solve this problem the enzyme starts its catalytic cycle, once the two substrates are bound, by adding the thiolate side chain of an active site cysteine residue into C_6 of bound dUMP (Figure 9.27). This covalent adduct now has a negative charge delocalized between the C_4 carbonyl and the adjacent C_5, solving the first problem of creating a carbon nucleophile at C_5 of dUMP.

That carbanion equivalent can attack the CH_2 group attached to THF, either directly on the N_5,N_{10} methylene adduct or on the N_5 iminium species that may be in equilibrium in the active site. This step covalently tethers the dUMP moiety, both to the enzyme-cysteine thioether link at C_6 and *via* the C_5–CH_2–N_5-THF cosubstrate in a covalent ternary complex (Figure 9.27, bottom of left hand column). One might call this the first half reaction. To this point the dUMP-C_5–CH_2 carbon–carbon bond has formed but there is still the redox problem of converting a CH_2 to a CH_3 group at C_5, breaking the CH_2–N_5 bond to THF, and forming the 7,8-dihydrofolate.

The second half reaction mechanism (right hand column in Figure 9.27) derives from the experimental finding that the C_6–hydrogen of THF ends up in the CH_3 group of dTMP product. This result suggests this ternary adduct can decompose in the front direction with an internal hydride transfer. It could happen in a concerted transition state or stepwise. For ease of discernment Figure 9.27 shows a two-step reaction manifold. The hydrogen attached to the sp^3 C_5 of the dUMP scaffold in this adduct is adjacent to the C_4 carbonyl group. Abstraction as a proton gives a C_5 transient carbanion that can be used to form an exomethylene $C_5{=}CH_2$ species while ejecting THF as the N_5 anion. The exomethylene group is electrophilic at its terminus by virtue of conjugation to the UMP C_4 carbonyl. Ejection of the C_6–H of THF as a hydride ion is driven by the negative charge on folyl N_5 as it forms the $C_6{=}N_5$ double bond that constitutes 7,8-dihydrofolate. The C_6-derived hydride adds in (as hydrogen nucleophile) to the exomethylene group creating the CH_3 attached to C_5 in dTMP. Now, one needs only to use the C_4, C_5 enolate anion to eject the initially added active site cysteine thiolate. This re-forms the 5-6 olefinic link as dTMP (5-CH_3-dUMP) is released.

A quick aside on the DNA methyltransferases that convert dCMP residues in dGdC islands of DNA to 5-methyl-dCMP residues that implement maternal patterning for gene expression[30] (Figure 9.28). Those enzymes use the same starting strategy, adding an active site cysteine residue across the 5-6-double bond of a cytidine ring. As seen above, that generates the resonance stabilized C_5-carbanion. Now, the cosubstrate is *S*-adenosyl-methionine for an S_N2 type transfer of the SAM methyl group as a nascent $[CH_3{}^+]$ equivalent, directly forming the C_5–CH_3 bond. Subsequent ejection of the cysteine residue thiolate gives the methylated DNA residue. The use of SAM in place of N_5,N_{10}–CH_2-THF obviates the need for an internal redox step (Figure 9.28). Indeed, SAM with its methyl sulfonium group is chemically a more active methyl donor than the N_5–methyl group of methyl-THF. The only context where cells use methyl THF rather than SAM as methyl donor is methionine synthase. The hundreds of other methyltransferases use SAM.

Together, biotin and tetrahydrofolate ferry around one-carbon units at four oxidation states in cellular metabolism, ready for transfer out to nitrogen and carbon nucleophiles for biosynthesis of the building blocks for RNA, DNA, proteins (*via* methionine) and for the premier cellular methyl transfer substrate *S*-adenosylmethionine.

9.2.5 Riboflavin

The sole vitamin with a heterotricyclic framework is riboflavin[1] (vitamin B2). The tricyclic isoalloazine ring system is composed of a dimethylbenzene fused to a pyrazine fused to a pyrimidine ring. The pyrazino-pyrimidine bicyclic moiety is reminiscent of the similar pyrazino-pyrimidine rings system of the pterin ring in the folate coenzymes above. Conversion of the ingested

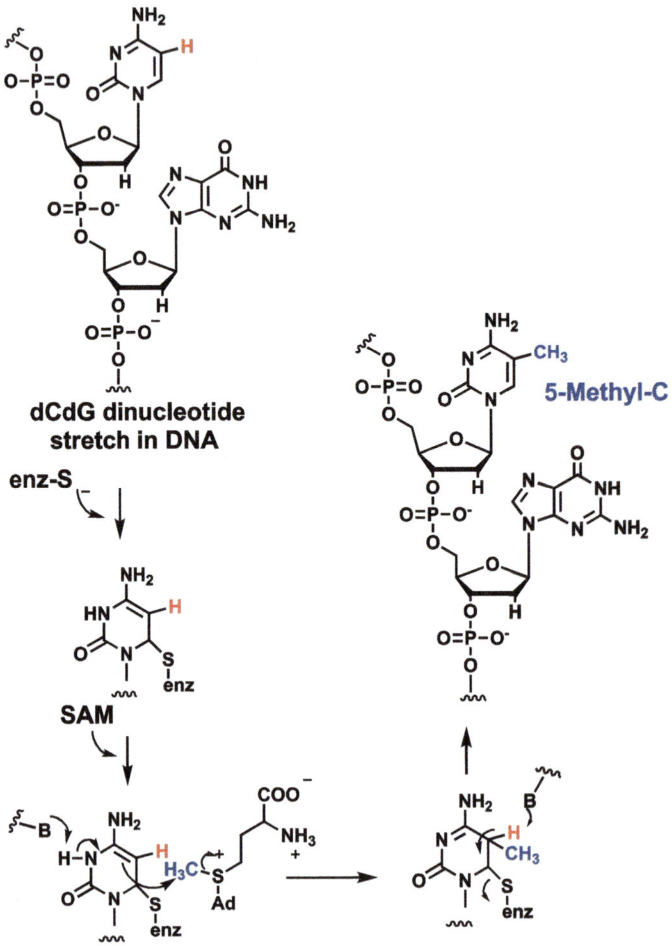

Figure 9.28 Comparative logic for formation of 5-methyl cytosine residues in DNA. *S*-adenosylmethionine as methyl donor.

riboflavin vitamin to two active coenzyme forms involves enzymatic phosphorylation at the ribityl C_5–OH to give FMN and then adenylyl transfer to the FMN phosphate oxygen to give FAD, noted early in Chapter 2[2] (Figure 2.3). Like the discussion on NAD^+ as a key redox cofactor that has already surfaced in so many examples in the previous chapters, the flavin coenzymes have also appeared in specific contexts in previous chapters. Flavins, tightly bound to particular apoenzymes, participate in specific redox transformations, from hydride transfers to and from NAD(P)H (Chapter 2), from redox reactions with active site disulfides and dithiols (Chapter 4), to epoxidation of squalene (Chapter 3), or premonensin (Chapter 8).

The redox active end of FMN and FAD coenzymes is the tricyclic isoalloxazine ring system. In oxidized flavins the central pyrazine ring at N_5 is an electrophilic nitrogen, accepting a hydride ion from such substrates as

NADH, NADPH (Figure 9.29). Hydride transfer to N_5 and proton transfer to N_1 generates the dihydro form of FMN or FAD. In addition to N_5, the adjacent bridgehead carbon is a site for covalent formation of both C4a-thiol flavin adducts (Chapter 4) and C_{4a}–OOH adducts, providing a second point of entry of two electrons at a time (Figure 9.29).

Once the two electron-reduced $FMNH_2$ or $FADH_2$ forms have been achieved, they are labile to unwanted autoxidation by O_2 and thus >98% of flavoproteins keep both oxidized and reduced flavin coenzymes tightly sequestered, not released into free solution. The dihydroflavins can be reoxidized two electrons at a time or one electron at a time (Figure 9.30) because the one electron reoxidized species the flavin semiquinone radical ($FlH^{\bullet}$) is both kinetically and thermodynamically accessible due to delocalization of the odd electron density.[31]

Indeed, reoxidation of $FMNH_2$/$FADH_2$ by one electron at a time is a central attribute of flavin coenzyme chemistry, allowing it to function as the molecular connector between obligate two electron *vs.* one-electron redox partners. In addition to enabling flavins to function as step down two electron to one electron transformers at many cellular redox intersections, the one-electron reactivity unlocks the thermodynamic potential of O_2 by one-electron reduction routes.[32] Dihydroflavins transfer one electron to O_2 to give superoxide anion and flavin semiquinone radical (Figure 9.30). In the active sites of flavoenzyme oxidases and oxygenases, but not dehydrogenases, the radical

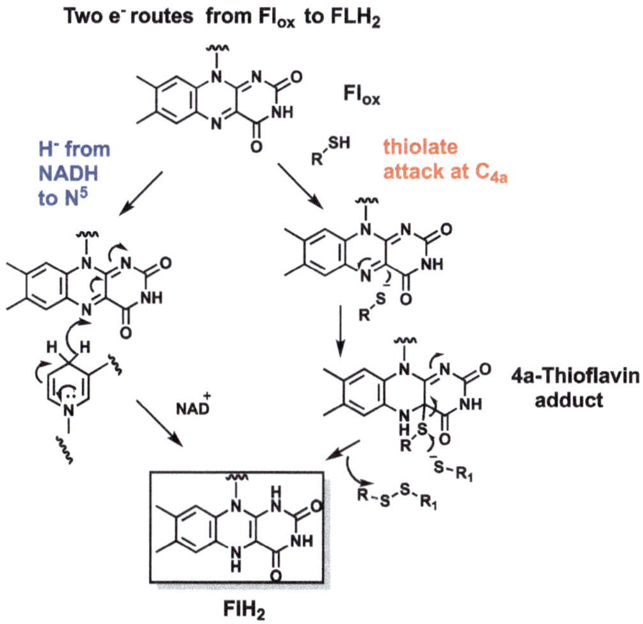

Figure 9.29 Two electron routes from Fl_{ox} to FlH_2. Hydride transfer to N_5 of flavin *vs.* thiol addition to bridgehead carbon C_{4a}.

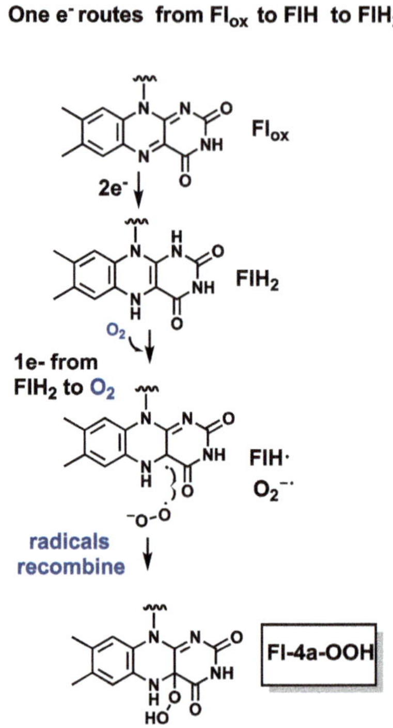

Figure 9.30 One-electron transfer to FlH_2 generates flavin semiquinone radical. Pairing with superoxide anion radical generates the covalent flavin-4a-OOH adduct.

pair undergoes regiospecific recombination to give the C_{4a}-hydroperoxyflavin (Figure 9.30) (the same C_{4a} captured in two electron chemistry by thiolate addition).

That peroxyflavin can be vectored to three different fates (Figure 9.31). In oxidase active sites, the C_{4a} peroxyflavin eliminates HOOH as oxidized flavin is formed, a net two-electron reduction of O_2 to HOOH by two one-electron steps. Instead, in flavin-containing monooxygenase active sites such as squalene epoxidase, a nucleophilic cosubstrate atom attacks the distal oxygen of the flavin-O–OH with resultant transfer of an electrophilic OH equivalent $[OH^+]$ to yield the epoxide. The residual flavin-4a-OH adduct then eliminates water as it reoxidizes (Figure 9.31). A third fate, actually a second oxygenation variant, involves initial dissociation of the flavin-4a-OOH to a proton and flavin-4a-peroxy anion that now behaves as a nucleophilic oxygen, for example toward the cyclic five membered ketone in Figure 9.31. That leads to conversion of a co-substrate ketone to a ring-expanded cyclic lactone as explored in both Chapter 3 on C–O group chemical biology and in Chapter 8 on heterocycles in biology.

Coupled with the redox active nicotinamide coenzymes, flavin coenzymes provide the complementary redox partner, serving as two-electron/one-electron

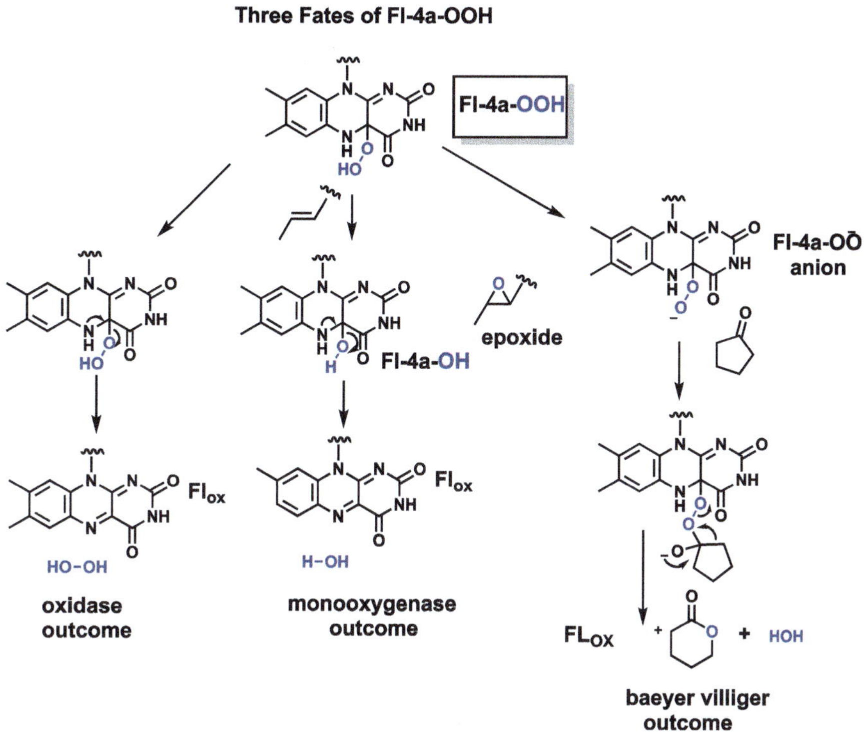

Figure 9.31 Three fates of the enzyme-bound 4a-OOH-flavin adducts.

step down redox intersection to enable cells to carry out the great bulk of electron transfers that are the keys to energy metabolism.

9.2.6 Vitamin B12

Vitamin B12 is one of a small family of tetrapyrrolic macrocycles, with different divalent cations embedded in the equatorial plane of the macrocycles.[22] The divalent cation insertion is mediated by different chelatases, ferrochelatases for heme family members, magnesium chelatases for chlorophylls, nickel chelatases for methanogen coenzyme F430, and cobalt ion chelatases for B12. Although humans can make the heme tetrapyrrole macrocycle scaffolds (see Chapter 15), they cannot biosynthesize the one-carbon contracted macrocyclic ring system of B12.

Viatmin B12, as isolated, has a cobalt ion in the cobalt^{III} oxidation state with the cobalt embedded in the equatorial plane of the ring-contracted tetrapyrrole macrocycle known as the corrin ring system.[33] As isolated, it typically has the top axial ligand as an OH group, aquo B12. This is the substrate for one of the two human enzymes requiring B12, methionine synthase. That enzyme generates the organometallic methyl B12 as a reactive intermediate as we shall note below. In that enzyme the cobalt shuttles between redox state Co^{I} and the two-electron oxidized, air-stable Co^{III}. The second organometallic form of B12 depicted in Figure 9.32 is adenosyl B12. It is the starting form of B12 used in human methyl CoA mutase and shuttles between starting Co^{III} and one-electron-reduced Co^{II}. We examine adenosyl B12 further as a carbon radical generator in Chapter 12.

B12 is required for two different types of reactions in human metabolism. Both involve organocobalt species, where the organic moiety provides the top axial ligand to B12.[34] We noted in Figure 9.25 that human methionine synthase uses methyl B12 as a transient reaction intermediate in forming the methylthioether linkage in methionine.[28] In methylmalonyl CoA mutase adenosyl B12 is the starting and ending cofactor[34] (Figure 9.33).

The methionine synthase at rest binds the Co^{III} oxidation state of aquocobalamin. An activase reduces the embedded Co^{III} ion to the powerfully nucleophilic Co^{I} ion. This is a strong enough nucleophile to capture the methyl group from N_5-CH_3-THF in a S_N2 type reaction where the CH_3 group gets transferred as a formal $[CH_3^+]$ species (Figure 9.25). This forms the organocobalt bond in methyl B12. The second half reaction involves a comparable transfer of the CH_3 group from the B12-cobalt to the thiolate

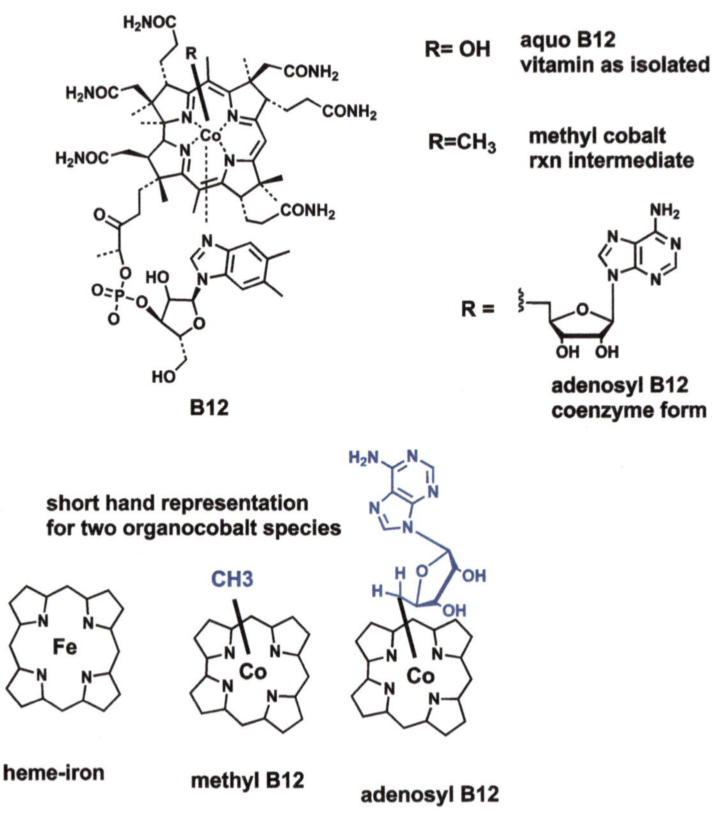

Figure 9.32 Three forms of B12: aquo (vitamin); methyl-B12, and adenosyl-B12.

anion of homocysteine. The two electrons in the Co–CH_3 bond remain with the Co^I ion, leaving the two-electron reduced B12 ready for the next catalytic cycle. The central cobalt ion in B12 cycles between Co^I and Co^{III} in two-electron steps in each half of a catalytic cycle. The activase is only required if adventitious air oxidation of Co^I intervenes every *n*th catalytic cycle.

As examined further in Chapter 12 on carbon radical chemical biology, the other organometallic B12 metabolite adenosyl B12 instead engages in one-electron redox chemistry, cycling between Co^{III} in the starting adenosyl B12 and Co^{II} in a homolyzed reaction intermediate. Methyl malonyl CoA mutase interconverts the indicated branched chain four carbon acyl thioester substrate to the straight chain succinyl CoA that feeds into the citrate cycle (Figure 9.33). Carbon radical-based framework rearrangements are in play, initiated by homolysis of the Co^{III}–$C_{5'}$ adenosyl bond in the adenosyl B12. The resultant 5′-deoxyadenosyl radical (dA•) is a robust enough oxidant to abstract a hydrogen atom (H•) from nearby bound methylmalonyl CoA. That step generates stable 5′-deoxyadenosine (dA–H) and a methylmalonyl CoA carbon radical (Figure 9.33). After radical based rearrangements to the succinyl CoA carbon radical, the reaction is quenched by reabstraction of an H• from dA–H to create the neutral succinyl CoA product. The resultant dA• is recombined at the active site with the Co^{II} form of B12 to regenerate adenosyl B12 for the next catalytic cycle.[34]

B12 is unique among the vitamins as a metalloporphyrin tailored macrocycle, a corrin ring. The central Co^{III} ion engages in organometallic chemistry and redox chemistry. The resting Co^{III} oxidation state can cycle between two-electron-reduced Co^I (perhaps the strongest biologic nucleophile) or between one-electron-reduced Co^{II}. In that latter case it is the transient unleashing of the 5′-deoxyadenosyl carbon-based radical that sets in motion subsequent substrate to product radical rearrangement enzyme chemistry.

9.3 Chemical Roles of the Other Five Vitamins

The remaining five vitamins A, C, D, E, K are not included within the B vitamin molecular diaspora[2] (Figure 9.1). However, vitamins C and E do utilize accessible one-electron chemistry in their heterocyclic frameworks to intercept radicals and thereby interrupt damaging radical cascades. Vitamin K has a carbacyclic menaquinone ring rather than a heterocyclic scaffold but does act as a cosubstrate for posttranslational carboxylation of glutamyl side chains in a set of proteins essential for blood coagulation Vitamin A, in the form of the aldehyde retinal, undergoes light driven *cis–trans* isomerization essential for visual perception. Vitamin A as retinoic acid acts as a nuclear hormone as does the steroid-based trihydroxy vitamin D3. We will note in Chapter 13 that the two steps in the light-initiated conversion of 7-dehydrocholesterol to vitamin D3 are biologic examples of two pericyclic reactions: a retro-electrocyclic step on the steroid carbacyclic system and then a 1,7-sigmatropic rearrangement.

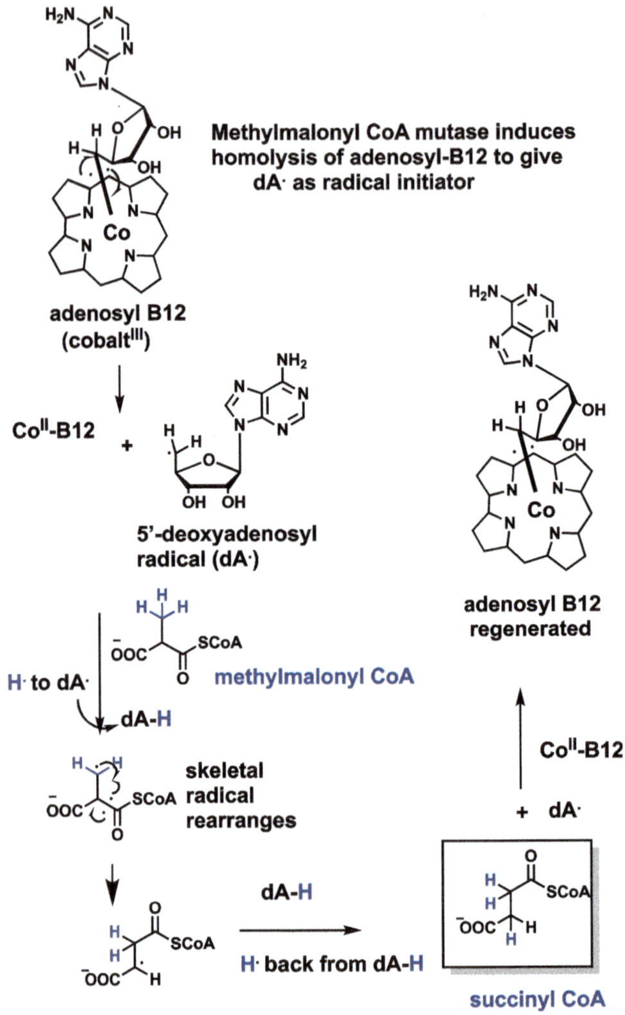

Figure 9.33 The methlymalonyl CoA mutase enzyme homolyzes the adenosyl B12 carbon–cobalt$^{(III)}$ bond to cobalt$^{(II)}$ and the 5′deoxyadenosyl radical. That radical abstracts a hydrogen atom from the CH_2 of bound substrate to set off a radical rearrangement to a product succinyl CoA radical before return of a hydrogen atom from dA–H and reformation of the starting adenosyl B12 Co$^{(III)}$–CH_2 adenosyl B12 organometallic bond.

9.3.1 Vitamin C and Vitamin E as Radical Scavengers

Both vitamin C and vitamin E have oxygen-containing heterocyclic rings that scavenge peroxy radicals and interrupt damaging radical chain reactions in cells. Vitamin C is the enol of the sugar 2-keto-L-gluconolactone, existing predominantly as the enediol (a 3,4-dihydroxyfuranone part structure)[35] (Figure 9.1). Vitamin C is perhaps the most celebrated (certainly the one with the highest sales) of the 13 human vitamins. In the absence of dietary vitamin C, high in fresh fruits (oranges, lemons, limes), the disease scurvy sets in and can be fatal. Vitamin C was first isolated as the antiscorbutic (scurvy = scorbutic disease) agent. For this reason the vitamin has the common name ascorbate ( ). Ascorbate is highly water soluble and stays in the aqueous sub-compartments of cells. It therefore can intercept and trap water-soluble but not lipid soluble radicals.

The complement to vitamin C as a lipid radical scavenger is vitamin E, a mixture of α-, β-, and γ-tocopherols (Figure 9.1). The term tocopherol drives from Greek words meaning live and bearing, to denote the initial nutritional deficiency assay amelioration by the vitamin was to enable successful full term pregnancy in rats.[36] The scaffold of vitamin E isomers contains a di- or trimethyl-phenol fused to a pyran cyclic ether: the bicyclic ring system is termed a hydroxychromanol. Anchored to the hydroxychromane business end of the molecule is a C_{20} phytyl chain, derived biosynthetically from reduction of the three triple bonds in geranylgeranyl-PP, that imparts membrane solubility.

The phenolic-OH of the hydroxychromane bis-heterocycle can dissociate to the phenolate anion in the vitamin E membrane-anchored populations to give an electron rich oxy anion. This electron rich membrane phenoxy anion is mirrored functionally by the electron rich enolate anion of ascorbate (vitamin C) (Figure 9.34).

Each of these electron rich anions can give up one electron to peroxy radicals diffusing through cell compartments, generating the vitamin C enolate anion radical or the vitamin E phenoxy anion (chromanolate) anion radical.[1] In both intracellular and extracellular aqueous compartments vitamin radicals break radical chain reactions. For example, in membrane compartments vitamin E phenoxy anions act as lipid radical interceptors. Vitamin C may similarly intercept polar radicals such as hydrogen peroxy radicals (or even rare hydroxyl radicals), converting them to much less reactive peroxide or hydroxide anions (Figure 9.36). In the membrane compartment the vitamin E anion radical's main target are lipid peroxy radicals arising from one electron transfer and capture of O_2 by unsaturated fatty acyl chains ( ). Lipid peroxy radicals participate in radical chain reactions, leading to amplified destruction of unsaturated fatty acyl chains and net oxidative cleavages and consequent membrane integrity disruptions.

Both the water soluble vitamin C enediolate anion radical and lipid soluble vitamin E chromanolate anion radical have in common the feature that the radical anions have low energy barriers to formation, because each radical gains stability by resonance delocalization (Figure 9.34). The targeted radicals that received the one electron have in turn been converted from species with reactive unpaired electrons to more stable anions, breaking peroxidative radical chain reactions. That same resonance stabilization ensures that neither the ascorbate radical anion radical nor the vitamin E anion radical are so reactive that they propagate an alternative set of radical chain reactions. Nature has titrated its water soluble and lipid soluble

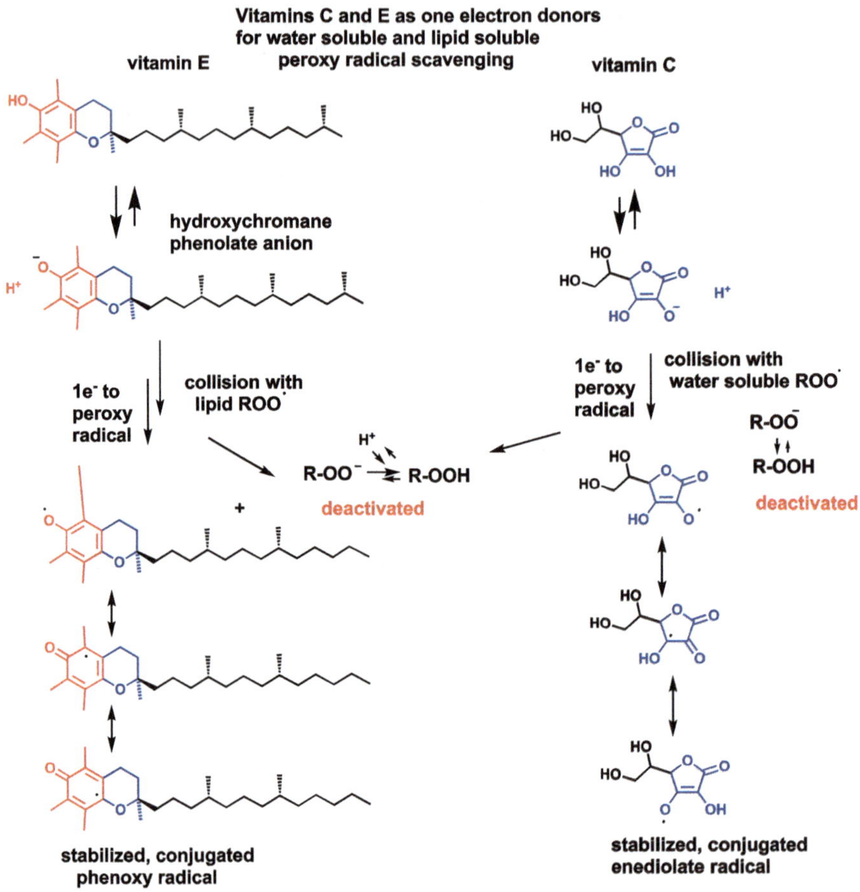

Figure 9.34 Vitamins C and E as one electron donors to water soluble and lipid soluble peroxy radicals for radical interception/scavenging.

radical scavengers such that they are stable enough to form by interrupting radical chains and themselves are not prone to propagate the chain.[1] The one electron transfers from the two vitamins, interrupts and turns off the radical chain propagations in different subcellular compartments. This is carbon/oxygen radical chemistry used for defensive, protective functions.

9.4 Three Carbacyclic Vitamins with Diverse Functions

The remaining three lipid soluble human vitamins, A, D, and K do not contain heterocyclic part structures but do contain carbacyclic ones (Figure 9.1), including a monocyclic ring in the isoprene-derived 20 carbon vitamin A aldehyde isoforms, the bicyclic menaquinone (naphthalene parent system) in vitamin K isoforms, and a tricyclic, steroid-derived scaffold in vitamin D3.[2]

This is not the place for a full description of the chemical biology of each of these three vitamins[1] but they do bring up three aspects of carbon chemical biology. These entail carbanion and redox reactions in vitamin K, vinylogous carbocation-driven polyolefin isomerization (vitamin A), and two distinct types of pericyclic reactions for endogenous conversion of 7-dehydrocholesterol to vitamin D3.

9.4.1 Vitamin K and O_2-dependent Posttranslational Carboxylation of Protein Glutamyl Side Chains

We briefly examine the vitamin K-dependent posttranslational modification of blood coagulation proteases by carboxylation of up to a dozen glutamyl residues at C_4 of the side chains, creating γ-carboxy-Glu (=GLA) residues as bivalent high affinity calcium ion binders (Figure 9.35). These posttranslational carboxylations and consequent Ca^{2+} chelations drive large conformational changes in the half dozen protein substrates that lead to assembly of coagulation protease complexes at a wound site and up to a millionfold amplification in laying down insoluble polymerized fibrin strands to act as clots.[37]

The posttranslational formation of each of these malonic acid side chains on a nested set of Glu side chains require a molecule of O_2, of dihydrovitamin K (KH_2), a molecule of CO_2 to be fixed at C_4 of a glutamyl residue,

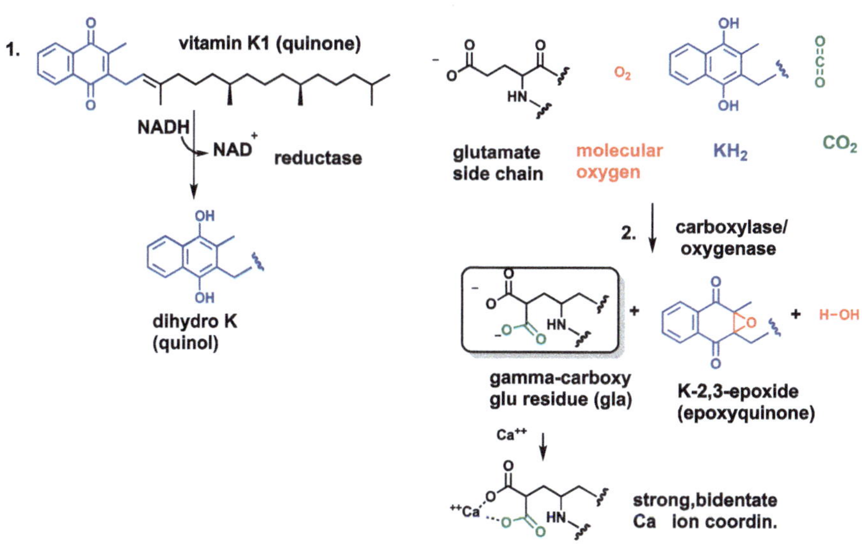

++

Figure 9.35 The dihydro form of vitamin K (KH_2) reacts with molecular oxygen and glutamyl side chain residues in a small pre-protease enzyme family to create γ-carboxy glutamyl side chains as bidentate calcium on chelators. Concomitantly O_2 is reduced to HOH and the 2,3-epoxide of vitamin K.

and the folded protein substrate with an appropriate array of glutamate residues to be carboxylated. The requirement for O_2 is that for every C–C bond formed (GLA residue formed), a molecule of vitamin K-2,3-epoxide is produced. The dihydro vitamin K is the hydroquinol form of the menaquinone ring and is anticipated to be a one electron donor to O_2 because one electron chemistry of benzoquinone/quinols and menaquinone/quinols operates by way of the accessible semiquinone radicals. Recombination of the one-electron reduced superoxide radical and the semiquinone radical would yield the menaquinone hydroperoxide anion. One of its fates would be capture by the 2,3-double bond to produce the 2,3-epoxide as the peroxy O–O bond cleaves.

However, that does not explain why O_2 reduction and one of the methylene hydrogens (CH_2) at C_4 of the several glutamyl residues become activated for removal as a proton. The equivalent question is how is the carbanionic transition state, stabilizable as the *aci*-form of the carboxylate, generated to serve as carbon nucleophile. It is CO_2 not HCO_3^- that is the cosubstrate. A mechanistic proposal, not yet proven, is that the vitamin K-peroxy anion acts as sufficiently strong base to generate the C_4 glutamyl carbanion,[38] as depicted in Figure 9.36.

The posttranslational cascade of glutamyl side chain carboxylations to the bidentate GLA side chains enables the defensive response to wound formation and possible exsanguination in animals (humans). In plants a perhaps earlier evolutionary function of the lipid soluble menaquinones occurs in the packed thylakoid membranes of chloroplasts. In that photosynthetic organelle, the menaquinone vitamin K_1 functions as one-electron redox carriers in photosystem I, receiving an electron from chlorophyll A and passing one to the ferredoxin protein acceptor.[39] Nature has found very different redox niches for both one-electron and two-electron redox chemistry of the naphthoquinone carbacycle in carbon chemical biology.

9.4.2 13-*cis*-All-*trans* Isomerization of Vitamin A Aldehyde in the Visual Cycle

Among the many remarkable aspects of the mammalian visual cycle is the use of 11-*cis*-retinal as the light absorbing pigment while in imine linkage to a specific lysine side chain of the apoprotein opsin[40] (forming the holoprotein rhodopsin) (Figure 9.37). The ionone carbacyclic ring is characteristic of the C_{40} polyene carotenoids and plant β-carotene is the likely metabolic precursor of the all-*trans*-penta-enal retinal (vitamin A aldehyde) by action of a dioxygenase (the extensive conjugation in the five olefins accounts for the absorption of light in the visible spectrum by retinal).

The 11-*cis*-aldehyde is substrate for light absorption while bound in iminium ion linkage to an active site lysine side chain in the protein opsin (holo protein = rhodopsin). Absorption of a photon of light promotes the chromophore into a set of excited states. One consequence is isomerization of the less stable-*cis*-11 retinal to all *trans*-retinal. This drives a rhodopsin

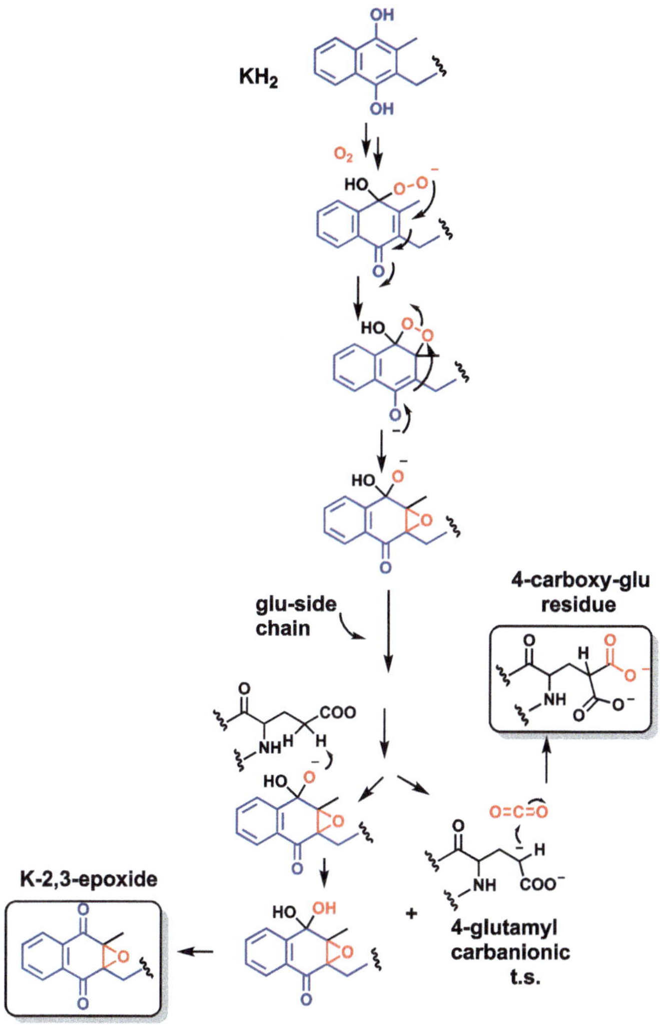

Figure 9.36 Mechanistic proposal for action of vitamin K-dependent carboxylase with O_2 and KH_2 to generate a γ-glutamyl carbanion equivalent to attack CO_2.

conformer change that is read out by the transducing signaling partner in a set of steps leading to perception of vision, beyond the scope of this chemistry-centered discussion.

Indeed, the redox and geometric isomer chemistry of retinoids in the rod outer segment and the rod cell body is a lesson in the kinds of C–O functional group transformations noted for olefins in Chapter 2 and alcohols in Chapter 3. The *trans*-retinal (aldehyde, capable of imine linkage) is released from imine linkage to opsin where its binding fit is less optimal than the 11-*cis*-retinal geometric isomer. That conjugated reactive *trans*-aldehyde can either be reduced by an NADH-driven enzymatic reduction to alcohol or oxidized (irreversibly to the *trans*-retinoate (Figure 9.37). Thus, all three C–O bond oxidation states, alcohol, aldehyde, and acid are available in cells.[41] The pool of retinoid molecules in the retinoic acid oxidation state are not active in vision. Instead, retinoic acid serves as transcription factor when bound to and presented by a cognate nuclear hormone receptor.

The *trans*-retinol pool, less promiscuously reactive than the conjugate retinal, is recycled to the *cis*-11-retinal. This must involve geometric isomerization of the all *trans* conjugated tetraene side chain to 11-*cis*-, and the reoxidation of alcohol to aldehyde able to form an iminium ion linkage to opsin. The strategy used by rod cells is first to convert the *trans*-retinol to a membrane-embedded retinyl-palmitate oxoester.[1] The retinol alcohol acts as oxygen nucleophile attacking the electrophilic carbonyl of palmityl CoA to form the oxoester bond. Note, in this oxoester, the retinol is supplying the alcohol component not the acid component.

Retinyl palmitate is then substrate for an iron-containing protein, known as retinyl pigment epithelial protein of 65 kDa molecular weight, RPE65, that has both isomerase and esterase activity.[42] The net result is conversion of all *trans*-retinyl-palmitate to 11-*cis*-retinol and free palmitate (Figure 9.37).

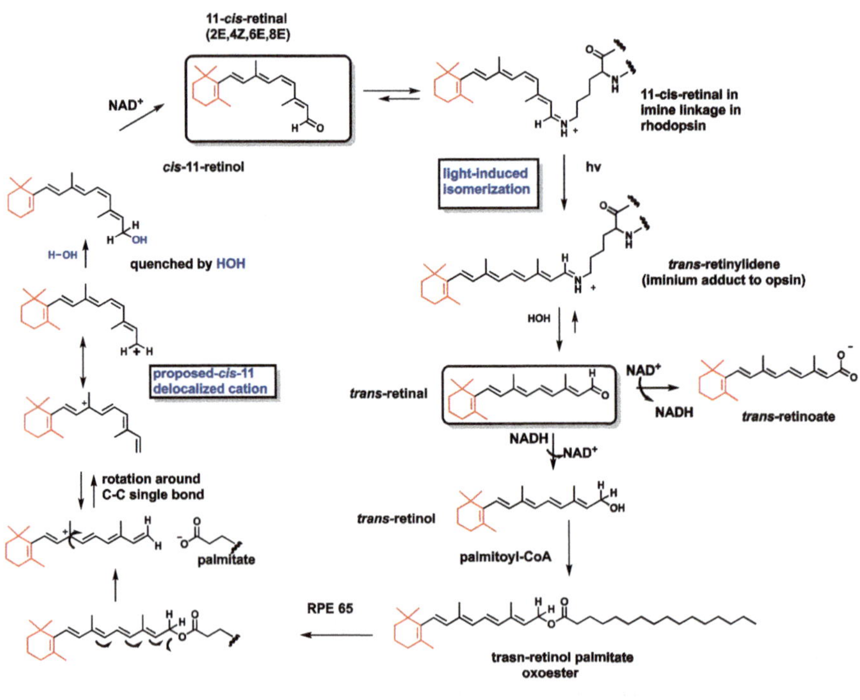

Figure 9.37 Metabolic cycle for interconversion of all-*trans* and 11-*cis*-retinyl isomers at the oxidation states of alcohols and aldehydes. 11-*cis*-Retinal in aldimine linkage to a lysine side chain in rhodopsin is the visual pigment.

A putative mechanism involves participation of three of the four conjugated alkene groups in the *trans*-retinol side chain as an internal nucleophile by triply vinylogous neighboring group participation to cleave the C_1–O alcohol bond of the retinol side chain. The resultant carbocation can be written in the structure shown to note the possibility of rotation around the single bond in the tertiary carbocation resonance contributor. Among the resonance contributors of the *cis*-11-cation is positive charge at the terminal methylene group. Quenching this species by cosubstrate water accounts for the net esterase activity, releasing the *cis*-11-retinol. This is an NAD^+ oxidation away from the *cis*-11-retinal with an aldehyde group that can participate in imine formation with opsin. Absent mechanistic study of this intrinsic pigment epoithelial cell iron-containing RPE65 enzyme, one could also write a one-electron mechanism involving conjugated retinyl radical isomers rather than carbocations (not shown). In a sense this is a prelude to the topic of Chapter 12.

By either mechanism the triply vinylogous participation of three of the four olefin groups in the retinoid side chain are testimony to their chemical reactivity, harking back to Chapter 2 and olefins as C=C functional groups. The full traverse of oxidation states from alcohol to aldehyde to acid to oxoester also reinforces the message of Chapter 3 and analysis of metabolic logic by redox interconversion of functional groups.

9.4.3 Two Pericyclic Reactions on the Way to Vitamin D3

The last vitamin to note that has either a heterocyclic ring system or a carbocyclic ring system is vitamin D (Figure 9.1). Analogous to the retinoic acid form of vitamin A, most of the biologic effects of vitamin D as a hormone are effected by presentation of vitamin D in complex with its cognate nuclear hormone receptor as a transcription factor for up to hundreds of target genes. The conditional biosynthesis of vitamin D in human skin with adequate impinging UV light involves two pericyclic reactions to convert 7-dehydrocholesterol to vitamin D3. We note those two uncatalyzed pericyclic reaction types in Chapter 13.

9.5 Overview

This brief tour of some of the major roles of 12 of the 13 human vitamins emphasizes the pervasive reach of heterocyclic chemistry into many areas of central metabolism. On the one hand, it is testament to how the heterocycles and carbacyclic ring systems augment the inventory of protein-based catalysis. On the other hand, it emphasizes how extensively the several variants of heterocyclic chemistry embedded in vitamin scaffolds enable central reaction types of carbon chemical biology.

All four types of C–C bond forming/cleaving reactions (Chapters 10–13) are enabled by the vitamins and/or their coenzymic forms: carbanions, carbocations, carbon radicals, pericyclic transitions states. The coenzyme forms of B1, B6 and K lower barriers to substrate-derived carbanion chemistry. Vitamin A aldehyde visual pigment regeneration uses polyene-driven carbocation isomerization chemistry. Vitamin B12 initiates carbon radical chemistry. The conversion of 7-dehydrocholesterol to vitamin D3 involves two types of pericyclic reactions.

Vitamins C and E use carbon and oxygen-centered radical chemistry billions of times per day to interrupt damaging radical chain reactions. Vitamins B2 and B3 carry much of the electron transfer flux in energy metabolism, two electrons at a time *via* hydride transfer from/to nicotinamides or $2e^-$ to $1e^-$ redox switching by flavins. Vitamin KH_2 joins flavin H_2 coenzyme forms as oxygen-reactive organic scaffolds. Vitamin K joins vitamins B7 and B9 in mediating one-carbon unit transfers at the several oxidation states of the transferring C_1 unit.

In Chapter 6 we observed that the foundation of nucleic acid structure and information transfer in both RNAs and DNAs rest on the purine and pyrimidine heterocycles. Add in the heterocyclic chemistry of the human vitamins summarized in this chapter: one could assert that heterocyclic chemical biology is in many ways the logical centerpiece for dissecting the chemical logic that underpins life.

References

1. C. T. Walsh and Y. Tang, *The Chemical Biology of Human Vitamins*, Royal Society of Chemistry, London, 2018.
2. G. Combs Jr. and J. McClung, *The vitamins fundamental aspectes in nutrition and health*, Elsevire, 2022.
3. J. K. Carpenter, The discovery of thiamin, *Ann. Nutr. Metab.*, 2012, **61**, 219–223.
4. R. Breslow, Rapid deuterium exchange in thiazolium salts, *J. Am. Chem. Soc.*, 1957, **79**, 1762–1763.
5. R. Perham, Swinging arems and swinging domains in mutlifunctional enzymes: catalytic machines for multistep reactions, *Annu. Rev. Biochem.*, 2000, **69**, 961–10004.
6. L. Mitschke, C. Parthier, K. Schröder-Tittmann, J. Coy, S. Lüdtke and K. Tittmann, The crystal structure of human transketolase and new insights into its mechanism of action, *J. Biol. Chem.*, 2010, **285**, 31559–31570.
7. J. A. West, The chemistry of enamines, *J. Chem. Educ.*, 1963, **40**, 194.
8. Z. H. Zhou, D. B. McCarthy, C. M. O'Connor, L. J. Reed and J. K. Stoops, The remarkable structural and functional organization od rthe eukaryotic pyruvate dehydrogenase complexes, *Proc. Natl. Acad. Sci. U. S. A.*, 2001, **98**, 14802–14807.
9. G. Magni, A. Amici, M. Emanuelli, N. Raffaelli and S. Ruggieri, Enzymology of NAD+ synthesis, *Adv. Enzymol. Relat. Areas Mol. Biol.*, 1999, **73**, 135–182.

10. M. Wojcik, H. F. Seidle, P. Bieganowski and C. Brenner, Glutamine-dependent NAD synthetase. How a two domain, three substrate enzyme avoids waste, *J. Biol. Chem.*, 2006, **281**, 33395–33402.
11. R. S. Cahn, C. Ingold and V. Prelog, Specification of molecular chorality, *Angew. Chem., Int. Ed.*, 1966, **5**, 385–415.
12. K.-S. You, L. J. Arnold, W. S. Allison and N. O. Kaplan, Enzyme stereospecificities for nicotinamide nucleotides, *Trends Biochem. Sci.*, 1978, **3**, 265–268.
13. I. J. Okazaki and J. Moss, Structure and function of eukaryotic mono-ADP ribosyltransferases, *Rev. Physiol., Biochem. Pharmacol.*, 1966, **129**, 51–104.
14. A. A. Suave, Sirtuin chemical mechanisms, *Biochim. Biophys. Acta*, 2010, **1804**, 1591–1603.
15. A. Wilkinson, J. Day and R. Bowater, Bacterial DNA ligases, *Mol. Microbiol.*, 2001, **40**, 1241–1248.
16. R. A. John, Pyridoxal-P dependent enzymes, *Biochim. Biophys. Acta*, 1995, **1248**, 81–95.
17. W. M. Rabeh and P. F. Cook, Structure and mechanism of *O*-acetylserine sulfhydrylase, *J. Biol. Chem.*, 2004, **279**, 26803–26806.
18. M. D. Toney, Controlling reaction specificity in pyridoxal phoisphate enzymes, *Biochim. Biophys. Acta*, 2011, **1814**, 1407–1418.
19. T. Yoshimura and N. Esak, Amino acid racemases: functions and mechanisms, *J. Biosci. Bioeng.*, 2003, **96**, 103–109.
20. J. Ayling and E. E. Snell, Mechanism of action of pyridoxamine pyruvate transaminase, *Biochemistry*, 1968, **7**, 1616–1625.
21. L. Zheng, R. H. White, V. L. Cash, R. F. Jack and D. R. Dean, Cysteine desulfurase activity indicates a role for NIFS in metallocluster biosynthesis, *Proc. Natl. Acad. Sci. U. S. A.*, 1993, **90**, 2754–2758.
22. A. R. Battersby, Tetrapyrroles: the pigments of life, *Nat. Prod. Rep.*, 2000, **17**, 507–526.
23. G. A. Hunter and G. C. Ferreira, 5-Aminolevulinate synthase: catalysis of first step of heme biosynthesis, *Cell. Mol. Biol.*, 2009, **55**, 102–110.
24. L. Tong, Structure and fuction of biotin-dependent carboxylases, *Cell. Mol. Life Sci.*, 2013, **70**, 863–891.
25. C. T. Walsh, B. P. Tu and Y. Tang, Eight kinetically stable but thermodynamically activated molecules that power cel metabolism, *Chem. Rev.*, 2018, **118**, 1460–1494.
26. L. Schirch and M. Ropp, Serine transhydroxymethylase: affinity of tetrahydrofolate for the enzyme and the enzyme-glycine complex, *Biochemistry*, 1967, **6**, 253–257.
27. B. D. Gueenther, C. A. Sheppard, P. Trran, R. Rozen, R. G. Matthews and M. L. Ludwig, The structure and properties of methylenetetrahydrofilate reductase from Escherici coli suggest how folate ameliorates human hyperhomocysteinemia, *Nat. Struct. Biol.*, 1999, **6**, 359–365.
28. J. C. Gonzalez, R. V. Banerjee, S. Huang, J. S. Sumner and R. G. Matthews, Comparison of cobalamin-independent and cobalamin-dependent

methionine synthases from Eschericia coli: two solutions to the same proboem, *Biochemistry*, 1992, **31**, 6045–6056.
29. C. W. Carreras and D. V. Santi, The catalytic mechanism and structure of thymidylate synthase, *Annu. Rev. Biochem.*, 2005, **64**, 721–762.
30. P. Weigele and E. A. Raleigh, Biosynthesis and function of modified bases in bacteria and their viruses, *Chem. Rev.*, 2016, **116**, 12655–12687.
31. C. T. Walsh and T. A. Wencewicz, Flavoenzymes in natural product biosynthesis, *Nat. Prod. Rep.*, 2013, **30**, 175–200.
32. C. Walsh, Flavin coenzymes: at the crossorads of biological redox chemistry, *Acc. Chem. Res.*, 1980, **13**, 148–155.
33. D. C. Hodgkin, J. Kamper, M. McKay, J. Pickworth, K. N. Trueblood and J. G. White, Structure of vitamin B12, *Nature*, 1956, **178**, 64–66.
34. R. Bannerjee, Radical carbon skeletal rearrangements: catalysis by coenzyme B12-dependent mutases, *Chem. Rev.*, 2003, **103**, 2083–2094.
35. K. J. Carpenter, The discovery of vitamin C, *Ann. Nutr. Metab.*, 2012, **61**, 259–264.
36. R. Brigelius-Flohé and M. G. Traber, Vitamin E: function and metabolism, *FASEB J.*, 1999, **13**, 1145–1155.
37. B. Furie and B. C. Furie, Molecular basis of vitamin K-dependent carboxylation, *Blood*, 1990, **75**, 1753–1762.
38. P. Dowd, S. W. Ham and S. J. Geib, Mechanism of action of vitamin K, *J. Am. Chem. Soc.*, 1991, **113**, 7734–7743.
39. J. H. Golbeck, Structure, function, and organization of the Photosystem I reaction center omplex, *Biochim. Biophys. Acta*, 1987, **895**, 167–204.
40. P. D. Kiser, M. Golczak and K. Palczewski, Chemistry of the retinoid (visual) cyclr, *Chem. Rev.*, 2014, **114**, 194–232.
41. N. Y. Kedishvili, Enzymology of retinoic acid biosynthesis and degradation, *J. Lipid Res.*, 2013, **54**, 1744–1760.
42. X. Cai, S. M. Conley and M. I. Nash, RPE65: role in the visual cycle, human retinal disease, and gene therapy, *Ophthalmic Genet.*, 2009, 57.

Section IV: Four Types of Reaction Modalities in Carbon Chemical Biology

Sections I–II provide a sense of the scope of organic functional groups found in molecules in biologic pathways. A deep look at the kinds of oxygen, nitrogen, and sulfur containing heterocycles in Section III informs on both the major heterocycle types in primary metabolism and the key energy-lowering functions of the coenzyme forms of vitamins. In particular seven of the eight B vitamins use their nitrogen heterocyclic ring systems and functional groups to increase nature's catalytic inventory of the kinds of chemical transformations accessible to cells and organisms.

Now in Section IV we examine the four major types of reaction profiles for organic substrates as they break and make chemical bonds in the active sites of specific enzymes to yield specific covalent connectivities in reaction products. As a C–H or C–C bond breaks or forms, abiotically or biotically, one of four possible transition state structures will be obtained. In biologic systems, the inventory of available carbon-based functional groups will limit/bias the mechanisms for bond breaking and bond making.

The four chapters, Chapters 10–13, of this section focus on carbanion-driven mechanisms (Chapter 10), carbocation-driven mechanisms (Chapter 11), carbon radical generation (Chapter 12), typically at un-activated/remote carbon centers, and on concerted reactions without any finite-lived intermediates (Chapter 13).

Chapter 10 assesses the inventory of energetically accessible substrate carbanions within enzyme active sites. The most obvious categories are enolate anions adjacent to aldehydes or ketones, and thioester enolates that can react as transient carbon nucleophiles for C–C bond formations by aldol and Claisen thioester C–C bond forming or cleaving steps. The enolate anion of pyruvate is often accessed by O–P bond cleavage by phosphatase action on the glycolysis pathway reactant phosphoenolpyruvate, a covalently trapped

C_3 carbanion precursor. Another major category of functional groups that can behave as carbon nucleophiles in the C–C bond forming reaction are π-electrons of alkene groups. The participation of four C=C alkene bonds in the cyclization of squalene epoxide to tetracyclic or pentacyclic sterols is a powerful reminder of how the alkene potential nucleophilic reactivity is marshalled in biology.

Chapter 11 looks at the other perspective for C–C bond formation in two electron pathways, the availability of the requisite, and corresponding carbon electrophiles. One abundant category of carbon electrophiles contains substrate carbonyl groups, both as internal ketones, terminal aldehydes, and the carbonyl groups of oxoesters and thioesters. These are stable, ground-state electrophilic functional groups.

Two other accessible carbocation species are accessible but evanescent. One is the delocalized allylic carbocation available from Δ^2-prenyl-diphosphates in enzyme active sites as early C_1–OPP bond cleavage gives the stabilized allylic cation. These carbocations are capturable (at both C_1 and C_3 of the allyl moiety) both intermolecularly by alkenes acting as nucleophiles in C–C bond formation (prenyl chain elongations). They are also capturable intramolecularly by internal olefins in the active site of cyclases that lead to cyclic cations, many of which can propagate by rearrangements before quenching occurs.

A third set of carbocationic intermediates/transition states encompass glycosyl C_1–O_5 oxocarbenium ion species that are formed in glycosyl transferase active sites when early C_1–O-UDP bond cleavage occurs. The delocalized oxocarbenium ion is stable enough to form and yet reactive enough then be captured by a variety of cellular nucleophilic atoms, including oxygens, nitrogens, thiolate sulfurs, and even carbanions.

One might include then hundreds of redox enzyme reactions where some alcohol or amine group (as examples) of a specific substrate is oxidized (or in the other direction a carbonyl or imine group reduced) by hydride transfer to (from) NAD. The hydrogen transfer is always direct from substrate to C_4 of the pyridinium ring of $NAD(P)^+$, interpretable as hydride ion transfer, consistent with the substrate having a transiently polarized electrophilic carbon before the electron pair on oxygen or nitrogen swoop in to form the carbonyl group of product.

The above two reaction alternatives have been essentially two-electron reaction, heterolytic pathways. A transient carbanion retains both electrons that had been in a C–X bond. Likewise, a transient carbocation has retained neither of the pair of electrons in the bond that has cleaved. As a carbon nucleophile (*e.g* carbanion) attacks an electrophilic carbon partner, both electrons move in the same transition state.

A mechanistic alternative is two one-electron steps. In bond making or breaking at a carbon center, the electron pair in a bond can be distributed equally: one electron remaining/going to the carbon, the other electron departing with the leaving group. This is a homolytic mechanism.

Some 30 years ago the prevailing consensus was that organic radical chemistry was rare to very rare in *in vivo* metabolism. That view has turned around with further exploration of two major families of enzymes, the radical *S*-adenosylmethionine (SAM) enzymes and the iron-based oxygenases. Radical SAM enzymes operate in anaerobic microniches. Oxygenases, consuming O_2 as a reducible cosubstrate, work in air. These two routes to radicals at unactivated carbon centers are summarized in Chapter 12.

Radical SAM enzymes and also the two dozen or so adenosyl B12 enzymes both converge on homolytic fragmentation of either SAM or adenosyl B12. That homolysis generates a transient enzyme bound 5′-deoxyadenosyl radical as the initiator of cosubstrate radical chemistry by substrate C–H bond homolysis. This yields a substrate carbon radical with chemical reactivity dependent on its particular functional group content. Concomitantly, the deoxyadenosyl radical, having gained a substrate-derived hydrogen atom, is converted to a stable CH_3 group in 5′-deoxyadenosine.

Iron-containing oxygenases create a different one-electron oxidant able to break bound cosubstrate C–H bonds homolytically. They generate transient substrate carbon radicals, typically at otherwise unreactive carbon centers of those substrate. The active site oxidant in hundreds of heme oxygenases is thought to be a high valent $Fe^{V}=O$ oxoiron. A comparable $Fe^{IV}=O$ oxidant is generated in nonheme iron oxygenases. One electron transfer of a hydrogen atom from a C–H bond of cosubstrate yields a carbon radical and an Fe^{IV}–OH or an Fe^{III}–OH, capable of delivering an [OH•] equivalent by Fe–O bond homolysis. The radical recombination of carbon radical and hydroxyl radical creates a product alcohol group. The one electron left in that homolyzing Fe–O bond goes back to iron to create Fe^{III} or Fe^{II} starting oxidation sites. (We note in Chapter 15, that 18 oxygenative steps occur in the conversion of lanosterol to the female sex hormone estradiol. That implies 18 different carbon radical species in the late stages of steroid hormone biosynthesis.)

Chapter 13 concludes the analysis of types of reaction mechanisms for C–C bond formations and cleavages in biology by raising the question of cascade reactions and the intersecting question of whether there are concerted mechanisms. After some initial consideration of stepwise *vs.* concerted transformations in primary metabolism, the chapter turns to four classes of abiotic pericyclic reactions that have *in vivo* counterparts.

Of the four groups of cycloaddition reactions, electrocyclic reactions, sigmatropic reactions and Alder-ene reactions, pericyclase enzymes have been detected for all but the electrocyclic reaction case. That category occurs nonenzymatically in the siphoning of some of the flux of 7-dehydrocholesterol away from cholesterol and down a pericyclic pathway to previtamin D3. The most celebrated of three classes of pericyclases, each of which catalyzes concerted chemistry *via* a cyclic transition state, is exemplified by chorismate mutase (3,3-sigmatropic rearrangement) in aromatic ring biosynthesis, Several Diels Alderases have been observed to catalyze (4+2π-electron) cycloadditions. Other Diels Alderases have been shown to carry out hetero Diels Alder cyclizations where aldehydes and ketones or imines, respectively, accelerate oxo-DA and aza-DA cycloadditions. Finally, some of the hetero-Diels Alderases partition pericyclic transition states between oxo-Diels Alder and Alder-ene product outcomes are presented.

CHAPTER 10

Routes That Make and Break C–H and C–C Bonds In Vivo *I: Carbanion Generation*

One estimate of the potential complexity of low molecular weight (<1000 Daltons) organic molecules that act as biologic metabolites approaches 100 000 compounds. That inventory of carbon-based molecular frameworks is not static in living cells. To reach a full molecular inventory the dynamics of metabolic pathways encompasses myriad transformations and interconversions *in vivo*.

It is useful to remember that there are actually only a small number of energetically accessible reaction paths with the limited set of carbon-based functional groups for the bulk of the organic chemistry of life.[1] The reactivity of organic metabolites *in vivo* conform to the reactivity rules of molecules in abiotic organic chemistry. Indeed, there are biologic examples of many of the "name reactions" of abiotic organic chemistry.[2] Two of the simplifying themes of this monograph that analyze patterns in the organic chemistry of life have been to categorize the common organic functional groups and to analyze many enzyme-catalyzed reactions involving changes in redox state of substrates as they are converted to products. A third integrating theme in the four chapters of this section are to analyze how the functional groups in organic substrate structures enable enzyme-catalyzed transformations.

10.1 A Short Aside into Types of Bonds Made and Broken in Anabolic *vs.* Catabolic Pathways

Before delving into the ways in which biologic systems accelerate reactions that make and break C–H bonds and C–C bonds, we offer a brief comment

Chemical Biology No. 21
The Chemical Biology of Carbon
By Christopher T. Walsh

Published by the Royal Society of Chemistry, www.rsc.org

about the division of metabolism into two major categories. Anabolic metabolism is about biosynthesis. Catabolic metabolism is about breaking down dietary sources of molecular fuels for cellular energy production.[1]

The latter set of reactions takes commonly ingested, partially reduced organic molecules such as glucose and fatty acids and oxidize them all the way to CO_2. Along the convergent paths of fatty acid or glucose catabolism, the electrons that are stripped out from each carbon on the way to CO_2 are stored in thermodynamically activated, kinetically stable molecules that cells use to do all manner of chemical and mechanical work. These molecules include ATP, NAD(P)H, acetyl CoA, carbamyl phosphate, UDP glucose and several others. First among equals is ATP production during catabolic metabolism to create the cellular currency that can be spent to drive otherwise unfavorable equilibria in biosynthetic reactions.

The metabolic reactions on the catabolic side of the ledger, key to energy generation from reduced organic substrates, utilize one of the four reaction modalities in Chapters 10–13. When we turn to the anabolic side of the cell metabolism ledger, one can divide biosynthesis into two broad parts. Perhaps most obvious is the assembly of dehydrated long biopolymers, including nucleic acids, proteins, and polysaccharides. Those biopolymer assemblies do not involve C–C bond formation.

RNA and DNA chains are made from nucleoside triphosphate building blocks that undergo dehydrative chain extensions in every catalytic cycle. The only type of covalent bond that links together the nucleotides in RNA and DNA are phosphodiester bonds, generated by attack of an oxygen nucleophile (3′-OH of a ribose or deoxyribose moiety) on the electrophilic α-phosphorus of the incoming monomeric NTP/dNTP. Neither the nucleophile (3′oxyanion, nor the electrophile (electrophilic phosphorus at Pα of the incoming NTP/dNTP) are carbon atoms (Figure 10.1).

Analogously, in protein biosynthesis the covalent bond connecting all amino acid residues is the amide bond. An amine nucleophile captures the carbonyl carbon of an oxoester linkage in the incoming aminoacyl-*O*-tRNAS in the ribosomal peptidyl transferase center. The carbonyl carbon participant is electrophilic but there are no C–C bonds formed in polymerization, only N–C bonds.

In oligosaccharide and polysaccharide biopolymer biosynthetic elongations, acetal C–O–C bonds are the iterative linkages that hold sugar monomer units together. The nucleophile is an oxygen atom of a terminal sugar in the growing polymer. The electrophile is the C_1–O_5 oxocarbenium ion of the incoming sugar-NDP monomer.

Thus, in one out of three of the major biopolymer classes (nucleic acids) no carbon nucleophiles or electrophiles are involved. In both protein biosynthesis and polysaccharide biosynthesis, carbon electrophiles are involved but no carbon nucleophiles (Figure 10.1). For these two biopolymer classes Chapter 11 on carbon electrophilic groups is clearly relevant.

The second part of anabolic metabolism, beyond the thermodynamically unfavorable accumulation of the major classes of biopolymers, does involve

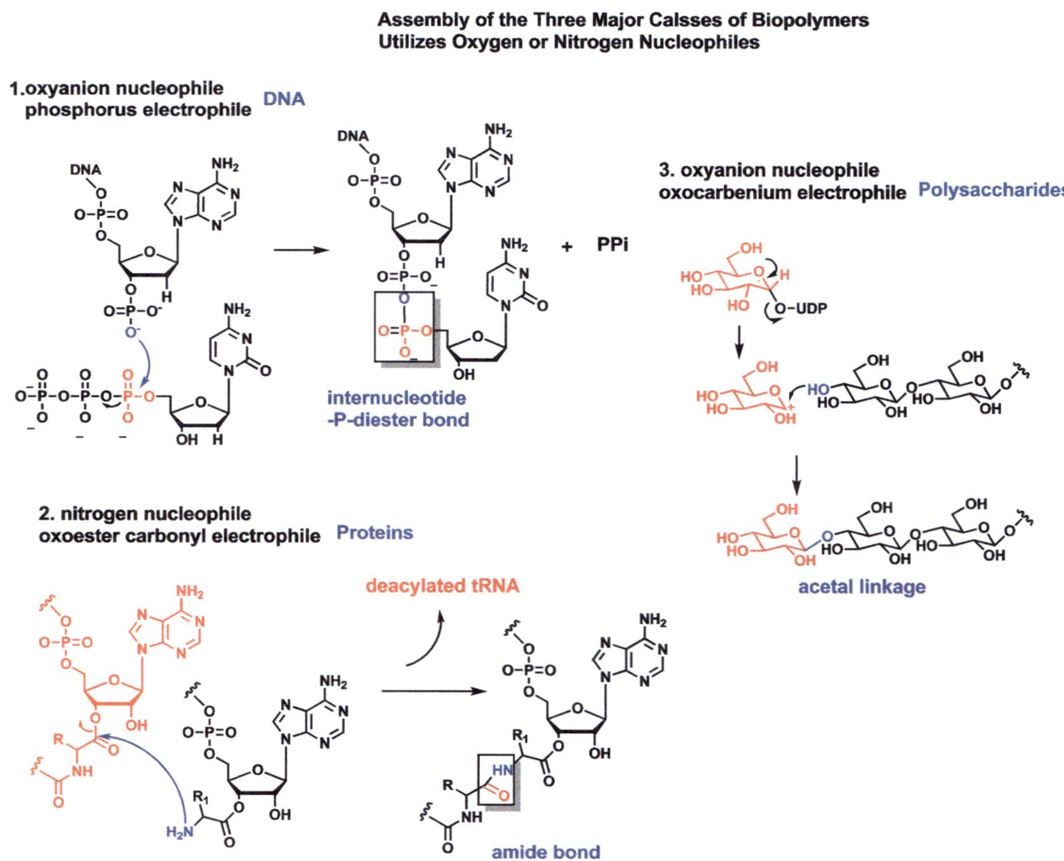

Figure 10.1 Oxygen and nitrogen nucleophiles, but no carbon nucleophiles, in an assembly of the three major classes of biopolymers.

C–C bond formation. Five pathways, among others, that come to mind are (1) fatty acid biosynthesis (and related polyketide biosynthesis in plants and microbes), (2) the ∼70 000 isoprenoid molecules, large and small including all the steroid metabolites, (3) aromatic ring biosynthesis exemplified, among other routes, by the shikimate/chorismate/phenylpyruvate pathway, (4) tetrapyrrole biosynthesis where the four pyrrole rings are connected by intervening CH_2 groups, and (5) radical SAM-based C–C bond couplings (Figure 10.2).

In each of these biosynthetic aspects of anabolic metabolism carbon–carbon bonds need to be made. Concomitantly, some C–C bonds need to be broken, most often by way of decarboxylations[3] to release CO_2. For example, each molecule of C_{27} cholesterol, biosynthesized from 18 acetyl CoA starter molecules, involves removal of nine carbons as 9 CO_2 (Chapter 15). Analogously, animal biosynthesis of heme, a key pigment of life, from glycine and succinyl CoA releases 14 CO_2 molecules while 48 input carbons end up in the 34 carbon framework of protoporphyrin-IX (Chapter 16).

Understanding the mechanistic alternatives for C–C bond-forming and bond-breaking reactions available to enzymes as they carry out catabolic and anabolic transformations of organic metabolites identifies the four possible ways that reaction profiles can proceed.

For heterolytic reactions, this means identification of accessible carbon nucleophiles (Chapter 10) and their electrophilic carbon partners (Chapter 11). For homolytic reaction profiles, it means identification of the kinds of one electron oxidants transiently available in enzyme active sites, whether from O_2

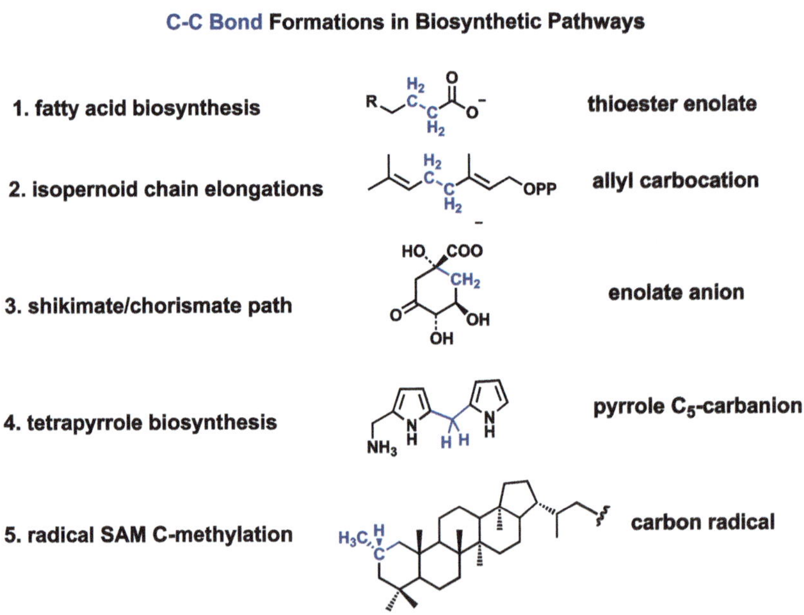

Figure 10.2 Five sets of C–C bond formations in biosynthetic pathways.

or from *S*-adenosylmethionine as potential radical generators (Chapter 12). Additionally, there are cases where bonds between/among substrates break and form products apparently without any dip in the energy surface that would indicate finite-lived intermediates. Chapter 13 takes up the four types of pericyclic concerted reactions observed to date for C–C and/or C–H bond formations in biologic systems.

10.2 C–H Bond Cleavage and Formation: Three Possibilities

As a prelude to accessible C–C bond formation chemistry *in vivo*, we focus first on C–H bonds and how they are formed or cleaved. The options devolve into three mechanistic routes. There are two possible heterolytic paths and one homolytic path to C–H bond cleavage. In the heterolytic paths the C–H bond breaks (or in the reverse direction, forms) with unequal partition of the two electrons that make up the C–H single bond (Figure 10.3). One outcome is for the electron pair to stay with the carbon as the H departs as a positively charged proton. Formally, those two electrons that remain behind place a negative charge on carbon as a carbanion. We examine the major accessible carbanion equivalents that arise by such proton abstraction routes in this chapter.

Alternatively, the departing hydrogen could move with both the electrons from the C–H bond. In this polarity extreme, the hydrogen in flight is a hydride ion and the carbon center becomes an electron deficient carbocation. Based on fundamental electronegativity considerations one expects the less electronegative hydrogen is more likely to depart as a proton rather than a hydride ion from a C–H bond. Indeed, proton abstraction steps are featured in the catalytic cycles of thousands of enzymes that carry out the constituent reactions of metabolomics, including essentially all hydrolases. Nonetheless and, surprisingly, hydride ion chemical biology turns out to be

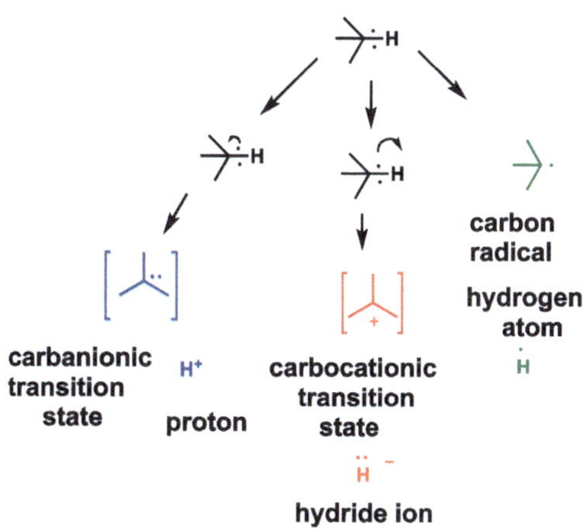

Figure 10.3 Three routes to break a C–H bond.

the predominant mode of oxidation in both glycolysis, fatty acid β-oxidation, and the citrate cycle, when nicotinamide coenzymes are the recipients of the hydride ions in flight from oxidizable substrates.

Another set of biologic contexts where hydride ions are implicated are in internal rearrangements of carbon metabolite scaffolds. The enzyme oxidosqualene synthase,[4] the entry point catalyst to all animal steroids, converts 2,3-epoxysqualene to lanosterol by way of a nascent bound protosterol cation. To get to the released tetracyclic steroid framework of lanosterol, two methyl groups and two hydrogens all migrate intramolecularly.[5] All four substituents are believed to migrate with their electron pairs. In particular, the two blue hydrogens indicated in Figure 10.4 move across the carbon scaffold face as hydride ions, one of them quenching the carbocation positive charge at C_{20} as lanosterol is formed and released into solution. A fifth hydrogen, at C_9 (shown in red), is lost as a proton to form the Δ^9 double bond in lanosterol.[6]

A somewhat related context of an internal group migration occurs in the ring opening aromatization of cyclohexadiene epoxide metabolites to the corresponding phenols (Figure 10.4). Driven by the gain in aromatic resonance energy, a direct 1,2-migration of one of the carbon bound hydrogens is

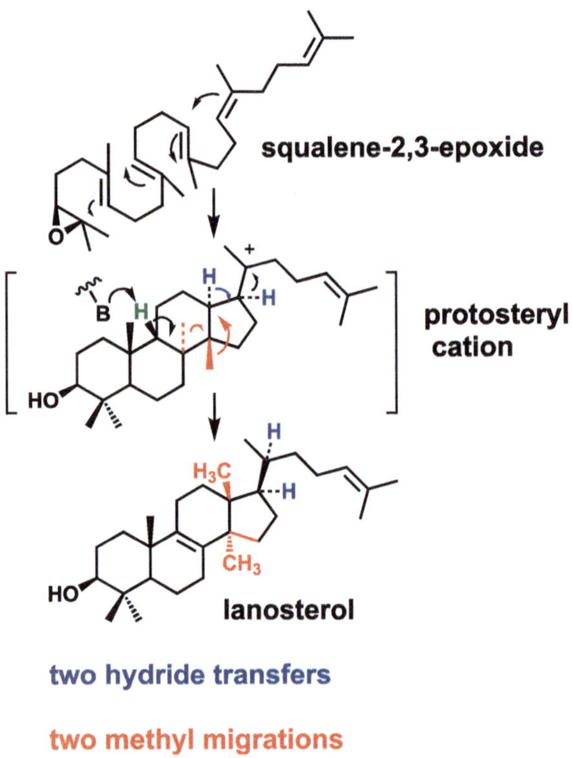

Figure 10.4 Two hydride transfers along with two methyl migrations convert bound protosteryl cation to lanosterol as released cyclic sterol product.

observed as the epoxide opens to the cyclohexadienone.[7] One can formulate this as a hydride ion migration. Proton abstraction then yields the more stable phenol tautomer. This two-step metabolic conversion illustrates the two possible heterolytic routes for C–H bond cleavages: first a net 1,2-hydride transfer and then a proton transfer. The isomerization of the cyclohexadienone to aromatized phenol involves removal of one of two prochiral methylene hydrogens. If enzyme-catalyzed, that could be stereospecific (proR *vs.* proS, see Figure 3.11).

In the mechanistic analysis of thymidylate synthase in Chapter 9, the net migration of the C_6-hydrogen in the tetrahydrofolate coenzyme to the new C_5–CH_3 group in product 5-methyl-2-dUMP is featured as an internal hydride transfer (Figure 9.27).

In addition to the above examples of internal (intramolecular) hydride transfers within reacting substrate scaffolds, we have repeatedly noted that the coenzyme forms of nicotinamide (vitamin B2) and riboflavin (vitamin B3) can each mediate hydride transfers in the active sites of oxidoreductases as cosubstrates undergo oxidation or reduction (see Chapters 2, 3, 4, and 9 for many specific examples). The nicotinamide coenzymes NAD^+, $NADP^+$ are obligate hydride acceptors at C_4 of the oxidized pyridinium ring while oxidized flavins are permissive for either hydride transfers to N_5 or one electron transfers during coenzyme redox cycles (Figure 10.5).

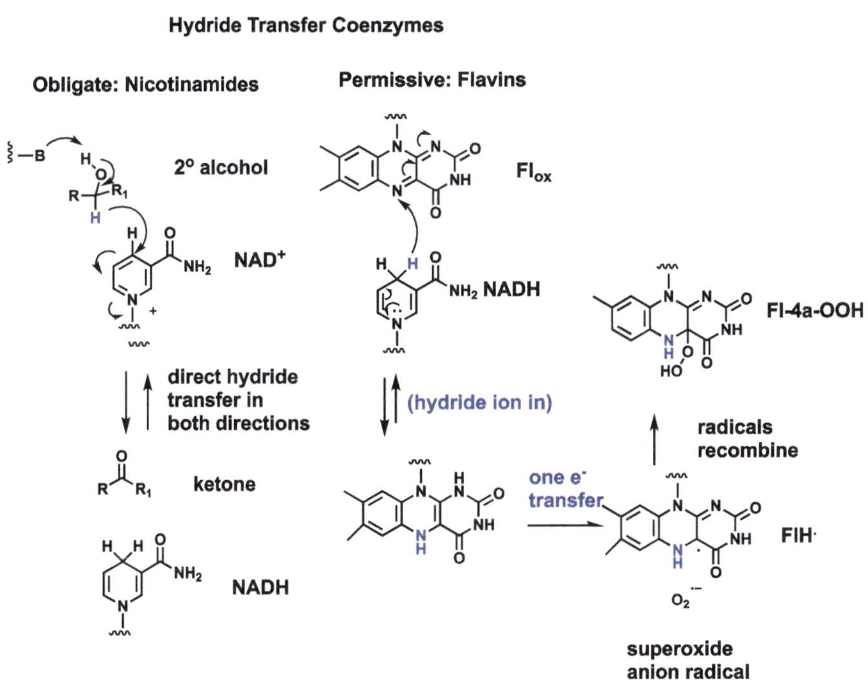

Figure 10.5 Hydride transfer coenzymes: nicotinamides as obligate hydride transfer reagents, flavins as permissive ones.

The third route for C–H bond cleavage (formation in the reverse direction) is homolytic.[8] By definition the word homolysis means an equal lysis. One electron stays with the carbon center, one electron departs with the hydrogen. A pair of radicals, a carbon radical and a hydrogen atom (hydrogen with one electron) (Figure 10.3), are formed. Although heterolytic routes for C–H bond cleavage/formation predominate over homolytic routes in chemical biology, there are two contexts where the homolytic route is prevalent. One context is in the action of hundreds of iron-based and copper-based oxygenase enzymes.[9–11] The second context is at the opposite end of the aerobic–anaerobic spectrum where radical *S*-adenosylmethionine enzymes[12] generate a sufficiently robust oxidant (the 5′-deoxyadenosyl radical) that initiates homolytic C–H bond cleavage in bound cosubstrates. These parallel radical topics are the subject of Chapter 12.

10.2.1 C–H *vs.* O–H, N–H, and S–H Bonds

The behavior of hydrogens as protons in X–H bonds in biologic contexts is related to pH and pK_a values,[13] and also the kinetics of exchange. The pH of biologic solutions tend to cluster around neutral pH (pH = 7; blood pH = 7.4) with homeostatic buffering equilibria that keep the pH values within narrow ranges. There are membrane bound proton pumping ATPases that do acidify specific compartments. Among them are the ATPases of gastric epithelial membranes[14] that can pump protons into gastric fluid to get to pH values as low as pH = 2, ($[H^+] = 0.01$ M), and ATPases associated with vacuoles in yeast cells and lysosomes in animal cells where pH values may be as low as pH 4–5. In those acidified compartments such acid pK_a values are consequential for the partition between neutral and carboxylate anion species.

The (thermodynamic) tendency of an X–H bond (X = C, O, N, S in cells) to dissociate to X^- and H^+ can be measured and calibrated on an acidity scale, the pK_a scale, where the pK_a value[13] is equal to the pH at which there is an equal concentration of X–H and the ions X^- and H^+. On this scale strong acids such as nitric acid and HCL may have very low pK_a values (0 or −3, for example) indicating that they are essentially fully dissociated at all physiologic pH values. Figure 10.6 lists some pK_a values for common functional groups in metabolites. Alcohols such as the side chain of serine are close to the pK_a for water itself (p$K_a \approx 14$, meaning there are 10^{-7} protons and 99.99999% HX at pH 7). Acid side chains of glutamate are present as >99% carboxylate anions at pH 7. The thiol side chain from cysteine is present as 10% thiolate anion at pH 7.2. The imidazolium cation is 90% abundant at pH 7, while serine alkoxide is only present as 1 part in $\sim 10^7$ at pH 7. The phenolic-OH of tyrosine with pK_a of ~10 is some 10^4 more acidic than the serine-OH, driven largely by resonance delocalization and stabilization of the phenolate anion. The equilibrium values say nothing about the kinetics of proton exchange, the rates at which protons come on or off the X–H bonds.

Selected pKa Values

Alcohols: ROH to RO^-+ H^+	
CH_3OH	15.5
CH_3CH_2OH	15.5
Phenol	10

Carboxylic Acids: RCOOH to RCOO + H^+	
CH_3COOH	4

Amines: NH_3. to NH_2^- + H^+	
NH_3	35

Protonated Amines: RNH_3^+ to RNH_2 + H^+	
RNH_3^+	10
Imidazole	6.9

Thiols: RSH to RS^- + H^+	
CH_3SH	13
Cysteine	8.2

Carbon: RCH to RC^- + H^+	
CH_4	50
acetone	24
malonate	13
alkyne	25
methyl acetate	25

Figure 10.6 Selected pK_a values (pH where the ionizable group is 50% ionized).

In terms of pK_a values, the carboxylic acids, simple alkyl or hydroxy or amino acids will be present as the carboxylate anions at 99% or higher percentage of those metabolites. Carboxylates as resonance-stabilized anions are not very nucleophilic in physiologic aqueous solutions, but in desolvated enzyme active sites they are typically used in nucleophilic attack on phosphoric anhydride side chains. From the imidazole pK_a of 6 for dissociation to the neutral heterocyclic ring, one would expect the deprotonated form of the imidazole ring of histidine to be the highest fraction of an unprotonated biologic amine at neutral pH. That neutral imidazole can react as a base to capture back a proton or react at an alternate electrophilic center as a nucleophile. Further dissociation of amines to the amine anions is energetically unfavorable (p$K_a \approx 35$). Amine anions are not thought to play any significant roles in carbon functional group chemistry *in vivo*. The two most likely anionic nucleophiles are thiolate anions and transition states with evanescent oxyanion equivalents as nucleophiles.

10.3 Carbon pK_a Suggests Most Carbanions Are Not Accessible in Neutral Aqueous Solutions

None of the biologically prevalent carbon scaffolds with C–H bonds have pK_a values close to neutral pH (Figure 10.6). That is, the C–H bonds are much less acidic than O–H, N–H, and S–H bonds: the resultant carbanions are higher in energy, more difficult to access kinetically and thermodynamically, than the corresponding $-O^-$, $-N^-$, and $-S^-$ anions. The pK_a for C–H bond dissociation in a β-ketone C_2 methylene group of acetone ($pK_a = 19$) is more acidic than most C–H bonds by many orders of magnitude, due to the lowered energy of the resultant carbanion by delocalization as the enolate anion(s). Still, less than one part in 10^{12} would form at equilibrium at pH 7. A simple alkane such as ethane has an estimated pK_a value of 62; requiring a formal pH of 62 before half ionization would be achieved at equilibrium. The size of the pK_a value and its distance from pH = 7 is an indication of lowered acidity and higher energy to form the alkane-derived carbanion.

Given the vast difference in acidity scales between O–H, N–H, and S–H bonds compared to C–H bonds in metabolites, it is not surprising that there is also a vast difference in the rates at which the X–H bonds lose protons. In general, hydrogen bonds to oxygen, nitrogen, and sulfur atoms in physiologic metabolite structures (alcohols, amines, thiols) undergo exchange with solvent water hydrogens within millisecond to seconds.

Thus, in deuterated media, the O–H, N–H, and S–H bonds would rapidly exchange with deuterons to becomes O–D, N–D, and S–D bonds. By contrast, C–H bonds in carbon frameworks of biologic metabolites are kinetically slow to exchange (half-lives could be measured in years). When fast exchange is detected, as in the insightful studies of Breslow on thiamin function[15] (Figure 10.7, also see Chapter 8), they constitute indirect evidence for reversible carbanion formation on a fast time scale. Those unusually fast exchange rates indicate enhanced acidity at that carbon center. They suggest accessible energetics for carbanion formation, meriting mechanistic study for the structural and electronic basis for enhanced access to carbanion intermediates. If a carbanion is

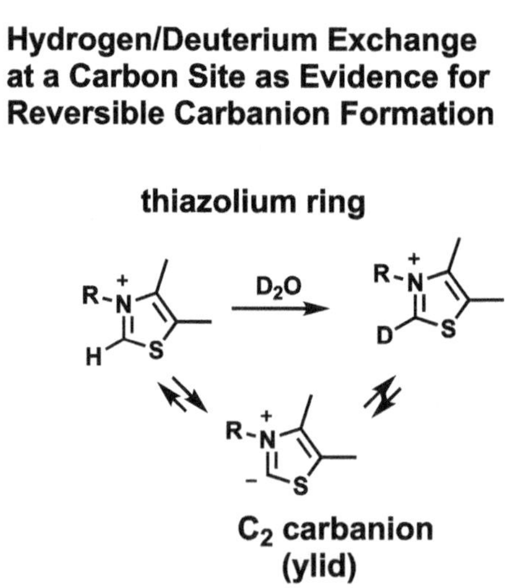

Figure 10.7 Hydrogen/deuterium exchange at a carbon site as evidence for reversible carbanion formation.

formed reversibly before its forward capture, and the active site is accessible to solvent protons, then one may see a preequilibrium solvent isotope exchange. If the carbanion is captured as fast as it forms, even though such an intermediate forms, it will not be detectable by solvent exchange.

One conclusion is the realization that C–H bonds are typically of very low acidity without some structural feature that lowers the energy for carbanion formation and stabilization.[16] Therefore, to carry out carbanion chemistry for C–C bond formation in hundreds to thousands of metabolites, organisms must have evolved structural and mechanistic answers to the need for generating thermodynamically and kinetically accessible carbanions. Those are the carbon nucleophiles needed for C–C bond creation/cleavage in metabolite chemical transformations. This holds for both catabolic and anabolic pathways. Before taking up nature's strategies for accessing both carbon nucleophiles and carbon electrophiles for C–C bond formations/cleavages, we remind that we use C–H bond content as a marker for different oxidation states of carbon-based functional groups, as a clue to whether redox has occurred and, if so, to define the interrelationship between functional groups in substrates (reactants) and products.

10.4 Oxidation/Reduction Status and C–H Bond Content

As noted briefly in Chapter 1, much of carbon-based functional group chemistry is based on electron transfers to and from carbon centers. One simple single carbon axis that we used to illustrate the formal change in carbon from -4 to $+4$ is the set of interconversions of methane to methanol to formaldehyde to formate to CO_2 (Figure 1.11). The traverse from methanol oxidation state to formaldehyde oxidation state to formate oxidation state is also carried out when those one carbon units are covalently attached to tetrahydrofolate cofactors (Chapter 9, Figure 9.23).

In the progression from methane to higher alkanes, such as propane and beyond, regiospecificity considerations also come into play. If a terminal carbon is oxidized to a 1-alkanol, the next oxidation step up is an aldehyde functional group. If an internal CH_2 is oxidized, then a secondary alcohol is formed and is progenitor to a ketone rather than an aldehyde functional group. Aldehydes and ketones are at the same formal oxidation step, four electrons up from the alkane. Each of the four two-electron steps outlined in Figure 1.11 involves a loss of a C–H bond in the oxidation direction and a gain of a C–H bond in the direction of increased reduction.

All the metabolite interconversions that involve transformations of one of the four C–O based functional groups (carbon dioxide, carboxylate, aldehyde/ketone, alcohol) to the alkane or to any of the three other C–O functional groups (*e.g.* carboxylate to aldehyde) involves C–H bond cleavage or formation and thereby entails a redox step. Analogously, for C–N functional groups (amines, imines, amides) the addition or removal of C–H bonds during interconversions (Figure 2.5) reflect net two-electron oxidations or reductions. This is also the case for the parallel but much rarer C–S redox interconversions of thiols to thioaldehydes to thioacids (Figure 2.5).

Note, in contrast, interconversions at the same carbon oxidation state do not involve any formal electron transfers between reactants. For example, progressive conversions of a carboxylate to an acyl-phosphate to a thioester to an amide (Figure 3.17) do not involve any redox changes at the acyl group.

10.5 C–C Bond Forming and Cleaving Strategies

Equally important to the organic chemistry of life as the underlying strategies for C–H bond cleavage/formation mechanisms are the comparable chemical biology logics for forming and breaking C–C bonds by low energy routes. We will take up the homolytic paths, *via* action of oxygenases or anaerobic radical SAM enzymes in Chapter 12, but stipulate here that in other transformations, not involving SAM or O_2 as specific cosubstrates, most routes are heterolytic for C–C bond formation or cleavage. This holds especially in primary metabolic pathways (on in all cells of an organism).

In turn, that requires organisms to find low energy routes to carbon nucleophiles (nucleophile and electrophile terminology introduced by Ingold in 1933[17]), most often carbanion equivalents but also the π-electrons of some olefinic groups. Carbon electrophiles act as partners in C–C bond formations or as coproducts in C–C bond cleavages. We take up the strategy for carbon nucleophiles here and then turn to carbon electrophiles as reaction partners in Chapter 11. Aldehydes, ketones, and acyl thioesters are stable metabolites that circulate in cells and provide carbon electrophiles for C–C bond formation. By contrast, there are few if any ground state stable carbanion functional groups, as requisite carbon nucleophiles, other than cases where π-electrons of alkenes are induced to act as carbon nucleophiles.

10.6 Accessing Carbanion Equivalents as Carbon Nucleophiles

Figure 10.8 lists some of the common carbon-based functional groups that can be converted in enzyme active sites to carbanionic transition states or high energy (unstable) intermediates that function as carbon nucleophiles in C–C bond formations.[18] All of them have been mentioned, sometimes in extensive mechanistic analysis, in previous chapters. Here they are all collected together to give a sense of scope of structure and electronics that lower energy barriers to allow timely access to different variants of carbanion equivalents. In turn, different types of C–C bonds are generated or cleaved.

Of the 14 entries in Figure 10.8, enolate anions arise from carbons with one or more hydrogens adjacent to an aldehyde or ketone. The thioester enolate anions are comparable in adjacency to the carbonyl of a thioester. Carbons adjacent to oxoesters are intermediate, while amides and carboxylates are substantially less likely to have energetically accessible, stable carbanions adjacent. A special case is found in the phosphorylated form of

Some kinetically accessible carbon nucleophiles *in vivo*

1. enolate anions (carbanion contributors)
2. thioester enolate anions (carbanion contributors)
3. enediolate anions (carbanion contributors)
4. dienolate anions (carbanion contributors)
5. enamines (carbanion contributors)
6. phenolate anions (cyclohexadienyl carbanion contributors)
7. aminoacid-PLP delocalized carbanions
8. glutamyl residue C_4-carbanions in vitamin K-carboxylations
9. Vitamin B1-keto acid or ketol carbanionic adducts
10. doubly and triply vinylogous double bond isomerizations
11. C3 enamine carbon of indole ring
12. C5 carbon of pyrrole ring
13. SAM- CH_2=ylid
14. Alkene groups, singly and in concert

Figure 10.8 Fourteen kinetically accessible carbanion equivalents *in vivo*.

the enolate anion of pyruvate, phosphoenolpyruvate (see Chapter 15 for its formation) (Figure 10.9). The enol form in PEP is trapped as a phosphate ester,[19] that arises by enzymatic dehydration of 2-phosphoglycerate. Any phosphatase subsequently acting on PEP with O–PO_3^{2-} bond cleavage releases the enolate anion as a ground state carbanion.

Two prior examples that use the enolpyruvate C_3 carbanion equivalent as a carbon nucleophile are (1) in the shikimate pathway where that carbanion attacks the aldehyde carbonyl of erythrose-5-P to create the linear seven-carbon framework of DAHP (Figure 7.6) and (2) in the second half of the biotin-dependent pyruvate carboxylase reaction where electrophilic CO_2 is captured to form the four-carbon oxaloacetate product (Figure 9.21).

Kinetically accessible enediolate anions are a central chemical feature of α-hydroxy-carbonyl groups that predominate at carons 1 and 2 of both common aldohexose (glucose, galactose, mannose) and aldopentose sugars (ribose). Perhaps the most famous of the enzymes working *via* carbanionic forms of enediolate anions is ribulose-1,5-bis phosphate carboxylase[20] (Figure 3.35) that globally fixes an estimated 250 trillion tonnes of CO_2 each year in plants.

The reverse regiochemistry in a 2-keto-1-hydroxy group likewise enables enediolate chemistry from such sugars as fructose and ribulose sugars. The related dienolate anions as carbanion equivalents are less common. However, they are the key carbanion equivalents in the bacterial isomerization of reducible 2,3-*trans*-enoyl thioesters to unreducible *cis*,3,4-enoyl thioesters on pathway to monounsaturated fatty acyl chains, noted back in Chapter 2 (Figure 2.24).

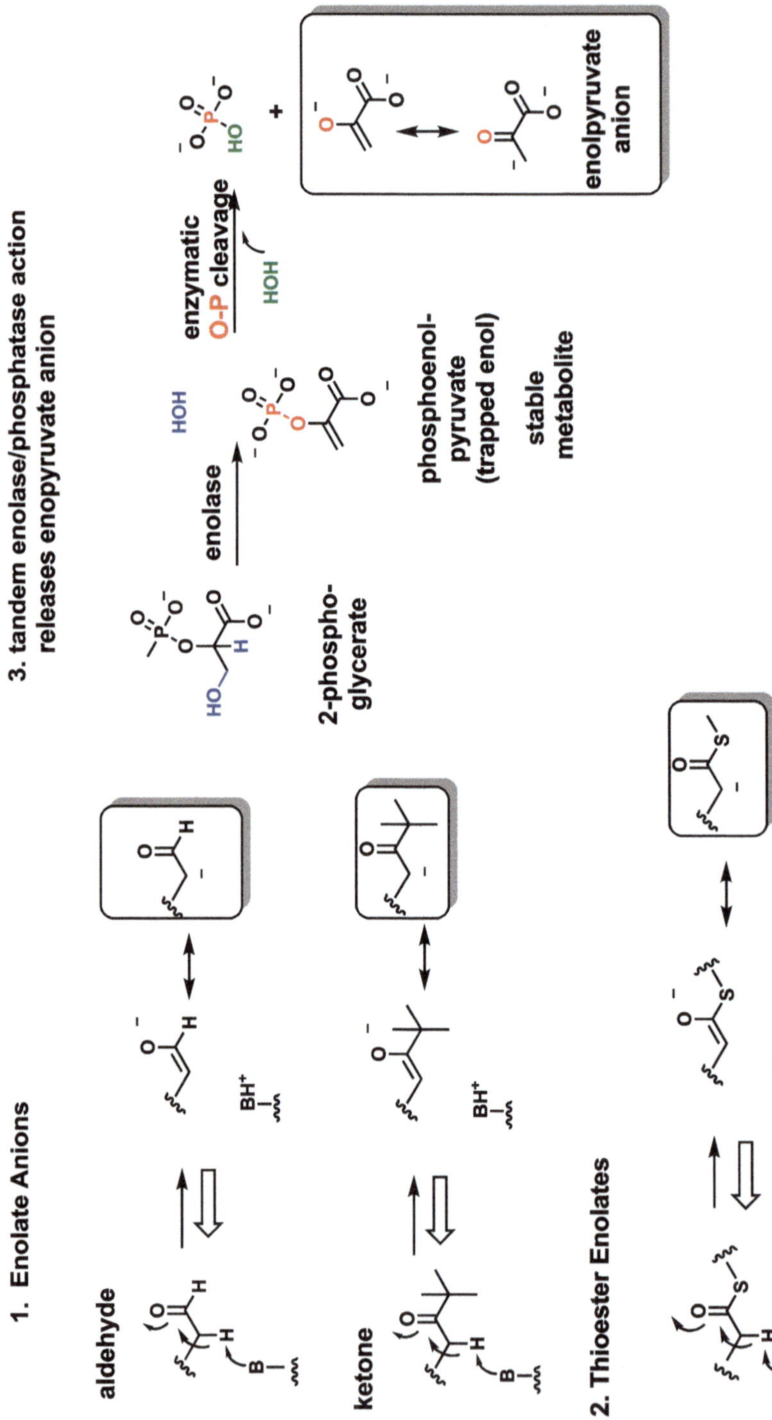

Figure 10.9 Enolate ions and thioester enolate ions as carbon nucleophiles.

Carbon–nitrogen metabolic scaffolds can also yield two types of carbanion equivalents. One is through simple enamine chemistry. Given that the side chains of the 20 proteinogenic amino acids lack any ketone or aldehyde electrophilic groups, protein-based enamine adducts derive from substrate aldehyde or ketone groups that start in initial iminium ion[21] linkages to the side chains of lysine residues. If the substrate carbon adjacent to the carbonyl group has a C–H bond, then isomerization to the neutral enamine often provides a lower energy pathway than the equivalent charged enolate anion.

Many aldolase enzymes offer side chain lysines for iminium formation with substrate carbonyl groups to set up low energy paths to carbanion formation. For the fructose-1,6-diphiosphate aldolase, acting in the cleavage direction, C_3–C_4 bond cleavage leads initially to the low energy dihydroxy-acetone phosphate enamine adduct. In the opposite, C–C bond-forming direction aldolases generate the enamine as low energy carbon nucleophile (Figure 10.10).

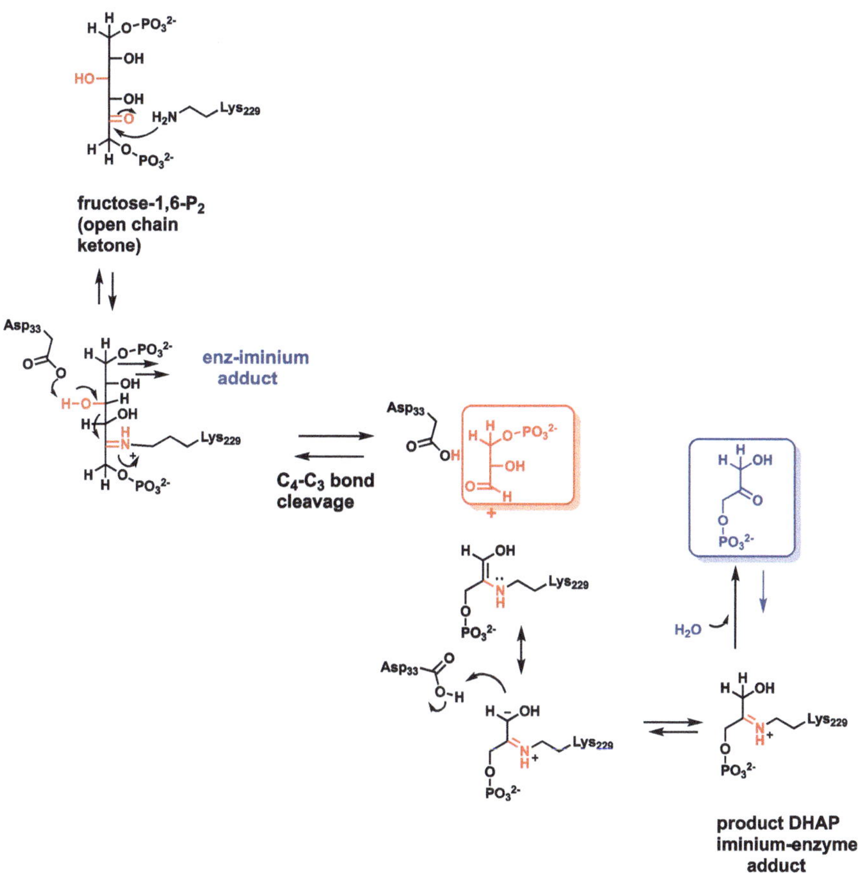

Figure 10.10 Iminium/enamine as a low energy carbanion regime in fructose-bis phosphate aldolase catalysis.

Nature has invented a route to achieve the opposite polarity of imines[21] by providing an electrophilic aldehyde carbonyl group in the coenzyme form, pyridoxal-P, of vitamin B6.[22] Amine groups of more than 200 natural amino acids can engage in aldimine adducts with PLP as prelude to a range of amino acid-based carbanion chemistry, exemplified in previous chapters.

We note here three routes that different PLP-enzymes have used to access distinct variants of delocalized, stabilized-PLP carbanion equivalents. One is by abstraction of the C_2–H hydrogen as a proton to yield carbanion adduct-1. A second is employed by PLP-decarboxylases where it is cleavage of the C_2–C_1OO^- bond that leads to a carbanionic-PLP adduct-2. A third variant arises in PLP-aldolase type enzymes as in threonine aldolase where a nascent glycyl-PLP carbanion adduct-3 is formed (Figure 10.11).

Two other features of PLP as an amino acid carbanion-stabilizing electron sink have surfaced in the previous chapters. One is the ability of transaminases to generate and use pyridoxamine-benzyl carbanions in the ketimine-PNP adducts in the second half of transaminase reactions. The second

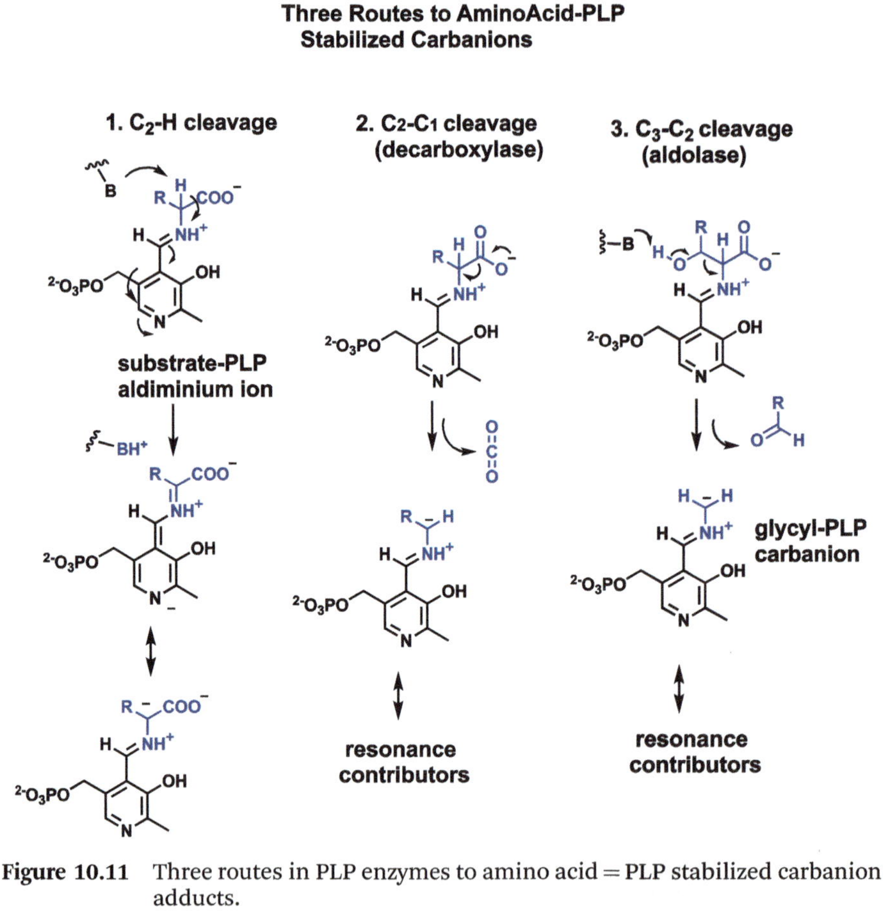

Figure 10.11 Three routes in PLP enzymes to amino acid = PLP stabilized carbanion adducts.

feature is that in some enzyme active sites, double carbanionic-PLP equivalents can be formed transiently. These are alternatively viewed as amino acid-PLP adducts with four electrons transiently stored in the system, as in the cystathionine-γ-lyase catalytic cycle.[23]

Two other vitamin coenzyme forms lower energy barriers for specific types of organic substrates to access carbanion equivalents. Most famous is the cationic thiazolium ring of thiamin-PP (Chapter 9) where either 2-keto acid substrates (Figure 9.4) or structurally dissimilar 2-ketose substrates (Figure 9.6) are processed to accessible substrate carbanion-thiazolium adducts where the $C_2{=}N^+$ moiety of the thiazolium ring of thiamin is a β-electron sink to form delocalized carbanion equivalents. The 2-ketoacids are decarboxylated: either one-carbon shorter aldehydes or acyl thioester products released. The ketoses undergo fragmentation to release two-carbon shorter aldoses while the two-carbon ketol group is retained for subsequent transfer out to an alternative aldose carbonyl acceptors.

The hydroquinol form (KH_2) of the vitamin K menaquinone nucleus is reactive with O_2 by one-electron redox chemistry. In a mechanistically still mysterious manner,[24] as each KH_2 is oxidatively epoxidized, a set of up to a dozen glutamyl side chains, one by one, in a family of 10 or so blood coagulation protease zymogens become carboxylated at C_4. These dicarboxyglutamate side chains are strong bidentate chelates with divalent calcium ions, effectively neutralizing local patches of anionic glutamate residues. This posttranslational modification of the zymogen forms of the proteases leads to conformational changes. The now, net neutral carboxylated domains bury into the outside surfaces of cell membranes to initiate clotting cascades. At the heart of these transformations are vitamin K-mediated transient formation of glutamyl C_4-carbanion equivalents that capture CO_2 covalently (Figure 9.38).

Two additional less common carbanionic forms of substrates that are accessible are cyclohexadienyl carbanion contributors to phenolate anions and the $CH_2{=}$ylid form of *S*-adenosylmethionine. The cyclohexadienyl carbanions account for *C*-methylation or *C*-glycosylation outcomes. The stabilized zwitterionic ylid of SAM[25] is thought to be a progenitor of CH_2 groups in some cyclopropanated enzymatic products.

Two heterocyclic ring systems have minor resonance contributors that explain ground state carbanion behavior. The pyrrole ring in the bicyclic indole group, both as the rare free indole molecule and in the abundant tryptophan and tryptamine scaffolds, is known to behave as a carbon nucleophile both abiotically and biotically (Figure 8.21). The enamine part structure makes C_3 of the indole ring the most reactive carbon a nucleophile, but prenylated tryptophan metabolites, substituted at each carbon position around the indole ring are known, consistent with carbanionic/carbon nucleophilic character at those ring sites as well. Although C_3 carbon nucleophilic behavior is suppressed in all known tryptophan residues in proteins it is the reactivity portal to thousands of prenylated tryptophan alkaloid secondary metabolites.

The other metabolite with equivalent built in inducible carbon nucleophilic reactivity is the stand-alone pyrrole ring in the aminomethyl-bis-carboxy pyrrole metabolite porphobilnogen. The behavior of porphobilinogen as both a carbon nucleophile at C_5 and an electrophile after elimination of the C_2-aminogroup explains the self-condensation properties that led to assembly of four new C–C bonds in tetrapyrrolic pigments of life[26] (Figure 10.12). This self-condensation

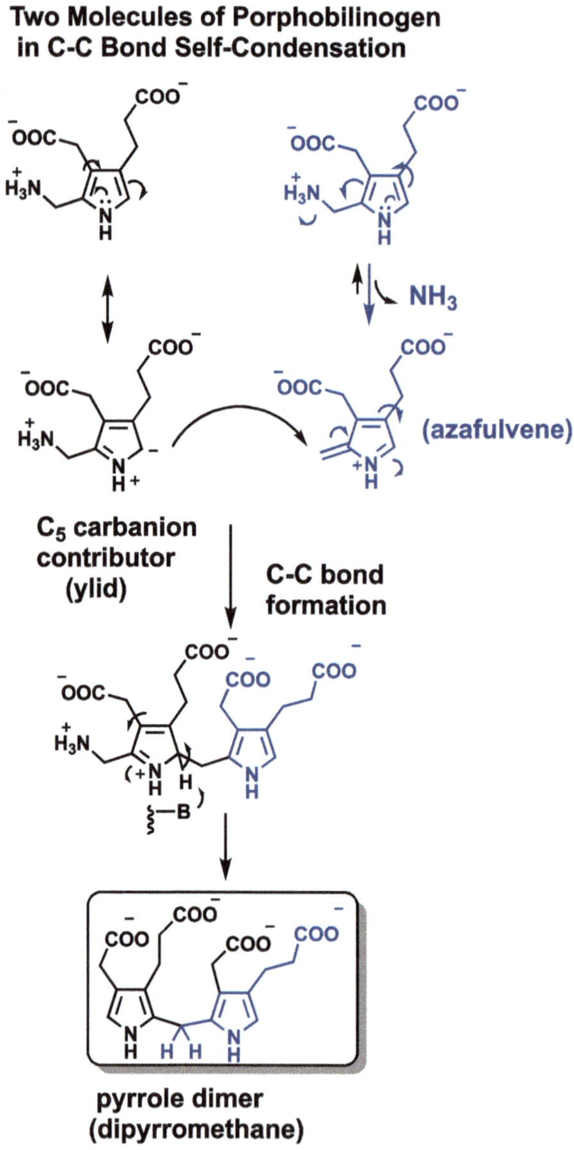

Figure 10.12 Two molecules of porphobilinogen in a net self-condensation for C–C bond formation.

tendency, *via* intermolecular carbon nucleophile and carbon electrophile contributors, is a property also shared with acetyl CoA.

10.7 Alkenes as Carbon Nucleophiles

The last of the categories of carbon nucleophiles listed in Figure 10.8 are alkene groups, particularly alkenes out of conjugation, as explored previously in Chapter 2. The reactivity of the alkene group of Δ^3-IPP as carbon nucleophile in all the prenyltransferase-catalyzed head to tail isoprenoid chain elongations is one prominent manifestation. So is the celebrated formation of many sterol variants with four or five fused carbacyclic rings in squalene cyclizations. The degree of concertedness of participation of all the unconjugated double bonds as carbon nucleophiles may vary between one triterpene cyclase to the next.

There are many cases where conjugated olefins can participate as neighboring groups that act as electron rich complex functional groups. The enzymatic conversion of hydroperoxide group in LTA_4 to the epoxide in LTB_4 discussed in Chapter 4 (Figure 4.11) is viewed as a simultaneous participation of the diene, a vinylogous neighboring group effect, where the proximal end of the diene forms a C–O bond as the epoxide forms. In a related context the isomerization/hydrolysis of *trans*-retinyl palmitate to *cis*-11-retinol (Figure 9.37) is suggested to involve triply vinylogous participation of the 4, 6, 8 triene in the unusual regiospecificity of C–O bond cleavage of the retinol component of the oxoester substrate. (Both of these vinylogous olefin neighboring group reactions cited make or break C–O bonds not C–C bonds.) Many other examples of single and polyolefins as carbon nucleophiles point to the widespread use of the π-electrons of carbon–carbon double bonds in many kinds of substrates.

Most of the first 9 of the 13 functional groups listed in Figure 10.8 are not stable ground state molecules, likely only populated in useful concentrations in the active sites of particular enzyme types. In contrast, the olefinic groups of examples 10 and 13 are ground state substrate forms. So are the enamine moiety in indole and to a significantly lessened extent the C_5 carbanion contributor to pyrrole ring structures.

In all the C–C bond formations represented by the accessible carbon nucleophiles of Figure 10.8, there must always be an electrophilic carbon partner to enable C–C bond formation. We turn to elaboration of those accessible carbon electrophilic groups in the next chapter. A completely orthogonal set of reactivity routes to C–C bond formation and cleavage is by homolytic pathways. That reactivity *in vivo* is examined in Chapter 12.

References

1. C. T. Walsh, B. P. Tu and Y. Tang, Eight kinetically stable but thermodynamically activated molecules that power cel metabolism, *Chem. Rev.*, 2018, **118**, 1460–1494.

2. C.-I. Lin, R. M. McCarty and H.-W. Liu, The enzymology of organic transformations: a survey of name reactions in biological systems, *Angew. Chem., Int. Ed.*, 2017, **56**, 3446–3489.
3. C. T. Walsh, Biologically generated carbon dioxide: nature's versatile chemical strategies for carboxy lyases, *Nat. Prod. Rep.*, 2015, **37**, 100–135.
4. K. U. Wendt, Enzyme mechanisms for triterpene cyclization: new pieces of the puzzle, *Angew. Chem., Int. Ed.*, 2005, **44**, 3966–3971.
5. E. J. Corey, K. Lin and H. Yamamoto, Separation of the cyclization and rearrangement processes of lanosterol biosynthesis. Enzymic conversion of 20,21-dehydro-2,3-oxidosqualene to a dehydroprotosterol, *J. Am. Chem. Soc.*, 1969, **91**(8), 2132–2134.
6. H. Diao, N. Chen, K. Wang, F. Zhang, Y.-H. Wang and R. Wu, Biosynthetic mechanism of lanosterol: a completed story, *ACS Catal.*, 2020, **10**, 2157–2168.
7. D. R. Boyd, J. W. Daly and D. M. Jerina, Rearrangement of (1-2 H)- and (2-2 H)naphthalene 1,2-oxides to 1-naphthol. Mechanism of the NIH shift, *Biochemistry*, 1972, **11**(10), 1961–1966.
8. S. J. Blanksby and G. B. Ellison, Bond dissociation energies of organic molecules, *Acc. Chem. Res.*, 2003, **36**(4), 255–263.
9. E. I. Solomon, U. M. Sundaram and T. E. Machonkin, Multicopper Oxidases and Oxygenases, *Chem. Rev.*, 1996, **96**(7), 2563–2606.
10. M. D. White and E. Flashman, Catalytic strategies of the non-heme iron dependent oxygenases and their roles in plant biology, *Curr. Opin. Chem. Biol.*, 2016, **31**, 126–135.
11. A. J. Komor, A. J. Jasniewski, L. Que and J. D. Lipscomb, Diiron monooxygenases in natural product biosynthesis, *Nat. Prod. Rep.*, 2018, **35**(7), 646–659.
12. H. J. Sofia, G. Chen, B. G. Hetzler, J. F. Reyes-Spindla, N. E. Miller and S. A. M. Radical, a novel protein superfamily linking unresolved steps in familiar biosynthetic pathways with radical mechanisms: functional characterization using new analysis and information visualization methods, *Nucleic Acids Res.*, 2001, **29**, 1097–1106.
13. A. Kutt, S. Selberg, I. Kaljurand, S. Tshepelevitsh, A. Heering, A. Darnell, K. Kaupmees, M. Piirsalu and I. Leito, pK_a values in organic chemistry-making maximum use of the available data, *Tetrahedron Lett.*, 2018, **59**, 3738–3748.
14. J. M. Shin, K. Munson, O. Vagin and G. Sachs, The gastric HK-ATPase: structure, function, and inhibition, *Pflugers Arch.*, 2009, **457**(3), 609–622.
15. R. Breslow, Rapid deuterium exchange in thiazolium salts, *J. Am. Chem. Soc.*, 1957, **79**, 1762–1763.
16. T. L. Amyes and J. P. Richard, Substituent Effects on Carbon Acidity in Aqueous Solution and at Enzyme Active Sites, *Synlett*, 2017, **28**(12), 2407–2421.
17. C. K. Ingold, Significance of tautomerism and of the reactions of organic compounds in the electronic theory of organic reactions, *J. Chem. Soc.*, 1933, 1120.

18. E. Anslyn and D. A. Dougherty, *Modern Physical Organic Chemistry*, Univerist Science Books, 2006.
19. C. T. Walsh, T. E. Benson, D. H. Kim and W. J. Lees, The versatility of phosphoenolpyruvate and its vinyl ether products in biosynthesis, *Chem. Biol.*, 1996, **3**, 83–91.
20. I. Anderson, Catalysis and regulation in RUBISCO, *J. Exp. Bot.*, 2008, **59**, 1555–1568.
21. A. Erkkila, I. Majander and P. M. Pihko, Iminium catalysis, *Chem. Rev.*, 2007, **107**, 5416–5470.
22. P. K. Mehta and P. Christen, The molecular evolution of pyridoxal-P dependent enzymes, *Adv. Enzymol. Relat. Areas Mol. Biol.*, 1999, **74**, 129–184.
23. C. T. Walsh and Y. Tang, *The Chemical Biology of Human Vitamins*, Royal Society of Chemistry, London, 2018.
24. P. Dowd, S. W. Ham and S. J. Gelb, Mechanism of action of vitamin K, *J. Am. Chem. Soc.*, 1991, **113**, 7734–7743.
25. B. J. Landgraf and S. J. Booker, The ylide has landed, *Nature*, 2013, **498**, 45–47.
26. A. R. Battersby, Tetrapyrroles: the pigments of life, *Nat. Prod. Rep.*, 2000, **17**, 507–526.

CHAPTER 11

Routes to Make and Break C–C Bonds II: Carbon Electrophiles

The bulk of the ~25 pounds of carbon in an adult human (amount for a male of 70 kg weight) is in dynamic flux as metabolites with carbon scaffolds are formed and transformed. Given that most of the enzyme-catalyzed C–C bond-forming and/or C–C bond-cleaving reactions *in vivo* occur in two-electron, heterolytic steps, each of those C–C bonds formed or broken involves carbon nucleophiles (*e.g.* carbanions or π-electron pairs of some olefins) and functional groups behaving as carbocation equivalents coming together or breaking apart. The preceding chapter delineated many of the commonly accessible carbanion equivalents acting as carbon nucleophiles in metabolism. Here, we turn to electrophilic partners in C–C bond formation/consumption chemical biology and look at the kinds of molecular electrophilic groups that are transferred to cellular nucleophiles.

11.1 Electrophilic Hydrogen, Oxygen, Sulfur, Phosphorus, and Sulfur Metabolites

As an aside, we iterate that only six chemical elements, carbon, hydrogen, oxygen, nitrogen, sulfur, and phosphorus comprise 99% of the body mass of humans. These five elements other than carbon can also function as electrophilic (electron-deficient) groups at one or more stages in metabolism, relevant to the themes in these chapters.

The most obvious are likely the positively charged protons, by far the major form of hydrogen in biologic systems. Hydrogen as the least electronegative element in the periodic table tends to dissociate from C–H, O–H, N–H, or S–H bonds as protons (the big surprise is how prevalent hydride ion transfers are in so many redox metabolic pathways). Nitrogen when present in imine or

Chemical Biology No. 21
The Chemical Biology of Carbon
By Christopher T. Walsh

Published by the Royal Society of Chemistry, www.rsc.org

positively charged iminium ion linkages is clearly electrophilic. So, too, is N_5 of the oxidized flavin isoalloxazine ring system (Chapter 9), to and from which hydride transfer can occur (Chapter 9).

Phosphorus in biologic systems is almost redox inert, presenting mostly as pentavalent phosphate compounds.[1] Elsewhere we have emphasized the remarkable property of phosphate groups as electrophilic at the central phosphorus atom but electronegative at the peripheral oxyanions (a mix of phosphate monoanions and dianions at neutral pH).[2] We have observed in previous chapters, particularly Chapters 3 and 5 (*e.g.* protein biosynthesis or glutamine synthetase catalytic logic), that activation of carboxylate groups almost always starts with attack of the carboxylate oxyanion on one of the electrophilic phosphorus atom in the sidechain of ATP.

In Chapter 4 during analysis of important categories of C–S bonds we noted both nucleophilic and electrophilic forms of sulfur that impact on carbon chemical biology. On one hand, both thiol and thioether oxidation states of carbon–sulfur compounds can behave as sulfur nucleophiles. On the other hand, the most obvious sulfur electrophile[3] is the sulfonium cation of *S*-adenosylmethionine as we note in a section below. Oxidation of C–S bonds to convert nucleophilic thiolate anions to electrophilic forms of sulfur occurs in formation of sulfenates, sulfinates, and sulfonates. Also, the generation of disulfides, *e.g.* in glutathione metabolism, generate a pair of electrophilic sulfur atoms needing reduction to regain nucleophilic capability. Finally, the oxidation of cysteine sulfur to the persulfide adducts in the active sites of a subgroup of cysteine lyases create the C–SS–H persulfide group, analogous to the C–SS–C disulfide group.

Molecular oxygen, O_2, and its one electron reductive metabolism is a main topic of radical chemistry in the next chapter. While it is appropriate to think of oxygen chemical biology as almost exclusively oxygen radical generation and capture, there are metabolic intersections where an O_2-derived –OH group acts as an electrophilic $[OH^+]$ equivalent. One such locus is the capture of HO–OH by glutathione in the detoxification carried out by glutathione peroxidase (Chapter 4). Another is in flavoprotein oxygenase active sites where the weak O–O bond in the flavin hydroperoxide adduct (Chapter 9) is broken with net $[OH^+]$ transfer to attacking cosubstrate nucleophiles (N, O, C atoms).

One should view the ability of the partner five chemical elements that make up our bodies to exist in electrophilic oxidation states and or functional groups as complementary to much of the electrophilic carbon chemistry accessible to organisms, described in this chapter.

11.2 Types of Accessible Carbon Electrophiles

In parallel to the list of accessible common carbon nucleophiles in Figure 10.8. Figure 11.1 provides a comparable list of biologically accessible carbon electrophiles. The combinations can provide the partners needed to generate the major types of C–C-bonded product functional groups.

Common Carbon Electrophiles for C-C Bond Formations *in vivo*

1. aldehyde and ketone carbonyl groups
2. acyl thioester carbonyl groups
3. acyl oxoester carbonyl groups
4. acyl amides (hydrolysis)
5. aldimine and ketimine groups
6. allylic cations and derived carbocations
7. oxocarbenium ions
8. strained small ring carbacycles and heterocycles
9. conjugated olefins
10. carbons adjacent to the sulfonium ion in SAM
11. Phosphate esters as precursors to electrophiles
12. Carbon dioxide
13. dehydroalanyl and dehydrobutyryl residues

Figure 11.1 Thirteen common carbon electrophiles for C–C bond formations *in vivo*.

11.3 Aldehyde, Ketone, and Thioester Carbonyls as Common Carbon Electrophiles

The most abundant and stable carbon electrophiles in primary metabolites are the carbonyl groups at two oxidation states. The first are the electrophilic carbonyl carbons in aldehydes and ketones, as we emphasized back in Chapter 3. The carbonyl groups of these primary metabolites are stable in aqueous solutions at physiologic pH values.

When paired with enolizable aldehyde or ketone partners to access enolate anions, aldehydes react in aldolase active sites. An intramolecular version of aldol condensation is provided by the conversion of the 5-keto-1-aldehyde metabolite from glucose-6-P where the C_6 contributor to the enolate carbanion attacks the C_1 aldehyde, converting the linear hexose scaffold to the hydroxy cyclohexane framework of inositol-P (Figure 7.18). An intermolecular aldol condensation occurs in gluconeogenesis when the enolate anion of dihydroxyacetone-P acts on the electrophilic aldehyde carbon of glyceraldehyde-3-P (Figure 11.2).

The second commonly available electrophilic groups in the carbon inventory of cells are acyl group carbonyls, typically when esterified or preferably thioesterified.[2] The free carboxylates are nearly fully dissociated ($pK_a s \approx 4–5$) at physiologic pH values (pH 7.2–7.4). Those negatively charged carboxylate anions are both resonance-stabilized and, if anything, nucleophilic rather than electrophilic. When a rare free carboxylate serves as a functional electrophile, as in the dihydrorotase-mediated cyclic amide bond formation in pyrimidine biosynthesis (Figure 6.4), it is likely that carboxylate is protonated in the relevant enzyme's active site to behave as a neutral electrophile for subsequent water elimination. Acyl oxoesters serve as the electrophilic partner in all ribosomal-mediated peptide bond formations (see Figure 5.9).

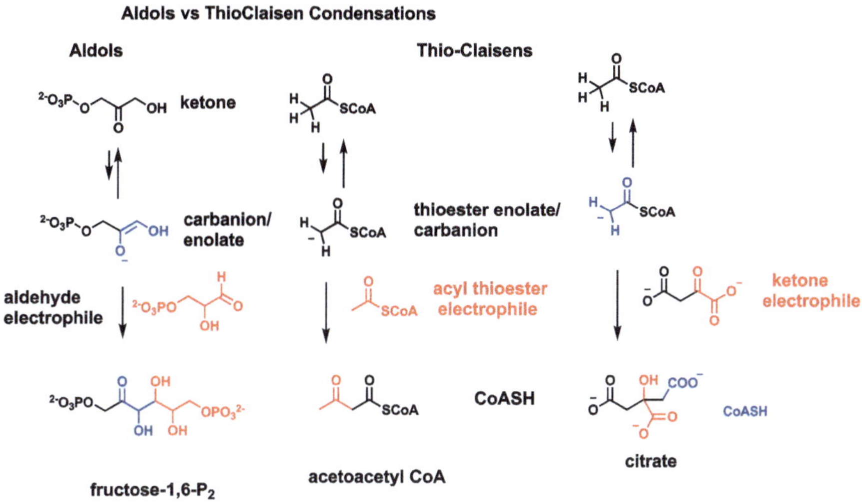

Figure 11.2 Aldol *vs.* thio-Claisen condensations: ketones *vs.* thioester.

The predominant form of esterified carboxyl groups in primary metabolites are thioesters,[4] largely acyl CoA thioesters (Chapters 1 and 3) (Figure 11.2). Because the radius of sulfur atoms are larger than oxygen atoms, the carbonyl carbon is further away in thioesters compared to oxoesters. Thioester carbonyls behave more like ketone carbonyls as electrophiles. The aldol condensations/cleavages *in vivo* use aldehydes and enolizable ketones as both electrophilic and nucleophilic partners. The thio-Claisen condensations/cleavage in fatty acid biosynthesis/degradations instead utilize the thioester carbonyls as electrophiles and the thioester enolates as nucleophiles. Two thio-Claisen condensation examples are shown in Figure 11.2. The first is the self-condensation of two acetyl CoA molecules, one acetyl group as C_2 acyl carbanion, the other as C_1 acyl electrophile. The four-carbon acetoacetyl acyl product is the first committed metabolite in biosynthesis of all isoprenoid biosynthesis (Figure 11.2). The second example uses the ketone carbonyl of the four carbon dicarboxylate metabolite oxaloacetate as electrophile and the acetyl CoA thioester enolate as nucleophile. This is the first reaction in the citric acid cycle, driven in the condensation direction by hydrolysis of the acyl thioester group before citrate release (see Chapter 14).

A variant of the chemical logic of using aldehyde and ketone carbonyls as common cellular electrophiles is to preform imines (usually as an equilibrium mixture) by attack of amines on those carbonyl carbons[5] (Figure 11.3). Because nitrogen is more basic than oxygen the pK_a values of those imines are such that protonated iminium ions can accumulate as improved electron sinks[6] (improved electrophilicity due to the full positive charge).

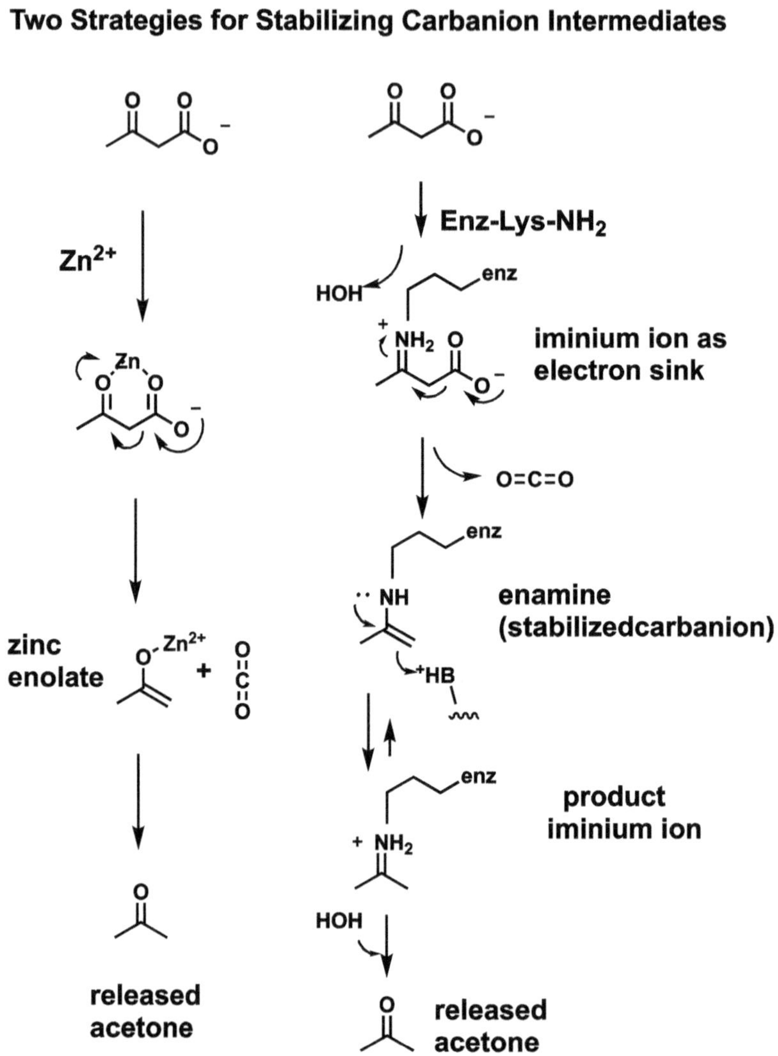

Figure 11.3 Two strategies for stabilizing carbanion intermediates during production of CO_2 from carboxylate substrates.

11.4 Iminium Ion Catalysis Logic in Aldolases and PLP Enzymes

Aldolases and acetoacetate decarboxylases can tether their ketone substrates as covalent iminium adducts. Those electrophilic centers are potent electron sinks for electron flow into them during aldol cleavage or β-keto acid decarboxylation, respectively. Figure 11.3 shows that microbial acetoacetate decarboxylases have evolved two different strategies to stabilize the incipient product, acetone enolate, before ketonization and release of acetone. An active site zinc ion in the left hand column accelerates catalysis in type I acetoacetate decraboxylases by chelation of both substrate and the nascent enolate anion as the zinc complex.

The more general solution to stabilization of high energy substrate-derived or nascent product enols is to use the side chain of an active site lysine as initial nucleophile, converting enolizable aldehdye or ketone to the corresponding aldimine or ketimine (type II aldolases). Because nitrogen is more basic than oxygen, the pK_a for converting imines to iminium ions is generally accessible in enzyme active sites. This makes for a more effective electron sink. Equally, though, is the fact that the carbanion fragment is now not a charged enolate anion but a neutral, more stable enamine.

Enamines as stabilized carbanion equivalents provide an overarching catalytic advantage in all the PLP-enzymes that were discussed in the last chapter and in earlier ones (Chapter 5). The above discussion on aldolases involves substrates bringing the carbonyl group and the enzyme providing the amine side chain to create the iminium ion/enamine axis.

The reverse is true in all PLP-enzymes that act on amino acid substrates. The substrates bring the nucleophilic amine groups. The pyridoxal-P co-factor provides the electrophilic aldehyde. We have examined the various kinds of substrate-PLP/PNP adducts that represent accessible, stabilized carbanions in Chapter 10.

However, in all those PLP-enzymes the vitamin B6 coenzyme is acting as a heterocyclic electron sink conjugated to the aldehyde group: a potent electrophilic functionality. The pyridine-aldehyde framework of PLP is a more effective electron sink than the simple iminium ions between enzyme lysine side chains and substrate aldehyde or ketones noted above for aldolases and acetoacetate decarboxylases. Thus, nature can build aldimines/ketimines with either polarity: substrate as nucleophile (amino acids) or substrate as electrophile (substrate aldehydes and ketones).

11.5 Acyl Oxoester and Amide Bonds as Electrophiles: Acyl Transfers to Water or Amines

In Chapter 3 during the discussion on different types of acyl groups, depending on what atom had been the attacking nucleophile to create that particular acyl group product, we noted that acyl thioesters were more activated as electrophiles than oxoesters.[4] After the thioesters and oxoesters came amides, largely deactivated for any synthetic acyl transfers, and finally the resonance-stabilized carboxylate anions which had nucleophilic tendencies but no electrophilic acyl group transfer reactivity (Figure 3.17). Acyl oxoesters can play notable roles in biology, the most prominent being conversion to the amide(peptide) bonds during protein biosynthesis at the ribosome. Clearly, they balance kinetic stability with enough retained thermodynamic activation for acyl transfer to amines (Figure 5.9). We have observed chemical stability of all the oxoesters links in membrane phospholipids and the requirement for an enzyme, acetylcholinesterase, to hydrolyze the oxoester link in the neurotransmitter acetyl choline to inactive acetate and choline.

Compared to oxoesters the amide bonds are even less activated due to resonance contributors to the planar amide bond. The major acyl transfer reaction observed *in vivo* for amide groups is enzyme catalyzed hydrolysis, conversion of amides to resonance-stabilized carboxylate anions and the amine cations. There are over 500 such amide bond hydrolases in human proteomes showing distinction in mechanisms of hydrolysis and in selectivity for the myriad protein and peptide substrates. Amides are effectively dead for C–C bond forming reactions in biology.

11.6 Allylic Cations

A distinct type of enzyme-generated carbocation underlies the formation of the 70 000–100 000 known isoprenoid natural products, including linear isoprenoid chains and cyclic steroidal scaffolds. We have catalogued several types of C–C bond-forming logic in converting 18 molecules of acetyl CoA to the C_{30} sterol lanosterol in Chapter 2 and also later in Chapter 15 in the next section of this book. Here, we call out the facts that three of the four phases that build C–C bonds on the way to tetracyclic and pentacyclic sterols utilize carbocationic transition states or transient high energy intermediates as carbon electrophiles.

The key building block as carbocation generator in the elongation phase of isoprenoid chain buildup is the Δ^2-isomer of isopentenyl-PP.[7,8] We noted in Chapter 2 that the other isoprenoid building block, the Δ^3-isopentenyl-PP with a terminal olefin, is repeatedly utilized as carbon nucleophile *via* the π-electrons of the double bond (Figure 2.10 and Figure 11.4), the latter an example of olefins as nucleophiles, delineated in the preceding chapter.

The key to initiation of C–C bond formation by the family of prenyltransferases is to catalyze early C–O bond cleavage in the Δ^2-isopentenyl-PP. The resulting isopentenyl C_1 cation is delocalized *via* the C_2–C_3 olefin as contributing resonance structure that puts the positive charge at carbon 3, a tertiary carbon. This allyl cation strategy for generating nascent electrophilic carbon centers at either C_1 or C_3 of the five-carbon scaffold arising from Δ^2-isopentenyl-PP is used throughout chain elongations that add five carbons at a time to build C_{10}, C_{15}, C_{20}, C_{40}–C_{60}, and even C_{110}–C_{120} long prenyl chains (see Chapter 2).

In contrast to the stable ground state carbonyl groups of aldehydes, ketones, and acyl thioesters, these allyl cations, and the rearranged cations that arise by internal capture by olefins elsewhere in the substrate scaffold, are transient high energy intermediates. Those cations get quenched by three common routes: proton abstraction from an adjacent carbon to form an olefin in the product, water addition to form a product alcohol, or C–C single bond migration, setting up rearranged frameworks that still have to be quenched, as noted back in Chapter 2, Figure 2.15. The C_{20} metabolite

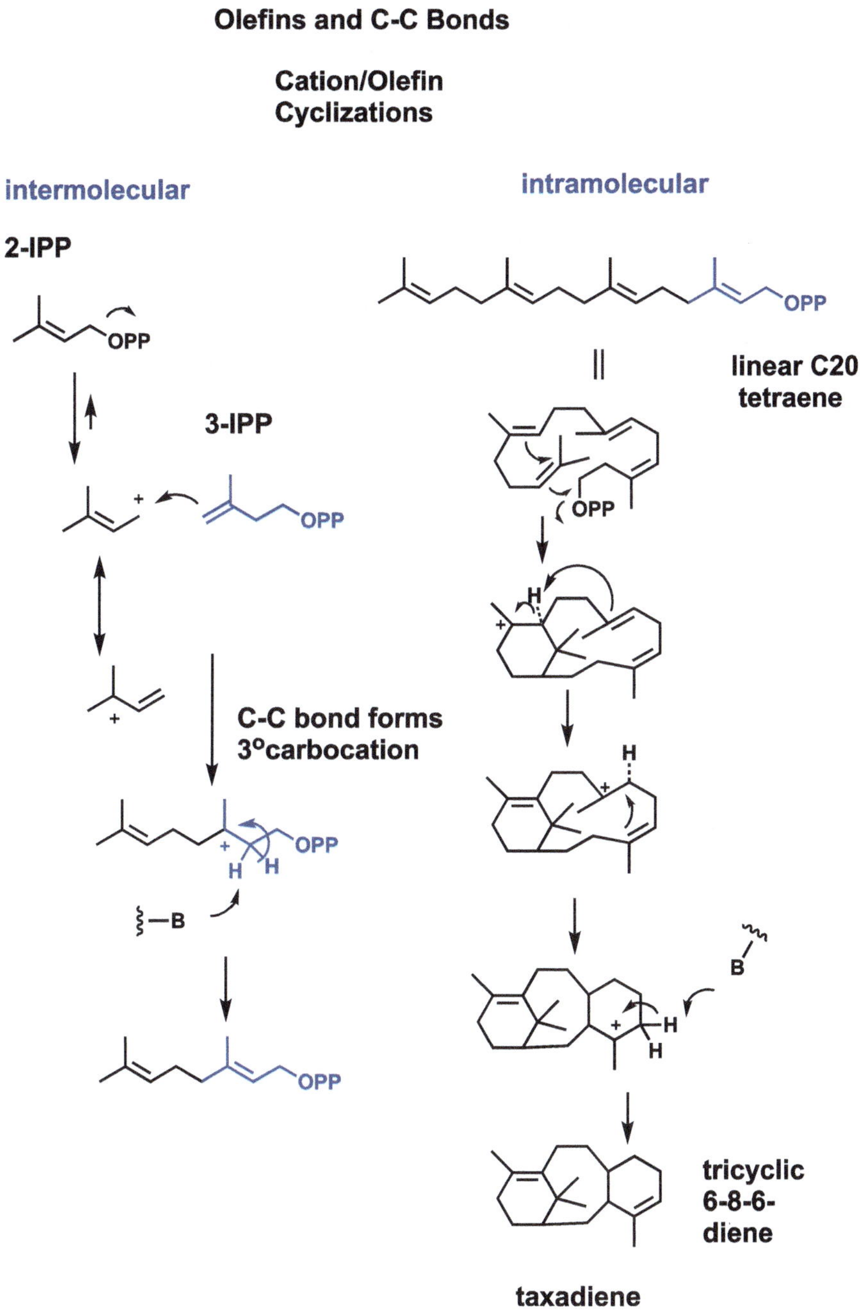

Figure 11.4 Olefins and C–C bonds.

geranyl-PP is progenitor to a large array of products *via* carbocation scaffold rearrangements. One such product, taxadiene, is shown in Figure 11.4.

One can distinguish between intermolecular C–C bond formations by prenyltransferases that extend the growing isoprenoid chains five carbons at a time and hundreds of intamolecular cylases.[8] Both are driven by olefin capture of carbocations that derive from the initial linear allylic cation. The right hand column of Figure 11.4 shows such an intramolecular fate for the linear C_{20} tetraene geranylgeranyl-PP. Not counting the initial allyl cation (not shown), one can write a series of three cyclic cations. The first two result in C–C bond formations creating the two six-carbon carbacyclic rings in the final tricyclic taxadiene product. The last cyclic cation has presumably reached a portion of the active site where proton abstraction leads to quenching of the cationic cascade. Note that the linear tetraene goes to a tricyclic C_{20} diene scaffold. Each missing olefin has gone to create one of the C–C bonds that create the cyclohexene rings in taxadiene. This cyclization logic is followed in hundreds to thousands of cyclase active sites on initially linear allyl cations generated from $C_{10,}$ C_{15}, C_{20}, C_{25}, and longer isoprenyl-PPs formed by the several prenyltransferases in plant cells.

Later stages of isoprenoid biosynthetic pathways (see Chapter 15) involve a change from head to tail coupling by prenyltransferases to head to head couplings of two C_{15}-farnesyl-PPs to the C_{30} hexaene squalene and two C_{20} geranylgeranyl-PPs to the C_{40}-nonaene phytoene.

The epoxidation of squalene sets up the subsequent cascade of cyclizations to tetracyclic and pentacyclic sterols that show additional attributes of olefins as C–C bond-forming carbon nucleophiles. It may be fair to argue that this olefin/carbocation capture chemistry regime, practiced in plant secondary metabolism, is the most abundant form of C–C bond formation in nature in terms of molecule output.

Isoprenoid chemical biology is essentially olefin functional group chemistry. The starting alcoholic-PP in the 2-IPP isomer is lost in the allyl cation-forming initial step. This is a nitrogen free arena. The only oxygen functionality comes from an occasional quenching of a cyclic cation by water or by an O_2-dependent double bond epoxidation (squalene). We will see in Chapter 12 and in Chapter 15 that in the late stages of sterol scaffold maturations, many oxygen substituents are introduced, creating new functional groups and new chemistry by oxygenases that switch to carrying out carbon radical chemistry.

11.7 Oxocarbenium Ions

Switching from isoprenoid molecular scaffolds to sugar-based metabolic chemistry brings in a different type of carbon electrophile. Sugar-based oxocarbenium ion transition states[9,10] feature prominently in the chemical biology of phosphorylated hexoses such as glucose-1-P, galactose-1-P, and mannose-1-P as well as in the key five-carbon sugar 5-P-ribose-1-PP metabolite, respectively. The hexose-1-P metabolites are all precursors to nucleoside diphosphohexoses that in turn are the proximal donors of glucosyl-, galactosyl-, and mannosyl-groups to cosubstrate nucleophiles (Figure 11.5). Nucleoside diphosphohexoses are formed by directed enzymatic attack of the hexose-1-phosphate oxyanion on Pγ of a nucleoside triphosphate cosubstrate, typically UTP for glucose-1-P and galactose-1-P, but GTP for mannose-1-P (Figure 11.5).

The $C_{1'}$ of the hexoses in UDP-glucose, UDP-galactose, GDP-mannose, and related NDP-sugar metabolites are rendered electrophilic by way of early-C_1'-O-UDP/GDP) bond cleavage in glycosyltransferase active sites. The incipient carbocationic transition state[11] is delocalized and stabilized to the point of ready accessibility by the adjacent oxygen (O_5) in the pyranose ring

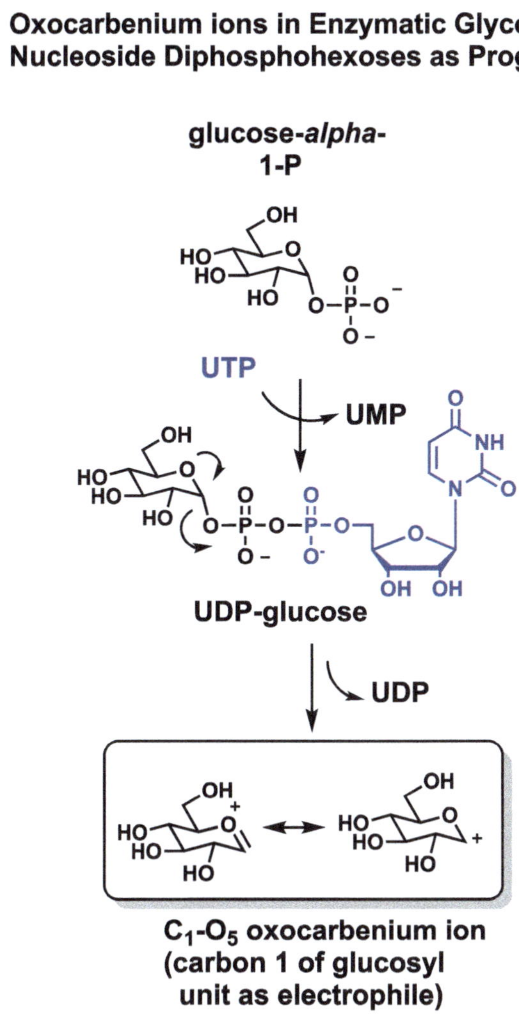

Figure 11.5 Oxocarbenium ions in glycosyl transfers: nucleoside diphophohexoses as progenitors.

structures of the NDP-hexose (Figure 11.5). Because oxygen is more basic than carbon, the resonance contributor with positive charge on oxygen may be the predominant contributor to the resonance hybrid structure but $C_{1'}$ is sufficiently electron deficient to be capturable by a range of cellular nucleophiles.

These are almost always oxygen or nitrogen nucleophiles. Figure 11.6, left hand partition, shows that the glycosyl group can be transferred from UDP-glucose as an electrophile at C_1 to the terminal 4′-OH of growing polysaccharide chains. That creates a new C_1–O_4 link as the chain is extended by one glycosyl unit. Glycosyltransferases exist that can use cellulose chains (cellulose synthases) that are repeat polymers of β-1,4-glycosyl units. There are also glycogen synthases that instead add the glucosyl unit to an α-1,4-linked glycogen chain. Both types of transferases, cellulose synthases and glycogen synthases, capture the planar C_1–O_5 oxocarbenium intermediate/transitions state with the stereochemistry inherent in the substrate chains,

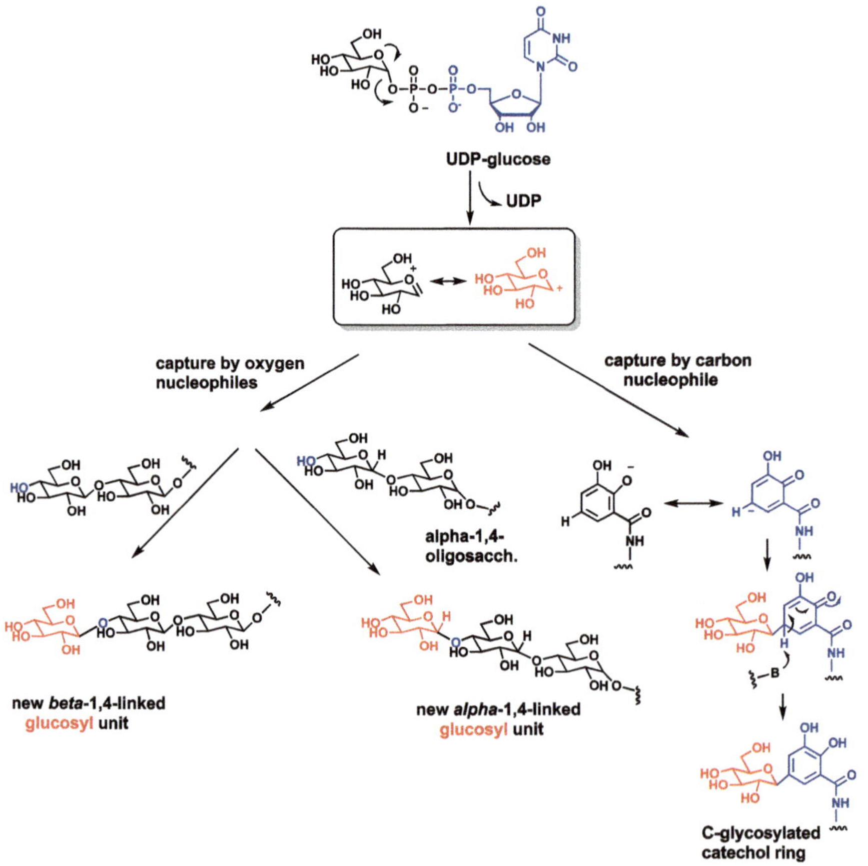

Figure 11.6 Capture of glycosyl oxocarbenium ions by oxygen *vs.* carbon nucleophiles.

β-1,4- for cellulose and α-1,4- for glycogen. When oxyanions are the nucleophiles in the enzymatic transition state, the products have acetal linkages, configurationally stable C–O–C bonds.

The right hand partition in Figure 11.6 shows a rare instance of an enzyme that makes a *C*-glucoside, using the cyclohexadienyl contributor to a catecholate oxyanion as carbon nucleophile. On re-aromatization of the glycosylated cyclohexadienone, the enzyme gives a β-*C*-glucoside. This is part of a *C*-glucosylated microbial iron chelator isolated from salmonella bacteria and is known as salmochelin.

Oxocarbenium ion transition states are also postulated in pentose metabolism as well as hexose metabolism. In the pentose case the metabolic precursor is typically not a nucleoside diphospho pentose but instead the central metabolite 5-phospho-ribose-α-1-pyrophosphate (Figure 11.7).

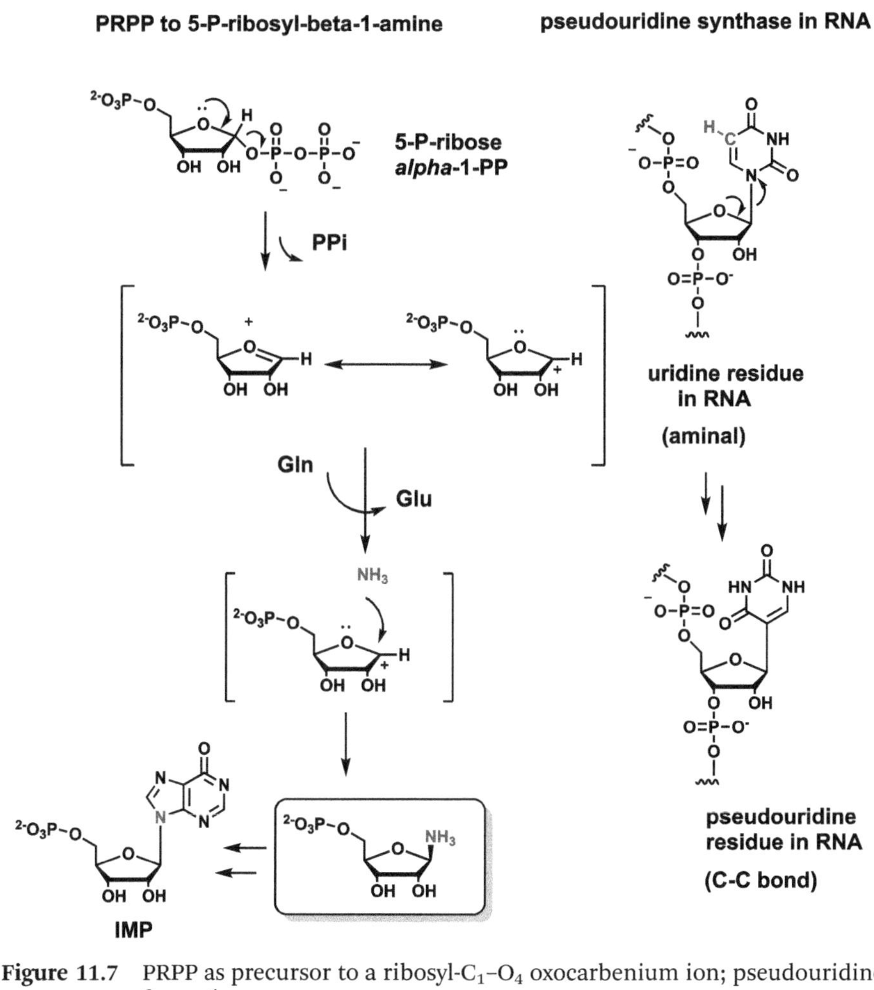

Figure 11.7 PRPP as precursor to a ribosyl-C_1–O_4 oxocarbenium ion; pseudouridine formation.

Early C_1–OPP bond cleavage generates a comparable C_1–O_4 oxocarbenium ion. The major fate of this transient ribosyl-C_1-electrophile is capture by nascent NH_3 (a nitrogen nucleophile not a carbon nucleophile) as the first step in purine biosynthesis, elaborated back in Chapter 6. The product has inverted stereochemistry at C_1 of the 5-P-ribosyl unit. It is 5-phosphoribosyl-1-β-amine.

The most well studied physiologic *C*-riboside found in RNAs from different biologic sources is pseudouridine (Figure 11.7). As shown, pseudouridine[12] synthase converts UMP (uridine) residues in RNAs into pseudouridine residues with a C–C riboside linkage to nucleobase instead of the original N–C linkage. Among the mechanistic constraints is a requirement for a reversible disconnection of the N_1–C_1 uracil-ribose bond and a reconnection as uracil C_5–C_1-ribose C–C bond. An oxocarbenium ion transition state at ribose C_1 is a possible participant in this remarkable posttranscriptional N to C migration reaction in RNAs (Figure 11.7).

11.8 Strained Small Ring Carbacycles and Heterocycles as Carbon Electrophiles

In previous chapters we have noted the formation of strained small ring carbacycles and heterocycles, among them cyclopropanes and cyclobutanes as carbacycles. Epoxides, oxetanes, and β-lactams are comparable cyclic strained heterocyclic ring systems. Due to the ring strain relief on ring opening, these various functional groups are potentially susceptible to ring-opening covalent capture by nucleophiles. The oxidosqualene cyclase reaction clearly features intramolecular capture of the squalene 2,3-β-epoxide by a cascade of internal olefins acting as carbon nucleophiles, anywhere from concerted, partially concerted, or stepwise. Analogously, the expansion of epoxides to tetrahydropyrans and tetrahydrofurans that we noted in Chapter 7 is a set of examples of intramolecular capture of the electrophilic strained epoxides by oxygen nucleophiles.

The strained beta lactam ring is a paradigm in both antibiotics and covalent modification of bacterial cell wall target enzymes when the electrophilic lactam ring is opened in protein active sites. The canonical nucleophiles are the oxyanions of serine side chains as these proteins are members of the super family of active site serine hydrolases.[13] There are no known cases where beta lactam antibiotics target carbon nucleophiles in their target enzymes.

11.9 Carbons Attached to the Sulfonium Sulfur of *S*-adenosylmethionine

We examine the role of *S*-adenosylmethionine (SAM) in the next chapter for its ability to undergo homolytic cleavage in the active sites of members of the radical SAM enzyme family. SAM is also one of the key metabolites that engage in electrophilic group transfers to a wider range of cellular nucleophiles by a two-electron reaction manifold.[14] Here, we note only that the permanent positive charge on the trivalent sulfonium cation of SAM activates all three of the carbon substituents at that trivalent sulfur for transfer to nucleophiles as net electrophilic fragments (Figure 11.8) in S_N2 transition states.

The transfer of the CH_3 substituent to cellular nucleophiles is the predominant role for SAM as cellular methyl donor. Stereochemical studies on several SAM methyltransferases using CHDT-chiral methyl group methodologies have verified inversion of configuration for the methyl group in flight, consistent with a S_N2 reaction manifold.[15] In those transfers the CH_3 group is transferred to cosubstrate nucleophiles without the electron pair in the starting CH_3–S^+ bond. We treat such reactions as akin to the transfer of a [CH_3^+] equivalent (Figure 11.8).

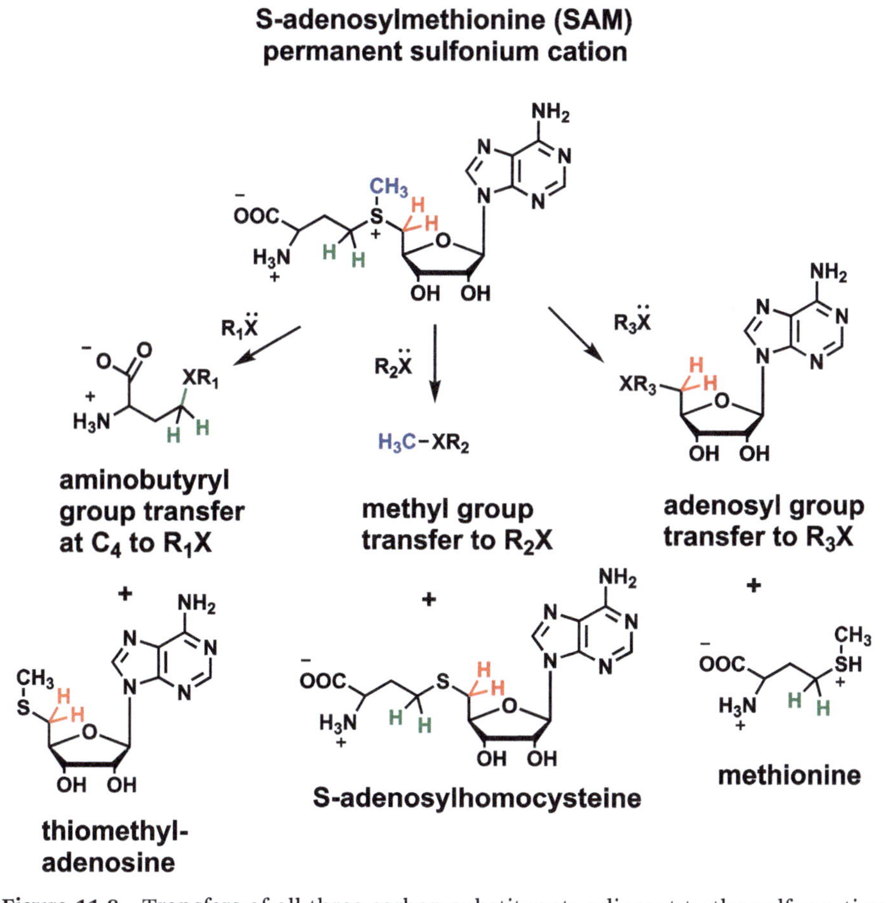

Figure 11.8 Transfers of all three carbon substituents adjacent to the sulfur cation in *S*-adenosylmethionine enzyme reactions.

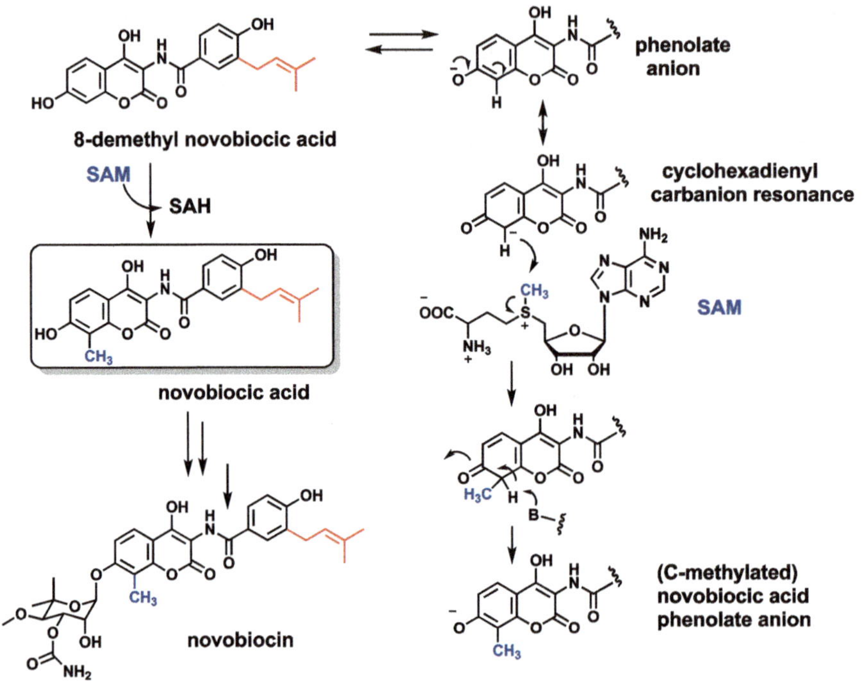

Figure 11.9 *C*-Methylation *ortho* to phenol oxygen *via* cyclohexadienone anion resonance contributor in novobiocin biosynthesis.

While oxygen, nitrogen and sulfur ground state nucleophiles are more common than carbon nucleophiles in cells (Figure 4.28),[14] Figure 11.9 shows the *C*-methylation step in novobiocin biosynthesis where the attacking carbon nucleophile is the cyclohexadienyl carbanion contributor to a phenolate anion structure.

C-Methylation occurs at C_8 adjacent to the phenol before that phenol is glycosylated. The free phenolate anion has a very minor resonance contributor that has cyclohexadienyl carbanion character at C_8. That carbanion is the carbon nucleophile attacking the methyl group on SAM. The transfer of the electrophilic CH_3 group to C_8 of the substrate quenches the positive charge on the sulfonium cation of SAM as it is converted to the neutral thioether linkage in coproduct *S*-adenosylhomocysteine (SAH). In passing, one would argue that a comparable cyclohexadienyl carbanion contributor to the phenolate anion of 4-hydroxybenzoate had acted earlier in the novobiocin biosynthetic pathway as carbon nucleophile for prenylation at carbon 3, *ortho* to the phenol.

Three examples of intramolecular transfer of the aminobutyryl side chain of SAM involve its internal capture by the carboxylate anion as oxygen nucleophile, by the deprotonated amino group acting as nitrogen nucleophile, or as carbon 2 acting as a carbon nucleophile. As depicted in Figure 11.10, these three different nucleophiles respectively create three different cyclic

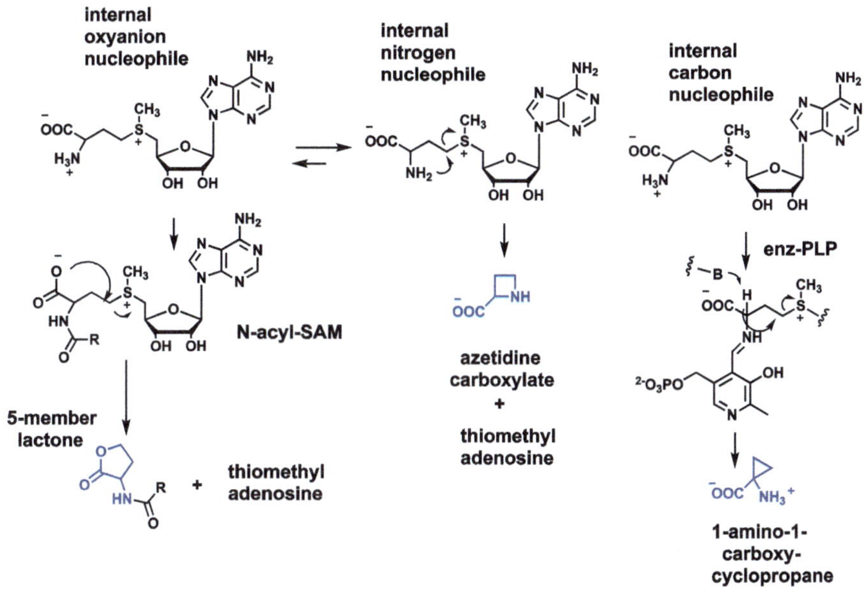

Figure 11.10 SAM can fragment internally to a five member lactone, a four member azetidine, or a three member cyclopropane ring.

products of differing ring sizes. Carboxylate capture of the electrophilic carbon 4 of the methionyl group of SAM gives the five member homoserine lactone scaffold. Capture by the amino group creates the four member azetidine heterocycle. Capture by the C_2 carbon acting a carbanion in a SAM = PLP aldimine adduct instead yields the three member carbacyclic cyclopropane ring of amino-carboxy-cyclopropane. This nested series of three, four, and five member rings (carbacycles and heterocycles) shows the remarkable features of SAM with three potential nucleophiles built into to its methionyl side chain in the presence of an electrophilic carbon adjacent to the sulfonium cation group.

11.10 Conjugated Olefins as Electrophiles

While the π-olefins in C=C double bonds can act as nucleophiles in C–C bond formations, in contexts noted in the preceding chapter, such as in oxidosqualene cyclizations, when olefins are conjugated to electron sinks such as aldehydes, ketone, acyl thioesters and even free carboxylates (*e.g.* fumarate), they can instead behave as electrophiles at the olefinic terminus furthest from the electron sink (Figure 11.11). In organic chemistry these are termed conjugate addition or Michael reactions.[16] We noted in Chapter 2 that this reactivity is observed in two common metabolites: α,β (= 2,3) enoyl thioesters in fatty acid biosynthesis (Figure 2.3), and in the citric acid cycle step of fumarate hydration to L-malate (Figure 2.29). In fatty acid biosynthesis, the 2,3-enoyl-thioesters are subjected to hydride addition at C_3 (hydrogen

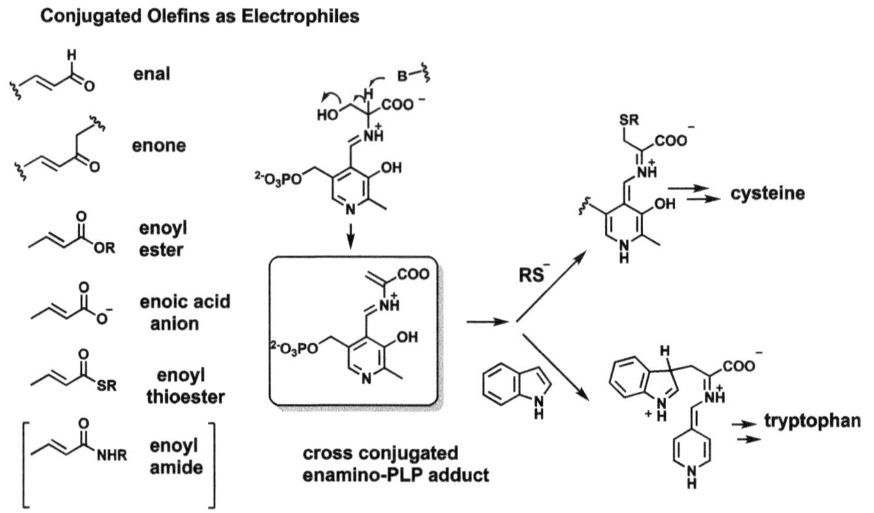

Figure 11.11 Conjugated olefins as electrophiles.

transferred from NADPH as nucleophile) (Figure 2.3), while in the citrate cycle conversion of fumarate to L-malate, an OH^- equivalent from water is the incoming nucleophile.

Conjugated enoyl-imines as electrophiles can be generated as intermediates by enzymes that require pyridoxal-P to metabolize β- or γ-substituted amino acids for either substitution or elimination of the β-substituent (see Chapters 9 and 10). The reaction paths start with Cα–H abstraction as a proton from initial amino acid = PLP aldimines to give the delocalized C_2 carbanion. Then, that electron lone pair can displace certain β-substituents (*e.g.* the –OH group of threonine). Inspection of the resultant enamino-PLP adducts reveal that the pyridine aldehyde component functions as an electron sink for the Cα, Cβ-olefin, polarizing Cβ (C_3) to act as a carbon electrophile.[17]

Figure 11.11 schematizes the addition of a thiolate nucleophile (as in cysteine synthase or cystathionine beta synthase of Chapter 4). This figure also shows the capture of that electrophilic enamino-PLP adduct by indole acting as a carbon nucleophile to form a new C_3–C_4 carbon–carbon bond (note that a free enamine would act as a nucleophile, not as electrophile, at C_3; it is the PLP as electron sink at C_2 that reverses olefinic terminus polarity). Intriguingly, tryptophan synthase couples a nucleophilic enamine (indole) to an electrophilic enamine (PLP adduct) for C–C bond formation.

11.11 The Interplay of Phosphorylation, C–O Bond Cleavages, and the Generation of Carbon Electrophiles

Organisms utilize the oxygens in a variety of carbon-based alcohol or carboxylate anion groups to act as nucleophiles towards any of the three electrophilic side chain phosphorus atoms of ATP.[1] The phosphoryl transfers result in phosphate esters from attack at Pγ, pyrophosphate esters from attack at Pβ (rare), and AMP phosphodiesters by attack at Pα of ATP. Phosphorylations or pyrophosphorylations can have the net effect of labilizing the C–O bonds for elimination. Prior to phosphorylation, C–OH bond cleavage in alcohols, hemiacetals, hemiaminals have high energy barriers (pK_a of the leaving group water ~14). Cleavage of C–O–PO_2^-R generates a resonance-stabilized inorganic phosphate or pyrophosphate groups ($pK_a \approx 6$–7), much more stable anions at physiologic pH values, with lower energy barrier for C–O bond cleavage.

Three enzymatic examples of C–O–P cleavages reveal this logic as they set up three types of electrophilic chemical groups[1] (Figure 11.12). The first case is the familiar C_1–OPP bond cleavage of the alcohol–phosphodiester bond in Δ^2-prenyl diphosphates. Low energy C_1–O bond cleavage releases that oxygen as part of the pyrophosphate leaving group *en route* to the electrophilic allyl cation, as we have noted many times in previous chapters.

The second and third cases generate C_1-oxocarbenioum ions in both the pentose and hexose metabolite series. For ribose the key metabolite is 5-P-ribose-1-PP (Figure 11.11). The phosphoryl group at the 5-OH derives ultimately from decarboxylation of a 6-P-gluconate metabolite. The pyrophosphate at C_1 of ribose arises by a direct pyrophosphoryl group transfer from ATP to the attacking C_1-α-OH of the hemiacetal form of 5-P-ribose.

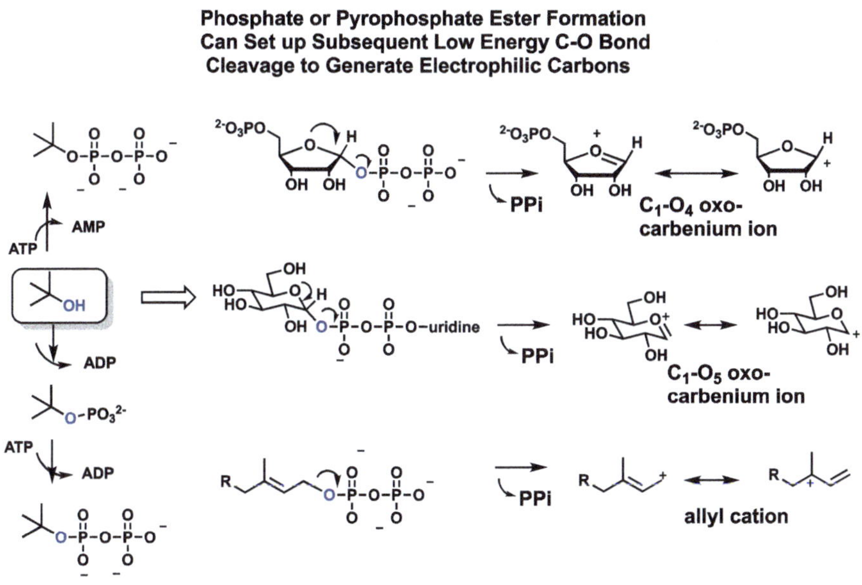

Figure 11.12 Phosphorylation or pyrophosphorylation can set up low energy pathways to C–O bond cleavage generating electrophilic carbocation equivalents.

The cognate 1-pyrophosphate ester in glucose metabolism is UDP-glucose. This is a substituted pyrophosphate leaving group but otherwise tracks the PRPP activation logic to facilitate C_1–O bond cleavage. Early dissociation of the C_1–OPP bonds in both the ribose and glucose series have lower energy since the leaving oxygen is not free $-OH^-$ but instead part of the resonance stabilized pyrophosphate polyanion. The electrons in the C_1–O bond depart with the pyrophosphate, creating the oxocarbenium ion centered at C in both the cyclic ribose and cyclic glucose scaffolds. We have noted the different fates for the ribosyl oxocarbenium ion as the starting point for purine ring biosynthesis (Chapter 6), while glucosyl transfers are predominantly to growing oligosaccharide and polysaccharide chains (Figure 11.5).

Addition of just one phosphoryl group to an alcohol to a phosphate monoester, rather than net pyrophosphorylation, can also achieve net C–O bond cleavage and generation of a carbon electrophile. This is the strategy for microbial posttranslational modification of serine or threonine alcohol side chains in some protein targets.[18] The modifications occur in two steps. First, the alcohol is converted to the phosphate ester in an ATP-utilizing phosphoryl transfer (Figure 11.13). Then, elimination of a C_2–H and the C_3-phosphoryl group follows (to the olefin in dehydroalanyl (from serine) or dehydrobutyrine (from threonine) residues).

These newly formed enamide C=C double bonds (Figure 11.13) are electrophilic by virtue of conjugation to the backbone amide carbonyl group. The double bonds may persist but in most cases are captured by the thiolate anion side chains of cysteine residues close by in primary sequence or in folding space. The crosslinked structures contain stable, nonreducible thioether bonds linking cysteine to the electrophilic C_3 amino acid (dehydroalanine *via* serine) or cysteine to the C_4 electrophile (dehydrobutyrine from threonine).[18] These thioether-containing double headed amino acids are termed lanthionine and methyllanthionine respectively. Unlike reducible disulfide bonds (Chapter 4), thioether bonds are stable to redox conditions, acting as conformational constraints in the crosslinked mature peptides.

11.12 Carbon Dioxide as Electrophile

From an animal as opposed to a plant perspective, the major flux of organic carbon is oxidation to CO_2 as the end product metabolite from which all metabolically capturable electrons have been removed. On the other hand, this one-carbon metabolite is perhaps the simplest organic electrophile that participates in metabolism. In Chapter 3 on C–O bonds, we noted several different strategies for decarboxylases to generate CO_2 as well as some complementary ones for CO_2 fixation back into organic scaffolds. This analysis featured the strategy that plants employ to fix 250 billion tonnes a year globally by the catalytic action of ribulose-bis-phosphate carboxylase (Figure 3.35). The requisite carbanion, accessible because of the 2-keto group in the ribulose framework, arises from the 2,3-enediolate anion.

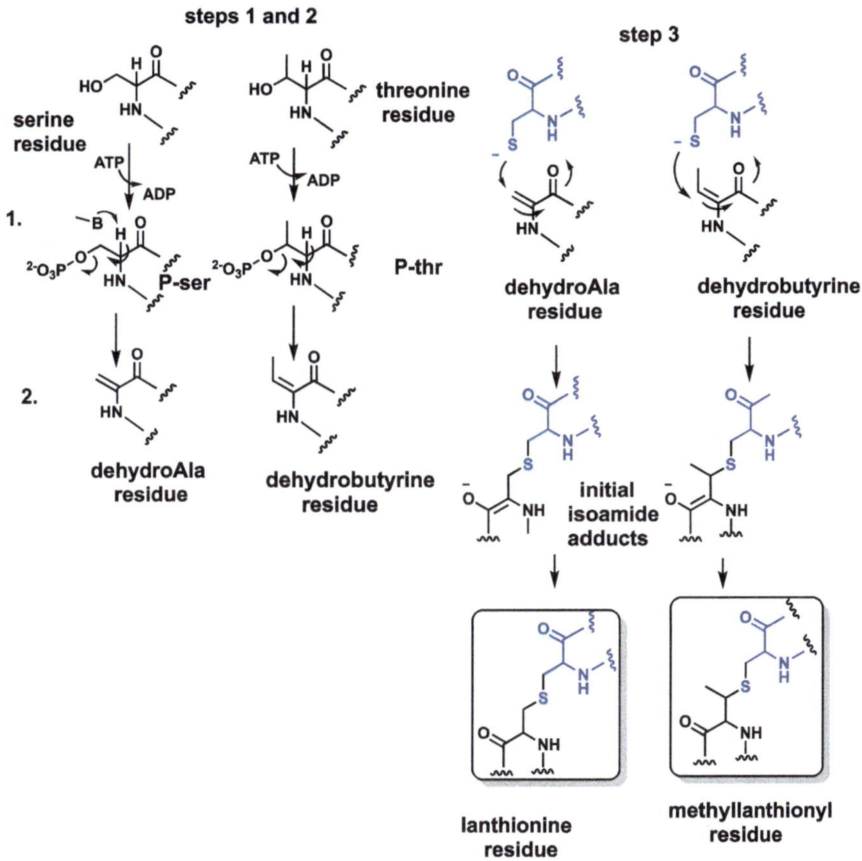

Figure 11.13 Phosphorylation of side chain ser or thr residues can lead to elimination to electrophilic dehydroalanine and dehydrobutyrine side chains, capturable by neighboring cysteine thiolate anions to give the thioether bonds in lanthionine and methyllanthionine residues.

CO_2 is preconcentrated in the enzyme active site by equilibration as a carbamate adduct with an active site lysine. Analogously, CO_2 acts as electrophilic cosubstrate for C–C bond formation to carbon 4 of glutamyl residues in a family of proteases (Figure 9.36) involved in blood coagulation.

11.13 Overview of Accessible Carbon Electrophiles

For organisms to both build and break carbon–carbon bonds in the framework of tens of thousands of metabolites, carbon nucleophiles and carbon electrophiles need to be generated when the reaction pathways are heterolytic. Organisms have evolved an itinerary of largely stable, spin-paired organic metabolites in the C_1 to C_6 range that serve as both building blocks and chemical intersection points for the transformation of one-carbon-based functional group to another.

There is an equal need for carbon electrophiles to be paired up with carbon nucleophiles in the active sites of a cell's enzyme catalysts. In this chapter we summarized the stable electrophilic carbon groups that circulate as diffusible metabolites compared to the functional groups (methyl-, prenyl-, glycosyl-) that move as fleeting enzyme-bound intermediates between nucleophilic acceptors (Figure 11.1). The carbonyl groups at the oxidation state of aldehydes and ketones are stable enough to exist in dozens of primary metabolites but reactive enough to be subjected to nucleophilic additions in the active sites of various C–C bond-forming enzyme catalysts. The aldehydes are more reactive than ketones. That propensity to nucleophilic addition may show up as measurable amounts in reversible equilibria as hydrates, 1,1-diol adducts from water addition to the aldehyde carbonyls, noted back in Chapter 3.

Ketones in primary metabolites do not accumulate appreciably as tetrahedral water addition products when free in neutral aqueous solutions. When there are intramolecular hydroxyl groups four or five carbons away from the carbonyl groups, as in fructose and glucose open chain forms, the cyclization to strain free hemiacetals and hemiketals may be dramatically favored, reflecting both the propensity of the aldehyde and ketone carbonyls to react as electrophiles and the stability of the strain-free furanose and pyranose hemiacetal rings. Aldolase enzymes catalyze often reversible condensation and cleavage of electrophilic aldehydes with enolate/enamine carbanions.

The second major electrophilic carbonyl group in enzymatic C–C bond formations are thioester carbonyls. Diffusible acyl-*S*-coenzyme A thioesters constitute the major acylation inventory of cells. They transfer the electrophilic acyl groups to cosubstrate nucleophiles, including to carbanion equivalents. This is the province of thio-Claisen condensation enzyme catalysts in fatty acid synthesis and oxidation, in polyketide chain assemblies in microorganisms and plants, and in the key first steps in both the citric acid cycle and the pathway to 3-ketobutyryl CoA for steroid biosynthesis.

The other three electrophilic common carbon group transfers are alkyl transfers from *S*-adenosylmethionine, allyl cation equivalent transfers from Δ^2-prenyl diphosophates, and transfers of $C_{1'}$ glycosyl and ribosyl oxocarbenium ion equivalents in both polysaccharide and nucleic acid biosynthetic routes. These groups may not form frank carbocations with measurable independent lifetimes. Describing them as electrophilic group transfers speaks to the polarity of the transition states as they are captured by nucleophiles. For instance, the enzymatic methyl transfers from *S*-adenosylmethionine proceed with inversion of stereochemistry when chiral methyl groups are used, consistent with an S_N2 processes. Yet, the methyl groups in flight leave the pair of electrons behind on the residual carbon center. We posit the central metabolic theme that electrophilic group transfer from a small set of central metabolites to nucleophilic cosubstrates is the dominant chemical logic that undergirds many of the metabolically central chemical transformations in cells. We now turn to the chemical logic and enzymatic strategies by which cells and organisms generate transient radicals in C–H and C–C bond homolytic reaction manifolds.

References

1. C. T. Walsh, *The Chemical Biology of Phosphorus*, Royal Society of Chemistry, London, 2021.
2. C. T. Walsh, B. P. Tu and Y. Tang, Eight kinetically stable but thermodynamically activated molecules that power cel metabolism, *Chem. Rev.*, 2018, **118**, 1460–1494.
3. C. T. Walsh, *The Chemical Biology of Sulfur*, Royal Society of Chemistry, London, 2020.
4. W. P. Jencks, *Catalysis in Chemistry and Enzymology*, McGraw Hill, 1969.
5. A. Erkkila, I. Majander and P. M. Pihko, Iminium catalysis, *Chem. Rev.*, 2007, **107**, 5416–5470.
6. V. M. Dong, D. Fiedler, B. Carl, R. G. Bergman and K. N. Raymond, Molecular recognition and stabilization of iminium ions in water, *J. Am. Chem. Soc.*, 2006, **128**(45), 14464–14465.
7. C. A. Lesburg, J. M. Caruthers, C. M. Paschall and D. W. Christianson, Managing and manipulating carbocations in biology: terpenoid cyclase structure and mechanism, *Curr. Opin. Struct. Biol.*, 1998, **8**(6), 695–703.
8. D. W. Christianson, Structure and chemical biology of terpenoid cyclases, *Chem. Rev.*, 2017, **117**, 1157–11648.
9. H. Elferink, M. E. Severijnen, J. Martens, R. A. Mensink, G. Berden, J. Oomens, F. Rutjes, A. M. Rijs and T. J. Boltje, Direct experimental characterization of glycosyl cations by infrared ion spectroscopy, *J. Am. Chem. Soc.*, 2018, **140**, 6034–6038.
10. A. Francenetti, A. Ardá, J. L. Asensio, Y. Bleriot, S. Thibaudeau and J. Jimenz-Barbero, Glycosyl oxocarbenium ions: structure, conformation, reactivity, and interactions, *Acc. Chem. Res.*, 2021, **54**, 2552–2564.
11. H. Elferink, M. E. Severijnen, J. Martens, R. A. Mensink, G. Berden and J. Oomens, *et al.*, Direct Experimental Characterization of Glycosyl Cations by Infrared Ion Spectroscopy, *J. Am. Chem. Soc.*, 2018, **140**(19), 6034–6038.
12. T. Hamma and A. R. Ferre-D'Amare, Pseudouridine synthases, *Cell Chem. Biol.*, 2006, **13**, 1125–1135.
13. E. Sauvage, F. Kerff, M. Terrak, J. A. Ayala and P. Charlier, The penicillin-binding proteins: structure and role in peptidoglycan biosynthesis, *FEMS Microbiol. Rev.*, 2008, **32**, 234–258.
14. M. Fontecave, M. Atta and E. Mulliez, S-adenosylmethionine: nothing goes to waste, *Trends Biochem. Sci.*, 2004, **29**, 243–249.
15. D. O'Hagan and J. W. Schmidberger, Enzymes that catalyse SN2 reaction mechanisms, *Nat. Prod. Rep.*, 2010, **27**(6), 900–918.
16. J. Clayden, N. Greeves and S. Warren, *Organic Chemistry*, Oxford University Press, Oxford, UK, 2nd edn, 2012.
17. R. A. John, Pyridoxal-P dependent enzymes, *Biochim. Biophys. Acta*, 1995, **1248**, 81–95.
18. L. M. Repka, J. R. Chekan, S. K. Nair and W. van der Donk, Mechanistic understanding of lanthipeptide biosynthetic enzymes, *Chem. Rev.*, 2017, **117**, 5457–5520.

CHAPTER 12

Carbon Radicals

12.1 Biologic Carbon Radical Generators

The previous two chapters have analyzed how organisms use organic molecules with functional groups that enable low energy routes to carbon nucleophiles and carbon electrophiles for C–C bond formations/cleavages in metabolic pathways. Carbanion equivalents as enolates or thioester enolates are generated transiently in enzyme active sites and used to attack electrophilic carbonyl carbons including aldehydes, ketone, oxoesters, thioesters, protonated carboxylic acids, and CO_2. Alternatively, in tens of thousands of isoprenoid examples allyl carbocations are generated from Δ^2-prenyl diphosphates. Enzymes pair them with olefinic groups acting as nucleophilic C–C bond-forming partners, both intermolecularly (for prenyl chain elongations) and intramolecularly in cascades of cyclizations and cation-driven rearrangements (see Figures 10.8 and 11.1).

However, there are hundreds to likely thousands of metabolic transformations, many in conditional metabolic pathways in plants and microbes, where the carbon centers undergoing reaction are devoid of obvious functional groups that can be processed readily to carbon nucleophiles or electrophiles. These can be terminal $-CH_3$ groups in alkyl chains or $-CH_2-$, or –CH– groups (Figure 12.1), far from any of the kinds of C–O, C–S, or C–N activating functional groups noted in Chapters 3–5.

These unactivated carbon centers are the typical sites of carbon-centered radical reactions. Precisely because the reactivity of these carbons is intrinsically low by lack of influence of any nearby activating functionality, they require strong oxidants to break any of the C–H bonds. Those strong one electron biologic oxidants come in two flavors, one anaerobic, one aerobic. Both catalyze homolytic cleavages of unactivated C–H bonds, yielding hydrogen atoms ($H^•$) and transient carbon radicals (Figure 12.1).

Chemical Biology No. 21
The Chemical Biology of Carbon
By Christopher T. Walsh

Published by the Royal Society of Chemistry, www.rsc.org

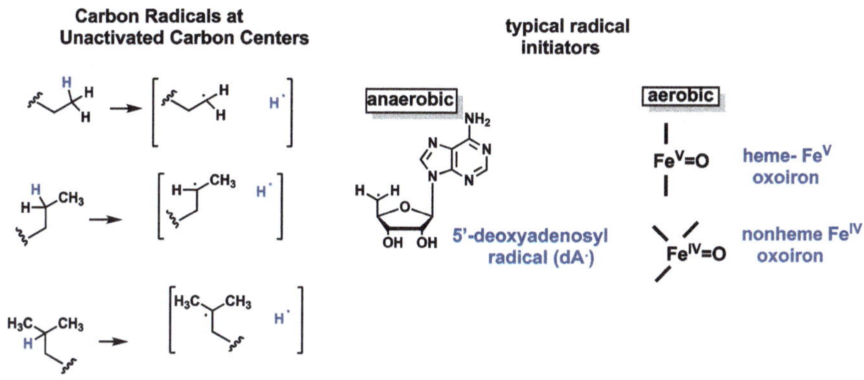

Figure 12.1 Carbon radicals at unactivated carbon centers.

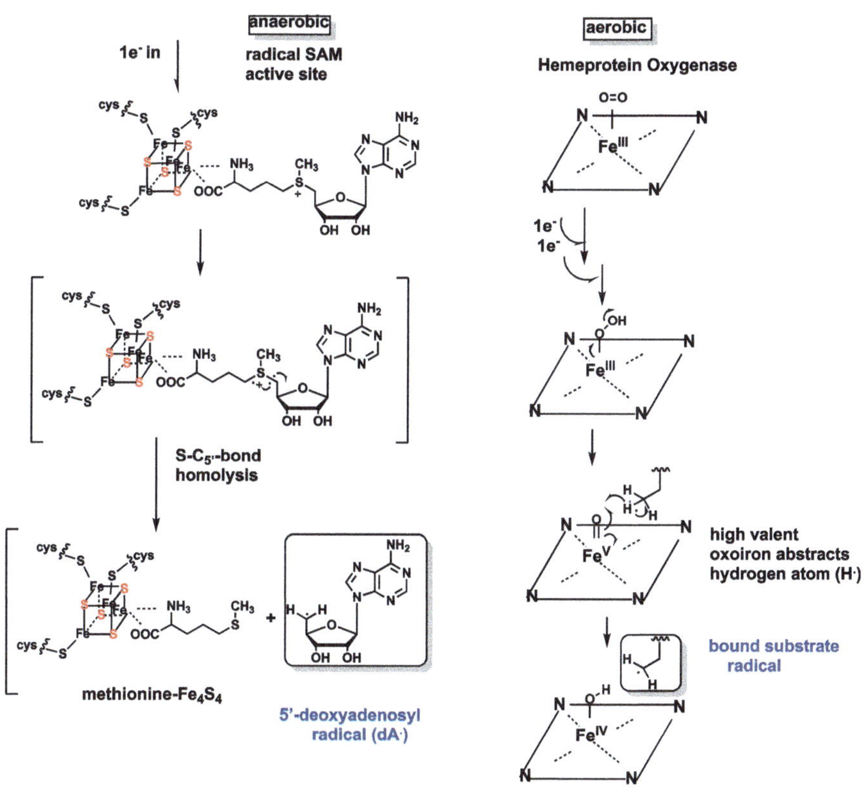

Figure 12.2 Radicals from anaerobic radical SAM enzymes and aerobic oxygenases.

The anaerobic reactions are largely initiated by nascent 5′-deoxyadenosyl radicals[1] transiently generated in SAM-cleaving enzyme active sites. The aerobic radical chemistry at unactivated carbon centers is typically propagated by cosubstrate O_2, processed to high valent oxoiron species that act as powerful one electron oxidants[2] (Figure 12.2).

12.2 Anaerobic Generation of 5′-Deoxyadenosyl Radical as Initiator of Substrate Carbon Radical Chemistry

The biologic one-electron oxidant generated in the active site of enzymes in anaerobic microenvironments is the 5′-deoxyadenosyl radical.[1] It is a strong enough oxidant to abstract a hydrogen atom from a C–H bond in a bound organic cosubstrate, resulting in the formation of the stable bound intermediate 5′-deoxyadenosine and the cosubstrate carbon radical[3] (Figure 12.3). Such carbon radicals, often high energy species, may last as little as one molecular vibration, undergoing a variety of one-electron reaction manifolds.

12.2.1 Adenosyl B12 as Progenitor of Bound 5′-Deoxyadenosyl Radicals

Organisms have evolved two vastly different chemical progenitors[4] to the 5′-deoxyadenosyl radical (dA•). One such metabolic precursor is the covalent adenosyl form of coenzyme B12[5] (Figure 9.34 and Figure 12.4). Adenosyl B12

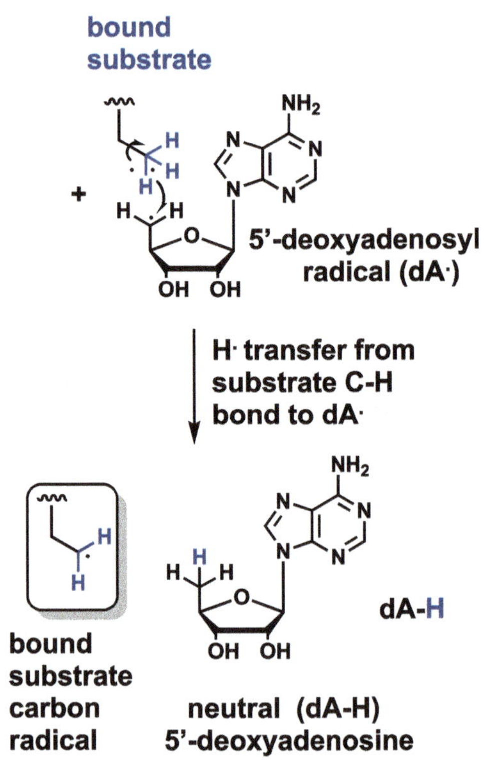

Figure 12.3 5′-Deoxyadenosyl radicals scavenge hydrogen atoms from bound adjacent C–H bonds of substrates to initiate radical catalysis.

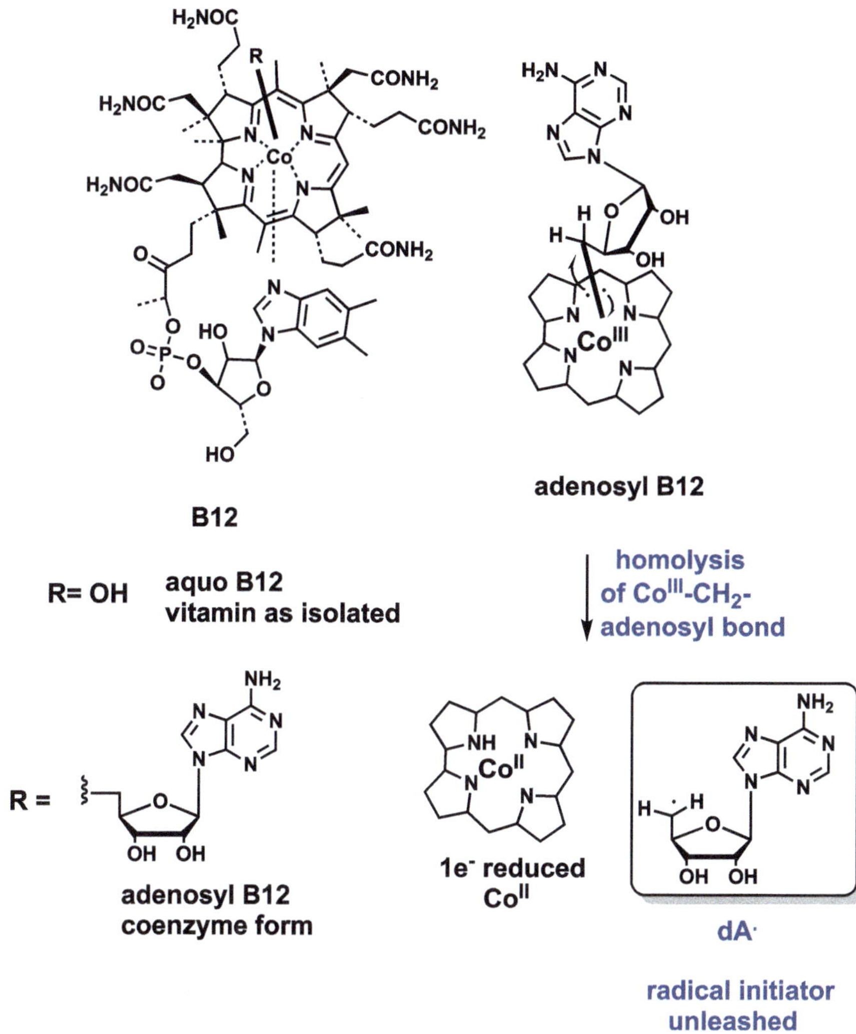

Figure 12.4 Homolysis of adenosyl-B12 generates Co^{II}-B12 radical and 5′-deoxy-adensoyl radical as initiator of substrate radical catalysis.

is a bacterial metabolite generated enzymatically from vitamin B12 in an unusual enzymatic adenosyl transfer reaction. ATP is attacked at the $C_{5'}$–CH_2 group of the ribose ring, by the cobalt (Co^I) ion sitting in the equatorial cavity of the corrinoid tetrapyrrole macrocycle. Inorganic triphosphate is split out from ATP (Chapter 9). To get to the supernucleophilic oxidation stable Co^I form of vitamin B12, the air stable Co^{III} form is first reductively converted to the Co^I form by an anaerobic activase enzyme system. Only in that reduced Co^I oxidation state is vitamin B12 able to form the rare organometallic metabolite adenosyl B12 (see Chapter 9).

In about two dozen known enzymes, the carbon–cobalt bond in adenosyl B12 is the immediate progenitor to the 5′-deoxyradical after a cosubstrate is bound. This occurs by enzyme-induced homolysis of the Co^{III}–CH_2 bond, leaving one electron from that bond at cobalt, reducing it from Co^{III} to Co^{II}. The other electron from that Co–CH_2 bond departs with the adenosyl moiety as 5′-deoxyadenosyl carbon-centered radical (dA•), ready to abstract a hydrogen atom (H•) from a cosubstrate C–H bond (Figure 12.4). Thus, dA• is viewed as a transiently formed radical initiator of carbon radicals in bound cosubstrates. The energetics of quenching dA• to dA-H as a substrate C–H goes to radical C• must drive the initial hydrogen atom transfer.[4]

The fate of the transient carbon radicals in bound substrates depends on the structural features of those specific cosubstrates, both in the small cadre of adenosyl B12 enzymes and the much larger radical SAM family discussed next. In Chapter 9 we discussed the sole human adenosyl-B12 requiring enzyme methylmalonyl CoA mutase mediating a carbon skeletal rearrangement of the four carbon branched acyl chain in L-methylmalonyl CoA to the citric acid cycle core metabolite succinyl CoA[6] (Figure 9.35).

Another example is the bacterial enzyme glutamate mutase, interconverting L-glutamate and β-glutamate.[7] The amine group has migrated to carbon 3 while one of the two methylene hydrogens at C_3 of glutamate has migrated to carbon 2 of the β-glutamate (Figure 12.5).

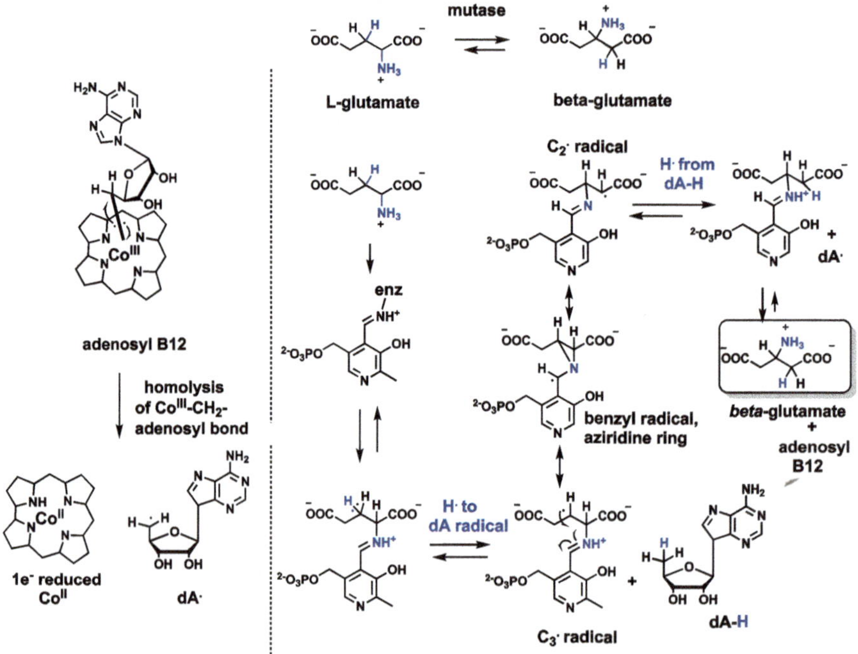

Figure 12.5 Radical intermediates in the adenosyl-B12- and PLP-dependent glutamate mutase reaction cycle.

The enzyme requires both pyridoxal-P and adenosyl-B12 as cofactors. The pyridoxal serves to form the typical aldiminium adduct with the amino group of substrate glutamate, or β-glutamate as the reverse reaction also proceeds. During a catalytic cycle PLP acts as an anchor to hold onto the amino group while substrate C–N bond formation and cleavage is occurring. Once the glutamate-PLP aldimine is in place, homolysis of B12 creates the Co^{II} corrin and the 5′-deoxyadenosyl radical. That radical initiator cleaves a specific C_3–H bond of glutamyl-PLP as a hydrogen atom. That hydrogen atom transfer quenches the dA• radical as dA-H, while generating the C_3 carbon radical of the glutamyl-PLP aldimine. One route forward would be capture of one of the two electrons in the adjacent aldimine C=N bond. This would form a new C_3–N bond, creating a transient aziridine ring and a benzyl radical from the PLP coenzyme. That cyclic three member amine can have radical density at C_3 or C_2, as depicted in Figure 12.5. The C_2 radical can abstract a hydrogen atom back from dA-H. This step creates the product β-glutamyl=PLP aldimine and regenerates dA• for coupling back to the Co^{II}-corrin to regenerate starting adenosyl cobalamin. Release of β-glutamate from aldimine linkage completes a radical-based catalytic cycle.[7]

Note that for adenosyl B12 enzymes to function for many catalytic turnovers the organometallic bond in adenosyl B12 must be fragmented reversibly. That organometallic bond must be regenerated each time. In turn the generation of dA•, and its action as a radical generator during conversion to dA-H, must also be reversible. A bound coproduct radical must be able to break one of the 5′-methyl C–H bonds to recreate dA• as a prelude to its one-electron recombination with Co^{II}-B12. By the way, the return of a hydrogen atom from the $C_{5'}$–CH_3 group in dA-H to product radical to reform dA• has a statistical probability of one third for returning the actual hydrogen that was initially transferred to dA•. This probability arises from the fact that the three C–H bonds in a methyl group are equivalent and the C–CH_3 bonds in substrates, even when enzyme bound, tend to have free rotation around the C–C bond axis. Thus, the methyl group becomes functionally torsiosymmetric and the three hydrogens cannot be distinguished. This holds for the many enzymes below where the equivalent dA• radical is generated from one electron homolysis of SAM rather than homolytic fragmentation of adenosyl B12.

All in all, a baroque and unique combination of organometallic chemical biology with transient high energy primary carbon radicals that rearrange to products while regenerating the organometallic cofactor.[7] The outcome from the other progenitor to 5′-deoxyadenosyl radicals, *S*-adenosylmethionine, can be more variable with regard to the fate of dA-H.

12.2.2 The Large Superfamily of Radical SAM Enzymes: An Alternate Route to dA• Radical Initiators

Marvelous as the organometallic adenosyl-B12 is in its role as a 5′deoxadenosyl radical generator, it is restricted to only about two dozen known rearrangement enzyme reactions. By contrast, *S*-adenosylmethionine (SAM) is so widely used in biology[3] that its usage lags behind only ATP, the paradigm of thermodynamically activated, kinetic stable cellular energy currency. We noted the bioconstruction of the remarkable trivalent sulfonium cation in the inventory of C–S bond functional group arrays in Chapter 4. We also discussed the two electron pathways for the three different carbon substituents (methyl, aminobutyryl, adenosyl) to the sulfonium cation to act as electrophiles in enzyme catalyzed group transfer reactions in Chapter 1 and in Chapter 11. For the first half century after its initial characterization in the 1950s as the prime cellular methylating alkylation reagent, SAM was analyzed exhaustively for its two-electron S_N2 alkylation reaction manifolds.[8]

However, from 2001 on,[9] the reactivity of SAM in an orthogonal one-electron, radical mode has mushroomed to some 700 000 currently predicted radical SAM enzymes.[3] Almost all these predicted radical catalysts are proposed to be progenitors of the 5′-deoxyadenosyl radical (dA•), the identical radical initiator discussed above in adenosyl B12 radical chemistry (Figure 12.6). It is anticipated that members of the radical SAM superfamily will likewise cleave bound substrate C–H bonds homolytically[10] to form dA-H and initiate a variety of carbon-radical chemistries at otherwise unactivated carbon sites of substrates.[3]

We turn to the mode of SAM one-electron homolytic cleavage below. We note in passing the question of whether adenosyl B12 or SAM would be favored as generators of deoxyadenosyl radicals. it appears that SAM is ahead in the evolutionary competition, with a ratio of ~700 000 extant enzymes to ~20–30 for adenosyl B12.[4] That could reflect the complex biosynthetic pathway to the B12 tetrapyrrolic macrocycle, its sensitivity to light and air, the aerobic stability of SAM and its two in one ability to serve in both two-electron (methyl group) and one electron (dA•) reaction modes. In the radical SAM reaction mode, those enzymes contain an air-sensitive Fe_4S_4 redox cluster so they are typically inactivated by air.

12.2.3 Multiple Fates of Carbon Radicals in Radical SAM Enzyme Active Sites

12.2.3.1 Generation of dA• as an Initiator

All radical SAM enzymes, both those characterized experimentally[4] and those as yet predicted bioinformatically,[3] (will) contain an Fe_4S_4 obligate one-electron redox active center. Three cysteine side chains from the enzyme provide thiolate ligands to three of the four iron atoms. The fourth iron atom instead has the amine and carboxylate of bound SAM as ligands (Figure 12.2). On one electron input from an external reductant, the Fe_4S_4 cluster is transiently reduced as it then passes one electron to the sulfonium cation of the ligated SAM molecule, inducing homolytic cleavage at the $-S^+-CH_2$-adenosyl bond of SAM in the active sites of radical SAM enzymes (Figure 12.2).

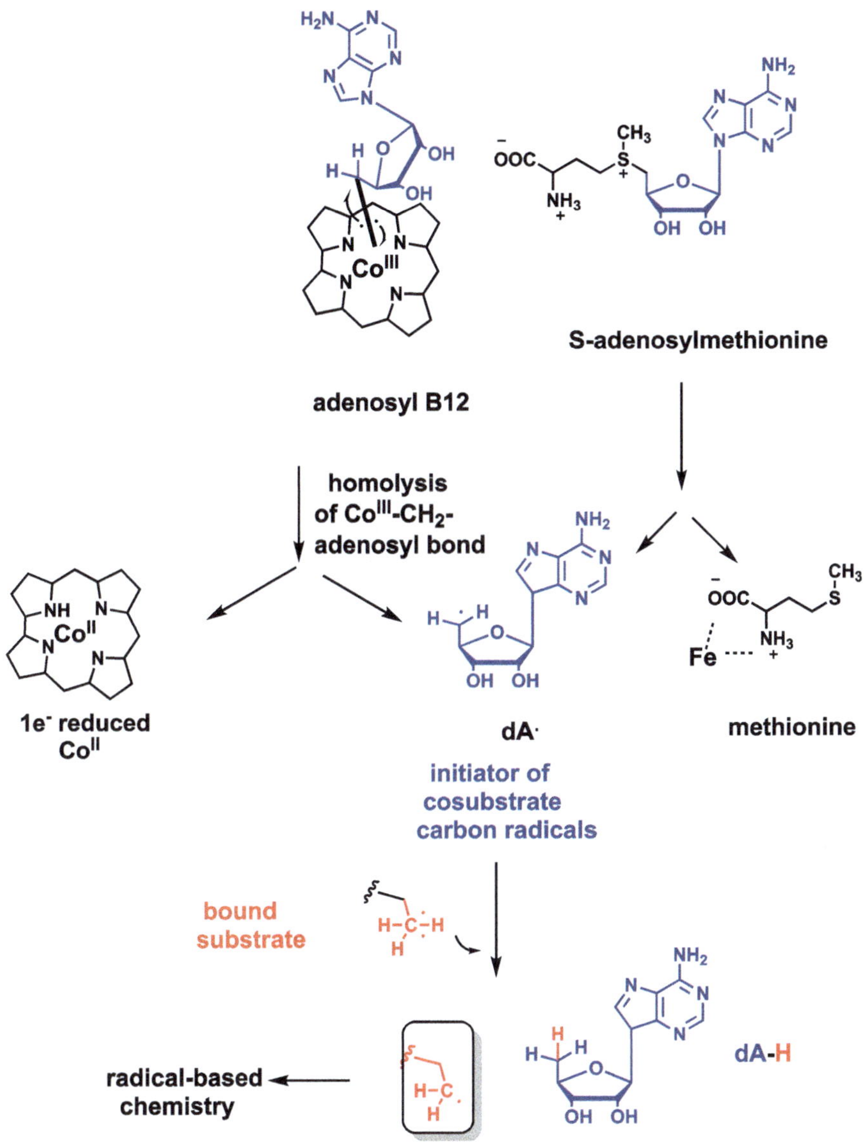

Figure 12.6 Homolysis of both adenosyl-B12 and the Fe_4S_4-S-adenosylmethionine complexes lead convergently to the 5′-deoxyadenosyl radical as initiator of carbon radical intermediates.

One electron stays with the sulfur atom and joins up with the electron coming from the FeS cluster to give coordinated methionine. The other electron remains at C_5' of the liberated adenosyl fragment as dA•. In keeping with the lessons from adenosyl B12 homolysis, the nascent dA• then acts as radical initiator by abstraction of a hydrogen atom from a nearby cosubstrate's unactivated C–H bond (Figure 12.6). The fate of the cosubstrate carbon radical and of the starting SAM molecule differs depending on the electronic demands of the specific cosubstrate and the particular catalytic properties of a given radical SAM enzyme.[3]

12.2.3.2 *Lysine to β-Lysine: SAM Cleavage is Occult*

One outcome is found in the reaction catalyzed by the microbial lysine-2,3-aminomutase, interconverting L-lysine and *S*-β-lysine, a net movement of the amine group between carbons 2 and 3 of the lysine scaffold.[11] This is a PLP- and Fe_4S_4-dependent radical SAM enzyme (Figure 12.7).

Otherwise, the chemical logic is fully parallel to the glutamate mutase strategy just analyzed in Figure 12.5, once the 5′-deoxyadenosyl radical has been generated *in situ* from SAM (Figure 12.6). The L-lysine substrate forms the typical aldimine with bound PLP. The dA• generated on one-electron-mediated homolysis of SAM, abstracts a hydrogen atom from C_2 of the Lys-PLP aldimine. That substrate radical with the amine at C_2 can interconvert through an aziridine ring with a benzyl radical to the corresponding carbon radical at C_3. Then, back abstraction of one of the three C–H hydrogens as a hydrogen atom from the C_5–CH_3 of dA-H completes formation of the β-Lys-PLP aldimine. The reversible isomerization of substrate to product has occurred with re-formation of the dA• radical *in situ.* Recombination with the methionine ligated to the Fe_4S_4 must require exit of the initially added electron back to the external reductant.

This reaction path, regenerating SAM in every turnover, resembles the adenosyl B12-dependent rearrangement enzymes that also regenerate the starting Co^{III}–carbon bond in each catalytic cycle. The reversible cleavage of SAM does not show up in the reaction stoichiometry. SAM is used catalytically. As is adenosyl B12. By contrast, most radical SAM enzymes studied for reaction stoichiometry consume one or two SAM molecules irreversibly in each catalytic cycle as exemplified in the next subsections.

12.2.3.3 *Net Cleavage of SAM to dAH and Methionine*

One way to categorize radical SAM enzymes is by the stoichiometry of SAM consumed during a reaction (Figure 12.8). The lysine aminomutase just noted consumes zero SAM molecules because the SAM sulfonium cation is regenerated in the enzyme active site before the next catalytic cycle. The underlying redox steps are cryptic.

A second outcome is one SAM consumed, by net fragmentation to methionine and dA-H in each turnover (Figure 12.8). For example, the enzyme completing formation of the thiane ring in biotin is such a catalyst.[12] One CH_3 group and one CH_2 group in desthiobiotin are converted to the bridging sulfide –CH_2–S–CH– as the cyclic thiane ring is formed (Figure 12.9). First the pendant CH_3 is converted to a CH_2–S– linkage, one SAM is fragmented to dA-H and methionine, indicating the intermediacy of a 5′-deoxyadenosyl radical intermediate (dA•) (Figure 12.9). The source of the sulfido sulfur in the thiane

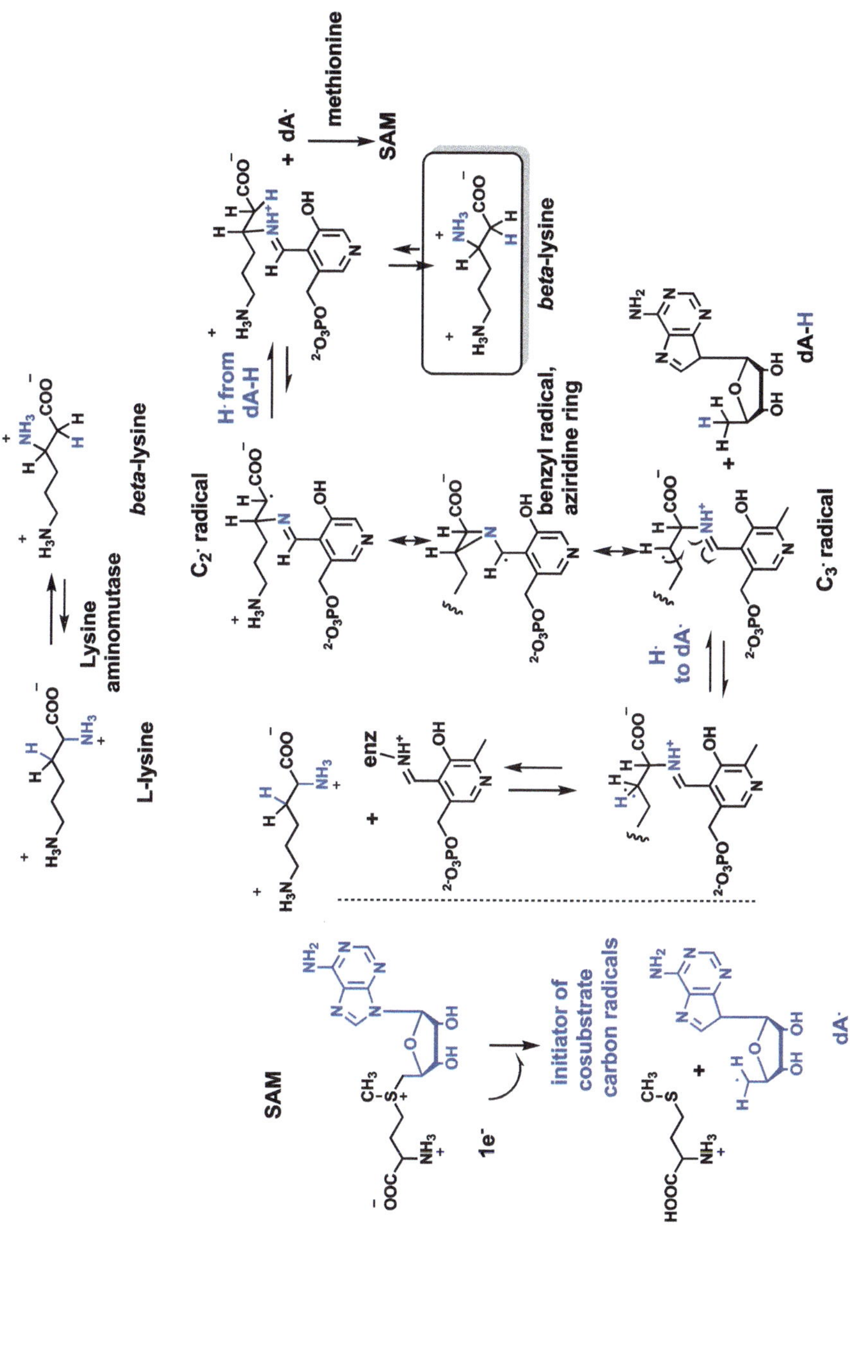

Figure 12.7 The SAM- and PLP-dependent lysine mutase works by radical pathways reminiscent of Figure 12.5.

Radical SAM Enzymes Categorized by Stoichiometry of SAM Usage

1. **Zero consumption of SAM molecules: SAM cleavage to dA· is reversible; SAM acts catalytically, not stoichiometrically**
2. **One SAM fragmented to methionine and dA-H in each catalytic cycle**
3. **Two SAMs consumed, one to generate dA·, One to transfer methyl group: distinct products from each SAM**
4. **One SAM cleaved in each catalytic cycle: dA· captured by cosubstrate olefin**

Figure 12.8 Radical SAM enzymes categorized by stoichiometry of SAM usage.

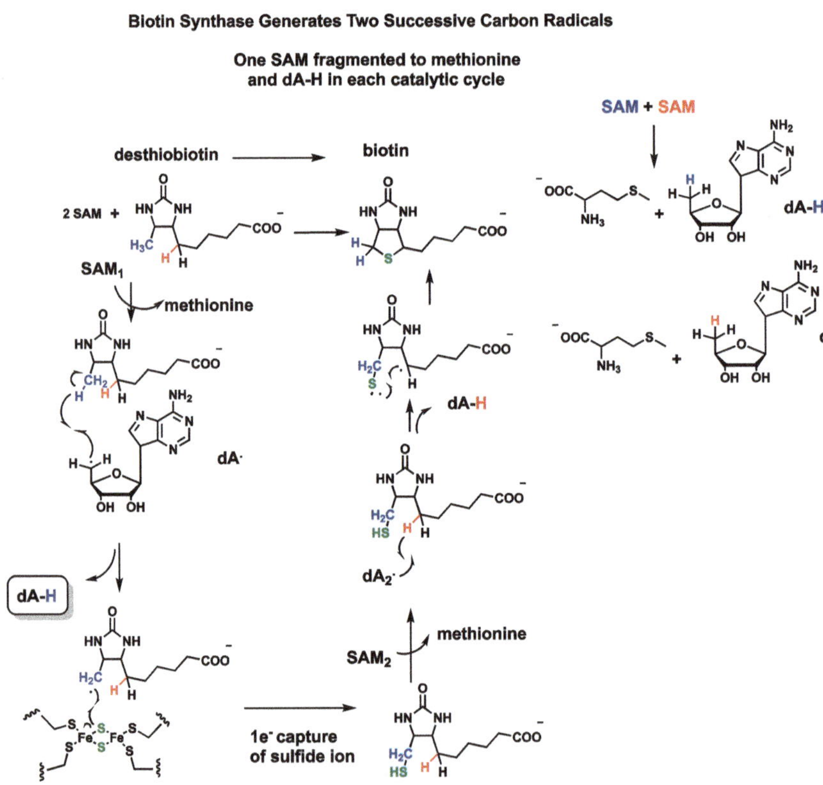

Figure 12.9 Biotin synthase generates two successive carbon radicals during closure of the thiane ring in biotin.

ring is an Fe_2S_2 center in biotin synthase, cannibalized during the reaction.[12] Note that conversion of desthiobiotin to biotin breaks two C–H bonds, makes two new C–S bonds. Two SAMs are fragmented along the way, each to dA-H, each converting a cosubstrate C–CH_3/C–CH_2 to a carbon transient radical.

12.2.4 Consumption of Two SAMs to Distinct Products

In some radical SAM enzymes a cosubstrate molecule undergoes methylation at an unactivated carbon site.[3] The stoichiometry typically involves two SAM molecules cleaved, but in these instances to two different types of products (Figure 12.8). The first SAM is fragmented to dA-H and methionine, emblematic of a dA• radical intermediate for C–H bond cleavage in the cosubstrate.[13]

The second SAM molecule is fragmented to *S*-adenosylhomocysteine as the methyl group of that SAM ends up in the coproduct (Figure 12.10).

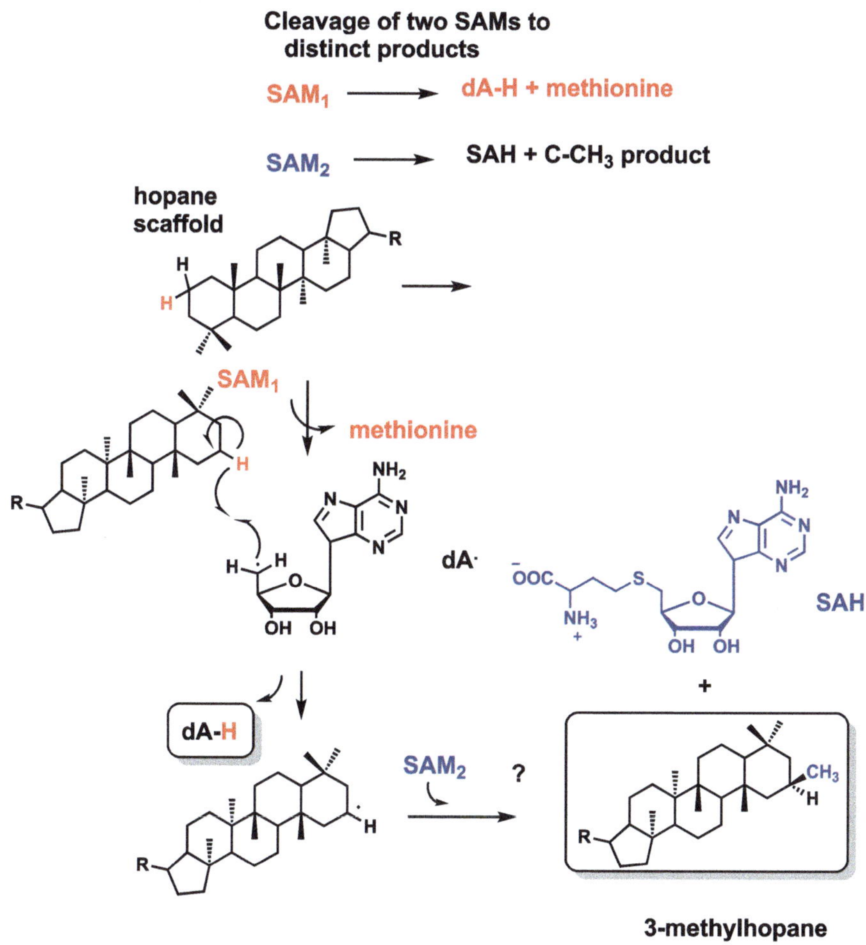

Figure 12.10 Bacterial C-methylation of hopane at the unactivated C_2 position of the A ring. Two SAM molecules cleaved into two distinct products.

The second SAM molecule could be donating a methyl radical equivalent to the cosubstrate carbon radical. The resultant SAH radical would need to pick up one electron, most likely from the Fe_4S_4 center[3] (or the transient substrate radical could pick up a second electron from the Fe/S cluster to give a transient carbanion to attack the CH_3 group in the second molecule of SAM). The enzymatic conversion of bacterial hopane metabolite to the C_3-methyl hopane[14] is presumed to involve utilization of two SAM substrate molecules with fragmentation to two distinct products (Figure 12.10).

12.2.5 Alternate Partitioning of dA• in Radical SAM Enzyme Active Sites

In the few dozen radical SAM enzymes examined for mechanism to date[15] there is evidence for alternate partitioning of the strong radical oxidant dA• as it undergoes quenching while acting as radical initiator. Figure 12.11

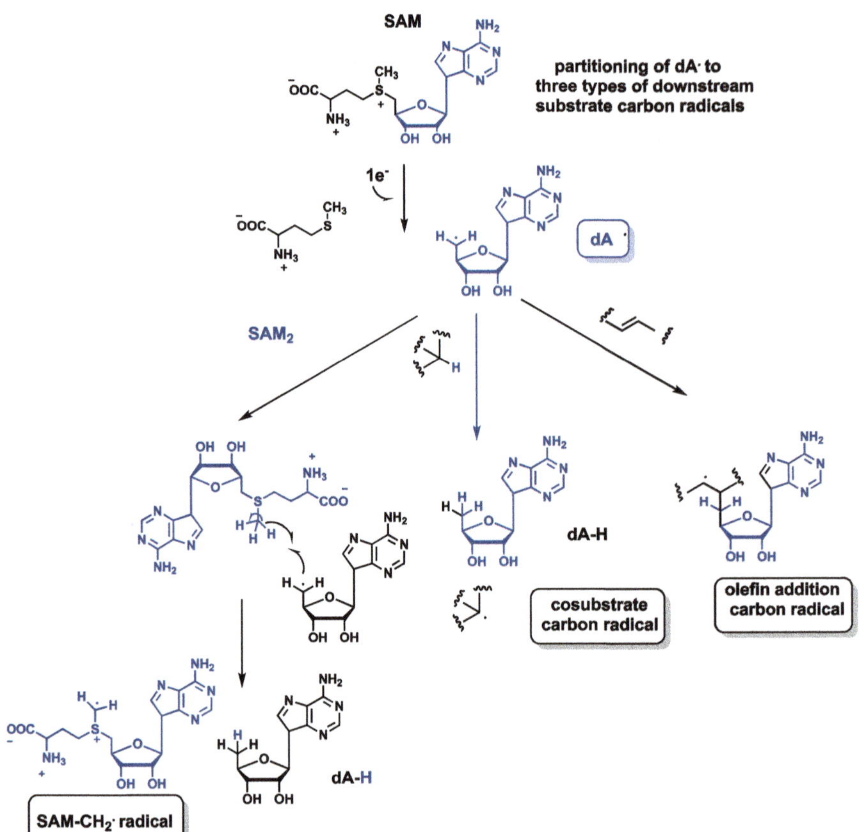

Figure 12.11 Partitioning of 5′-deoxyadenosyl radical (dA•) to three types of downstream carbon radicals.

suggest three partitioning modes, two of which create dA-H, the third reflecting covalent addition of dA• to an alkene linkage in a bound cosubstrate.

While a cosubstrate C–H bond is the most common source of H•, abstracted as dA• is quenched, an alternate source is the methyl group of a second SAM, yielding the methylene radical of SAM. That SAM-CH_2• radical is thought to be the proximal radical source of the cyclopropyl group in the DNA-targeting antitumor antibiotic CC1065 (Figure 12.12).

The third active site fate of transient dA• radical oxidant is addition to an olefin. This fate is illustrated in the net adenosylation of the terminal double bond in diplotene (Figure 12.13). The ensuing secondary carbon radical in the nascent product adduct gains an electron and proton to yield 22-*R*-adenosylhopane.[16]

All in all, a plethora of carbon-based radicals can be propagated and then quenched by several routes in the active site of radical SAM enzymes analyzed to date. These all ensue from the early common 5′-deoxyadenosyl radical

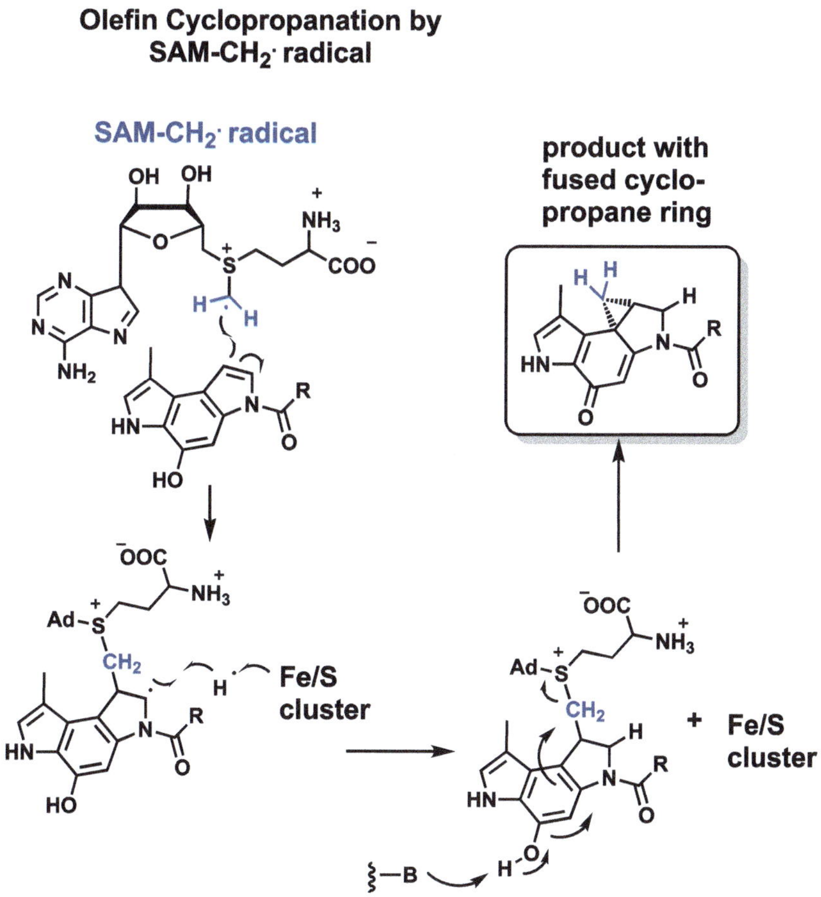

Figure 12.12 Olefin cyclopropanation by SAM-CH_2• radical.

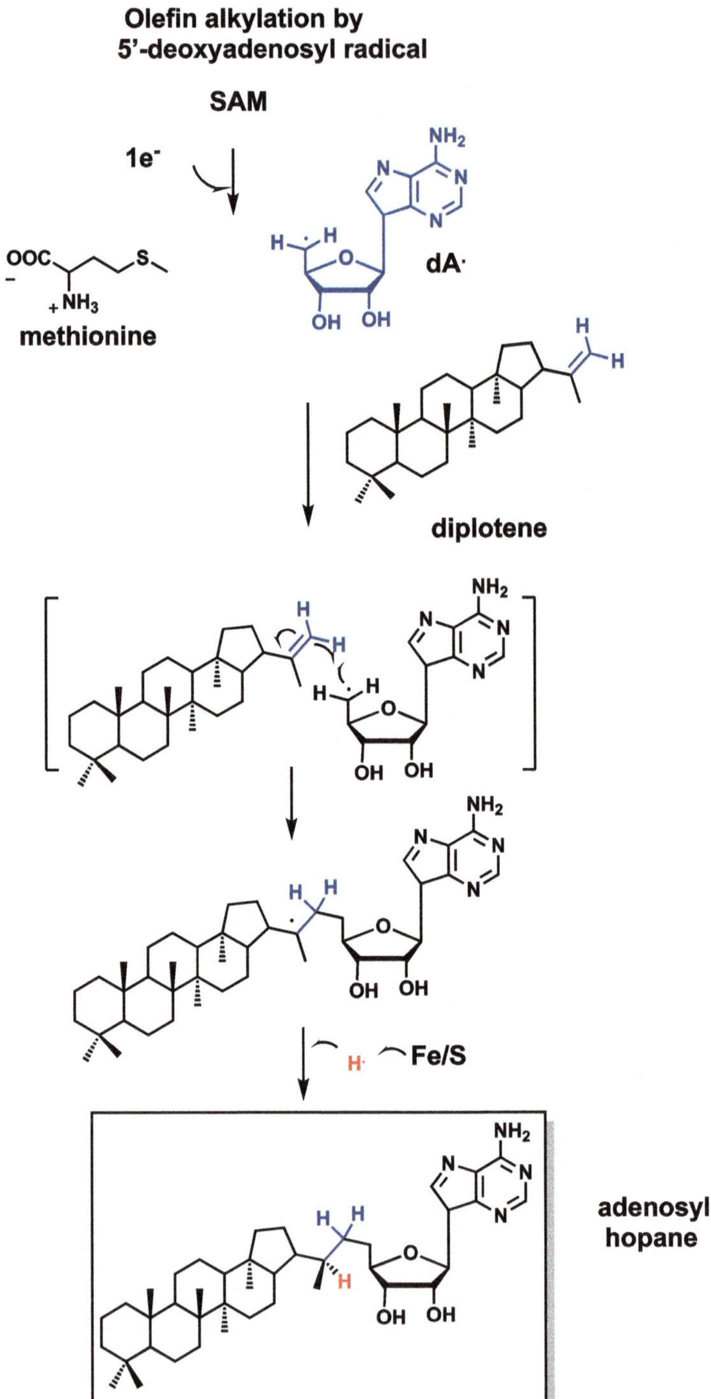

Figure 12.13 Olefin alkylation by 5′-deoxyadenosyl radical.

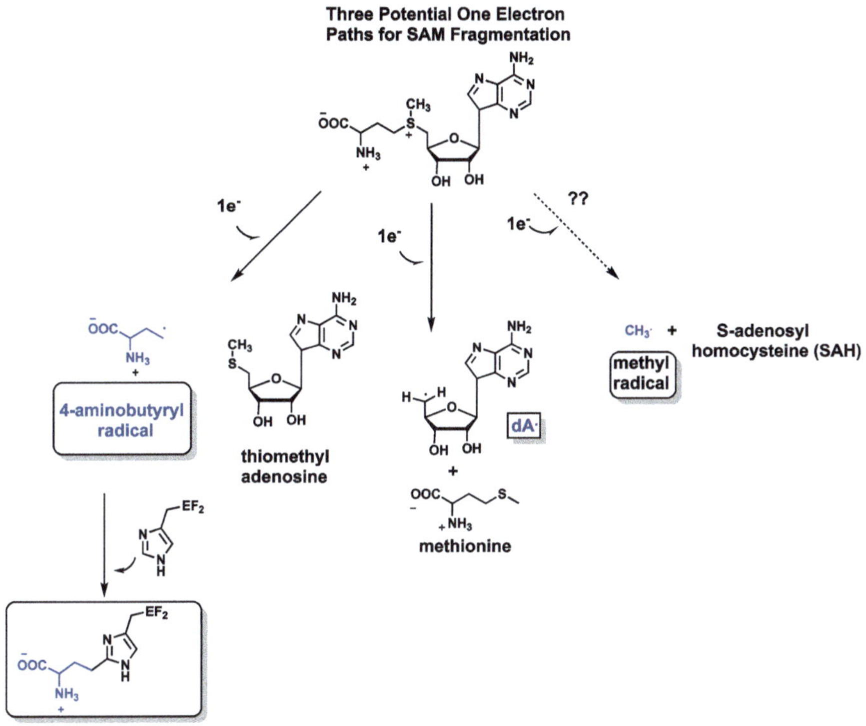

Figure 12.14 Three potential paths for SAM one-electron fragmentation.

fragment liberated from SAM (Figure 12.11). A variety of substrate-based carbon radicals at otherwise unactivated carbon centers can propagate. Finally, given that any of the three C–S^+-bonds in SAM can be independently cleaved in two-electron transfer modes, it is reasonable to anticipate enzymes operating in the orthogonal one-electron/radical mode can also cleave, for example, the carbon–sulfur bond to the methionyl side chain (Figure 12.14). This homolytic cleavage regiospecificity to the transient 4-aminobutyryl radical equivalent does occur in the active site of the enzyme in diphthamide assembly[17] in the maturation of protein synthetic elongation factor 2 (Figure 12.14). All this radical chemistry for C–H cleavages, C–C bond formations, and organic scaffold rearrangements testify to the high oxidizing power of the 5′-deoxyadenosyl radical.

12.3 Aerobic Carbon Radical Generation: High Valent Oxoiron Species in Oxygenase Catalyses

The second pervasive route to carbon radical chemistry in biologic systems depends on enzymes that reductively activate O_2 to generate high valent oxoiron[18] (occasionally oxocopper) species that abstract H• atoms from bound cosubstrate C–H bonds (Figure 12.2).

This is not the place for a detailed exposition on the chemical biology of O_2, but it is fair to state that molecular oxygen (O_2) can be considered one of those thermodynamically activated but kinetically stable molecules that play such key roles in the chemistry of aerobic life.[19] The kinetic stability in the presence of typical spin paired organic molecules found in biologic systems is manifest by the fact that our global atmosphere contains 21% O_2 and yet people are stable organic entities. On the other hand, aerobic organisms have learned how to funnel substrate level electrons down membrane electron transport chains to O_2 as terminal acceptor. One measure of the favorable energetics of passage of four electrons from two NADH to reduce O_2 to $2H_2O$ is the calculated equilibrium constant of 10^{38} in favor of this direction of electron flow. That is a lot of thermodynamic activation locked in the O_2 molecule as electron acceptor.

The main chemical feature contributing to the kinetic stability of O_2 in the presence of stable, spin paired organic molecules is that molecular oxygen is a ground state triplet ($^3O_2$)[20] (the lowest energy configuration for atomic oxygen has one electron each, with spins parallel, in the two 2p orbitals). O_2 has hugely favorable potential for accepting electrons, undergoing reduction, but only by one electron at a time. As an obligate one electron transfer reagent, O_2 does not have a readily available mechanism to react by the two-electron reaction pathways of organic functional groups noted in the preceding two chapters. Hence, the kinetic stability.

To unlock the thermodynamic driving force of reductive oxygen metabolism, nature serves up one-electron redox partners. These come in two distinct types. One set subsumes redox active transition metals, such as Fe^{II} or Cu^{I}, that also react in obligate one-electron redox reactions. Thus, O_2 in iron oxygenase or copper oxygenase active sites can undergo a sequential pair of one-electron transfers to first yield the one-electron-reduced superoxide and then the two-electron reduced metal peroxide complexes (Figure 12.15).

The complementary strategy for one electron transfer to O_2 are the coenzyme forms of the organic redox active vitamin riboflavin (see Chapter 8). We have noted that flavin enzymes differ from the two electron-only nicotinamide redox coenzymes in having access to both low energy two-electron-redox and one-electron-redox routes.[21] Specifically, in the two-electron reduced FlH_2 oxidation states, the flavin can transfer one electron to O_2 to generate the superoxide anion radical and the flavin semiquinone radical (a delocalized nitrogen tris-heterocyclic radical) (Figure 12.16). Radical recombination generates the covalent flavin-4a-hydroperoxide (Fl-OOH) adduct. O_2 has been reduced by two electrons, as occurs also in both the Fe^{III}–OOH forms of heme iron oxygenases and in the Fe^{II}–OOH forms of nonheme iron (di)oxygenases.

Although some iron-oxygenases may utilize Fe^{III}–OOH to initiate oxygen transfers to substrates, most of the heme and nonheme iron enzymes are thought to first generate more powerful oxidants by prior cleavage of the weak –O–OH peroxide bond.[18] This can be formulated using an electron pair on Fe^{III} or Fe^{II} to break the O–OH bond, creating the high valent heme $Fe^{V}{=}O$ or $Fe^{IV}{=}O$, respectively (Figure 12.2).

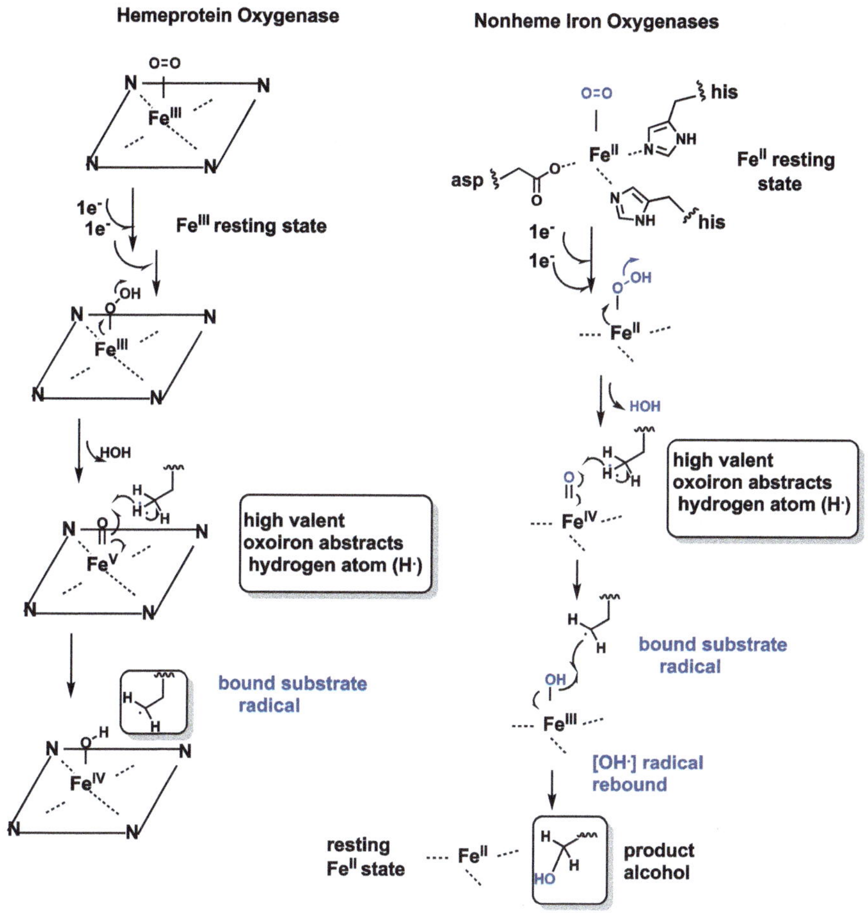

Figure 12.15 Related high valent oxoiron species in heme *vs.* nonheme iron-containing oxygenases.

Now, these oxoiron species are strong enough oxidants to break unactivated C–H bonds of bound cosubstrates homolytically. For example, hydrogen atom transfer to $Fe^{V}{=}O$ involves one-electron transfer from the Fe=O double bond to a hydrogen in flight to create an Fe–O–H bond, with the other electron back into the iron atom to reduce it from Fe^{V} to Fe^{IV} in the transiently formed Fe^{IV}–OH (Figure 12.2). The hydrogen atom removal from substrate C–H bond has generated a substrate carbon radical. The fate of that high energy carbon radical is typically radical rebound, perhaps within the lifetime of a molecular vibration, with an [OH•] from the Fe^{IV}–OH. One electron is left to convert Fe^{IV} back to the starting Fe^{III} oxidation state. The substrate C–H bond has become an alcohol.

The reaction manifold is sketched out for a nonheme iron oxygenase[22] in Figure 12.15, illustrating the parallels to the high valent heme iron oxygenase strategy. Most nonheme iron enzymes have an Fe^{II} rather than Fe^{III} in

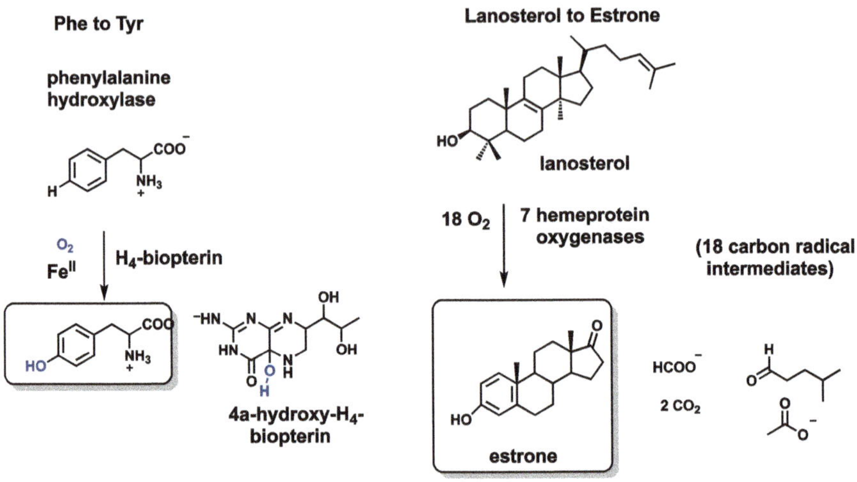

Figure 12.16 Aromatic oxygenation (phenylalanine to tyrosine) by phenylalanine hydroxylase. Eighteen molecules of O_2 consumed in aliphatic oxygenations from lanosterol to estrone.

resting state so the high valent oxoiron species are shown as Fe^{IV}=O species. Hydrogen atom abstraction yields an Fe^{III}–OH and a bound substrate carbon radical as in the hemeprotein oxygenases. Collapse of the Fe-liganded -OH as an [OH•] equivalent is a radical coupling step that produces the product alcohol. In the nonheme iron oxygenase iron is returned to the starting Fe^{II} oxidation state. For hemeprotein oxygenases it is a return to starting Fe^{III}.

This oxygenative, carbon-radical route from alkanes to alcohols was noted early in Chapter 3 as one of the ways aerobic organisms create alcohol functional groups for further metabolic processing. The scope of carbon oxygenation at unactivated sp^3 carbon centers can be immense. Aliphatic carbons in every type of secondary metabolic pathway in plants and microbes are oxygenated to create polar OH groups[2] capable of further chemistry such as subsequent acylation or glycosylation.

Aromatic and heteroaromatic ring metabolites are also oxygenase substrates. Aromatic ring net hydroxylation typically goes through an arene oxide intermediate.[23] A mammalian example of *de novo* aromatic alcohol group formation is the *para*-hydroxylation of the aromatic amino acid phenylalanine to tyrosine noted below. A second set of mammalian biosynthetic oxygenases involve a cascade of 18 sequential oxygenations by a family of heme protein oxygenases that catalyze the oxygenative carving up of the C30 hydrocarbon lanosterol to the C18 female steroid hormone estrone[13] (Figure 12.16). Phenylalanine hydroxylase uses nonheme oxoiron oxidants while the steroid hydroxylases all use the heme-based Fe^{V}=O oxidant strategy.

12.3.1 Aromatic Carbon Hydroxylation Route

Phenylalanine hydroxylase in humans catalyzes oxygen insertion/hydroxylation regiospecifically at the *para* position of the aromatic ring to provide tyrosine, a partner aromatic amino acid and progenitor to many alkaloid metabolites in plants. The enzyme contains one essential equivalent of mononuclear heme iron and also utilizes tetrahydrobiopterin, oxidizing it to the quinoid isomer of H_2 biopterin by way of a 4a-peroxy-H_4 biotin adduct.[24] The cosubstrate O_2 undergoes net four-electron reduction, two electrons from the *para*-C–H bond of phenylalanine and two from the biopterin oxidation (Figure 12.17). Peroxide level oxygen is likely an adduct between Fe^{II}–O–O-biopterin on the way to an $Fe^{IV}{=}O$ proximal oxidant.

One can formulate a route to an intermediate arene oxide by one-electron transfer from the 3,4-double bond in the phenylalanine aromatic ring to

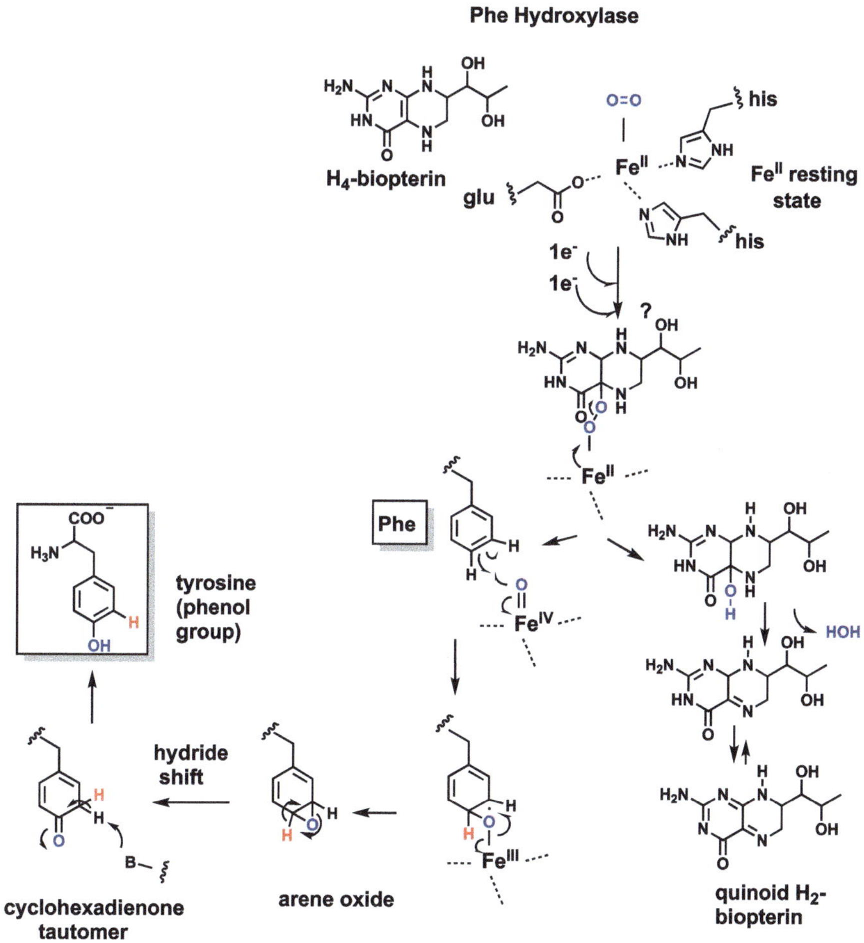

Figure 12.17 Phenylalanine hydroxylase, a mononuclear nonheme iron oxygenase uses O_2 and tetrahydrobiopterin as cosubstrates with L-phenylalanine. The quinoid form of H_2-biopterin is a coproduct with L-tyrosine and a molecule of water. An arene oxide intermediate is the nacent oxygenated product that undergoes a 1,2-hydride shift during ring opening to the product phenol.

create a transient oxy adduct with an adjacent carbon radical (Figure 12.17). That can collapse by convergent one-electron transfer to form the second C–O single bond in the nascent arene oxide, a cyclohexadienyl epoxide (Figure 12.17). The rearrangement of cyclohexadienyl epoxides to aromatic phenols is well studied. The epoxide opens intramolecularly as a hydride ion undergoes a net 1,2-shift to generate the cyclohexadienone. This is the unstable tautomer of the phenol in tyrosine.[25]

12.3.2 Aliphatic Carbon Hydroxylation Route

Of the many examples that one could cite in human steroid oxygenative maturation to steroid hormones, both the first and last step between lanosterol and estrone (Figure 12.16) may give insight into comparable tandem oxygenation strategies. *Inter alia*, the first enzyme pair, acting at both C_4–CH_3 groups and then at the C_{14}–CH_3, reminds us of the redox connections between methyl group, alcohol, aldehyde and carboxylic acid functional groups, as well as decarboxylation to release CO_2. Overall, this is a remarkably efficient eight-electron oxidation to convert each of two C–CH_3 groups to the C–H products and CO_2. The proximal substrate is the 3-keto, 4,4-dimethyl steroid (Figure 12.18). This pre-oxidation to the ketone at C_3 will eventually serve as the electron sink for low energy decarboxylation three oxygenations hence.

Each of the two C_4 geminal dimethyl groups arising from action of oxidosqualene cyclase are removed oxygenatively,[26] the β-, and then the α-*C*-methyl group at C_4. In retrospect, the strategy seems straightforward if not simple. Use the same hemeprotein oxygenase three times in succession, each time creating a high valent Fe^V=O as radical oxidant. Concomitantly, the C_4–CH_3 group of substrate forms the C–$CH_2^\bullet$ radical, captured in [OH•] radical rebound to the C–CH_2OH product. The C–CH_2OH is substrate for the next catalytic cycle to the •C-CHOH carbon radical, captured in [OH•] rebound to the gem diol (Figure 12.18). That gem diol is the less stable hydrate adduct of an aldehyde carbonyl. So, after two cycles, the C–CH_3 has become a C–CH=O functional group. A third O_2 consumed leads to one more Fe^V=O oxidant, one more carbon radical, the transient C–[C=O] radical, capturable by [OH•] rebound to the carboxylate anion. Now, the product is a β-keto acid with a low energy enolate pathway to loss of CO_2 and protonation to the steroid C–H product[27] (Figure 12.18).

Note that the triple oxygenation sequence, consuming three O_2 molecules (generating the oxygen atoms in four HOH product molecules and one CO_2) uses nature's iron-oxo heavy artillery *ad seriatim* to create (1) a C–CH_2 primary radical, (2) an alcohol-based carbon radical, and (3) an aldehyde-based carbon radical (an acyl radical) to be captured by [OH•] rebound chemistry (Figure 12.18).

At the C_{14}–CH_3 group the first two oxygenations to alcohol and then aldehyde are paralleled by a regiospecific hemeprotein oxygenase.[27] Then the third oxygenation converts the aldehyde to an *O*-formyl ester. That oxoester is subsequently hydrolyzed to release formate instead of CO_2.

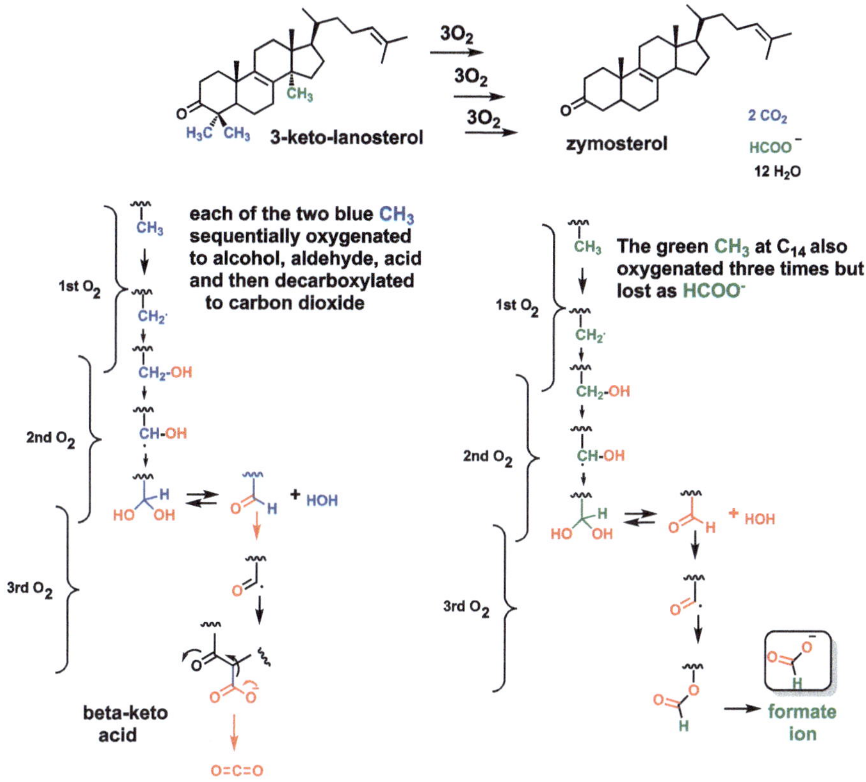

Figure 12.18 Oxygenative removal of three C–CH_3 groups in lanosterol (to zymosterol) consumes 9 O_2, releasing 2 CO_2, $HCOO^-$, and 12 H_2O. Nine carbon radical species are generated, one for each O_2 converted to a high valent oxoiron oxidant.

At the other end of the steroid oxygenation cascade, in the last step of steroid hormone maturation, the C19 human male sex hormone androstenedione is converted to the C18 human female sex hormone estrone.[28] The enzyme aromatase carves out the angular C_{19}-methyl group oxygenatively, transiently forming the A ring cyclohexadienone that aromatizes to estrone nonenzymatically (Figure 12.19).

Three O_2 molecules are consumed by the hemeprotein aromatase *via* three high valent oxoiron oxidants utilized as radical oxygen transfer agents. The one carbon oxidized fragment is released as formate, not CO_2, following the product pattern for the lanosterol C14 demethylase in Figure 12.18 above. The first two oxygenations follow the logic shown above for conversion of C–CH_3 to CH_2OH and CH_2OH to $CH(OH)_2 = C–CH{=}O + HOH$. The third oxygenation results in net insertion of the itinerant oxygen into the C–CHO bond to create a formate ester. Elimination of the formate fragment then creates the cyclohexadienone moiety, a nonenzymatic isomerization away from the aromatic phenol A ring of female sex hormones (Figure 12.19).

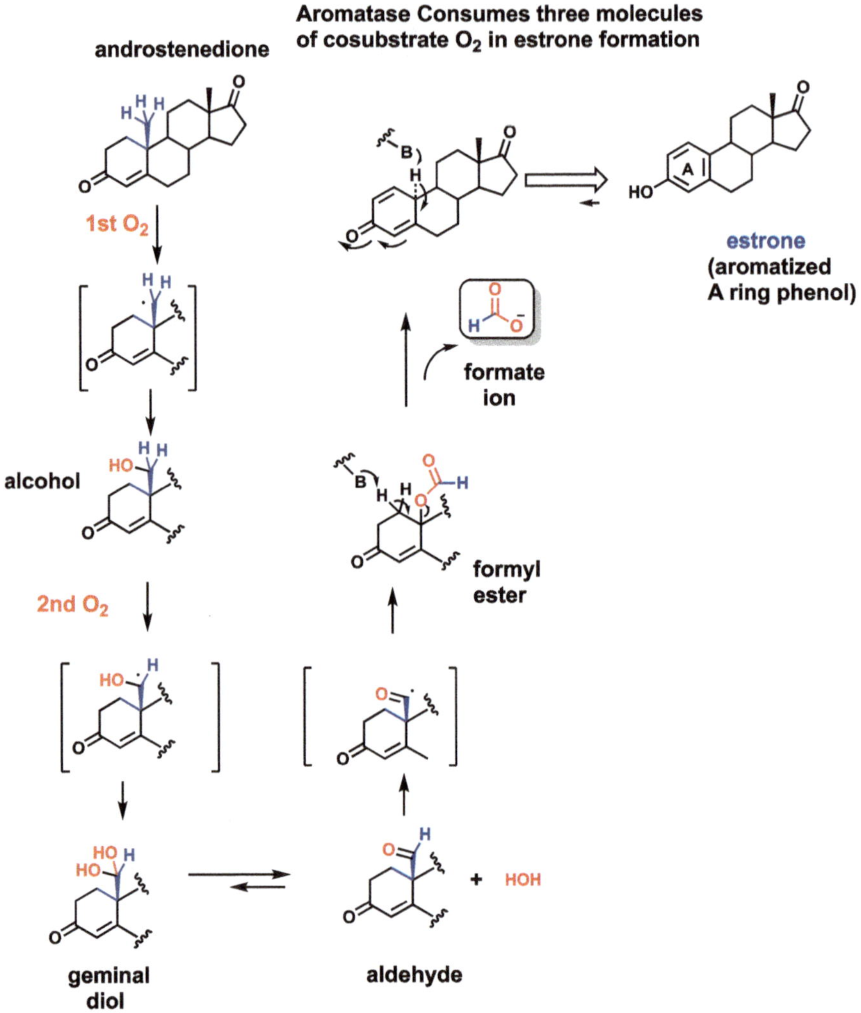

Figure 12.19 Three molecules of O_2 reduced as the C_{19} angular methyl group in androstenedione is oxygenatively removed as formate ion during aromatization of the steroid A ring to estrone by the enzyme aromatase.

Again, three carbon radical species are generated in the three catalytic cycles of aromatase action[29] (Figure 12.19).

One can view the overall steroid biosynthetic pathway as involving three types of carbon-based intermediates as the pathway matures. The first three enzymes use acetyl CoA as thioester enolate/carbanion chemistry to build in the branching of hydroxymethyl CoA (Chapter 3). Then, the isoprenyl-PP pair of isomers employ allyl cation-initiated chain elongations and cyclizations. O_2 enters the pathway only at the C30 stage, epoxidizing squalene for the subsequent cyclization cascade. The subsequent consumption of 18 O_2 molecules at a late stage, as C18 estrone/estradiol is oxygenatively carved out from the C30 lanosterol, putatively involves 18 different carbon radical species (see Chapter 15).

12.3.3 Radical Rearrangements Before Oxygen Rebound

The question arises in consideration of the wide range of oxygenations of carbon centers by high valent oxoiron enzyme-bound oxidants as to the lifetimes of various high-energy carbon-centered radicals proposed on the reaction pathways. Thus, the primary C–$CH_2^•$ radicals seem forbiddingly unstable to be more than a transition state and not a finite intermediate. Two kinds of observations bear on the prospect that transiently formed carbon radicals in enzyme bound substrates can live long enough for oxygen-independent intramolecular radical rearrangements to occur, competitive with the intermolecular proposed rebound step of [OH•] transfer. One is observed scaffold rearrangement before completion of oxygen transfer. The second is a growing realization of many "thwarted" oxygenases[30] where carbon radicals react intramolecularly and appear to escape completely from any oxygen quenching events.

One delayed hydroxylation outcome, after carbon radical formation and rearrangement, is seen in the enzymatic ring contraction of *ent*-kaurenic acid during three tandem hydroxylation at carbons 7 and 6 on the way to gibberellic acid 12 by a heme iron-containing oxygenase[31] (Figure 12.20).

The 7-hydroxylation proceeds first without incident *via* a carbon radical at C_7. Then, the same enzyme creates a high valent oxoiron in the next catalytic cycle and generates the C_6 carbon radical. Before oxygen transfer can capture this carbon radical, the tetracyclic 6-6-6-5 scaffold rearranges with extrusion of the carbon that becomes the C_6 hydroxy radical during a net ring contraction. The carbon radical has moved in the process to the adjacent carbon, bearing the initial-OH group. Delivery of the second [OH•] equivalent to that rearranged radical creates the gibberellic 12 aldehyde scaffold. Thus, a tandem hydroxylation occurs, but the second carbon radical first rearranges from a bridging to a terminal position. Completion of the second hydroxylation creates the geminal diol that is the gibberellic aldehyde. One more oxygenation converts the aldehyde to the acid in gibberellic acid 12. If the scaffold had not rearranged at the stage of the second carbon radical extrusion, a 6,7 diol without skeletal rearrangement would have been formed.[32] The scaffold rearrangement suggests a finite lifetime for the substrate carbon radical before its capture in an [OH•] rebound.

12.3.4 Thwarted Oxygenase Outcomes: Carbon Radicals Without Oxygenation

If there were an infinite delay between an oxoiron-generated carbon radical formation and then its oxygenative capture in the active site of an oxygenase family member, the oxygen transfer would never be completed

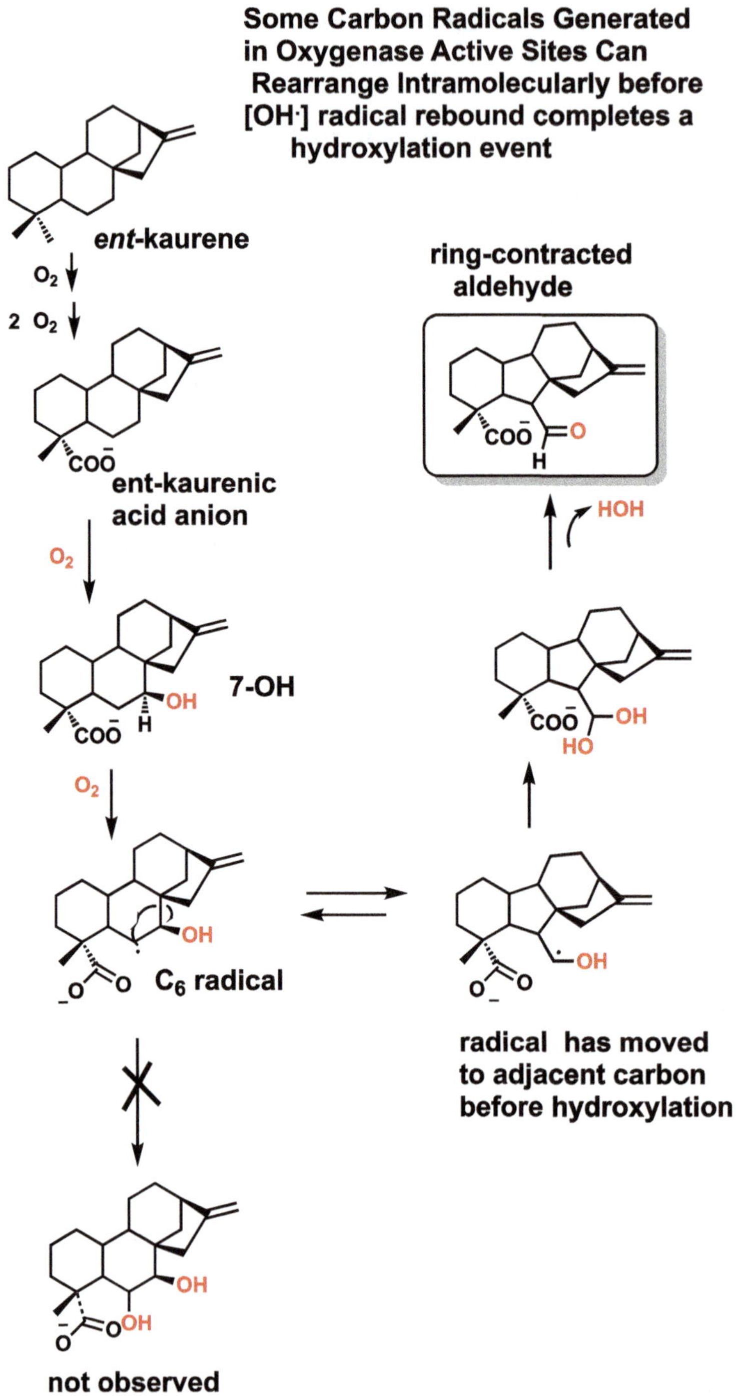

Figure 12.20 Some carbon radicals generated in iron-oxygenase active sites can rearrange intramolecularly before capture by iron-bound [OH•] equivalents. An example in the conversion of *ent*-kaurenic acid anion to a ring contracted aldehyde is shown.

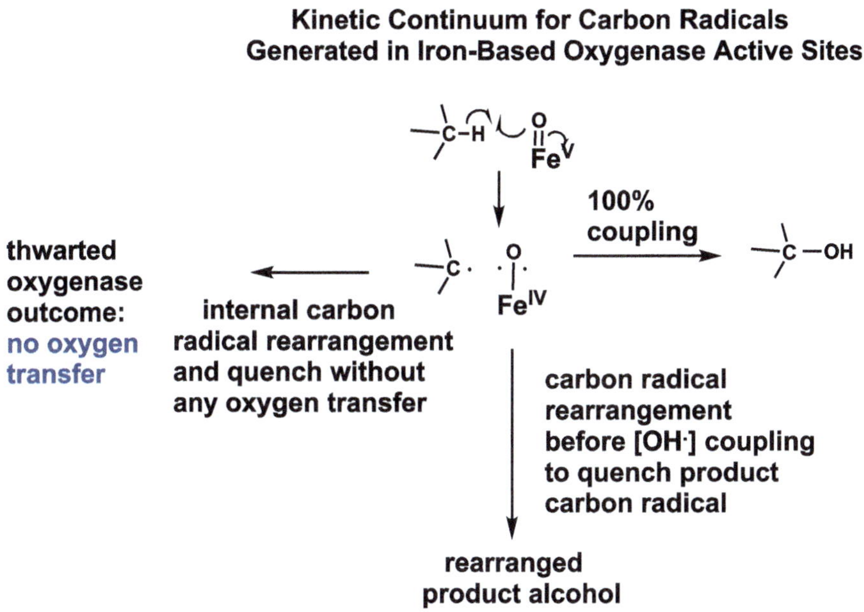

Figure 12.21 Kinetic continuum for substrate radicals generated in oxygenase active sites from 0% to 100% coupling with [OH•].

(Figure 12.21). Some intramolecular reaction of the carbon radical would outcompete any oxygenation. We have termed the growing number of these oxygenase subsets as "thwarted" oxygenases.[30] O_2 is absolutely required as cosubstrate to create the high valent oxoiron oxidant that in turn creates the carbon radical. In all such thwarted oxygenases, O_2 is ultimately reduced by four electrons to two molecules of product water.

We note below a small subset of examples that illustrate that thwarted oxygenases can create carbon radical-derived products that are of significant value to the producer organisms. One might suggest that these are actually "evolved" oxygenases, evolved to form carbon radicals at otherwise unactivated carbon centers and let them rearrange to novel product scaffolds. The failure to capture and sequester the carbon radicals by hydroxylation may unlock novel reaction outcomes that confer some advantage to the producer organisms.

Among the most intriguing nonheme iron-containing oxygenases is clavaminate synthase (CAS) in the bacterial biosynthetic pathway to the 4,5-fused bicyclic β-lactam antibiotic clavulanate.[33] Clavulanate is itself a weak antibiotic but a potent acylating agent of β-lactamases. CAS catalyzes three separate steps in the clavulanate pathway (Figure 12.22). The first appears to be a conventional hydroxylation at a methylene carbon center, presumably *via* an $Fe^{IV}{=}O$ radical oxidant. Then, an arginase enzyme intervenes to convert the guanidinium side chain to a primary amine, with hydrolytic release of urea.

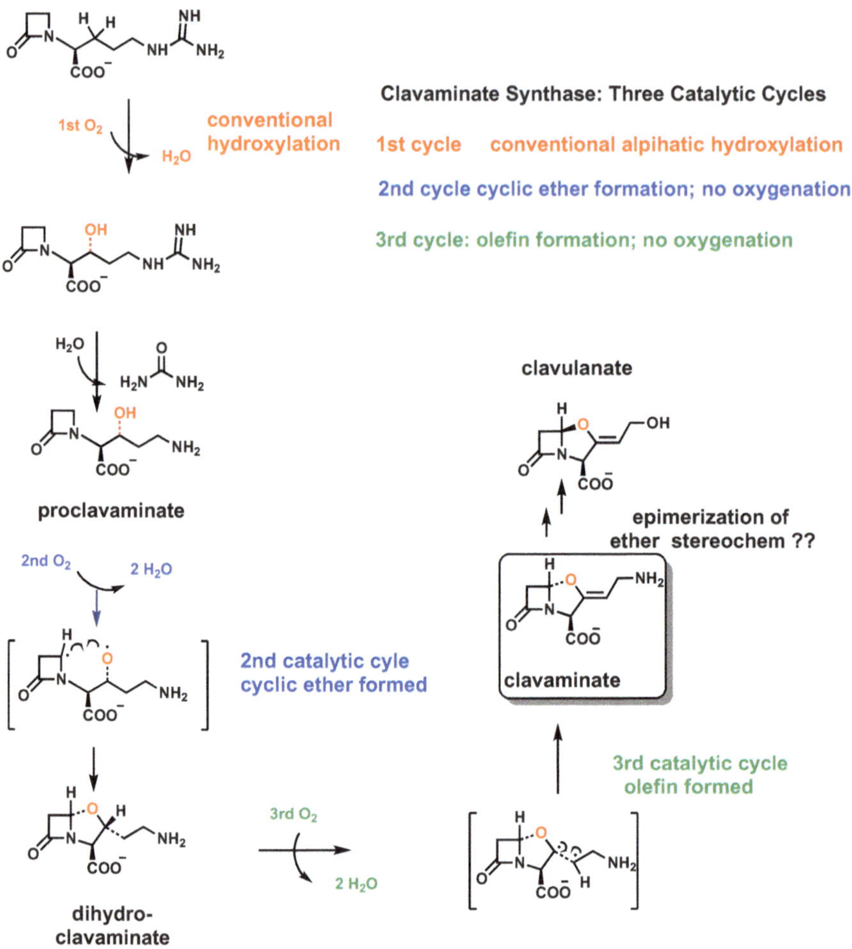

Figure 12.22 The nonheme iron enzyme clavaminate synthase is a successful oxygenase in one out of three catalytic cycles. The other two cycles use carbon radical chemistry for cyclic ether and exocyclic olefin formation on the way to clavaminate.

The second catalytic cycle of CAS reduces O_2 to $2H_2O$. The alcohol group in substrate is converted to the α-fused cyclic ether, creating the 4,5-bicyclic framework of proclavaminate. One can write a putative biradical pair (Figure 12.22) that can close to the cyclic ether and out compete any intermolecular oxygen transfer from an oxygenated iron species. Analogously, the third catalytic cycle of CAS follows, again without any net oxygenation from the third O_2 substrate molecule. Presumably, a $CH^{\bullet}$ radical is generated but may collapse with a second hydrogen atom abstraction to form the exocyclic double bond of clavaminate. Subsequent oxidation of the amine to aldehyde and reduction to alcohol must be accompanied by a still mysterious inversion

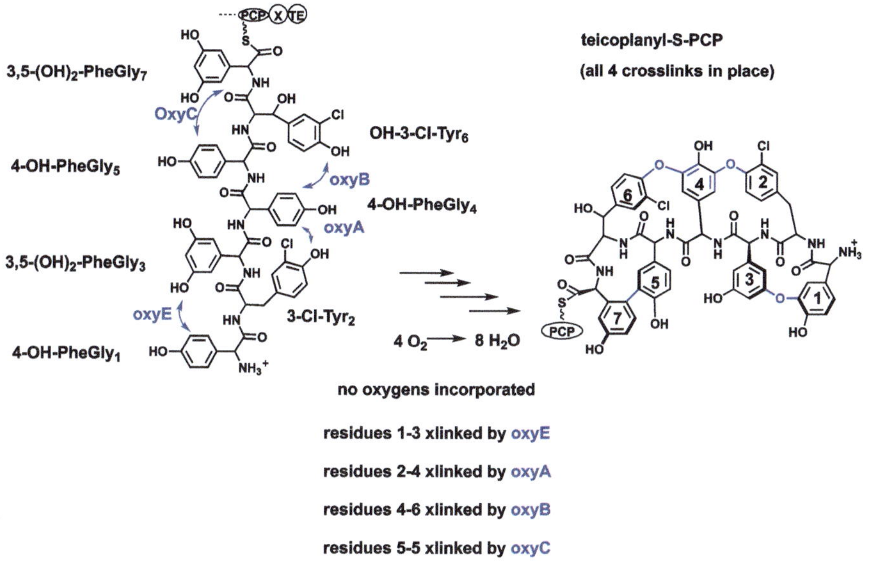

Figure 12.23 Crosslinking P450s act by carbon radical chemistry but without oxygen incorporation.

of cyclic ether configuration from α- to β- in the final clavaminate framework[34] (Figure 12.22).

One can define clavaminate synthase as an oxygenase that only once out of three cycles is actually an oxygenase. In the other two cycles it is a thwarted oxygenase, putatively reflecting faster intramolecular kinetic fates for substrate carbon radical species (cyclic ether formation, then olefin formation) that outcompete intermolecular [OH•] transfers.

Two examples of thwarted oxygenases in the natural antibiotic literature involved in the creation of the mature heptapepide antibiotic teicoplanin (Figures 12.23 and 12.24) and the bicyclic beta lactam antibiotics, respectively, show two kinds of oxygen-independent radical couplings.

In Figure 12.23 the linear heptapeptidyl chain on the seventh peptidyl carrier protein domain in the teicoplanin synthetase assembly line undergoes four sets of "thwarted" oxygenase couplings that link all seven residues in a macrocyclic framework. The set of four tandemly acting thwarted hemeprotein oxygenases that crosslink the electron rich side chains of the heptapeptide antibiotic teicoplanin (OxyB, OxyE, OxyA, OxyC in that order) convert a floppy acyclic heptapeptide to a highly rigidified cup-shaped antibiotic that targets bacterial cell walls and prevents their covalent crosslinking[35] (Figure 12.23). The four oxygenases work in tandem on the linear heptapeptidyl chain while still attached in acyl thioester linkage to the most downstream peptidyl carrier protein (PCP) domain.[36] In teicoplanin all seven electron-rich tyrosyl and hydroxyphenylglycyl residues have been crosslinked (Figure 12.23). Four molecules of O_2 are consumed, each undergoing four-electron reduction to two water molecules. Eight electrons have been removed from the linear teicoplanin heptapeptide on the

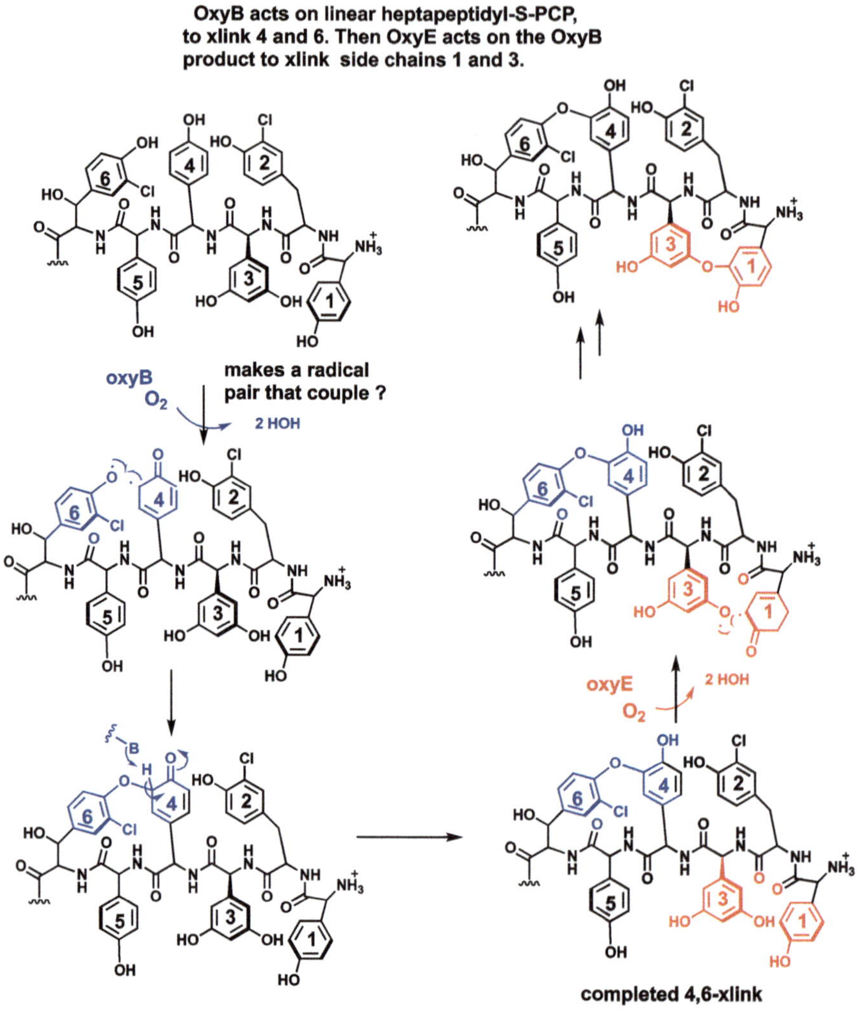

Figure 12.24 The hemeprotein oxygenases act regiospecifically in sequence. The first side chains crosslinked through carbon radicals as aryl ethers are residues 4 and 6 by OxyB. Then OxyE performs O_2-dependent radical C–O–C coupling of the phenoxy radical side chains of residues 1 and 3.

way to the mature full crosslinked product. None of those eight oxygens are incorporated into mature crosslinked teicoplanin. All four enzymes act as thwarted oxygenases.

The mechanistic presumption is that high valent $Fe^{V}{=}O$ heme oxidants are generated in each catalytic cycle and create substrate radicals in the electron rich phenolic side chains. These phenolic radicals, as both phenolate radicals and cyclohexadienyl radicals, couple intramolecularly, forming three aryl ethers and one direct aryl-aryl coupling (residues 5 and 7).

The first thwarted oxygenase OxyB is proposed to generate phenolic radicals at Tyr_6 and HO-$PheGly_4$. They go on to form crosslinked aryl ethers between residues 4, and 6 (Figure 12.24). The next enzyme, OxyE, acts regiospecifically to couple radical species between 3,5-hydroxy-PhenylGly$_3$ and 4-OH-Phegly$_1$ (Figure 12.24). Then, the subsequent two enzymes OxyA and OxyC finish the conversion of an acyclic, linear, biologically inactive heptapeptidyl precursor to the constrained cup-shaped architecture of the active, mature, crosslinked teicoplanin antibiotic.[36]

A second example of directed intramolecular scaffold rearrangements from transient radicals that outcompete intermolecular [OH$^•$] capture is in the enzymatic cyclization of the linear tripeptide aminoadipyl-cysteinyl-valine to isopenicillin N by isopenicillin N synthase (IPNS) (Figure 12.25). This nonheme iron "thwarted oxygenase" reduces obligate cosubstrate O_2 to two water molecules as it constructs both the four member β-lactam ring and the five member fused thiane ring[37] (also see Chapter 8).

This is the same kind of high valent oxoiron mediated enzyme chemistry, using radicals to create the four membered β-lactam warhead first before generating the fused five member thiane heterocyclic ring (Figure 12.25). The highly strained β-lactam ring is thermodynamically activated for ring opening acylation of nucleophiles but is kinetically stable enough to undergo catalyzed ring-opening acylation only in the active sites of its bacterial cell wall target enzymes. Both sulfur and carbon radicals are proposed for the steps forming the two heterocyclic rings.

The very next enzyme in this antibiotic biosynthetic pathway is another thwarted iron-containing oxygenase. It is known colloquially as expandase for its ability to expand the sulfur five ring to a sulfur six ring while leaving the β-lactam warhead intact on the way to the 4,6-fused ring system of cephem antibiotics[38] (Figure 12.26). The ring expansion steps are again featured to involve carbon-centered, sulfur, and again carbon-centered transient radicals.

Expandase not only reduces O_2, it also decarboxylates 2-ketoglutarate to succinate and CO_2 to generate the high valent $Fe^{IV}{=}O$ oxidant[39] (Figure 12.26). On binding of substrate, the $Fe^{IV}{=}O$ abstracts a hydrogen atom from one of the two prochiral methyl groups of the valyl residue of penicillin. Rather than be captured by the [OH$^•$] equivalent now bound to Fe^{III}, the substrate radical rearranges internally to two additional radicals (Figure 12.26). The first is a strained three member episulfide radical, with unpaired electron density on sulfur. The second arises from opening of that strained episulfide. This is the expansion step of five member to six member ring. This third substrate radical is quenched by adjacent hydrogen atom abstraction, forming the C=C double bond in the new six member cephem ring. The electron gets transferred to the iron, reducing it from Fe^{III}–OH back to resting Fe^{II}, and the iron-bound OH is no longer an [OH$^•$] equivalent but instead protonates to the product water molecule.

These two examples of thwarted oxygenases demonstrate how the high valent oxoiron reagents allow entry into carbon radical chemistry without immediately quenching those radicals by (deactivating) hydroxylation. Instead, both the teicoplanin and the penicillin and cephalosporin cases show how these carbon radicals can be used for shape- and reactivity-changing scaffold rearrangements.

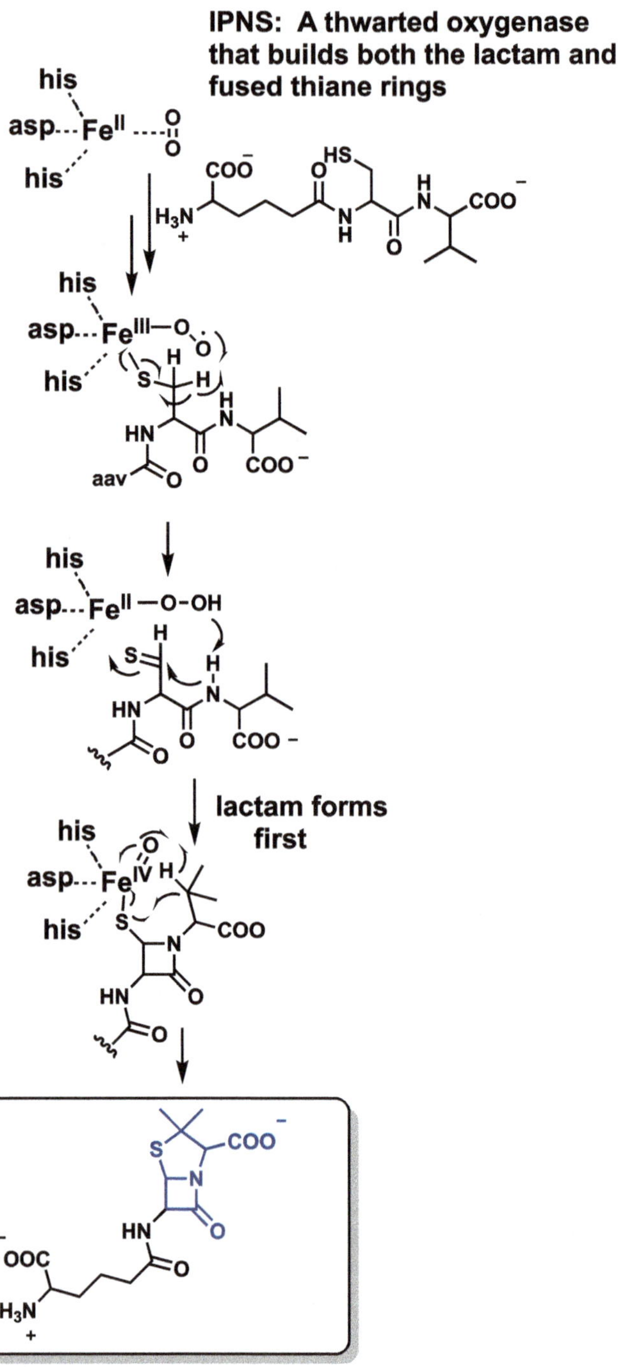

isopenicillin N

Figure 12.25 Isopenicillin N synthase: a thwarted oxygenase that first builds the beta-lactam ring then the fused thiane ring.

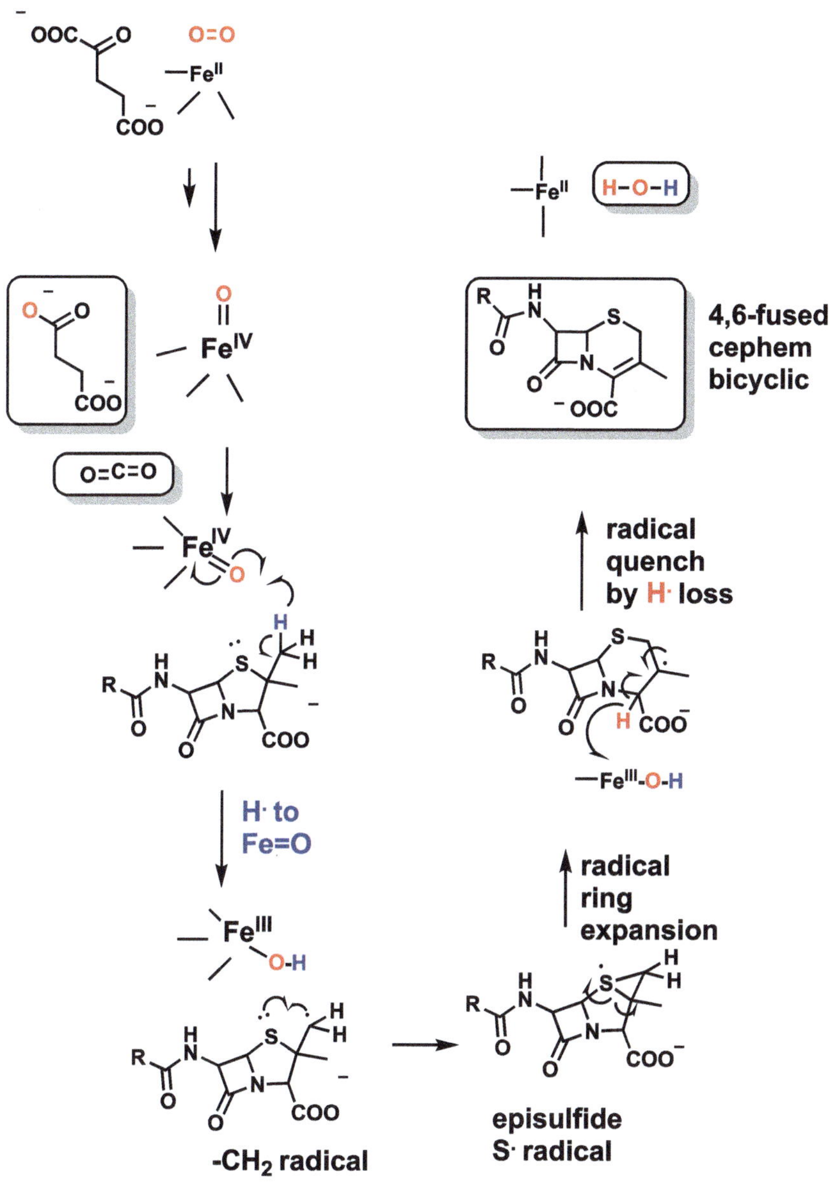

Figure 12.26 Tandem action of iron enzymes IPNS and DAOCS converts acyclic tripeptide to 4,5-fused ring of penicillin and then to fused 4,6-bicyclic scaffold of cephem antibiotics *via* radical chemistry.

In teicoplanin the radical polymerization of electron-rich aryl side chains builds a three dimensional shape that is a high affinity surface for its biologic cell wall target. For the β-lactam antibiotics, the transient radicals are directed to the construction of the key fused nitrogen and sulfur heterocycles. These are illuminating cases of enzyme-directed reactivity of carbon radicals without oxygenative capture.

12.4 Carbon Radical Overview

Carbon-centered radicals, as transition states or short-lived high energy intermediates, play significant roles in the chemical biology of carbon. As a strategy to carry out homolytic chemistry at sets of otherwise unactivated carbons in organic metabolites, nature has developed two kinds of carbon radical initiator strategies in the active sites of two major enzyme families. Radical chemistry is at play both in anaerobic, oxygen free niches and in oxygenated microenvironments under different chemical constraints.

Anaerobic radical chemistry, probably the early strategy evolutionarily, involves transient generation of 5′-deoxyadenosyl radicals in enzyme active sites. Two kinds of molecules are stable precursors of the carbon-centered 5′-deoxyadenosyl radical (dA•). One is adenosyl-B12. Homolysis of its unusual organometallic Co^{III}–CH_2-adenosyl bond generates the Co^{II} oxidation state of B12 and the dA• radical. The second metabolically stable precursor is *S*-adenosylmethionine, operating in its one-electron radical SAM mode. The enormous size of the superfamily of radical SAM enzymes suggests nature has duplicated this strategy over and over to homolyze SAM to Fe_4S_4-bound methionine and dA•.

From the convergent homolytic generation of dA• in either adenosyl-B12 or radical SAM enzyme active sites, the dA• radicals are powerful enough one-electron oxidants to initiate radical chemistry on bound cosubstrate carbon centers by homolytic C–H bond cleavage. The formation of the bound substrate radical along with the CH_3-group of dA-H is often, but not always, reversible. An enormous range of primary, secondary, and tertiary carbon radicals as well as nitrogen, sulfur and oxygen radicals (from C–N, C–S, or C–O) are likely available by this route. The radical enzyme superfamily enzymes always contain an Fe_nS_n cluster, typically sensitive to air autoxidation, so radical SAM enzymes are typically inactive in aerobic microenvironments due to FeS cluster oxidative decomposition.

The second major route to carbon radical-based reaction profiles in biology occurs only in aerobic niches, complementing the anaerobic requirement for radical SAM enzyme activity. The favorable thermodynamic potential of O_2 by one-electron reductive pathways is unlocked either by the dihydro form of flavin coenzymes or the reduced forms of redox active transition metals. In biology the most common such metal ions are Fe and Cu, with iron enzymes dominant. Among the family of iron-containing enzymes that interact with molecular oxygen, both heme proteins and nonheme mononuclear iron coordination sites have evolved to reduce O_2 and incorporate one or both oxygen atoms into oxygenated coproducts.

We note the ability of both heme and nonheme iron oxygenases to make transient high valent oxoiron species as proximal radical oxidants of C–H and C=C bonds in bound cosubstrates. The generation of substrate radicals and their capture by hydroxy radical equivalents from the oxoiron centers can be almost instantaneous for generation of alcohols from sp^3 carbons and epoxides from C=C bonds in substrates. We also note that in the back end of mammalian steroid pathways, converting lanosterol to estrone, 18 cycles of heme iron-based oxygenases occur, indicating 18 separate carbon radicals in the transforming steroid substrates. Particularly notable are the oxygenative strategies for conversion of unactivated C–CH_3 groups to the C–H and CO_2 or $HCOO^-$ products.

Finally, in the oxygenase lineages we call a few examples, out of many, thwarted oxygenases. O_2 is consumed, used to make the standard high valent oxoiron oxidant, cosubstrate carbon radicals generated, but the intermolecular [OH•] transfer is never completed. No oxygen-derived oxygen atoms end up in products, Instead the intramolecular reaction/rearrangement of the nascent carbon radicals outcompete oxygen transfer.

The rationale for these thwarted oygenases may precisely be to slow down or totally withhold the oxygen rebound. Let the rearranging carbon radicals go on to build scaffolds that may offer selective advantage to those producer organisms. One can appreciate the crosslinking of five of the seven side chains to give teicoplanin its active antibiotic shape. Analogously, the ability of carbon and sulfur radicals in isopenicillin N synthase and expandase active sites to live long enough (without oxygen quenching) allows them to rearrange the linear tripeptide substrate to the fused 4,5 (penicillins) and 4,6-bicyclic (cephalosporin) β-lactams ring systems. Clearly, these thwarted oxygenases represent evolutionary positive chemical advantages to the producing microorganisms.

In sum, between anaerobic routes *via* transient enzyme generated 5′-deoxyadenosyl radicals and aerobic O_2-consuming high valent oxoiron oxidants, the chemical biology of carbon at unactivated carbon sites in organic metabolites, both cryptic and observed in stoichiometry, are likely to proceed *via* radical-based reaction paths.

References

1. H. Yang, E. C. McDaniel, S. Impano, A. S. Byer, R. J. Jodts, K. Yokoyama, W. E. Broderick, J. B. Broderick and B. M. Hoffman, The elusive 5′-deoxyadenosyl radical: captured and characterized by electron paramagnetic resonance and electron double resonance spectroscopies, *J. Am. Chem. Soc.*, 2019, **141**, 12139–12146.
2. A. J. Mitchell and J.-K. Weng, Unleashing the synthetic power of plant oxygenases: from mechanism to application, *Plant Physiol.*, 2019, **179**, 813–829.
3. Y. Nicolet, Structure-function relationships of radical SAM enzymes, *Nat. Catal.*, 2020, **3**, 337–350.
4. J. Bridewell-Rabb, T. A. J. Grell and C. L. Drennan, A rich man, poor man story of S-adenosylmethionine and cobalamin revisitied, *Annu. Rev. Biochem.*, 2018, **87**, 555–584.
5. J. Stubbe, Binding site revealed of nature's most beautiful cofactor, *Science*, 1994, **266**, 1663–1664.
6. R. Bannerjee, Radical carbon skeletal rearrangements: catalysis by coenzyme B12-dependent mutases, *Chem. Rev.*, 2003, **103**, 2083–2094.
7. W. Buckel and B. T. Golding, Glutamate ans 2-methyleneglutarate mutase: from microbial curiosities to paradigms for coenzyme B12-dependent enzymes, *Chem. Soc. Rev.*, 1996, **25**, 329–337.

8. M. Fontcave, M. Atta and E. Mulliez, *S*-Adenosylemthionine: nothing goes to waste, *Trends Biochem. Sci.*, 2004, **29**, 243–249.
9. H. J. Sofia, G. Chen, B. G. Hetzler, J. F. Reyes-Spindola and N. E. Miller, Radical SAM, a novel protein superfamily linking unresolved steps in familiar biosynthetic pathways with radical mechanisms: functional characterization using new analysis and information visualization methods, *Nucleic Acids Res.*, 2001, **29**, 1097–1106.
10. P. A. Frey, A. D. Hegeman and F. J. Ruzicka, The radical SAM superfamily, *Crit. Rev. Biochem. Mol. Biol.*, 2008, **43**, 63–88.
11. P. A. Frey, Lysine-2,3-aminomutase: Is S-Adenosylmethionine a poor man's adenosylcobalamin?, *FASEB J.*, 1993, 7, 662–670.
12. S. J. Booker, R. M. Cicchillo and T. Y. Grove, Self sacrifice in radical S-adenosylmethionine proteins, *Curr. Opin. Chem. Biol.*, 2007, **11**, 543–552.
13. C. T. Walsh and Y. Tang, *Natural Product Biosynthesis: Chemical Logic and Enzymatic Mechanisms*, Royal Society of Chemistry, London, 2nd edn, 2022.
14. P. V. Welander and R. E. Summons, Discovery, distribution, and phenotyoic characterization of a gene required for 3-methylhopanoid production, *Proc. Natl. Acad. Sci. U. S. A.*, 2012, **109**, 12905–12910.
15. S. Joshi, D. Fedoseyenko, N. Mahanta and T. P. Begley, Aminofutalosine Synthase (MqnE): a new catalytic motif in radical SAM enzymology, *Methods Enzymol.*, 2018, **606**, 179–198.
16. S. Sato, F. Kudo, M. Rohmer and T. Egichi, Characterization of radical SAM adenosylhopane synthase, HpnH, which catalyzes the 5′-deoxyadenosyl radical addition to diplotene in the biosynthesis of the C35 bacterialhopanepolyols, *Angew. Chem., Int. Ed.*, 2020, **59**, 237–241.
17. M. Dong, Y. Zhang and H. Lin, Noncanonical radical SAM enzyme chemistry learned from diphthamide biosynthesis, *Biochemistry*, 2018, **57**, 3454–3459.
18. J. Hollenberger, K. Ray and K. Meyer, The biuology and chgemistry of high valent iron-oxo and iron-nitrido complexes, *Nat. Commun.*, 2012, **3**, 720.
19. C. T. Walsh, B. P. Tu and Y. Tang, Eight kinetically stable but thermodynamically activated molecules that power cel metabolism, *Chem. Rev.*, 2018, **118**, 1460–1494.
20. W. T. Borden, R. Hoffmann and T. Stuyver, Oxygen: what makes this triplet diradical kinetically persistent?, *J. Am. Chem. Soc.*, 2017, **139**, 9010–9018.
21. C. T. Walsh and T. A. Wencewicz, Flavoenzymes versatile catalysts in biosynthetic pathways, *Nat. Prod. Rep.*, 2013, **30**, 175–200.
22. C. Krebs, D. G. Fujimori, C. T. Walsh and J. M. Bollinger Jr., Non-heme Fe(IV)-Oxo intermediates, *Acc. Chem. Res.*, 2007, **40**, 484–492.
23. J. W. Daly, D. M. Jerina and B. Witkop, Arene oxides and the N.I.H. shift: the metabolism, toxicity, and carcinogenicity of aromatic compounds, *Biochemistry*, 1972, **11**, 1961–1966.
24. P. F. Fitzpatrick, The aromatic amino acid hydroxylases, in *Advances in Enzymology - and Related Areas of Molecular Biology*, 2000, vol. 74, pp. 235–294.

25. E. Olsson, K. Teigen, A. Martinez and V. R. Jensen, The aroimatic amino acid hydroxylase mechanism: a perspective from computational chemistry, in *Advances in Inorganic Chemistry*, 2010, pp. 437–500.
26. S. Pascal, M. Taton and A. Rahier, Identification and charcaterization of two distinct microsomal oxidatrive enzymatic systems involved in sterol C4 demethylation, *J. Biol. Chem.*, 1993, **268**, 11639–11654.
27. W. D. Nes, Biosynthesis of cholesterol and other sterols, *Chem. Rev.*, 2011, **111**, 6423–6451.
28. H. J. Chan, K. Petrossian and S. Chen, Structural and functional characterization of aromatase, etstrogen receptor, and their genes in endocrine responsive and -resistant breast cancer cells, *J. Steroid Biochem. Mol. Biol.*, 2016, **161**, 73–83.
29. F. K. Yoshimoto and F. P. Guengerich, Mechanism of the third oxidative step in the convertsion of androgens to estrogens by cytochrom P450 19A1 steroid aromatase, *J. Am. Chem. Soc.*, 2014, **136**, 15016–15025.
30. C. T. Walsh and B. S. Moore, Enzymatic cascade reactions in biosynthesis, *Angew. Chem., Int. Ed.*, 2019, **58**, 6846–6879.
31. C. Helliwell, P. M. Chandler, A. Poole, E. S. Dennis and J. W. Peacock, The CYP88A cytochrome P450 ent-kurenic acid oxidase catalyzes three steps of the gibberellin biosynthesis pathway, *Proc. Natl. Acad. Sci. U. S. A.*, 2001, **98**, 2065–2070.
32. M. Mizutani and F. Sato, Unusual P450 reactions in plant secondary metabolism, *Arch. Biochem. Biophys.*, 2011, **507**, 194–203.
33. S. P. Salowe, W. J. Krol, D. Iwata-Reuyl and C. A. Townsend, Elucidiation of the order of oxidations and identification of a multistep clavaminate synthase reaction, *Biochemistry*, 1991, **30**, 2281–2292.
34. Z. Zhang, J. Ren, D. K. Stammers, J. E. Baldwin, K. Harlos and C. J. Schofiled, Structural origins of the selectivity of the trifunctional oxygenases clavaminate synthase, *Nat. Struct. Biol.*, 2000, **7**, 127–133.
35. B. Hadatsch, D. Butz, T. Schmiederer, J. Steudle, W. Wohlleben, R. Sussmuth and E. Stegmann, The buoisynthesis of teicoplanin-type glycopeptide antibiotics: assignment of P450 mono-oxygenases to side chain cyclizations of glycopeptide A47934, *Chem. Biol.*, 2007, **14**, 1078–1089.
36. Y. Zhao, R. J. A. Goode, R. B. Schittelheim, J. Tailhades and M. J. Cryle, Exploring the tetracylcization of teicoplanin precursor peptides through chemoenzymatic synthesis, *J. Org. Chem.*, 2020, **85**, 1535–1547.
37. N. I. Burzlaff, P. J. Rutledge, I. J. Clifton, C. M. H. Hensgens, M. Pockford, R. M. Adlington, P. L. Roach and J. E. Baldwin, The reaction cycle of isopenicillin N synthase observed by X-ray diffraction, *Nature*, 1999, **401**, 721–724.
38. X. Niu, J. Zhang, X. Xue, D. Wang, L. Wang and Q. Gao, Deacetoxycephalosporin C synthase (expandase): research progress and application potential, *Synth. Syst. Biotechnol.*, 2021, **6**, 396–401.
39. K. Valegard, A. C. T. van Scheltinga, A. Dubus, G. Ranghino and L. M. Oster, The structural basis of cephalosporin formation in a mon onuclear ferrous enzyme, *Nat. Struct. Mol. Biol.*, 2003, **11**, 95–101.

CHAPTER 13

Cascades vs. *Concerted Reactions*

13.1 Cascades *vs.* Concertedness

Synthetic chemists often use cascade reactions, where more than one chemical step is catalyzed in a one pot reaction without isolation of any intermediates, as practical and theoretical measures of efficiency of functional group or scaffold transformations.[1] Those abiotic cascade reactions can be driven by carbanion chemistry, by carbocation chemistry, or by carbon radical chemistry, three stepwise chemical processes. As a fourth alternative, one or more of the reactions could be concerted.

In concerted processes multiple bonds are broken and made in a single transition state.[2] There are no intermediates and no species of finite lifetime along the energy surface of the reaction coordinates. There are reactions where the transition state has a cyclic geometry, bonds forming as other bonds are breaking. These are defined as pericyclic reactions (Figure 13.1). While originally silent about how synchronous each bond cleavage and formation would be, later analyses predicted that concerted pericyclic reactions could be completely synchronous or have some asynchronicity where one or more bond breaking or forming steps could lag behind the others. Houk and colleagues proposed that asynchrony would be less than 60 femtoseconds, the time for a molecular vibration in a transition state.[3]

An enormous collection of apparently disparate abiotic concerted reactions were codified by the insights of Woodward and Hoffmann in the 1960s with their elaboration of *The Principles of Orbital Symmetry* that defined what pericyclic reactions were allowable thermally *vs.* those allowed photochemically.[4] Over the past 60 years[5] the question has arisen repeatedly[6] whether concerted reactions of organic compounds occur *in vivo.*

Chemical Biology No. 21
The Chemical Biology of Carbon
By Christopher T. Walsh

Published by the Royal Society of Chemistry, www.rsc.org

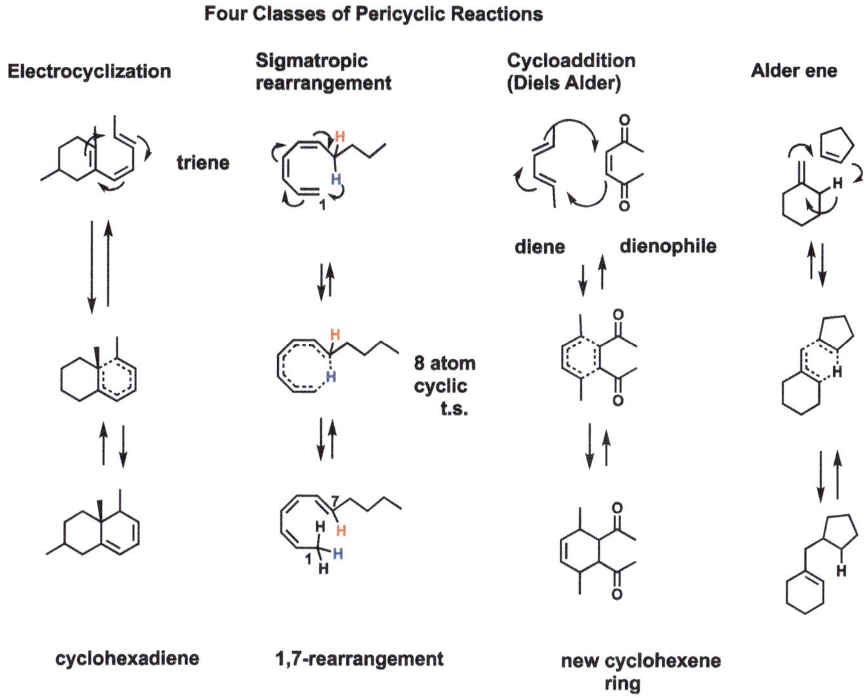

Figure 13.1 Four classes of pericyclic reaction that occur *in vivo*.

For example, pericyclic reactions, catalyzed by pericyclase categories of enzymes, would constitute a fourth alternative, to the carbanion-driven, carbocation-driven, and carbon-radical driven enzymatic reactions noted in preceding Chapters 10–12.

The twin issues of cascade reactions[7] and concertedness in types of enzymatic reactions are the subject of this chapter. We take up a brief reminder that we have encountered enzymatic cascades driven by carbanions, carbocations, and radicals. Then, we revisit three enzymes, squalene and oxidosqualene cyclases, nicotinamide dependent oxidase-decarboxylases, and flavin-dependent proton-hydride type dehydrogenases for analyses of stepwise *vs.* concerted reaction profiles before turning to four categories of pericyclase enzymes (Figure 13.1).

13.1.1 Enzymatic Cascades Driven by Carbanions, Carbocations, or Carbon Radicals

The polyketide synthase assembly lines that generate macrocyclic lactones such as deoxyerythronolide B or tylactone (Chapters 3 and 15), use iterated thioester enolates for chain elongations, carbanions derived from *in situ* decarboxylation of malonyl CoA equivalents. These are cascade reactions because only the full length cyclized polyketidyl chains are released.[1]

One can also argue for undecaprenyl-PP synthase which uses farnesyl-PP as substrate and then adds eight equivalents of Δ^3-isopentenyl-PP iteratively to chain extend by eight five-carbon prenyl units with *Z*-olefin geometry (Figure 2.17). This is cascade enzymology driven by allyl cation equivalents.

A third type of cascade is catalyzed by the enzyme that removes both of the C–CH_3 groups from carbon 4 of 3-keto-lanosterol. Each of the one carbon CH_3 groups are successively oxygenated, to alcohol, to aldehyde, and to carboxylate before release as CO_2 (Chapter 12). In all six of the oxygenation cycles in the oxygenase active site, carbon radicals are transient odd electron species captured by [OH•] equivalents.

Clearly enzymatic cascades are efficient in iterative modification of initial substrates and intermediates without release.[8] Doubtless, this increases both the probability of each catalytic event, such as chain elongation or bound substrate oxygenation, and the catalytic flux to the final released product. The three types of carbon-based intermediates discussed in Chapters 10–12 can clearly function in enzymatic cascades. In Section 13.3.2 on the bifunctional chorismate mutase-prephenate dehydrogenase, the two enzyme activities could be viewed as catalyzing a tandem pair of concerted transformations.[9]

13.2 Challenges for Establishing Stepwise *vs.* Concerted Reactions in Multistep Enzymatic Reactions

Enzymes always have one or more physical binding steps for substrate(s) before any chemical steps, followed by one or more physical steps leading to release of product(s). The kinetics of binding and release for multiple substrates and products may be strictly ordered or partially or fully random. There are often multiple chemical steps in the catalytic cycle. The step under examination for its stepwise or concerted nature may not be readily accessible or separable chemically or energetically from preceding or following steps.[3] All of these factors may well complicate the analysis of a single chemical step within the several steps of a catalytic cycle in any enzyme's active site.

13.2.1 Squalene and Oxidosqualene Cascades

One enzymatic arena examined for stepwise *vs.* concerted chemistry are the family of triterpene cyclases,[10] both the subclass of squalene/hopene cyclases[11] and the oxidosqualene cyclases noted at several points in previous chapters (*e.g.* Chapters 2, 3, and 11) as well as in Chapter 15. These enzymes are famous for the cascades of C–C bonds that they catalyze, as linear polyenes are converted to tetracyclic and pentacyclic carbon frameworks in bacterial, fungal, plant, and animal steroids.

Our previous focus has been on the nature of the tandem C–C bond constructions in the carbocyclic ring formations, with olefins acting as carbon nucleophiles that capture transient carbocations. We have characterized them as carbocation-driven cyclizations, similar to the types of cyclizations that occur in the smaller diterpenes, sesquiterpenes, and monoterpene scaffolds.

We also comment later in Chapter 15 on how distinct acyclic folded conformers of squalene or 2*S*-oxidosqualene in specific cyclase active sites can lead to chair–boat–chair or chair–chair–chair geometries that control stereochemistries, for example *en route* to lanosterol *vs.* euphol (Figure 13.2).

The reaction from squalene to hopene catalyzed by the hopene cyclases constructs 13 covalent bonds and sets 9 stereocenters at carbons[10] (Figure 13.3). There are some 57 natural products known in the hopene/hopanol series that reflect distinct enzyme guidance to the foldamers and cationic species in various squalene cyclase active sites.

Here we ask a different question. Are the cyclization cascades, olefins capturing carbocations, stepwise or concerted?[12] The question has been posed many times in the literature with some divided opinion, mostly on the question of whether formation of, say, the first three carbacyclic rings on the way to steroid cyclic scaffolds are concerted (Figure 13.4). After that point, there are so many branch point minor metabolites that indicate quenching of different carbocations that the later reactions to form the fourth or fifth

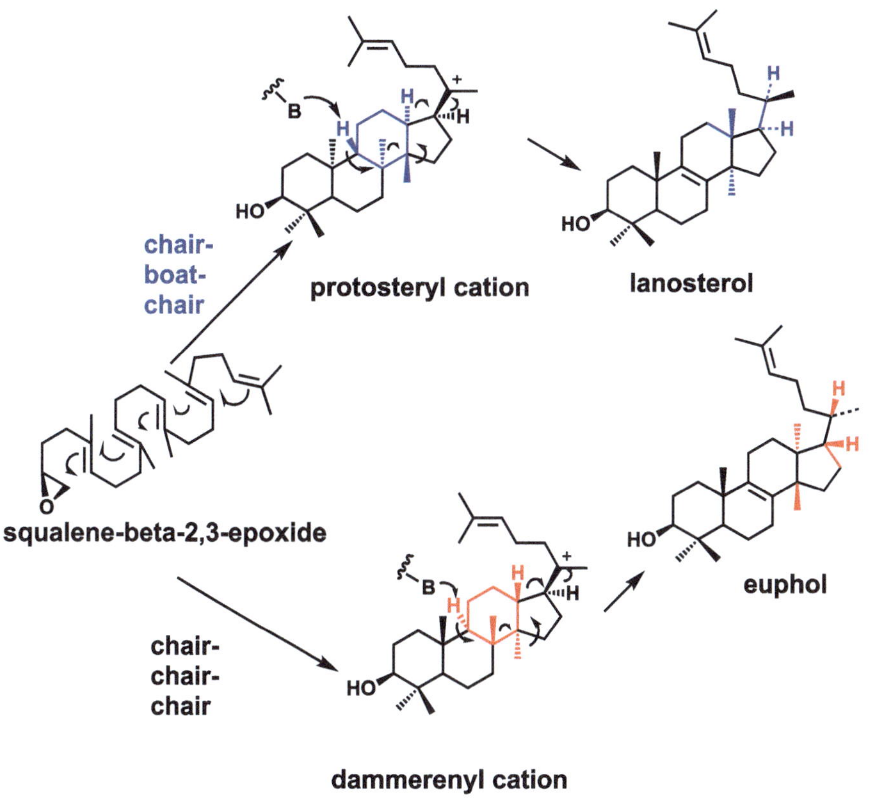

Figure 13.2 Partition of squalene-2,3-epoxide to protosteryl cation or dammerenyl cation through chair–boat–chair *vs.* chair–chair–chair foldamers on the way to lanosterol *vs.* euphol. Lanosterol and euphol are diastereomers at the four carbons indicated in blue or red respectively.

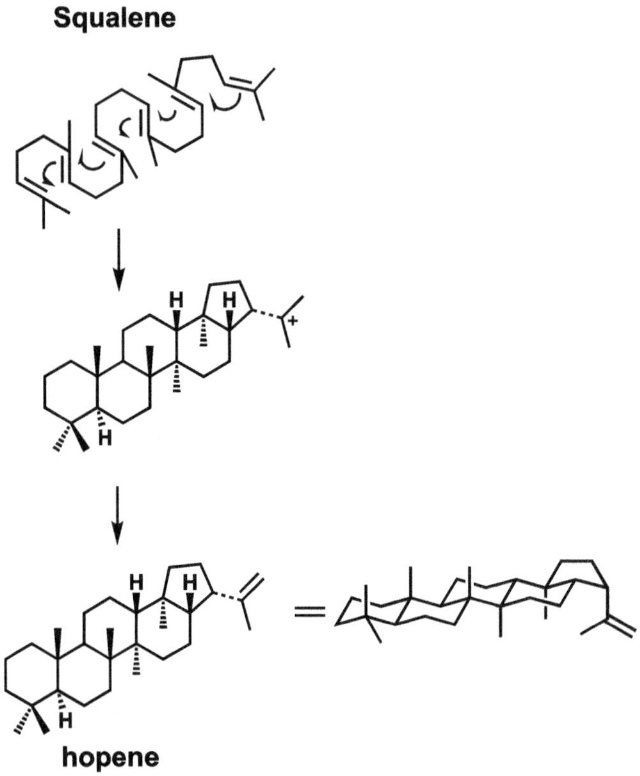

Figure 13.3 Squalene hopene cyclase generates new covalent bonds and sets stereochemistry at multiple carbon atoms.

carbacyclic rings seem clearly stepwise. In Chapter 15, Figure 15.11, we note that oxidosqualene cyclases can generate either tetracyclic protosteryl cations or the diastereomeric dammerenyl cation on the way to lanosterol or plant product euphol. To go from the dammerenyl cation to the major released plant sterol β-amyrin requires stepwise passage through another four rearranged carbocations.[13]

One can frame this triterpene cyclase issue from a different angle: the promiscuity of product formations, reflecting the release, anticipated or unanticipated quenching, of intermediate carbocations. In that regard baruol synthase directs ~90% of the flux to the major tetracyclic baruol product but also generates 22 minor products in amounts from 0.02% to 3% (Figure 13.5). The authors interpret the product profile as resulting from 14 different carbocations around the nascent cyclic scaffold in the cyclase's active site.[14]

In addition to a path to evolution of sterol diversity, these minor products suggest the inability of the enzyme active site to selectively lower energy barriers for only one product formation. In turn, these minor sterols tri- and tetracyclic versions presumably reflect diversion of some, if not all, of the stepwise carbocations generated during the cascade of squalene oxide to baruol.

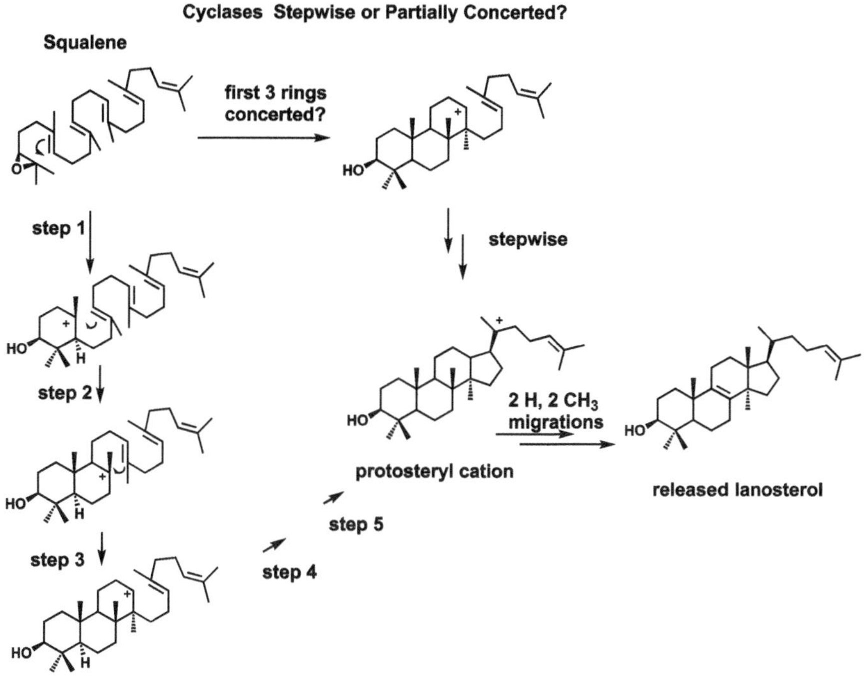

Figure 13.4 Oxidosqualene/lanosterol cyclase. Fully stepwise or partially concerted cascade?

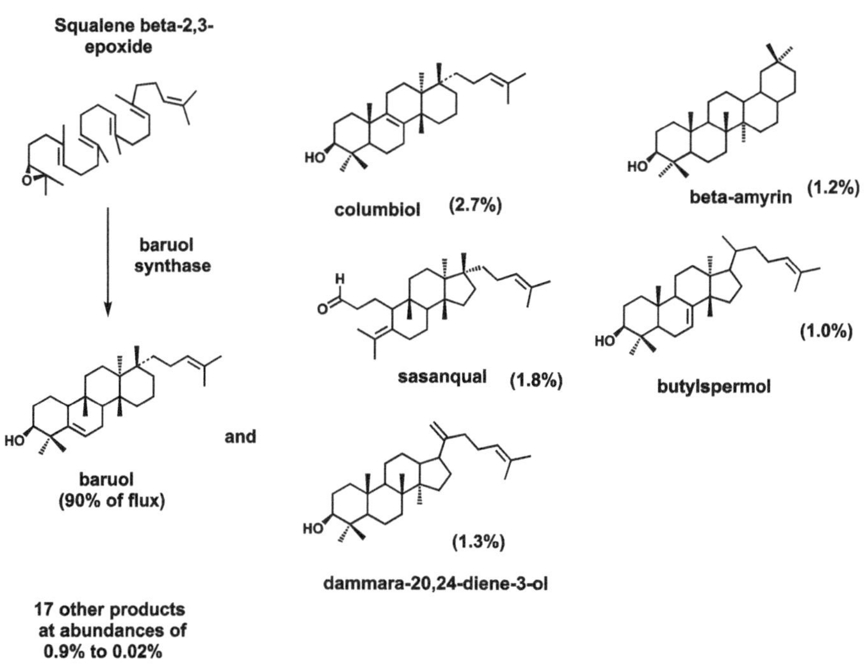

Figure 13.5 Baruol synthase: 90% of the flux to baruol. The next five of 22 minor products are shown with their relative abundances.

The later stage stepwise routing of carbocations does not speak to whether the first three rings (rings ABC) formed during action of steroid cyclase are themselves concerted or stepwise (Figure 13.4).

13.3 Double Isotope Effect Tests for Concerted *vs.* Stepwise Chemistry in Enzyme Active Sites

One kinetic approach developed in the 1980s, and recently reviewed, to get at stepwise *vs.* concerted catalysis during reactions that may be occurring at some stage in enzyme active sites was to examine isotope effects.[15] In particular, examination of the magnitude and effect on the size of a kinetic isotope effect by a second isotope at a second reacting center might inform whether both isotope-sensitive bonds were broken/formed (rehybridized) in the same transition state.

Figure 13.6 shows two options for the mechanistic course of isocitrate dehydrogenase. One is a concerted route where CO_2 formation and hydride transfer occur in the same transition state. The other is stepwise with initial hydride transfer during oxidation of the alcohol to the ketone in

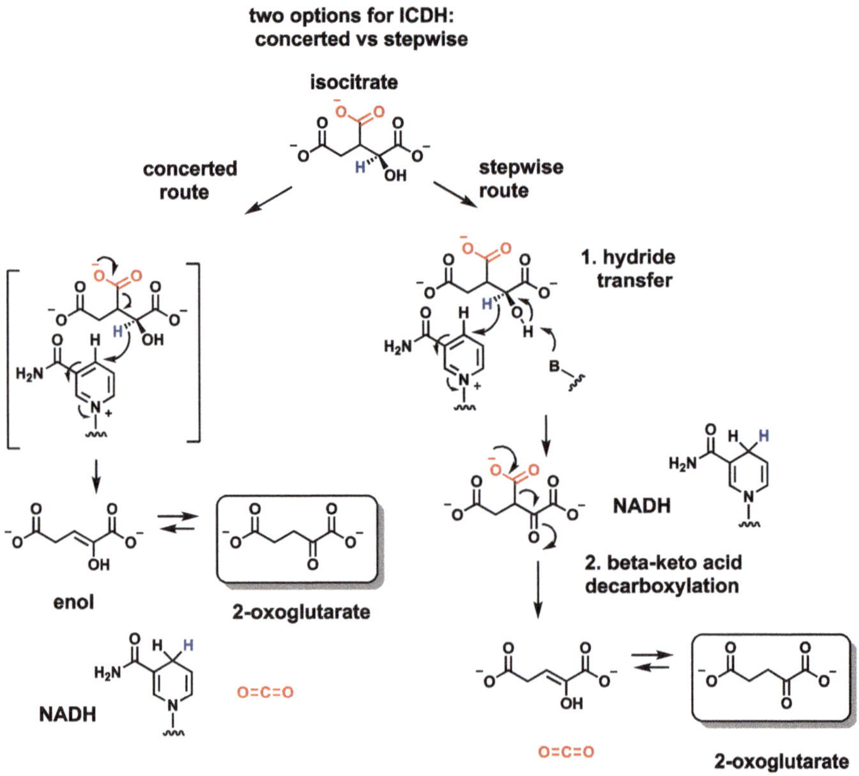

Figure 13.6 Concerted *vs.* stepwise possible routes for enzymatic oxidative decarboxylation of isocitrate to 2-oxoglutarate.

oxalosuccinate as intermediate. Then decarboxylation, specifically, of the red carboxylate now beta to the new ketone, would occur.

One approach to distinguish the stepwise from the concerted path was to examine the double isotope effects[16] where deuterium substitution was one isotope anchor that typically labels all molecules in a population and reports on V_{max}. Additionally, isotope effects on low abundance isotopes such as ^3H, ^{13}C, and ^{18}O that are minor species in the bulk population report on isotopically sensitive steps in V_{max}/K_m ratios[17] (Figure 13.6).

In a simple case, say for a desaturation of an alkane to an alkene by a flavin-linked dehydrogenase as examined in Chapter 2, one could prepare substrates deuterated at either of the C–H bonds. For dihydroorotate dehydrogenase, the sole redox enzyme in pyrimidine biosynthesis, that would be the 5-[^{2}H] and the 6-[^{2}H] or the doubly labeled 5,6-[$^2H_2$]-dihydroorotate to examine the mechanism of dihydroorotate dehydrogenase[18] (Chapter 6) (Figure 13.7).

Substitution of a C–H by a C–^{2}H bond lowers the zero point energy in the deuterated bond, provides a larger energy barrier to the transition state where the C–^{2}H bond breaks compared to the C–^{1}H bond and so retards the rate. Primary deuterium isotope effects can be up to a seven-fold reduction in rate. That full isotope effect may be expressed on V_{max} if that C–H bond cleavage step is fully rate-determining. If some other step is slow, the transmission of the intrinsic isotope effect may be anywhere from the full intrinsic effect down to 1.0. If a deuterium isotope effect on V_{max} is seen for a

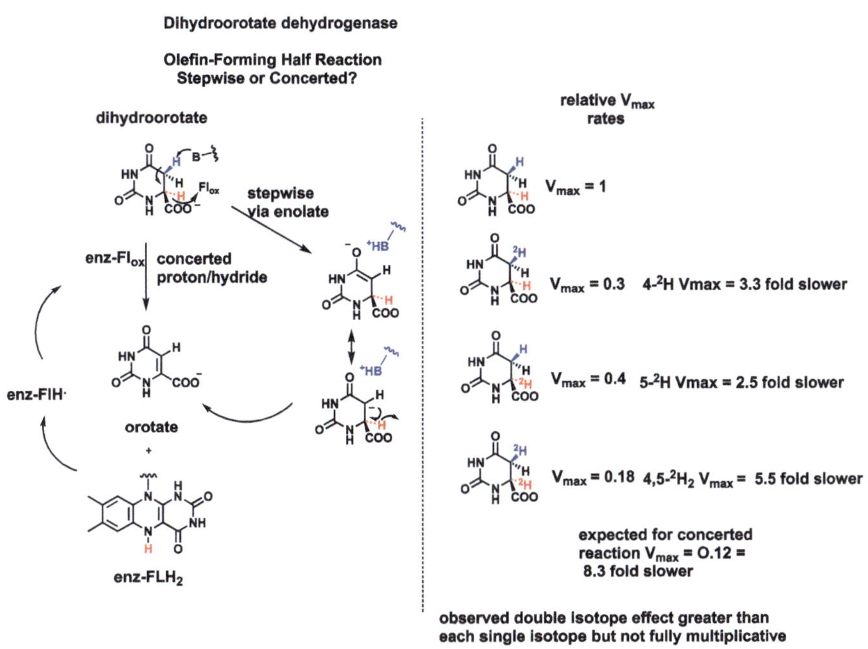

Figure 13.7 Mechanistic considerations for the oxidation of dihydrororotate to orotate (alkane to alkene group).

4-[^{2}H]-dihydroorotate and also for the 5-[^{2}H]-orotate, and the double isotope effect for 4,5-[$^2H_2$] dihydroorotate is close to the full multiplicative isotope effect, one could conclude that both C–H cleavage steps were slow, and occurring in the same transition state. That is, the bonds are breaking at the same time, definitionally[17] a concerted transition state. Three enzyme examples are briefly summarized using the double isotope effect approach.

13.3.1 Malic Enzyme

An early example of the double isotope fraction method,[17] by Cleland's group, was a $^{13}C_4$, ^{2}H–C_2 kinetic study of malic enzyme, an enzyme known both to oxidize C_2 of malate with hydride transfer to NAD^+ and to cause decarboxylation of the C_4-carboxylate to give CO_2 and pyruvate (Figure 13.8). From our earlier discussion on decarboxylations in Chapter 3, we would

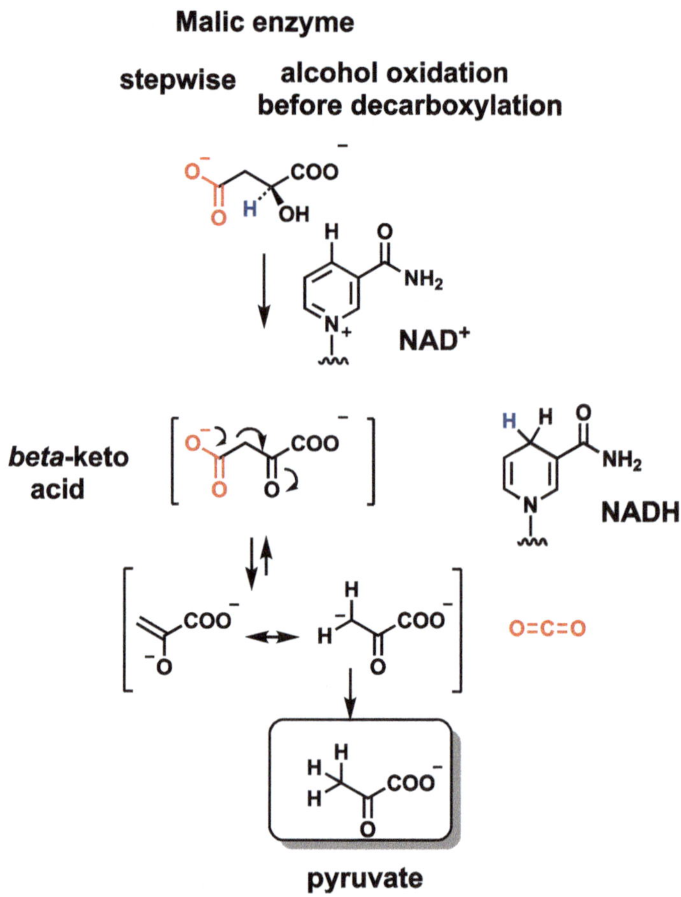

Figure 13.8 Malic enzyme catalyzes the oxidative decarboxylation of L-malate to pyruvate and CO_2 with hydride transfer to NADH. The *in situ* oxidation of malate to oxaloacetate precedes the decarboxylation of the C_4-carboxylate.

anticipate prior oxidation of L-malate to oxaloacetate in the enzyme active site. This would set up the new C_2 ketone β to the C_4 carboxylate to be lost as CO_2. The energy barrier for C_3–C_4 bond cleavage should be lowered because the incipient carbanion density at C_3 could be delocalized and stabilized by the C_3 enolate ion formed as nascent decarboxylation product. In the double isotope fractionation study. Hermes *et al.*[19] concluded the C_2–H and the C_3–C_4 bonds were cleaved in different steps, with C_2–H likely undergoing cleavage first. This is the exactly the insight one would gain from the organic chemistry analysis just noted.

13.3.2 Prephenate Decarboxylations: Three Variants

A second enzyme analyzed by the Cleland group[20] was the bifunctional chorismate mutase-prephenate dehydrogenase from *E.coli*. This bifunctional protein catalyzes two tandem metabolic reactions, converting chorismate all the way to *para*-hydroxyphenylpyruvate (Figure 13.9). We take up the first of these two enzyme, chorismate mutase in Section 13.4 below as a bona fide pericyclase, catalyzing a concerted 3,3-sigmatropic rearrangement of chorismate to prephenate. Here, the attention of Hermes *et al.* was on the second enzyme activity, the aromatizing decarboxylase of prephenate. As the C_2–C_1OO^- bond cleaves and carbanion density flows into the adjacent carbon, that carbon is part of the cyclohexadienyl ring of substrate prephenate. That electron density is used in the enzyme active site to expel a *para*-hydrogen as a hydride ion to cosubstrate NAD^+. The question was whether the C–COO bond cleavage was concerted with the hydride transfer at the other end of the developing phenyl ring (Figure 13.9).

The Cleland group concluded from a comparable [^{13}C,^2H] kinetic isotope study that the prephenate dehydrogenase reaction[20] was indeed concerted. By this analysis both the tandem 3,3-sigmatropic rearrangement of chorismate to prephenate and then the aromatizing decarboxylation/dehydrogenation of prephenate to *para*-OH-phenylpyruvate are concerted (Figure 13.9).

We note in passing there are two other structurally related enzymes that work on prephenate (Figure 13.10). One is the prephenate dehydratase, also an aromatizing decarboxylase, that expels the other substituent in the *para* position across the substrate's cyclohexadienyl ring, the 4-OH group. That ejection does not require a cosubstrate and gives phenylpyruvate, the precursor of L-phenylalanine, noted back in Chapter 7. One presumes the aromatizing mechanism for this enzyme may be similar to the dehydrogenase and if so, may also be a concerted decarboxylation, water elimination catalyst.

Yet a third prephenate decarboxylase in a small enzyme subfamily is related structurally but has an anomalous, non-aromatizing outcome.[21] The BacA enzyme in the bacterial pathway from prephenate to the antibiotic bacilysin decarboxylates prephenate but the product is an isomerized cyclohexadiene. The incipient carbanion from decarboxylation is trapped by delivery of a proton regio- and stereospecifically within the BacA active site to give the

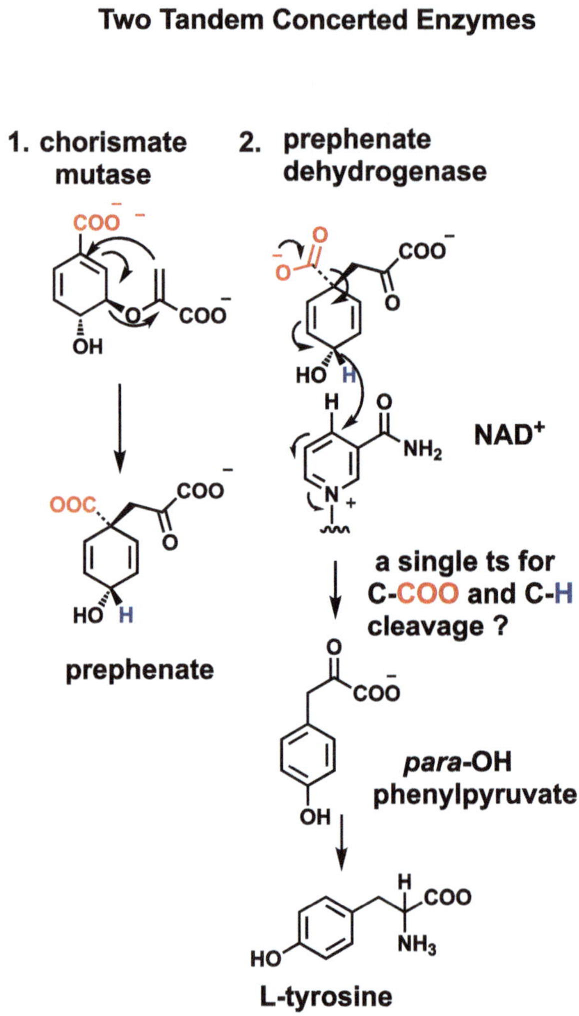

Figure 13.9 Chorismate mutase and prephenate dehydrogenase: two tandemly acting enzymes through concerted mechanisms.

endocyclic diene dihydro-4-hydroxyphenylpyruvate. That is unstable with respect to rearrangement to an exocyclic diene and subsequent reductive capture by NADH and BacB enzyme to give a stable H_2-hydroxyphenylpyruvate[21] (Figure 13.10).

The ability of the BacA to intercept a carbanion at C_6 and forestall aromatization suggests a finite lifetime for the initial carbanion from decarboxylation and raises the prospect that aromatization can be decoupled as a driving force for prephenate decarboxylation. This line of thought also raises the prospect of cyclohexadienyl carbanions of finite lifetime in this scaffold and thereby the possibility of stepwise reactions (Figure 13.10).

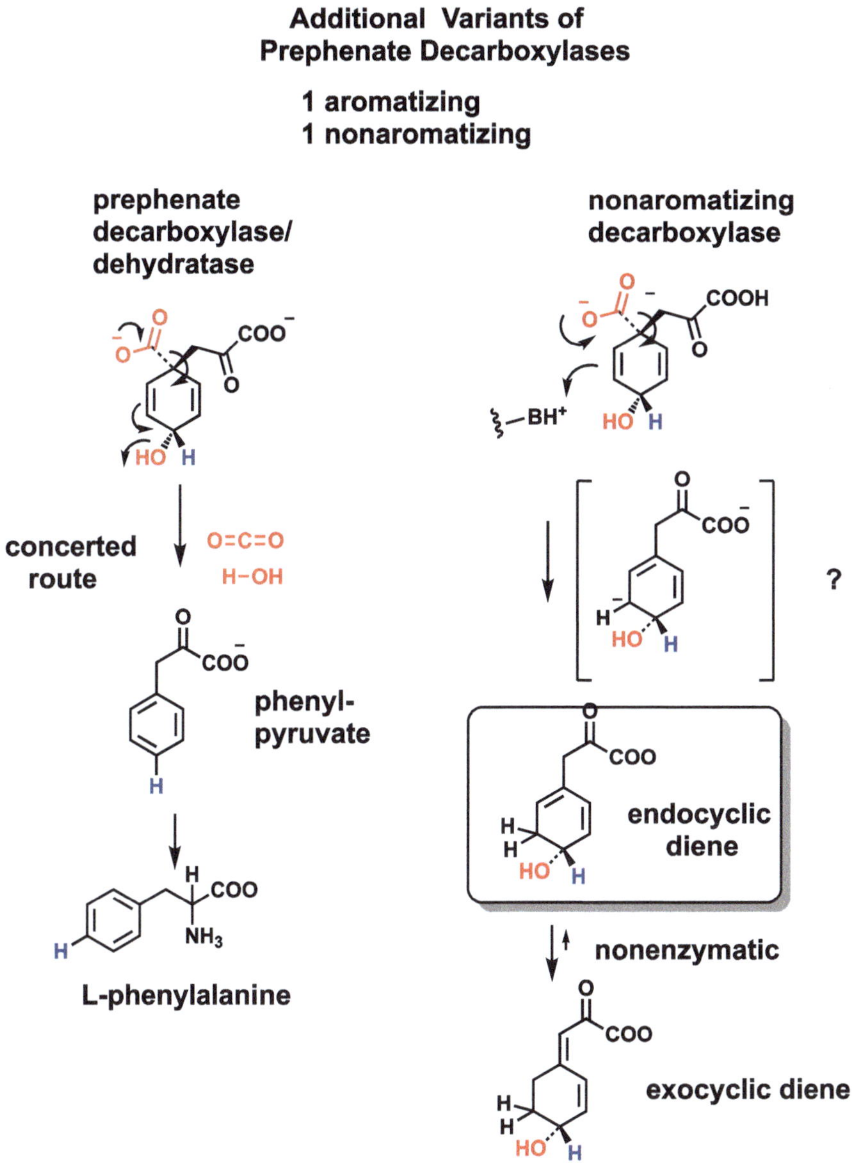

Figure 13.10 Two additional variants of prephenate decarboxylases: one aromatizing, one non-aromatizing.

13.3.3 Flavoprotein Dehydrogenases: Proton, Hydride Mechanisms

The third and final example to be discussed on the question of stepwise *vs.* concerted nonpericyclic reactions is the flavin-dependent dihydroorotate dehydrogenase example noted above (Figure 13.10). In Chapter 2 and subsequently we discussed conjugated olefin formation and saturation by proton hydride mechanistic routes. The proton would be removed from a C–H bond adjacent to a carbonyl group to give an energetically accessible enolate anion. The hydride would arise from cleavage of the adjacent C–H bond as it was ejected to N_5 of oxidized flavin coenzyme. It is reasonable to assume that these desaturations to conjugated enoyl groups are stepwise, with the polarity and timing as intuited from the organic chemistry analyses (Figure 13.11).

For dihydroorotate the reaction profile may be less straightforward. In the back reaction one would expect C_6 of orotate to be electrophilic, but the enzyme isolated from several sources does not run in the back direction from orotate to dihydroorotate. Determination of hydride transfer from C_5 of dihydroorotate to N_5 of FAD is experimentally difficult because the N5–H in

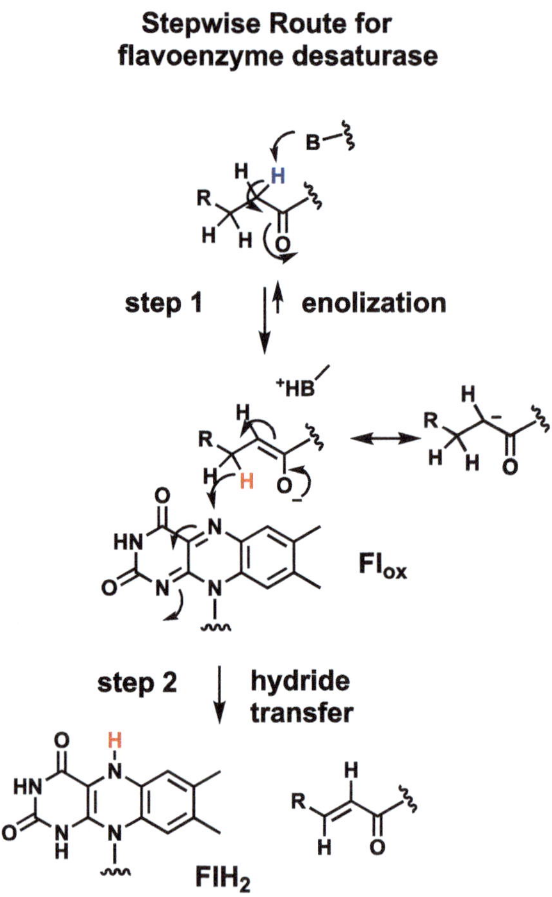

Figure 13.11 Desaturation to conjugated enones by stepwise enolization by proton abstraction and then ejection of the adjacent hydride by carbanion participation to form the olefin.

$FADH_2$ rapidly exchanges with solvent protons, obscuring a possible origin from substrate. In three separate cases, double deuterium isotope fractionation was performed with the enzyme isolated from the parasite *Crithidia fasciculata*, from bovine mitochondria, and from the bacterium *Clostridium oroticum*. The evidence for the parasite dihydroorotate dehydrogenase was interpreted in favor of a stepwise proton, hydride reaction coordinate.[18] From the latter two enzymes the interpretation was of a concerted proton-hydride desaturation mechanism. On the other hand, the human dihydroorotate dehydrogenase was classified as a stepwise catalyst.[22] These cases may suggest limits to the double isotope fractionation approach to give a unique answer and/or may indicate the same enzyme from three sources is poised to have different internal energetics.

13.4 Pericyclic Reactions

The most compelling arena for concerted mechanisms for C–C and C–H bond formation and cleavage may be the category of pericyclic reactions where bonds are made and broken in a cyclic single transition state. The exact timing of all bond formations/cleavages may be asynchronous but still within 60 femtoseconds of each other to avoid formation of an intermediate, a dip in the energy surface of the reaction coordinate.[3] Pericyclic reactions proceed with no finite intermediates.

Of the many abiotic variants of pericyclic reactions of organic molecules, four pericyclic reaction types, proceeding without participation of catalysts except heat to provide thermal energy for bond rearrangements, are of current relevance to carbon chemical biology.[3] These are sigmatropic rearrangements, electrocyclic reactions, Diels Alder type cyclizations of dienes and dienophiles and their heteroatom variations, and Alder-ene reactions (Figure 13.1).

The best known may be the cycloaddition reactions, pioneered by Diels and Alder[5] almost a century ago. As depicted in Figure 13.1, two new C–C σ-bonds are formed. In the reverse direction, cycloreversion, two σ-bonds are broken. By this metric cycloaddition/reversion reactions have a $\Delta\sigma = 2$. We note examples of Diels Alderase enzymes below.

Also famous in aromatic biosynthesis pathways is the enzyme chorismate mutase. This is a sigmatropic rearrangement catalyst. In sigmatropic rearrangements, both abiotic and biotic, one carbon sigma bond breaks as another forms, so the $\Delta\sigma$ count $= 0$. In electrocyclization reactions one C–C σ bond is made or broken (reverse direction). The $\Delta\sigma$ count $= \pm 1$. The counting of how many C–C sigma bonds are made or broken therefore allows chemists to distinguish among these three types of pericyclic reactions. This sigmatropic rearrangement is also an example of a Claisen rearrangement.[23]

The fourth type of pericyclic transformation of relevance to *in vivo* organic chemistry was also discovered by the Alder[24] of Diels and Alder. This is the Alder-ene reaction. As shown in Figure 13.1, in the Alder-ene reaction one C=C π-bond is cleaved while one new σ-bond is formed. No carbacyclic or heterocyclic rings are either formed or broken.

Examples of these four types of pericyclic reactions in carbon chemical biology are presented in the next four subsections. Pericylases have been identified and characterized for sigmatropic rearrangements, cycloadditions, and even Alder-ene reactions.[3] The sole example of a retro-electrocyclization is in the biosynthesis of the steroid framework of vitamin D3. No pericyclases have been found for this reaction. It is a photochemically driven nonenzymatic reaction driven by absorption of sunlight in skin epithelial cells.

13.5 Tandem Nonenzymatic Pericyclic Reactions in Vitamin D Biosynthesis

During the biosynthesis of endogenous cholesterol, the last enzymatic step is saturation of the double bond in the 5,7-endocyclic diene known as 7-dehydrocholesterol (Figure 13.12). In epithelial cells of the skin in humans exposed to sufficient sunlight, the cyclohexadienyl B-ring of 7-dehydrocholesterol can be opened by absorption of UV-B light in a slow, nonenzymatic, retro-form of a pericyclic electrocyclization.[25] For those individuals between the tropic of cancer and tropic of capricorn, the angle of incident sunlight is sufficient throughout the year for the endogenous production of steroid hormone precursor D3. For everyone else, D3 is a vitamin. Although beyond the scope of this discussion, D3 is a precursor form of the active D3 hormone known as calcitriol.[26] The first alcoholic group is from the C_3–OH of 7-dehydrocholesterol. The subsequent hydroxylations occur in liver and kidney by hemeprotein oxygenases and presumably involve transient carbon radical species at the site of hydroxylation (Figure 13.12).

Examination of the product from the nonenzymatic pericyclic ring opening of 7-dehydrocholesterol shows the product that is known as previtamin D3. The C_9–C_{10} sigma bond of 7-dehydrocholesterol has been broken ($\Delta\sigma = -1$) as the tetracyclic cyclohexadiene moiety in substrate has been converted to a tricyclic triene. All three of the C=C double bonds are conjugated.

Previtamin D3 is then subjected to a thermally allowed, still nonenzymatic, isomerization to vitamin D3[27] (Figure 13.12). This second pericyclic reaction qualifies as a sigmatropic rearrangement. One of the C–H bonds (a sigma bond) in the CH_3 attached to C_{10} in the still intact A-ring, is cleaved as the H moves to C_9 in the C-ring. Concomitantly, each of the three C=C double bonds moves in a single cyclic transition state. The result is conversion of the C–CH_3 in the starting A ring of previtamin D3 to the exocyclic methylene group and a diene connecting the A ring to the C, D bicyclic ring moiety. This is vitamin D3. One can define this second pericyclic rearrangement as a 1,7-sigmatropic shift. One C–H sigma bond is cleaved while a new C–H sigma bond is formed. The $\Delta\sigma$ count = 0, the metric for sigmatropic pericyclic reactions.

The endogenous conversion of 7-dehydrocholesterol to vitamin D3 is remarkable on several counts.[26] First, the process relies on ambient UV-B

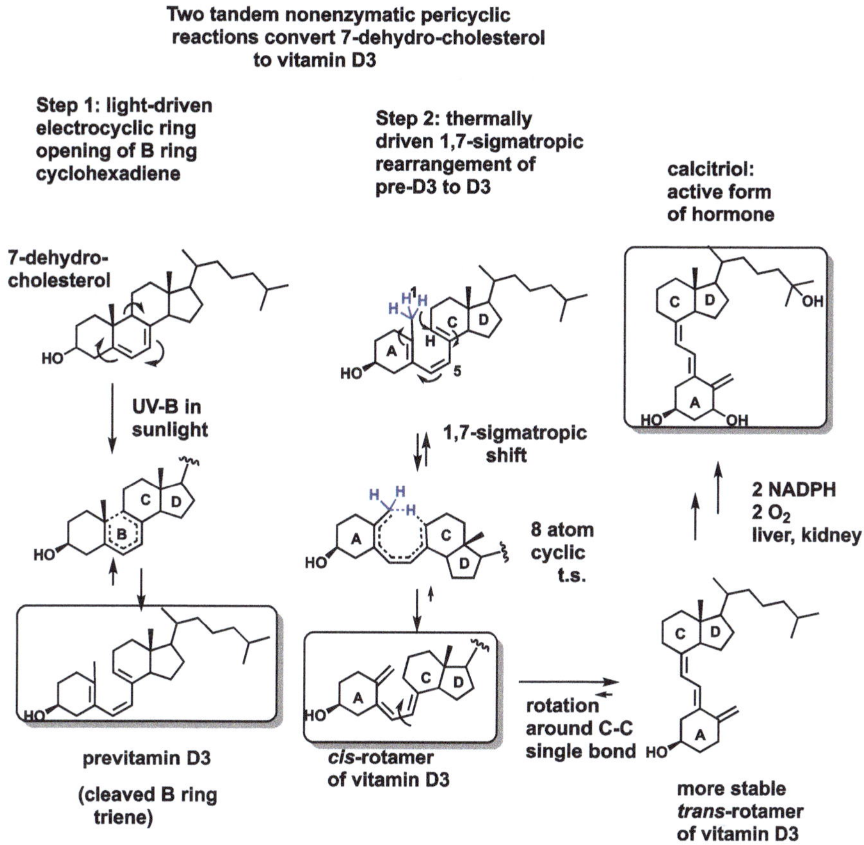

Figure 13.12 Two tandem nonenzymatic pericyclic reactions convert 7-dehydrocholesterol to vitamin D3. An initial retro-electrocyclization cleaves the B ring of the sterol to generated preD3. Then a 1,7-sigmatropic rearrangement forms vitamin D3.

light in sunlight to provide enough absorbed energy by the cylohexadienyl chromophore in 7-dehydrocholesterol to undergo the photochemically allowed pericyclic ring opening of the B-ring of the steroid. Second, the next reaction is another pericyclic reaction, also nonenzymatic, that is a thermally allowed (rather than the photochemically allowed first pericyclic reaction). That second reaction is a 1,7-sigmatropic rearrangement.[3]

The further downstream metabolite 1,3,25-trihydroxy vitamin D acts in both short and long term signaling. In the short term it accelerates calcium and phosphate uptake from the GI tract, essential for bone mineralization. In the longer term calcitriol acts as a steroid nuclear hormone turning on hundreds of genes in a variety of target tissues. Given the pleiotropic physiologic roles of D3 as a hormone, it is not surprising that many of the world's human populations, that cannot make enough vitamin D3 *via* the tandem pericyclic reactions, need dietary supplements of D3 as a vitamin.

13.6 Enzyme-catalyzed Sigmatropic Reactions

The nonenzymatic sigmatropic rearrangement of previtamin D3 to vitamin D3 is joined by at least three examples of enzyme-catalyzed sigmatropic rearrangements.[3] The best known is the 3,3-sigmatropic rearrangement of chorismate to prephenate[28] catalyzed by chorismate mutase[9] (Figure 13.13). We have already noted this path in Chapter 7 as the major biosynthetic flux to phenyl rings in aromatic metabolites in plant and microbial biosynthetic pathways. This example of conversion of an allyl vinyl ether to a $\gamma\delta$-unsaturated carbonyl system is the famous abiotic Claisen rearrangement, named after Rainer Ludwig Claisen[23] who discovered the rearrangement in 1912.

A second celebrated 3,3-sigmatropic rearrangement in abiotic organic chemistry is the Cope rearrangement,[29] named after the American chemist Arthur C. Cope (Figure 13.1). These are thermally allowed rearrangement of 1,5-dienes. Decades of study have validated a pericyclic reaction path *via* an aromatic transition state. Two C=C π-bonds are broken while two π-bonds are formed in the same cyclic transition state. The best studied example of a pericyclic Copase is in the biogenesis of hapalindoles noted in Figure 13.14.

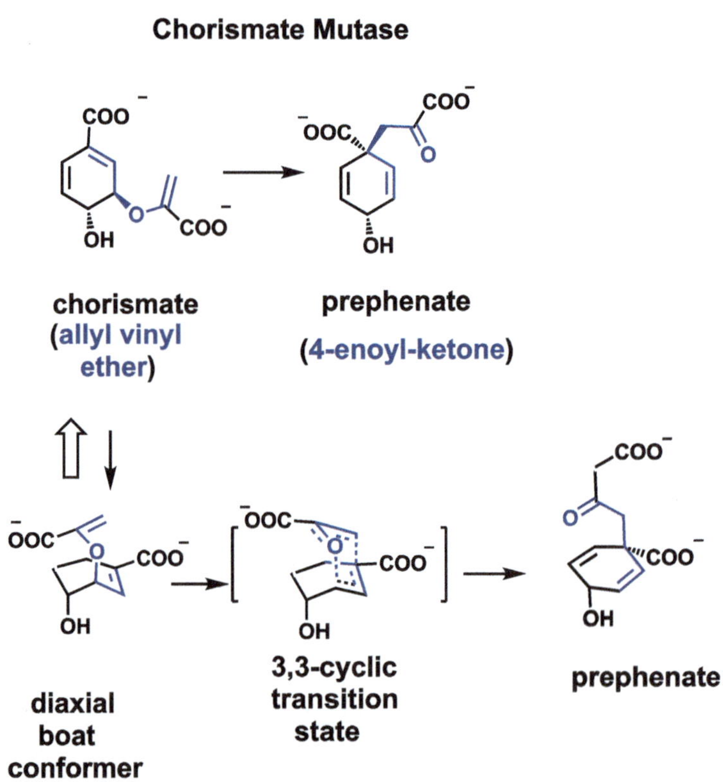

Figure 13.13 Chorismate mutase catalyzes a 3,3-sigmatropic rearrangement known as a Claisen rearrangement.

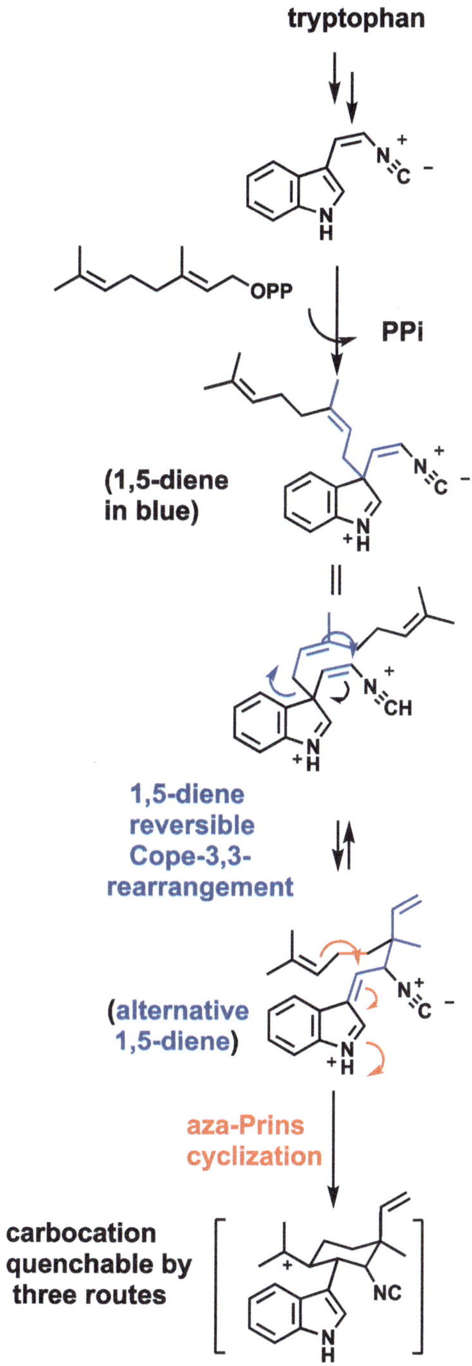

Figure 13.14 In hapalindole biosynthesis a 1,5-di-olefin part structure in a geranylated indole isonitrile is converted to the an isomeric 1,5-bis-olefin in a proposed 3,3-sigmatropic rearrangement in the Cope rearrangement variety. Then an aza-Prins coupling produces an isopropyl cationic advanced intermediate.

Proposals for 1,5-sigmatropic shifts have been advanced for isochorismate pyruvate lyase[30] and for CobH, the precorrin-8X mutase[3] in vitamin B12 biosynthesis, but we do not discuss those cases further here.

13.6.1 Chorismate Mutase – A Claisen Rearrangement Catalyst

Chorismate mutase has been characterized as a 3,3-sigmatropic rearrangement catalyst for 50 years. The substrate has both an allyl ether and a vinyl ether substituent proximal to the 1,2-endocyclic double bond in the carbacyclic ring. Mechanistic studies are consistent with the cyclic chair-like pericyclic transition state shown in Figure 13.13. A three million-fold rate acceleration over the nonenzymatic pericyclic 3,3-rearrangement has been estimated[31] *via* a pseudo-diaxial substrate conformer that can access a chair-like geometry in the cyclic transition state.

13.6.2 An Enzymatic Cope-3,3-rerrangement in Hapalindole Framework Maturations

Evidence interpreted in favor of a comparable enzymatic 3,3-sigmatropic Cope-type-1,5-diene interconversion has been accrued recently in a secondary metabolite pathway[32] involving a prenylated-*cis*-3-cyanoindole metabolite.[33] Early C_1–O bond cleavage as PPi departs would generate a familiar C_1-allyl cation, capturable by an enamino nitrogen to give the indolene depicted in Figure 13.14. This is the substrate for a proposed Cope isomerase, converting a 1,5-diene substrate to an alternative product diene (shown in blue[34] in Figure 13.15). This isomeric diene is not released from the active site but is then transformed in a doubly vinylogous aza-Prinz cyclization,[35] as a new cyclohexanyl-isopropyl cation forms (Figure 13.16, red arrows).

Three potential routes to carbocation quenching appear operant (Figure 13.15). One is by proton removal to 12-*epi*-hapalindole C with its terminal olefin and tricyclic scaffold. Alternatively, capture of the tertiary carbocation by from C_4 of the indole ring is on path to 12-*epi*-hapalindole U. If, instead, C_2 of the indole quenches the tertiary carbocation, the resultant framework is on path to 12-*epi*-fischerindole U. Both of the indole capture routes generate a new carbacyclic ring, creating a fused tetracyclic product scaffold. Each indole capture route reemphasizes the ability of C_2 and C_4 of indole to act as carbon nucleophile (Chapter 8).

In Section 13.2 we noted how difficult it can be to determine mechanistic details precisely within the cone of enzymes that hold onto substrates and intermediates tightly. The chemical step of interest may be occurring within a series of both preceding and subsequent chemical steps. This is the case for the hapalindole biosynthetic pathway. Investigators believe the embedded Cope rearrangement route is most reasonable but have not unambiguously proven the embedded 3,3-sigmatropic rearrangement proposal.[32]

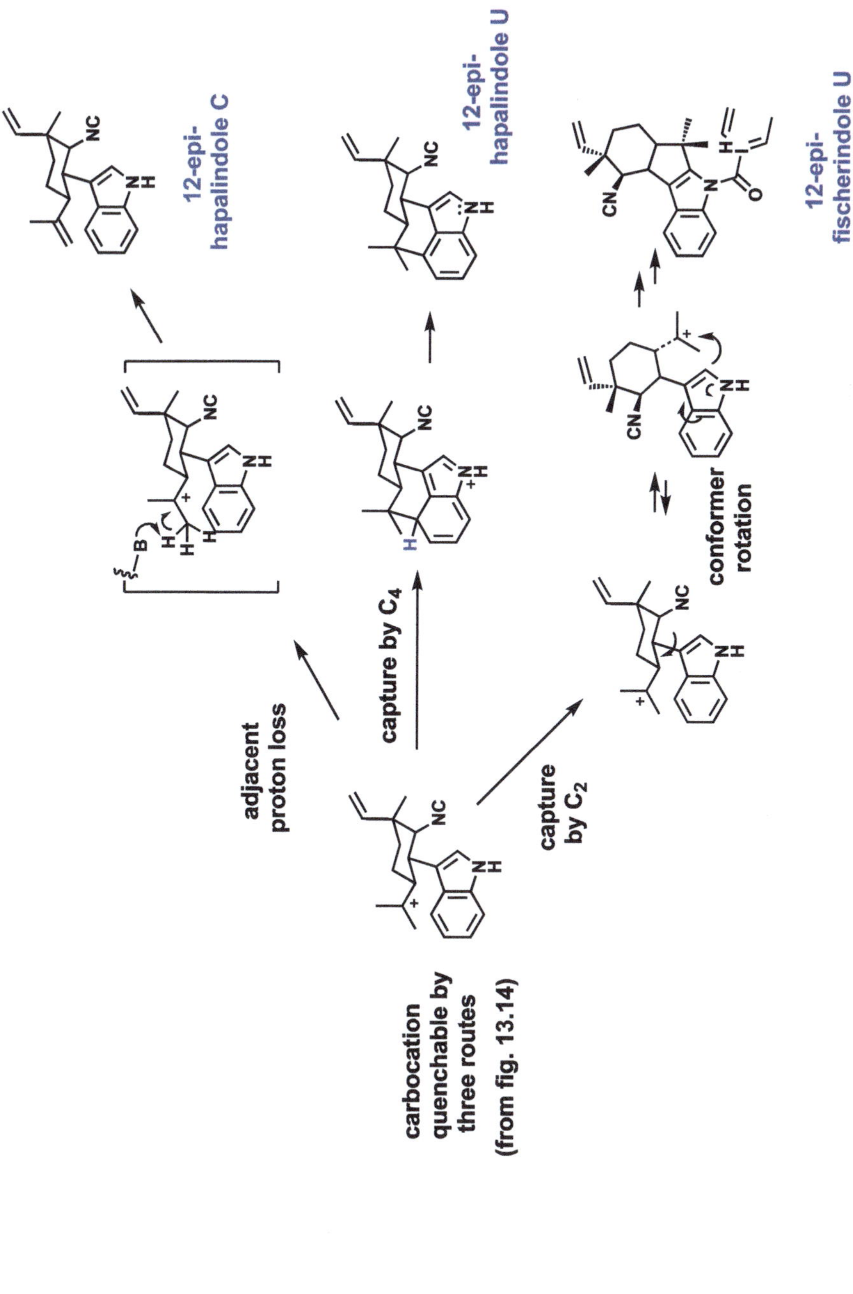

Figure 13.15 The isopropyl cationic intermediate can be quenched by three routes. Proton abstraction yields a terminal olefin. Internal capture by an indolyl carbanion equivalent at C_4 or C_2 are the other two quenching modes.

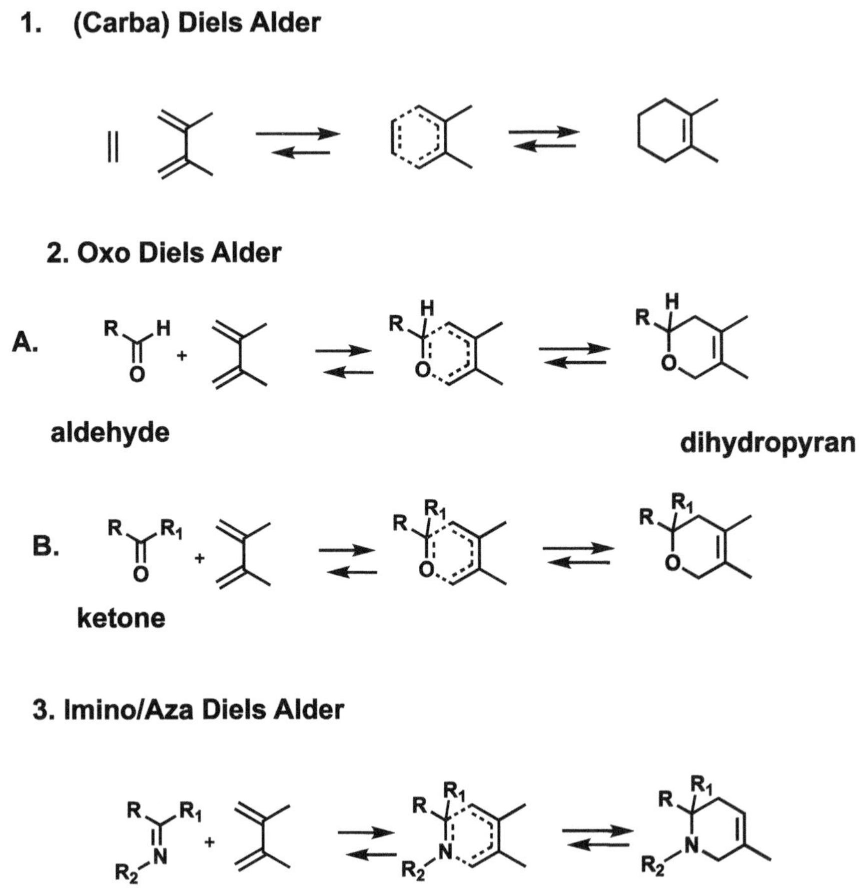

Figure 13.16 Diels Alder *vs.* hetero-Diels Alder cycloadditions. Normal (carba) Diels Alder 4 + 2π-electron cycloadditions yield cyclohexene rings. Oxo-Diels Alder reactions produce cyclic dihydropyrans. Aza-Diels Alder cycloadditions produce cyclohexenylamines.

13.7 Enzymatic Cycloadditions: Diels Alderases

The eponymous Diels Alder (4+2) pericyclizations of diene and dienophiles have drawn perhaps the most interest over the past half century to the question of whether there are biologic counterparts. That devolves to the question: are there Diels Alderase enzymes?[3] There appear to be no such examples in the primary metabolism of higher eukaryotes and megafauna, including humans. On the other hand, the explosion of genome sequencing, leading to more than 100 000 sequenced microbial genomes over the past two decades, has begun to provide new insights into the chemical logic of secondary metabolic pathways.

There are many natural product structures where Diels Alder cyclizations are chemically reasonable hypotheses for scaffold complexity generation. It has been possible in several cases to identify genes and encoded enzymes in those pathways and to characterize Diels Alderase catalysts.[3] Figure 13.16 sets out the essential components of Diel Alder cycloadditions: thermally allowed conversion of a diene and dienophile into a cyclohexene ring. Also depicted in Figure 13.16 are outcomes from oxo-Diels Alder cyclizations with either aldehydes or ketones that lead to substituted dihydropyrans. Similarly, aza Diels Alder reactions of dienes with imines create cylohexenylamine frameworks.

The regio- and stereochemistry of the abiotic Diels Alder cyclizations have been worked out in dozens of cases and all evidence is consistent with a pericyclic transition state and no intermediates of finite lifetimes.

The Diels Alderases classified to date carry out two kinds of cyclizations in natural product biosynthetic pathways.[36] One route leads to decalins, while the second route leads to spirotetronates or related spirotetramates, reflecting distinct functional group arrays in starting substrates (Figure 13.17). Decalin formation is the simplest outcome where the diene and dienophile react to create two new C–C sigma bonds ($\Delta\sigma = +2$).

The endocyclic olefin in the new cyclohexenyl ring of the decalin product arises from rearrangement of one of the starting diene olefinic pairs. The 4+2 concerted cycloaddition creates the new cyclohexene ring, *e.g.* in

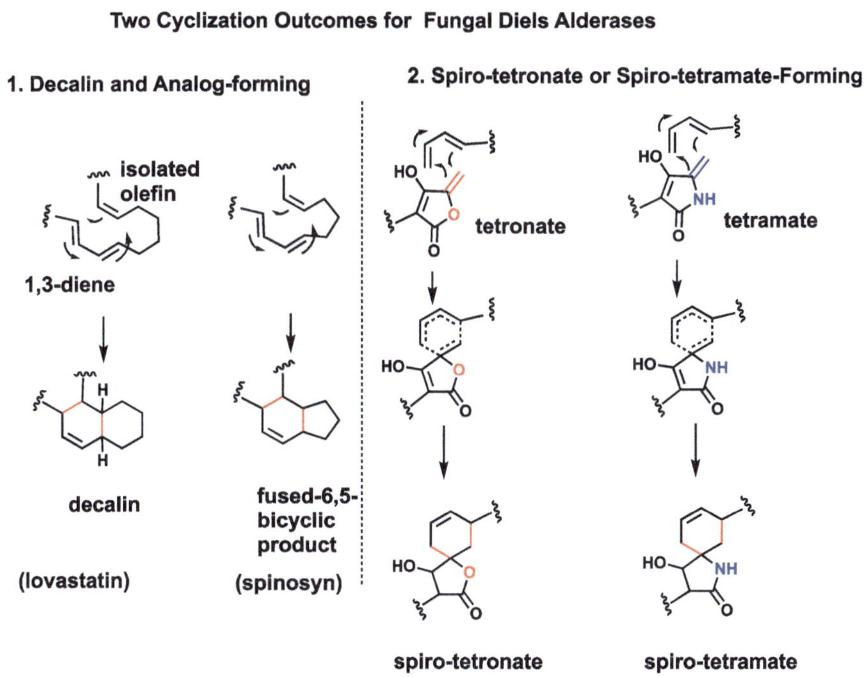

Figure 13.17 Two cyclization outcomes for fungal Diels Alderases. The new cyclohexene ring can be generated as part of a 6,6-fused bicyclic decalin or as part of a spirocycle with an exocyclic methylene group of tetronate or tetramate.

lovastatin[37] or mycellothermophilin.[38] A comparable cyclohexene ring is generated in spinosyn. However, the initial Diels Alder adduct in spinosyn is a fused 6,5 bicyclic ring system rather than the 6,6-decalin system because of the number of carbon atoms between diene and isolated double bond that acts as dienophile.

For spirotetronate or spirotetramate formation, typically the exocyclic olefin in the cyclic tetronate/tetramate is the dienophile, capturable by a diene moiety elsewhere in the intramolecular framework. The result is again a newly created cyclohexene ring, now in spiro connection to the preexisting tetronate/tetramate ring (tetramate five member heterocycles typically arise by Dieckman-type cyclizing release of a scaffold from a mixed polyketide/nonribosomal peptide synthetase[39]).

13.7.1 Spirotetronate Formation in Abyssomycin Biosynthesis

Biosynthesis of the polyketide antibiotic abyssomycin (Figure 13.18) may be the clearest example of participation of a spirotetronate-forming Diels Alderase enzyme AbyU.[40]

A polyketide synthase assembly line cascade transfers the growing polyketide chain to form an unusual glyceryl-*S*-thioester with subsequent cyclization to release a five member hydroxytetronate ring at one end and a linear triene at the other end. Tetronates are a subclass of unsaturated cyclic γ-lactones. Figure 13.18 also shows the atrop isomer of abyssomycin C also produced by the deep sea marine streptomycete bacterium. *Atrop* isomers are nonsuperimposable because of restricted rotation. In this case the indicated carbonyl group cannot freely equilibrate its geometry because of the size limitations in the 11-member ring. Altogether, the highly constrained architecture of abyssomycin C isomers arise from the interlaced tetracyclic framework. Finally, in Figure 13.18 the part structure of the initial Diels alder product from AbyU action is shown as the spiro cyclohexenyl-tetronate. Comparison with the structure of the final abyssomycin metabolites indicates additional enzyme chemistry must occur on the AbyU product.

The terminal diene of that linear triene moiety will be one of the -participants in an AbyU-catalyzed Diels Alder cyclization[40] (Figure 13.19). The requisite olefinic partner is not present in the PKS assembly line product. It must be generated by dehydration of the original hydroxytetronate ring to give the tetronate with exocyclic methylene group that becomes the dienophile.[40] AbyU is first a dehydratase and only then a Diels Alderase. It must also fold up that proximal intermediate to a conformer (Figure 13.19) where the terminal diene and the exocyclic methylene of the tetronate can undergo the $4+2\pi$-electron cycloaddition reaction. AbyU accelerates the Diels Alder cyclization by about a million fold over nonenzymatic background cycloaddition.[40] The newly formed cyclohexene ring has been created in spiro fusion to the preexisting tetronate ring (Figure 13.19). Note that an ancillary feature

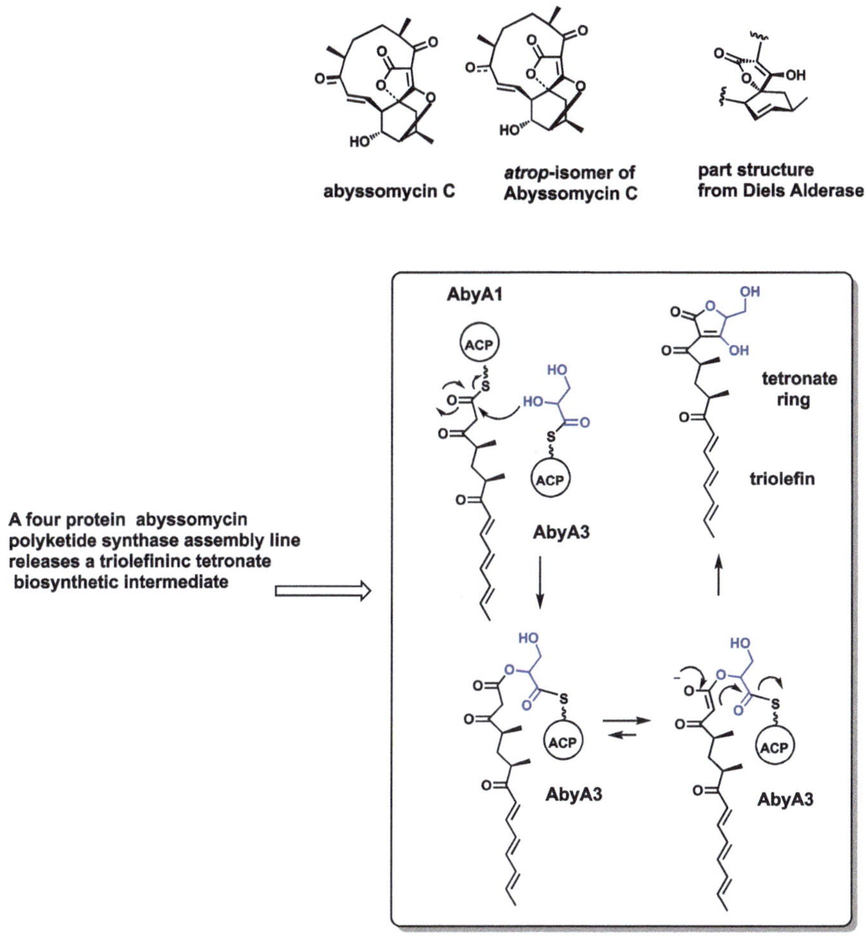

Figure 13.18 Abyssomycin C and *atrop*-isomer. Spirotetronate part structure from AbyU Diels Alderase. PKS assembly line releases a hydroxytetronate with an extended triene part structure.

to this enzymatic Diels Alder reaction is the creation of the macrocyclic 11-member ring in the AbyU product.

Two additional steps remain to convert the tricyclic AbyU product to the tetracyclic framework of abyssomycin (Figure 13.19). The first is enzymatic epoxidation of the new olefin in the just formed cyclohexene ring, example of an olefin as nucleophile, and generation of an electrophilic epoxide. As noted earlier in Chapter 3 epoxides can be opened intramolecularly by nearby hydroxyl groups acting as nucleophiles. The oxygen of the enol group in the tetronate is nucleophilic because that enol accumulates rather than the keto form to give the αβ unsaturated lactone conjugation. That enol oxygen opens the epoxide, forming the scaffold of the fourth ring in the constrained architecture of abyssomycin.

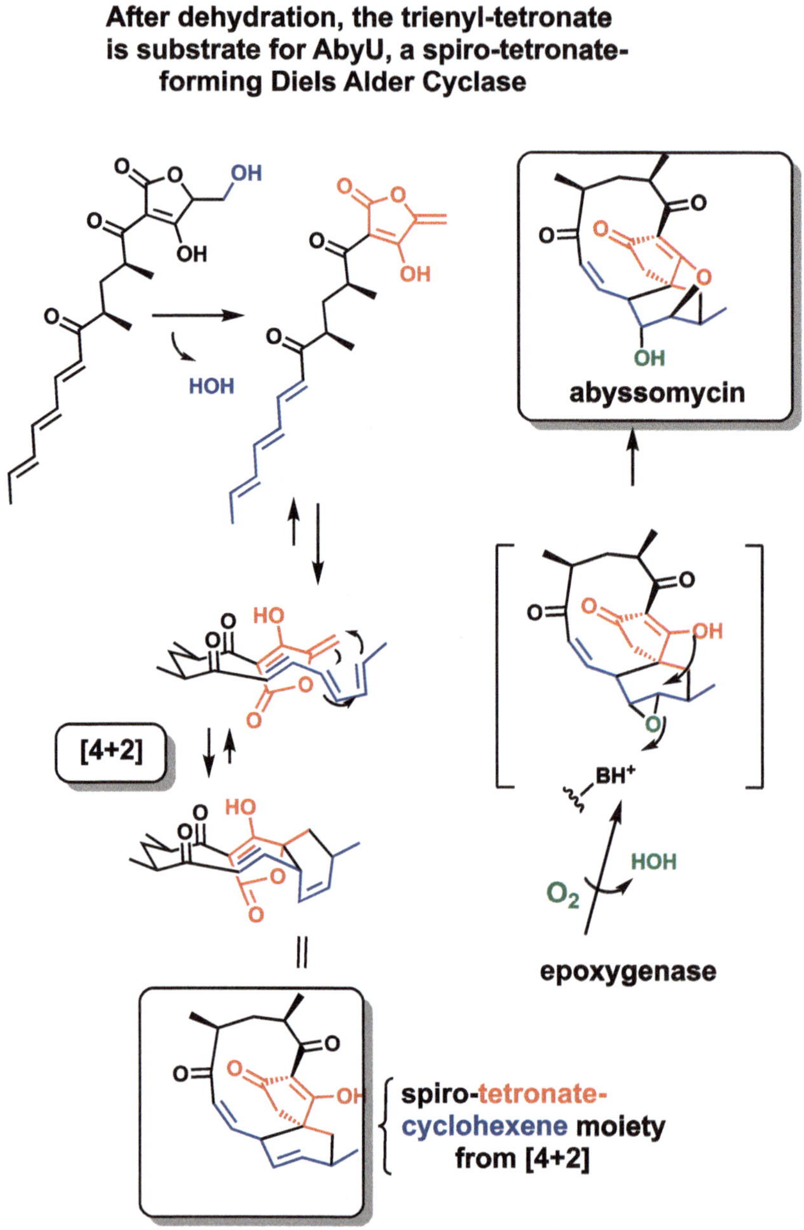

Figure 13.19 AbyU catalyzes dehydration on the hydroxytetronate substrate to create the exocyclic methylene group in the tetronate functional group to act as the dienophile in reaction with the terminal diene of substrate. The resulting spiro cyclohexenyl-tetronate part structure is further modified by epoxidation of the cyclohexenyl double bond and internal opening by the –OH group of the tetronate moiety.

13.7.2 Decalin Formation in Lovastatin Biosynthesis

The same underlying chemical logic, a $4\pi+2$ pericyclic cycloaddition, accounts for enzymatic construction of the fused decalin array in the cholesterol-lowering natural product lovastatin[41] (Figure 13.20). Semisynthetic versions of the lovastastin core have led to the statin class of cardiovascular drugs that lower cholesterol levels by inhibition of HMG CoA synthase, the rate-limiting enzyme in the cholesterol biosynthetic pathway (see Chapter 15). Many other decalins in secondary metabolites may be assembled by this enzymatic logic.

The lovastatin core is assembled by complementary action of two polyketide synthase type enzymes LovB and LovC. LovB is a classical cascade PKS, catalyzing eight chain elongation cycles, each time using decarboxylation of malonyl CoA to generate a two-carbon acetyl carbanion as carbon nucleophile (Figure 13.20). A flavoprotein enoylreductase (ER) domain is present in LovB but inactive. LovC is a standalone ER catalyst.

The first two elongation cycles of LovB bring a diene forward (no ER function), then the LovC ER works in *trans* for the next two elongation cycles to yield a pentaketidyl-*S*-enzyme, where the early diene has been carried through (Figure 13.20). In the next elongation cycle to the hexaketidyl chain, LovC does not recognize that new olefin, so now a diene and an isolated olefin persist.

The Diels Alder pericyclic ring forming step occurs at the hexaketidyl level as schematized in Figure 13.20. The formation of the cyclohexene ring occurs simultaneously with the bicyclic decalin.[42] The bicyclic decalin moiety is then carried through three additional elongation cycles from hexakide to full length nonaketidyl-*S*-enzyme. Hydrolysis at that stage yields dihydromonacolin L as the metabolite that is finally released from the cascade. The outlines of the lovastatin mature scaffold are visible in that metabolite.

Note that the Diels Alder 4+2 cyclization has occurred on the growing chain in the LovF polyketide cascade assembly line. The three olefins in the hexaketidyl intermediate have the spacing of a 1,3-diene and then four-carbon spacer before the isolated olefin (dienophile). That linear four-carbon spacing becomes the adjacent fused cyclohexane ring in the decalin product of the Diels Alder cyclization.

13.7.3 Tandem Diels Alderase Action

Both types of Diels Alder enzymatic cyclizations occur in two microbial natural product biosynthetic pathways, to pyrroindomycins and chlorothricins.[3] Figure 13.21 indicates that PyrE3 dehydrates the hydroxytetronate to the exomethylene ring but then acts on the other triene system noted in red. Diels Alder concerted cyclization creates the cyclohexene and fused cyclohexane that is the decalin system shown in red. Then, PyrI4 generates the spirotetronate functionality. By tandem action the two enzymes construct the pentacyclic core of pyrroindomycins. The PyrE3 prior dehydration to set up a C=C double bond for subsequent Diels Alder cyclization is a recurrent theme in how enzymes set up these pericyclic accelerations.

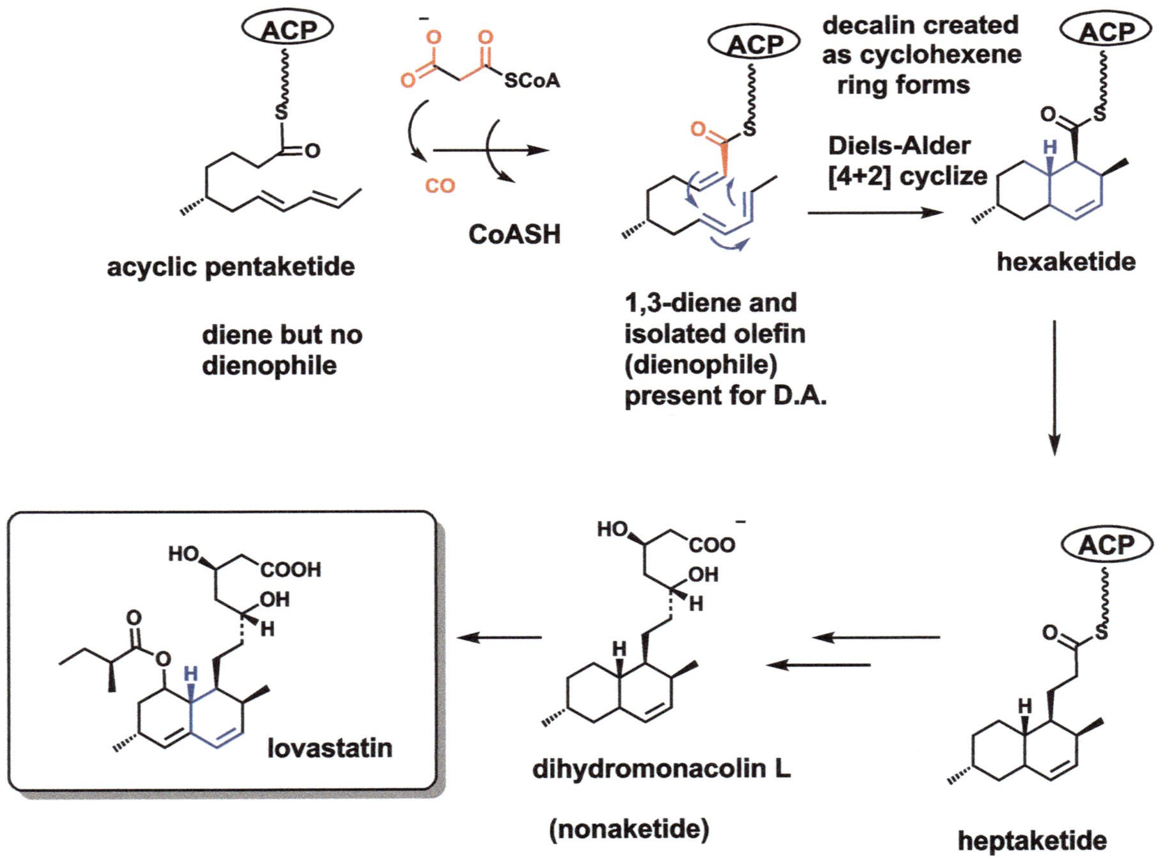

Figure 13.20 The decalin ring system in lovastatin is generated by LovF acting as a Diels Alderase at the level of a covalent hexaketidyl-*S*-enzyme intermediate. The decalin ring is carried through three additional PKS elongation cycles and released as dihydromonacolin L.

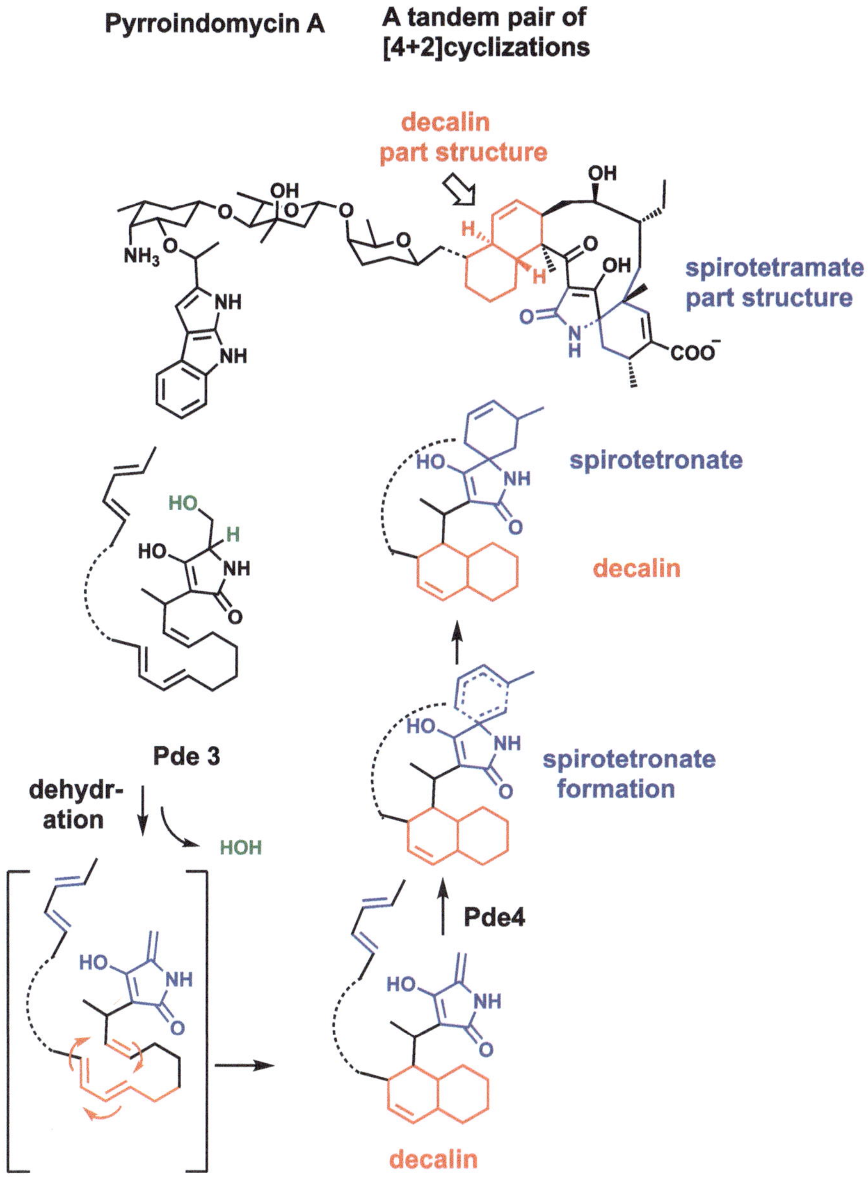

Figure 13.21 Two Diels Alderase enzymes work in tandem to first construct a decalin part structure and then the spirotetronate core of pyrroindomycin A.

13.7.4 Partitioning a Diene and Dienophile by Two Routes in Catharanthine and Tabersonine Alkaloids

The periwinkle plant turns out to be an expert and inventive alkaloid chemist, constructing a variety of complex indole alkaloids, including some molecules used as clinical antitumor agents. In the context of Diels Alder pericyclic enzyme logic the *O*-acetyl-stemmadenine alkaloid undergoes enzyme-catalyzed fragmentation to dehydrosecodine, an indolyl dihydropyridine structure.[43] The dihydropyridine ring has two double bonds and remarkably is utilized either as diene (to catharanthine) or dienophile (to tabersonine).

The ability of tabersonine synthase *vs.* catharanthine synthase to differentiate the function of the two double bonds in that nitrogen heterocycle must depend on different substrate conformers in the two Diels Alderase active sites[44] (Figure 13.22).

To get to catharanthine the pair of double bonds in the dihydropyridine would react as the diene component while the conjugated olefinic vinyl group of the side chain would be the dienophile. A 4+2 cyclization creates the

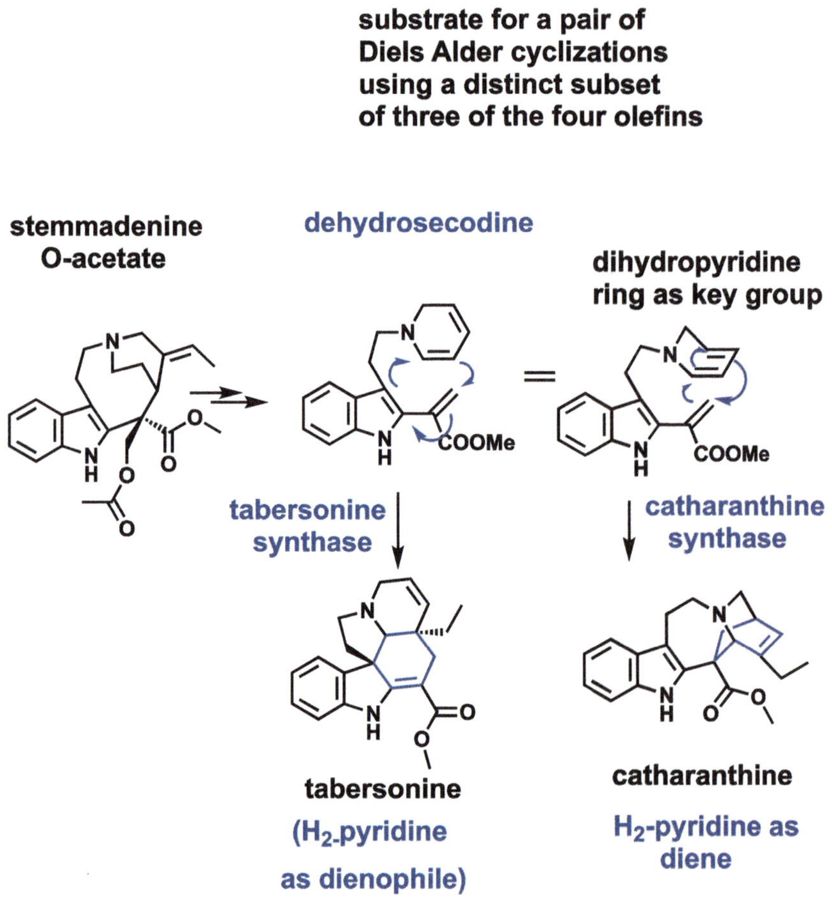

Figure 13.22 In biosynthesis of complex periwinkle indole alkaloids, stemmadenine-*O*-acetate is precursor to two different regiospecific Diels Alder enzymatic cyclizations. Tabersonine synthase uses only one of the two double bonds of the dihydropyridine as dienophile. Catharanthine synthase instead uses the dihydropyridine as diene to create a cyclohexene ring in a different pentacyclic product scaffold.

aza-bicycloctane ring with an endocyclic double bond in catharanthine. To get to tabersonine from the same tetraene substrate, a foldamer conformation in the active site is proposed that uses one of the two double bonds in the dihydropyridine ring as the dienophile. The diene moiety is provided by the exocyclic terminal olefin and the endocyclic 2,3-double bond in the pyrrole moiety of the indole ring. That cyclic transition state builds a new cyclohexene ring and a five member pyrrolidine ring fused to an indoline.

13.7.5 SpnF in Spinosyn Assembly: Ambimodal Transition States

A well-studied variant of the decalin-forming subclass of Diels Alderases is the enzyme spnF in the spinosyn biosynthetic pathway[41] (Figure 13.23). SpnF acts to create first the 6-5 fused carbacycle. Then, the next enzyme SpnL builds the third ring in a 5-6-5 tricyclic array embedded within the 12 member macrolactone. The nonenzymatic pericyclic reaction to form the 6–5 ring system occurs at detectable rates, but is accelerated some 500 fold by SpnF.[41]

The initial product released from the polyketide synthase assembly line (Figure 14.23) is the 12 member macrolactone. Further enzymatic maturation builds the fused 5-6-5 tricyclic part of the spinosyn aglycone. Lastly, the sugars are attached to create spinosyn A. In analogy with some of the previous cases of pericyclases, the dieneophile required for spnF to act as Diels Alderase is not present in the substrate and must be uncovered during catalytic cycles. As shown in Figure 13.24, several functional group chemical principles, as elaborated in earlier chapters, are in play. First the red alcohol is oxidized to the ketone, enabling adjacent enolization to a carbanion equivalent that acts as doubly vinylogous dienolate-carbon nucleophile to eliminate the indicated alcohol. This action creates a conjugated dienyl ketone part structure that has introduced the double bond that will serve as dienophile in the subsequent Diels Alder cyclization.

Further analysis of the SpnF cycloaddition reaction has led to intensive computational analysis. These efforts have suggested a possible ambimodal pericyclic transition state for Spnf[3] where the substrate could either be processed by the 4+2 pericyclic route denoted in Figures 13.23 and 13.24 or by a pericyclic 6+4 transition state (Figure 13.25). The [6+4] product is not detected, but energetics calculations suggest it could undergo a 3,3-sigmatropic rearrangement (Cope rearrangement of 1,5-dienes) to the fused 6,5-observed SpnF product. The Houk group has suggested ambimodal pericyclic transition states are possible in several proposed pericyclases,[3] indicating low energy barriers between different pericyclic outcomes.

13.8 Hetero-Diels Alderases

Among the many directions that synthetic chemists have explored in applications of abiotic Diels Alder cycloaddition chemistry have been the utilization of hetero-Diels Alder cyclizations.[45] One of the participating C=C bonds, either

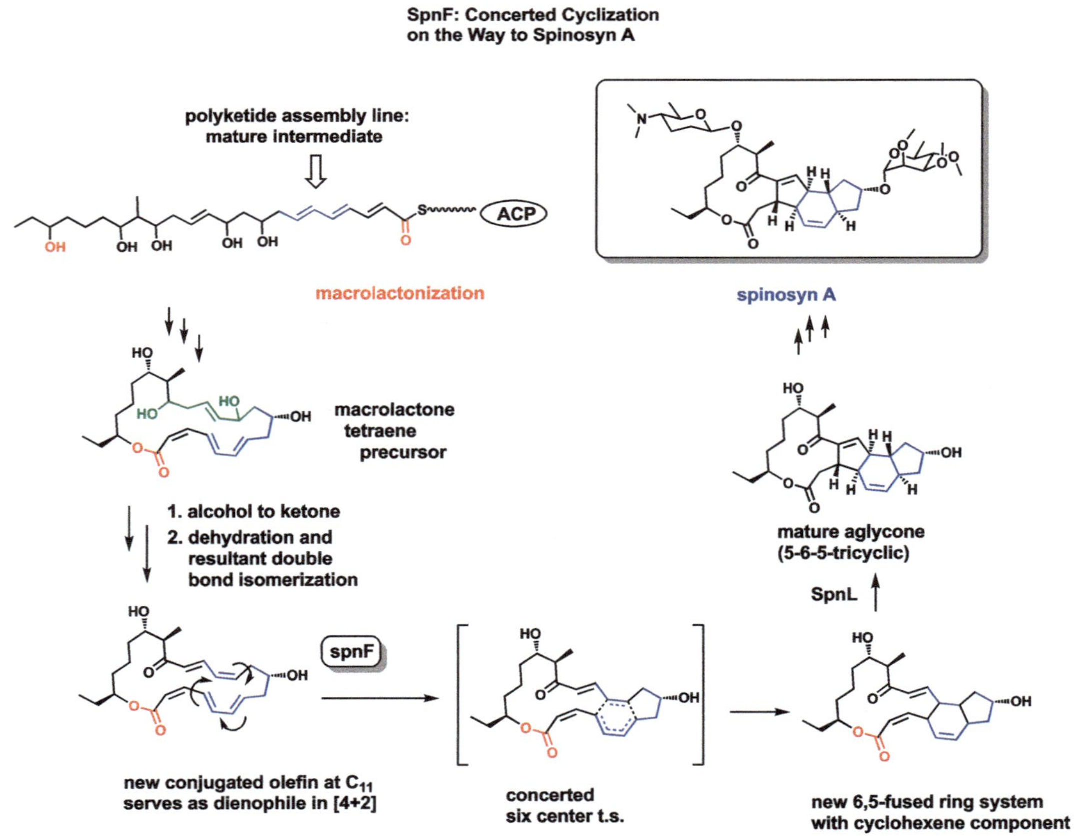

Figure 13.23 The SpnF enzyme catalyzes a concerted cyclization on the way to the 5-6-5 tricyclic moiety of spinosyn A, shown here as a 4+2 Diels Alder cyclization.

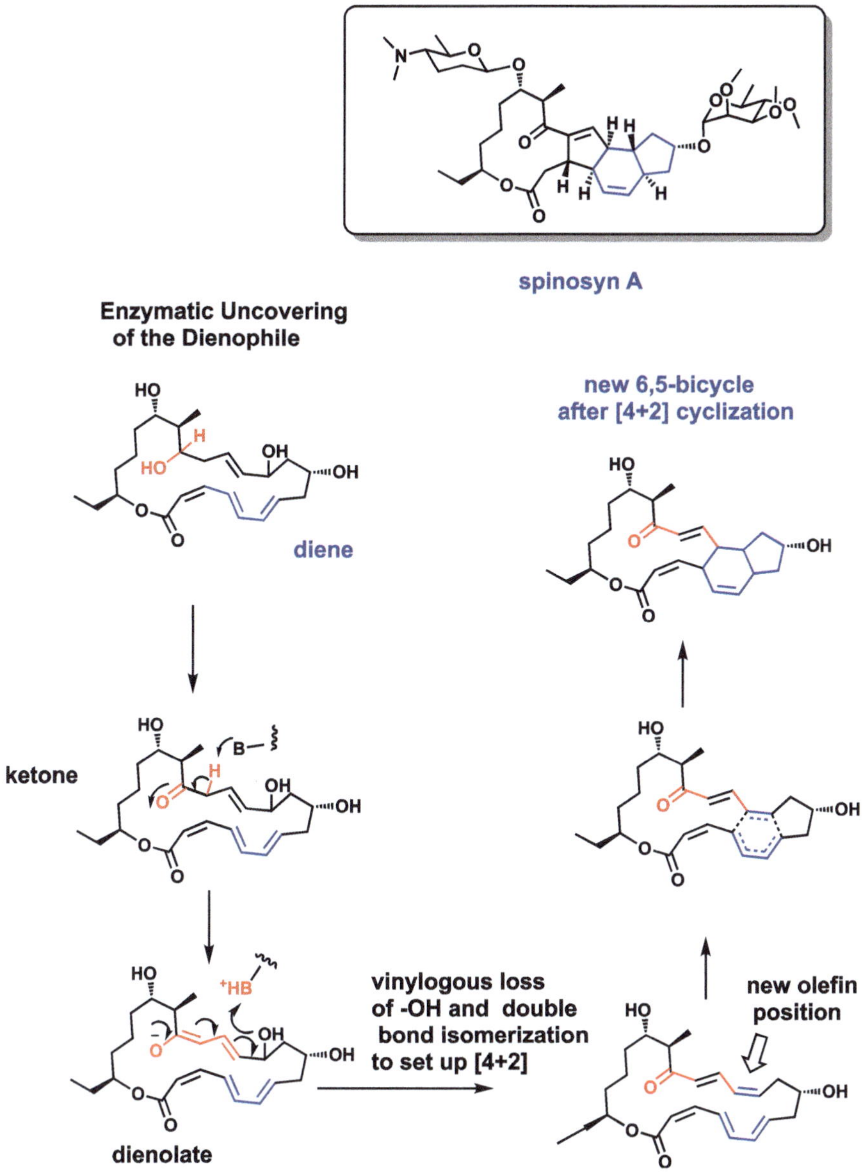

Figure 13.24 SpnF acts to uncover the required dienophile in substrate by an oxidation, dienolate-mediated dehydration to produce a conjugated diene where one of the double bonds is then utilized as dienophile in the cycloaddition. Uncovering the dienophile reveals the underlying strategy of alcohol oxidation, dienolate formation, and isomerization of the new diolefin moiety.

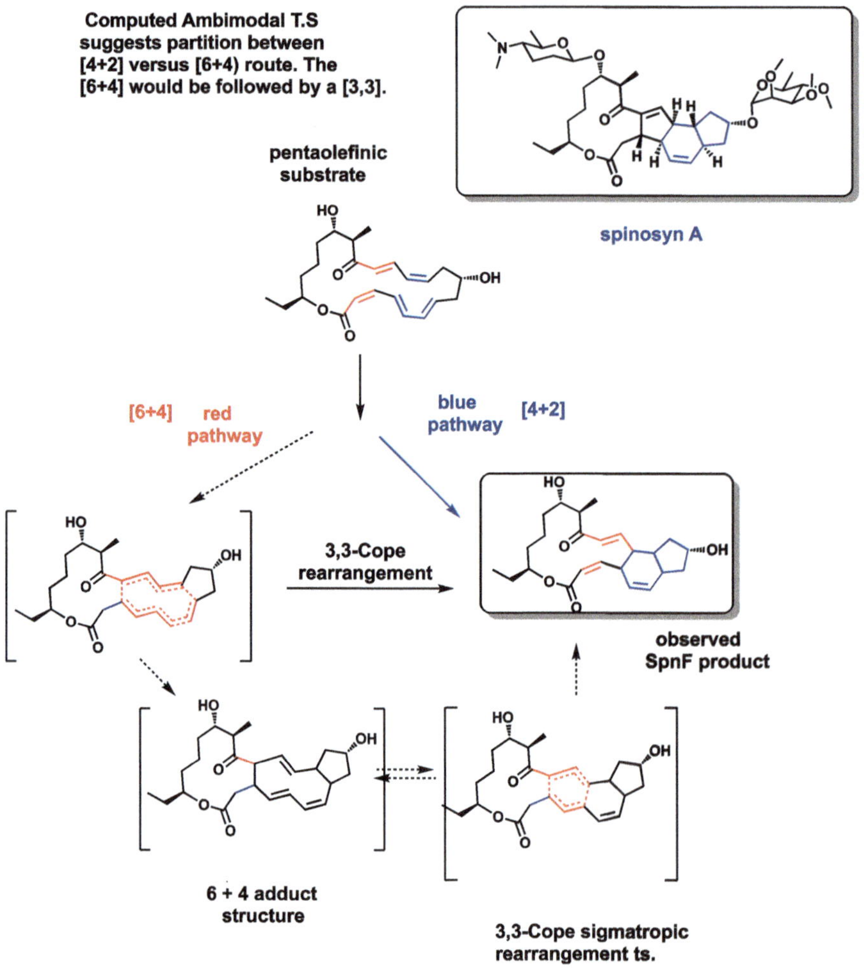

Figure 13.25 Computed ambimodal transition state.

in the dienophile or within the diene component, can be replaced by C=O or a C=NR functional group. Figure 13.16 shows examples of hetero-Diels Alder chemistry.

Recent studies on genome mining of fungal biosynthetic pathways have led to characterization of the enzyme LepI as a hetero-Diels Alderase in assembly of the lepoprin C scaffold, a toxin from Aspergillus molds.[46] Again, the starting substrate does not have the functional group array requisite for an oxo-Diel Alder cyclization. To get to an orthoquinone type quinone methide, LepI starts each catalytic cycle by catalyzing the dehydration indicated in Figure 13.26.

The nascent *ortho*-quinone methide takes the place of the diene in the normal Diels Alder cyclization.[47] The product has a fused tricyclic scaffold where the new six membered ring is the heterocyclic dihydropyran rather than a traditional cyclohexene. LepI could also produce a minor amount of

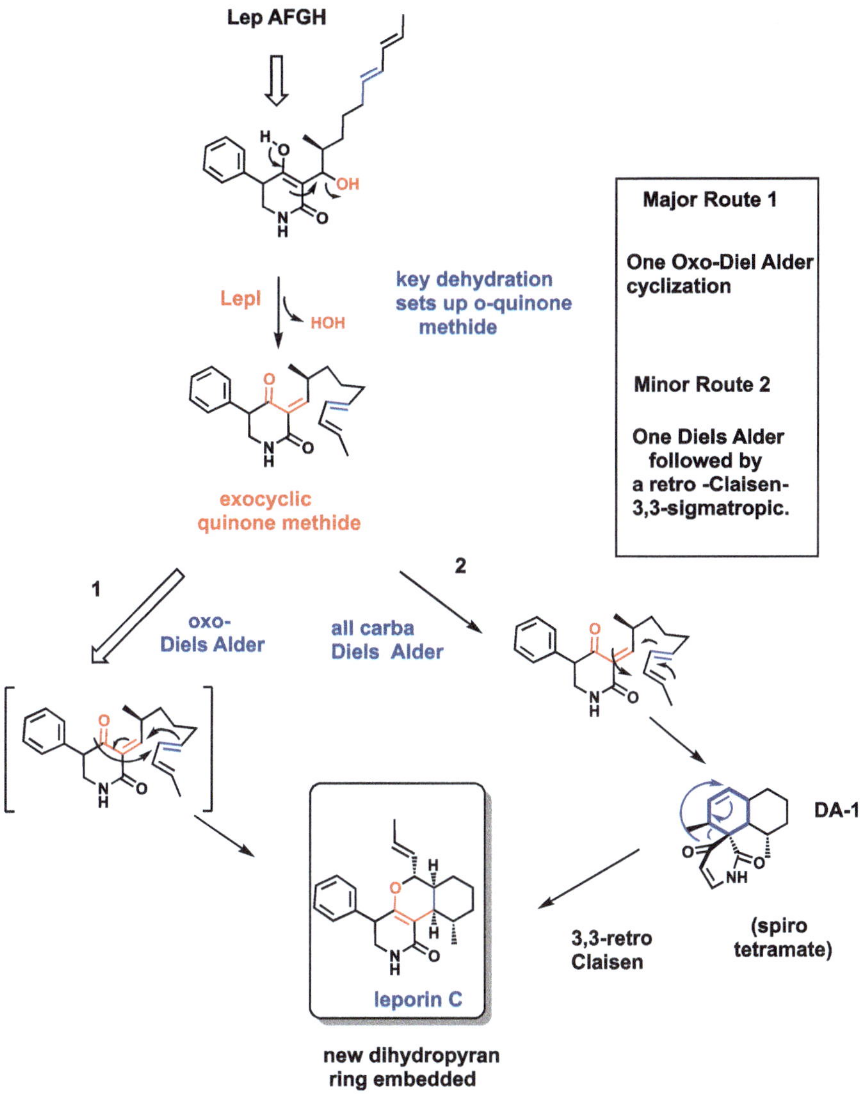

Figure 13.26 In the fungal biosynthesis of leporin C, the enzyme LepI catalyzes both an oxo-Diels Alder (major flux) and a normal Diels Alder (minor flux) competing cycloaddition on an *ortho*-quinone methide intermediate that arises by initial dehydration of substrate. The spirotetronate normal Diels Alder minor product is convertible by the enzyme to the oxo-Diels Alder product by a 3,3-sigmatropic rearrangement.

the tricyclic metabolite-DA1, that would result from a normal Diels Alder, all carbon electrocyclization. That minor product could be processed by LepI for a retro Claisen rearrangement to the major product.[47] In addition to the first example of a hetero-Diels Alderase, LepI also intimates closely balanced energetics between three types of pericyclic reactions, an oxo-Diels Alder cycloaddition, an all-carbon Diels Alder cycloaddition, and the retro-3,3-sigmatropic rearrangement.

13.9 Partition of an Intermediate Between Hetero-Diels Alder and Alder-ene Products

Consistent with balanced energetics between enzymatic transition states of bound substrates/intermediates that populate different pericyclic routes, a series of LepI fungal homologs could function either as oxo-Diels Alderases or, in retrospect perhaps not surprisingly, as Alder-ene catalysts (Figure 13.27). Alder-ene enzymes constitute the fourth example of an enzyme-accelerated pericyclic reaction[48] (Figure 13.9).

Again, the substrates can undergo initial dehydration to reveal a C=C double bond necessary for the oxo-Diels Alder route to proceed. The reactive *ortho*-quinone methide can undergo the oxo-Diels Alder pericyclic transition state shown on the left in Figure 13.27, or can undergo the Alder-ene transition state shown in the right hand column.[48] On one hand, the oxo-Diels Alder product is the now anticipated fused tricyclic framework where the central ring is the new dihydropyran. On the other hand, the Alder-ene reaction is not ring-forming; a C–H bond is broken and formed in a single cyclic transition state. The terminal methyl group of substrate becomes the olefin in the pyridoxatin product. The two product types are not interconverted by the enzymes. The partitioning between asperpyridone *vs.* pyridoxatin frameworks occurs in competing pericyclic transition states from the bound *ortho*-quinone methide intermediate.

13.10 Overview of Pericyclic Reactions in Carbon Chemical Biology

This chapter has dwelt briefly on enzymatic cascade reactions before turning to pericyclic reactions as the fourth route to C–C and C–H bond formation and cleavage reactions in carbon chemical biology. Enzymes can be highly efficient cascade catalysts, carrying out multiple catalytic cycles before any intermediates are released. Perhaps the epitome of such enzymatic cascades are found in a distinct realm of *in vivo* chemistry, in DNA and RNA polymerases where highly processive enzymes[49] may carry out thousands of catalytic cycles before intermediate or product dissociations. We have noted that iterative enzymatic cascades can be driven by any of the four reaction types available for carbon chemical biology: carbanion-driven, carbocation-driven, carbon-radical driven, and concerted chemistry-driven.

Pericyclic reactions, with concerted cyclic transition states for bond-making and breaking, are rare to nonexistent in primary metabolism of higher, multicellular eukaryotes. The sole example of two tandem nonenzymatic pericyclic reactions in skin fibroblasts to convert 7-dehydrocholesterol to previtamin D3 and then on to vitamin D3 are photochemically permitted and

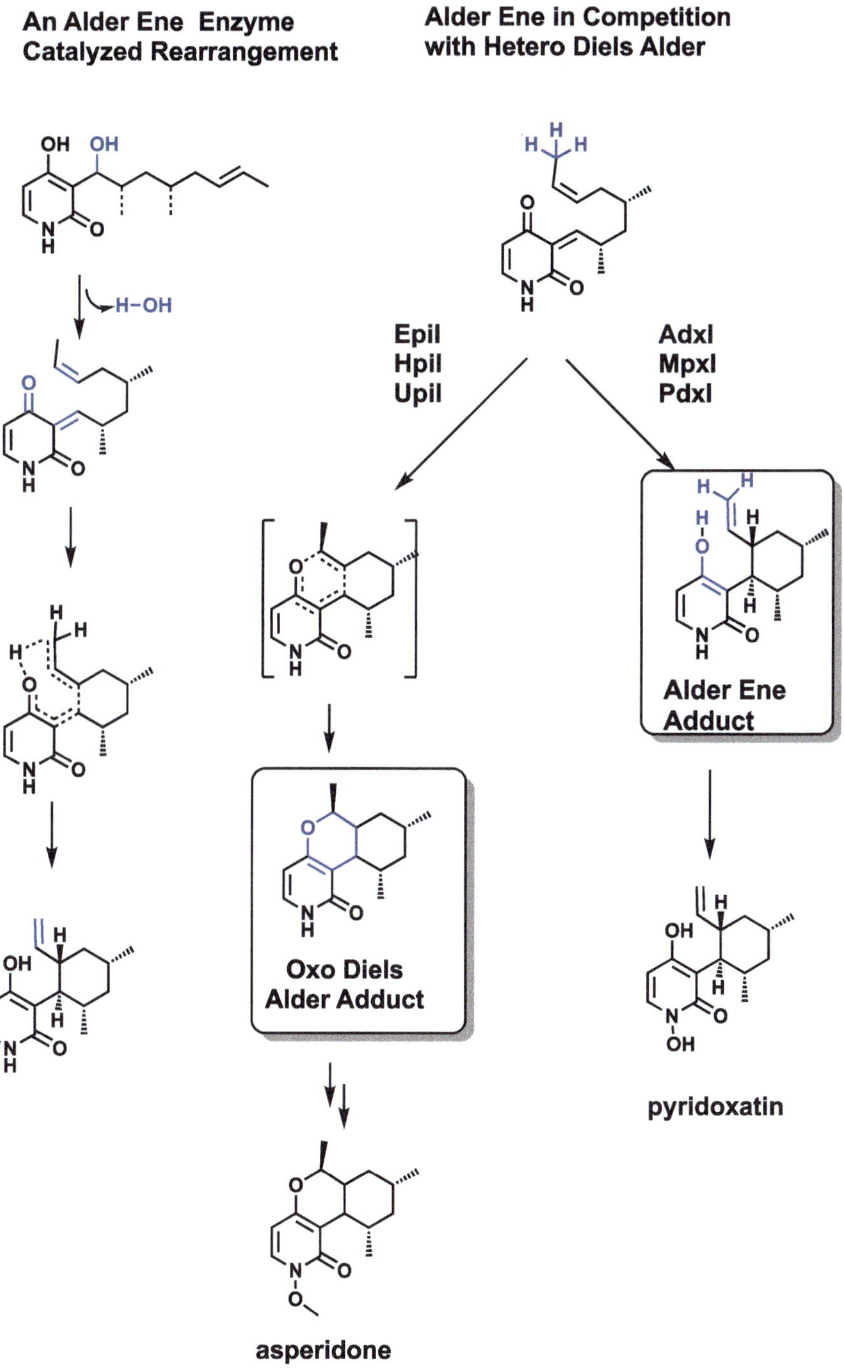

Figure 13.27 Alder-ene enzyme-catalyzed rearrangement occurs in competition with a hetero-Diels Alder pathway.

then thermally permitted, respectively. These tandem reactions introduce two of the four biologically relevant pericyclic reaction classes.

For half a century chorismate mutase was the prime and singular example of a 3,3-sigmatropic rearrangement catalyst at a key point in the flux of cell carbon to aromatic rings molecules in plant and microbial metabolism. That Claisen rearrangement enzyme turns out to be the tip of an iceberg floe of pericyclases. A comparable 3,3-rearangement of 1,5-dienes in marine microbial hapalindole scaffold assembly brings abiotic Cope rearrangement chemistry to the 3,3-sigmatropic rearrangement enzyme class.

In many ways the question of whether Diels Alderases, enzymes that catalyze cycloaddition of dienes and dienophiles in a single cyclic transition state, actually exist has been the focal point for investigators searching for concerted chemistry in biology. The past decade of fungal genomics has unlocked dozens, tending towards hundreds of such pericyclases. In addition to the all-carbon variants of intramolecular and intermolecular Diels Alderases, genome mining and candidate enzyme characterization has identified hetero-Diels Alderases and Alder-ene catalysts. Further information on how enzymes tip bound intermediates down different pericyclic transition states is sure to emerge in the near future.

This chapter, defining categories of pericyclases that accelerate four different subclasses of pericyclic transition states, sets the stage for examination of the fungal proteomes for additional classes of pericyclic reactions. Meanwhile, they validate the proposition that there are classes of organic reactions *in vivo* that are concerted, some of which react by cyclic transition states.

This chapter concludes Section IV of this volume where each of Chapters 10–13 has summarized one of the four known major routes by which C–C bonds and C–H bonds can be formed or broken by enzyme catalysts. While there are one-off mechanistic examples to be further studied, especially in secondary pathways generating complex organic scaffolds, these four chapters define the major reaction manifolds available for carbon chemical biology.

The final section of the book, Section V, includes three chapters, Chapters 14–16, that examine how three iconic metabolites are formed and their accessible chemistry utilized to determine the chemical logic of five metabolic pathways. Glucose, with its polyhydroxy-aldehyde-hemiacetal equilibria shows how the resulting carbonyl chemistry defines three metabolic pathways. Separately, squalene is a linear 30-carbon hexenyl olefin on the way to tetra- and pentacyclic reduced sterol scaffolds. Finally, porphobilinogen brings in nitrogen chemistry, heterocyclic chemistry, and both carbon nucleophile and carbon electrophile reactivity to construction of the tetrapyrrole pigments of life.

References

1. C. T. Walsh and B. S. Moore, Enzymatic cascade reactions in biosynthesis, *Angew. Chem., Int. Ed.*, 2019, **58**, 6846–6879.
2. E. Anslyn and D. A. Dougherty, *Modern Physical Organic Chemistry*, Univerist Science Books, 2006.

3. C. Jamieson, M. Ohashi, F. Liu, Y. Tang and K. Houk, The expanding world of pericyclases: cooperation of theory and experiment for discovery, *Nat. Prod. Rep.*, 2019, **36**, 698–713.
4. R. B. Woodward and R. Hoffmann, *The Conservation of Orbital Symmetry*, Elsevier, 2013.
5. K. N. Houk, Fang Liu, Z. Yang and J. I. Seeman, Evolution of the Diels-Alder reaction mechanism sinmce the 1930s: Woodward, Houk woth Woodward, and the influence of computational chemistry on understanding cycloadditions, *Angew. Chem., Int. Ed.*, 2020, **60**, 12660–12681.
6. K. Klas, S. Tsukamoto, D. H. Sherman and R. M. Williams, Natural Diels-Alderases: elusive and irresistable, *J. Org. Chem.*, 2015, **80**, 11672–11685.
7. B. T. Ueberbacher, M. Hall and K. Faber, Electrophilic and nucleophilic cascade reactions in biosynthesis, *Nat. Prod. Rep.*, 2011, **29**, 337–350.
8. E. Ricca, B. Brucher and J. H. Schrittweiser, Multi-enzymatic cascade reactions: overview and perspectives, *Adv. Synth. Catal.*, 2011, **353**, 2239–2262.
9. P. Sampathkumar and J. F. Morrison, Chorismate mutase-prephenate dehydrogenase from Eschericia coli. Purification and properties of the bifunctional enzyme, *Biochim. Biophys. Acta*, 1982, **702**, 204–211.
10. K. U. Wendt, Enzyme mechanisms for triterpene cyclization: new pieces of the puzzle, *Angew. Chem., Int. Ed.*, 2005, **44**, 3966–3971.
11. G. Siedenburg and D. Jendrossek, Squalene-hopene cyclases, *Appl. Environ. Microbiol.*, 2011, **77**, 3905–3915.
12. K. U. Wendt, G. E. Schulz, E. J. Corey and D. R. Liu, Enzyme mechanisms for polycyclic triterpene formation, *Angew. Chem., Int. Ed.*, 2000, **39**, 2812–2833.
13. T. Hoshino, Beta-Amyrin biosynthesis: catalytic mechanism and substrate recognition, *Org. Biomol. Chem.*, 2017, **15**, 2869–2891.
14. S. Lodeiro, Q. Xiong, W. K. Wilson, M. D. Kolesnikova, C. S. Onak and S. P. T. Matsuda, An oxidosqualene cyclase makes numerous products by diverse mechanisms: a challenge ro prevailing concepts of triterpene biosynthesis, *J. Am. Chem. Soc.*, 2007, **129**, 11213–11222.
15. P. F. Cook, Mechanism from isotope effects, *Isot. Environ. Health Stud.*, 1998, **34**, 3–17.
16. H. K. Schmidt and R. Medina, Possibilities and scope of the double isotope effect method in the elucidation of mechanisms of enzyme catalyzed reactions, *Isot. Environ. Health Stud.*, 1991, **27**, 1–4.
17. W. W. Cleland, The use of isotope effects to determine enzyme mechanisms, *Arch. Biochem. Biophys.*, 2005, **433**, 2–12.
18. R. A. Pascal and C. T. Walsh, Mechanistic studies with deuterated dihydrooroates on the dihydroorotae oxidase from Critihidia fasiculata, *Biochemistry*, 1984, **23**, 2745–2752.
19. J. D. Hermes, C. A. Roeske, M. H. O'Leary and W. W. Cleland, Use of multiple isotope effects to determine enzyme mechanisms and isotope effects. Malic enzyme and glucose-6-phosphate dehydrogenase, *Biochemistry*, 1982, **21**, 5106–5114.

20. J. D. Hermes, P. A. Tipton, M. A. Fisher, M. K. O'Leary, J. F. Morrison and W. W. Cleland, Mechanisms of enzymatic and acid-catalyzed decraboxylation of prephenate, *Biochemistry*, 1984, **23**, 6263–6275.
21. S. Mahlstedt, E. N. Fielding, B. S. Moore and C. T. Walsh, Prephenate decarboxylases: a new prephenate-utilizing enzyme family that performs non-aromatizing decarboxylation en route to diverse secondary metabolites, *Biochemistry*, 2010, **49**, 9021–9023.
22. C. N. Alves, J. R. A. Silva and A. E. Roitberg, Insights into the mechsnism of oxidation of dihydroorotate to orotate catalyzed by the human cllass 2 dihydroorotate dehydrogenase: a QM/MM free energy study, *Phys. Chem. Chem. Phys.*, 2015, **17**, 17990–17996.
23. A. Castro and M. Castro, Claisen rearrangement over the past nine decades, *Chem. Rev.*, 2004, **104**, 2939–3002.
24. K. Alder, F. Pascher and A. Schmitz, Uber die anlagerung von maleinsaure-anhydrid und azodicarbonsaure-ester an einfach ungesattingte koh an einfach ungesattigte kohlenwasserstoffe. ur kenntnis von substitutionsvorgangen in der allyl-steilung, *Ber. Dtsch. Chem. Ges.*, 1943, **76**, 27–53.
25. M. F. Holick, J. E. Frommer, S. C. McNeil, M. N. Richtand, J. W. Henley and J. T. Potts Jr., Photometabolism of 7-dehydrocholesterol to previtamin D3 in skin, *Biochem. Biophys. Res. Commun.*, 1977, **76**, 107–114.
26. D. D. Bikle, Vitamin D: production, metabolism, mechanism of action, in *Endotext*, 2021, www.endotext.org.
27. M. F. Holick and M. B. Clark, The photobiogenesis and metabolism of vitamin D, *Fed. Proc.*, 1978, **37**, 2567–2574.
28. W. J. Guilford, S. D. Copley and J. R. Knowles, The mechanism of the chirismate mutase reaction, *J. Am. Chem. Soc.*, 1987, **109**, 5013–5019.
29. A. C. Cope and E. M. Hardy, The introduction of substituted vinyl groups V. A rearrangement involving the migration of an alyl group in a three-carbon system, *J. Am. Chem. Soc*, 1940, **62**, 441–444.
30. M. S. DeClue, K. K. Baldridge, D. E. Kunzler, P. Kast and D. Hilvert, Isochoismate pyruvate lyase: a pericyclic reaction mechansism?, *J. Am. Chem. Soc.*, 2005, **127**, 1502–15003.
31. N. A. Khanjin, J. P. Snyder and F. M. Menger, Mechanism of chorismate mutase. Contribution of conformational retriction to catalysis in the claisnen rearrangement, *J. Am. Chem. Soc.*, 1999, **121**, 11831–11846.
32. R. M. Holman and D. H. Sherman, Recent advances in hapalindole type cyanobacterial alkaloids: biosynthesis, synthesis, and biological activity, *Nat. Prod. Rep.*, 2021, **38**, 1567–1588.
33. S. Li, A. N. Lowell, F. Yu, A. Raveh, S. A. Newmister, N. Bair, J. M. Schaub, R. M. Williams and D. H. Sherman, Hapalindole/Ambiguine biogenesis is mediated by a Cope rearrangement, C-C bond-forming cascade, *J. Am. Chem. Soc.*, 2015, **137**, 15366–15369.
34. S. A. Newmister, S. Li, M. Garcia-Borras, J. N. Sandrs, S. Yang, A. N. Lowell, F. Yu, J. L. Smith, R. M. Williams, K. N. Houk and

D. H. Sherman, Structural basis of rhe Cope rearrangement and cyclization in hapalindole biogenesis, *Nat. Chem. Biol.*, 2018, **14**, 345–351.
35. B. V. Subba Reddy, P. N. Nair, A. Anthony and N. Srivastava, Recent advances in Prins spirocyclization, *Eur. J. Org. Chem.*, 2017, **37**, 5484–5496.
36. C. T. Walsh and Y. Tang, *Natural Product Biosynthesis: Chemical Logic and Enzymatic Mechanisms*, Royal Society of Chemistry, London, 2nd edn, 2022.
37. K. Auclair, A. Sutherland, J. Kennedy, D. J. Witter, J. P. van den Heever, C. R. Hutchinson and J. C. Vederas, Lovastatin nonaketide synthase catalyzes an intramolecular Diels-Alder reaction of a substrate analogue, *J. Am. Chem. Soc.*, 2000, **122**, 11519–11520.
38. L. Li, P. Yu, M.-C. Tang, Y. Zhou, S.-S. Guo, Y.-S. Hung, M. Zhao, K. Watanabe, K. N. Houk and Y. Tang, Biochemical characterization of a eukaryotic decalin-forming Diels Alderase, *J. Am. Chem. Soc.*, 2016, **138**, 15837–15840.
39. M.-C. Tang, Y. Zou, K. Watanabe, C. T. Walsh and Y. Tang, Oxidative cyclization in natural product biosynthesis, *Chem. Rev.*, 2017, **117**, 5226–5333.
40. M. J. Byrne, N. R. Lees, L.-C. Han, M. W. van der Jamp, A. J. Mulholland, J. E. M. Stach, C. L. Willis and P. R. Race, The catalytic mechanism of a natural Diels-Alderase revealed in moleculafr detail, *J. Am. Chem. Soc.*, 2016, **138**, 6095–6098.
41. B.-S. Jeon, S.-A. Wang, M. W. Ruszczycky and H.-W. Liu, Natural [4+2] cyclases, *Chem. Rev.*, 2017, **117**, 5367–5388.
42. J. Wang, J. Liang, L. Chen, W. Zhang, L. Kong, C. Peng, C. Su, Y. Tang, Z. Deng and Z. Wang, Structural basis for the biosynthesis of lovastatin, *Nat. Commun.*, 2021, **12**, 867.
43. L. Caputi, J. Franke, S. C. Farrow, K. Chung, R. M. E. Payne, T.-D. Nguyen, T.-T. Dang, I. S. T. Carquijiero, K. Koudonas, T. Duge de Bernonville, B. Ameyaw, D. M. Jones, I. J. C. Viera, V. Courdavault and S. E. O'Connor, Missing enzymes in the biosynthesis of the anticancer drug vinblastine in Madagascar periwinkle, *Science*, 2018, **360**, 1235–1239.
44. L. Caputi, J. Franke, K. Bussey, S. C. Farrow, I. J. C. Viera, C. E. M. Stevenson, D. M. Lawson and S. E. O'Connor, Structural basis of cycloaddition in biosynthesis of iiboga and aspidosperma alkaloids, *Nat. Chem. Biol.*, 2020, **16**, 383–386.
45. *Comprehensive Organic Synthesis*, ed. P. Knochel and G. A. Molander, Elsevier, 2nd edn, 2014.
46. Z. Chang, T. Ansbacher, L. Zhang, Y. Yang, T.-P Ko, G. Zhang, W. Liu, J.-W Huang, L. Dai, R.-T. Guo, D. T. Major and C.-C. Chen, Crysta; structutrr of LepI, a mutlifunctiona enzyme which catalyzes pericyclic reactions in leporin biosynthesis, *Org. Biomol. Chem.*, 2019, **17**, 2070–2076.

47. M. Ohashi, F. Liu, Y. Hai, M. Chen, M.-C. Tang, Z. Yang, M. Sato, K. Watanabe, K. N. Houk and Y. Tang, SAM-dependent enzyme-catalyzed pericyclic reactions in natural product biosynthesis, *Nature*, 2017, **549**, 502–506.
48. M. Ohashi, C. S. Jamieson, Y. Cai, D. Tanm, D. Kannayama, M.-C. Teng, S. M. Anthony, J. V. Chari, J. S. Barber, E. Picazo, T. B. Kakule, S. Cao, N. K. Garg, J. Zhou, K. N. Houk and Y. Tang, An enzymatic Alder-ene reaction, *Nature*, 2020, **586**, 64–69.
49. Z. Zhuang and Y. Ai, Processivity factor of DNA polymerase and its expanding role in normal and translesion DNA synthesis, *Biochim. Biophys. Acta*, 2010, **104**, 1081–1093.

Section V: Metabolic Pathways Enabled by the Chemistry of Key Intermediates

In previous sections we have analyzed the underlying principles of chemical biology in three ways. Following an introductory chapter on properties of carbon in abiotic organic chemistry, Section II discusses four types of carbon-based functional groups that account for the reactions of most organic metabolites in biologic systems: C=C alkenes(and alkynes), C–O bonded functional groups, C–S functional groups, and C–N functional groups.

Section III broadens the scope to examine the kinds of carbacyclic and heterocyclic metabolites prominent in the carbon chemistry of life. The carbacycles include aromatic rings and tetracyclic steroids. Oxygen heterocycles include ethers, lactones, and cyclic sugar hemiacetals. Nitrogen heterocycles include purines and pyrimidines, indoles, imidazoles, pyrroles, and lactams. We also call out 10 of the 13 human vitamins as nitrogen-containing heterocycles that use distinct chemistries for central metabolic transformations.

Section IV contains four chapters, one each dedicated to the kinds of reaction mechanisms that make and break the inventory of carbon–carbon bonds in cells. The chapters focus in turn on (1) carbon nucleophiles, including accessible stabilized carbanions, (2) on accessible carbon electrophiles, including carbonyl groups and allyl cations, (3) on carbon radicals, generated both anaerobically and aerobically, and (4) on pericyclic transformations with concerted bond breaking and making in the same transition state.

We finish with three chapters in Section V that further examine the roles of three crucial metabolites in distinct metabolic pathways. The underlying chemical logic utilized at different stages of maturation of the metabolic pathways reveal how the chemistry open to the functional groups of those three metabolites dictate the types of chemical steps accessible for *in vivo* carbon chemistry.

The three key metabolites, out of several that could be chosen and analyzed, are glucose, squalene, and porphobilinogen. The hydroxy carbonyl groups in the open chain form of glucose and the rapid equilibrium with the cyclic hemiacetals as an aldehyde protecting group illustrate the many modes of carbonyl group and stabilized carbanion reactivity in three central metabolic pathways, one degradative and two biosynthetic.

Squalene is a thirty carbon hexa-olefin. Its functional groups are limited to those six alkenes, yet it is immediate progenitor to the four carbacyclic rings of all steroid frameworks. The logic of how that linear polyolefin is assembled from 18 molecules of acetyl CoA and how squalene goes on to use its olefins to form C–C bonds in the tetracyclic scaffolds of mature sterols reveals at least four types of bond-forming/cleaving logics along this carbacyclic pathway.

The third key metabolite, porphobilinogen, illustrates the chemical versatility of the 2-aminomethylpyrrole ring, as heterocyclic progenitor of both nucleophilic and electrophilic carbons. Those carbons are the reactants for elongation of the monopyrrole to linear tetrapyrrole and its directed cyclization to the tetrapyrrole macrocycle framework of hemes and chlorophylls.

Analyses of how each of these five metabolic pathway proceeds to use these key different organic molecules reveals the chemical logic played out through the distinct sets of carbon-based functional groups.

CHAPTER 14

Glucose Chemical Biology

14.1 The Accumulating Forms of Glucose in Cells

The six-carbon sugar D-glucose is the most abundant sugar in biologic systems. It is a major dietary component in animals and humans, released as the building block from enzymatic hydrolysis of polysaccharides (carbohydrates) in the GI tract (Figure 14.1). Glucose exported largely from liver and muscle into the blood serves as the main source for glucose transported into all the cells of different tissues.

The chemical formula for simple sugars is a stretch of hydroxylated carbons and typically a terminal aldehyde. Sugars common in central metabolic pathways range from the three-carbon D-glyceraldehyde, the four-carbon D-erythrose, the five-carbon D-ribose (the sugar building block for all nucleotide monomers in both RNA and DNA), and the six-carbon D-glucose (Figure 14.1).

Using the lens of the complement of carbon functional groups as predictor of reactivity traits, the sugars have two such groups: multiple alcohols and the electrophilic aldehyde group. It turns out that the free aldehyde form is a very minor species in a rapidly established equilibrium for glucose and for ribose that form cyclic hemiacetals. The three- and four-carbon sugars glyceraldehyde and erythrose do not form comparable intramolecular hemiacetals because the small rings would be strained compared to the linear aldose forms (Figure 14.1). D-ribose does form the comparable α- and β-hemiacetal isomers by attack on the C_4–OH group from below or above the plane of the electrophilic aldehyde carbonyl carbon. Those five-ring hemiacetal isomers are also called furanoses, after the parent furan ring system. Glucose can cyclize using either the 4-OH group to furan isomers or the 5-OH to six-ring pyranose hemiacetal forms. It turns out the pyranose forms dominate (Figure 14.1). The composition of glucose in neutral aqueous

Chemical Biology No. 21
The Chemical Biology of Carbon
By Christopher T. Walsh

Published by the Royal Society of Chemistry, www.rsc.org

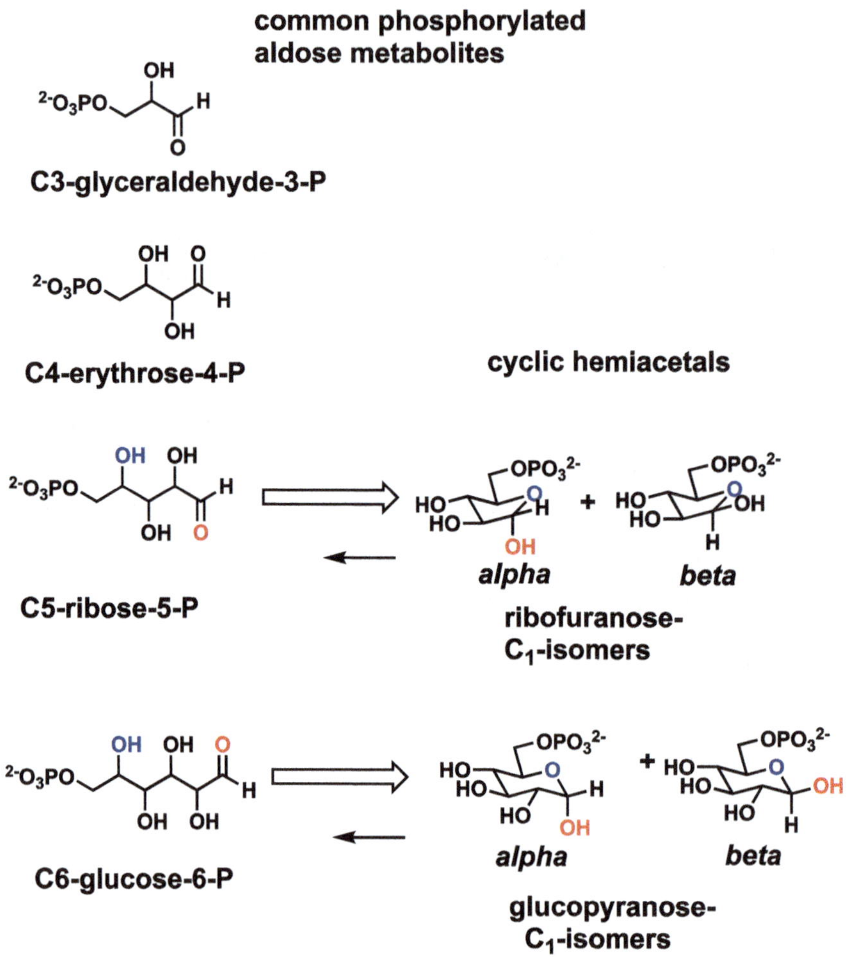

Figure 14.1 Common phosphorylated aldose metabolites.

solutions is ~63% β-hemiacetal, 37% α-hemiacetal, and less than 0.02% of the open chain aldehyde (also see Figure 8.1).

The presence of the alcohol groups in glucose and other sugars opens the possibility of isomers. Two additional common hexoses in metabolism are the C_4-isomer D-galactose, and the C_2-isomer D-mannose (Figure 14.2). One of the reasons that glucose may be the most abundant building block for oligo- and polysaccharides stored in plants, and then as the free metabolite and major dietary energy source in animals that consume plants, is that in the accumulating cyclic hemiacetal forms the alcohols are all equatorial substituents. The axial substituents are all the small hydrogen substituents. By comparison, the C_4–OH in the galactose hemiacetals is axial while in the mannose hemiacetals it is the C_2–OH that is axial, causing steric crowding.

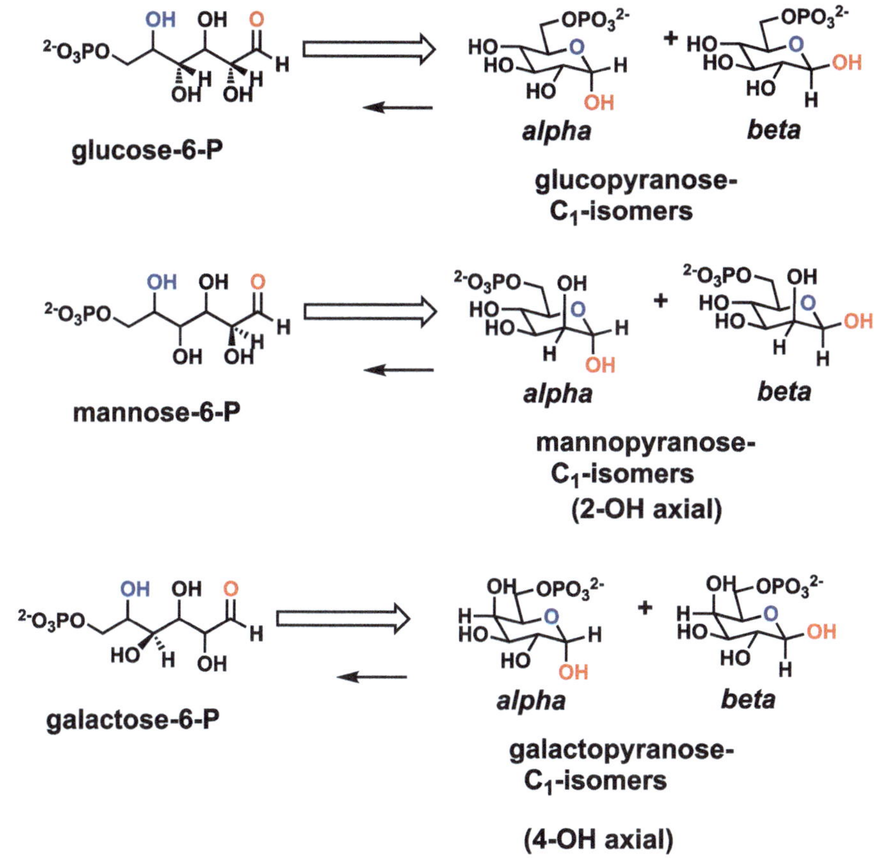

Figure 14.2 Open chain aldehyde forms in equilibrium with cyclic hemiacetal forms for glucose-6-P, mannose-6-P, and galactose-6-P.

An additional central feature of glucose chemical biology (and thereby of all glucose-derived sugar metabolites) is the fact that almost all the hexoses in cell have been phsophorylated enzymatically at the lone primary hydroxy group, C_6 in glucose[1] (Figure 14.1). As glucose diffuses from the blood down its concentration gradient into cells *via* glucose transporters, this neutral molecule is phosphorylated by a suite of ATP-utilizing glucokinases. The major rationale is to convert the neutral sugar into a dianionic glucose-6-phosphate, unable to diffuse back out of cells and therefore accumulate as a trapped metabolite. The equilibrium lies far in the direction of phosphoryl transfer, from attack of the C_6-oxygen nucleophile on the electrophilic γ-phosphoryl atom in the ATP side chain. This is a classic example of the thermodynamic activation of the phosphoric anhydride linkages being spent to drive a coupled equilibrium. Now there are three distinguishable functional groups in the minor open chain form of glucose-6-P: the C_6–OPO_3^{2-} phosphate ester, the C_2–C_5 alcohols and the C_1 aldehyde carbonyl. In the cyclic hemiacetal α- and β-isomers the C_1–OHs are isomeric, and the C_6-phosphate ester is chemically distinguishable from the C_{2-4} alcohol groups (Figure 14.2). The acyclic and the cyclic forms of glucose in particular display totally distinct reactivity, as electrophile or nucleophile, respectively.

14.2 Three Distinct Metabolic Pathways from Glucose-6-P for Three Different Purposes

Glucose-6-P is one of the *ur*-building blocks of chemical biology. Cells and organisms can direct it down three different pathways, utilizing distinct aspects of the chemistry available to the functional groups in the major cyclic hemiacetal *vs.* the minor open chain aldehyde forms. One route is degradative, while a second route leads biosynthetically both to D-ribose for nucleic acid assembly and to the phenyl ring in all metabolites of the phenylalanine diaspora. The third route builds oligosaccharides, polysaccharides, and the glycan chains in glycoproteins (Figure 14.3).

14.3 Strategies for the Oxidation of Glucose to Six Molecules of Carbon Dioxide: The Chemical Logic of Glycolysis and The Citric Acid Cycle

14.3.1 All 24 Electrons from Glucose-6-P Are Transferred as 12 Hydride Ions During Glycolysis and the Citric Acid Cycle

In the broad division between catabolic pathways of metabolism for energy production *vs.* anabolic pathways for biosynthesis,[2] including nucleic acid, protein and polysaccharide polymers, glucose is the primary starting point for abstracting energy from controlled room temperature oxidation of organic molecules from diet. The tandem action of the glycolytic (sugar lysis) pathway that feeds into the citric acid cycle converts all six carbons of glucose to six CO_2 molecules that are thrown away as escape gas. It is the 24 electrons removed from each glucose molecules that are captured and ultimately stored in cellular fuels such as NADH and ultimately ATP (Figure 14.3).

The molecular signposts of this tandem pair of energy yielding pathways are (1) enzymatic processing of glucose to two moles of pyruvate (by the end of the glycolysis route); (2) oxidative decarboxylation of two pyruvates to two acetyl CoAs and two CO_2: (3) acetyl CoA and oxaloacetate condensation to citrate. In each turn of that cycle two CO_2 are released from each acetyl CoA (Figure 14.4). An electron count shows four electrons removed in glycolysis, four more electrons as two pyruvates are decarboxylated, eight electrons during each turn of the citrate cycle. Two acetyl CoAs from each glucose mean 16 electrons at phase 3. Adding $4+4+16=24$, the total number of electrons removed.

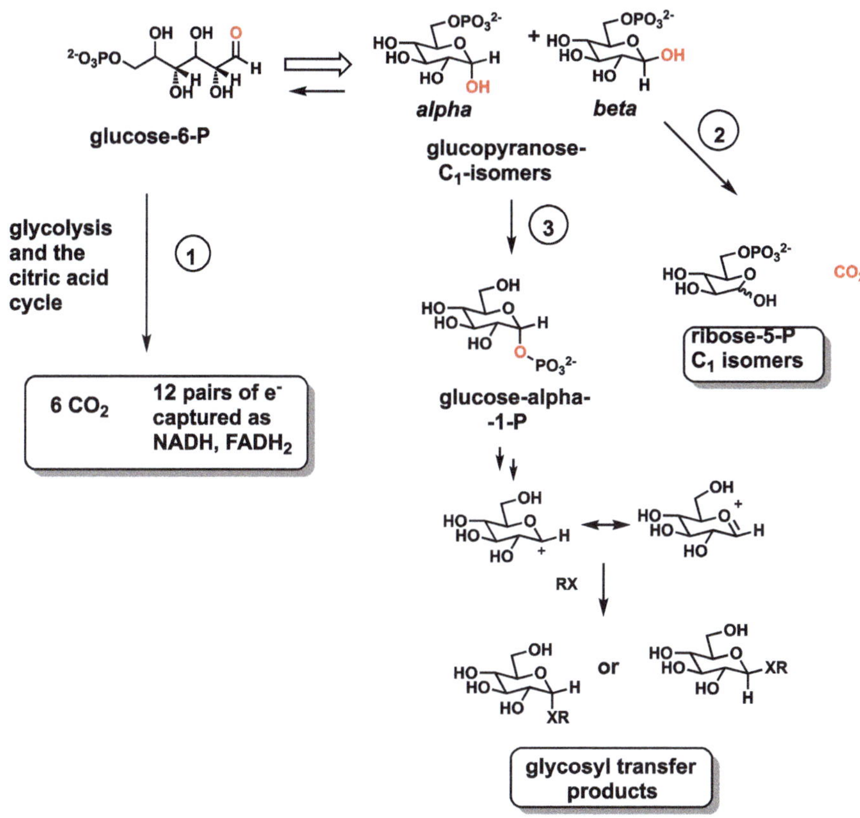

Figure 14.3 Three major metabolic fates for glucose-6-P: (1) glycolysis and citric acid cycle oxidation to 6 CO_2 and 24 electrons removed as 12 pairs of electrons in hydride ion transfers. (2) The pentose phosphate pathway provides ribose-5-P as nucleic acid building block and also carbons for the shikimate pathway to phenylalanine and tyrosine. (3) Glucose-1-P is the route to UDP-glucose and glycosyl transfers in polysaccharide biosynthesis.

Four thoughts on the above paragraph. As glucose is dehydrogenatively processed, all six carbons get jettisoned as CO_2. It is the electrons, not the carbon that are useful to cells. Also, both glycolysis and the citrate cycle are ancient, evolved before O_2 levels rose. No carbon radicals are involved in the workings of the 19 enzymes. Third, the major site of oxidative chemistry is the citric acid cycle, accounting for two-thirds of the electrons stripped away from glucose-derived carbons. Fourth, all 24 glucose electrons are removed as 12 hydride ions. In retrospect this oxidation by hydride removal strategy might not have been an *a priori* obvious chemical choice. Of those 12 hydride ions (and their associated 24 electrons), 10 end up in NADH molecules (20 electrons' worth), while the remaining four electrons end up as two $FADH_2$ equivalents bound to succinate dehydrogenase.

Finally, the six CO_2 molecules, accounting for all six carbons of each glucose molecule catabolized, are released by decarboxylase enzymes, as

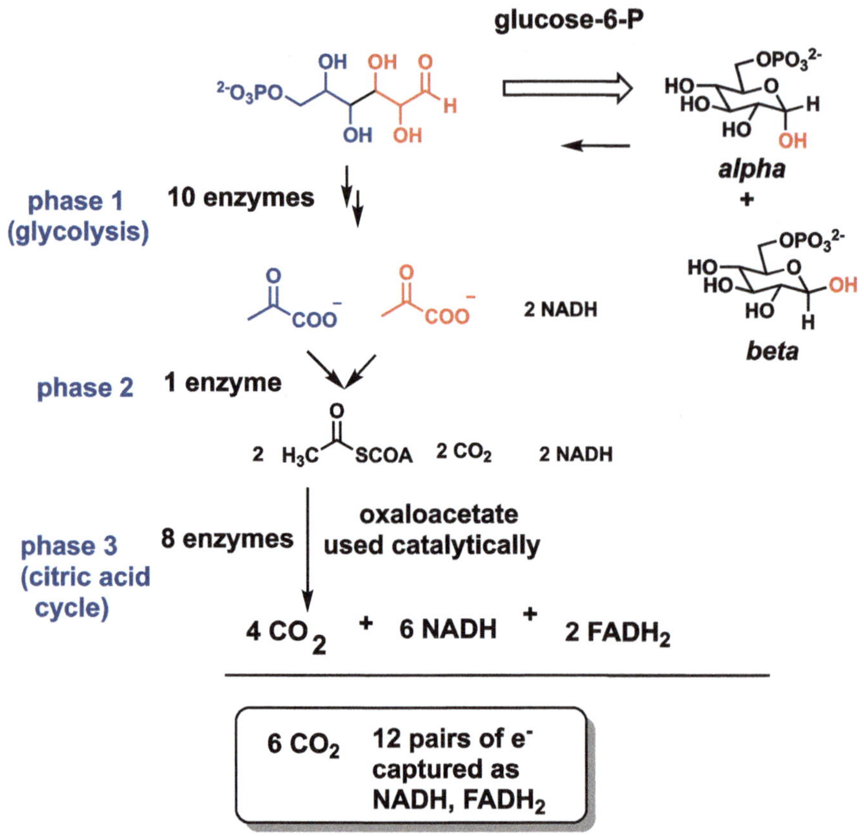

Figure 14.4 Three phases of glucose oxidation to six CO_2; the 10 enzymes of glycolysis comprise the first phase; the three-component pyruvate dehydrogenase is the second phase; the eight enzymes of the citric acid cycle constitute the third phase.

noted in Chapter 3. We have discussed low energy routes for decarboxylation of both α and β keto acids. The former requires thiamin-PP, the latter decarboxylate without cofactors by enolate/enamine pathways. One central strategy of the catabolism of glucose by the combined action of the enzymes of glycolysis and of the citric acid cycle is to form α or β keto acids to set up such decarboxylations. In the end four of the six molecules of product CO_2 come from α-keto acid decarboxylase action[3] while two CO_2 arise from β-keto acid decarboxylase action.

14.3.2 Glucose to Two Pyruvates: Net Four Electron Oxidation During Glycolysis

The eight enzymes that convert D-glucose-6-P to two equivalents of pyruvate constitute glycolysis (Figure 14.4). The predominant chemistry is carbanion-based, with one interposed aldehyde to acyl thioester redox step on each of two glyceraldehyde-3-P molecules. Glucose (and from it, fructose), in the minor open chain form but not the cyclic hemiacetal isomers, offers two types of hydroxycarbonyl groups, 1,2-hydroxycarbonyls (aldehyde) and 1,3-hydroxycarbonyls (ketone) that are converted in enzyme active sites into transient carbanion equivalents with two distinct fates (Figure 14.5).

The 1,2-hydroxycarbonyl groups come into play in the first enzyme (phosphoglucose isomerase), interconverting open chain 6-P-glucose to open chain 6-P-fructose. Then, again in the fourth enzyme in the pathway (triose-P isomerase), the 1,2-hydroxy carbonyls are equilibrated between hydroxy-aldehyde and keto-hydroxy groups at the level of the pair of three-carbon fragments glyceraldehyde-3-P and dihydroxyacetone-P.

Interspersed is the aldolase[4] working on fructose-1,6-P_2, recognizing the 1,3-hydroxy-ketone array at carbons 2–4 of fructose-1,6-P_2 substrate. The aldolase generates a low energy carbanion/enamine pathway for aldol

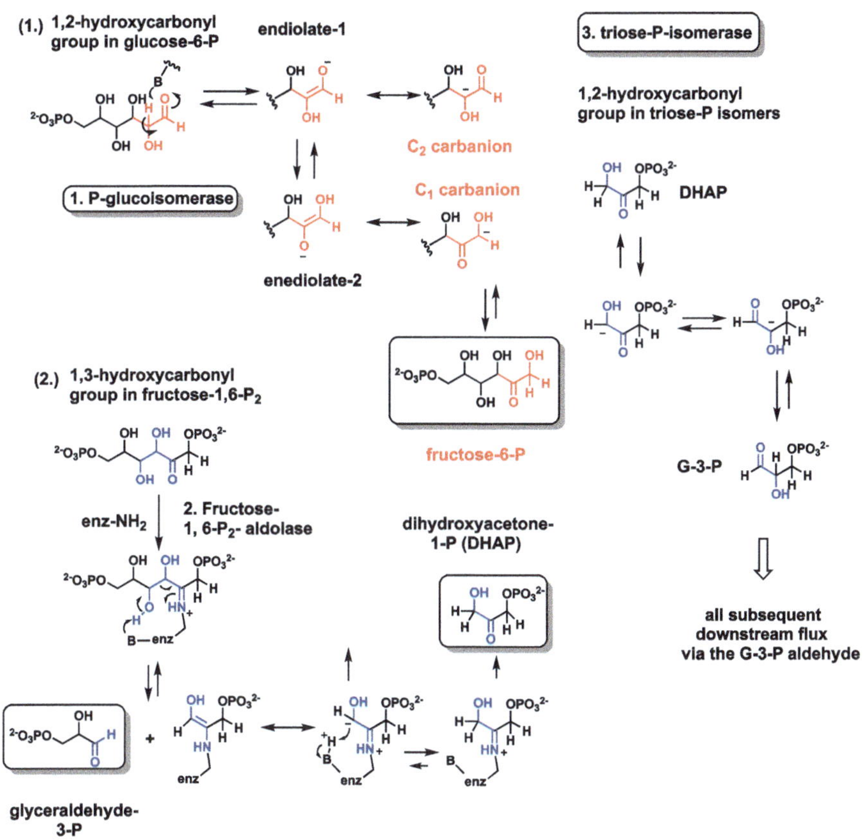

Figure 14.5 Carbanions from 1,2-hydroxycarbonyl *vs.* 1,3-hydroxycarbonyl groups in hexose and triose metabolites.

cleavage of the six-carbon fructose to a 1 : 1 mix of glyceraldehyde-3-P and dihydroxyacetone-P. The reactions catalyzed by these three enzymes are summarized in Figure 14.5.

The 1,2-hydroxycarbonyl group in glucose-6-P is progenitor *via* proton abstraction at the C_2–H to an enediolate carbanion form (negative charge at C_2) (Figure 14.5). Protonation at the C_1–oxygen of the enediolate anion-1 to the ene-diol, and then proton abstraction from O_2, creates the isomeric enediolate-2 (negative charge at C_1). Subsequent net protonation at C_1 creates the isomerized product fructose-6-P scaffold with its 1-hydroxy-2-keto group. Both at this step and later at the three-carbon fragment stage of catalyzed equilibration of dihydroxyacetone-P and glyceraldehyde-3-P by triose-P isomerase, the accessible 1,2-enediolate isomers mediate 2-OH-1-aldehyde interconversion with 1-OH-2-keto isomers (Figure 14.5). At the fructose level, it is the 2-keto isomer that is further metabolized. At the three-carbon level it is only the 3-P-glyceraldehyde isomer that is enzymatically processed further.

The second type of carbonyl group enabling low energy carbanion formation is the 1,3-hydroxy carbonyl group (Figure 14.5). This is the classic functional group arrangement for enolate anion formation in aldol cleavages and aldol condensations. As in the glucose-6-P isomer distributions, the cyclic forms of fructose-6-P and fructose-1,6-P_2 predominate over the open chain keto form. However, it is again the minor open chain keto form of fructose-1,6-P_2 that is the aldolase substrate. In that aldolase active site, the side chain of an amine engages in imine formation with the open chain ketone (Figure 14.5) (also see Chapter 10). C–C bond fragmentation occurs across C_3 and C_4 to yield glyceraldehyde-3-P and dihydroxyacetone-P after protonation and release of the initial DHAP enamine fragment.

The next enzyme triose-P isomerase uses the accessibility of the 1,2-enediolate route to equilibrate DHAP and glyceraldehyde-3-P, ensuring that both three-carbon fragments from the starting glucose scaffold are processed by the downstream enzymes.

In Chapter 3 we noted aldehyde dehydrogenase strategies where the dehydrogenase adds in a cysteine thiolate to the aldehyde carbonyl and then catalyzes hydride ejection to NAD^+ and the covalent acyl thioester-enzyme adduct (Figure 3.17).

In many such dehydrogenases, the activated acyl group would be transferred to water as an oxygen nucleophile. In the glyceraldehyde-3-P dehydrogenase active site an oxyanion of inorganic phosphate is specific cosubstrate to yield the released 3-P-glyceryl-1-phosphate, a mixed acyl-phosphoric anhydride (Figure 14.6). Each glucose-6-P molecule that starts down the glycolytic path yields two glyceraldehyde-3-P fragments. Thus, two NADH are produced by action of the glyceraldehyde-3-P-dehydrogenase, the only NADH molecules generated during glycolysis.

The remaining four enzymes serve to harvest energy both from this acyl phosphate (Figure 14.7), and, after isomerization and dehydration, from two

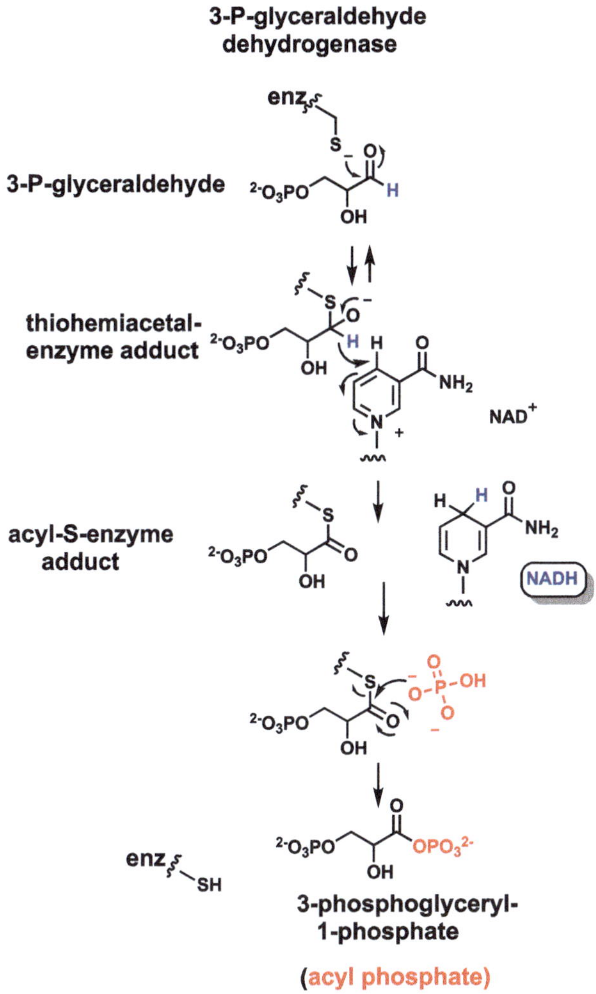

Figure 14.6 3-P-glyceraldehyde dehydrogenase oxidizes the substrate-enzyme thiohemiacetal adduct rather than the free aldehyde. Phosphorolysis rather than hydrolysis of the resultant 3-P-glyceryl acyl enzyme adduct gives 1,3-diphosphoglycerate.

phosphoenolpyruvates (PEP) to two pyruvates. At that point the second stage occurs. These final four enzymes-3-PGA kinase, 2,3-PGA mutase, enolase, and pyruvate kinase harvest 2-ATPs/pyruvate or net 4-ATPs per glucose molecule. Note that enolase, by dehydration of 2-PGA, generates a thermodynamically activated but kinetically stable enol phosphate ester. Any enzyme, *e.g.* hydrolase, catalyzing attack on the phosphorus of the PO_3 group will generate the pyruvyl enolate anion, resonance contributor the C_3 carbanion of pyruvate, and one of the canonically accessible cellular carbanion equivalents (Chapter 10).

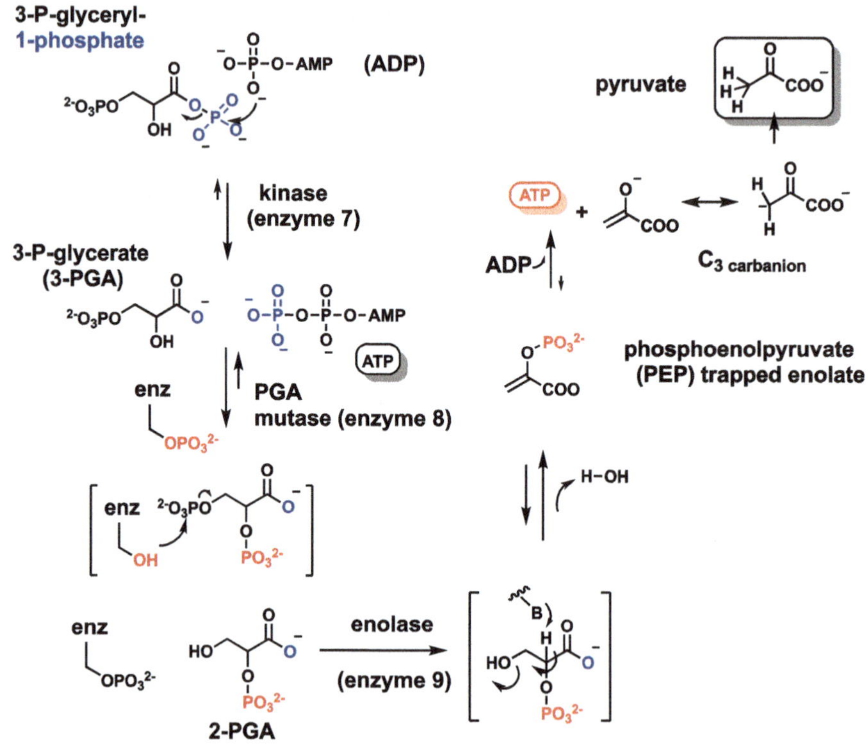

Figure 14.7 Enzymes 7–10 of glycolysis harvest energy trapped in the acyl phosphate linkage of 1,3-diphosphoglycerate, isomerize 3-PGA to 2-PGA and then dehydrate 2-PGA to the product enol trapped as a phosphate ester: phosphoenolpyruvate. That thermodynamically activated enol phosphate transfers the phosphoryl group to ADP, making ATP and the keto acid pyruvate, ending the glycolytic pathway: two pyruvates from one glucose.

14.3.3 Oxidative Decarboxylation of Pyruvate to Acetyl CoA and NADPH: Energy Trapping

Phase 2 of the three phase complete degradation of glucose to six CO_2 (Figure 14.4) starts with oxidative metabolism on each of the two pyruvates. The glycolytic enzymes thus far have generated two pyruvate molecules from glucose in the cytoplasm of cells. Pyruvate can be transported specifically into mitochondria where it is a substrate for a three-component pyruvate dehydrogenase complex[5] (Figure 14.8).

To this point all six carbons of glucose are still in organic form, albeit with oxidation states changed in the pairs of pyruvates to a methyl group, a ketone, and a carboxylate anion. In this next phase of catabolism each pyruvate is decarboxylated to release CO_2 with concomitant oxidation of the C_2-ketone to the acyl oxidation state at C_1 of the acetyl CoA product.

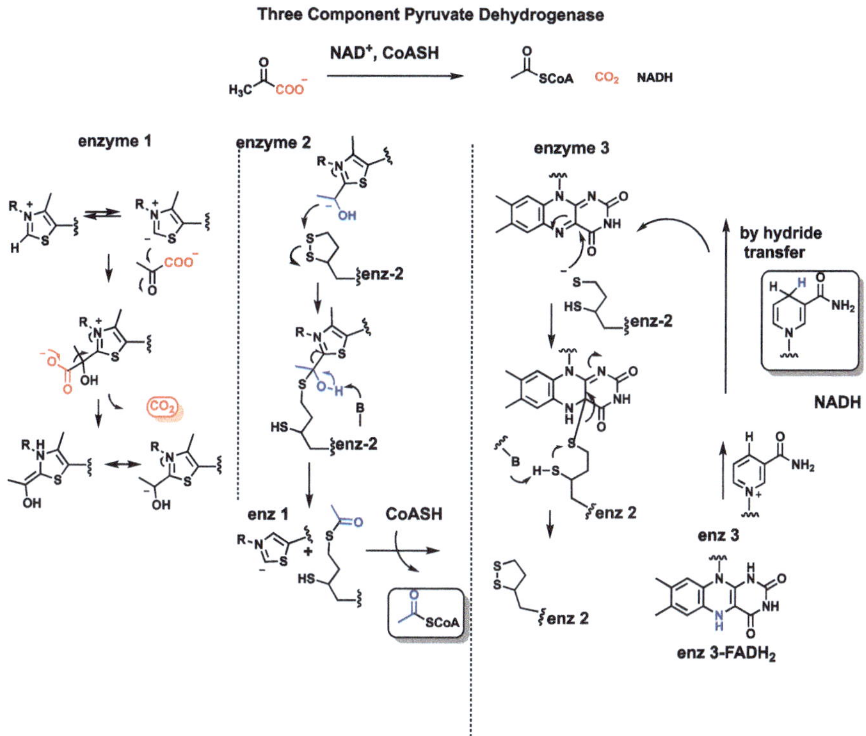

Figure 14.8 Three-component pyruvate dehydrogenase. Enzyme 1 has tightly bound thiamin-PP. Enzyme 2 has covalently tethered lipoamide in the oxidized disulfide form. Enzyme 3 has tightly bound FAD. In addition to pyruvate, NAD^+ and CoASH are substrates. The products from action of the three enzymes are CO_2, NADH, and acetyl CoA.

Each of the three protein components of the pyruvate dehydrogenase multisubunit enzyme complex contain a tightly bound coenzyme.[5] Enzyme 1 requires the heterocyclic thiamin-PP (TPP, Chapter 8); enzyme 2 has a molecule of covalently tethered oxidized lipoamide; enzyme 3 has tightly bound FAD (Chapters 2 and 8) with its tris-heterocyclic isoalloxazine ring.

From Chapter 8, we recall that TPP readily forms a C_2-thiazolium carbanion that adds reversibly into the electrophilic ketone of 2-keto acids. The resulting adduct now has the thiazolium cationic iminium group beta to the substrate carboxylate. A low energy path for decarboxylation proceeds, yielding the stabilized, enamine version of a carbanion (Figure 14.7). That carbanion on enzyme 1 is presented to the oxidized disulfide form of lipoamide on enzyme 2. The carbanion attacks one of the electrophilic sulfur atoms in the lipoamide disulfide to open it to the acyl thioester (Figure 14.8) and the free lipoamide thiol. Thioester interchange with external CoASH is rapid, releasing the acetyl CoA product. Thus far, CO_2 (derived from C_1 of pyruvate and ultimately C_3 and C_4 of glucose) has been released[3] along with acetyl CoA (this step represents removal of the first two of the six carbons from a starting glucose molecule as CO_2. The other four CO_2 molecules are released in the citric acid cycle, described below).

The function of enzyme 3 is to reoxidize the dithiol form of enzyme 2 for the next catalytic cycle (see Figure 9.29). One of the lipoamide thiolate anions adds in to C_{4a} of FAD in the enzyme 3 active site. That adduct is in turn attacked by the remining free thiol of the lipoamide cofactor to produce Enz-3-$FADH_2$ while regenerating the starting form of oxidized enzyme 2. Finally, enzyme 3 is in turn regenerated to its starting oxidized FAD form *via* hydride transfer to NAD^+ (Figure 14.8).

This tricomponent enzyme is an ingenious redox catalyst using the coenzyme forms of three nitrogen heterocyclic vitamins (see Chapter 9) as well as CoASH as external substrate and the disulfide form of lipoamide bound in enzyme 2. Carbanion chemistry is in play both in enzyme 1 and enzyme 2 while hydride transfer is the mode in enzyme 3. The substrate oxidation step can be identified as when the two-carbon hydroxyethyl carbanion-TPP adduct attacks the lipoamide disulfide (Figure 14.8). The resultant adduct, spanning both enzyme 1 and enzyme 2, can decay in the front direction to an acyl thioester. By this stage the C_2 ketone of pyruvate has been oxidized by two electrons to the acyl oxidation state and trapped as the thermodynamically activated acyl CoA.

In this step, two pyruvates to two acetyl CoA and two CO_2, another four electrons have been removed from the glucose scaffold. The total thus far of removal of 8 of 24 starting electrons means 16 of the electrons remain in two acetyl CoAs, 8 in each molecule. One can anticipate that further oxidation of each acetyl group to two CO_2 will formally remove 6 electrons from each methyl group and two from the acyl group, accounting for the 16 remaining electrons that were present in the starting glucose framework. The citric acid cycle is the covey of mitochondrial enzymes that carry out the net removal of those eight electrons, two at a time as four hydride transfers.

14.3.4 The Citric Acid Cycle as an Eight-electron Redox Cycle

The eight enzymes of the citric acid cycle constitute the third phase (Figure 14.4) of glucose oxidative breakdown, where all six carbons are released as six CO_2 and all 24 electrons removed have been stored in the form of 10 NADH and 2 Enz-$FADH_2$. The two acetyl CoA molecules, that trace back to two pyruvates and ultimately to one starting glucose molecule, each provide eight electrons, transferred as four hydride ions, in each turn of the cycle. The net outcome is shown across the top of Figure 14.9, depicting the reaction stoichiometry as eight-electron removal from the two acetyl carbons as the enzymes of the cycle produced the two CO_2 molecules. This is the overall goal of glucose catabolism and the citric acid cycle captures two thirds of the 24 electrons removed during the full oxidative conversion of each glucose molecule to six CO_2 molecules. We go through the strategy of each of the eight enzymes of the citric acid cycle in Figure 14.9.

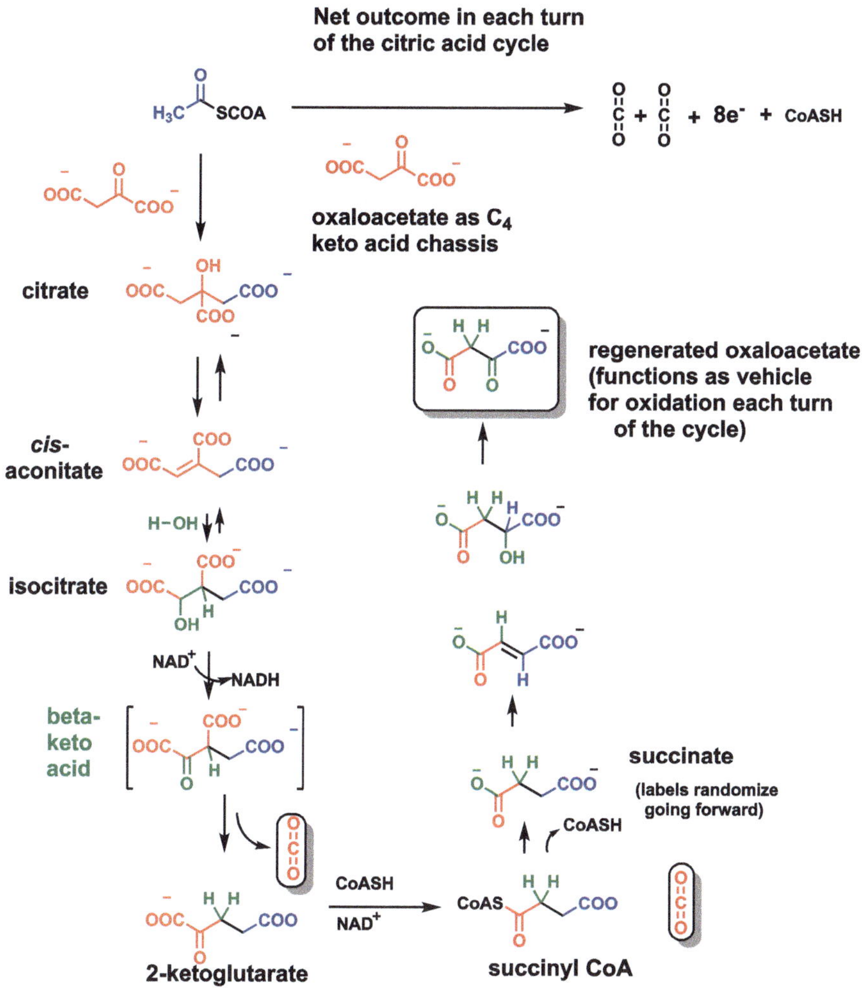

Figure 14.9 Net outcome in each turn of the citric acid cycle. The four-carbon 2-keto-dicarboxylate oxaloacetate is catalytic. Two molecules of CO_2 derive from each oxaloacetate in each turn of the cycle but those carbon are replaced by the acetyl carbons of acetyl CoA. The cycle is all about oxidative removal of four pairs of electrons as hydride ions.

First, however, it may be helpful to analyze what really goes in in the cycle that uses a molecule of the four-carbon α-keto acid oxaloacetate as the chassis for the acetyl group's net oxidation as charted in Figure 14.10. A central question is what mechanisms do cells possess to carry out low energy decarboxylations of carboxylic acids to CO_2? In Chapters 3, 9, and 10 we noted that β-keto acids represent one route by stabilizing incipient product carbanions by iminium/enamine formation in the nascent products. Second, we have just discussed the role of the thiazolium ring in TPP acting as C_2 carbon nucleophile on the α-keto acid pyruvate to create a β-iminium

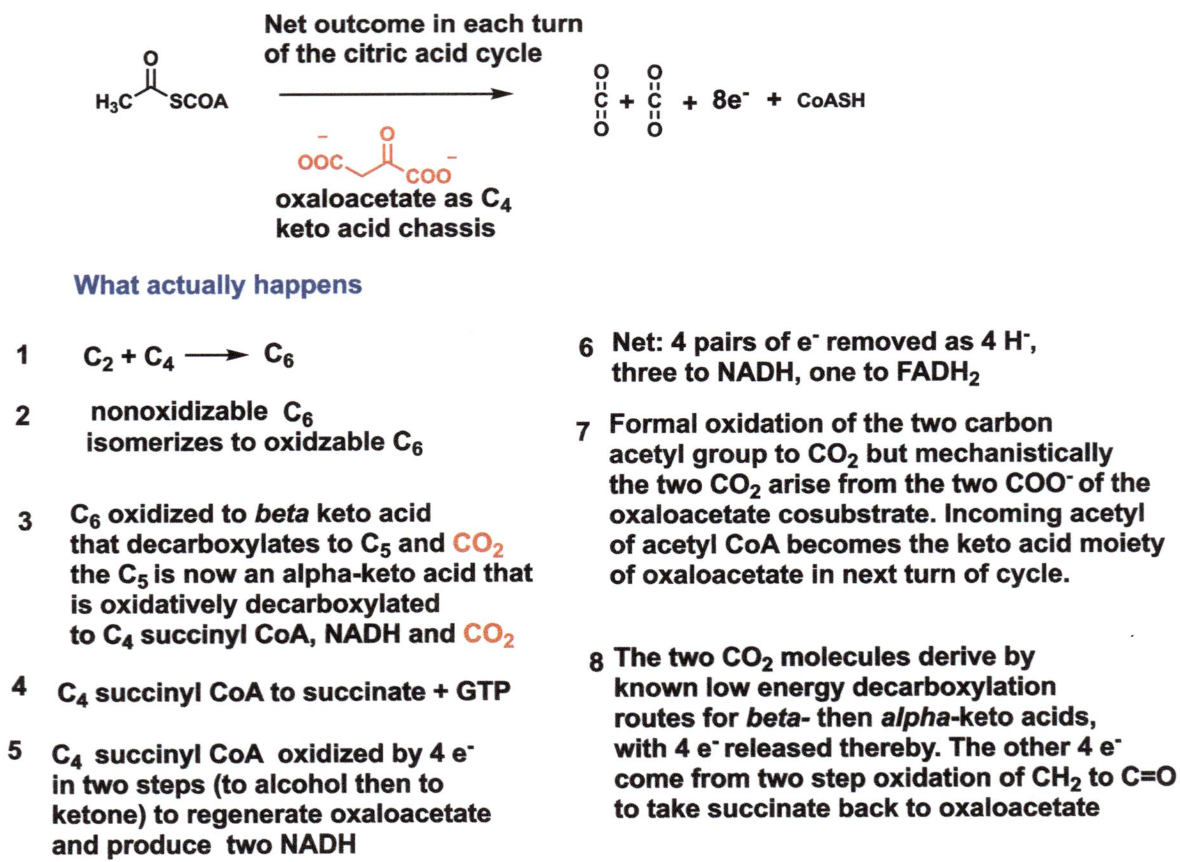

Figure 14.10 The eight chemical events that happen in each turn of the citric acid cycle.

ion functional group as the device for turning α-keto acids into β-iminium adducts, set up for a comparable low energy path to CO_2.

Indeed, this is the two-pronged chemical strategy for generating the two molecules of CO_2 from substrate carboxylate groups in the citric acid cycle, forming first a β- and then an α-keto acid and enabling low barrier decarboxylations. That strategy is not readily enacted on the two carbon acetyl fragment but is implementable on the oxaloacetate moiety used as a four-carbon chassis for the acetyl group.

Every reaction in the eight-enzyme citrate cycle is explicable in terms of the C–O functional group chemistry we discussed in Chapters 3 and 10, including C–C bond formation by thioester enolate as carbanion and then two C–C disconnections by the two decarboxylases[3] noted above. In between are dehydrations and rehydrations to set up alcohol functional groups for oxidation. While the first half of the eight enzyme cycle generates the two CO_2 molecules, the second half takes another four electrons out of the four-carbon framework of succinate to get back to oxaloacetate as a reusable carrier.

The first enzyme, citrate synthase,[6] uses the acetyl group of acetyl CoA as a C_2 carbanion (accessible thioester enolate, see Chapter 10) to attack the electrophilic C_2 ketone of the four-carbon oxaloacetate. The oxaloacetate molecules are the scaffolds on whose backs the cycle runs, as they undergo the initial citrate synthase C–C bond-forming reaction to bound *S*-citryl-CoA. Before product release the enzyme hydrolyzes the thioester linkage, providing the driving force for accumulating the six-carbon tricarboxylate citric acid, as the tricarboxylate anion citrate (Figure 14.9).

Two items to note about citrate. First, carbon three can be described as a Cabbc carbon center where the b groups are the identical CH_2COO^- groups) (see Chapter 3). Therefore, citrate C_3 is a prochiral carbon. The arm provided by the acetyl CoA substrate during citrate synthase action is specifically the proS arm (we will see below that the eight electrons removed actually come from the oxaloacetate framework not that acetyl moiety).

Second is the fact that citrate has a tertiary –OH group as one of the C_3 substituents. That is, C_3 bears no hydrogen substituents. Because the purpose of the citrate cycle is oxidative, the next enzyme, aconitase, converts the nonoxidizable tertiary –OH of citrate to the oxidizable secondary –OH in isocitrate (Figure 14.9). Aconitase logic involves dehydration of citrate to the olefinic *cis*-aconitate and then catalyzed rehydration with opposite regiochemistry *via* an ingenious "ferrous wheel" mechanism[7] (worth a mechanistic detour for interested chemists).

Enzymes 3 and 4 of the citrate cycle are the oxidative decarboxylases. The first, (isocitrate dehydrogenase), generates and then decarboxylates a β-keto acid intermediate, oxalosuccinate (Figure 14.9). The second enzyme 2-ketoglutarate dehydrogenase then works on the newly formed α-ketoacid substrate.

The C_2 secondary alcohol CHOH group of isocitrate is oxidizable to the ketone by the standard hydride transfer route to NAD^+ to give NADH (the first of four 2e redox steps in the cycle). The nascent 2-keto acid, oxalosuccinate is not released from ICDH.[8] Instead, it undergoes catalyzed decarboxylation. Only one of the three carboxylates (the one bonded to C_3) in oxalosuccinate is β to the newly introduced ketone. Only that carboxylate can undergo C–COO bond cleavage *via* a stabilized enolate intermediate, lowering the energy for its formation. The products from ICDH action are the linear five-carbon 2-ketoglutarate and CO_2.

Enzyme 4 of the cycle, 2-ketoglutarate dehydrogenase,[9] is presented with the same type of chemical challenge that pyruvate dehydrogenase faced back in Section 14.3.3: how to create a low energy path for α-keto acid decarboxylation. This enzyme uses the same strategy as the three component pyruvate dehydrogenase complex: TPP in the first enzyme, the disulfide form of lipoamide in the second enzyme component and FAD in the third protein (see Figure 14.8 for the strategy). The analogous mechanistic logic yields CO_2, NADH and succinyl CoA.

We can assess the results, four enzymatic steps into the citrate cycle, exactly halfway around the eight-enzyme circuit. Two molecules of CO_2 have been released. For two turns of the cycle to this point (processing the two acetyl CoA molecules that trace back to one starting glucose molecule), four CO_2 would be produced. Joined to the two CO_2 molecules released in two turns of pyruvate dehydrogenase action, that accounts for all six of the glucose carbons oxidized all the way to CO_2.

Thus far, two NADH molecules have been produced, the first from ICDH action, the second from 2-KGDH action, representing four electrons *via* transfer *via* two hydride ion pairs. Double that, to account for a net of two acetyl CoAs processed, gives eight electrons. Added to the eight electrons already released, totals 16 of the 24 electrons originally contained in glucose. To finish the cycle the remaining enzymes need to get four electrons each out of two succinyl CoA molecules as they are processed to oxaloacetate. Oxaloacetate then would be the catalytic backbone of the cycle, not consumed.

Enzyme 5 phosphorolyzes succinyl CoA to the mixed anhydride succinyl phosphate and then uses it to convert GDP to CTP to capture the thermodynamic activation in the acyl thioester bond as free succinate emerges. The remaining three enzymes, enzymes 6–8, of the eight-enzyme citrate cycle have precisely the role of turning the succinate scaffold back to starting oxaloacetate (Figure 14.9); nothing more complicated than a strategy for double oxidation of a CH_2 in a succinate backbone to a C=O group in an oxaloacetate backbone. This is the identical chemical challenge in fatty acid β-oxidation. Exactly the same chemical strategy is in effect.

Enzyme 6, succinate dehydrogenase,[10] converts succinate to the *trans*-olefinic diacid fumarate. The enzyme is a flavoprotein, which uses a proton, hydride mechanism, transferring the hydride ion to N_5 of the bound FAD. That constitutes the third hydride transfer in the citrate cycle, and the only one to FAD instead of NAD^+. The double bond in fumarate is then hydrated by fumarase, water acting as nucleophile towards the conjugated double bond as electrophile, to produce the β-hydroxy four-carbon dicarboxylate framework of L-malate. The eighth enzyme is the final one in the cycle, carrying out the prosaic oxidation of the β-OH group in malate to the β-ketone as oxaloacetate is regenerated for the next turn of the cycle. This final enzyme malate dehydrogenase once again uses NAD^+ as a cosubstrate. This is the fourth hydride transfer in the eight-enzyme citrate cycle. All eight electrons have been removed. The net result is as though the incoming acetyl group has been oxidized to two CO_2 and eight electrons, stored in the reduced forms of the two common redox coenzymes, NADH and $FADH_2$ (Figure 14.9).

Note from the color coding in Figure 14.9, and also point 7 of 14.10, that the acetyl moiety in the first turn of the citrate cycle has been incorporated throughout the four carbons of oxaloacetate, due to randomization at the level of the symmetric intermediate succinate. Without that randomization the incoming acetyl moiety at the start of the citric acid cycle would become the two-carbon α-keto acid moiety of oxaloacetate for the next turn of the cycle.

14.4 Glucose to Ribose and Beyond: Pentose-phosphate Pathway Logic

The other two metabolic pathways emanating from glucose-6-P are much shorter than the 19 tandemly acting enzymes of glycolysis and the citrate cycle. Those 19 enzymes are dedicated to "burning glucose" at physiologic temperature and pH completely to six CO_2 while harvesting all the electrons along the way in thermodynamically activated coenzyme forms. The NADH and enz-$FADH_2$ are either immediately available for cellular chemical work or serve by passage of electrons down a large potential drop to O_2 for coupled ATP synthesis.

The biosynthetic pentose phosphate pathway[11] that converts D-glucose-6-P to D-ribose-5-P is not engineered for the full degradation of glucose (pathway 2 of Figure 14.3). Instead, it is designed to set up low energy loss of carbon 1 of glucose-6-P on the way to ribose-5-P. That carbon is lost as CO_2. Thus, the strategy devolves again to implementation of a route to catalytic oxidation of carbon 1, converting the aldehyde oxidation state to an acid, and then setting up a low energy path for decarboxylation.[3]

In the pentose-P pathway the form of glucose-6-P recognized as substrate by the first enzyme is the predominant cyclic hemiacetal β-isomer, rather than the minor open chain aldehyde that was the starting substrate in the glycolytic pathway (Figure 14.11). The strain free cyclic hemiacetal is a protected cyclic form of the aldehyde. The first step is then enzymatic dehydrogenation of that C_1 hemiacetal group to the cyclic lactone. There are analogies to the glyceraldehyde-3-P dehydrogenase strategy of oxidation by hydride transfer from the hemithioacetal rather than the free aldehyde (Chapter 3, and Figure 14.6). Indeed, the G-6-P dehydrogenase transfers a hydrogen to $NADP^+$ to give the two-electron-reduced NADPH.

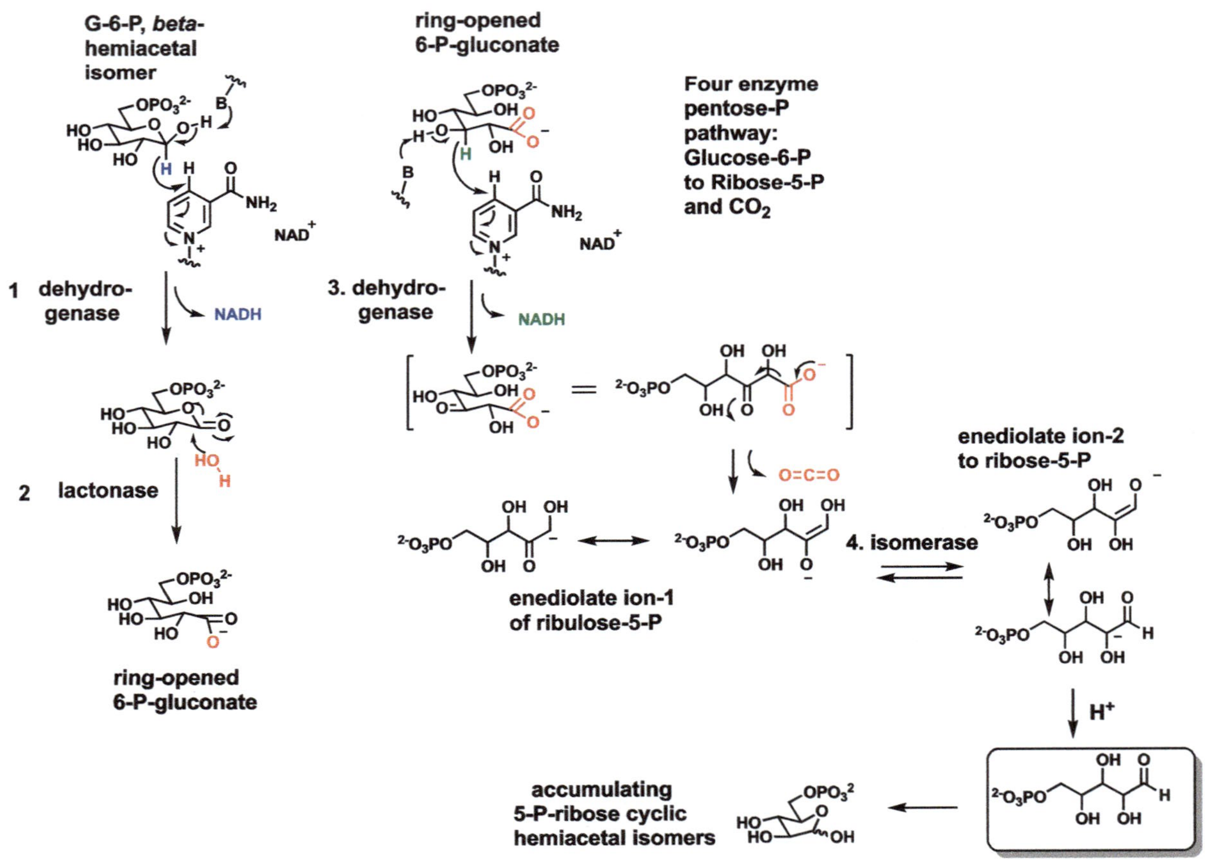

Figure 14.11 The pentose phosphate shunt: four enzymes convert glucose-6-P to CO_2 and ribose-5-P. This is carbonyl chemistry *in vivo*. Four electrons are removed from glucose-6-P, first at C_1 then at C_3, as two hydride ions to $NADP^+$. The lactone is opened hydrolytically by a lactonase. The 6-P-gluconate dehydrogenase uses the same logic as isocitrate dehydrogenase. The ribulose-5-P is isomerized to ribose-5-P by enediolate anion protonation equilibria.

The 6-P-gluconolactone is stable under physiologic conditions so organisms have evolved a specific lactonase that hydrolytically opens the electrophilic lactone to the linear 6-P-gluconic acid (Figure 14.11). The free carboxylate functionality, two electrons oxidized from the starting hemiacetal, has now been revealed. The subsequent strategy is to find a low energy path to disconnection of the C_2–C_1 bond by decarboxylation. One notable route would be to convert the 3-OH to a 3-keto group. Then, that ketone would be β to the C_1 carboxylate and provide a low energy enolate route for decarboxylation. In accord with this chemical logic the third enzyme in the pentose-P pathway is indeed an $NADP^+$-accepting hydride transfer catalyst, generating the intermediate 6-P-3-ketogluconate that is then decarboxylated before release. The observed products are CO_2, NADPH, and D-ribulose-5-P (Figure 14.11). This is exactly the same logic used by isocitrate dehydrogenase in the citric acid cycle above. To this point, four electrons, in two hydride pairs, have been removed from glucose-6-P.

Ribulose-5-P is the 2-keto isomer of D-ribose-5-P. The fourth enzyme in the pathway is the required isomerase. The chemistry involves an enzyme-generated 1,2-enediolate from abstraction of one of the carbon-bound hydrogens at the $-CH_2OH$ group as a proton. As noted for comparable 2-hydroxy-carbonyl chemistry in the glucose-6-P to fructose-6-P isomerization catalyzed by phosphoglucoisomerase at the start of glycolysis, this is low barrier carbanion formation (Figure 14.5). The enediolate anion can be readily isomerized to the alternate enediolate anion and be protonated at C_2 (Figure 14.11). That completes the four-enzyme pathway: two $NADP^+$-accepting hydride transfers, a hydrolytic lactonase, and a carbanion-generating 1,2-isomerase.

D-Ribose-5-P is the critical precursor for both purine and pyrimidine nucleotides as described in Chapter 6. The enzyme 5-P-ribose pyrophosphokinase[12] utilizes the α-hemiacetal's C_1-oxyanion of ribose-5-P to act as an oxygen nucleophile specifically towards the electrophilic β-phosphorus atom in the triphosphate side chain of ATP. This direct pyrophosphoryl transfer yields 5-P-ribose-α-1-PP (PRPP), activated at C_1 for capture by amine nucleophiles in both purine and pyrimidine ring assemblage (Figure 6.7). The pentose-P pathway must provide several billions of nucleotide triphosphates and deoxynucleotide triphosphates for DNA replication and its transcription to RNA in every cell cycle.

Beyond providing ribose-5-P from glucose-6-P, the pentose-P pathway encodes transketolases and transaldolases that carry out ketol transfers and aldol transfers between sugar substrates to provide adequate, equilibrative levels of the phosphorylated aldoses from the three-carbon 3-P-glyceraldehyde, the four-carbon 4-P-erythrose, and the seven-carbon-7-P-heptose. We noted back in Chapter 7 the chemistry that leads to the carbacyclic phenyl ring in phenylalanine and congeners. It is the pentose-P pathway that funnels erythrose-4-P and the glycolytic intermediate PEP by carbanion chemistry into the seven-carbon 2-keto-3-deoxyheptulose-7-P (Figure 7.8). In turn, this heptose becomes the committed building block to shikimate, chorismate, and phenylpyruvate in the way to phenylalanine and a suite of salicylate, isochorismate, dihydroxybenzoate and aminobenzoate congeners (Figure 7.11).

14.5 Glucose Activation at C_1 for Electrophilic Glycosyl Transfers

We commented earlier in this chapter that 6-P-glucose (and glucose itself) under physiologic conditions is a rapidly equilibrating mixture of a very minor amount of linear free aldehyde and 99% as a 2/1 mixture of β/α cyclic hemiacetal isomers. The predominant hemiacetal isomers are protected latent forms of the electrophilic aldehyde. Glycolysis utilizes the electrophilic properties of the minor linear aldehyde form in the first step catalyzed by 6-P-glucoisomerase. In contrast, the pentose-P pathway just summarized uses the cyclic hemiacetal β anomer as starting form of 6-P-glucose.

The third pathway, like the pentose-P pathway, involves chemistry at C_1. However, whereas the dehydrogenation of the hemiacetal of glucose-6-P to gluconolactone oxidizes C_1 as an electrophilic group (hemiacetal to lactone), the route from the *ur* metabolite 6-P-glucose to polysaccharides, such as glycogen and starch (energy storage forms for glucose) or cellulose (cell wall structural polymer in plants and microbes), requires different chemical logic.

Glcose-6-P flux is funneled down the polysaccharide biosynthetic pathway by enzymatic conversion of glucose-6-P to the α-isomer of glucose-1-P. The requisite enzyme, phosphoglucomutase,[13] acting reversibly, has an interesting strategy for effecting the apparent movement of the phosphoryl group between the 6-OH and the 1-OH of the glucose hemiacetal ring (Figure 14.12).

The enzyme starts each catalytic cycle with a serine side chain in the active site already converted to a phosphate monoester. When G-6-P binds, the C_1–OH of the α-anomer acts as nucleophile on the electrophilic phosphorus atom in the enz-Ser-OPO_3^{2-} group, forming a tightly bound 6-P-glucose-α-1-P. This intermediate must be able to rotate with regard to the now dephosphorylated serine residue and present itself for attack by the Enz-Ser-OH on the 6-PO_3^{2-} group, to effect an apparent net movement of the –PO_3^{2-} group between C_6–OH and C_1–OH. The enzyme will readily catalyze the reverse reaction but only from the α-isomer, not the β-isomer, of G-1-P.

The second enzyme[14] in the pathway utilizes one of the oxyanions of the phosphate group of the G-1-P as nucleophile, attacking the electrophilic Pα of cosubstrate UTP. The result is transfer of the UMP moiety as a (phosphorus) electrophile to give UDP-α-glucose[15] and the PPi fragment (Figure 14.12). This third pathway for use of the glycosyl scaffold by thousands of glycosyl transferases thus requires only two dedicated enzymes to create the glycosyl group donors.

UDP-glucose falls into that special category of core metabolites that are thermodynamically activated for group transfer but have sufficient kinetic stability that they can serves as diffusable substrates within cells.[15] In particular, UDP-glucose is the general source of glucosyl groups that are

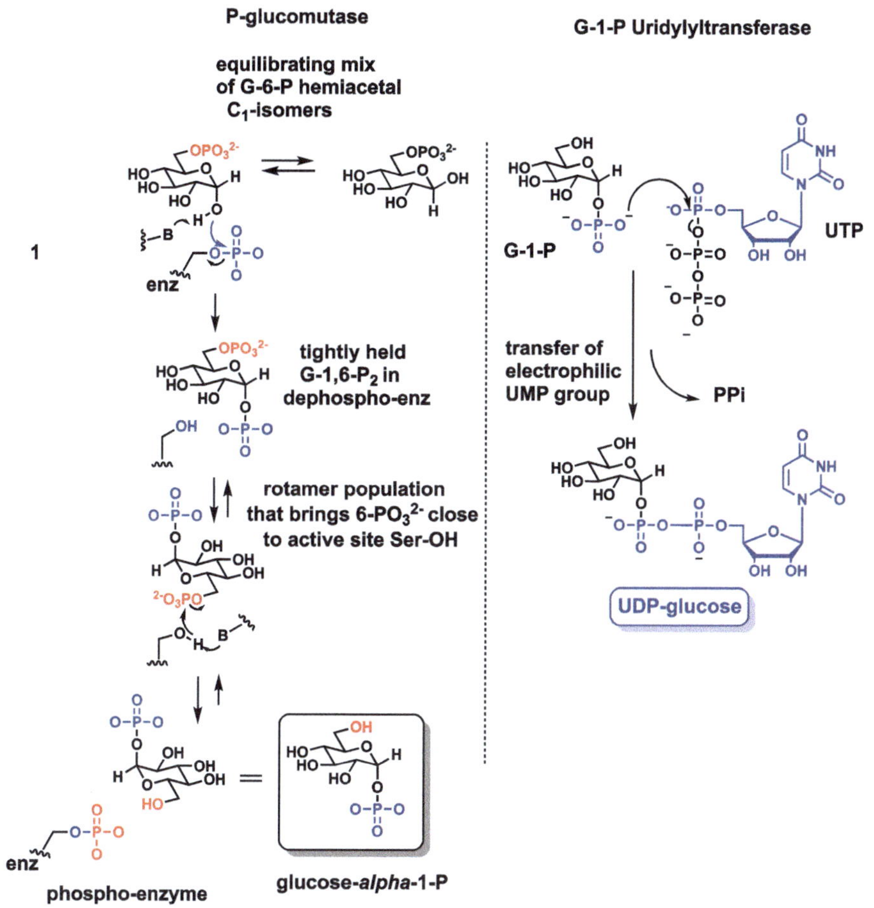

Figure 14.12 Phosphoglucomutase channels some of the glucose-6-P intracellular pool to glucose-alpha-1-P, that then attacks UTP to form UDP-glucose.

transferred as electrophilic fragments at C_1 to a variety of cosubstrate nucleophiles (Figure 14.13). For example, UDP-glucose (α-isomer only) is a substrate for chain elongation by glycogen synthetases of animals and humans, by starch synthases of plants and by cellulose synthases of plants and fungi.

As we featured in Chapter 11, many thousands of enzymatic glycosyl transfers proceed *via* early C_1–O bond cleavage of the bound UDP glucose, releasing UDP while creating an oxocarbenium glycosyl transition state or intermediate (Figure 14.13). That carbocation at C_1, as transition state or unstable finite lived intermediate, can be delocalized into the more basic oxygen, giving an oxocarbenium ion character to the transferring glycosyl group. Incoming nucleophiles can be vectored to approach from below (as in glycogen and starch synthases) or from above (cellulose synthases). In the former case, the transferring glycosyl unit ends up in an α-1,4-acetal linkage

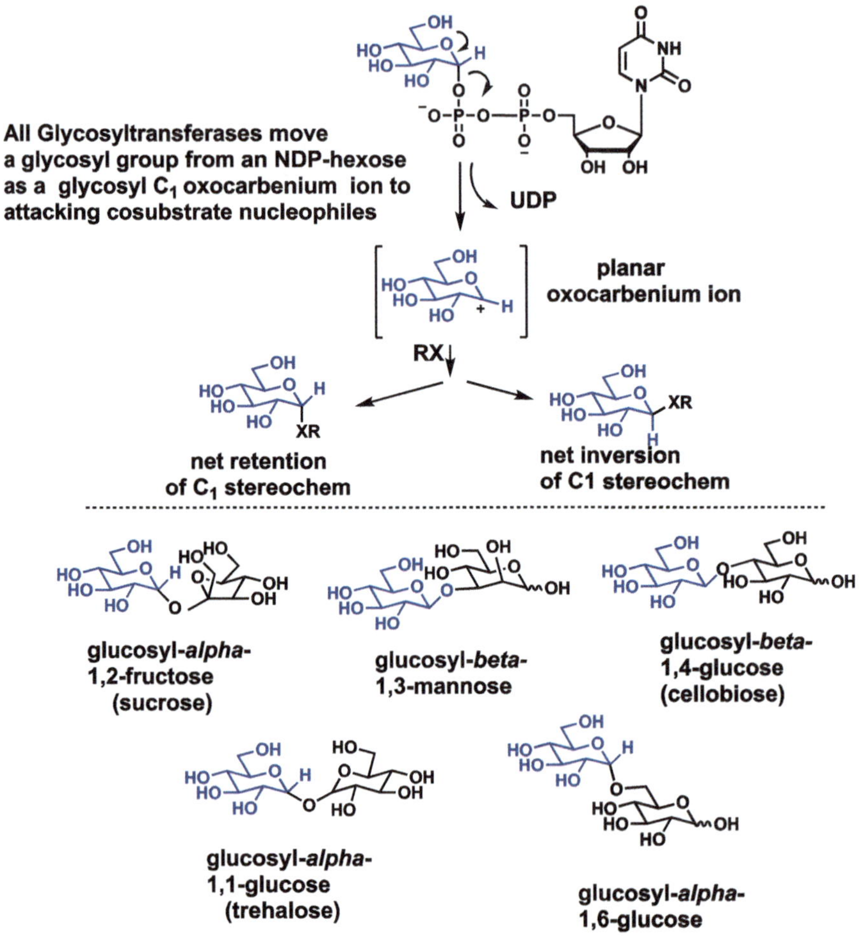

Figure 14.13 UDP glucose is progenitor to a glucosyl-C_1-oxcarbenium ion transition state, capturable by a series of biologic nucleophiles. When those are sugar oxygens a variety of sugar acetal linkages are shown with varying regiochemistry and α/β stereochemisry at C_1, dictated by particular glycosyl transferases.

(net retention of C_1 stereochemistry). In the latter case, the transferring glycosyl moiety ends up in a β-1,4-acetal (= glycoside; net inversion at C_1) linkage. Acetal linkages are both chemically and configurationally stable unlike the hemiacetals that can equilibrate through the linear aldehyde forms.

Thousands of glycosyl transferases use this strategy, with many variations in the nature of the capturing nucleophile and the regiochemistry of the nucleophile (*e.g.* to give 1,1- (trehalose)-, 1,2-, 1,3-, 1,4- and 1,6-, glucoside linkages to other glucose or hexose units (Figure 14.13) with either α or β stereochemistry in the acetyl link of the transferring glucosyl unit

(Figure 14.1). The strategy holds for transfer of other common hexosyl groups from nucleoside diphosphosugars, *e.g.* from UDP galactose or from GDP-mannose. As noted in Chapter 11, the glycosyl oxocarbenium ion strategy relies on the low energy barrier for C_1–O-UDP cleavage and the delocalization of the accumulating positive charge on C_1 of the glycosyl moiety.

14.6 Overview of the Chemical Biology of Glucose

The two types of functional groups in glucose, the five hydroxyl groups and the aldehyde, enable and dictate the several roles it plays as a central metabolite – as an energy source in degradative oxidation, and in two anabolic roles – as the building block for the ribose ring of nucleic acids, and as the monomer for oligo- and polysaccharide chain growth. For example, each of the five hydroxyl groups in glucose at some point in its metabolic diaspora are converted to carbonyl groups (all by hydride transfers to oxidized nicotinamides as cosubstrates) (Figure 14.14): the 2-keto in fructose-6-P, the 3-keto in 6-P-gluconate, the 4-keto in deoxyhexose biogenesis, the 5-keto in inositol-P synthase and even the 6-aldehyde in conversion of UDP glucose to UDP-glucuronate.

Those carbonyls give rise to low energy stabilized carbanions at all of adjacent carbons 1–6, so carbon electrophile and carbon nucleophile chemistry is enabled at every carbon. Thereby, much of glucose chemical biology is regiospecifically directed redox chemistry.

The hydroxy carbonyl pairings add additional functional group chemistry. The aldehyde in the open chain form of glucose allows entry, *via* the adjacent C_2 carbanion, to a pair of 1,2-enediolate isomers, and the conversion to the fructose scaffold. The 1,3 hydroxycarbonyl grouping sets up aldol cleavage in fructose-1,6-P_2 *via* enzyme enamine chemistry. The 1,4- and 1,5-hydroxycarbonyl pairings set up facile nonenzymatic cyclization to the furanose and pyranose hemiacetals forms.

The glucopyranose hemiacetals predominate in the population of glucose molecules. They serve as a readily reversible protecting group for the reactive aldehyde electrophile in the open chain form of glucose. Indeed, the open chain form is present at less than 1% of the population, blunting (but not fully abrogating) promiscuous reactivity as an electrophile. The two hemiacetal ring forms effectively reverse the polarity at C_1, from a carbon electrophile as aldehyde to an oxygen nucleophile in the hemiacetal isomers.

Remarkably, that very minor open chain aldehyde form of glucose-6-P carries all the flux down the tandem glycolysis and citric acid cycle ($\sim$50–60 grams per day $= \sim 2 \times 10^{23}$ molecules per person per day). The aldehyde carbon is the essential initiator functional group. On the other hand, the dominant cyclic hemiacetal forms are the starting substrates for the pentose-phosphate pathway and the glycosyl group transfers in polysaccharide chain elongations. The pentose-P pathway is also oxidative, but selectively at C_1 and C_3 of the glucopyranose to enable the low energy decarboxylation to get to the ribose scaffold.

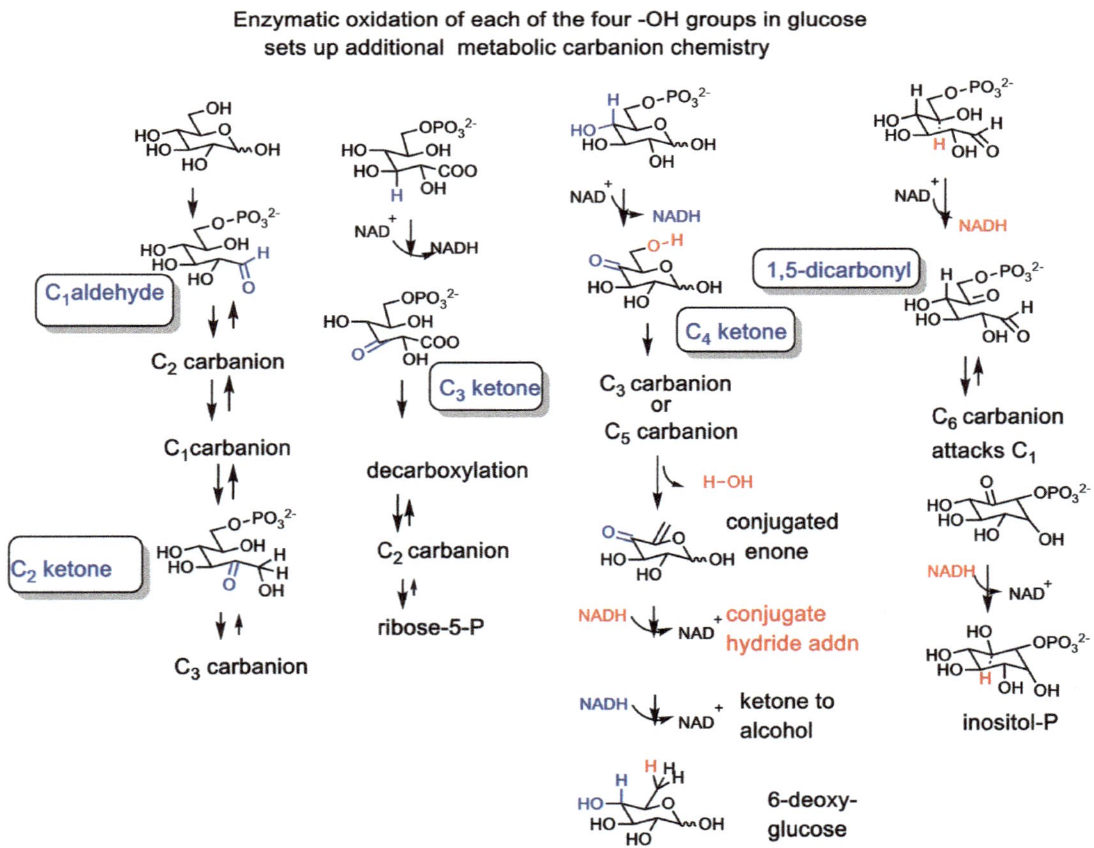

Figure 14.14 Each of the hydroxy groups in glucose can be enzymatically oxidized to the ketone, setting up adjacent 1,2- and 1,3-hydroxycarbonyl group chemistries that broaden the scope and reach of glucose metabolism.

In contrast, the third pathway is nonoxidative. The phosphoglucomutase engages in phosphoryl transfers to convert G-6-P to α-G-1-P. In that active site the oxygen at C_1 of the glucose hemiacetal acts as a nucleophile, not as the electrophilic carbonyl group of the open chain aldehyde. The glucose α-1-P oxygen then also acts as a nucleophile in UDP-glucose formation. That nucleotide diphosphosugar is precursor to oxocarbenium species at C_1 of the glucosyl group in transfer to a number of oxygen, nitrogen and even carbon cellular nucleophiles. Those are glycosyl transfers with breakage of the C_1–O bond, a novel reactivity compared to glucose chemistry in the other two pathways.

This is C_1 glycosyl carbocation chemistry but in the cyclic ring form not the open chain carbonyl form. When sugar hydroxyl oxygens are the attacking nucleophiles, the products have an acetal linkage. Unlike the rapidly equilibrating reversible hemiacetals, acetal linkages are configurationally and chemically stable. They are the only type of covalent bond connecting the hexose monomers in glycogen, starch, cellulose, and other polysaccharides.

In sum, the two types of carbon functional groups in glucose, the alcohols and the aldehyde, enable a wide range of carbon chemical biology, making glucose an almost perfect metabolic building block.

References

1. C. T. Walsh, *The Chemical Biology of Phosphorus*, Royal Society of Chemistry, London, 2021.
2. A. M. Helmenstine, *Anabolism and catabolism definition and examples*, 2020.
3. C. T. Walsh, Biologically generated carbon dioxide: nature's versatile chemical strategies for carboxy lyases, *Nat. Prod. Rep.*, 2015, **37**, 100–135.
4. E. Lorentzen, B. Siebers, R. Hensel and E. Pohl, Mechanism of the schiff base forming fructose-1,6-bis phosphate aldolase: structural analysis of reaction intermediates, *Bicohemistry*, 2005, **44**, 4222–4229.
5. R. N. Perham, Swinging arems and swinging domains in mutlifunctional enzymes: catalytic machines for multistep reactions, *Annu. Rev. Biochem.*, 2000, **69**, 961–10004.
6. D. Goodsell, *PDB Molecule of the month: citrate synthase*, 2007.
7. S. J. Lloyd, H. Lauble, G. S. Prasad and C. D. Stout, The mechanism of aconitase: 1.8A resolution crystal structure of the S642A: citrate complex, *Protein Sci.*, 1999, **8**, 2655–2662.
8. C. E. Quartararo, T. Hadi, S. M. Cahill and J. S. Blanchard, Solvent isotope-induced equilibrium perturbation for isocitrate lyase, *Biochemistry*, 2013, **52**, 9286–9293.
9. F. Qi, R. K. Pradhan, R. K. Dash and D. A. Beard, Detailed kinetics and regulation of mammalian 2-oxoglutarate dehydrogenase, *BMC Biochem.*, 2011, **12**, 53.
10. J. Rutter, D. R. Winge and J. D. Schiffman, Succinate dehydrogenase-assembly, regulatioin and role in human disease, *Mitochondrion*, 2011, **10**, 393–401.

11. N. J. Kruger and A. von Schaewen, The oxidative pentodse phosphate pathway: structure and organization, *Curr. Opin. Plant Biol.*, 2003, **6**, 236–246.
12. B. Hove-Jensen, K. R. Andersen, M. Kilstrup, J. Martinussen, R. L. Switzer and M. Willemoes, Phosphoribosly diphosphate (PRPP): biosynthesis, enzymology, utilization, and metabolic significance, *Microbiol. Mol. Biol. Rev.*, 2017, **81**, e00040-16.
13. M. D. Percival and S. G. Withers, Fluorine-19 NMR investigations of the catalytic mechamism of phosphoglucomutase using fluorinated substrates and inhibitors, *Biochemistry*, 1992, **31**, 505–512.
14. J. I. Fuhring, J. T. Cramer, J. Schneider, P. Baruch, R. Gerardy-Schahn and R. Fedorov, A quaternary mechanism enables the comoplex biological functions of octameric human UDP-glucose pyrophosphorylase, a key enzyme in cell metabolism, *Sci. Rep.*, 2015, **5**, 9618.
15. C. T. Walsh, B. P. Tu and Y. Tang, Eight kinetically stable but thermodynamically activated molecules that power cel metabolism, *Chem. Rev.*, 2018, **118**, 1460–1494.

CHAPTER 15

C–C Bonds in Biosynthesis: Squalene and the Sterol Biosynthetic Pathway

The second of the two iconic metabolites in this last section of the book is the 30-carbon hydrocarbon squalene, originally sourced from deep sea shark livers, but also abundant in the lipid extracts of olives. In contrast to glucose, the topic of Chapter 13, bristling with hydroxyls, and a latent aldehyde in the predominant cyclic hemiacetal forms, squalene has no C–O functional groups. It does have six alkene groups, each in a 1,4-relationship, each with *E*-geometry. The chemical logic that organisms employ to build up the 30-carbon hexaene scaffold of squalene and then to deploy it as immediate precursor to tetracyclic and pentacyclic steroids illuminate the several strategies of olefin chemical biology and multiple routes to C–C bond formation.

15.1 C–C Bonds in Two Biosynthetic Classes

In contrast to the use of heterocyclic purine and pyrimidine building blocks for nucleic acid assemblies, the 20 proteinogenic amino acids in ribosome-catalyzed protein biosynthesis, and in the assembly of tetrapyrrole macrocycles, the biosynthetic arms to fatty acids and isoprenoids, including the tetracyclic steroids, are effectively nitrogen free metabolic zones (so is much of the carbohydrate metabolism embodied by glucose processing in the previous chapter). There are no amine nucleophiles to build C–N bonds. Instead, these are two realms of distinct strategies to build C–C bonds.

Both the fatty acid and steroid realms in animal metabolism use as common biosynthon the thermodynamically activated acetyl group in the thioester acetyl CoA[1] (Figure 15.1). Both biosynthetic routes are highly reductive,

Chemical Biology No. 21
The Chemical Biology of Carbon
By Christopher T. Walsh

Published by the Royal Society of Chemistry, www.rsc.org

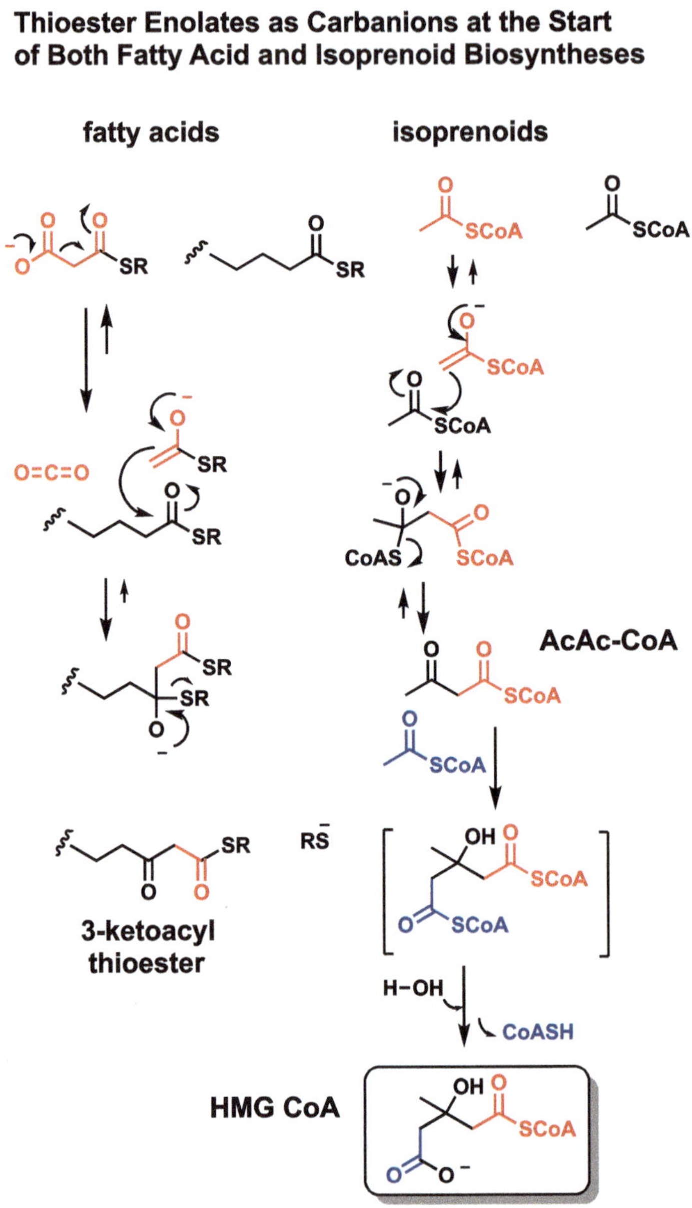

Figure 15.1 Thioester enolates as carbanion equivalents at the start of both fatty acid biosynthesis and isoprenoid biosynthesis pathways.

ultimately converting the acetyl carbonyl group to CH_2 groups in fatty acid and sterol frameworks. Each carbonyl group reduced to a methylene group is a four-electron reduction. In some senses this is *in vivo* chemistry where functional groups disappear into the background of methylene and methyl groups.

We note in Section 15.3.1 that some bacteria avoid acetyl CoA as starting point to isoprenoid metabolites. Instead, they couple two available, glycolytic three-carbon metabolic intermediates, glyceraldehyde-3-P and pyruvate, bypassing the hydroxymethylglutaryl CoA stage of the pathway entirely. This may have been a less evolutionarily favorable route as it is far less frequent than the acetyl CoA route in extant prokaryotes and eukaryotes.

For the great bulk of organisms, the two pathways commence with acetyl CoA (for isoprenoids) or its C_2-carbanion-generating daughter metabolite malonyl CoA (for fatty acids) as the initial building blocks for variants of thio-Claisen condensation chemistry (Figure 15.1). The stoichiometry for converting eight acetyl groups from acetyl CoA to the fully saturated C_{16} chain of palmitate consumes 14 hydride equivalents of NADPH. Seven ketone groups are each reduced by four electrons (28 electrons in from NADPHs), emphasizing that fatty acids with their array of methylene groups are sinks for metabolic electrons (see Chapter 2).

Analogously, the stoichiometry for converting the acetyl groups of 18 acetyl CoA molecules into the major mammalian sterol cholesterol indicates several precepts at work. First is the consumption of 46 molecules of NADPH, illustrating not only the profound reductive theme of sterol assembly but also the ubiquity of hydride transfers as the prime redox modality in cellular metabolism (Figure 15.2).

As in the fatty acid logic above, each of the 18 acetyl carbonyl groups undergo net four-electron reduction to 18 methylene groups. Each NADPH provide two electrons in a hydride ion packet, so 36 of the 46 NADPHs that are consumed are put to this direct reductive use. The remaining 10 NADPH are indirect external reductants for O_2 reduction by either a flavoenzyme

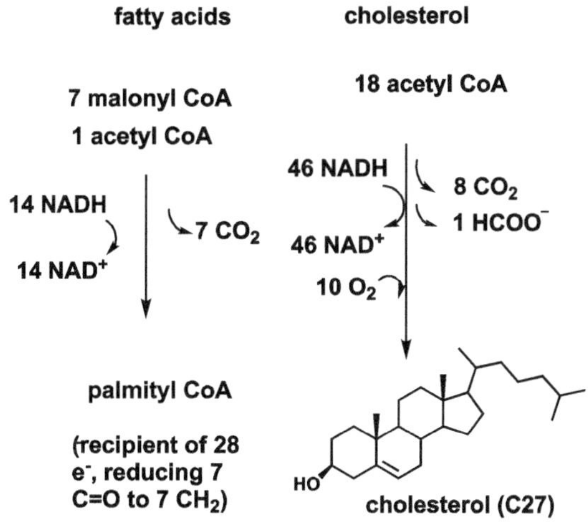

Figure 15.2 Fatty acid biosynthesis and cholesterol biosynthesis are profoundly reductive processes.

oxygenase[2] (squalene epoxidase) or three cytochrome P450 oxygenases[3] (converting three lanosterol C–CH_3 to C–H and CO_2, each acting for three consecutive oxygenations; see Chapter 12). Second, of the 36 carbons from 18 acetyl CoA building blocks, eight are released as CO_2[4] and one as a formate ion to leave a connected C_{27} framework in cholesterol (Figure 15.2). Six of the eight CO_2 are released early, as the biologic isoprenoid building blocks are constructed. The other two CO_2 are carved out late, during maturation of lanosterol to cholesterol. The single formate ion is also released during the late oxygenative phase of the pathway.

All 10 dioxygen molecules are utilized by four enzymes after the C_{30} framework of squalene has been built up in the isoprenoid phase of the pathway. Notably, while 10 O_2 molecules are reductively activated and fragmented by oxygenases, only one oxygen atom ends up in the cholesterol scaffold (as the 3-β-alcohol). It is instructive logically and mechanistically to observe the fate of the other 19 oxygen atoms (see Chapter 12).

15.2 Squalene as a Central Intermediate in the Biosynthesis of Cholesterol

Given all the above chemical operations, the intermediacy of the linear C_{30} hydrocarbon squalene, with six unconjugated olefins, each in a 1,4-relationship, as the key progenitor to the full tetracyclic scaffold of sterols, places it as a key metabolite (Figure 15.3).

As we examine the chemical logic of this highly reductive biosynthetic pathway, we pose several questions. How are the C–C bonds in the C_{30} hydrocarbon squalene set? How are the six olefins installed with the characteristic unconjugated spacing? How is the acyclic hexaene directed to cyclize? By what routes are the internal olefins converted to C–C single bonds? How is the initial product C_{30} lanosterol converted to major membrane sterol C_{27} cholesterol by oxidation of three C–CH_3 to three C–H groups and one formate ion and two CO_2?

15.3 Five Phases to the Bond-forming Strategies to Cholesterol

One can divide the C–C bond forming logic in cholesterol biosynthesis into five phases (Figure 15.4).

The first phase uses the carbanions available from thioester enolates to drive thio-Claisen condensations. The second phase depends on accessibility of allylic carbocations as carbon electrophiles as six C_5 units are stitched together. The third phase is the head to head joining of two C_{15} farnesyl-PPs reductively to the C_{30} hexaene squalene. The fourth phase is the carbocation-based cyclization of squalene-2,3-epoxide to lanosterol. The fifth phase can be viewed as oxygenative, carving out of three methyl groups as C_{30} lanosterol is processed to C_{27} cholesterol. One oxygenase cycle (*via* squalene epoxidase)

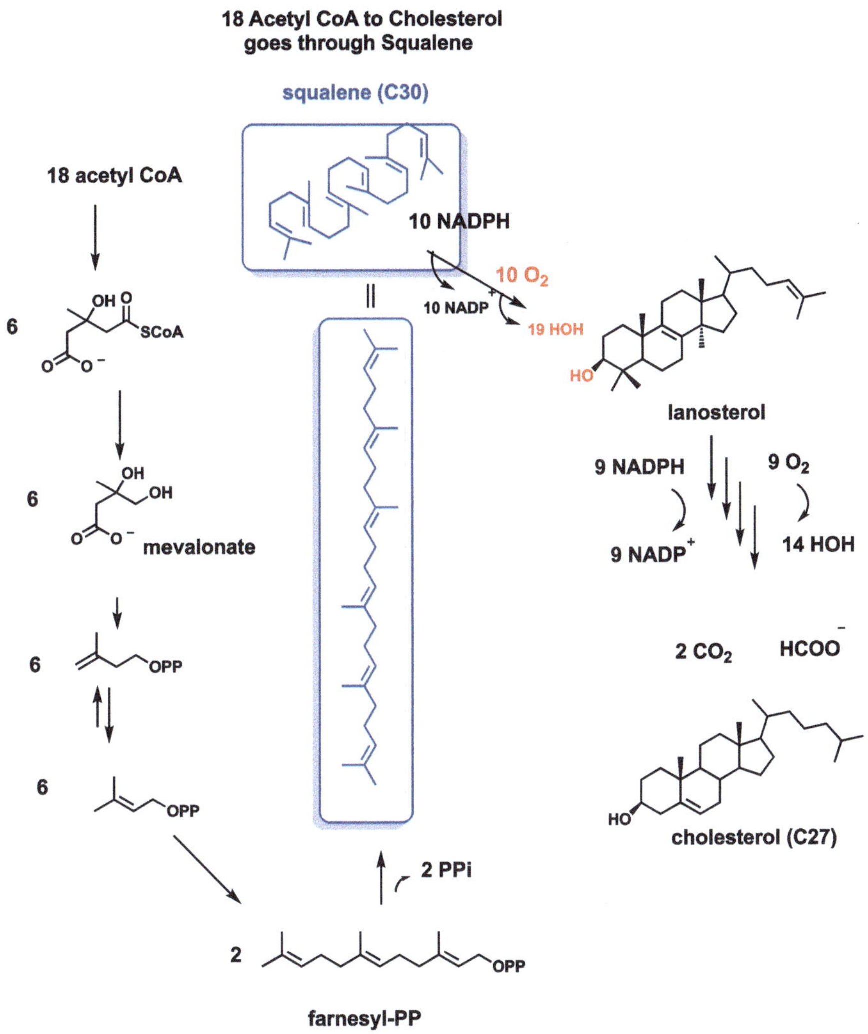

Figure 15.3 The cholesterol biosynthetic pathway goes through squalene, a linear C30 hydrocarbon with six olefins in 1,4-relationship.

initiates the fourth phase. Nine oxygenase cycles, from three cytochrome P450s each acting three times, characterize the fifth phase, a fundamental switch from cation-based hydrocarbon chemistry to radical-based oxygenation chemistry (see Chapter 12). Only the pericyclase enzyme class is missing from the steroid biosynthetic construction catalytic tool kits.

We have analyzed aspects of each of these five phases in previous chapters but we summarize the *ad seriatim* logic here. The full pathway illustrates the distinct and complementary roles that carbanions, carbocations, olefins as nucleophiles, epoxides, and oxoiron-generated carbon radicals play in maturation of the reduced tetracyclic steroid scaffold.

Five Distinct Chemistries in Cholesterol Biosynthesis

1. Thioester Enolates: 3 acetyl CoA ⟶ hydroxymethylglutaryl CoA

2. Head to tail C5 alkylations by prenyltransferases to farnesyl-PP

3. Head to head self alkylation of two farnesyl-PPs to squalene

4. Carbocation-based cyclization of squalene epoxide

5. Oxygenations with carbon radical species

Figure 15.4 Five distinct chemistries used in cholesterol biosynthesis.

15.3.1 Isoprenoid Biosynthesis: Phase 1. The Canonical Mevalonate Pathway Starts with Three Acetyl CoA Condensations by Thioester Enolate Carbanion Chemistry

As featured in Figure 15.5, the classical mevalonate pathway to isoprenyl-diphosphates starts with two molecules of acetyl CoA. The β-ketoacyl thiolase enzyme[5] generates the C_2 carbanion of the acetyl group in the active site, attacking the electrophilic thioester carbonyl in a second molecule of acetyl CoA, to produce the 3-ketoacetyl CoA four carbon product and free CoASH (Figures 2.8 and 15.1).

Then, instead of processing this 3-ketoacyl thioester for two-electron reduction as in fatty acyl thioester biosynthesis, the next enzyme catalyzes a second C–C bond forming reaction with another acetyl CoA[5] (Figure 15.1). This iterated thio-Claisen reaction sequence involves another generation of a thioester enolate/C_2 carbanion from the third equivalent of acetyl CoA. The electrophilic partner is the four-carbon 3-ketobutyryl-*S*-CoA. However, this time the enzyme directs attack at the 3-keto carbonyl group rather than the C_1 thioester carbonyl. The result is a nascent branched chain six-carbon glutaryl dithioester with a 3-OH group (Figure 15.1). This bound, nascent dithioester of 3-hydroxyglutaryl-bis-CoA undergoes regiospecific thioester group hydrolysis specifically on the thioester derived from the acetyl CoA reactant (Figure 15.1). This thermodynamically favored thioester hydrolysis provides a factor of 10^5 in favor of the condensation direction. It is the key step in drawing flux into the HMG-CoA pathway to isoprenoids. Note that the branched carbon chain that persists in the isoprenoid scaffolds has been created in this second C–C bond forming step and mandates the *C*-methyl group branching typical of isoprenoid scaffolds.

The rate-limiting step in the isoprenoid/steroid biosynthetic pathway in humans is catalyzed by the next enzyme, hydroxymethylglutaryl CoA reductase,[6] carrying out a tandem pair of hydride transfers from two NADPH molecules (a net four-electron reduction of acyl thioester to alcohol) to sequentially reduce the HMG CoA thioester carbonyl, first to the aldehyde, and then, before its dissociation, to the alcohol mevalonate (Figure 15.4). HMG-CoA reductase is the target for the family of statins used therapeutically as chronic blood pressure lowering agents.

15.3.2 Mevalonate to a Pair of Isopentenyl-PP Isomers

At this point, three successive ATP-cleaving kinase reactions occur[7] (Figure 15.5). The first two build up a phosphate ester and then a pyrophosphate ester at the newly generated C_1 alcohol of mevalonate (that pyrophosphate group will ultimately become the leaving group that leads to low energy cleavage of the C_1–O bond as the oxygen departs with inorganic pyrophosphate). The third enzyme-catalyzed phosphoryl transfer from ATP is to the 3-OH group, transiently generating the 3-phospho-mevalonate-1-pyrophosphate. This phosphoryl group transfer has converted the 3-OH to a 3-OPO_3^{2-} group, set up for low energy C_3–OP bond cleavage since the departing oxygen will be part of the resonance stabilized inorganic phosphate product (Figure 15.5).

The driving force for phosphate elimination from C_3 is decarboxylation at C_6. As the CH_2–COO bond cleaves and incipient CO_2 forms, the electrons that had been in the C–C bond undergoing cleavage flow into the C_4–C_3 bond. They act as a neighboring group to help eject the phosphate dianion from C_3, taking out the original C_3-tertiary-OH group as part of the departing phosphate dianion. Decarboxylation thus provides the electrons to function as intramolecular neighboring group assistance to C_3–O bond cleavage.

We have seen key decarboxylations in two of the three glucose pathways of Chapter 13 and will see others in the tetrapyrrole biosynthetic pathway. Each of these decarboxylations reflect strategies by which organisms sculpt substrate frameworks, usually employing incipient carbanion chemistry by C–COO^- bond disconnections.[4]

The result in the HMG CoA context is a branched five-carbon scaffold with a 1-pyrophosphate ester linkage and a newly formed olefin at $C_3{=}C_4$. This olefin-forming step provides the alkene functional group that shows up six times in the 30-carbon squalene metabolite (Figure 15.3). The product from the decarboxylation/inorganic phosphate elimination is Δ^3-isopentenyl-diphosphate (Figure 15.5).

Δ^3-Isopentenyl-diphosphate is one half of the pair of biologic isoprene isomers needed for the series of chain elongations characteristic of isoprenoid chain growth in chemical biology. The other isomer is Δ^2-isopentenyl-diphosphate available by isomerase action on the 3-IPP isomer (Figure 15.5). In this mevalonate pathway the Δ^2-isomer is derived biosynthetically from the initial Δ^3-isopentenyl-diphosphate isomeric scaffold by action of an isopentenyl diphosphate isomerase enzyme.[8] One mechanistic possibility is depicted in Figure 15.6 for equilibrative

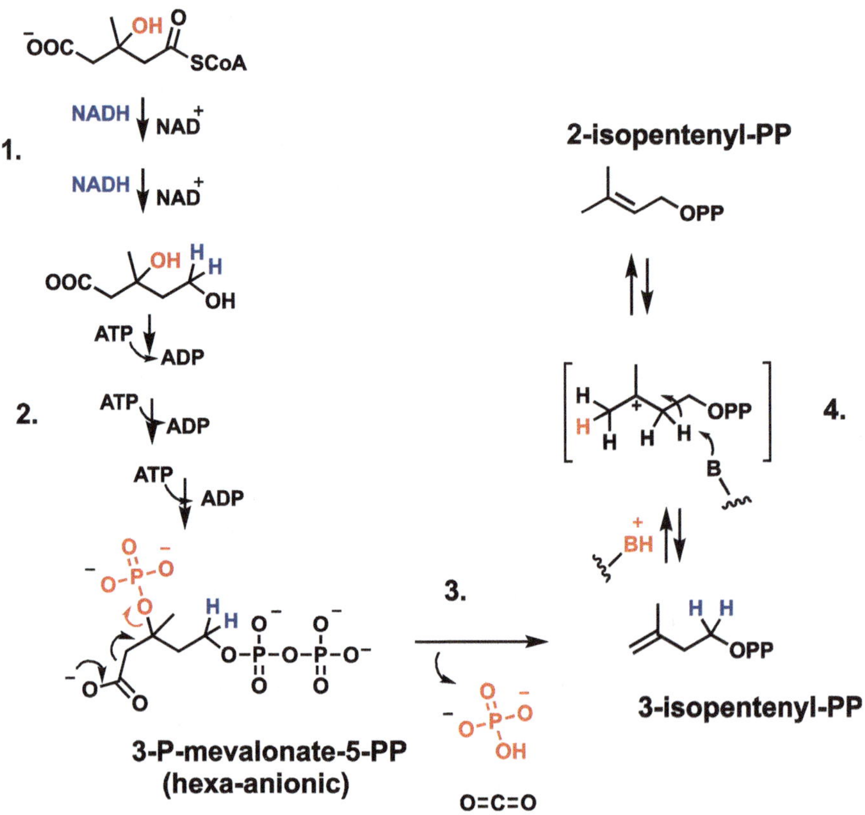

Figure 15.5 The stretch from hydroxymethylglutraryl CoA to the 3- and 2-isiomers of isopentenyl pyrophosphate involve four notable steps: (1) double reduction of thioester to alcohol in mevalonate; (2) three successive phosphorylation events *via* three ATPs; (3) decarboxylation and inorganic phosphate elimination create the C=C double bond in 3-isopentenyl-PP; (4) isomerization of 3-IP to 3-IPP by reversible protonation of the olefin group.

isomerization between the terminal double bond (Δ^3-isomer) and the internal double bond isomer (Δ^2-). It features protonation of the terminal olefin to give the tertiary carbocation. That cation is quenched in the front direction by proton abstraction of one of the two prochiral methylene hydrogens at C_2, creating the C_{2-3} olefin. This olefin/carbocation chemistry is a recurring motif in all the subsequent reactions up to squalene formation (Chapter 2).

15.3.2.1 *Isoprenoid Biosynthesis Alternate Phase 1: The MEP Pathway Alternative*

Before summarizing the carbon–carbon bond coupling mode of the isopentenyl isomers we turn to alternate logic for their formation used by some bacteria that are pathogenic to humans. This alternate route is termed the

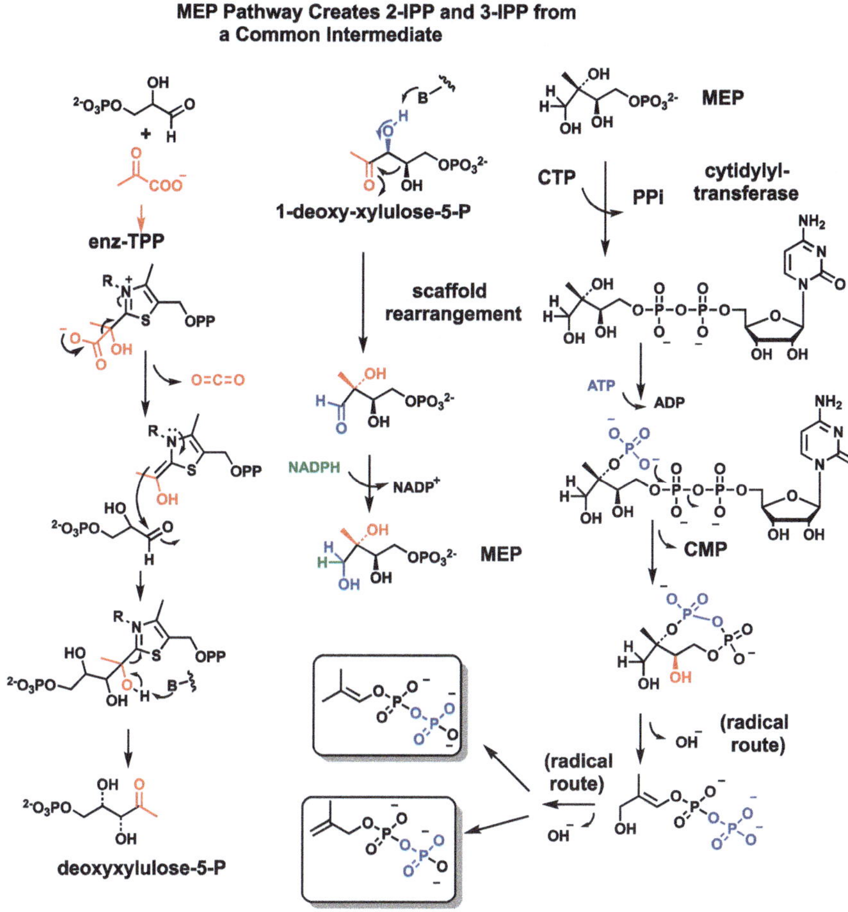

Figure 15.6 The alternative methylerythritol phosphate (MEP) pathway creates both 3-IPP and 2-IPP from a late stage common intermediate. The pathway starts with a TPP-mediated decarboxylation of pyruvate, with hydroxyethyl group transfer to glyceraldehyde-3-P to form deoxy-xylulose-1-P. A retro aldol, scaffold recombination gives the eponymous MEP metabolite. Multistep formation of a cyclic pyrophosphate precedes two radical chemistry steps eliminating an –OH group in each step on the way to fragmentation to the IPP isomers.

nonclassical methylerythritol-P (MEP) route[9] (Figure 15.6), compared to the classical mevalonate route.

Because it operates in bacterial pathogens and not humans the MEP pathway has gained interest for possible targeting by antibiotics. Note its distinctive chemical logic for getting to the five carbon branched chain olefinic pyrophosphate isomers.

The MEP pathway[9] begins with two abundant glycolytic three-carbon metabolites pyruvate and glyceraldehyde-3-P. The first enzyme decarboxylates pyruvate to a ketol carbanion equivalent to attack the electrophilic aldehyde carbonyl of glyceraldehyde-3-P. In keeping with nature's logic to lower the energy barrier for 2-keto acid decarboxylation (Chapter 9) this first enzyme utilizes tightly bound thiamin-PP to form a β-iminium adduct, inducing low energy barrier to decarboxylation. The enzyme then uses the resulting C_2-hydroxyethyl-TPP carbanion as carbon nucleophile (Figure 15.6). The released product is 1-deoxyxylulose-5-P.

The second enzyme is a reductoisomerase, consuming NADPH while generating methylerythritol-P. This is the catalyst that generates a branched carbon scaffold from the linear substrate scaffold by quite distinct chemical logic from the HMG CoA framework in the mevalonate pathway. Figure 15.6 shows that carbons 2 and 3 from starting pyruvate (colored red) end up in the rearranged scaffold. One proposed mechanistic route is for the deoxylulose-P substrate to undergo a retro aldol cleavage and then an aldol condensation on nascent rearranged product fragments in the active site to glycoaldehyde-2-P and the nascent C_1 carbanion of 3-methy-2-keto-1-propanol.

We have seen several previous examples, notably in the enzymatic conversion of glucose-6-P to fructose-6-P, and in ribulose-bis-phosphate carboxylase action, which enediolate anions can interconvert by proton transfer to isomeric enediolate anions. If that happens in this enzyme active site, the alternate enediolate anion isomer is the C_2-carbanion of lactaldehyde. Assuming the nascent glycoaldehyde-2-P is still in the active site, an aldol condensation yields the rearranged scaffold with a C_1 aldehyde. Reduction of that aldehyde to the alcohol by NADPH in the second half reaction of the isomeroreductase reaction gives methylerythritol-P. The first two enzymes of the MEP pathway to isoprenoids utilize several recognizable facets of carbonyl chemistry, carbanion generation and isomerization: a retro aldol and then an alternative aldol condensation within a single catalytic cycle. The reductoisomerase qualifies as a cascade catalyst.

Enzyme 3 is a cytidylyltransferase.[9] A phosphate oxyanion of MEP captures a CMP moiety as phosphorus electrophile from cosubstrate CTP to generate CDP-methylerythritol, a variant NDP-sugar metabolite (Figure 15.6). Enzyme 4 is an ATP-dependent kinase that transfers the gamma-PO_3^{2-} from ATP to the oxygen of the tertiary alcohol of CDP-methylerythritol (Figure 15.6). This CDP-O-phospho-methylerythritol is now set up to expel the CMP group as that newly added phosphoryl group acts as an intramolecular oxyanion on the distal phosphorus of the CDP moiety (Figure 15.6). This is a variant of precedented NDP-sugar enzyme chemistry and generates an unusual cyclic pyrophosphate scaffold. The purpose of the cytidylyl transfer two enzymes back is now clear along with the ATP-dependent alcohol kinase action of the previous enzyme. These steps set up intramolecular pyrophosphate (phosphoric anhydride) formation with CMP expulsion as low energy leaving group. The eight membered cyclic pyrophosphate is efficiently a precursor to both isomers of isopentenyl diphosphates.

The next enzyme, ispG, contains Fe_4S_4 clusters and carries out a net two-electron reduction on that cyclic pyrophosphate in two separate one-electron steps (Figure 15.6). Radical chemistry is the signature of iron-sulfur cluster-mediated electron transfers (Chapter 12). Three changes between substrate and product are: (1) directional opening of the cyclic pyrophosphate, (2) loss of the internal-OH group, (3) formation of a Δ^2-olefin in the acyclic olefinic pyrophosphate product.

The last enzyme in the pathway, ispY, likewise mediates a net two-electron reduction of substrate, in two $1e^-$ steps for a radical mechanism for loss of water (Figure 15.6). In the course of this reaction the double bond can end up either remaining at C_2 or moving to C_3, generating in the first case the dimethylallyl-PP (Δ^2-isopentenyl-PP) and in the second case the partner Δ^3-isopentenyl-diphosphate.

The final two enzymes[9] of the pathway have carried out a net four-electron reduction and removal of the two-OH groups present in the cyclic methylerythritol diphosphate. The iron–sulfur-based radical routes may indicate these were an early evolutionary pathway in anaerobic organisms to these key isoprenoid building blocks. The starting materials are pyruvate and glyceraldehyde-3-P. One of the six total carbons is jettisoned in the first step, as CO_2, to get to the five-carbon framework from which subsequent carbanionic rearrangement and then carbon radical reductive chemistry occurs.

The classical mevalonate pathway starts instead with three acetyl CoA molecules to give a six-carbon scaffold from which one carbon must also be jettisoned, also as CO_2, to get to the five-carbon isoprenoid isomers. The decarboxylation strategies are distinct and worth comparative analysis. There are two reductive steps in the mevalonate pathway, net four electrons by two hydride ions from NADPH, *versus* the six electrons pumped into the MEP pathway. The mevalonate pathway avoids any radical intermediates and may have been selected for robustness in aerobic organisms. Both types of underlying chemical logic lead to the common pair of isopentenyl-PP isomers through which all downstream flux occurs.

15.4 Isoprenoid Biosynthesis Phase 2: C–C Bond Chemistry Dominated by Allyl Carbocation Species Derived from Δ^2-Isopentenyl Diphosphate

A large family of prenyltransferase enzymes[10] carry out the next phase of isoprenoid biosynthesis: chain elongations by five carbons at a time. The iterative mechanism for chain growth utilizes a stabilized allyl cationic transition state derived from early cleavage of the C_1–OPP bond in the Δ^2-isopentenyl-PP substrate (Figure 15.7; also refer back to Chapter 11). Presumably, the conversion of that C_1-primary alcohol to the C_1-pyrophosphoryl ester back at the mevalonate stage was used ultimately to lower the

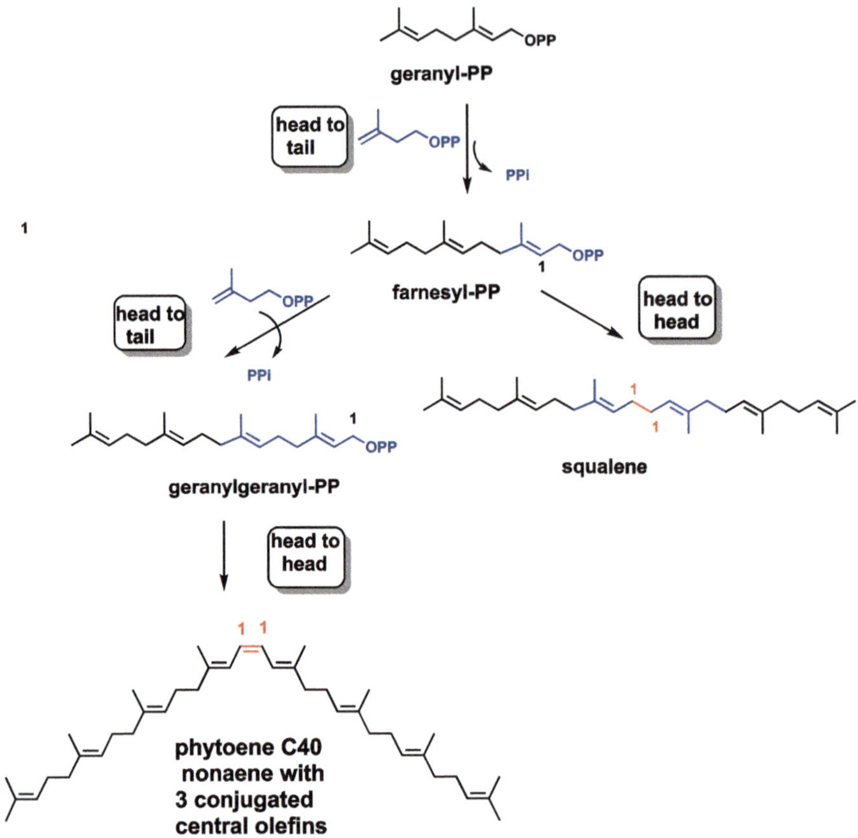

Figure 15.7 Head to tail *vs.* head to head alkylation reactions in isoprenoid linear chain elongations both involve allyl carbocations and capture by olefins acting as carbon nucleophiles.

energy for such C_1–O bond cleavages and enable access to the allyl cationic transition state/intermediate.

Inspection of the carbocation from early C_1–OPP (C–O) bond cleavage shows a C_1 carbocation that has a resonance contributor with the positive charge at the tertiary C_3 carbon, a more stable cation inductively. Indeed, a number of prenyltransferases[10] move the five-carbon cationic species as though it were the C_3-carbenium ion (Figure 15.7) rather than a C_1 carbenium ion. In forming C–C bonds one must pair a carbon electrophile (here the resonance-delocalized C_1-allyl cation) with a carbon nucleophile. The carbon nucleophiles in all the prenyl chain elongation steps are the π-electrons of the Δ^3-prenyl cosubstrate (Figure 15.7).

In the first chain elongation step, two five-carbon substrates are joined, the Δ^2-isomer of IPP as precursor to allyl cation electrophile, the Δ^3-IPP as carbon nucleophile. The new C–C bond formation generates a C_{10} nascent product with a tertiary carbocation at C_3 of the Δ^3-partner (Figure 15.7).

We noted in Chapters 2 and 11 that there are three common ways in which carbocations are quenched during isoprenoid biosynthesis. One route that happens in these elongations is loss of a proton from an adjacent carbon, here C_2, to (re)generate the Δ^2-double bond, now in a C_{10} framework. This molecule is geranyl-PP.

What has happened in this first chain elongation? The two five-carbon Δ^2- and Δ^3-IPP isomers have been enzymatically joined in a net carbon alkylation reaction. The new C–C bond joins what had been the terminal olefinic C_4 of the Δ^3-IPP to the C_1 cationic intermediate from the Δ^2-IPP isomeric substrate molecule (Figure 15.7). The 10-carbon geranyl-PP product has two double bonds: the initial one from the Δ^2-IPP isomer and a second one by net isomerization of the Δ^3-double bond to the Δ^2-position in the C_{10} linear product. Thus, geranyl-PP is a $\Delta^{2,6}$-diolefin with two methyl branches (carbons 9 and 10) occurring as methyl substituents at C_3 and C_7, respectively. The internal double bond geometry is *E*, controlled by the stereospecificity of which prochiral hydrogen is removed as a proton at C_2 in the C_3 cation quenching step. There are always four carbon spacings between the methyl branches as linear isoprenoid chains are elongated.

Note that the 3-IPP carbon nucleophile is transmuted into the 2-IPP moiety bearing the OPP group in the chain-extended product (Figure 15.7). That moiety serves as precursor to the allyl cation in the next elongation cycle, capturable by another five-carbon 3-IPP monomer. The elegant chemical logic at work: the 3-IPP building blocks serve first as carbon nucleophile for C–C bond formation. Then in the alkylated product, that moiety is progenitor to the carbon electrophile needed for the next elongation cycle. The 3-IPP building block thus cycles between nucleophile and captured electrophilic precursor in every chain elongation cycle. The chain elongated products always have a Δ^2-olefin so the partners for the next chain elongation cycle are always 3-IPP molecules (nucleophiles). This has been termed enzyme-catalyzed head to tail alkylation.

Subsequent isoprenoid chain elongations follow identical iterated logic, using the Δ^3-isopentenyl-diphosphate as carbon nucleophile to form a C–C bond with the allylic cationic species derived from a Δ^2-partner. For example, chain elongation between the 10-carbon geranyl-PP and a Δ^3-isopentenyl-diphosphate monomer leads to the next five-carbon chain elongation, a C_{15} triolefinic product known as farnesyl-PP (Figure 3.3 back in Chapter 3). The mechanism of C–C bond formation is the same as before. The initial tertiary carbenium ion adduct is quenched by adjacent proton removal at C_2 to create an *E*-C_2–C_3 double bond and set up the farnesyl-P as a Δ^2-prenyl diphosphate for a subsequent round of five-carbon chain elongation with another Δ^3-IPP. Farnesyl-PP is an *E*,*E*,*E*-$\Delta^{2,6,10}$ triolefin with methyl branches at C_3, C_7, and C_{10} (Figure 15.7). Note the placement of the three double bonds in the C15 isoprenoid chain are exactly those found in half of the C30 squalene molecule.

The third round of head to tail five-carbon additions, adding the third equivalent of Δ^3-isopentenyl-diphosphate monomer to the C_{15} farnesyl-PP, results in a C_{20} geranylgeranyl-PP (Figure 15.7). This is a tetra-olefin, $\Delta,^{2,6,10,14}$ with four methyl branches at $C_{3,7,11,15}$, respectively. The prenyl chain elongations, five carbons at a time, can go on to the C_{40} linear, the C_{60-80} linear, and the C_{110} linear isoprene chains, although partial saturation and/or isomerization of some *E*-olefinic links to *Z*-olefinic links may happen in the longer chains. Figure 2.18 back in Chapter 2 shows the structure of undecaprenyl-PP with eight isoprenyl units. The *trans* to *cis* olefinic isomerizations involve transient protonation of a 2,3-double bond to give the tertiary C_3 cation, free rotation, and then proton reabstraction (not shown).

In sum, these are carbon alkylations, five carbons at a time, with accrual of a methyl branch and an olefin functional group each time in the elongating chain. These are fundamentally different strategies from the thio-Claisen condensations in fatty acid saturated chain elongations and in the first three enzymes of the isoprenoid pathway.[5] We have noted in Chapters 2 and 11 some of the intramolecular capture of allyl cations by internal olefins that can then lead to a cascade of cation-driven scaffold rearrangements in cyclic monoterpene, sesquiterpenes, and diterpene products.

15.5 Isoprenoid Biosynthetic Logic: Phase 3. Head to Head Self-alkylations of Farnesyl-PP and Geranylgeranyl-PP

A third mode of C–C bond formation in the isoprenoid pathway, head to head alkylation, is directed enzymatically by homodimerization of two molecules of the C_{15} farnesyl-PP to the C_{30} linear hexaene squalene[11] (Figure 15.8). Analogously, two molecules of C_{20} geranylgeranyl-PP are enzymatically dimerized to the C_{40} nonane phytoene (Figure 15.7). The connectivity between the two building blocks in squalene or phytoene are dramatically distinct from the iterative prenyl chain elongations. We characterize those previous prenyl chain elongations, five carbons at a time, by sequential addition of Δ^3-IPP building blocks to the growing Δ^2-prenyl chains (C_5, C_{10}, C_{15}, C_{20} units), as head to tail carbon alkylations. The homodimerizations to form squalene or phytoene would then be head to head alkylations (Figure 15.7): new C_1 to C_1 carbon bonds, not new C_1–C_4 carbon bonds (prenyl transfers).

There is also a difference in the redox state of squalene *vs.* the C40 plant metabolite phytoene at the paired C_1 carbons.[12] Squalene has a pair of methylene groups while phytoene had a double bond (six double bonds in squalene, nine in phytoene). The reduction in squalene arises from consumption of an NADPH equivalent, absent in phytoene synthase as described below.

The distinct head to head regiochemistry of farnesyl-PP or geranylgeranyl-PP self-couplings argues for a different underlying chemical logic for C–C bond formations. This premise was borne out by subsequent experimental findings of cyclopropane-containing C_{30} (presqualene-PP) or C_{40} (prephytoene-PP)

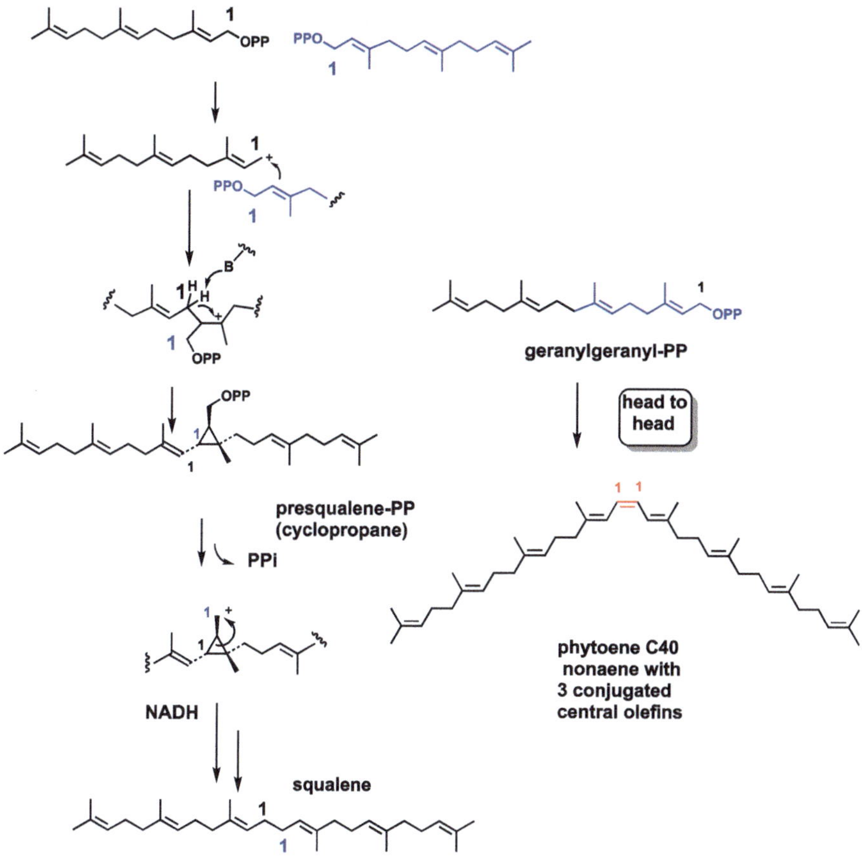

Figure 15.8 Squalene synthase involves formation and then opening of the cyclopropane ring-containing presqualene-PP from two molecules of farnesyl-PP.

intermediates that went on to squalene or phytoene respectively[13] (Figure 15.8). Deciphering the mechanism of formation of the homologous cyclopropane rings and their enzymatic breakdown to squalene and phytoene illuminated this third type of C–C bond forming logic. Those mechanistic analyses also indicated distinct routes to the six double bonds in squalene *vs.* the unanticipated ninth double bond (rather than the anticipated eight) in phytoene (Figure 15.8).

The initial C–C bond formation between two farnesyl-PPs or two geranylgeranyl-PPs follows a variant of the chemical logic of early C_1–OPP bond cleavage in one substrate partner and then capture by the π-electrons of the second molecule of substrate. This is an intermolecular C–C bond forming reaction, with the distinction that the double bond that acts as carbon nucleophile is in the Δ^2-, not the Δ^3-position. Also, the carbon nucleophile is in a 15-carbon partner molecule (squalene) or a 20 carbon partner (phytoene), not the five-carbon partner in the head to tail alkylations.

The nascent C_{30}-mer (or corresponding C_{40}-mer) has a 1′–2 single bond and a carbocation at C_3. Internal quenching by single bond migration yields the cyclopropane ring with three C–C bonds: 1′–2, a new 1′–3, and the starting 2–3 single bond. Cyclopropane formation has occurred by single bond migration to quench the initial C_3 cation (analogous to cyclopropane formation from the α-terpinyl cation back in Chapter 4). This is presqualene-PP (C_{30}), a bona fide pathway intermediate (Figure 15.8).

One of the two C_1–OPP bonds in the pair of farnesyl-PPs was cleaved on the way to forming presqualene-PP. In the catalyzed breakdown of the cyclopropane intermediates, the process again starts with C_1–O bond cleavage in the remaining pyrophosphate ester to again create the C_1-cation, in this case a cyclopropyl methyl cation.[12]

As shown in Figure 15.8, migration of the 1′–3 single bond of the cyclopropane ring to quench that primary cation creates a more stable cyclopropyl tertiary cation. It is believed that at this juncture the cosubstrate NADPH captures that intermediate by hydride transfer to $C_{1'}$ with migration/fragmentation of the 1′–2 single bond to quench the cation. That hydride ion reduction ensures that the 1–1′ bond connecting the two halves of the C_{15} substrate partners is a single bond with two sp^3 hybridized carbons (–CH_2–CH_2–) in the resultant linear squalene. Three double bonds from each of the farnesyl substrate scaffolds are preserved, as a result of the 2′–3′ olefin formation in that cyclopropane-cleaving hydride transfer step. Squalene is an acyclic C_{30} hexaene.

In sum, the switch from head to tail chain elongations to head to head dimerizations of farnesyl-PP to squalene represent a new mode of coupling of the allyl cation intermediates. First, in contrast to the head to tail five-carbon chain elongations where the π-electrons of Δ^3-IPP are used as the carbon nucleophile in elongating chain C–C bond formations, the head to head alkylations use only the pair of Δ^2-prenyl diphosphates. The polarity of the initial C–C bond forming step joining the two Δ^2-prenyl diphosphates is maintained: π-electrons of one Δ^2-prenyl substrate attacking the C_1-allyl cation derived from the other Δ^2-prenyl substrate (creating an initial $C_{3'}$ to C_1 bond). From there, the quenching of the resultant carbocation leads to cyclopropane rings in presqualene-PP (Figure 15.8).

Because this is head to head and not head to tail regiospecific alkylation chemical biology, the remaining $C_{1'}$–OPP on dissociation generates an allyl cation close enough to the first C–C bond center that cyclopropyl methyl cation chemistry comes into play to reach squalene. The different routes of presqualene-PP *vs.* prephytoene-PP devolution reveal two different strategies for quenching the rearranged cation frameworks. One rare quench mode is by hydride transfer from external NADPH on route to squalene. The other is by a classic proton abstraction to create an olefin. The first strategy generates the sp^3 hybridization state at C_1 and $C_{1'}$ (alkane link) in squalene. The second strategy creates the sp^2 hybridization state at C_1 and $C_{1'}$ (alkene link) in prephytoene. To this point we have seen allyl cations and cyclopropylemethyl cations as key intermediates in isoprenoid chain couplings.

15.6 Isoprenoid Biosynthetic Logic: Phase 4. Cation-mediated Cyclizations to Tetracyclic or Pentacyclic Steroidal Scaffolds

Yet another mechanistic phase of isoprenoid chemical biology can be delineated as the large set of triterpenoid reactions that convert the C_{30} acylic hexaene scaffold of squalene to the tetracyclic to pentacyclic frameworks characteristic of steroid metabolites[14] (Figure 15.9). As in the two previous phases of head to tail and head to head isoprenoid chain elongations, this carbacylization phase is carbocation-driven (see Chapter 11). The array of 1,4-C=C bonds act as a nested set of carbon nucleophiles. The cascade of six/five-ring carbacycles can be fully stepwise or partially concerted.

One can distinguish two parallel paths: (1) the enzymatic cyclization of squalene itself, exemplified by bacterial hopane hydrocarbon pentacyclic product (see Figure 2.16), *vs.* (2) prior formation of squalene-2,3-epoxide and cyclization to the tetracyclic alcohol lanosterol or the pentacyclic alcohol amyrin (Figure 15.9). Lanosterol is the major product from animal oxidocyclases, β-amyrin the major product from action of plant squalene oxidocyclases.

As described back in Chapter 2, hopane cyclases bind the hexaenyl hydrocarbon squalene in a folded conformation, presumably to place the terminal 2,3-double bond next to an acidic group in the active site that can protonate that double bond of bound substrate at C_3. That would generate a transient C_2 hopanyl acyclic carbocation. One can write a mechanism involving the participation of the remaining five double bonds to form four new C–C bonds as depicted in Figure 2.16 to reach the pentacyclic carbocation. Two of the now classic routes for quenching carbocations in the hydrocarbon class of metabolites include proton abstraction from an adjacent carbon (to yield hopene) or water addition to produce the alcohol (hopanol) (Figure 2.16). Both products can be observed in bacterial metabolism.

The process can be viewed as a concerted cascade of carbon–carbon single bond rearrangements or one can imagine finite short lifetimes from intermediate carbocations where one, two, or three C–C bonds have formed but not the fourth C–C bond. Triterpene cyclases are promiscuous for leaking minor amounts of partially cyclized (on *vs.* off pathway?) scaffolds consistent with finite lifetimes of intermediate cations that can be quenched along the way.[15]

To that end in Figure 15.10 the oxido-squalene to lanosterol conversion can have as few as one and as many as three on pathway cationic transition states/intermediates to reach the protosterol cation with positive charge at C_{20}. Then, there are still two hydride migrations and two methyl migrations before lanosterol is released.

15.6.1 Squalene Epoxidation and Cyclization

Enzymatic epoxidation of the squalene 2,3-double bond is a common prequel to squalene oxide cyclase actions. The epoxidase[14] is a flavin-dependent

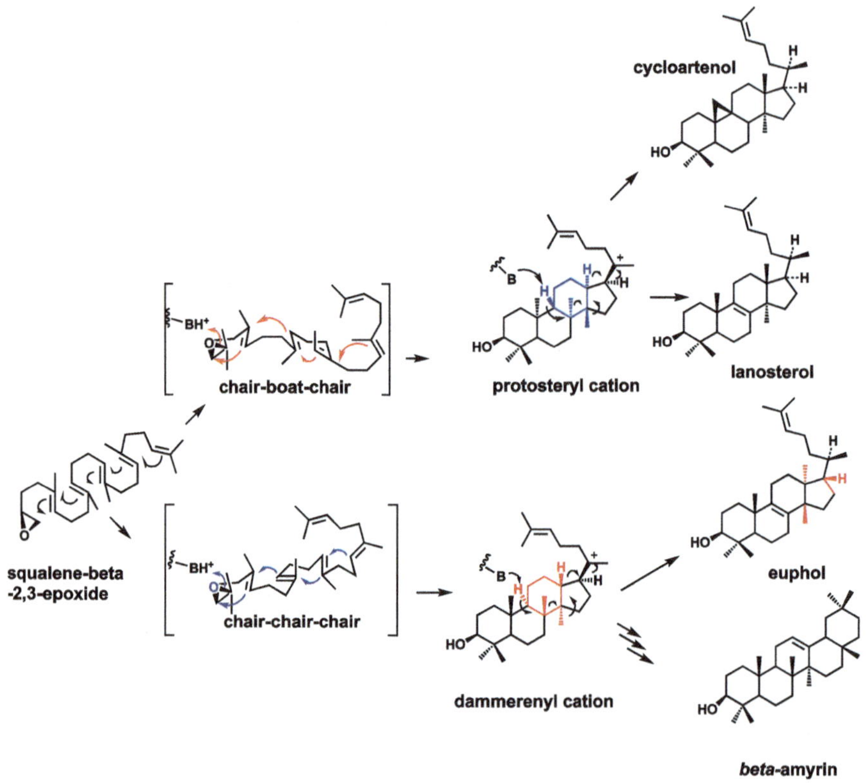

Figure 15.9 Squalene oxidocyclases fold up the linear squalene 2,3-epoxide either into chair–boat–chair-like acyclic conformers or chair–chair–chair-like acyclic conformers before the cyclization cascades. The chair–boat–chair array leads to the protosteryl cation that is progenitor to both lanosterol and cycloartenol. The all-chair array leads to the dammereny cation and then to euphol or, after a series of further cationic rearrangement to β-amyrin.

monooxygenase, using NADPH to generate the dihydroFADH$_2$ enzyme as the O$_2$-reactive species. Two one-electron transfers with O$_2$ are presumed to generate a covalent hydroperoxyflavin intermediate as the proximal oxygen donor to the squalene olefin, creating the β-epoxide geometry (Figure 15.10). This reaction epitomizes hundreds of examples where alkenes act as carbon nucleophiles to form strained cyclic ethers that then can in turn serve as electrophiles at either carbon of the epoxide ring.

Subsequent action of squalene oxidocyclases gives different sterol connectivities, depending on animal *vs.* plant sources.[16] The formation of lanosterol by initial participation of three C–C single bonds to form a tetracyclic protosterol cation is viewed as proceeding through a pre-chair–boat–chair bound conformer of squalene β-2,3-oxide (Figure 15.9).

The protosterol cation in the animal lanosterol synthase is proposed to undergo a series of intramolecular hydride and methyl group migrations as depicted as the blue substituents and blue arrows in the protosteryl cation

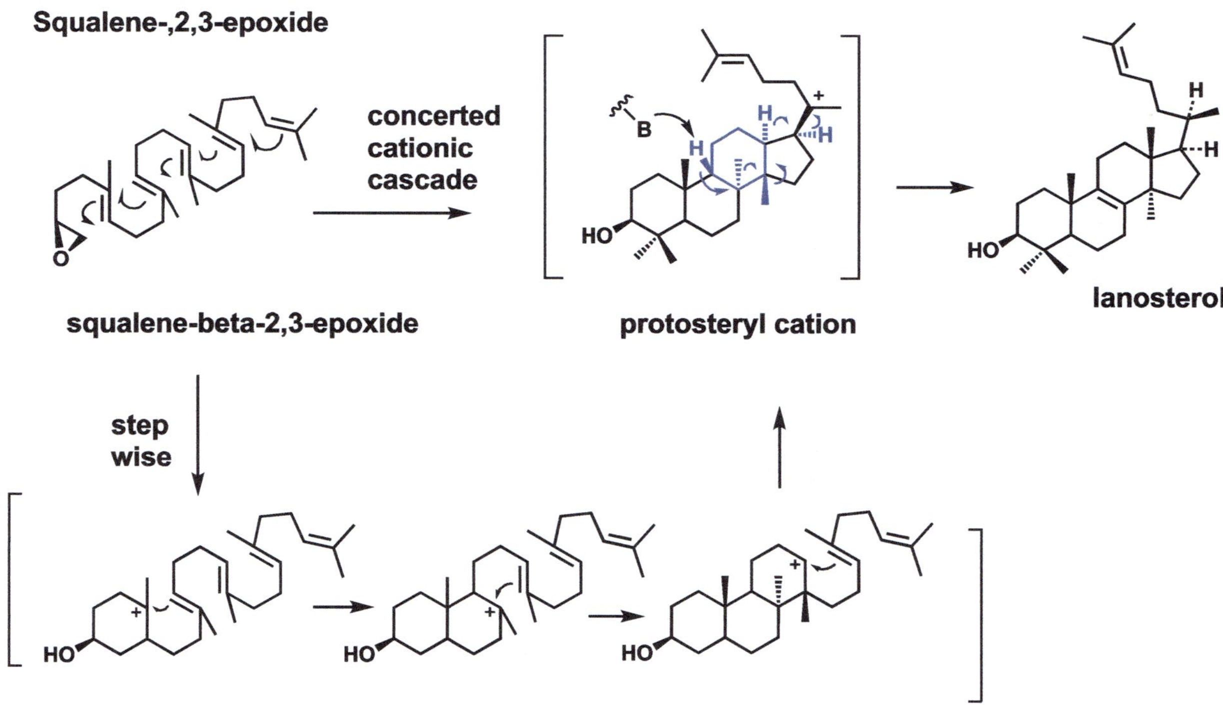

Figure 15.10 Concerted *vs.* stepwise alternatives to the protosteryl cation. A series of two hydride and two methyl group migrations shown in blue occur before quenching of the final cation with 9,10-olefin formation in released lanosterol.

structure in Figure 15.9. Removal of the C_9–H as a proton, generating the internal olefin, sets off the migrations that lead to the remote quenching of the carbocation, leading to release of neutral lanosterol.[16] The proposed migrations are two hydrogens as hydride ions and two methyl groups with the net polarity of methyl anions.

Two other notable divergent routes are formation of cycloartenol and β-amyrin by plant cyclases (Figure 15.9). For cycloartenol, the same initial protosterol cation undergoes the two hydride and the two methyl migrations to reach the C_8 carbocation. Then, instead of loss of the C_9–H as a proton to yield the 8,9-olefin of lanosterol, the C_9–H instead undergoes a 1,2 hydride transfer to C_8. This moves the as yet unquenched cation to C_9. Now the quench step constitutes proton abstraction from the C_{18} angular methyl group and use of the transient C_{18} methylene carbanion as carbon nucleophile, attacking the C_9 cation, generating the fused cyclopropane ring of cycloartenol.[16]

To get to the pentacyclic 6, 6, 6, 6, 6 ring scaffold of β-amyrin, the main triterpenoid metabolite from plant oxidocyclases, the all-pre-chair conformer of bound squalene epoxide would yield an initial tetracyclic cation, known as the dammerenyl cation (Figure 15.9). That cation differs from the protosteryl cation at the four carbon centers, shown in red for the dammerenyl cation and in blue for the protosteryl cation, arising in turn from a chair–chair–chair conformer *vs.* a chair–boat–chair conformer of epoxysqualene in the two cyclase active sites. If the same series of hydride and methyl group rearrangements occur before release, the product would be euphol (from the euphobia plant family). Note euphol differs from lanosterol only in the stereochemistry of the three (red) substituents.

To proceed from that dammerenyl cation to β amyrin requires an additional set of cationic rearrangements (Figure 15.11). Among them are rearrangement of the tetracyclic dammerenyl cation (6, 6, 6, 5) to the ring expanded D-ring in the baccharenyl cation (6, 6, 6, 6). That cation then cyclizes to form the E-ring in the pentacyclic lupenyl cation (6, 6, 6, 6, 5). In turn the lupenyl cation undergoes ring expansion of the E-ring known as the germanicenyl cation.[16]

Then, three C–H bonds are broken. The first two are 1,2-hydride migrations that result in the oleanyl cation (6, 6, 6, 6, 6) as the final bound carbocation. Cleavage of the last C–H bond occurs by loss of a proton. That quench route sets the 12–13 double bond in the pentacyclic scaffold of beta amyrin (Figure 15.11). Amyrin is the major sterol released by plant oxidosqualene cyclases. If all the steps pictured in Figures 15.10 and 15.11 are stepwise, then there are nine enzyme bound cationic species that are intermediates in these plant triterpene cyclases: a cascade of directed olefin/single bond/carbocation/hydride chemistry.

15.6.2 Many Triterpene Cyclases

Dozens of variant cyclases are known to make tetracyclic to pentacyclic triterpene hydrocarbons and alcohols that are then further metabolized

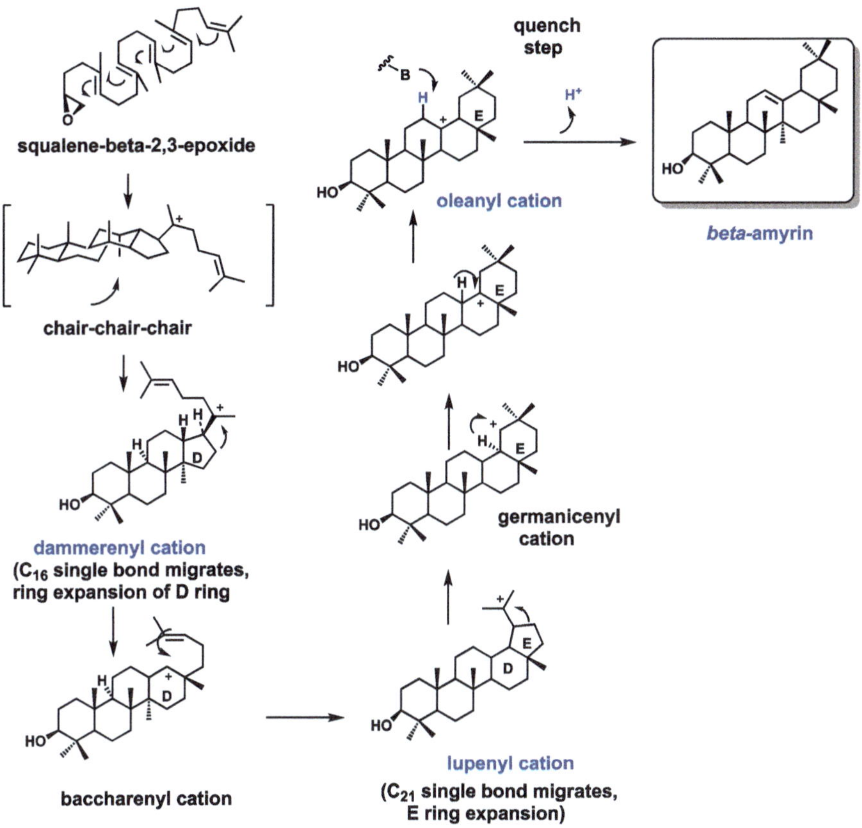

Figure 15.11 The dammerenyl cation is on a pathway, not only to euphol by the same kind of two-hydride, two-methyl group migrations shown in Figure 15.10 for the protosteryl cation, but also to β-amyrin. The route to the amyrin scaffold goes through the baccherenyl cation, lupenyl cation, germanicenyl cation and oleanyl cation, consistent with step-wise cascades in this portion of the reaction coordinate.

(*e.g.* a cascade of oxygenations to make the hydrocarbon frameworks more water soluble). It is clear that the capture of transient carbocations by carbon–carbon single bond migrations is the common currency of these enzyme-accelerated routes to steroidal carbacycles.

For several of the cyclases the leakage of minor, partially rearranged but quenched intermediates suggest low energy barriers between sets of enzyme-bound transition states and alternate placement of potential quenching groups within the active site.[15] Among potential causes for scaffold diversification are side chain acid residues, access to and orientation of bound waters, and the rearrangement potential of hydrogens, and methyl groups within the protosterol scaffolds themselves. For example, some triterpene cyclases quench the dammerenyl cation or the lupenyl cation or the oleanyl cation (Figure 15.10), with water to form corresponding alcohols.

The C_{30} acyclic hexaene triterpene squalene (and its 2,3-epoxide) is the poster child for conversion of acyclic polyalkene frameworks into tetracyclic and pentacyclic alkanes. Enzyme-catalyzed cyclizations are not unique to squalene; we saw that the C_{20} geranylgeranyl-PP can undergo cyclization and cation rearrangements to both ent-kaurene and taxadiene carbacyclic scaffolds back in Chapters 2 and 11. On the other hand, the C_{40} phytoene, with its additional, central *Z*-double bond does not fold into conformers that can undergo facile intramolecular cyclization. The downstream metabolites, such as lycopene, and many carotenoids are typically highly conjugated linear polyenes[17] (Figure 15.12).

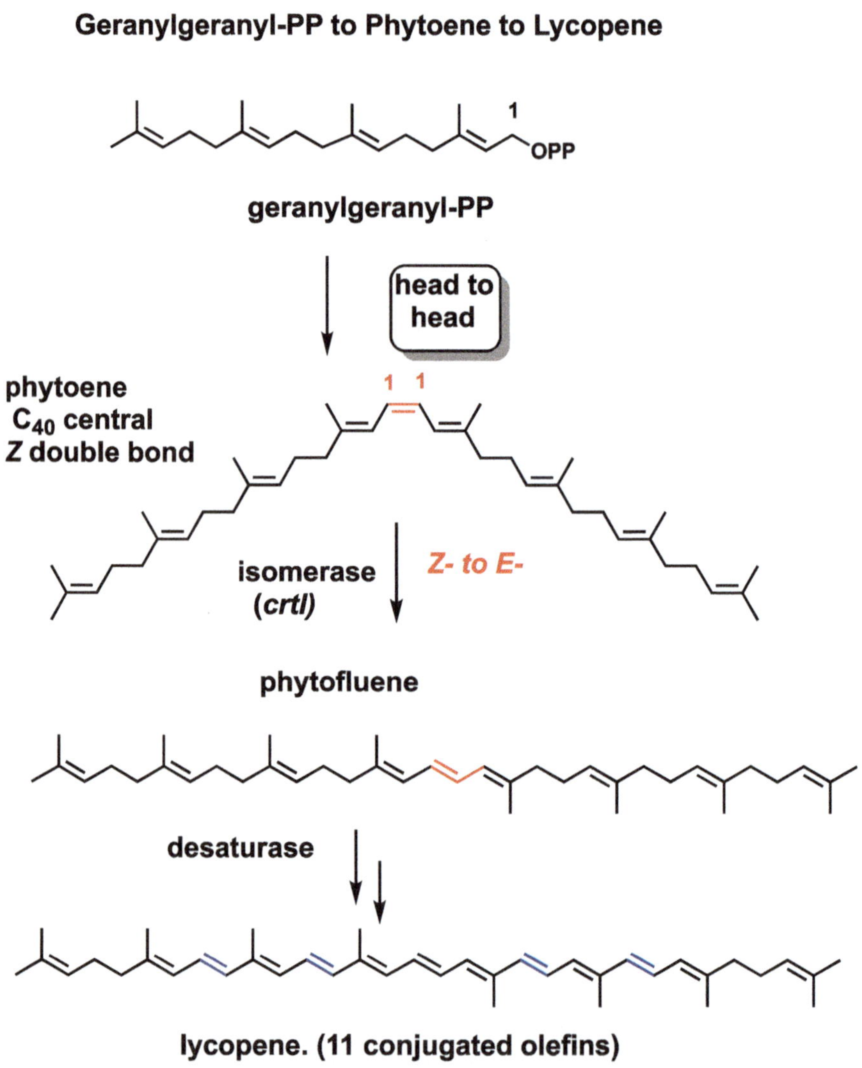

Figure 15.12 2 Geranylgeranyl-PP to phytoene and then lycopene.

Isoprenes are nature's preferred hydrocarbons in terms of architectural complexity and differentiated function. The generation and utilization of both Δ^2- and Δ^3-pyrophosphorylated alcohol forms of isoprene gives two functional groups to those five carbon building blocks: pyrophosphate ester and alkene as progenitors to carbon electrophiles and nucleophiles, respectively.

15.7 Lanosterol to Cholesterol: Phase 5. Consumption of Nine O_2 by Cytochrome P450 Oxygenases

The further processing of the initial tetracyclic C_{30} steroid metabolite lanosterol to the major mammalian sterol, the C_{27} cholesterol, and all the way to the C_{19} androgens and C_{18} estrogens brings changes from cation-driven chemistry to carbon radical chemistry (and oxygen–iron radical chemistry). Oyxgenative modifications convert the C_{30} hydrocarbon alcohol to the highly oxygenated steroid hormones (Figures 12.18 and 12.19).

Lanosterol is converted to cholesterol by iterative oxygenation of three *C*-methyl groups, two at carbon 4 and one at carbon 14, that ultimately lead to removal of the one carbon units as two CO_2 and one $HCOO^-$. In the process nine molecules of O_2 are reduced by four electrons each, *via* iterative action of a pair of cytochrome P450 monooxygenases. Also, nine NADPH are consumed as external reductants for the feeder flavoenzymes to perform their 2/1 electron step down function and deliver one electron at a time to the heme iron in the P450 active sites (Figure 12.18).

Clearly, the isoprenoid biosynthetic pathway to cholesterol has arrived at a new phase. The chemistry switches from carbocation-dominated C–C bond formations to carbon radicals and [OH•] equivalent transfers from high valent oxoiron species, as discussed back in Chapters 4 and 12.

15.7.1 C30 Lanosterol to C27 Cholesterol

In Chapter 12 on carbon radicals in iron-based oxygenase catalytic cycles, we noted that conversion of lanosterol to cholesterol involves the formal removal of two *C*-methyl groups at C_4 and one C–CH_3 group at C_{14} and their net replacement by hydrogens. Each oxygenation step by a cytochrome P450 oxygenase involves a carbon radical at the site of oxygenation (Figure 12.18). To set up the dimethyl groups at a carbon 4 system for decarboxylation at the end of three oxygen transfer cycles, the 3-β-alcohol group is first oxidized to the ketone by an NAD^+ oxidoreductase. The eventual decarboxylation then occurs from a β-keto acid grouping. The product enolate can isomerize the remaining methyl group from α to β. Then, the β-monomethyl metabolite in turn undergoes three oxidations and decarboxylation. Protonation of the enolate ion completes the oxygenative removal of the two *C*-methyl groups at carbon (Figure 12.18). The two CH_3 groups at C_4 are removed by three sequential oxygenations each to give

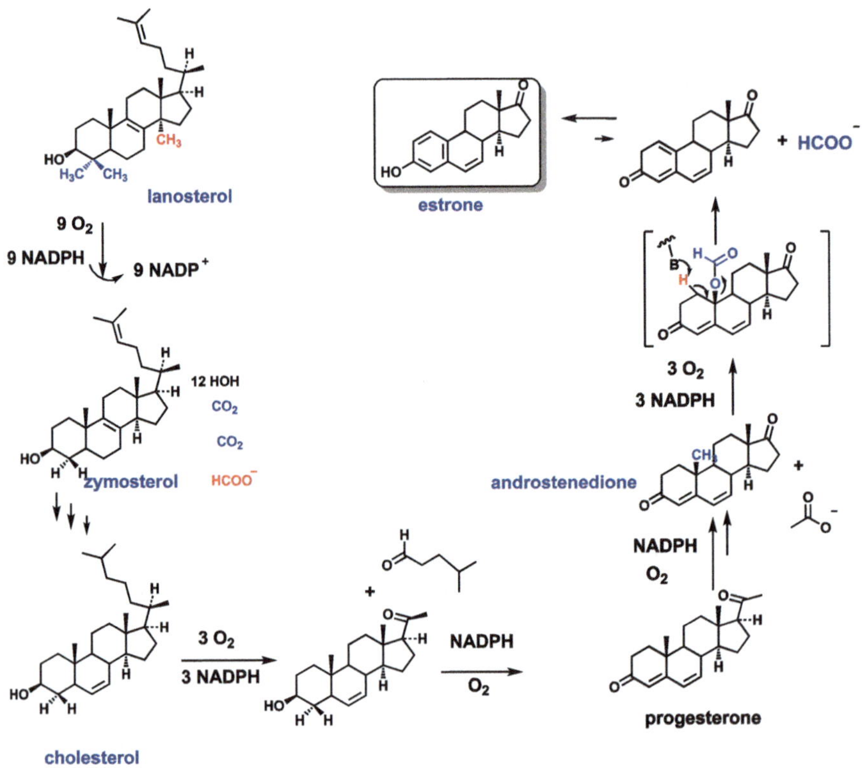

Figure 15.13 Lanosterol to estrone involves multienzyme oxygenases carving out 12 of the 30 carbons to reach the 18 carbon female steroid sex hormone. Along the way 18 O_2 molecules are consumed and 18 carbon radicals in steroid substrates are presumed intermediates, emphasizing the radical nature of the three last stages of steroid hormone biosynthesis.

two CO_2 molecules and four H_2Os as six O_2 molecules are each reduced by four electrons. Six carbon radical species are presumed to be formed and captured during the oxidative removal of the two C4-methyl groups (Figure 15.13).

Oxygenative removal of the *C*-methyl group at carbon 14 occurs by a comparable P450-mediated strategy,[18] actually preceding the C_4 demethylations noted above. However, the disconnection of the one carbon unit that starts as the methyl group at C_{14} requires a different strategy for C–C bond cleavage because there is no ketone β- to C_{14}. Therefore, the α-CH_3 group at carbon 14 is oxygenated twice by the oxygenase Cyp51, first to the alcohol and then to the aldehyde group (Figure 12.18). A third oxygenation does not give the carboxylate which would be a stranded intermediate. Instead, Cyp51 inserts an oxygen between the C_{14} and aldehyde carbon to yield the formyl ester transiently. This can be eliminated as the carboxylate anion of formic acid, initiated by proton abstraction at C_{15}. Proton removal and adjacent loss of formate creates to give the 14,15-olefin in the C–C bond cleavage step (Figure 12.18).

To complete the pathway to cholesterol two double bonds at $C_{14,15}$ and $C_{24,25}$ need to be enzymatically saturated by enzyme-catalyzed NAD(P)H hydride transfers, giving zymosterol (Figure 15.13). The remaining olefin at $C_{8,9}$ undergoes a net isomerization to the 5,6-olefin in cholesterol.[19]

Figure 15.13 summarizes the transformation from initial cyclization product lanosterol to the main sterol metabolite cholesterol.[20] Nine O_2 are reductively activated and split. Of the nine oxygen atoms transferred to carbon centers in the sterol intermediates, four end up in CO_2, two in formate ion, and three were released as water, from geminal diols in conversion to the equilibrating aldehydes. Overall, 12 of the 18 oxygen atoms end up as water molecules. Remarkably, cholesterol emerges with none of the 18 oxygen atoms in its scaffold. While tetracyclic carbon steroid scaffolds such as hopanes are ancient metabolites on an evolutionary scale, the branch point to oxygenated sterols must have evolved after oxygen levels rose. Squalene epoxidase requires O_2 to divert squalene down an oxygen-dependent path and then to bombard lanosterol with nine more oxygen molecules to oxygenate the three methyl groups removed on the way to cholesterol.

15.7.2 Cholesterol to Androgens and Estrogens. (C_{27}–C_{18}): Consumption of Another Eight O_2

Cholesterol, in addition to its major membrane lipid roles in fluidity control and membrane raft formation, is a metabolic precursor to both vitamin D3, *via* 7-dehydrocholesterol when there is sufficient sunlight (Chapter 12), and also to a variety of hormones.[21] The human steroid hormones include adrenal steroid hormones, mineralcorticoids, glucocorticoids and the androgen and estrogen sex hormones. The logic for just one of the pathways, to androgens and estrogens, involves another cascade of cytochrome P450 hemeprotein oxygenases.

The first catalyst is the cholesterol side chain cleavage enzyme[22] ($P450_{scc}$), another oxygenase, like the three in the lanosterol to cholesterol arm, that uses up three molecules of substrate O_2 (Figure 15.13). Two such oxygenations yield the 20,22-dihydroxycholesterol as intermediate before the third oxygenation step effects the C_{20}–C_{22} bond cleavage generating the side chain aldehyde as the C_6 methyl branched aldehyde and the methyl ketone on the steroid framework.[23] This is the C_{21} metabolite pregnenolone. The net fragmentation of the C_{20}–C_{22} pair of unactivated carbons into an aldehyde and a ketone fragment emphasizes the remarkable power of iron-based oxygenases to perform radical chemistry/oxygenation that can cleave unactivated C–C bonds. This set of actions trims away another six carbons, in the form of the branched aldehyde, from the C_{27} cholesterol framework to yield the C_{21} pregnenelone.

The C_3–OH that has been reversibly oxidized and reduced during lanosterol 4-demethylase action is again oxidized by an NAD^+ dependent oxidoreductase. Isomerization of the B-ring double bond into conjugation (to a 4,5-double bond) then creates the electrophilic enone in progesterone (Figure 15.13) (that enone is precursor to the aromatized ring in estrone – see below). Progesterone is now at a branch point, as substrate for two different regioselective P450 oxygenases. The 17-hydroxylase is on a pathway to androgens and estrogens.[24] The 21-hydroxylase (not shown) is instead on pathway to cortisol and corticosterone hormones. The two hydroxylases presumably make carbon radicals at C_{17} or C_{21} respectively in bound progesterone.

The 17-α-hydroxyprogesterone is subjected to the next P450 that cleaves the 17,20 bond (the 17,20 lyase) oxygenatively,[25] releasing the two-carbon fragment as oxidized acetate and producing the 19 carbon androstenedione (Figure 15.13). The pathway has now entered into the realm of androgen hormones. Add this two-carbon acetate fragment to the six-carbon branched isocaproic aldehyde released three enzymes earlier and the starting cholesterol C_{27} framework has been whittled down to a C19 androgen, all by a family of radical-based P450 monooxygenases, carving out the six-carbon and then the two-carbon fragments. Five oxygen molecules have been reductively consumed by three P450s.

If one also adds in the three methyl groups carved out of lanosterol on the way to cholesterol, a total of eleven carbons have been removed by action of five different P450s, three of which act iteratively four times to consume 12 of the 14 O_2 molecules to get to androgens. And there is one last oxygenative removal of a carbon from the C_{19} androgens.

That carbon is the angular methyl group at C_{18}. We have seen C–CH_3 oxygenative removal strategies in the lanosterol 4- and 14-demethylases. Indeed, the aromatase[26] is another one of those methyl oxygenating P450s that act iteratively. Two such oxygen transfers *via* two successive radicals at C_{18} of androstenedione yield the by now anticipated aldehyde[27] (Figure 12.19). To cleave the C–C bond the enzyme has to resort to a variation of the strategies used by the previous C–C bond cleaving P450s in the pathway. Figure 12.19 suggests addition of the Fe^{III}–O–O^- form of aromatase acting as oxyanion nucleophile on the C_{18} aldehyde carbonyl. The tetrahedral adduct can be expelled by proton removal at C_1, perhaps at the same instant that the tetrahedral one-carbon fragment is collapsing to the observed formate ion and Fe^{III}–OH enzyme. The nascent C_{18} steroid product is the cyclohexadienone which quickly isomerizes to gain the resonance energy of the aromatic A-ring phenolate ion (Figure 15.13). The C_{18} aromatized A-ring accumulating product is estrone oxidoreductase-mediated reduction of the C_{17} ketone to the alcohol yields estradiol. Both estrone and estradiol are human female sex hormones.

The regiospecific barrage of hemeprotein oxygenases, acting *via* cascades of carbon-based radical intermediates, have added water solubility to the steroid hydrocarbon backbone and created a diaspora of differentially oxygenated steroids of differing molecular weight to function as distinct hormonal entities in the evolution of steroid metabolism.

15.8 Overview

The multistep route for funneling 18 acetyl group equivalents from 18 acetyl CoA molecules to the carbon scaffold of cholesterol and its downstream oxygenated steroid hormone derivatives reveals a changing set of underlying chemical logics brought into play at different stages of structural and functional group morphing. As three initial two-carbon acyl fragments are assembled into the branched six-carbon 3-hydroxyglutaryl acyl thioester scaffold, carbanion-based thioester enolate chemistry builds the two C–C bonds. Ultimately six such hydroxyglutaryl units are assembled into the acyclic C30 hexaene of squalene. From a functional group perspective, the acyl carbonyl in each 3-hydroxyglutaryl group must undergo four-electron reduction to the alcohol group. Four-electron reductions in six hydroxyglutaryl CoA building blocks means 24 electrons consumed, as 12 hydride ions from 12 molecules of NADPH. The reductive nature of the steroid pathway shows up early, then again at the squalene formation stage, and finally in NADH consumption in each oxygenation cycle.

Iterative utilization of six 3-hydroxyglutaryl units (36 carbons) to the C30 scaffold of squalene and then tetracyclic lanosterol mean six carbons have to be jettisoned, one from each HMG unit. That chemistry also occurs early, in the decarboxylation/phosphate elimination step that creates both the branched five-carbon backbone and the all-important C=C double bond in the Δ^3-isopentenyl-PP.

Isomerization of the double bond from the 3 to the 2 position reveals one of the reactivities of that olefin group and crucially generates the Δ^2-IPP partner for all the subsequent head to tail prenyl transfers. At the isomerase stage the pathway has already entered the olefinic/carbocation phase of reactivity. The pair of IPP isomers are perfectly designed to function as carbon nucleophile (3-olefin) and allyl cation electrophile (2-olefin on early C_1–OPP cleavage). Two such alkylation reactions, utilizing two 3-IPP units to alkylate the growing 2-allyl cation species gives the triolefinic 15 carbon farnesyl-PP.

While such head to tail five caron unit alkylations can proceed for as many as 20–24 iterations, the route to squalene instead uses head to head carbon alkylation from a pair of farnesyl-PPs. Instead of a 3-olefin capturing the carbocation from a 2-olefin partner in head to tail alkylations in 1,4-bond formations, the head to head route utilizes two Δ^2-olefin partners. This directed route leads to cation quenching by C–C bond capture to the cyclopropane ring in presqualene-PP. C_1–OPP cleavage then sets off the cyclopropane ring opening and hydride quenching from NADPH reduces the new C_1–C_1 bond as a pair of sp3 methylene groups in squalene.

The linear hexaene squalene with its array of six olefins in 1,4-relationship is perfectly suited for folding up in a cyclase enzyme site to orient the π-electrons of the double bonds as potential carbon nucleophiles. Each of those double bonds originated back in the decarboxylative/inorganic phosphate elimination to form Δ^3-IPP.

Two reactivity modes of the squalene C=C double bonds are utilized. In the presumably primordial bacterial cyclases, the folded squalene presents one of its two terminal double bonds to an active site base to create a transient cation at C_3. Alternatively, once global O_2 levels rose to useful levels, the formal epoxidation of a squalene terminal double bond created the 2,3-epoxide as a kinetically stable, diffusible electrophilic group.

By either mode of conversion of terminal olefin to electrophile, the internal double bonds in a cyclase/oxidocyclase active site then act as nucleophiles. As many of four of them in animals, in concert or stepwise, create the C–C single bonds that form the A, B, C, and D rings of the protosterol cation. In plants the remaining five double bonds of squalene epoxide all react as carbon nucleophiles in the amyrin cyclase active site. Different routes of cyclized carbocation quenchings can yield product olefins or alcohols (in the hopane series). In the lanosterol-generating squalene oxidocyclase, a remarkable migration of two hydrides and two methyl groups constitute the quenching regime as the presumptive bound nascent protosterol cation is converted to released lanosterol or to euphol.

At this juncture, O_2 reduction, regiospecific carbon radical generation on the bound sterol scaffold, and [OH•] radical rebound becomes the dominant reactivity mode. An astonishing number of nine O_2 molecules are employed for the otherwise extremely difficult cleavage of three unactivated C–CH_3 carbon–carbon bonds to convert lanosterol to the major pathway end product cholesterol. The oxygenase onslaught continues as cholesterol is converted to more polar steroid hormones, consuming another eight O_2 molecules in the oxygenase-rich pathway to estrone and estradiol. Carbanions early, carbocations in the middle, carbon radicals at the end: the steroid biosynthetic pathway uses all three types of accessible carbon intermediates along the way. The concomitant oxidation of 17 NADPH molecules illustrates the use of 34 electrons (as 17 hydride ions *via* flavoprotein intermediates) to reduce all those O_2 molecules on the way to oxoiron species that can generate a presumptive 17 carbon radicals in the steroid frameworks of specific cosubstrates.

References

1. C. T. Walsh, B. P. Tu and Y. Tang, Eight kinetically stable but thermodynamically activated molecules that power cel metabolism, *Chem. Rev.*, 2018, **118**, 1460–1494.
2. B. P. Laden, Y. Tang and T. D. Porter, Cloning, heterologous expression, and enzymological characterization of human squalene monooxygenase, *Arch. Biochem. Biophys.*, 2000, **374**, 381–388.
3. S. G. Sligar, Nature-s universla oxygenases: the cytochromes P450, *Essays Biochem.*, 1999, **34**, 71–83.

4. C. T Walsh, Biologically generated carbon dioxide: nature's versatile chemical strategies for carboxy lyases, *Nat. Prod. Rep.*, 2015, **37**, 100–135.
5. R. J. Heath and C. O. Rock, The Claisen condensation in biology, *Nat. Prod. Rep.*, 2002, **19**, 581–596.
6. J. A. Friesen, VWR. The 3-hydroxy-3-methyl glutaryl coenzyme-A (HMG-CoA) reductases, *Genome Biol.*, 2004, **5**, 248.
7. C. T. Walsh, *The Chemical Biology of Phosphorus*, Royal Society of Chemistry, London, 2021.
8. J. E. Reardon and R. H. Abeles, Mechanism of action of isopentenyl pyrophosphate isomerase: evidence for a carbonium ion intermediate, *Bicohemistry*, 1986, **25**, 5609–5616.
9. L. Zhao, W.-C. Chang, Y. Xiao, H.-W. Liu and P. Liu, Methylerythritol phosphate pathway of isoprenoid biosynthesis, *Annu. Rev. Biochem.*, 2013, **82**, 497–530.
10. C.-C. Chen, S. R. Malwal, X. Huan, I. W. Liu, L. Ma, C. Zhai, L. Dai, J.-W. Huang, A. Shillo, J. Desai, X. Ma, Y. Zhang, R.-T. Guo and E. Oldfield, Terpene cyclases and prenyltransferases: structures and mechanisms of action, *ACS Catal.*, 2021, **11**, 290–303.
11. C.-I. Liu, W.-Y. Jeng, W.-J. Chang, M.-F. Shih, T.-P. Ko and A. J.-H. Wang, Structural insights into the catalytic mechanism of hiuman squalene synthase, *Acta Crystallogr., Sect. D: Biol. Crystallogr.*, 2014, **70**, 231–241.
12. C. T. Walsh and Y. Tang, *Natural Product Biosynthesis: Chemical Logic and Enzymatic Mechanisms*, Royal Society of Chemistry, London, 2nd edn, 2022.
13. N. B. Myant, *The biology of cholesterol and related steroids*, Butterworth-Heinemann, 2014, p. 924.
14. A. K. Padyana, S. Gross, L. Jin, G. Cianchetta, R. Narayanaswamy and F. Wang, *et al.*, Structure and inhibition mechanism of the catalytic domain of human squalene epoxidase, *Nat. Commun.*, 2019, **10**(1), 97.
15. S. Lodeiro, Q. Xiong, W. K. Wilson, M. D. Kolesnikova, C. S. Onak and S. P. T. Matsuda, An oxidosqualene cyclase makes numerous products by diverse mechanisms: a challenge ro prevailing concepts of triterpene biosynthesis, *J. Am. Chem. Soc.*, 2007, **129**, 11213–11222.
16. K. U. Wendt, Enzyme mechanisms for triterpene cyclization: new pieces of the puzzle, *Angew. Chem., Int. Ed.*, 2005, **44**, 3966–3971.
17. Z. Wang, L. Zhang, C. Dong, J. Guo, L. Jun, P. Wei, F. Li, X. Zhang and R. Wang, Characterization and functional analysis of phytoene synthase gene family in tobacco, *BMC Plant Biol.*, 2021, 32.
18. G. I. Lepesheva and M. R. W. Sterol, 14-alpha demethylase cytochrome P450 (CYP51), a P450 in all biologic kingdoms, *Biochim. Biophys. Acta, Gen. Subj.*, 2007, **1770**, 467–477.
19. J. M. Risley, Cholesterol biosynthesis: from lanosterol to cholesterol, *J. Chem. Educ.*, 2002, **79**, 377.
20. L. Miao, M. Nielsen, J. Thewalt, J. J. Ipsen, M. Bloom, M. J. Zuckerman and O. G. Mouritsen, From lanosterol to cholesterol: structural evolution and differential effects on lipid bilayers, *Biophys. J.*, 2002, **82**, 1429–1444.

21. M. Haggstrom and D. Richfield, Diagram of the pathways of human steroidogenesis, *WikiJournal Med.*, 2014, **1**, DOI: 10.15347/WJM/2014.005.
22. J. Hanukoglu, Steroidogenic enzymes: structure, function, and role in regulation of steroid hormone biosynthesis, *J. Steroid Biochem. Mol. Biol.*, 1992, **43**, 779–804.
23. N. Mast, A. J. Annalora, D. T. Lodowski, K. Palczewski, C. D. Stour and I. A Pikuleva, Structural basis for three setp sequential catalysis by the cholesterol side chain cleasvage enzyme CYP11A, *J. Biol. Chem.*, 2011, **286**, 5607–5613.
24. S. M. Kim and J. H. Rhee, A case of 17 alpha hydroxylase deficiency, *Clin. Exp. Reprod. Med.*, 2015, **42**, 72–76.
25. S. D. van Rooyen, R. Yadav, E. E. Scott and A. C. Stewart, CTP17A1 exhibits 17 alpha-hydroxylase/17,20-lyase activity tpwards 11-beta-hydroxyprogesterone and 11-ketoprogesterone metabolites in the C11-oxy backdoor pathway, *J. Steroid Biochem. Mol. Biol.*, 2020, **199**, 105614.
26. R. J. Santen, H. Brodie, E. R. Simpson, P. K. Siiteri and A. Brodie, History of aromatase: saga of an important biological mediator and therapeutic target, *Endocr. Rev.*, 2009, **30**, 343–375.
27. H. J. Chan, K. Petrossian and S. Chen, Structural and functional characterization of aromatase, etstrogen receptor, and their genes in endocrine responsive and -resistant breast cancer cells, *J. Steroid Biochem. Mol. Biol.*, 2016, **161**, 73–83.

CHAPTER 16

Chemical Biology of the Nitrogen Heterocycle Porphobilinogen in Tetrapyrrole Macrocycle Biosynthesis

16.1 The Metalated Tetrapyrrole Pigments of Life

The set of tetrapyrrolic macrocycles with embedded metal ions have been termed the pigments of life[1] (Figure 16.1). The first to be recognized were the intensely red brown-colored cytochromes (pigments in the cytoplasm) of which the heme B cofactor in heme and the cytochrome P450s in families of oxygenases are members.[2] They are found in metalated Fe^{2+} or Fe^{3+} starting oxidation states, respectively. When Mg^{2+} instead is inserted into the equatorial cavity framed by the four pyrrole rings, chlorophylls A or B result with the characteristic green color imparted to the plant world. Bacteria make vitamin B12 (see Chapter 9), one of the 13 essential human vitamins. The corrin macrocycle of B12 is contracted by one carbon bridge from the heme tetrapyrrole macrocycle. The color of cobalt^{III}-B12 is salmon pink. Some anaerobic bacteria and archaeal strains can reduce CO_2 all the way to CH_4. The coenzyme for the last reductive step is a yellow nickel tetrapyrrolic macrocycle known as Factor 430 (for its absorbance maximum at 430 nM). A fifth variant of note is siroheme, an octacarboxy scaffold, an intermediate on the way to corrin and F430 scaffolds (not shown). Siroheme is purple in color and is also the redox cofactor for microbial sulfite reductases and nitrite reductases.[3]

The iron-containing heme macrocycles and the cobalt-containing corrins are redox active, as noted in previous chapters, as is the nickel-F_{430}. The photosynthetic chlorophyll pigments are the primary receptors for light absorption in

Chemical Biology No. 21
The Chemical Biology of Carbon
By Christopher T. Walsh

Published by the Royal Society of Chemistry, www.rsc.org

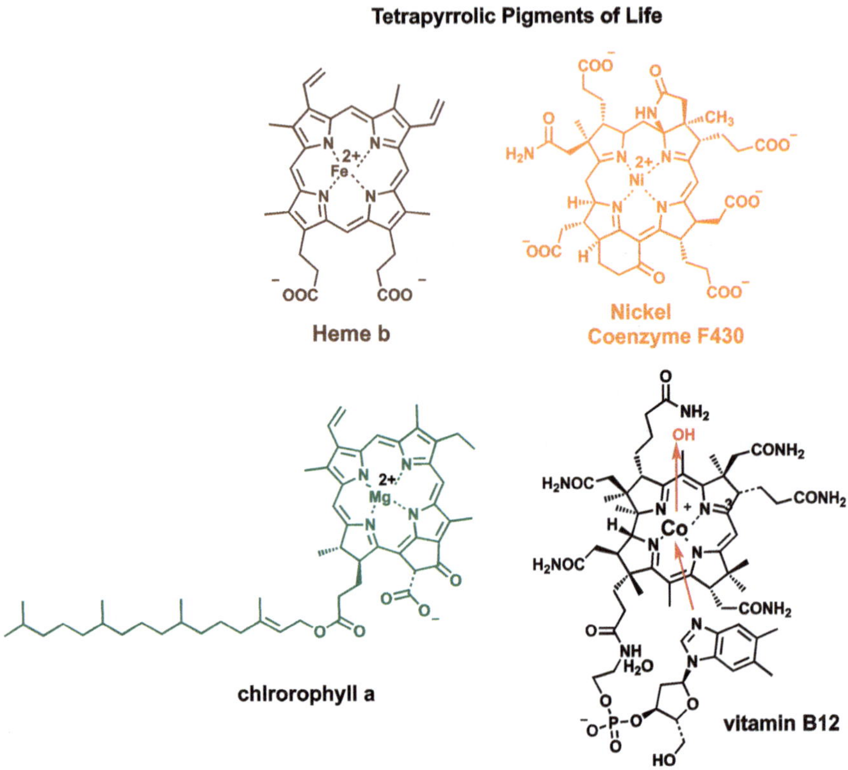

Figure 16.1 Four tetrapyrrole macrocyclic pigments of life.

plants that start the dual processes of NADPH formation and oxidation of $2H_2O$ to O_2. Taken together, between chlorophyll-mediated photosynthesis, microbial production of methane, and both the electron transport and the O_2 binding and or reduction activities of heme proteins, one cannot see how life can function without their absorptive and subsequent functional properties.

The array of those functions is beyond the scope of the focus of this chapter, which is analysis of what kinds of chemical logic are employed to build the pyrrole ring from primary metabolites, to polymerize it to a linear tetrapyrrole, to control regiospecific cyclization, side chain maturation and metal cation insertions.

16.1.1 Porphobilinogen: An Aminopyrrole Heterocycle Is the Key Metabolite

The strategies to build, extend, and cyclize tetrapyrroles separated by a methylene bridge that has cavities that can fit their specific metal ions are examples of how heterocycle chemical biology plays a central role in the pathway to the pigments of life. The pathway starts with generation of an unusual 4-keto-5-amino-1-carboxy pentane, known historically as

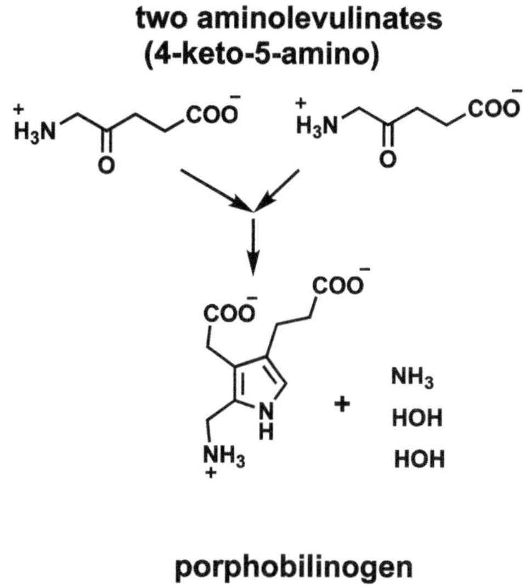

Figure 16.2 Porphobilinogen is formed by dehydrative aldol condensation from two molecules of aminolevulinate.

δ-aminolevulinate[4] and its subsequent enzyme-catalyzed dimerization to the pyrrole monomer porphobilinogen.[5] Porphobilinogen, a dicarboxy-2-aminomethyl pyrrole is the star molecule in this pathway (Figure 16.2). This aminopyrrole can behave as both a nucleophile and an electrophile for controlled oligomerization and then cyclization to form the tetrapyrrole macrocycle framework.

16.2 Aminolevulinate Formation Strategies

All organisms that make tetrapyrrole macrocycles *de novo*, including humans, start with δ-aminolevulinate as the first committed framework (Figures 16.2 and 16.3). During evolution two pathways to this key five-carbon building block have evolved. Plants conscript the available five-carbon amino acid glutamate and morph it into the 4-keto-5-amino-valerate framework (Figure 16.3). This occurs in two steps.

First is a reduction of the C_1 acyl group of glutamate to the glutamic-1-semialdehyde in an NAD^+ hydride transfer. However, as we have noted in previous chapters, the deprotonated glutamate carboxylate anion is, if anything, nucleophilic, not electrophilic, not a ready recipient of a hydride ion. To that end plants divert some of the glutamyl-$tRNA^{glu}$, destined for protein synthesis at ribosomes, as an electrophilic glutamyl C_1 oxoester for hydride reduction to give the C_1 glutamyl aldehyde[6] (Figure 16.3).

Second, is an unusual pyridoxal-P dependent internal transaminase reaction: reductive amination at C_1 while C_2 undergoes oxidative deamination.[7] This is an internal redox reaction. The result of switching the positions

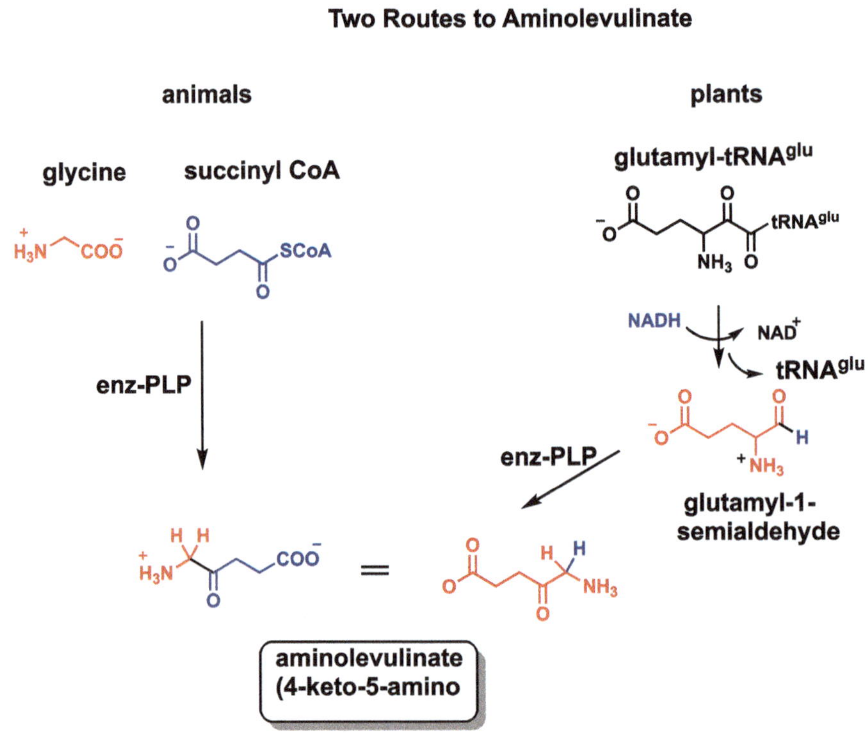

Figure 16.3 Two routes to aminolevulinate.

of the C_1 aldehyde and C_2 amine in glutamic semialdehyde to the terminal amine and adjacent ketone in the product is to form δ-aminolevulinate (Figure 16.4).

The numbering of the framework changes. The remaining carboxylate has precedence and becomes C_1. Then, the ketone is at C_4 and the amino group at C_5. The reaction mechanism involves initial aldimine formation between the C_2 amine of glutamic semialdehyde and the bound PLP aldimine. The typical net 1,3-proton transfer scheme for transaminases (Chapters 5 and 8) yields the ketimine-pyridoxamine adduct. Dissociation to the bound 1,2-dicarbonyl nascent product and pyridoxamine is prelude to formation of the isomeric C_1 aldimine. Now, reversal of the initial 1,3-proton transfer yields the 4-keto-5-aldimine-PLP adduct. Release gives free δ-aminolevulinate.

The alternate route that has evolved in animals including humans instead starts with C_2 and C_4 common metabolites, generates one C–C bond to create a transient C_6 adduct, and then decarboxylates the carboxylate group specifically from what had been the glycine carboxylate to yield the C_5 product scaffold before release by aldimine cleavage[4] (Figure 16.5).

The C_2 metabolite is glycine, the C_4 metabolite is succinyl CoA. The enzyme requires one equivalent of tightly held pyridoxal-P. Glycine can form the aldimine adduct with PLP to generate the C_2 carbanion as carbon

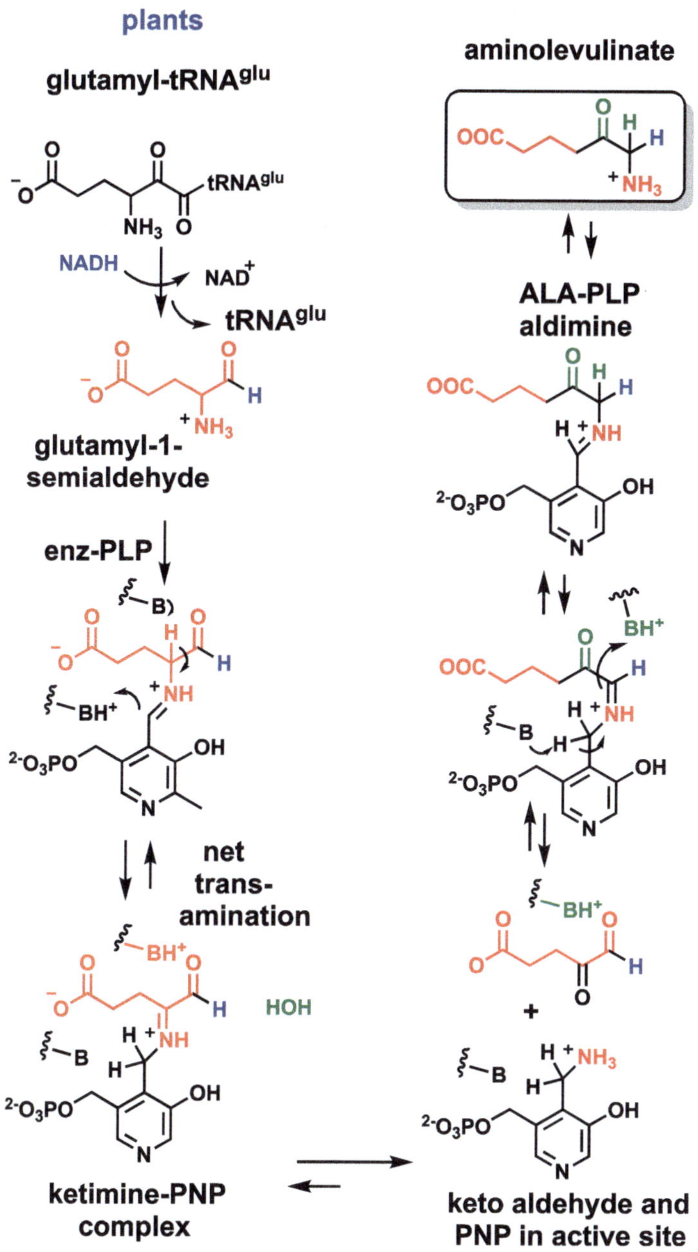

Figure 16.4 The mechanism of the plant transaminase that converts glutamyl-1-semialdehyde to the aminoketone group in aminolevulinate.

nucleophile. The thioester carbonyl of succinyl CoA is electrophilic so C–C bond formation with release of CoASH proceeds (Figure 16.5). That six-carbon branched adduct can then undergo facilitated decarboxylation of the

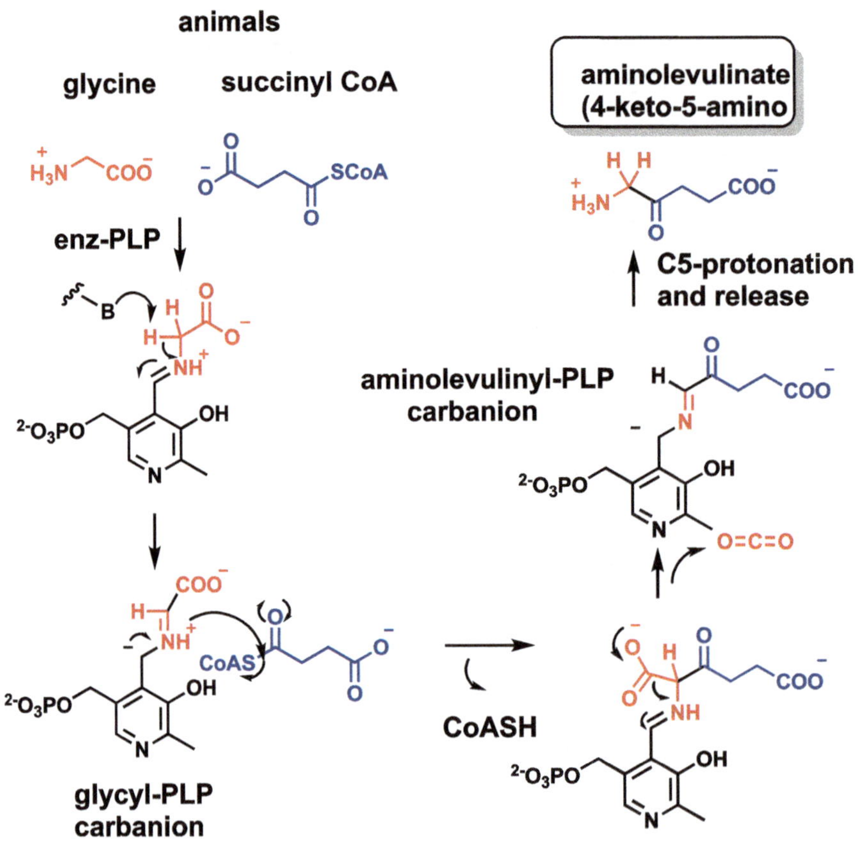

Figure 16.5 The mechanism of the animal aminolevulinate synthase.

COO^- that had been the glycine carboxylate. Of the two acyl carbons from succinyl CoA, the thioester carbonyl has undergone C–C bond formation to become the C_4 ketone. The other acyl carbon of succinyl CoA remains untouched as the C_1 carboxylate of aminolevulinate. On release of the product, the glycine C_2–NH_2 group has become the C_5 amine of aminolevulinate.[4] The rest of product framework is from the succinate starting moiety.

These two quite distinct, chemically logical routes to the 4-keto-5-aminovaleryl-1-carboxylate scaffold of δ-aminolevulinate are reminiscent of the two routes to isopentenyl diphosphates noted in Chapters 2 and 15. The plant route from five-carbon glutamate does not involve any formation or cleavage of C–C bonds but does require an unusual PLP-dependent internal transamination catalyst and diverts some of the glutamyl-tRNA pool. The animal route, starting from separate C_2 and C_4 metabolites, must both make and break separate C–C bonds. PLP is again the perfect coenzyme for this oxosynthase reaction. The two routes show the versatility of amino acid-PLP accessible carbanion chemistry.

16.2.1 Dimerizing Dehydration of Aminolevulinate Creates the Pyrrole Ring of Porphobilinogen

The second enzyme in the pathway takes two of the δ-aminolevulinate monomers and condenses them together, making a C–C and a C–N bond, with double dehydrations to create a heteroaromatic, trisubstituted pyrrole product[5] (Figure 16.6). This aldol condensation yields porphobilinogen, with three substituents on the five membered heterocycle: 2-methylamino-3-carboxymethyl, 4-carboxypropyl substituents. The condensation between the

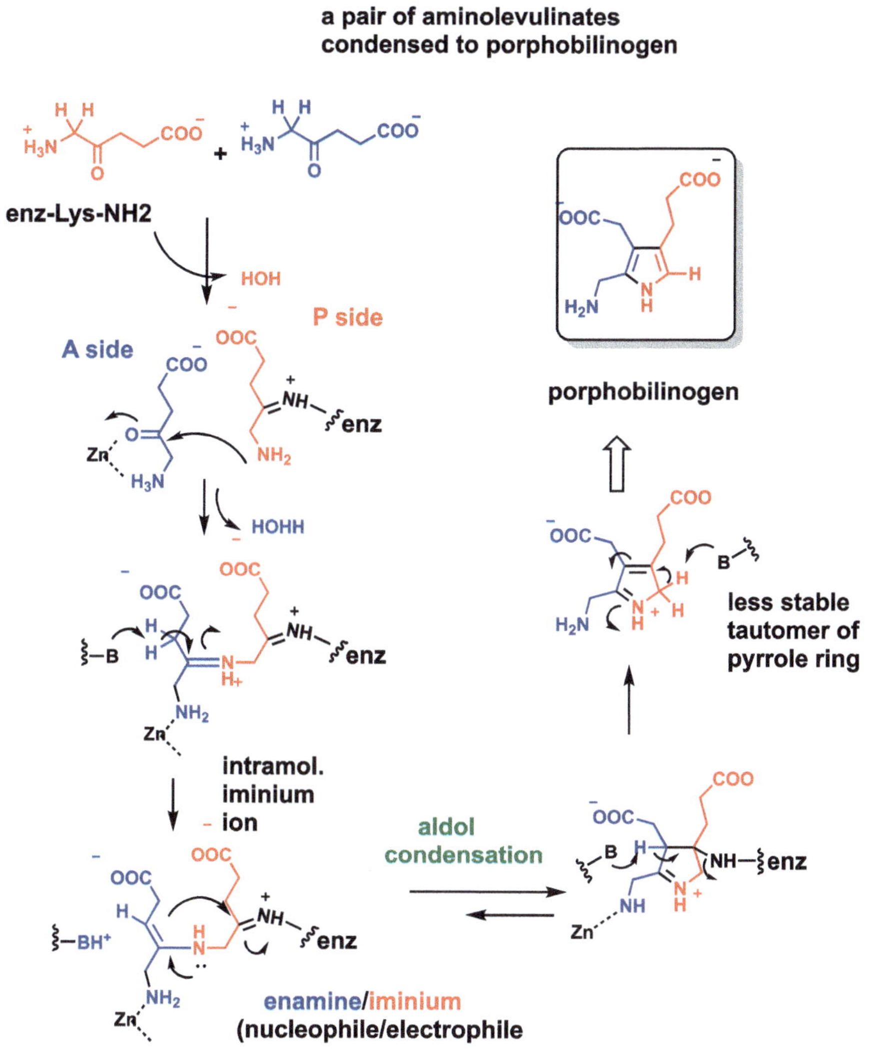

Figure 16.6 Porphobilinogen synthase uses both enamine/iminium ion catalysis and zinc chelation to effect the dimerizing dehydrative aldol condensation of two aminolevulinates to porphobilinogen.

two monomers of aminolevulinate is out of register, accounting for the two carbon and three carbon lengths of the carboxymethyl and carboxypropyl substituents.

The enzyme is dimeric and uses an active site lysine side chain amine in each subunit to engage in imine linkage with one of the aminolevulinate monomers (red becomes propionate-bearing side) through its C_4 ketone group at the intersubunit active site.[5] The other aminolevulinate monomer (blue) will become the pyrrole half that bears the aminomethyl group and the carboxymethyl side chain (Figure 16.6). The five-amino group, in the particular ALA unit that is tethered as an iminium ion to the lysine residue in one subunit, is proposed as the nucleophile in an imine forming reaction That action creates a first covalent bond (N–C) between the two ALA units. That intramolecular iminium ion can be isomerized to the enamine as shown. That isomerization sets up an intramolecular aldol condensation between the blue C_3 carbanion equivalent attacking the red C_4 iminium enzyme adduct. That forms a second intramolecular covalent bond (this time a C–C bond) and thereby creates the five member heterocycle, still tethered to the active site lysine. Elimination of that lysine enzyme group creates a second double bond (C=C) to go with the iminium ion that formed during the aldol condensation. The product is the unstable tautomer of the pyrrole. There is a large driving force for isomerization to the heteroaromatic ring tautomer of porphobilinogen.

The pyrrole ring construction has been engineered by imine/enamine (carbanion/carbocation equivalents) coupling chemistry: a straightforward but effective route to a five-ring nitrogen heterocycle (see Chapter 7). Formation of the pyrrolic porphobilinogen is atom efficient in use of the two identical building aminolevulinate blocks.

16.3 Chemical Ode to Porphobilinogen as a Bifunctional Building Block

Porphobilinogen has three kinds of functional groups relevant to its downstream metabolism. (1) There are the two carboxylate anions as carboxymethyl and carboxyethyl substituents, respectively, at C_3 and C_4. (2) The second functional group is the side chain aminomethyl group at C_2. (3) The third is the pyrrolic NH. The pK_a for that N–H is 16.5 so it effectively never dissociates to the N anion under physiologic conditions.

Most relevant to the immediate downstream enzyme chemistry are each of the two nitrogens. The lone pair on the pyrrole-NH group can function in two ways in porphobilinogen. One resonance contributor places a positively charged iminium group at $C_2{=}NH^+$, with concomitant carbanion density at C_5 (Figure 16.7). This resonance form (blue) suggests C_5 can behave as a carbon nucleophile, as will be evident one enzyme reaction hence. The pyrrole NH lone pair can also participate in expulsion of the side chain amine at C_2. That elimination would lead to an azafulvene isomer of a

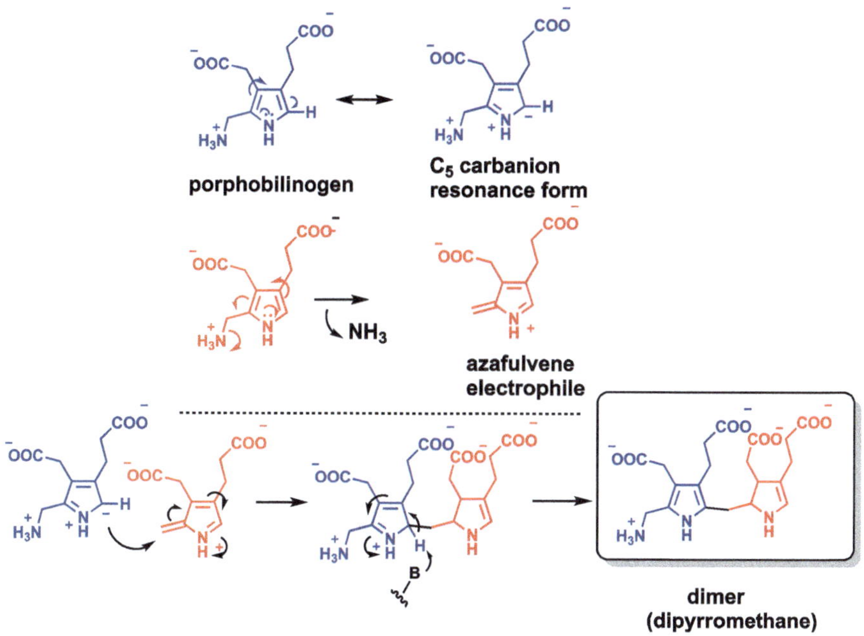

Figure 16.7 Porphobilinogen as both C_5 carbanion and as progenitor to an exomethylene group in an azafulvene tautomer leading to C–C bond formation in a dipyrromethane dimer.

2-methylpyrrole (red). The exocyclic double bond of the azafulvene form is electrophilic at its terminus (Figure 16.7). Addition of a nucleophile would regenerate the heteroaromatic pyrrole, now tethered *via* the 2-methylene group to the incoming nucleophile.

These two differential reactivities, a carbanion nucleophile at C_5 and an azafulvene exocyclic methylene group as electrophile at C_2, show the chemical versatility of porphobilinogen as both a potential carbon nucleophile and a carbon electrophile. This chemistry is realized in the growth of hexapyrrolic oligomers derived from porphobilinogen in the active site of the next enzyme in the pathway, hydroxybilane synthase.[8] Figure 16.7 shows the formation of a pyrrole dimer, a dipyrromethane from the intermolecular capture of the azafulvene form of one porphobilinogen by the C_5 carbanion of the second porphobilinogen.

16.4 The Remarkable Bilane Synthase/ Porphobilinogen Deaminase

The bilane synthase, aka porphobilinogen deaminase to emphasize its release of free NH_4^+, is a remarkable catalyst in at least two ways.[9] The enzyme, after it emerges from ribosomes and folds to its native three dimensional conformation, is inactive. It self-activates by using the inherent propensity of the porphobilinogen molecule to undergo NH_4^+ elimination to the azafulvene structure.[10] That electrophile is captured by the thiolate anion of an active site cysteine residue (Figure 16.8), restoring the electronic configuration to the pyrrole ring, while tethering the pyrrole covalently through a thioether linkage to the enzyme. The enzyme is halfway to its mature active state.

Now, the resonance contributor of the C_5 pyrrole carbanion acts as carbon nucleophile to attack a second azafulvene form of another porphobilinogen molecule, forming a C–C bond between C_5 of the first tethered pyrrole and the exocyclic methylene at C_2 of the second pyrrole. This covalent dipyrromethane-enzyme adduct (Figure 16.8) is now the active form of the enzyme.[9] It is the form that is returned at the end of each catalytic cycle.

This dipyrromethane enzyme adduct then proceeds to repeat the self-priming covalent capture of azafulvene forms of four porphobilinogen

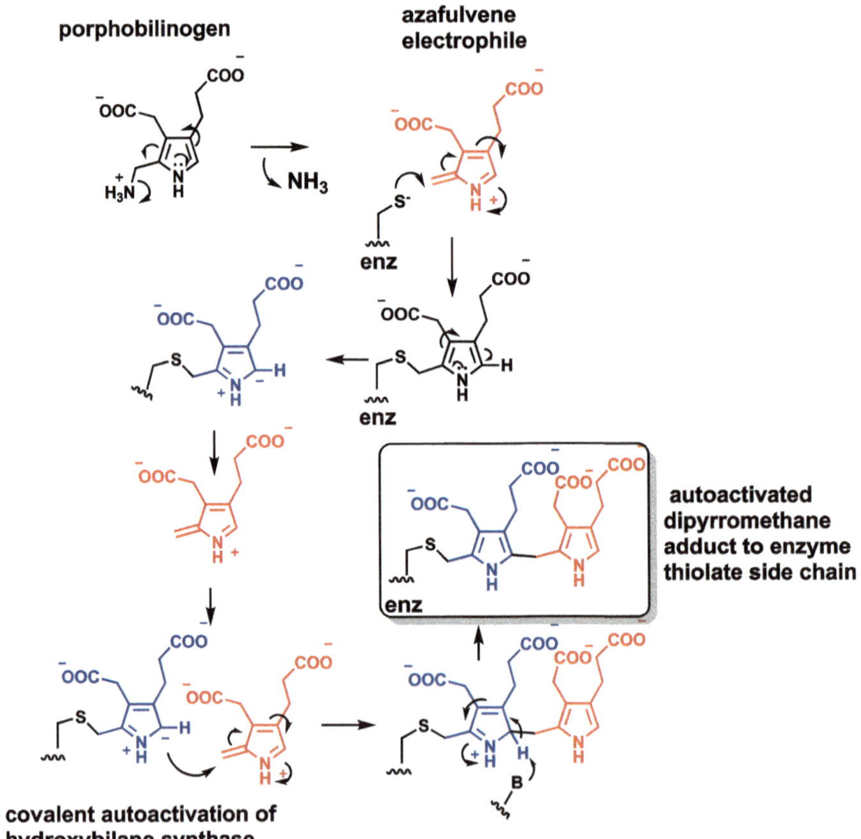

Figure 16.8 Covalent autoactivation of hydroxybilane synthase involves conversion of an active site cysteinyl thiolate anion side chain to the dipyrromethene enzyme adduct *via* the porphobilinogen nucleophile/electrophile behavior schematized in Figure 16.7.

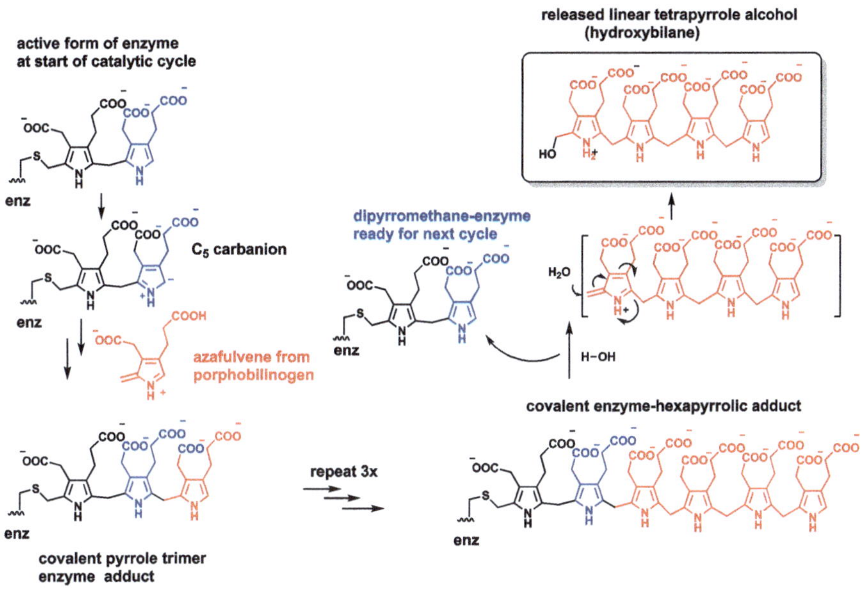

Figure 16.9 The active form of hydroxybilane synthase extends the dipyrromethane covalent adduct to a covalent hexapyrrole chain, then catalyzes the hydrolytic cleavage between pyrroles-2 and -3 to release the linear tetrapyrrole alcohol hydroxybilane.

bound scaffolds sequentially.[8] These are chain extensions, respectively, to a covalent trimer, tetramer, pentamer and hexamer (Figure 16.9). The C_5 carbanion nucleophile of the previously added pyrrole ring is the carbanion that captures the azafulvene electrophile. Each time a molecule of ammonia is liberated as the 2-carboxymethy-3-carboxypropyl-pyrroles are added by a methylene bridge connection to the previous incorporated monomer.

Chain elongation stops at the covalently tethered hexapyrrole adduct (Figure 16.9), perhaps by filling the volume of the active site cavity. At that point the geometry of the connection between the second and third pyrrolic ring must be affected to undergo fragmentation to cleave the connection between C_5 of pyrrole #2 and the exomethylene carbon of pyrrole #3 as the azafulvene isomer of pyrrole #3 forms (Figure 16.9). This fragmentation/water capture of the transient azafulvene #3 releases a linear hydroxy-tetrapyrrole product and regenerates the dipyrromethane form of the covalently tethered dipyrrole adduct to the enzyme, the active form for the next catalytic cycle. This linear tetrapyrrole alcohol product is known as hydroxybilane.

The chain growth of a hexapyrrolic enzyme covalent adduct and then fragmentative release of the linear tetrapyrrolic alcohol may be the most remarkable step in the heme biosynthetic pathway.[11] It further reveals the dual role of porphobilinogen as both a C_5 carbanion and a precursor to the C_2 azafulvene electrophile, the complementary reactivities for oligopyrrole assembly *via* connecting methylene groups. This $-CH_2$ bridge joining each pyrrole ring is the signature connectivity of tetrapyrrole macrocycles. The chemical duality built into the porphobilinogen pair of nitrogens biosynthetically enshrines the pyrrole ring as one of the key heterocycles of life's chemistry.

16.5 Uroporphyrinogen III, Not Uroporphyrinogen I

The linear tetrapyrrolic alcohol – hydroxybilane – once released from its synthase has a very short half-life in cells, estimated to be as short as four minutes. The default fate is ejection of the –OH group, leading to the aza-fulvene form of the A-ring pyrrole and then intramolecular capture by the C_6 carbanion equivalent of the D-ring pyrrole. That macrocyclization generates a cyclic regioisomer known as uroporphyrinogen I (Figure 16.10). However, that symmetric nonenzymatic default tetrapyrrole macrocycle uroporphyrinogen I is a dead end metabolite. It is not on a pathway, and is not recognized for processing by downstream enzymes.

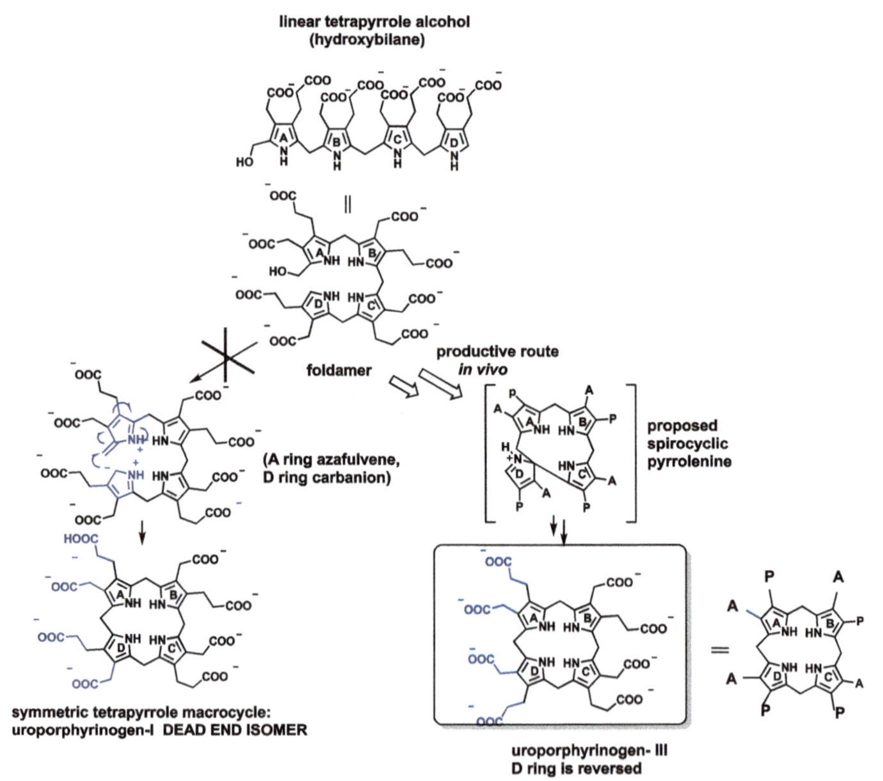

Figure 16.10 Cyclization of linear hydroxybilane is directed enzymatically to the unsymmetric (with regard to P and A side chains) uroporphyrinogen III rather than the dead-end symmetric uroporphyrinogen I.

Instead, an enzyme known as uroporphyrinogen III synthase intercepts the short-lived linear hydroxybilane and macrocyclizes it to a distinct isomer, uroporphyriniogen III. The III isomer has an asymmetric distribution in rings C and D of the carboxymethyl and carboxypropyl (abbreviated as A for acetate and P for propionate, respectively) substituents. This is in contrast to the alternating symmetry of A and P side chains around the periphery of all four pyrrole rings (A, B, C and D) in uroporphyrinogen I (Figure 16.10). Specifically, in the III isomer, the substituent order in the pyrrole D-ring has been reversed. The C-, D-ring substituent pattern of uroporphyrinogen III is P-A-A-P, instead of the P-A-P-A pattern in nonenzymatically formed isomer I.

The exact route to uroporphyrinogen III rather than isomer I in the active site of the isomer III synthase is not fully resolved.[12] The leading hypothesis is initial capture of an A-ring azafulvene by the D-ring to form a transient spiropyrrolenine (Figure 16.10). That adduct would have to unravel at the methylene between rings D and C with a developing azafulvene in ring C. Now the D-ring can rotate around the single bond between ring D and the bridging methylene to ring A. Recapture of the C-ring exomethylene group by C_5 of ring D, acting as carbon nucleophile, yields uroporphyrinogen III (this mechanism proposes formation of the spirocyclic intermediate where the A-ring is the carbon nucleophile and the D-ring the azafulvene. Collapse of the spirocycle back to an acyclic intermediate is proposed with opposite polarity: now the D-ring is a nucleophile and the A-ring is an electrophile). Why evolution chose the uroporphyrinogen III isomer rather than uroporphyrinogen I is not clear. The terminology of uroporphyrinogens derives from their initial isolation from urine as precursors of porphyrins.

16.6 Decarboxylation of Six Out of Eight Side Chain Carboxylates

The next two enzymes in the heme biosynthetic pathway act *ad seriatim* to remove six of the eight side chain carboxylate ions on the periphery of uroporphyrinogen III (Figure 16.11). The result is conversion of the highly anionic octacarboxy tetrapyrrole macrocycle to the dicarboxylate dianion of hemes.[13] The first enzyme chemoselectively decarboxylates all four carboxymethyl (acetates) side chains (Figure 16.11). The second enzyme regioselectively decarboxylates two of the carboxypropyl side chains (propionates), specifically at the A and B rings, leaving the remining propionate side chains in the C and D pyrrole rings (Figure 16.11).

Despite the chemical similarities of the acetate and propionate side chains, the two decarboxylases employ distinct mechanisms for effecting the four and then the two $C–COO^-$ bond cleavages. The acetate side chain decarboxylations are heterolytic,[14] leaving transient carbanionic density at the incipient four methyl groups (Figure 16.11). One can write iminium tautomers of the pyrrole ring, surely minor contributors to pyrrole

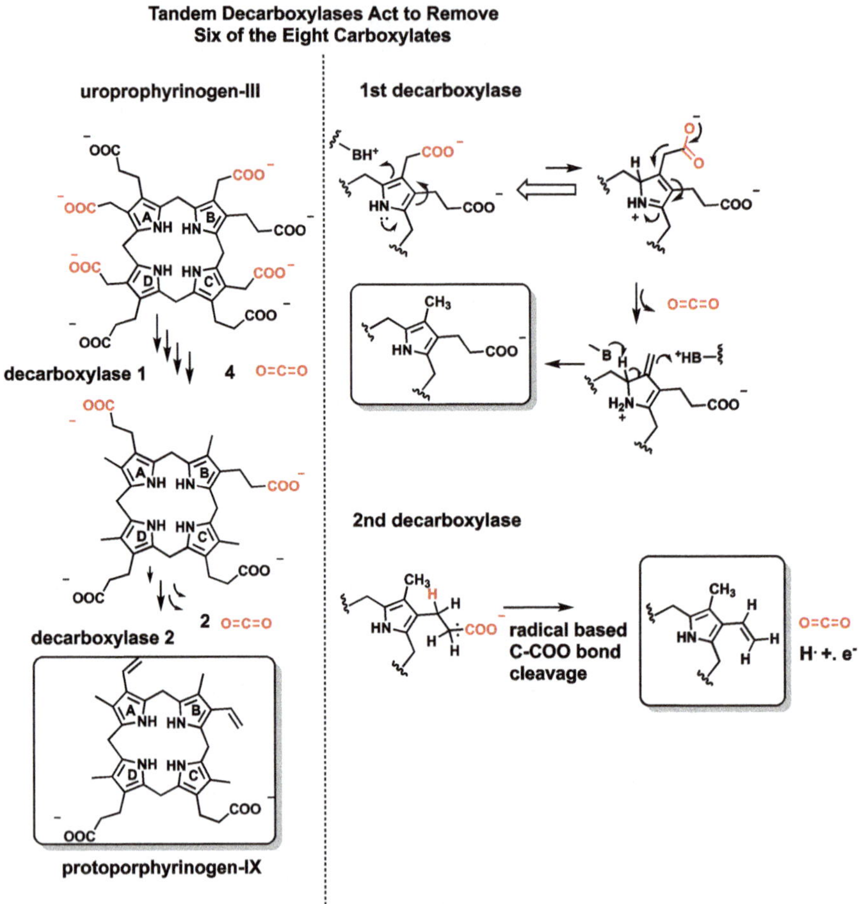

Figure 16.11 Tandem decarboxylases act to remove six of the eight peripheral carboxylate groups. The four acetate side chains are converted to four methyl groups; two of the four propionate side chains are oxidatively decarboxylated to vinyl groups, leaving protoporphyrinogen IX as a dianion.

structures, as potential electron sinks to stabilize the incipient negative charge on what becomes an exocyclic C_2 methylene anionic transition state prior to protonation to the methyl group on each of the four pyrrole rings in coproporphyrinogen III (the prefix copro reflects initial detection in feces).

The next enzyme, decarboxylates the propionate side chains in A- and B-rings only, leaving the propionate negative charges on rings C and D, by a homolytic, radical mechanism, rather than a heterolytic carbanion mechanism (not shown). There are both aerobic oxygen dependent enzymes and anaerobic radical SAM enzyme catalysts (see Chapter 12) that can each carry out the propionate decarboxylation to a vinyl side chain plus CO_2.

The anaerobic C–COO$^-$ bond fragmentation is proposed to cleave to give the $^\bullet CO_2$ radical and the incipient side chain ethyl radical. Each of the two side chains in A- and B-rings quickly devolves to a vinyl side chain by $H^\bullet$ abstraction. At this stage the pathway has reached protoporphyrinogen IX. It is colorless and has 14 π-electrons in the macrocycle.

16.7 Desaturation to the 20 π-Electron Tetrapyrrole Macrocycle

The next enzyme converts protoporphyrinogen IX to protoporphyrin IX, the fully conjugated tetrapyrrole macrocycle into which iron atoms will be inserted (Figure 16.12). Three double bonds are introduced, adding 6 π-electrons to the existing 14 π-electrons to reach the fully conjugated 20 π-electrons in the tetrapyrrole macrocycle. The full conjugation is essential for the electronic properties of the iron or magnesium-embedded tetrapyrrole macrocycle behavior for light absorption or/and subsequent chemistry.

The responsible catalyst is a flavin-containing desaturase known as protoporphyrinogen oxidase,[15] encoded by an mRNA from which 18 introns have been spliced out (!). The desaturase terminology refers to the first half reaction converting a saturated carbon system to an olefin. In this case three of the methylene bridges to methine carbon atoms (Figure 16.12) are generated by (blue) hydride transfer to N_5 of oxidized FAD in the enzyme active site (see Chapter 2 for olefin chemical biology). The driving force for hydride ejection from the methylene bridges are presumed to be the prototropic rearrangements of the pyrrole rings. When the third methylene hydrogen from three of the four methylene bridges has been ejected, the remaining (fourth) methylene hydrogen is removed as a proton not as a hydride ion. The 10 double bonds of the macrocycle can be written as a number of contributing resonance structures to account for the macrocycle stability.

In each of the three second half reactions the reduced flavin is reoxidized in one electron steps by O_2. The resultant FAD-4a-hydroperoxy adduct breaks down intramolecularly to expel HOOH (see Chapters 8 and 11). A total of three O_2 molecules are reduced to HOOH, accounting for the six electrons removed from protoporphyrinogen IX.

One might compare this tandem reduction of three O_2 molecules by two electrons to H_2O_2, to the P450-mediated aromatase reduction of three O_2 molecules by four electrons each (total 6 electrons *vs.* 12 electrons removed, in the two systems) as androstenedione is converted to estrone – Chapter 12).

Protoporphyrin IX is the initial end product metabolite for heme biosynthesis. It has 20 carbon atoms that make up the macrocycle and another 14 carbons as substituents on the four pyrrole rings for a total of 34 carbon atoms. The mature porphyrin scaffold also has the four pyrrolic nitrogens. Looking back to the start of the pathway, *e.g.* in human cells, eight glycines

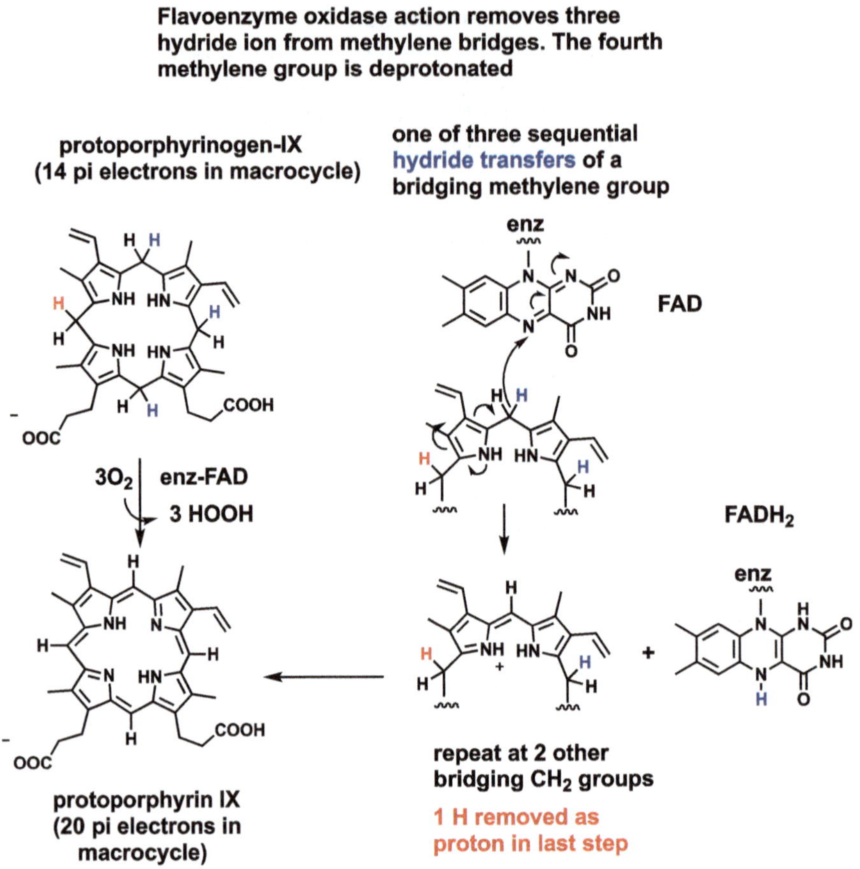

Figure 16.12 Protoporphyrinogen IX has 14 π-electrons. To get to the fully conjugated 20 π-electrons of protoporphyrin IX, a flavoenzyme oxidase removes a hydride ion from three of the bridging CH_2 groups, reducing O_2 to HOOH in each cycle. The fourth –CH_2 bridge undergoes proton loss to complete the conjugation.

and eight molecules of succinyl CoA serve as the building blocks for a total of 48 carbon atoms. Along the way 14 carbons are released as CO_2, consistent with $48 - 14 = 34$ carbons in protoporphyrin IX. Of those 14 CO_2 molecules pared away, eight happen in the first step on the way to eight aminolevulinates. Four are lost from acetate side chains and two are lost from propionate side chains late in the pathway. Four of the eight nitrogen atoms are lost as ammonium ions, all at the stage of hydroxybilane formation, *via* azafulvene isomeric structures of the pyrroles. In Chapter 15 we noted that 18 acetyl CoA molecules yield C_{27} cholesterol and ultimately C_{18} estradiol. To get to cholesterol, eight of the nine original carbons jettisoned are CO_2 and one is formate, emphasizing the trimming power of decarboxylases in tailoring carbon frameworks in nature.

16.8 Metal-specific Chelatases

The final step in heme biosynthesis is insertion of ferrous iron (Fe^{2+}) by the enzyme ferrochelatase, forming heme B[16] (Figure 16.13). The ferochelatase has an initial binding site for Fe^{2+}, allowing discrimination against Zn^{2+} for subsequent selective insertion of the divalent iron cation into the equatorial plane of the macrocycle to be chelated by the pyrrole ring nitrogens. It is thought that chelatases need to bind and transiently distort the macrocycle out of planarity to allow Fe^{2+} to be inserted. In plants this same protoporphyrin IX in chloroplasts is partitioned between Fe^{2+} chelation to heme B

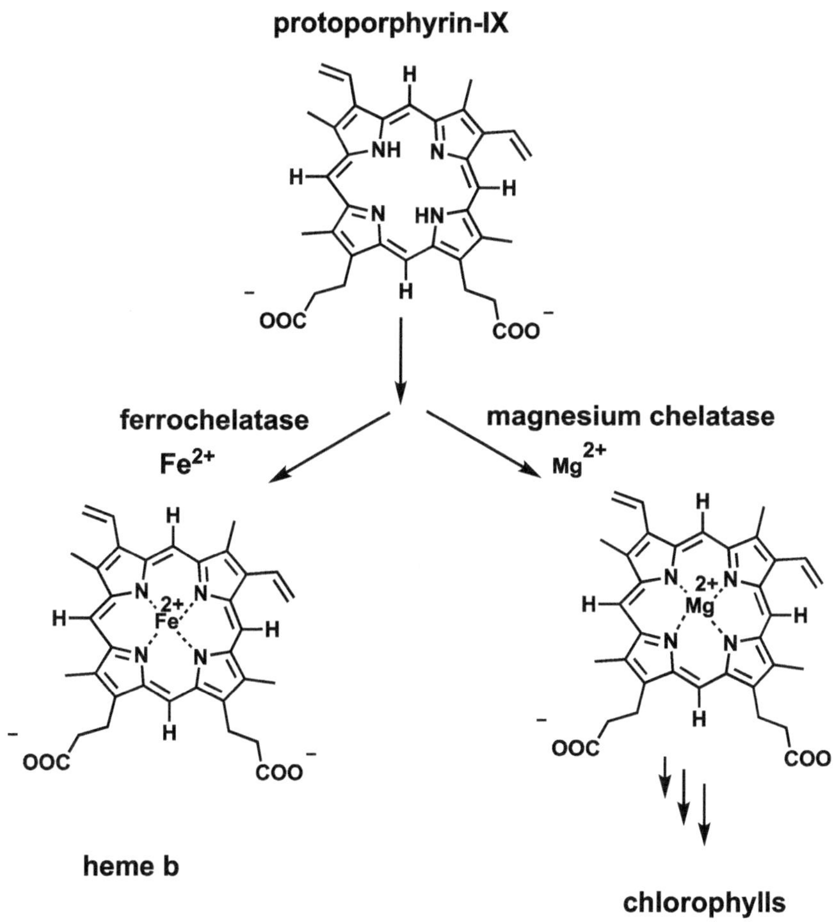

Figure 16.13 Ferrochetalases insert an Fe^{2+} ion into the equatorial plane of protoporphyrin IX with the four pyrrolic nitrogens as ligands, completing the formation of heme B. Plants compete for the open cavity of protoporphyrin IX with a magnesium chetalase that instead inserts Mg^{2+} to give the magnesium-metalated tetrapryrrole that is an intermediate for plant-mediated further modifications to chlorophylls (Figure 16.1).

and Mg^{2+} chelation as a prelude for maturation to chlorophyll. The Mg^{2+} chelatase[17] requires ATP for its insertion ability, as distinct from the ferrochelatases.

16.9 Different Branch Points to Siroheme Chlorophyll, B12 and F430

Although we have focused on the enzymatic steps to protoporphyrin IX in this chapter we stop for a moment to note two major branch points.[3] One branching occurs at the level of uroporphyrinogen III, to lead to coenzyme F430 and ultimately the ring-contracted corrin system for B12. The second branch point, at the end stage of protoporphyrin IX, is at the stage of Fe^{II} *vs.* Mg^{II} by competing metal chelatases in photosynthetic organisms (Figure 16.13). The Mg-tetrapyrrole molecules are then further modified to chlorophyll A and B.

16.10 Overview of Tetrapyrrole Biosynthetic Logic

Biosynthetic assembly of the 34 carbon scaffold of heme B is at a complexity level equal to construction of the 30 carbon framework of lanosterol (Chapter 15). The chemical logic underpinning tetrapyrrole macrocycle assembly is quite distinct since pyrrole heterocyclic chemistry is the core set of transformations. The repeating building block is porphobilinogen with its 2-aminomethylpyrrole core.

In the chain elongation phase of isoprenoid assembly (Chapter 15), it was the isomeric pair of Δ^2- and Δ^3-isopentenyl diphosphates that cooperate as intermolecular paired electrophile and nucleophile. In chain elongations in the hydroxybilane synthase/porphobilinogen deaminase active site, the pyrrole core can also act intermolecularly both as C_2 azafulvene electrophile and C_5 carbon nucleophile. Both pathways generate intermediates that are used iteratively in chain growth. For steroid assembly it is two chain elongations to get to the key C_{15} farnesyl-PP that dimerizes to squalene. For tetrapyrroles it is first dimerization of two aminolevulinates to porphobilinogen, then four inter-pyrrole methylene groups that are built up as linear frameworks (hydroxybilane). The cyclization mechanisms also utilize very different chemistries: oxidosqualene cationic cyclization path *vs.* the complex head to tail (A-ring to D-ring) joining of the linear tetrapyrrole azafulvene species to uroporphyrinogen III. Late stage maturations involve release of carbons as CO_2 in both cases, but lanosterol utilizes nine O_2 molecules (oxygenatively) *vs.* up to three O_2 spent (oxidatively) in the aerobic route to convert coproporphyrinogen III to protoporphyrn IX.

On the way to lanosterol six decarboxylations happen early, at the stage of 3-phospho-mevalonate-1-diphosphate. On the way to protoporphyrin IX, δ-aminolevulinate synthases act eight times to decarboxylate an enzyme bound intermediate to produce the eight required aminolevulinates.

Six more CO_2 one-carbon units are released by tandem action of carboxylases on uroporphyrinogen III late in the pathway for a total of 14 CO_2.

Also, in the contrast and compare mode for the two pathways, the conversion of lanosterol to the female sex hormone estradiol consumes 17 molecules of O_2 by action of a cascade of cytochrome P450 oxygenases. There is no involvement of O_2 up to the coproroporphyrinogen decarboxylase and protoporphyrinogen oxidase enzyme, the last two catalysts on the way to the protoporphyrin IX. Even then, O_2 is used as an electron acceptor not as an oxygenation reagent. These may be a clue to earlier evolution of the tetrapyrrole biosynthetic pathway in anaerobes before global oxygenation selected for aerobes.

Tetrapyrrole biosynthesis, to heme, chlorophyll, vitamin B12, and coenzyme F430, was most likely an early pathway on the evolution of organisms, given the key roles those four pigments play in redox metabolism, in photosynthesis, and in methane production.[3] The construction of the aminomethylpyrrole metabolite porphobilinogen appears to be the central invention among several unusual chemical steps in protoporphyrin IX assembly. The core chain elongation chemistry, enabled by the pyrrole ring acting consecutively both as electrophile and nucleophile, emphasizes how important nitrogen heterocycles are in the chemical biology of carbon atoms.

Indeed, the carbon chemical biology and nitrogen chemical biology are inextricably linked in nitrogen heterocycles, whether porphobilinogen, purines and pyrimidines, tetrahydrofolate, pyridoxal, thiamin, flavins, or nicotinamides, are the topic.

References

1. A. R. Battersby, Tetrapyrroles: the pigments of life, *Nat. Prod. Rep.*, 2000, **17**, 507–526.
2. H. A. Dailey, T. A. Dailey, S. Gerdes, D. Jahn, M. R. O'Brian and M. J. Warren, Prokaryotic heme biosynthesis: multiple pathways to a common essential product, *Microbiol. Mol. Biol. Rev.*, 2017, **81**, e00048.
3. D. A. Bryant, C. N. Hunter and M. J. Warren, Biosynthesis of the modified tetrapyrroles-the pigments of life, *J. Biol. Chem.*, 2020, **295**, 6888–6925.
4. G. A. Hunter and G. C. Ferreira, 5-Aminolevulinate synthase: catalysis of first step of heme biosynthesis, *Cell. Mol. Biol.*, 2009, **55**, 102–110.
5. E. K. Jaffee, The porphobilinogen synthase-catalyzed reaction mechanism, *Bioorg. Chem.*, 2004, **32**, 316–325.
6. J. Moser, W.-D. Schubert, V. Beier, I. Bringemeier, D. Jahn and D. W. Heinz, V-shaped structure of glutamyl-tRNA reductase, the first enzyme of tRNA-dependent tetrapyrrole biosynthesis, *EMBO J.*, 2001, **20**, 6583–6590.
7. M. A. Smith, C. G. Kannangara and B. Grimm, Glutamate 1-semialdehyde aminotransferase; anomalous enantiomeric reaction and enzyme mechanism, *Biochemistry*, 1992, **31**, 11249–11254.

8. P. Pluta, P. Roversi, G. Bernardo-Seisdedos, A. L. Tojas, J. B. Cooper, S. Gu, R. W. Pickersgill and O. Millet, Structural basis of pyrrole polymerization in human porphobilinogen deaminase, *Biochim. Biophys. Acta*, 2018, **1862**, 1948–1955.
9. N. Bung, A. Roy, B. Chen, D. Das, M. Pradhan, M. Yasuda, M. I. New, R. J. Desnick and G. Bulsu, Human hydroxybilane synthase: molecular dynamics of the pyrrole chain elongation identifies site-specific residues thst cause acute intermittent porphyria, *Proc. Natl. Acad. Sci. U. S. A.*, 2018, **115**, E4701–E4708.
10. P. Eswar, L. S. Krishna, V. Srinivasadesikan and R. K. Kottalanka, Unprecedented structural evidence for pyrrole-imine to azafulvene amine tautomerization in iminopyrrolyl Schiff's base ligand: coexistence of both tautomers in a single crystal, *J. Mol. Struct.*, 2021, **1244**, 131030.
11. R. Gill, S. E. Kolstoe, F. Mohammed, A. A. D. Bass, J. E. Moseley, M. Sarwar, J. B. Cooper, S. P. D. Wood and P. M. Shooligan-Jordan, Structure of human porphobilinogen deaminase at 2.8 angstroms: the molecular basis of acute intermittent porphyria, *Biochem. J.*, 2009, **420**, 17–25.
12. M. A. Matthews, H. L. Schubert, F. R. G. Whitby, K. J. Alexander, K. Schadick, H. A. Bergonia, J. D. Phillips and C. P. Hill, Crystal structur of human uroprophyrinogen III synthase, *EMBO J.*, 2001, **20**, 5832–5839.
13. C. T Walsh, Biologically generated carbon dioxide: nature's versatile chemical strategies for carboxy lyases, *Nat. Prod. Rep.*, 2015, **37**, 100–135.
14. B. G. Miller and R. A. Wolfenden, Catalytic proficiency: the unusual case of OMP decarboxylase, *Annu. Rev. Biochem.*, 2002, **71**, 847–885.
15. T. A. Dailey and H. A. Dailey, Human protoporphyrinogen oxidase: expression, purification, and characterization of the cloned enzyme, *Protein Sci.*, 1996, **5**, 98–105.
16. D. Lecerof, M. Fodje, M. Hansson, A. Hansson and S. Al-Karadaghi, Structural and mechanistic basis of prophyrin metallation by ferrochetalase, *J. Mol. Biol.*, 2000, **297**, 221–232.
17. D. Zhang, E. Chang, X. Yu, Y. Chen, Q. Yang, Y. Cao, W. Li, Y. Wang, A. Fu and M. Xu, Molecular characterization of magnesium chelatase in soybean [Glycine max (L.) Merr.], *Front. Plant Sci.*, 2018, **9**, 720.

Subject Index